Principles of Financial Engineering

Principles of Financial Engineering

Third Edition

Robert L. Kosowski

Oxford-Man Institute of Quantitative Finance
Oxford University

Salih N. Neftci

Department of Finance
Imperial College Business School
Imperial College London, UK

AMSTERDAM • BOSTON • HEIDELBERG • LONDON
NEW YORK • OXFORD • PARIS • SAN DIEGO
SAN FRANCISCO • SINGAPORE • SYDNEY • TOKYO
Academic Press is an imprint of Elsevier

Academic Press is an imprint of Elsevier
32 Jamestown Road, London NW1 7BY, UK
525 B Street, Suite 1800, San Diego, CA 92101-4495, USA
225 Wyman Street, Waltham, MA 02451, USA
The Boulevard, Langford Lane, Kidlington, Oxford OX5 1GB, UK

First published 2004
Second edition 2008
Third edition 2015

Notices

Knowledge and best practice in this field are constantly changing. As new research and experience broaden our understanding, changes in research methods, professional practices, or medical treatment may become necessary.

Practitioners and researchers must always rely on their own experience and knowledge in evaluating and using any information, methods, compounds, or experiments described herein. In using such information or methods they should be mindful of their own safety and the safety of others, including parties for whom they have a professional responsibility.

To the fullest extent of the law, neither the Publisher nor the authors, contributors, or editors, assume any liability for any injury and/or damage to persons or property as a matter of products liability, negligence or otherwise, or from any use or operation of any methods, products, instructions, or ideas contained in the material herein.

ISBN: 978-0-12-386968-5

British Library Cataloguing-in-Publication Data
A catalogue record for this book is available from the British Library

Library of Congress Cataloging-in-Publication Data
A catalog record for this book is available from the Library of Congress

For information on all Academic Press publications
visit our website at **http://store.elsevier.com**

Typeset by MPS Limited, Chennai, India
www.adi-mps.com

Printed and bound in the United States

Dedicated to
Salih Neftci and my family.

Contents

Preface to the Third Edition

This book is an introduction. It deals with a broad array of topics that fit together through a certain logic that we generally call *Financial Engineering*. The book is intended for beginning graduate students and practitioners in financial markets. The approach uses a combination of simple graphs, elementary mathematics, and real-world examples. The discussion concerning details of instruments, markets, and financial market practices is somewhat limited. The pricing issue is treated in an informal way, using simple examples. In contrast, the engineering dimension of the topics under consideration is emphasized.

Like Salih, I learned a great deal from technically oriented market practitioners who, over the years, have taken my courses. The deep knowledge and the professionalism of these brilliant market professionals contributed significantly to putting this text together. I first met Salih at Hong Kong University of Technology in 2006 where I gave a research seminar. Salih struck me as a very knowledgable finance professional and charismatic teacher. It was with sadness that I learned of Salih's passing in 2009. I was asked to teach his course at HEC Lausanne in Switzerland in 2009 and 2010. I based the course on his *Principles of Financial Engineering* book, since I could relate to the pedagogical approach in the book. I found the opportunity to revise the book for the third edition a humbling and enjoyable experience. The world of financial engineering and derivatives has changed significantly after the Global Financial Crisis (GFC) of 2008–2009 with a bigger emphasis on simplicity, standardization, counterparty risk, central clearing, liquidity, and exchange trading. But only 5 years after the GFC, new complex products such as contingent convertibles (CoCos) have been sold by banks to investors and prices of risky assets are again at all time highs. Understanding the principles of financial engineering can help us not only to solve new problems but also to understand hidden risks in certain products and identify risky and inappropriate financial engineering and market practices early enough to take action accordingly.

My main objective was to update the book and keep it topical by discussing how existing markets and market practices have changed and outline new financial engineering trends and products. In 2009 and 2010, I served as specialist advisor to the UK House of Lords as part of their inquiries into EU legislation related to alternative investment funds and Over-The-Counter (OTC) derivatives and I have continued to follow regulatory changes affecting derivatives markets and alternative investment funds with interest. I also benefitted greatly from my conversations with Marek Musiela and Damiano Brigo on various topics included in the book. Several colleagues and students read the original manuscript. I especially thank Damiano Brigo and Dimitris Karyampas and several anonymous referees who read the manuscript and provided comments. The book uses several real-life episodes as examples from market practices. I would like to thank Thomson Reuters International Financing Review (IFR), Derivatives Week (now part of GlobalCapital), Futuresmag, Efinancialnews, Bloomberg and Risk Magazine for their kind permission to use the material. I would like to thank Aman Kesarwani for excellent assistance with the creation of additional new end-of-chapter exercises.

What is new in the third edition?

Financial engineering principles can be applied in similar ways to different asset classes and therefore the third edition is structured in the form of different chapters on the application of financial engineering principles to interest rates, currencies, commodities, credit, and equities.

The material has been reorganized and streamlined. Since duration and related concepts are referred to repeatedly in the book, a new section that introduces duration and other measures of interest rate risk has been added to Chapter 3. The new Chapter 7 (on commodities and alternative investments) now contains a new expanded and updated section on the hedge fund industry which has grown in importance in recent years. The section on commodities introduces the spot-futures parity theorem and applications such as the cash and carry arbitrage. A section on callable bonds has been added to Chapter 16 (on option applications in fixed-income and mortgage markets). Chapter 18 on credit default swaps now contains material on CDS pricing and recent developments in sovereign CDS markets. In Chapter 19, the discussion of discounted cash flow approaches to equity valuation has been replaced by a financial engineering perspective in the form of the Merton model which views equity as an option on the firm's assets. Reverse convertibles were added to the list of equity structured products discussed in Chapter 20. We introduce securitization, ABS, CDOs in Chapter 21 and apply our financial engineering toolkit to the valuation and critical analysis of CoCos, a new post-GFC hybrid security. Chapter 22 discusses default correlation trading including hedging and risk management of such positions. Market participants and many academics were aware of the importance of counterparty risk before the GFC, but one of the biggest revolutions in financial engineering and derivatives practice has been how comprehensively counterparty risk is now being incorporated into derivatives pricing. We no longer assume as before that the counterparties in a derivatives transaction will honor their payment obligations. Chapter 24 is one of the new chapters and deals with how counterparty risk adjustments such as CVA, DVA, and FVA affect the pricing of derivatives and the choice of the risk free rate proxy.

All the remaining errors are, of course, mine. Solutions to the exercise and supplementary material for the book will be available on the companion website. A great deal of effort went into producing this book. Several more advanced issues that I could have treated had to be omitted, and I intend to include these in the future editions.

Robert L. Kosowski
July 31, 2014
London

INTRODUCTION

Market professionals and investors take long and short positions on *elementary assets* such as *stocks, default-free* bonds, and debt instruments that carry a *default* risk. There is also a great deal of interest in trading currencies, commodities, and, recently, inflation, volatility, and correlation. Looking from the outside, an observer may think that these trades are done overwhelmingly by buying and selling the asset in question *outright*, for example, by paying "cash" and buying a US Treasury bond. This is wrong. It turns out that most of the *financial* objectives can be reached in a much more convenient fashion by going through a proper *swap*. There is an important logic behind this and we choose this as the first principle to illustrate in this introductory chapter.

1.1 A UNIQUE INSTRUMENT

First, we would like to introduce the equivalent of the integer *zero*, in finance. Remember the property of zero in algebra. Adding (subtracting) zero to any other real number leaves this number the same. There is a unique financial instrument that has the same property with respect to market and credit risk. Consider the cash flow diagram in Figure 1.1. Here, the time is continuous and the t_0, t_1, t_2 represent some specific dates. Initially we place ourselves at time t_0. The following deal is struck with a bank. At time t_1 we borrow 100 US dollars (USD100), at the *going* interest rate of time t_1, called the *LIBOR* and denoted by the symbol L_{t_1}.[1] We pay the interest and the principal back at time t_2. The loan has no default risk and is for a period of δ units of time.[2] Note that the contract is written at time t_0, but starts at the future date t_1. Hence this is an example of *forward* contracts. The actual value of L_{t_1} will also be determined at the future date t_1.

Now, consider the time interval from t_0 to t_1, expressed as $t \in [t_0, t_1]$. At *any* time during this interval, what can we say about the value of this forward contract initiated at t_0?

It turns out that this contract will have a value *identically* equal to zero for all $t \in [t_0, t_1]$ regardless of what happens in world financial markets. Perceptions of future interest rate movements may go from zero to infinity, but the value of the contract will still remain zero. In order to prove this assertion, we calculate the value of the contract at time t_0. Actually, the value is obvious in one sense. Look at Figure 1.1. No cash changes hands at time t_0. So, the value of the contract at time t_0 must be zero. This may be obvious but let us show it formally.

To value the cash flows in Figure 1.1, we will calculate the time t_1 value of the cash flows that will be exchanged at time t_2. This can be done by discounting them with the *proper* discount factor.

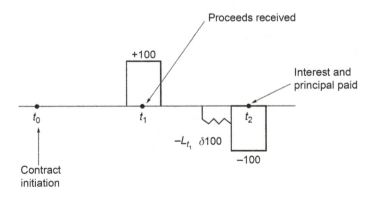

FIGURE 1.1

A forward loan.

---

[1]The London Interbank Offered Rate (LIBOR) is the average interest rate reported by a group of banks in London for borrowing from other banks. Traditionally LIBOR played a key role in financial engineering and derivatives pricing. As we discuss in Chapter 4, during the Global Financial Crisis (GFC) LIBOR rose significantly due to increased counterparty risk. This and recent scandals surrounding LIBOR underreporting and manipulation have changed its role in derivatives pricing as discussed in Chapters 3 and 24.

[2]The δ is measured in proportion to a year. For example, assuming that a "year" is 360 days and a "month" is always 30 days, a 3-month loan will give $\delta = 1/4$.

The best discounting is done using the L_{t_1} itself, although at time t_0 the value of this LIBOR rate is not known. Still, the time t_1 value of the future cash flows are

$$PV_{t_1} = \frac{L_{t_1} \delta\, 100}{(1 + L_{t_1} \delta)} + \frac{100}{(1 + L_{t_1} \delta)} \tag{1.1}$$

At first sight, it seems we would need an estimate of the random variable L_{t_1} to obtain a numerical answer from this formula. In fact, some market practitioners may suggest using the corresponding *forward rate* that is observed at time t_0 in lieu of L_{t_1}, for example. But a closer look suggests a much better alternative. Collecting the terms in the numerator

$$PV_{t_1} = \frac{(1 + L_{t_1} \delta)100}{(1 + L_{t_1} \delta)} \tag{1.2}$$

the unknown terms cancel out and we obtain:

$$PV_{t_1} = 100 \tag{1.3}$$

This looks like a trivial result, but consider what it means. In order to calculate the value of the cash flows shown in Figure 1.1, we don't *need* to know L_{t_1}. Regardless of what happens to interest rate expectations and regardless of market volatility, the value of these cash flows, and hence the value of this contract, is *always* equal to *zero* for *any* $t \in [t_0, t_1]$. In other words, the *price volatility* of this instrument is identically equal to zero.

This means that given any instrument at time t, we can add (or subtract) the LIBOR loan to it, and the value of the original instrument will *not* change for all $t \in [t_0, t_1]$. We now apply this simple idea to a number of basic operations in financial markets.

1.1.1 BUYING A DEFAULT-FREE BOND

For many of the operations they need, market practitioners do not "buy" or "sell" bonds. There is a much more convenient way of doing business.

The cash flows of buying a default-free coupon bond with par value 100 *forward* are shown in Figure 1.2. The coupon rate, set at time t_0, is r_{t_0}. The price is USD100, hence this is a *par bond* and the maturity date is t_2. Note that this implies the following equality:

$$100 = \frac{r_{t_0} \delta\, 100}{(1 + r_{t_0} \delta)} + \frac{100}{(1 + r_{t_0} \delta)} \tag{1.4}$$

which is true, because at t_0, the buyer is paying USD100 for the cash flows shown in Figure 1.2.

Buying (selling) such a bond is inconvenient in many respects. First, one needs cash to do this. Practitioners call this *funding*, in case the bond is purchased.[3] When the bond is sold short it will generate new cash and this must be managed.[4] Hence, such outright sales and purchases require inconvenient and costly *cash management*.

---

[3] Following the GFC the issue of how to account for funding costs in valuing OTC derivatives has generated an intense controversy. The arguments for and against the so-called FVA (funding value adjustment) will be discussed in Chapter 24.
[4] Short selling involves *borrowing* the bond and then selling it. Hence, there will be a cash management issue.

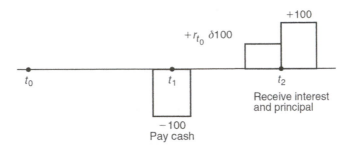

FIGURE 1.2

Buying default-free bond.

Second, the security in question may be a *registered* bond, instead of being a *bearer* bond, whereas the buyer may prefer to stay anonymous.

Third, buying (selling) the bond will affect *balance sheets*, called *books* in the industry. Suppose the practitioner borrows USD100 and buys the bond. Both the asset and the liability sides of the balance sheet are now larger. This may have *regulatory* implications.[5]

Finally, by securing the funding, the practitioner is getting a loan. Loans involve *credit risk*. The loan counterparty may want to factor a default risk premium into the interest rate.[6]

Now consider the following operation. The bond in question is a contract. To this contract "add" the forward LIBOR loan that we discussed in the previous section. This is shown in Figure 1.3a. As we already proved, for all $t \in [t_0, t_1]$, the value of the LIBOR loan is identically equal to zero. Hence, this operation is similar to adding *zero* to a risky contract. This addition does not change the *market* risk characteristics of the original position in any way. On the other hand, as Figures 1.3a and b show, the resulting cash flows are significantly more convenient than the original bond.

The cash flows require *no upfront* cash, they do not involve buying a *registered* security, and the *balance sheet* is not affected in any way.[7] Yet, the cash flows shown in Figure 1.3 have *exactly* the same market risk characteristics as the original bond.

Since the cash flows generated by the bond and the LIBOR loan in Figure 1.3 accomplish the same market risk objectives as the original bond transaction, then why not package them as a *separate* instrument and market them to clients under a different name? This is an interest rate swap (IRS). The party is paying a *fixed* rate and receiving a *floating* rate. The counterparty is doing the reverse.[8] IRSs are among the most liquid instruments in financial markets.

---

[5]For example, if this was an emerging market or corporate bond; the bank would be required to hold additional capital against this purchase. Regulatory capital or capital requirement is the amount of capital a bank or other financial institution has to hold as required by its financial regulator.

[6]If the Treasury security being purchased is left as collateral, then this credit risk aspect mostly disappears.

[7]Here, for simplicity, we ignore potential cash flows related to margin requirements, discussed in Chapter 2, and adjustments related to counter-party risk such as CVA (credit valuation adjustment) and DVA (debit valuation adjustment), discussed in Chapter 24.

[8]By market convention, the counterparty paying the fixed rate is called the "payer" (while receiving the floating rate), and the counterparty receiving the fixed rate is called the "receiver" (while paying the floating rate). The fixed rate payer (floating rate payer) is often referred to as having bought (sold) the swap or having a long (short) position.

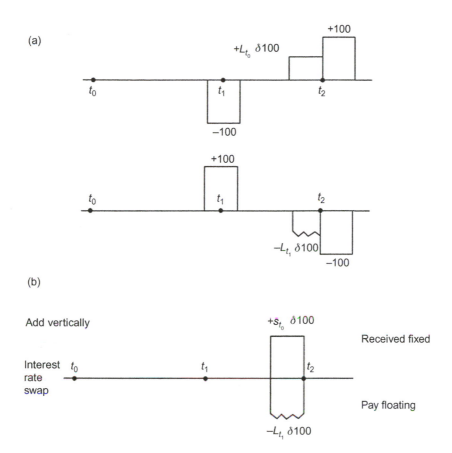

FIGURE 1.3

Engineering a simple IRS.

1.1.2 BUYING STOCKS

Suppose now we change the basic instrument. A market practitioner would like to buy a stock S_t at time t_0 with a t_1 delivery date. We assume that the stock does not pay dividends. Hence, this is, again, a forward purchase. The stock position will be liquidated at time t_2. Also, assume that the time t_0 perception of the stock market gains or losses is such that the markets are demanding a price

$$S_{t_0} = 100 \qquad (1.5)$$

for this stock as of time t_0. This situation is shown in Figure 1.4a, where ΔS_{t_2} is the unknown stock price appreciation or depreciation to be observed at time t_2. Note that the original price being 100, the time t_2 stock price can be written as

$$S_{t_2} = S_{t_1} + \Delta S_{t_2}$$
$$= 100 + \Delta S_{t_2} \qquad (1.6)$$

Hence the cash flows shown in Figure 1.4a.

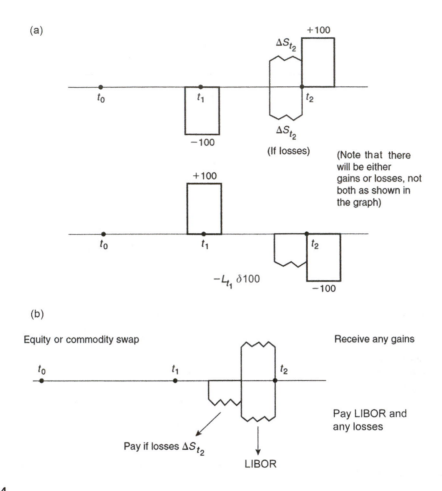

FIGURE 1.4

Engineering an equity or commodity swap.

It turns out that whatever the purpose of buying such a stock was, this *outright purchase* suffers from even more inconveniences than in the case of the bond. Just as in the case of the Treasury bond, the purchase requires *cash*, is a *registered* transaction with significant *tax* implications, and immediately affects the balance sheets, which have *regulatory* implications. A fourth inconvenience is a very simple one. The purchaser may not be allowed to own such a stock.[9] Last, but not least, there are regulations preventing highly leveraged stock purchases.

Now, apply the same technique to this transaction. Add the LIBOR loan to the cash flows shown in Figure 1.4a and obtain the cash flows in Figure 1.4b. As before, the market risk characteristics of the portfolio are identical to those of the original stock. The resulting cash flows can be

---

[9]For example, only special foreign institutions are allowed to buy Chinese A-shares that trade in Shanghai.

marketed jointly as a separate instrument. This is an *equity swap* and it has none of the inconveniences of the outright purchase. But, because we added a zero to the original cash flows, it has exactly the same market risk characteristics as a stock. In an equity swap, the party is receiving any stock market gains and paying a floating LIBOR rate plus any stock market losses.[10]

Note that if S_t denoted the price of any *commodity*, such as oil, then the same logic would give us a *commodity swap*.[11]

1.1.3 BUYING A DEFAULTABLE BOND

Consider the bond in Figure 1.1 again, but this time assume that at time t_2 the issuer can *default*. The bond pays the coupon c_{t_0} with

$$r_{t_0} < c_{t_0} \tag{1.7}$$

where r_{t_0} is a risk-free rate. The bond sells at *par* value, USD100 at time t_0. The interest and principal are received at time t_2 *if* there is no default. If the bond issuer defaults the investor receives nothing. This means that we are working with a *recovery rate* of zero. Figure 1.5a shows this characterization.

This transaction has, again, several inconveniences. In fact, all the inconveniences mentioned there are still valid. But, in addition, the defaultable bond may not be very *liquid*.[12] Also, because it is defaultable, the regulatory agencies will certainly impose a capital charge on these bonds if they are carried on the balance sheet.

A much more convenient instrument is obtained by adding the "zero" from Figure 1.1 to the defaultable bond and forming a new portfolio. Figures 1.5a and b visualized the cash flows of a defaultable bond together with those of a forward LIBOR loan. The combination of the defaultable bond and the LIBOR loan is show in Figure 1.5c, in which we assume $\delta = 1$. But we can go one step further in this case. Assume that at time t_0 there is an IRS, as shown in Figure 1.3, trading actively in the market. Then we can subtract this IRS from Figure 1.5c and obtain a much clearer picture of the final cash flows. This operation is shown in Figure 1.6. In fact, this last step eliminates the unknown future LIBOR rates L_{t_i} and replaces them with the known swap rate s_{t_0}.

The resulting cash flows don't have any of the inconveniences suffered by the defaultable bond purchase. Again, they can be packaged and sold separately as a new instrument. Letting s_{t_0} denote the rate on the corresponding IRS, the instrument requires receipts of a known and constant premium $\mathrm{Sp}_{t_0} = c_{t_0} - s_{t_0}$ periodically. Against this a *floating* (contingent) cash flow is paid. In case of default, the counterparty is compensated by USD100. This is similar to buying and selling default insurance. The instrument is called a *credit default swap* (CDS). Since their initiation during the

---

[10]If stocks decline at the settlement times, the investor will pay the LIBOR indexed cash flows *and* the loss in the stock value.

[11]To be exact, this commodity should have no other payout or storage costs associated with it; it should not have any *convenience yield* either. Otherwise, the swap structure will change slightly. This is equivalent to assuming no dividend payments and will be discussed in Chapter 7.

[12]Many corporate bonds do not trade in the secondary market at all.

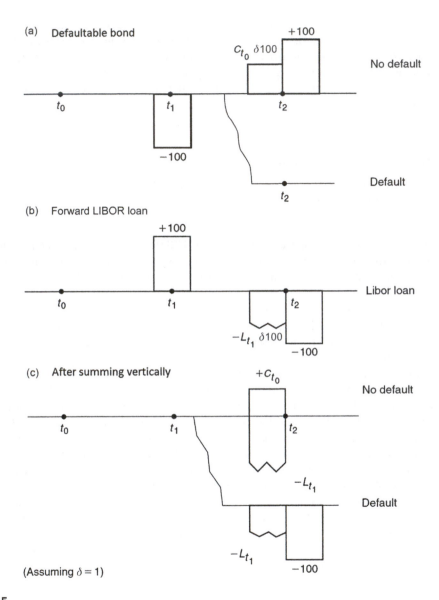

FIGURE 1.5

A risky bond and a LIBOR loan.

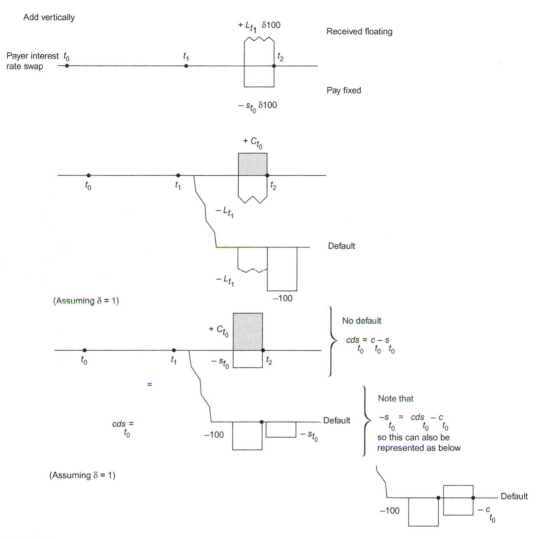

FIGURE 1.6

A credit default swap.

1990s CDSs have become very liquid instruments and completely changed the trading and hedging of credit risk. The insurance premium, called the *CDS spread*, cds_{t_0}, is given by

$$cds_{t_0} = Sp_{t_0} = c_{t_0} - s_{t_0} \tag{1.8}$$

This rate is positive since c_{t_0} should incorporate a default risk premium, which the default-free bond does not have.[13]

1.1.4 FIRST CONCLUSIONS

This section discussed examples of the first method of financial engineering. Switching from cash transactions to trading various swaps has many advantages. By combining an instrument with a forward LIBOR loan contract in a specific way, and then selling the resulting cash flows as separate swap contracts, the financial engineer has succeeded in accomplishing the same objectives much more efficiently and conveniently. The resulting swaps are likely to be more efficient, cost effective, and liquid than the underlying instruments. They also have better regulatory and tax implications.

Clearly, one can sell as well as buy such swaps. Also, one can reverse engineer the bond, equity, and the commodities by combining the swap with the LIBOR deposit. Chapter 4 will generalize this swap engineering. In the next section, we discuss another major financial engineering principle: the way one can build *synthetic* instruments.

We now introduce some simple financial engineering strategies. We consider two examples that require finding financial engineering solutions to a daily problem. In each case, solving the problem under consideration requires creating appropriate *synthetics*. In doing so, legal, institutional, and regulatory issues need to be considered.

The nature of the examples themselves is secondary here. Our main purpose is to bring to the forefront the *way* of solving problems using financial securities and their derivatives. The chapter does not go into the details of the terminology or of the tools that are used. In fact, some readers may not even be able to follow the discussion fully. There is no harm in this since these will be explained in later chapters.

1.2 A MONEY MARKET PROBLEM

Consider a Japanese bank in search of a 3-month money market loan. The bank would like to borrow USD in *Euromarkets* and then on-lend them to its customers. This *interbank* loan will lead to cash flows as shown in Figure 1.7. From the borrower's angle, USD100 is received at time t_0, and then it is paid back with interest 3 months later at time $t_0 + \delta$. The interest rate is denoted by the symbol L_{t_0} and is determined at time t_0. The *tenor* of the loan is 3 months. Therefore,

$$\delta = \frac{1}{4} \tag{1.9}$$

and the interest paid becomes $L_{t_0}(1/4)$. The possibility of default is assumed away.[14]

---

[13]The connection between the cds_{t_0} and the differential $c_{t_0} - s_{t_0}$ is more complicated in real life. Here we are working within a simplified setup.

[14]Otherwise at time $t_0 + \delta$ there would be a conditional cash outflow depending on whether or not there is default.

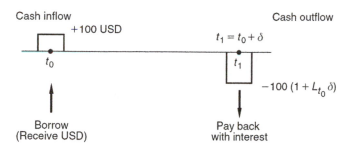

FIGURE 1.7

A USD loan.

The money market loan displayed in Figure 1.7 is a fairly liquid instrument. In fact, banks purchase such "funds" in the wholesale interbank markets, and then on-lend them to their customers at a slightly higher rate of interest.

1.2.1 THE PROBLEM

Now, suppose the above-mentioned Japanese bank finds out that this loan is not available due to the lack of appropriate *credit lines*. The counterparties are unwilling to extend the USD funds. The question then is: Are there other ways in which such *dollar funding* can be secured?

The answer is yes. In fact, the bank can use *foreign currency* markets judiciously to construct exactly the same cash flow diagram as in Figure 1.7 and thus create a *synthetic* money market loan. The first cash flow is negative and is placed *below* the time axis because it is a *payment* by the investor. The subsequent sale of the asset, on the other hand, is a receipt, and hence is represented by a positive cash flow placed *above* the time axis. The investor may have to pay significant taxes on these capital gains. A relevant question is then: Is it possible to use a strategy that postpones the investment gain to the next tax year? This may seem an innocuous statement, but note that using currency markets and their derivatives will involve a completely different set of financial contracts, players, and institutional setup than the money markets. Yet, the result will be cash flows identical to those in Figure 1.7.

1.2.2 SOLUTION

To see how a synthetic loan can be created, consider the following series of operations:

1. The Japanese bank first borrows *local* funds in yen in the onshore Japanese money markets. This is shown in Figure 1.8a. The bank receives yen at time t_0 and will pay yen interest rate $L_{t_0}^Y \delta$ at time $t_0 + \delta$.
2. Next, the bank sells these yen in the *spot* market at the current exchange rate e_{t_0} to secure USD100. This spot operation is shown in Figure 1.8b.
3. Finally, the bank must eliminate the currency mismatch introduced by these operations. In order to do this, the Japanese bank buys $100(1 + L_{t_0}^Y \delta)f_{t_0}$ yen at the known forward exchange rate f_{t_0},

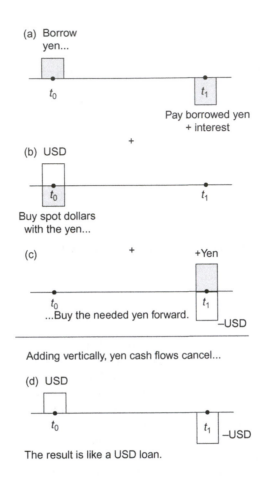

FIGURE 1.8

A synthetic USD loan.

in the *forward* currency markets. This is the cash flow shown in Figure 1.8c. Here, there is no exchange of funds at time t_0. Instead, forward dollars will be exchanged for forward yen at $t_0 + \delta$.

Now comes the critical point. In Figure 1.8, *add vertically* all the cash flows generated by these operations. The yen cash flows will cancel out at time t_0 because they are of equal size and different sign. The time $t_0 + \delta$ yen cash flows will also cancel out because that is how the size of the forward contract is selected. The bank purchases just *enough* forward yen to pay back the local yen loan and the associated interest. The cash flows that are left are shown in Figure 1.8d, and these are exactly the *same* cash flows as in Figure 1.7. Thus, the three operations have created a *synthetic* USD loan. The existence of the FX-forward played a crucial role in this synthetic.

1.2.3 **SOME IMPLICATIONS**

There are some subtle but important differences between the actual loan and the synthetic. First, note that from the point of view of Euromarket banks, lending to Japanese banks involves a principal of USD100, and this creates a *credit risk*. In case of default, the 100 dollars lent may not be repaid. Against this risk, some capital has to be put aside. Depending on the state of money markets and depending on counterparty credit risks, money center banks may adjust their credit lines toward such customers.

On the other hand, in the case of the *synthetic* dollar loan, the international bank's exposure to the Japanese bank is in the forward currency market only. Here, there is *no* principal involved. If the Japanese bank defaults, the burden of default will be on the domestic banking system in Japan. There is a risk due to the forward currency operation, called *counterparty risk* and since the Global Financial Crisis (GFC) this has received much more attention since it can be economically important. Thus, the Japanese bank may end up getting the desired funds somewhat easier if a synthetic is used.

There is a second interesting point to the issue of credit risk mentioned earlier. The original money market loan was a Euromarket instrument. Banking operations in Euromarkets are considered *offshore* operations, taking place essentially outside the jurisdiction of national banking authorities. The local yen loan, on the other hand, would be subject to supervision by Japanese authorities, obtained in the onshore market. In case of default, there may be some help from the Japanese Central Bank, unlike a Eurodollar loan where a default may have more severe implications on the lending bank.[15]

The third point has to do with pricing. If the actual and synthetic loans have identical cash flows, their values should also be the same *excluding* credit risk issues. If there is a value discrepancy the markets will simultaneously sell the expensive one, and buy the cheaper one, realizing a windfall gain. This means that synthetics can also be used in *pricing* the original instrument.[16]

Fourth, note that the money market loan and the synthetic can in fact be each other's *hedge*. Finally, in spite of the identical nature of the involved cash flows, the two ways of securing dollar funding happen in completely different markets and involve very different financial contracts. This means that legal and regulatory differences may be significant.

1.3 **A TAXATION EXAMPLE**

Now consider a totally different problem. We create synthetic instruments to restructure taxable gains. The legal environment surrounding taxation is a complex and ever-changing phenomenon; therefore, this example should be read only from a financial engineering perspective and not as a tax strategy. Yet the example illustrates the close connection between what a financial engineer does and the legal and regulatory issues that surround this activity.

---

[15]Chapter 2 provides more information on Euromarkets.

[16]However, the credit risk issues mentioned earlier may introduce a wedge between the prices of the two loans.

1.3.1 THE PROBLEM

In taxation of financial gains and losses, there is a concept known as a *wash-sale*. Suppose that during the year 2014, an investor realizes some financial gains. Normally, these gains are taxable that year. But a variety of financial strategies can possibly be used to postpone taxation to the year after. To prevent such strategies, national tax authorities have a set of rules known as wash-sale and *straddle* rules. It is important that professionals working for national tax authorities in various countries understand these strategies well and have a good knowledge of financial engineering. Otherwise some players may rearrange their portfolios, and this may lead to significant losses in tax revenues. From our perspective, we are concerned with the methodology of constructing synthetic instruments.

Suppose that in September 2014, an investor bought an asset at a price $S_0 = \$100$. In December 2014, this asset is sold at $S_1 = \$150$. Thus, the investor has realized a capital gain of $50. These cash flows are shown in Figure 1.9.

One may propose the following solution. This investor is probably holding assets *other* than the S_t mentioned earlier. After all, the right way to invest is to have diversifiable portfolios. It is also reasonable to assume that if there were *appreciating* assets such as S_t, there were also assets that lost value during the same period. Denote the price of such an asset by Z_t. Let the purchase price be Z_0. If there were no wash-sale rules, the following strategy could be put together to postpone year 2014 taxes.

Sell the Z-asset on December 2014, at a price Z_1, $Z_1 < Z_0$, and, the next day, buy the same Z_t at a similar price. The sale will result in a loss equal to

$$Z_1 - Z_0 < 0 \tag{1.10}$$

The subsequent purchase puts this asset back into the portfolio so that the diversified portfolio can be maintained. This way, the losses in Z_t are recognized and will cancel out some or all of the capital gains earned from S_t. There may be several problems with this strategy, but one is fatal. Tax authorities would call this a wash-sale (i.e., a sale that is being intentionally used to "wash" the 2014 capital gains) and would disallow the deductions.[17]

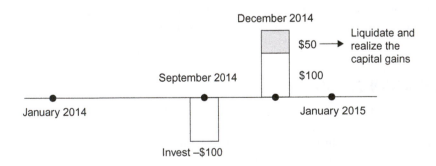

FIGURE 1.9

An investment liquidated on December 2014.

---

[17]The equivalent of the US wash-sale rule is evocatively called a *bed and breakfast* rule in the UK tax code.

1.3.1.1 Another strategy

Investors can find a way to sell the Z-asset without having to sell it in the *usual* way. This can be done by first creating a *synthetic* Z-asset and then realizing the implicit capital losses using this synthetic, instead of the Z-asset held in the portfolio.

Suppose the investor originally purchased the Z-asset at a price $Z_0 = \$100$ and that asset is currently trading at $Z_1 = \$50$, with a paper loss of \$50. The investor would like to recognize the loss without directly selling this asset. At the same time, the investor would like to retain the original position in the Z-asset in order to maintain a well-balanced portfolio. How can the loss be realized while maintaining the Z-position *and* without selling the Z_t?

The idea is to construct a proper synthetic. Consider the following sequence of operations:

- Buy another Z-asset at price $Z_1 = \$50$ on November 26, 2014.
- Sell an at-the-money call on Z with expiration date December 30, 2014.
- Buy an at-the-money put on Z with the same expiration.

The specifics of call and put options will be discussed in later chapters. For those readers with no background in financial instruments we can add a few words. Briefly, options are instruments that give the purchaser a *right*. In the case of the call option, it is the right to *purchase* the underlying asset (here the Z-asset) at a prespecified price (here \$50). The put option is the opposite. It is the right to sell the asset at a prespecified price (here \$50). When one sells options, on the other hand, the seller has the *obligation* to deliver or accept delivery of the underlying at a prespecified price.

For our purposes, what is important is that short call and long put are two securities whose expiration payoff, when added, will give the synthetic short position shown in Figure 1.10. By selling the call, the investor has the *obligation* to deliver the Z-asset at a price of \$50 if the call holder demands it. The put, on the other hand, gives the investor the *right* to sell the Z-asset at \$50 if he or she chooses to do so.

The important point here is this: When the short call and the long put positions shown in Figure 1.10 are added together, the result will be equivalent to a *short* position on stock Z_t. In fact, the investor has created a *synthetic* short position using options.

Now consider what happens as time passes. If Z_t appreciates by December 30, the call will be exercised. This is shown in Figure 1.11a. The call position will lose money, since the investor has to deliver, at a *loss*, the original Z-stock that cost \$100. If, on the other hand, the Z_t decreases, then the put position will enable the investor to sell the original Z-stock at \$50. This time the call will expire worthless.[18] This situation is shown in Figure 1.11b. Again, there will be a loss of \$50. Thus, no matter what happens to the price Z_t, either the investor will deliver the *original* Z-asset purchased at a price \$100, or the put will be exercised and the investor will sell the original Z-asset at \$50. Thus, one way or another, the investor is using the original asset purchased at \$100 to close an option position at a loss. This means he or she will lose \$50 while keeping the same Z-position, since the *second* Z, purchased at \$50, will still be in the portfolio.

The timing issue is important here. For example, according to US tax legislation, wash-sale rules will apply if the investor has acquired or sold a *substantially identical* property within a

---

[18]For technical reasons, suppose both options can be exercised only at expiration. They are of European style.

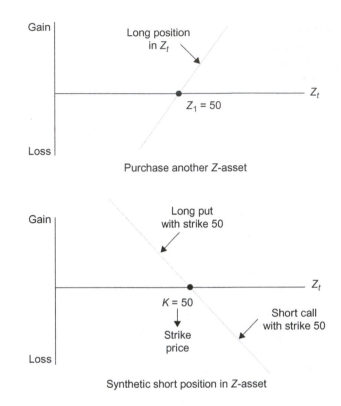

FIGURE 1.10

Two positions that cancel each other.

31-day period. According to the strategy outlined here, the second Z is purchased on November 26, while the options expire on December 30. Thus, there are more than 31 days between the two operations.[19]

1.3.2 IMPLICATIONS

There are at least three interesting points to our discussion. First, the strategy offered to the investor was *risk-free* and had *zero cost* aside from commissions and fees. Whatever happens to the new long position in the Z-asset, it will be canceled by the synthetic short position. This situation is shown in the lower half of Figure 1.10. As this graph shows, the proposed solution has no market risk, but may have counterparty, or operational risks. The second point is that, once again, we have created a *synthetic*, and then used it in providing a solution to our problem. Finally, the example

---

[19]The timing considerations suggest that the strategy will be easier to apply if over-the-counter (OTC) options are used, since the expiration dates of exchange-traded options may occur at specific dates, which may not satisfy the legal timing requirements.

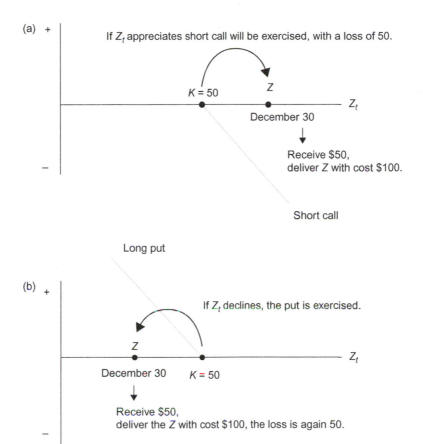

FIGURE 1.11

The strategy with the Z initially at 50. Two ways to realize loss.

displays the crucial role *legal* and *regulatory* frameworks can play in devising financial strategies. Although this book does not deal with these issues, it is important to understand the crucial role they play at almost every level of financial engineering.

1.4 SOME CAVEATS FOR WHAT IS TO FOLLOW

A newcomer to financial engineering usually follows instincts that are *harmful* for good understanding of the basic methodologies in the field. Hence, before we start, we need to lay out some basic rules of the game that should be remembered throughout the book.

1. This book is written from a market practitioner's point of view. Investors, pension funds, insurance companies, and governments are *clients*, and for us they are always on the *other side* of the deal. In other words, we look at financial engineering from a trader's, broker's, and dealer's angle. The approach is from the *manufacturer's* perspective rather than the viewpoint of the *user* of the financial services. This premise is crucial in understanding some of the logic discussed in later chapters.

2. We adopt the convention that there are *two* prices for every instrument unless stated otherwise. The agents involved in the deals often quote *two-way* prices. In economic theory, economic agents face the law of one price. The same good or asset cannot have two prices. If it did, we would then buy at the cheaper price and sell at the higher price.

    Yet for a market maker, there *are* two prices: one price at which the market participant is willing to *buy* something from you, and another one at which the market participant is willing to *sell* the same thing to you. Clearly, the two cannot be the same. An automobile dealer will buy a used car at a *low* price in order to sell it at a *higher* price. That is how the dealer makes money. The same is true for a market practitioner. A swap dealer will be willing to buy swaps at a low price in order to sell them at a higher price later.[20] In the meantime, the instrument, just like the used car sold to a car dealer, is kept in inventories.

3. It is important to realize that a financial market participant is not an investor and never has "money." He or she has to secure *funding* for any purchase and has to *place* the cash generated by any sale. In this book, almost no financial market operation begins with a pile of cash. The only "cash" is in the investor's hands, which in this book is *on the other side* of the transaction.

    It is for this reason that market practitioners prefer to work with instruments that have zero-value at the time of initiation. Such instruments would not require funding and are more practical to use.[21] They also are likely to have more *liquidity*.

4. The role played by regulators, professional organizations, and the legal profession is much more important for a market professional than for an investor. Although it is far beyond the scope of this book, many financial engineering strategies have been devised for the sole purpose of dealing with them.

Remembering these premises will greatly facilitate the understanding of financial engineering.

1.5 TRADING VOLATILITY

Practitioners or investors can take positions on expectations concerning the *price* of an asset. Volatility trading involves positions taken on the volatility of the price. This is an attractive idea, but how does one buy or sell volatility? Answering this question will lead to a *third* basic methodology in financial engineering. This idea is a bit more complicated, so the argument here will only be an introduction. Chapter 9 will present a more detailed treatment of the methodology.

In order to discuss volatility trading, we need to introduce the notion of *convexity gains*. We start with a *forward* contract. Let us stay within the framework of the previous section and

---

[20]The price at which a market participant such as a swap dealer is willing to buy is called the *bid* price and the price at which the market participant is willing to sell is called his *ask* price. The difference between the two is called the bid/ask spread and discussed further in Chapter 2.

[21]Although one could pay bid-ask spreads or commissions during the process.

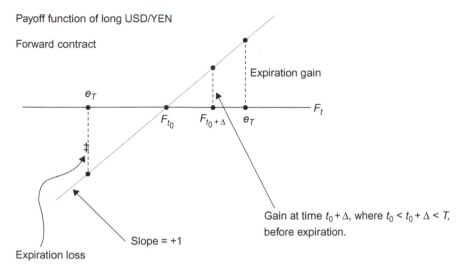

FIGURE 1.12

Payoff function of long USD/YEN forward contract.

assume that F_{t_0} is the forward dollar–yen exchange rate.[22] Suppose at time t_0 we take a long position in USD as shown in Figure 1.12. The upward sloping line is the so-called *payoff function*.[23] For example, if at time $t_0 + \Delta$ the forward price becomes $F_{t_0+\Delta}$, we can close the position with a gain:

$$\text{Gain} = F_{t_0+\Delta} - F_{t_0} \tag{1.11}$$

It is important, for the ensuing discussion, that this payoff function be a *straight line* with a constant slope.

Now, suppose there exists *another* instrument whose payoff depends on the F_T. But this time the dependence is nonlinear[24] or *convex* as shown in Figure 1.13 by the convex curve $C(F_t)$. It is important that the curve be smooth, and that the derivative

$$\frac{\partial C(F_t)}{\partial F_t} \tag{1.12}$$

exist at all points.

Finally, suppose this payoff function has the additional property that as time passes the function changes shape. In fact as expiration time T approaches, the curve becomes a (piecewise) straight line just like the forward contract payoff. This is shown in Figure 1.14.

---

[22]The e_{t_0} denotes the spot exchange rate USD/JPY, which is the value of one dollar in terms of Japanese yen at time t_0. On April 11, 2014, for example, USD/JPY = 101.60.
[23]Depending on at what point the *spot* exchange rate denoted by e_T, ends up at time T, we either gain or lose from this long position.
[24]Options on the dollar–yen exchange rate will have such a pricing curve. But we will see this later in Chapter 9.

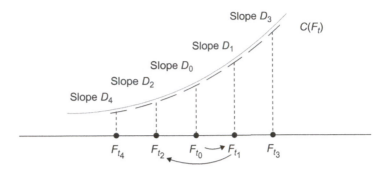

FIGURE 1.13

Payoff function of a nonlinear (convex) instrument.

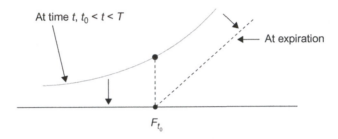

FIGURE 1.14

Payoff function before and at expiration.

1.5.1 A VOLATILITY TRADE

Volatility trades depend on the simultaneous existence of two instruments, one whose value moves linearly as the underlying risk changes, while the other's value moves according to a convex curve.

First, suppose $\{F_{t_1}, \ldots F_{t_n}\}$ are the forward prices observed successively at times $t < t_1, \ldots,$ $t_n < T$ as shown in Figure 1.13. Note that these values are selected so that they *oscillate* around F_{t_0}.

Second, note that at every value of F_{t_i} we can get an approximation of the curve $C(F_t)$ using the tangent at that point as shown in Figure 1.13. Clearly, if we know the function $C(.)$, we can then *calculate* the slope of these tangents. Let the slope of these tangents be denoted by D_i.

The third step is the crucial one. We form a *portfolio* that will eliminate the risk of directional movements in exchange rates. We first buy one unit of the $C(F_t)$ at time t_0. Note that we *do* need cash for doing this since the value at t_0 is nonzero.

$$0 < C(F_{t_0}) \tag{1.13}$$

Simultaneously, we sell D_0 units of the forward contract F_{t_0}. Note that the forward sale does not require any upfront cash at time t_0.

Finally, as time passes, we recalculate the tangent D_i of that period and adjust the forward sale accordingly. For example, if the slope has increased, sell $D_i - D_{i-1}$ units more of the forward

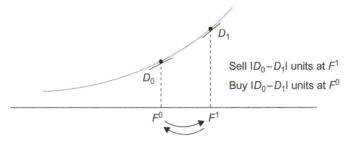

Ignore the movement of the curve, assumed small.
Note that as the curve moves down slope changes

FIGURE 1.15

Oscillations in the forward price and rebalancing the hedge.

contracts. If the slope has decreased cover $D_i - D_{i-1}$ units of the forwards.[25] As F_{t_i} oscillates continue with this *rebalancing*.

We can now calculate the net cash flows associated with this strategy. Consider the oscillations in Figures 1.13 and 1.15,

$$(F_{t_{i-1}} = F^0) \rightarrow (F_{t_i} = F^1) \rightarrow (F_{t_{i+1}} = F^0) \tag{1.14}$$

with $F^0 < F^1$. In this setting if the trader follows the algorithm described above, then at every oscillation, the trader will

1. First sell $D_i - D_{i-1}$ additional units at the price F^1.
2. Then, buy the same number of units at the price of F^0.

For each oscillation, the cash flows can be calculated as

$$\text{Gain} = (D_1 - D_0)(F^1 - F^0) \tag{1.15}$$

Since $F^0 < F^1$ and $D^0 < D^1$, this gain is *positive* as summarized in Figure 1.15. By hedging the original position in $C(.)$ and periodically rebalancing the hedge, the trader has in fact succeeded to *monetize* the oscillations of F_{t_i}.

1.5.2 **RECAP**

Look at what the trader has accomplished. By holding the convex instrument and then trading the linear instrument *against* it, the trader realized *positive* gains. These gains are bigger, the bigger the oscillations. Also they are bigger, the bigger the changes in the slope terms D_i. In fact, the trader gains whether the price goes down or up. The gains are proportional to the realized volatility.

---

[25]This means buy back the units.

Clearly, this dynamic strategy has resulted in extracting cash from the volatility of the underlying forward rate F_t. It turns out that one can package such expected volatility gains and sell them to clients. This leads to volatility trading. It is accomplished by using options and, lately, through volatility swaps.[26]

1.6 CONCLUSIONS

This chapter uses some examples in order to display the use of synthetics (or *replicating portfolios* as they are called in formal models). The main objective of this book is to discuss *methods* that use financial markets, instruments, and financial engineering strategies in solving practical problems concerning pricing, hedging, risk management, balance sheet management, and product structuring. The book does *not* discuss the details of financial instruments, although for completion, some basics are reviewed when necessary. The book deals even less with issues of corporate finance. We assume some familiarity with financial instruments, markets, and rudimentary corporate finance concepts.

Finally, the reader must remember that regulation, taxation, and even the markets themselves are "dynamic" objects that change constantly. Actual application of the techniques must update the parameters mentioned in this book.

SUGGESTED READING

There are excellent sources for studying financial instruments, their pricing, and the associated modeling. An excellent source for instruments and markets is **Hull** (2014). For corporate finance, **Brealey and Myers** (2013) and **Ross et al.** (2012) are two well-known references. **Bodie et al.** (2014) is recommended as background material on investments. **Wilmott** (2006) is a comprehensive and important source. **Duffie** (2001) provides the foundation for solid asset pricing theory.

EXERCISES

1. Which two instruments can be used to replicate the payoffs of an IRS?

2. What are Eurodollars?

3. Which two instruments can be used to create the payoffs of an equity swap?

4. How can a synthetic foreign currency loan be engineered?

---

[26]Volatility swaps will be discussed in Chapter 16.

CASE STUDY: JAPANESE LOANS AND FORWARDS

You are given the Reuters news item below. Read it carefully. Then answer the following questions.

1. Show how Japanese banks were able to create the dollar-denominated loans synthetically using cash flow diagrams.
2. How does this behavior of Japanese banks affect the balance sheet of the Western counterparties?
3. What are *nostro accounts*? Why are they needed? Why are the Western banks not willing to hold the yens in their nostro accounts?
4. What do the Western banks gain if they do that?
5. Show, using an "appropriate" formula, that the negative interest rates can be more than compensated by the extra points on the forward rates. (Use the decompositions given in the text.)

NEW YORK, (Reuters) - Japanese banks are increasingly borrowing dollar funds via the foreign exchange markets rather than in the traditional international loan markets, pushing some Japanese interest rates into negative territory, according to bank officials.

The rush to fund in the currency markets has helped create the recent anomaly in short-term interest rates. For the first time in years, yields on Japanese Treasury bills and some bank deposits are negative, in effect requiring the lender of yen to pay the borrower.

Japanese financial institutions are having difficulty getting loans denominated in U.S. dollars, experts said. They said international banks are weary of expanding credit lines to Japanese banks, whose balance sheets remain burdened by bad loans.

"The Japanese banks are still having trouble funding in dollars," said a fixed-income strategist at Merrill Lynch & Co.

So Japan's banks are turning to foreign exchange transactions to obtain dollars. The predominant mechanism for borrowing dollars is through a trade combining a spot and forward in dollar/yen.

Japanese banks typically borrow in yen, which they have no problem getting. With a three-month loan, for instance, the Japanese bank would then sell the yen for dollars in the spot market to, say, a British or American bank. The Japanese bank simultaneously enters into a three-month forward selling the dollars and getting back yen to pay off the yen loan at the stipulated forward rate. In effect, the Japanese bank has obtained a three-month dollar loan.

Under normal circumstances, the dealer providing the transaction to the Japanese bank should not make anything but the bid-offer spread.

But so great has been the demand from Japanese banks that dealers are earning anywhere from seven to 10 basis points from the spot-forward trade.

The problem is that the transaction saddles British and American banks with yen for three months. Normally, international banks would place the yen in deposits with Japanese banks and earn the three-month interest rate.

But most Western banks are already bumping against credit limits for their banks on exposure to troubled Japanese banks. Holding the yen on their own books in what are called NOSTRO accounts requires holding capital against them for regulatory purposes.

So Western banks have been dumping yen holdings at any cost—to the point of driving interest rates on Japanese Treasury bills into negative terms. Also, large Western banks such as Barclays Plc and J.P. Morgan are offering negative interest rates on yen deposits—in effect saying no to new yen-denominated deposits.

Western bankers said they can afford to pay up to hold Japanese Treasury bills—in effect earning negative yield—because their earnings from the spot-forward trade more than compensate them for their losses on holding Japanese paper with negative yield.

Japanese six-month T-bills offer a negative yield of around 0.002 percent, dealers said. Among banks offering a negative interest rate on yen deposits was Barclays Bank Plc, which offered a negative 0.02 percent interest rate on a three-month deposit.

The Bank of Japan, the central bank, has been encouraging government-lending institutions to make dollar loans to Japanese corporations to overcome the problem, said [a market professional]. (Reuters, November 9, 1998).

INSTITUTIONAL ASPECTS OF DERIVATIVE MARKETS

CHAPTER OUTLINE

2.1 INTRODUCTION

This chapter takes a step back and reviews in a nutshell the prerequisite for studying the methods of financial engineering. Readers with a good grasp of the conventions and mechanics of financial markets may skip it, although a quick reading would be preferable.

Financial engineering is a *practice* and can be used only when we define the related environment carefully. The organization of markets, and the way deals are concluded, confirmed, and carried out, are important factors in selecting the right solution for a particular financial engineering problem. This chapter examines the organization of financial markets and the way market practitioners interact. Issues related to settlement, to accounting methods, and especially to *conventions* used by market practitioners are important and need to be discussed carefully.

In fact, it is often overlooked that financial practices will depend on the conventions adopted by a particular market. This aspect, which is relegated to the background in most books, will be an important parameter of our approach. Conventions are not only important in their own right for proper pricing, but they also often reside behind the correct choice of theoretical models for analyzing pricing and risk management problems. The *way* information is provided by markets is a factor in determining the model choice. While doing this, the chapter introduces the mechanics of the markets, instruments, and who the players are. A brief discussion of the syndication process is also provided.

2.2 MARKETS

The first distinction is between *local* and *Euromarkets*. Local markets are also called *onshore markets*. These denote markets that are closely supervised by regulators such as central banks and financial regulatory agencies. There are basically two defining characteristics of onshore markets. The first is *reserve requirements* that are imposed on onshore deposits. The second is the formal *registration process* of newly issued securities. Both of these have important cost, liquidity, and taxation implications.

In money markets, reserve requirements imposed on banks increase the *cost* of holding onshore deposits and making loans. This is especially true of the large "wholesale" deposits that banks and other corporations may use for short periods of time. If part of these funds is held in a noninterest-bearing form in central banks, the cost of local funds will increase.

The long and detailed registration process imposed on institutions that are issuing stocks, bonds, or other financial securities has two implications for financial engineering. First, issue costs will be higher in cases of registered securities when compared to simpler *bearer* form securities. Second, an issue that does not have to be registered with a public entity will disclose less information.

Thus, markets where reserve requirements do not exist, where the registration process is simpler, will have significant cost advantages. Such markets are called Euromarkets.

2.2.1 EUROMARKETS

We should set something clear at the outset. The term "Euro" as utilized in this section does not refer to Europe, nor does it refer to the Eurozone currency, the Euro. It simply means that, in terms of reserve requirements or registration process we are dealing with markets that are outside the formal control of regulators and central banks. The two most important Euromarkets are the Eurocurrency market and the Eurobond market.[1]

2.2.1.1 Eurocurrency markets

Start with an onshore market. In an *onshore* system, a 3-month retail deposit has the following life. A client will deposit USD100 cash on date *T*. This will be available the same day. That is to say, "days to deposit" will equal zero. The deposit-receiving bank takes the cash and deposits, say, 10% of this in the central bank. This will be the *required reserves* portion of the original 100.[2] The remaining 90 dollars are then used to make new loans or may be lent to other banks in the interbank overnight market.[3] Hence, the bank will be paying interest on the entire 100, but will be receiving interest on only 90 of the original deposit. In such an environment, assuming there is no other cost, the bank has to charge an interest rate around 10% higher for making loans. Such supplementary costs are enough to hinder a liquid *wholesale market* for money where large sums are moved. Eurocurrency markets eliminate these costs and increase the liquidity. Let's briefly review the life of a Eurocurrency (offshore) deposit and compare it with an onshore deposit. Suppose a US bank deposits USD100 million in another US bank in the New York Eurodollar (offshore) market. Thus, as is the case for Eurocurrency markets, we are dealing only with banks, since this is an interbank market. Also, in this example, all banks are located in the United States. The Eurodeposit is made in the United States and the "money" never leaves the United States. This deposit becomes usable (settles) in 2 days—that is to say, *days to deposit* is 2 days. The entire USD100 million can now be lent to another institution as a loan. If this chain of transactions was happening in, say, London, the steps would be similar.

2.2.1.2 Eurobond markets

A bond sold publicly by going through the formal registration process will be an onshore instrument. If the same instrument is sold without a similar registration process, say, in London, and if it is a *bearer* security, then it becomes essentially an offshore instrument. It is called a Eurobond. Again the prefix "Euro" does not refer to Europe, although in this case the center of Eurobond activity happens to be in London. But, in principle, a Eurobond can be issued in Asia as well.[4]

---

[1]For further information about Eurocurrency and Eurodollar markets.

[2]In reality the process is more complicated than this. Banks are supposed to satisfy reserve requirements over an average number of days and within, say, a 1-month delay.

[3]In the United States, this market is known as the federal funds market.

[4]In their proposal on November 21, 2011, designed to tackle the financial crisis, the European Union Commission referred to European bonds backed collectively by all eurozone countries as *Eurobonds*. Although the same word is used, the onshore bonds mentioned in the EU proposal do not have anything to do with the offshore Eurobond Market which started in 1963 and is discussed in this chapter.

A Eurobond will be subject to less regulatory scrutiny, will be a *bearer* security, and will not be (as of now) subject to withholding taxes. The primary market will be in London. The secondary markets may be in Brussels, Luxembourg, or other places, where the Eurobonds will be *listed*. The settlement of Eurobonds will be done through Euroclear or Clearstream.

2.2.1.3 Other Euromarkets

Euromarkets are by no means limited to bonds and currencies. Almost any instrument can be marketed offshore. There can be Euro-equity, Euro-commercial paper (ECP), Euro medium-term note (EMTN), and so on. In derivatives, we have onshore forwards and swaps in contrast to offshore nondeliverable forwards and swaps.

2.2.2 ONSHORE MARKETS

Onshore markets can be organized over the counter or as formal exchanges. *Over-the-counter* (OTC) markets have evolved as a result of spontaneous trading activity. An OTC market often has no formal organization, although it will be closely monitored by regulatory agencies and transactions may be carried out along some precise *documentation* drawn by professional organizations, such as ISDA and ICMA.[5] Some of the biggest markets in the world are OTC. A good example is the interest rate swap (IRS) market, which has the highest notional amount traded among all financial markets with very tight bid-ask spreads. OTC transactions are done over the phone or electronically and the instruments contain a great deal of flexibility, although, again, institutions such as ISDA draw standardized documents that make traded instruments homogeneous.

In contrast to OTC markets, organized *exchanges* are formal entities. Most exchanges now use *electronic* trading, but some *open-outcry* exchanges with *pits* remain.[6] The distinguishing characteristic of an organized exchange is its formal organization. The traded products and trading practices are homogeneous while, at the same time, the specifications of the traded contracts are less flexible.

Stock markets are organized exchanges that deal in equities.[7] Futures and options markets process derivatives written on various underlying assets. In a *spot deal*, the trade will be done and confirmed, and within a few days, called the *settlement period*, money and securities change hands. In futures markets, on the other hand, the trade will consist of taking positions, and settlement will be after a relatively longer period, once the derivatives *expire*. The trade is, however, followed by depositing a "small" guarantee, called an *initial margin* (IM).

---

[5]The International Capital Market Association (ICMA) is a professional organization that among other activities may, after lengthy negotiations between organizations, homogenize contracts for OTC transactions. ISDA, the International Swaps and Derivatives Association, and NASD, the National Association of Securities Dealers in the United States, are two other examples of such associations.

[6]According to a January 3, 2014, CME press release, around 90% of trading on CME in 2013 was electronic. However, open-outcry trading is still used for certain complex (options) trades and some commodity futures contracts such as those traded on the LME and Nymex. For recent developments in floor trading, see the Reuters article "Insight: Chicago pits going quiet, 165 years after shouting began," 5/8/2013, by Tom Polansek.

[7]Stocks are also sometimes referred to as *cash* equities to distinguish them from equity derivatives.

2.2.2.1 *Futures and options exchanges*

CME Group (including CME, CBOT, NYMEX, COMEX), ICE (including ICE futures, Liffe futures, NYSE and Euronext), EUREX, National Stock Exchange of India, and BM/FBovespa are some of the major futures and options exchanges in the world.[8] The exchange provides three important services:

1. A physical location (i.e., the *trading floor* and the accompanying *pits*) for such activity, if it is a historical open-outcry system. Otherwise, the exchange will supply an electronic trading platform (such as CME's Globex). Such platforms use the internet as the underlying network which implies that the location becomes less relevant. Many electronic trading platforms provide Application Programming Interfaces (APIs) that facilitate algorithmic trading.

2. An exchange *clearing house* that becomes the real counterparty to each buyer and seller once the trade is done. A clearing house stands between two clearing firms (also known as *member* firms or *clearing* participants) and its purpose is to reduce the risk of one (or more) clearing firm failing to honor its trade settlement obligations. Why does central clearing reduce risk compared to bilateral trading? In a bilateral trade, two parties A and B enter into an agreement to exchange financial flows.[9] Each of the parties bears counterparty risk or the risk that the other party defaults and is unable to fulfill its obligations. In order to reduce such risks a central counterparty (CCP) approach can be used. The clearing house becomes the CCP by splitting the bilateral trade into two trades involving the CCP and party A and the CCP and party B, respectively. The offsetting positions imply that the CCP does not have any market risk but it faces counterparty risk. A clearing house addresses the counterparty risk by requiring collateral deposits or *margin* payments from each of the counterparties. It reduces settlement risks by netting offsetting transactions between multiple counterparties. Other functions carried out by the clearning house that reduce risk involve providing independent valuation of trades and collateral, monitoring the credit worthiness of the clearing firms, oftentimes, providing a guarantee fund that can be used to cover losses that exceed a defaulting clearing firm's collateral on deposit.

3. The service of *creating* and *designing* financial contracts that the trading community needs and, finally, providing a transparent and reliable trading environment.

The mechanics of trading in futures (options) exchanges is as follows. Two *traders* trade directly with each other according to their client's wishes. One sells, say, at 100; the other buys at 100. Then the *deal ticket* is signed and stamped.[10] Until that moment, the two traders are each other's counterparties. But once the deal ticket is stamped, the clearinghouse takes over as the counterparty. For example, if a client has bought a futures contract for the delivery of 100 bushels

---

[8]See FIA Annual Survey 2013, http://www.futuresindustry.org/files/css/magazineArticles/article-1612.pdf.

[9]The agreement will specify terms and conditions regarding early termination of transactions, settlement, and collateral calculations (in the Credit Support Annex (CSA)). The ISDA Master agreement is typically used for this purpose. See http://www.isda.org/educat/faqs.html, for more details.

[10]A deal ticket is similar to a trading receipt. It records key information regarding a trade agreement (such as price, volume, names, and dates of a transaction, along with all other important information). Deal tickets are important for internal control systems since they represent the transaction history. Deal tickets can be kept in either electronic or physical form.

of wheat, then the entity eventually responsible (they have agents) for delivering the wheat is not the "other side" who physically sold the contract on the pit, but the exchange clearinghouse. By being the *only* counterparty to all short and long positions, the clearinghouse will lower the counter-party risk dramatically.[11] The counterparty risk is actually reduced further, since the clearinghouse will deal with *clearing members*, rather than the traders directly.[12] The *open interest* in futures exchanges is the number of outstanding futures contracts. It is obtained by totaling the number of short or long positions that have not yet been closed out by *delivery*, *cash settlement*, or *offsetting* long/short positions.

2.2.2.2 Futures compared with forward contracts

Forwards are OTC contracts, designed according to the needs of the clients and negotiated between two counterparties. They are easy to price and *almost* costless to purchase. Futures are different from forward contracts in this respect. Some of the differences are minor; others are more impor-tant, leading potentially to significantly different forward and futures prices on the same underlying asset with identical characteristics. Most of these differences come from the design of futures con-tracts. Futures contracts need to be homogeneous to increase liquidity. The way they *expire* and the way *deliveries* are made will be clearly specified, but will still leave some options to the players. Forward contracts are initiated between two specific parties. They can state exactly the delivery and expiration conditions. Futures, on the other hand, will leave some room for last-minute adjust-ments and these "options" may have market value.

We consider two contracts in order to review the main parameters involved in the design of a futures and contrast it with forward contracts. The key point is that most aspects of the transaction need to be pinned down to make a homogeneous and liquid contract. This is relatively easy and straightforward to accomplish in the case of a relatively standard commodity such as soybeans. Consider the following soybeans futures contract.[13]

EXAMPLE: *CBOT (CME GROUP) SOYBEANS FUTURES*

1. **Contract size.** 5000 bushels (~136 metric tons).
2. **Deliverable grades.** No. 2 yellow at par, No. 1 yellow at 6 cents per bushel over contract price, and No. 3 yellow at 6 cents per bushel under contract price. (Note that in case a trader accepts the delivery, a special type of soybeans will be delivered to him or her. The trader may, in fact, procure the same quantity under better conditions from someone else. Hence, with a large majority of cases, futures contracts do not end with delivery. Instead, the position is unwound with an opposite transaction sometime before expiration.)
3. **Tick size.** Quarter-cent/bushel ($12.50/contract).
4. **Price quote.** Cents and quarter-cent/bushel.

---

[11]We discuss the role of counter-party risk in financial engineering in detail in Chapter 24.

[12]In order to be able to trade, a pit trader needs to "open an account" with a clearing member, a private financial company that acts as a clearing firm that deals with the clearinghouse directly on behalf of the trader.

[13]The parameters of futures contracts are sometimes revised by Exchanges; hence the reader should consider the information provided here simply as examples and check the actual contract for specifications.

5. **Contract months.** January, March, May, July, August, September, and November. (Clearly, if the purpose behind a futures transaction is delivery, then forward contracts with more flexible delivery dates will be more convenient.) On June 23, 2014, the *most distant* contract month available was November 2015.

6. **Last trading day.** The business day prior to the 15th calendar day of the contract month.

7. **Last delivery day.** Second business day following the last trading day of the delivery month.

8. **Trading hours.** Open outcry: 9:30 a.m. to 1:15 p.m. Chicago time, Monday through Friday. Electronic, 8:30 p.m. to 6:00 a.m. Chicago time, Sunday through Friday. Trading in expiring contracts closes at noon on the last trading day.

9. **Daily price limit.** 50 cents/bushel ($2500/contract) above or below the previous day's settlement price. No limit in the spot month. (Limits are lifted two business days before the spot month begins.)

An important concept that needs to be reviewed concerning futures markets is the process of *marking to market.* When one "buys" a futures contract, a margin is put aside, but no cash payment is made. This leverage greatly increases the liquidity in futures markets, but it is also risky. To make sure that counterparties realize their gains and losses daily, the exchange will reevaluate positions every day using the settlement price observed at the end of the trading day.[14]

EXAMPLE

A soybeans futures contract has a price of $1470.2 on day T. At the end of day $T + 1$, the settlement price is announced as $1469.4. The price fell by 80 cents, and this is a loss to the long position holder. The position will be marked to market, and the clearinghouse—or more correctly the clearing firm—will lower the client's balance by the corresponding amount.

The above example illustrates one of the differences between futures and forwards. Futures contracts are always marked to market, whereas this requirement does not apply to all forwards, as we explain below.

We consider the cash flows generated by a futures contract and compare them with the cash flows on a forward contract on the same underlying. It turns out that, unlike forwards, the effective maturity of a futures position is, in fact, *1 day*. This is due to the existence of marking to market in futures trading. The position will be marked to market in the sense that every night the exchange will, in effect, close the position and then reopen it at the new settlement price. It is best to look at this with a precise example. Suppose a futures contract is written on one unit of a commodity with spot price S_t. Let F_t be the *futures price* quoted in the exchange. Suppose t is a

---

[14]The settlement price is decided by the exchange and is not necessarily the last trading price.

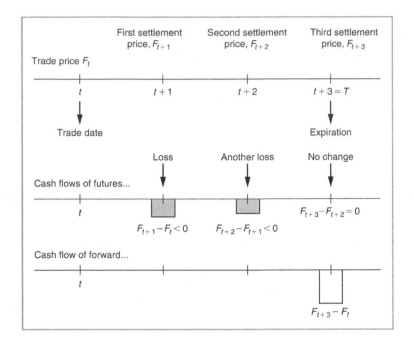

FIGURE 2.1

Marking to market.

Tuesday (March 11, 2014, for example), and that the expiration of the contract is within *3* trading days, that is on Friday, March 14, 2014:

$$T = t + 3 \tag{2.1}$$

where T = March 11, 2014 and t = March 14, 2014. Suppose further that during these days, the *settlement prices* follow the trajectory

$$F_t(=\$1470.2) > F_{t+1}(=\$1469.4) > F_{t+2}(=\$1468.4) = F_{t+3}(=\$1468.4) \tag{2.2}$$

where t = March 11, 2014, $t + 1$ = March 12, 2014, $t + 2$ = March 13, 2014, and $t + 3$ = March 14, 2014. What cash flows will be generated by a *long* position in one futures contract if at expiration date T the position is closed by taking the offsetting position?[15]

The answer is shown in Figure 2.1. Marking to market is equivalent to forcing the long (short) position holder to close his position at that day's settlement price and reopen it again at the same price. Thus, at the end of the first trading day after the trade, the futures contract that was "purchased" at f_t will be "sold" at the f_{t+1} shown in Eq. (2.2) for a loss:

$$F_{t+1} - F_t = 1469.4 - 1470.2 = -0.8 < 0 \tag{2.3}$$

---

[15]Instead of taking the offsetting position and canceling out any obligations with respect to the clearinghouse, the trader could choose to accept delivery.

Similarly, at the end of the second trading day, marking to market will lead to another loss:

$$F_{t+2} - F_{t+1} = 1468.4 - 1469.4 = -1.0 < 0 \tag{2.4}$$

This is the case since, according to trajectory in Eq. (2.2), prices have declined again. The expiration date will see no further losses, since, by chance, the final settlement price is the same as the previous day's settlement.

In contrast, the last portion of Figure 2.1 shows the cash flows generated by the forward prices F_t. Since there is no marking to market (in this case), the only capital loss occurs at the expiration of the contract. Clearly, this is a very different cash flow pattern.[16] Marking to market may significantly alter the implied cash flows and result in some moderate *convexities*.

Interest rate futures are some of the most liquid futures. The following example illustrates the typical quoting convention for interest futures. The futures price is quoted as 100 minus the interest rate. This implies that a falling interest rate futures price indicates a rise in the underlying interest rate.

EXAMPLE: *LIFFE 3-MONTH EURIBOR INTEREST RATE FUTURES*

1. **Unit of trading.** €1,000,000.
2. **Delivery months.** January, February, March, April, May, June, July, August, September, October, November, and December. December 2019 is the last contract month available for trading, as of April 24, 2014.
3. **Price quotes.** 100 minus rate of interest. (Based on the European Bankers Federations' Euribor Offered Rate (EBF Euribor) for 3-month Euro deposits at 11.00 Brussels time (10:00 London time) on the Last Trading Day. The settlement price will be 100.00 minus the EBF Euribor Offered Rate rounded to three decimal places.)
4. **Minimum price movement.** (Tick size and value) 0.005 (€12.50).
5. **Last trading day.** 10.00—Two business days prior to the third Wednesday of the delivery month.
6. **Delivery day.** First business day after the last trading day.
7. **Trading hours.** 01:00—06:45, 07:00—21:00.
8. **Clearing.** ICE Clear Europe Limited.

Eurocurrency futures contracts will be discussed in the next chapter and will be revisited several times later. In particular, one aspect of the contract that has not been listed among the parameters noted here has interesting financial engineering implications. Eurocurrency futures have a *quotation convention* that implies a *linear* relationship between the forward interest rate and the price of the futures contract. This is another example of the fact that conventions are indeed important in finding the right solution to a financial engineering problem.

---

[16]For details about differences in collateral and margin requirements for futures and forwards, see the CME document, Clearing and Bookkeeping Processing for Forwards, 18/9/2013, http://www.cmegroup.com/clearing/files/Clearing-Forwards.pdf.

Although, for simplicity, we assume in some of the financial engineering applications in this book that forward contracts are costless to enter, this is not the case in reality since forward contracts and other OTC derivatives require the posting of collateral. However, there are some subtle differences in collateral requirements between forward and futures contracts. First, according to the latest BCBS-IOSCO guidelines, financial and nonfinancial counterparties will be required to exchange variation margin (VM) for all new contracts entered into after December 1, 2015, and two-way IM in several phases starting in December 2015.[17] This requirement also applies to derivatives that are not centrally cleared, including certain forward contracts. There are exemptions for physically settled FX OTC derivatives which require VM, but not IM.

Second, there are differences in interest payments on the collateral in OTC and exchange-traded markets. Exchange-traded futures are settled daily and requirement daily margin that does not earn interest. If OTC forwards are cleared through a CCP or bilaterally, interest is paid on the VM if it is held in the form of cash.

2.2.3 CHANGES TO THE INFRASTRUCTURE OF DERIVATIVES MARKETS FOLLOWING THE GFC

Prior to the financial crisis, many derivatives contracts such as IRSs were mostly traded OTC and were often not centrally cleared. Oftentimes, OTC derivative contracts were not standardized and trade repositories did not exist for contracts such as credit default swaps, for example. In the years following, the GFC regulation has been implemented that saw a move towards central clearing, exchange trading, standardization and trade repositories for an increasing number of derivative contracts.

In September 2009, the G20 set an objective of introducing mandatory clearing for standardized derivatives, including OTC IRSs. The United States (through Dodd–Frank), the European Union (through EMIR), and other jurisdictions in the G20 developed and implemented regulations to achieve this objective. US regulations are driving reporting, clearing and settlement functions to much more tightly regulated Swap Execution Facilities (SEFs).[18] LCH.Clearnet cleared $282.6 trillion in notional amount of IRSs and $446.1 billion in cleared nondeliverable FX forwards volume in 2013. ICE Clear Credit's annual cleared CDS volume rose to $7.7 trilllion in 2013.[19]

In Europe, the derivatives market infrastructure is defined and supervised by the European Securities and Markets Authority (ESMA) under the European Markets Infrastructure Regulation (EMIR) regulation. Following the Dodd–Frank Act, the US Commodity Futures Trading Commission (CFTC) has developed the regulation, under which Swap Data Repositories are regulated.

The hope is that recent regulation and moving most privately negotiated, bilateral contracts in the OTC market to central clearing and requiring real-time trade reporting will allow regulators to see the volume of contracts trading in the market, to assess the trading activity in different asset classes, to monitor derivatives trading data and thus, to oversee risk exposures and reduce systemic

---

[17]See "The Future: Margin Requirements for Uncleared Derivatives" Citi OpenInvestor report 01/2014.

[18]As of April 2014, four categories of interest rate swaps and two categories of credit default swap are currently subject to CFTC clearing mandates.

[19]See http://www.futuresindustry.org/files/css/magazineArticles/article-1613.pdf.

risk in the derivatives market. From a financial engineering perspective, the effect of regulatory changes is of great importance. Regulation affects margin and funding costs of different derivatives instruments and strategies. It affects the profitability and permissibility of certain structured products and trades. In later chapters, we will discuss the effect of recent regulation on financial engineering practice in different asset classes.

2.3 PLAYERS

Market makers make markets by providing days to delivery, notice of delivery, warehouses, etc. Market makers provide liquidity and must, as an obligation, buy and sell at their quoted prices. Thus for every security at which they are making the market, the market maker must quote a bid and an ask price. A market maker does not warehouse a large number of products, nor does the market maker hold them for a long period of time. Different exchanges have different structures and use different approaches in liquidity provision or *market making*. For example, at the New York Stock Exchange (NYSE), market making is based on the Designated Market Maker (DMM) system.[20] DMMs have the primary responsibility of guaranteeing a fair and orderly market. Sometimes, this may involve taking the other side of trades when there are short-term buy-and-sell-side imbalances in customer orders. In return, the DMM is granted various informational and trade execution advantages.

Traders buy and sell securities. They do not, in the pure sense of the word, "make" the markets. A trader's role is to execute clients' orders and trade for the company given his or her *position limits*. Position limits can be imposed on the total capital the trader is allowed to trade or on the risks that he or she wishes to take.

A trader or market maker may *run* a portfolio, called a *book*. There are "FX books," "options books," "swap books," and "derivatives books," among others. Books run by traders are called "trading books"; they are different from "investment portfolios," which are held for the purpose of investment. Trading books exist because during the process of buying and selling for clients, the trader may have to warehouse these products for a short period of time. These books are hedged periodically. Hedge funds are an important type of arbitrageur in financial markets and we discuss hedge funds in detail in Chapter 7.

Brokers do not hold inventories. Instead, they provide a platform where the buyers and sellers can get together. Buying and selling through brokers is often more *discreet* than going to bids and asks of traders. In the latter case, the trader would naturally learn the identity of the client. In open-outcry options markets, a *floor-broker* is a trader who takes care of a client's order but does not trade for himself or herself. (On the other hand, a market maker does.)

Dealers quote two-way prices and hold large *inventories* of a particular instrument, maybe for a longer period of time than a market maker. They are institutions that act in some sense as market makers.

---

[20]DMMs were formerly known as *specialists* and run *books* on stocks that they specialize in.

Risk managers are relatively new players. Trades, and positions taken by traders, should be "approved" by risk managers. The risk manager assesses the trade and gives approvals if the trade remains within the preselected boundaries of various risk managers.

Regulators are important players in financial markets. Practitioners often take positions of "tax arbitrage" and "regulatory arbitrage." A large portion of financial engineering practices is directed toward meeting the needs of the practitioners in terms of regulation and taxation.

Researchers and analysts are players who do not trade or make the market. They are information providers for the institutions and are helpful in sell-side activity. Analysts in general deal with stocks and analyze one or more companies. They can issue buy/sell/hold signals and provide forecasts. Researchers provide macrolevel forecasting and advice.

2.4 THE MECHANICS OF DEALS

What are the mechanisms by which the deals are made? How are trades done? It turns out that organized exchanges have their own clearinghouses and their own clearing agents. So it is relatively easy to see how accounts are opened, how payments are made, how contracts are purchased, and positions are maintained. The clearing members and the clearinghouse do most of these. But how are these operations completed in the case of OTC deals? How does one buy a bond and pay for it? How does one buy a foreign currency?

Turning to another detail, *where* are these assets to be kept? An organized exchange will keep positions for the members, but who will be the custodian for OTC operations and secondary market deals in bonds and other relevant assets?

Several alternative mechanisms are in place to settle trades and keep the assets in custody. A typical mechanism is shown in Figure 2.2. The mechanics of a deal in Figure 2.2 are from the point of view of a market practitioner. The deal is initiated at the *trading* or dealing room. The trader writes the *deal ticket* and enters this information in the computer's front office system. The *middle office* is the part of the institution that initially verifies the deal. It is normally situated on the same floor as the trading room. Next, the deal goes to the *back office*, which is located either in a different building or on a different floor. Back office activity is as important for the bank as the trading room. The back office does the final verification of the deal, handles settlement instructions, releases payments, and checks the incoming cash flows, among other things. The back office will also handle the messaging activity using the SWIFT system, to be discussed later. Following the Global Financial Crisis, pre- and posttrade operational processes are increasingly merging to support new trading and clearing procedures required under financial regulation such as the Dodd−Frank Act and EMIR. The merging of pre- and posttrade processes is especially crucial to support the clearing certainty requirement. New requirements to execute certain OTC derivatives via SEFs, coupled with mandated clearing via a CCP means there is now a two-stage process to the execution of some derivatives. Execution and clearing via a CCP are now two intrinsic components of a derivatives trade, while in the past they were separate.

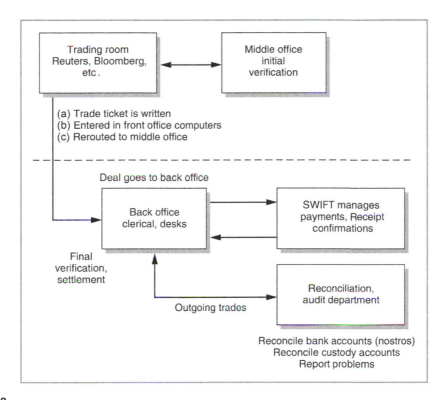

FIGURE 2.2

How trades are made and confirmed.

2.4.1 **ORDERS**

There are two general types of orders investors or traders can place. The first is a *market order*, where the client gets the price the market is quoting at that instant.

Alternatively, parties can place a *limit order*. Here a derived price will be specified along the order, and the trade will go through only if this or a better price is obtained. A limit order is valid only during a certain period, which needs to be specified also. A *stop loss* order is similar. It specifies a target price at which a position gets liquidated automatically.

Processing orders is by no means error free. For example, one disadvantage of traditional open-outcry exchanges is that in such an environment, mistakes are easily made. Buyer and seller may record different prices. This is called a "price out." Or there may be a "quantity out," where the buyer has "bought 100" while the seller thinks that he has "sold 50." In the case of options exchanges, the recorded expiration dates may not match, which is called a "time out." *Out-trades* need to be corrected after the market close. There can also be *missing trades*. These trades need to be negotiated in order to recover positions from counterparties and clients.[21] Electronic trading systems are also vulnerable to outages. During one such outage for agricultural contracts in April 2014, the CME temporarily resorted to open-outcry trading.

2.4.2 CONFIRMATION AND SETTLEMENT

Order confirmation and *settlement* are two integral parts of financial markets. Order confirmation involves sending messages between counterparties, to confirm trades verbally agreed upon between market practitioners. Settlement is exchanging the cash and the related security, or just exchanging securities.

The *SWIFT system* is a communication network that has been created for "paperless" communication between market participants to this end. It stands for the Society for Worldwide Financial Telecommunications and is owned by a group of international banks. The advantage of SWIFT is the *standardization* of messages concerning various transactions such as customer transfers, bank transfers, foreign exchange (FX), loans, and deposits. Thousands of financial institutions in more than 100 countries use this messaging system.

Another interesting issue is the relationship between settlement, clearing, and custody. *Settlement* means receiving the security and making the payment. The institutions can settle, but in order for the deal to be complete, it must be *cleared.* The orders of the two counterparties need to be matched and the deal terminated. *Custody* is the safekeeping of securities by depositing them with carefully selected depositories around the world. A *custodian* is an institution that provides custody services. Clearing and custody are both rather complicated tasks. *FedWire, Euroclear*, and *Clearstream* are three international *securities clearing firms* that also provide some custody services. Some of the most important custodians are banks.

Countries also have their own clearing systems. The best known bank clearing systems are CHIPS and CHAPS. CHAPS is the clearing system for the United Kingdom, CHIPS is the clearing system for payments in the United States. Payments in these systems are cleared multilaterally and are netted. This greatly simplifies settling large numbers of individual trades.

Spot trades settle according to the principle of DVP—that is to say, *delivery versus payment*—which means that first the security is delivered (to securities clearing firms) and *then* the cash is paid.

Issues related to settlement have another dimension. There are important conventions involving normal ways of settling deals in various markets. When a settlement is done according to the convention in that particular market, we say that the trade settles in a *regular way.* Of course, a trade can settle in a special way. But special methods would be costly and impractical.

EXAMPLE

Market practitioners denote the trade date by T, and settlement is expressed relative to this date.

---

[21]As an example, in the case of a "quantity out," the two counterparties may decide to split the difference.

US Treasury securities settle regularly on the first business day after the trade—that is to say, on $T + 1$. But it is also common for efficient clearing firms to have cash settlement—that is to say, settlement is done on the trade date T.

Corporate bonds and international bonds settle on $T + 3$.

Commercial paper (CP) settles the same day.

Spot transactions in stocks settle regularly on $T + 3$ in the United States.

Euromarket deposits are subject to $T + 2$ settlement. In the case of overnight borrowing and lending, counterparties may choose cash settlement.

FX markets settle regularly on $T + 2$. This means that a spot sale (purchase) of a foreign currency will lead to two-way flows 2 days after the trade date, regularly. $T + 2$ is usually called the spot date.

It is important to expect that the number of days to settlement in general refers to business days. This means that in order to be able to interpret $T + 2$ correctly, the market professional would need to pin down the corresponding *holiday convention*.

Before discussing other market conventions, we can mention two additional terms that are related to the preceding dates. The settlement date is sometimes called the *value date* in contracts. Cash changes hands at the value date. Finally, in swap-type contracts, there will be the *deal date* (i.e., when the contract is signed), but the swap may not begin until the *effective date*. The latter is the actual start date for the swap contract and will be at an agreed-upon later date.

2.4.2.1 Regulatory update following the GFC

On April 23, 2014, The Depository Trust & Clearing Corporation (DTCC) released a white paper outlining the rationale for supporting a move to shorten the settlement cycle (SSC) in the US financial markets for equities, corporate and municipal bonds, and unit investment trust (UIT) trades. The rationale for the reduction from $T + 3$ (trade date plus 3 days) to $T + 2$ (trade date plus 2 days) is that a shortened settlement cycle will reduce the following: counterparty risk exposure, procyclicality, NSCC's liquidity need, and overall operational and industry risk.[22]

2.5 MARKET CONVENTIONS

Market conventions often cause confusion in the study of financial engineering. Yet, it is *very* important to be aware of the conventions underlying the trades and the instruments. In this section, we briefly review some of these conventions.

---

[22]The reduction would enable funds to be freed up faster for reinvestment, thus, reducing credit and counterparty exposure. A shortened cycle would further reduce the liquidity requirement of DTCC's subsidiary, National Securities Clearing Corporation (NSCC), freeing up capital for broker dealers by reducing the NSCC Clearing Fund requirement. For details, see http://www.dtcc.com/~/media/Files/Downloads/WhitePapers/T2-Shortened-Cycle-WP.ashx.

Conventions vary according to the location and the type of instrument one is concerned with. Two instruments that are quite similar may be quoted in very different ways. *What* is quoted and *the way* it is quoted are important.

As mentioned, in Chapter 1 in financial markets there are always *two* prices. There is the price at which a market maker is willing to *buy* the underlying asset and the price at which he or she is willing to *sell* it. The price at which the market maker is willing to buy is called the *bid* price. The *ask* price is the price at which the market maker is willing to sell. In London financial markets, the ask price is called an *offer*. Thus, the bid-ask spread becomes the bid-offer spread.

As an example consider the case of deposits in London *money and FX markets*, where the convention is to quote the *asking* interest rate first. For example, a typical quote on interest rates would be as follows:

Ask (Offer)	Bid
$5\frac{1}{4}$	$5\frac{1}{8}$

In other money centers, interest rates are quoted the other way around. The first rate is the bid, the second is the ask rate. Hence, the same rates will look as such:

Bid	Ask (Offer)
$5\frac{1}{8}$	$5\frac{1}{4}$

A second characteristic of the quotes is decimalization. The Eurodollar interest rates in London are quoted to the nearest $\frac{1}{16}$ or sometimes $\frac{1}{32}$. But many money centers quote interest rates to *two* decimal points. Decimalization is not a completely straightforward issue from the point of view of brokers/dealers. Note that with decimalization, the bid-ask spreads may narrow all the way down to zero, and there will be no *minimum* bid-ask spread. This may mean lower trading profits, everything else being the same.

2.5.1 WHAT TO QUOTE

Another set of conventions concerns *what* to quote. For example, when a trader receives a call, he or she might say, "I sell a bond at a price of 95," or instead, he or she might say, "I sell a bond at yield 5%." Markets prefer to work with conventions to avoid potential misunderstandings and to economize time. Equity markets quote individual stock prices. On most stock exchanges including the NYSE, the quotes are to decimal points.

Most *bond markets* quote prices rather than yields, with the exception of short-term T-bills. For example, the price of a bond may be quoted as follows:

Bid Price	Ask (Offer) Price
90.45	90.57

The first quote is the price a market maker is willing to pay for a bond. The second is the price at which the market maker dealer is willing to sell the same bond. Note that according to this, bond prices are quoted to *two* decimal points, out of a par value of 100, regardless of the true denomination of the bond.

It is also possible that a market quotes neither a price nor a yield. For example, caps, floors, and swaptions often quote "volatility" directly. Swap markets prefer to quote the "spread" (in the case of USD swaps) or the swap rate itself (Euro-denominated swaps). The choice of what to quote is not a trivial matter. It affects pricing as well as risk management.

2.6 INSTRUMENTS

This section provides a list of the major instrument classes from the perspective of financial engineering. A course on markets and instruments along the lines of Hull (2014) is needed for a reasonable understanding.

The convention in financial markets is to divide these instruments according to the following sectors:

1. *Fixed income instruments.* These are interbank certificates of deposit (CDs), or deposits (depos), CP, banker's acceptances, and Treasury bill (T-bills). These are considered to be *money market* instruments.

   Bonds, notes, and floating rate notes (FRNs) are bond market instruments
2. *Equities.* These are various types of stock issued by public companies
3. *Currencies*
4. *Commodities*
5. *Derivatives*, the major classes of which are interest rate, equity, currency, and commodity derivatives
6. *Credit instruments*, which are mainly high-yield bonds, corporate bonds, credit derivatives, CDSs, and various guarantees that are early versions of the former
7. Structured products MBS, CDO, ABS.

We discuss these major classes of instruments from many angles in the chapters that follow.

2.7 POSITIONS

By buying or short-selling assets, one takes *positions*, and once a position is taken, one has *exposure* to various risks.

2.7.1 LONG AND SHORT POSITIONS

A *long* position is easier to understand because it conforms to the instincts of a newcomer to financial engineering. In our daily lives, we often "buy" things, we rarely "short" them. Hence, when we buy an item for cash and hold it in inventory, or when we sign a contract that obliges us to buy something at a future date, we will have a long position. We are long the "underlying instrument," and this means that we benefit when the value of the underlying asset increases.

A *short* position, on the other hand, is one where the market practitioner has sold an item without really owning it. For example, a client calls a bank and buys a particular bond. The bank may not have this particular bond on its books, but can still *sell* it. In the meantime, however, the bank has a short position.

2.7.1.1 Payoff diagrams

One can represent short and long positions using *payoff diagrams*. Figure 2.3 illustrates the long position from the point of view of an investor. The investor has savings of 100. The upward-sloping vertical line *OA* represents the value of the investor's position given the price of the security. Since its slope is $+1$, the price of the security P_0 will also be the value of the initial position. Starting from P_0 the price increases by ΔP; the gain will be equal to this change as well.

In particular, if the investor "buys" the asset when the price is 100 using his or her *own* savings, the net worth at that instant is represented by the vertical distance *OB*, which equals 100. A market professional, on the other hand, has no "money." So he or she has to *borrow* the OB (or the P_0) first and then buy the asset. This is called *funding* the long position.

This situation is shown in Figure 2.4. Note that the market professional's total *net* position amounts to zero at the time of the purchase, when $P_0 = 100$. In a sense, by first borrowing and then buying the asset, one "owns" not the asset but some *exposure*. If the asset price goes up, the position becomes profitable. If, on the other hand, the price declines, the position will show a loss.

Figure 2.5 shows a short position from a market practitioner's point of view. Here the situation is simpler. The asset in the short position is borrowed anyway at $P_0 = 100$. Hence, when the price

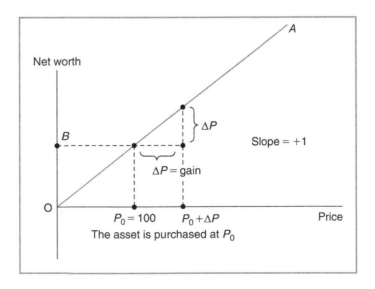

FIGURE 2.3

Long position of a market professional.

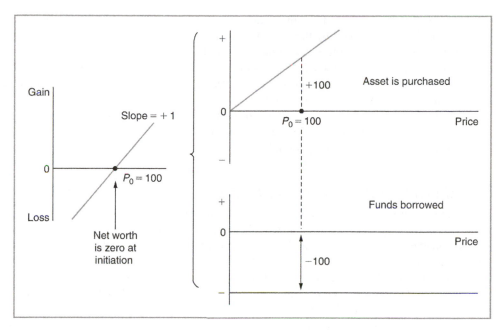

FIGURE 2.4

Funding a long position.

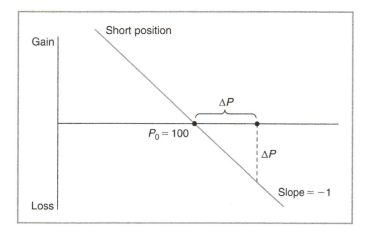

FIGURE 2.5

Short position of a market professional.

is 100 at the time of the sale, the net worth is automatically zero. What was sold was an asset that was worth 100. The cash generated by the sale just equals the value of the asset that was borrowed. Therefore, at the price $P_0 = 100$, the position has zero value. The position will gain when the price *falls* and will lose when the price goes *up*. This is the case since what is borrowed is a security and not "money." The asset is sold at 100; and, when P increases, one would have to return to the original owner a security that is worth *more* than 100.

Similarly, when P falls, one *covers* the short position by buying a new security at a price lower than 100 and then returning this (less valuable) asset to the original owner. Overall, the short position is described by a downward sloping straight line with slope -1.

2.7.1.2 Real-world complications and short selling

When we discuss synthetics and replicating portfolios for derivatives in this book, for simplicity, we assume that borrowing costs associated with shorting the underlying assets are negligible. This may be a reasonable assumption for liquid forward and futures markets, but it is not the case for other assets such as *cash* equities and bonds. Here, we briefly discuss real-world complications and the mechanics and costs of shorting stocks. Consider a hedge fund that would like to short a share such as Coca-Cola (KO), for example, in the hope of buying it back later at a lower price. The fund could contact a broker dealer such as Citibank, for example, who would normally *locate* the shares by borrowing them from a major custody bank or fund management company such as State Street, for example. The incentive for State Street to lend the shares lies in the fact that it could charge a fee for lending the shares. If the hedge fund was to borrow a share of KO and sell it for, say $41, it could however not keep the $41. The reason is that the security lender—State Street in this example—needs to protect itself against the potential inability of the hedge fund to return the share. This is achieved by asking for cash collateral. In an institutional US stock loan, the borrower is required to put up cash collateral, typically 102% of the value of the stock or $41.82 in the case of KO. The $0.82 represents the margin requirement. Hence, in contrast to stylized academic accounts, short selling in practice does not free up capital, but instead it uses capital. If the price of KO was to increase, to say $50, then the hedge fund would receive a *margin call* and would have to put up additional margin, of say $51. The opposite would happen if the stock price decreased.

Once the hedge fund returns the KO share, State Street returns the cash collateral plus interest. If Street can invest the cash collateral and a market interest rate that is higher than the *rebate rate*, that is interest rate paid to the hedge fund, this represents an additional return source for State Street. The difference between the rebate rate and market interest rates is the *loan fee* that the hedge fund has to pay. Securities lending fees for most stocks are small, that is around 1% per year in the United States, for most stocks, but sometimes they can be as high as 50% per year or even higher for individual stocks.[23] The market for securities lending is after all subject to the same laws of supply and demand as other markets and *hard to borrow* securities can only be borrowed at substantial cost. Most securities lending transactions are *open loans*. An open loan has an overnight tenor, but continues until one of the counterparties decides to cancel it. The short seller faces *recall risk* if the lender decides to recall or cancel the loan.

---

[23]The above example is just for illustration. For a thorough introduction to the securities lending market, see http://www.eseclending.com/pdfs/Data_Explorer_Intro_to_Sec_Lending.pdf.

In the above example, before the hedge fund closes the short position in KO it must buy back the share. Sometimes a lot of shortsellers try to do so at the same time and this can drive up the price and lead to a *short squeeze*, as the following recent example illustrates.

EXAMPLE

Great cornering and eye-popping acceleration make Porsche's cars popular among thrill-seeking bankers and hedge fund managers. Now its clients are discovering that the carmaker itself has an unexpected talent for cornering markets. [...]

[...] on October 26th it executed a handbrake turn, saying that it owned nearly 43% of VW's shares outright and had derivative contracts on nearly 32% more [...] Hedge funds quickly did the maths, concluding that they could be caught in an "infinite squeeze"

Their frenzied buying sent VW's share price soaring [...]. After languishing below €200 last year, it jumped to more than €1,005 at one point on October 28th [...]

The Economist, October 30, 2008

In the three years preceding October 2008 Porsche had secretly used derivatives to accumulate a stake in VW and repeatedly denied that it was doing so. In the process Porsche had accumulated most of the freely available shares of VW. The remainder of the shares was owned by the state government and passive funds which implied that hedge funds were short VW risked not being able to close their short positions in time. The above illustrates that shorting is not simply the opposite of going long. Not only are there borrowing fees, but the risk profile of shorting stocks is also different from that of going long. Significant borrowing costs also have implications for financial engineering and derivatives pricing. It can affect put-call parity and the early exercise of American options, which we will discuss in later chapters. The price discontinuities introduced by short squeezes make continuous hedging impossible and violate assumptions about the smooth stochastic processes that some financial engineering models rely on. Financial engineers and traders need to be aware of such model limitations in turbulent times and anticipate them in the pricing of products and execution of arbitrage strategies.

2.7.2 PAYOFF DIAGRAMS FOR FORWARDS AND FUTURES

Forwards and futures contracts are the most basic type of derivatives and involve some of the simplest cash flow exchanges. We discussed institutional differences between forwards and futures markets in Section 2.2. Next we extend our discussion of positions and payoff diagrams to forwards. A *forward* is a contract written at time t_0, with a commitment to accept delivery of (deliver) N units of the underlying asset at a future date t_1, $t_0 < t_1$, at the *forward price* F_{t_0}. At time t_0, nothing changes hands; all exchanges will take place at time t_1. The current price of the underlying asset S_{t_0} is called the spot price and is not written anywhere in the contract, instead, F_{t_0} is used during the settlement. Note that F_{t_0} has a t_0 subscript and is fixed at time t_0. An example of such a contract is shown in Figure 2.6.

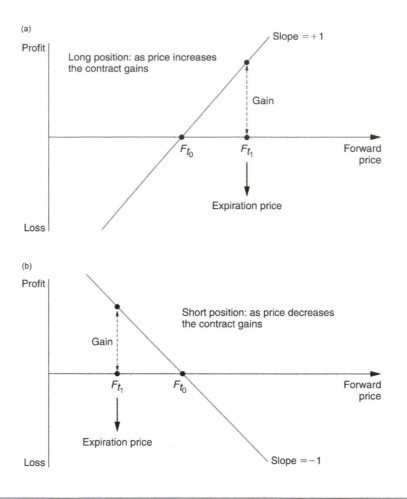

FIGURE 2.6

Long and short forward position.

Forward contracts are written between two parties, depending on the needs of the client. They are *flexible* instruments. The size of contract N, the expiration date t_1, and other conditions written in the contract can be adjusted in ways the two parties agree on.

To recap our earlier discussion: if the same forward purchase or sale is made through a *homogenized contract*, in which the size, expiration date, and other contract specifications are preset, if the trading is done in a *formal exchange*, if the counterparty risk is transferred to a clearinghouse, and if there is formal *mark-to-market*, then the instrument is called *futures*.

Positions on forward contracts are either *long* or *short*. A *long position* is a commitment to accept delivery of the contracted amount at a future date, t_1, at price F_{t_0}. This is displayed in Figure 2.6a. Here F_{t_0} is the contracted forward price. As time passes, the corresponding price on newly written contracts will change and at expiration the forward price becomes F_{t_1}. The difference, $F_{t_1} - F_{t_0}$, is the profit or loss for the position taker. Note two points. Because we assume, for simplicity, that the forward contract does not require any cash payment at initiation, the time t_0 value is *on* the x-axis. This implies that, at initiation, the market value of the contract is zero. Second, at time t_1 the spot price S_{t_1} and the forward price F_{t_1} will be the same (or very close).

A *short position* is a commitment to deliver the contracted amount at a future date, t_1, at the agreed price F_{t_0}. The short forward position is displayed in Figure 2.6b. The difference $F_{t_0} - F_{t_1}$ is the profit or loss for the party with the short position.

EXAMPLES

Elementary forwards and futures contracts exist on a broad array of underlyings. Some of the best known are the following:

1. Forwards on interest rates. These are called forward rate agreements (FRA). An FRA is an OTC contract designed to fix the interest rate that will apply to either borrowing or lending a certain amount during a specific time period in the future.
2. Forwards on *currencies*. These are called FX forwards and consist of buying (selling) one currency against another at a future date t_1.
3. Futures on *loans and deposits*. Here, a currency is exchanged against itself, but at a later date. We call these *forward loans* or *deposits*. Another term for these is *forward–forward*. Futures provide a more convenient way to trade interest rate commitments; hence, forward loans are not liquid. Futures on forward loans are among the most liquid.
4. Futures on commodities, e.g., be oil, corn, pork bellies, and gold. There is even a thriving market in futures trading on weather conditions.
5. Futures and forwards on individual stocks and *stock indices*. Given that one cannot settle a futures contract on an index by delivering the whole basket of stocks, these types of contracts are *cash settled*. The losers compensate the gainers in cash, instead of exchanging the underlying products.
6. Futures contracts on (interest rate) swaps. These are relatively recent and they consist of future swap rate commitments. They are also settled in cash. Compared to futures trading, the OTC swap and OTC forward market are more dominant here.[24]
7. Futures contracts on volatility indices.
8. Futures contracts on swaps.

It is interesting to note some technical aspects of the graphs in Figures 2.3−2.7. First, the payoff diagrams that indicate the value of the positions taken are *linear* in the price of the asset. As the price P changes, the payoff changes by a constant amount. The sensitivity of the position to price changes is called *delta*. In fact, given that the change in price will determine the gains or losses on a one-to-one basis, the *delta* of a long position will be 1. In the case of a short position, the *delta* will equal -1.

---

[24]Liquidity in futures on swaps may increase in the future if they are seen as having advantages over SEFs in terms of costs and reporting requirements. See "Buy-side welcomes swap futures as OTC market undergoes structural reform," HedgeWeek, 6/11/2013.

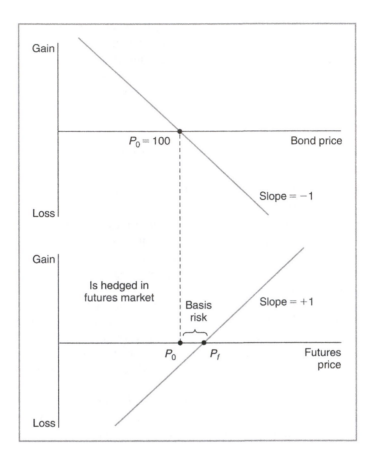

FIGURE 2.7

A hedge.

One can define many other sensitivity factors by taking other partial derivatives. Such sensitivities are called *Greeks* and are extensively used in option markets.[25] Financial derivatives with a delta of 1 have no optionality and are called Delta One products and the trading desks that pursue strategies related to them are called Delta One desks.[26]

Forwards and futures contracts are ideal for creating synthetic instruments for many reasons. Forwards and futures are, in general, linear permitting static replication. They are often very liquid and, in case of currency forwards, have homogeneous underlying. Many technical complications

---

[25]Note that bid-ask spreads are not factored in the previous diagrams. The selling and buying prices cannot be the *same* at 100. The selling price P^{ask} will be larger than the buying price P^{bid}. The $P^{ask} - P^{bid}$ will be the corresponding *bid-ask spread*. The original point is not zero but bid-ask.

[26]Two recent rogue trading scandals (those of Jérôme Kerviel at Société Générale in 2008 and Kweku Adoboli at UBS in 2011) involved Delta One traders.

are automatically eliminated by the homogeneity of a currency. Forwards and futures on interest rates present more difficulties and we will discuss them in the next chapter.

2.7.3 TYPES OF POSITIONS

Positions can be taken for the purposes of *hedging*, *arbitrage*, and *speculation*. We briefly review these activities.

Let us begin with *hedging*. Hedging is the act of eliminating the exposures of existing positions without unwinding the position itself. Suppose we are *short* a bond (i.e., we borrowed somebody's bond and sold it in the market for cash). We have cash at hand, but at the same time, we owe somebody a bond. This means that if the bond price goes *up*, our position will have a *mark-to-market* loss.

In order to eliminate the risk we can buy a "similar" bond. Our final position is shown in Figure 2.7. The long and short positions "cancel" each other except for some remaining *basis risk*.[27] At the end, we will have little exposure to movements in the underlying price P. To hedge the same risk we can also take the long position not in the *cash* or *spot* bond markets, but in a *futures* or *forward* market. That is to say, instead of buying another bond, we may write a contract at time t promising that we *will* buy the bond at a prespecified price P^f after δ days. This will not require any cash disbursement until the settlement time $t + \delta$ arrives, while yielding a gain or loss given the way the market prices move until that time. Here, the forward price P^f and the spot price P will not be identical. The underlying asset being the same, we can still anticipate quite *similar* profits and losses from the two positions. This illustrates one of the basic premises of financial engineering. Namely that as much as possible, one should operate by taking positions that do not require new funding.

The difference between the spot price of a given cash market asset and the price of its related futures contract, $P - P^f$, is called the *basis*.[28] The purchase of a particular financial instrument or commodity and the sale of its related derivative (e.g., the purchase of a particular bond and the sale of a related futures contract) may sometimes not be motivated by a hedging motive, but by speculation, in which case it is also called *basis trading*.

2.7.3.1 Arbitrage

The notion of arbitrage is central to financial engineering. It means two different things, depending on whether we look at it from the point of view of market *practice* or from the *theory* of financial engineering.

We begin with the definition used in the theory of financial engineering. In that context, we say that given a set of financial instruments and their prices, $\{P_1, P_2, \ldots, P_k\}$, there is no arbitrage opportunity if a portfolio that costs nothing to assemble now, with a nonnegative return in the future is ruled out. A portfolio with negative price and zero future return should not exist either.

If prices P_i have this characteristic, we say that they are *arbitrage free*. In a sense, arbitrage-free prices represent the *fair* market value of the underlying instruments. One should not realize

---

[27]Basis risk is the risk that offsetting investments in a hedging strategy will not experience price changes that are perfectly correlated.

[28]We discuss the basis in commodity markets in some detail in Chapter 7.

gains without taking some risk and without some initial investment. Many arguments in later chapters will be based on the no arbitrage principle.

In market practice, the term "arbitrage," or "arb," has a different meaning. In fact, "arb" represents a position that *has* risks, a position that *may* lose money but is still highly likely to yield a "high" profit.

2.7.3.2 Comparing performance

There are two terms that need to be defined carefully in order to understand the appendix to Chapter 12 and several examples. An asset A is set to *outperform* another asset B, if a long position in A and a simultaneous short position in B makes money. Otherwise A is said to *underperform B*. According to this, outperform indicates relative performance and is an important notion for spread trading.

2.8 THE SYNDICATION PROCESS

A discussion of the syndication process will be useful. Several contract design and pricing issues faced by a financial engineer may relate to the dynamics of the syndication process. Stocks, bonds, and other instruments are not sold to investors in the primary market the way, say, cars or food are sold. The selling process may take a few days or weeks and has its own wrinkles. The following gives an indicative time table for a syndication process.

2.8.1 SELLING SECURITIES IN THE PRIMARY MARKET

Time tables show variations from one instrument to another. Even in the same sector, the timing may be very different from one issuer to another, depending on the market psychology at that time. The process described gives an example. The example deals with a Eurobond issue. For *syndicated loans*, for facilities, and especially for IPOs, the process may be significantly different, although the basic ideas will be similar.

1. The week of D-14: Manager is chosen, *mandate* is given. Issue strategy is determined. Documentation begins.
2. The week of D-7: *Documentation* completed. Comanagers are determined.
3. D-Day: "Launch" date. Contacting *underwriter* and *selling group* members. Issue is published in the press.
4. D + 8: Preliminary allotment done by lead manager.
5. D + 9: Pricing day.
6. D + 10: Offering day. Allotment messages are sent to group members.
7. D + 24: Payment day. Syndicate members make payments.

In other markets, important deviations in terms of both timing and procedure may occur during actual syndication processes. But overall, the important steps of the process are as shown in this simple example.

2.8.1.1 Syndication of a bond versus a syndicated loan

We can compare a bond issue with processing a syndicated loan. There are some differences. Syndicated loans are instruments that are in *banking books* or *credit departments* of banks. The follow-up and risk management is done by the banking credit departments with methodologies similar to standard loans. For example, information in the offering circular is not as important.

Bonds, on the other hand, are handled by investment or in *trading books*, and the analysis and information in the circular are taken seriously. Documentation differences are major.

The syndicated loan tries to maintain a relationship between the bank and its client through the agent. But in the bond issue, the relationship between the lender and the borrower is much more distant. Hence, this type of borrowing is available only to good names with good credit standing. Banks have to continuously follow lesser names to stay aware of any deterioration of credit conditions. The maturities can also be very different.

2.9 CONCLUSIONS

This chapter reviewed some basic information the reader is assumed to have been exposed to. The discussion provided here is sketchy and cannot be a substitute for a thorough course on conventions, markets, and players. Also, market conventions, market structure, and the instrument characteristics may change over time.

SUGGESTED READING

It is important for a financial engineering student to know the underlying instruments, markets, and conventions well. This chapter provided only a very brief review of these issues. Fortunately, several excellent texts cover these further. **Hull** (2014) and **Wilmott** (2006) are first to come to mind. Market-oriented approaches to instruments, pricing, and some elementary financial market strategies can be found in **Steiner** (2012) and **Roth** (1996). These two sources are recommended as background material.

EXERCISES

1. What is the difference between initial and variation margin?

2. How do collateral requirements differ between futures and OTC derivative markets.

3. What are the differences between centrally and bilaterally cleared derivatives?

4. What does open interest refer to?

5. Suppose the following stock prices for GE and Honeywell were observed before any talk of merger between the two institutions:
   Honeywell (HON) 27.80
   General Electric (GE) 53.98

Also, suppose you "know" somehow that GE will offer 1.055 GE shares for each Honeywell share during any merger talks.

a. What type of "arbitrage" position would you take to benefit from this news?

b. Do you need to deposit any of your funds to take this position?

c. Do you need to and can you borrow funds for this position?

d. Is this a true arbitrage in the academic sense of the word?

e. What (if any) risks are you taking?

6.   Read the market example below and answer the following questions that relate to it.

Proprietary dealers are betting that Euribor, the proposed continental European-based euro money market rate, will fix above the Euro BBA LIBOR alternative... The arbitrage itself is relatively straightforward. The proprietary dealer buys the Liffe September 1999 Euromark contract and sells the Matif September 1999 Pibor contract at roughly net zero cost. As the Liffe contract will be referenced to Euro BBA LIBOR and the Matif contract will be indexed to Euribor, the trader in effect receives Euribor and pays Euro BBA LIBOR.

The strategy is based on the view that Euribor will generally set higher than Euro BBA LIBOR. Proprietary dealers last week argued that Euribor would be based on quotes from 57 different banks, some of which, they claimed, would have lower credit ratings than the eight LIBOR banks. In contrast, Euro BBA LIBOR will be calculated from quotes from just 16 institutions. (From Thomson Reuters IFR, December 18, 1998)

a. Show the positions of the proprietary dealers using position diagrams.

b. In particular, what is on the horizontal axis of these diagrams? What is on the vertical axis?

c. How would the profits of the "prop" dealers be affected at expiration, if in the meantime there was a dramatic lowering of all European interest rates due, say, to a sudden recession?

CASH FLOW ENGINEERING, INTEREST RATE FORWARDS AND FUTURES

CHAPTER OUTLINE

3.1 INTRODUCTION

All financial instruments can be visualized as bundles of cash flows. They are designed so that market participants can trade cash flows that have different *characteristics* and different *risks*. This chapter uses interest rate *forwards* and *futures* to discuss how cash flows can be replicated and then repackaged to create synthetic instruments.

It is easiest to determine replication strategies for linear instruments. We show that this can be further developed into an analytical methodology to create synthetic equivalents of complicated instruments as well. This analytical method will be summarized by a (contractual) equation. After plugging in the right instruments, the equation will yield the synthetic for the cash flow of interest. Throughout this chapter, we assume that there is *no default risk* and we discuss only *static replication methods*. Positions are taken and kept unchanged until expiration, and require no rebalancing. Dynamic replication methods will be discussed in Chapter 8.

This chapter develops the financial engineering methods that use forward loans, *forward rate agreements* (FRAs), and Eurocurrency futures. We first discuss these instruments and obtain contractual equations that can be manipulated usefully to produce other synthetics. The synthetics are used to provide pricing formulas.

3.2 WHAT IS A SYNTHETIC?

The notion of a synthetic instrument, or replicating portfolio, is central to financial engineering. We would like to understand how to price and hedge an instrument, and learn the risks associated with it. To do this, we consider the cash flows generated by an instrument during the lifetime of its contract. Then, using other *simpler, liquid* instruments, we form a portfolio that replicates these cash flows exactly. This is called a *replicating portfolio* and will be a synthetic of the original instrument. The constituents of the replicating portfolio will be easier to price, understand, and analyze than the original instrument.

In this chapter, we start with synthetics that can be discussed using *forwards* and *futures* and *money market products*. At the end, we obtain a *contractual equation* that can be algebraically manipulated to obtain solutions to practical financial engineering problems.

3.2.1 CASH FLOWS

We begin our discussion by defining a simple tool that plays an important role in the first part of this book. This tool is the graphical representation of a cash flow.

By a cash flow, we mean a payment or receipt of cash at a specific *time*, in a specific *currency*, with a certain *credit risk*. For example, consider the default-free cash flows in Figures 3.1 and 3.2. Such figures are used repeatedly in later chapters, so we will discuss them in detail.

EXAMPLE

In Figure 3.1 we show the cash flows generated by a default-free loan. Multiplying these cash flows by -1 converts them to cash flows of a deposit, or depo. In the figure, the horizontal axis represents time. There are two time periods of interest denoted by symbols t_0 and t_1. t_0 represents the time of a USD100 cash *in*flow. It is shown as a rectangle *above* the line. At time t_1, there is a cash *out*flow, since the rectangle is placed *below* the line and thus indicates a debit. Also note that the two cash flows have different sizes.

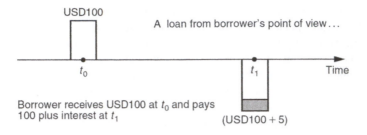

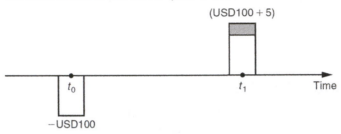

FIGURE 3.1

Cash flows generated by a default-free loan.

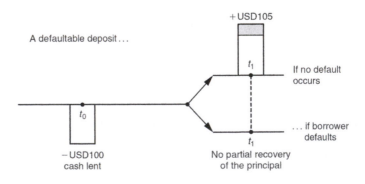

FIGURE 3.2

Cash flows generated by a risky loan.

We can interpret Figure 3.1 as cash flows that result when a market participant borrows USD100 at time t_0 and then pays this amount back with interest as USD105, where the interest rate applicable to period $[t_0, t_1]$ is 5% and where $t_1 - t_0 = 1$ year.

Every financial transaction has at least two counterparties. It is important to realize that the top portion of Figure 3.1 shows the cash flows from the borrower's point of view. Thus, if we look at the same instrument from the lender's point of view, we will see an inverted image of these cash flows. The lender lends USD100 at time t_0 and then receives the principal and interest at time t_1. The bid-ask spread suggests that the interest is the asking rate.

Finally, note that the cash flows shown in Figure 3.1 do not admit any uncertainty, since, both at time t_0 and time t_1 cash flows are represented by a single rectangle with known value. If there was uncertainty about either one, we would need to take this into account in the graph by considering different states of the world. For example, if there was a *default* possibility on the loan repayment, then the cash flows would be represented as in Figure 3.2. If the borrower defaulted, there would be no payment at all. At time t_1, there are *two* possibilities. The lender either receives USD105 or receives nothing.

Cash flows have special characteristics that can be viewed as *attributes*. At all points in time, there are market participants and businesses with different needs in terms of these attributes. They will exchange cash flows in order to reach desired objectives. This is done by trading financial contracts associated with different cash flow attributes. We now list the major types of cash flows with well-known attributes.

3.2.1.1 Cash flows in different currencies

The first set of instruments devised in the markets trade cash flows that are identical in every respect except for the *currency* they are expressed in.

In Figure 3.3, a decision maker pays USD100 at time t_0 and receives $100e_{t_0}$ units of Euro at the *same* time. This is a *spot FX deal*, since the transaction takes place at time t_0. The e_{t_0} is the *spot* exchange rate. It is the number of Euros paid for one USD.[1]

FIGURE 3.3

A spot FX deal.

---

[1]We will discuss financial engineering applications involving different currencies in Chapter 6.

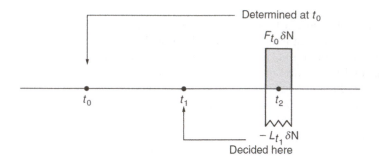

FIGURE 3.4

A swap.

3.2.1.2 Cash flows with different market risks

If cash flows with different market risk characteristics are exchanged, we obtain more complicated instruments than a spot FX transaction or deposit. Figure 3.4 shows an exchange of cash flows that depend on different market risks. The market practitioner makes a payment proportional to L_{t_1} percent of a *notional amount N* against a receipt of F_{t_0} percent of the same N. Here L_{t_1} is an unknown, *floating* LIBOR rate at time t_0 that will be learned at time t_1. F_{t_0}, on the other hand, is set at time t_0 and is a *forward* interest rate. The cash flows are exchanged at time t_2 and involve two different types of risk. Instruments that are used to exchange such risks are often referred to as *swaps*. They exchange a *floating* risk against a *fixed* risk. Swaps are not limited to interest rates. For example, a market participant may be willing to pay a floating (i.e., to be determined) oil price and receive a fixed oil price. One can design such *swaps* for all types of commodities.

3.2.1.3 Cash flows with different credit risks

The probability of default is different for each borrower. Exchanging cash flows with different credit risk characteristics leads to *credit instruments.*

In Figure 3.5, a counterparty makes a payment that is contingent on the *default* of a decision maker against the *guaranteed* receipt of a fee. Market participants may buy and sell such cash flows with different credit risk characteristics and thereby adjust their credit exposure. For example, AA-rated cash flows can be traded against BBB-rated cash flows.[2]

3.2.1.4 Cash flows with different volatilities

There are instruments that exchange cash flows with different volatility characteristics. Figure 3.6 shows the case of exchanging a fixed volatility at time t_2 against a realized (floating) volatility observed during the period, $[t_1, t_2]$. Such instruments are called *volatility swaps.*[3]

---

[2]Chapter 18 introduces financial engineering applications and derivatives involving credit risk.
[3]Recent developments regarding volatility derivatives are discussed in Chapter 16.

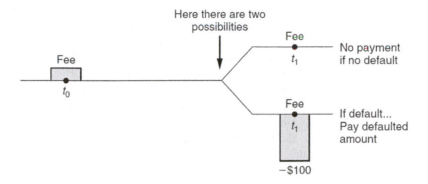

FIGURE 3.5

Example of payment contingent on default.

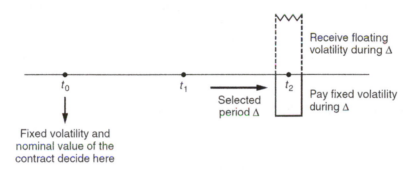

FIGURE 3.6

A volatility swap.

3.3 ENGINEERING SIMPLE INTEREST RATE DERIVATIVES

Forwards, *futures contracts*, and the underlying interbank *money markets* involve some of the simplest cash flow exchanges. They are ideal for creating synthetic instruments for many reasons. The *money market* refers to a segment of the financial market which is used by participants as a means for borrowing and lending in the short term. Money markets are very liquid and maturities range from several days to just under a year. We can contrast the money market with the *capital market* for longer-term funding, which is supplied by bonds and equity.

A forward or a futures contract can fix the future selling or buying price of an underlying item. This can be useful for hedging, arbitraging, and pricing purposes. They are essential in creating synthetics. Consider the following interpretation.

Foreign currency and commodity forwards (futures) are the simplest types of derivative instruments. Forwards and futures are, in general, linear permitting static replication. They are often very liquid and, in case of currency forwards, have homogeneous underlying. Many technical complications are automatically eliminated by the homogeneity of a currency.[4] Forwards and futures on interest rates present more difficulties, and we discuss them in this chapter. The chapter discusses financial engineering methods that use *forward loans, Eurocurrency futures*, and FRAs. The discussion prepares the ground for the next chapter on swap-based financial engineering. In fact, the FRA contracts considered here are precursors to plain vanilla swaps.

Interest rate strategies, hedging, and risk management present more difficulties than FX, equity, or commodity-related instruments for at least two reasons. First of all, the payoff of an interest rate derivative depends, by definition, on some interest rate(s). In order to price the instrument, one needs to apply discount factors to the future payoffs and calculate the relevant present values. But the discount factor itself is an interest rate-dependent concept. If interest rates are *stochastic*, the present value of an interest rate-dependent cash flow will be a nonlinear random variable; the resulting expectations *may* not be as easy to calculate. There will be *two sources* of any future fluctuations—those due to future cash flows themselves and those due to changes in the *discount factor* applied to these cash flows. When dealing with equity or commodity derivatives, such non-linearities are either not present or have a relatively minor impact on pricing. Second, every interest rate is associated with a maturity or *tenor*. This means that, in the case of interest rate derivatives we are not dealing with a single random variable, but with vector-valued stochastic processes. The existence of such a vector-valued random variable requires new methods of pricing, risk management, and strategic position taking. We start our discussion with interest rates and simple interest rate derivatives since interest rates are the foundation for pricing derivatives in other asset classes.

3.3.1 A CONVERGENCE TRADE

Before we start discussing replication of elementary interest rate derivatives we consider a real-life example.

For a number of years before the European currency (euro) was born, there was significant uncertainty as to which countries would be permitted to form the group of euro users. During this period, market practitioners put in place the so-called *convergence plays*. The reading that follows is one example.

EXAMPLE

Last week traders took positions on convergence at the periphery of Europe.

Traders sold the spread between the Italian and Spanish curves. JP Morgan urged its customers to buy a 12 × 24 Spanish FRA and sell a 12 × 24 Italian FRA. According to the bank, the spread, which traded at 133 bp would move down to below 50 bp (basis points).

The logic of these trades was that if Spain entered the single currency, then Italy would also do so. Recently, the Spanish curve has traded below the Italian curve. According to this logic, the Italian yield curve would converge on the Spanish yield curve, and traders would gain.

(Episode based on Thomson Reuters IFR issue number 1887).

---

[4]Foreign currency and commodity derivatives are discussed in Chapters 6 and 7, respectively.

Traders *buy* and *sell spreads* in order to benefit from a likely occurrence of an event. In this episode, these spreads are sold using the FRAs, which we discuss in this chapter. If the two currencies converge, the *difference* between Italian and Spanish interest rates will decline.[5] The FRA positions will benefit. Note that market professionals call this *selling* the spread. As the spread goes *down*, they will profit—hence, in a sense they are *short* the spread. The example makes references to the *yield curve*.

3.3.2 YIELD CURVE

The yield curve is a curve that plots several interest rates or *yields* across different maturities for a given borrower in a given currency. The maturity can range from several days to 50 years. The contract length of the debt is typically referred to as the *maturity* or the *term*. Therefore, the yield curve is also sometimes referred to as the term structure of interest rates. Figure 3.7 provides a real-world example. It plots the US dollar yield curve on July 13, 2014.[6] For comparison the figure also shows the yield curve on June 15, 2007, that is, before the GFC. Short-term and

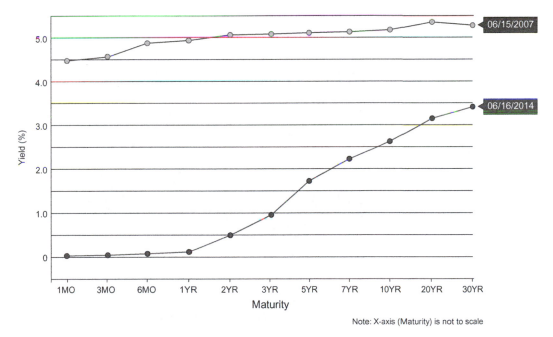

Note: X-axis (Maturity) is not to scale

FIGURE 3.7

US Treasury yield curve.

---

[5]Although each interest rate may go up or down individually.
[6]The chart and the underlying data are available from www.treasury.gov/resource-center/data-chart-center.

long-term yields exhibit significant variation over time and represent important risk factors for fixed income instruments.

There are several ways of calculating the yield curve.[7] One approach to the calculation of the yield curve is based on zero-coupon bonds. We discuss alternative approaches in the Appendix to this chapter. We can calculate a yield curve using zero-coupon bonds with par value 100 based on the following definition of the yield to maturity:

$$B(t,T) = \frac{100}{(1+y_t^T)^{T-t}} \tag{3.1}$$

where $B(t,T)$ is time t price of default-free zero-coupon bond that pays 100 at time T. The y_t^T will correspond to the $(T-t)$-maturity zero-coupon yield.

Broadly speaking, the yield curve can be calculated from prices available in the *bond market* or the *money market* and different methods lead to different results. Of course, there is no single yield curve describing the cost of money for every market participant since borrowing costs depend on the currency in which the securities are denominated as well as the creditworthiness of the borrower. Yield curves corresponding to the bonds issued by governments in their own currency are called government bond yield curves. Figure 3.8 shows Treasury yield curve for the United States, United Kingdom, Japan, and Germany on June 16, 2014. On that day, the spreads between yield

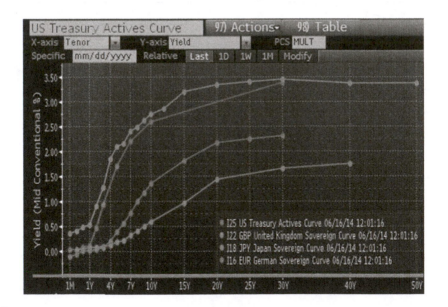

FIGURE 3.8

Treasury yield curves for US, Japan, and Germany on 16 June 2014.

---

[7]The methodology underlying the Treasury's yield curve represented in the above figure is detailed here http://www.treasury.gov/resource-center/data-chart-center/interest-rates/Pages/yieldmethod.aspx.

curves were as large as 200 bps in the case of the 10-year US and JPY treasuries. Yield curves based on money market prices may combine information on short-term LIBOR rates, futures prices, and interest rate swaps to plot the curve.

3.4 LIBOR AND OTHER BENCHMARKS

We first need to define the concept of LIBOR rates. The existence of such reliable *benchmarks* is essential for engineering interest rate instruments. Banks with high credit ratings borrow money from each other at the LIBOR rates. The LIBOR yield curves are typically a little higher than government curves and are known as the LIBOR curves or the swap curve. Figure 3.9 shows the USD LIBOR curve on June 13, 2014. It is apparent that on this day the LIBOR curve was above the US government bond yield curve. Later in this chapter, we discuss the construction of the yield curve from bond data, but first let's look at the precise definition of LIBOR.

LIBOR is an arithmetic average interest rate that measures the cost of borrowing from the point of view of a panel of preselected *contributor* banks in London. It stands for London Interbank Offered Rate. It is the ask or offer price of money available only to banks. It is an unsecured rate in the sense that the borrowing bank does not post any collateral. The ICE−LIBOR (formerly known as BBA LIBOR) is obtained by polling a panel of preselected banks in London.[8]

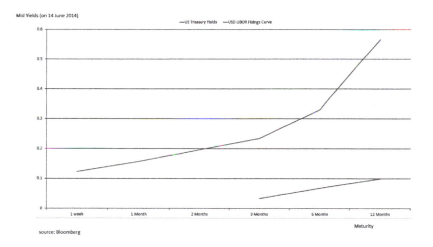

FIGURE 3.9

USD LIBOR versus Treasury curve.

[8]On February 1, 2014 ICE Benchmark Administration Limited took over the administration of LIBOR from the British Bankers Association (BBA). This followed the acquisition of NYSE Euronext by Intercontinental Exchange (ICE) Group on November 13, 2013. See https://www.theice.com/iba_libor.jhtml for further details.

LIBOR interest rates are published daily at 11:00 London time for nine currencies. In 2012, a scandal engulfed LIBOR as the US Department of Justice started a criminal investigation into the manipulation of LIBOR through communication between bankers before the rates were set. The investigations led to large fines by the US Commodity Futures Trading Commission (CFTC) and the UK Financial Services Authority (FSA) in 2012 and the investigations are continuing. The manipulation scandal in addition to evidence of understating of borrowing costs reported for LIBOR in 2008 has led to a revision in the LIBOR framework and a new code of conduct. To make it more likely that the submitted rates are underpinned by real trades, the number of rates has been reduced. As of July 2013, only five currencies and seven maturities are quoted every day (35 rates). This is a significant drop from the 150 different LIBOR rates (15 maturities for each of 10 currencies) which were quoted before the GFC. Euribor is a similar concept determined in Brussels by polling a panel of banks in continental Europe. These two benchmarks will obviously be quite similar. London banks and Frankfurt banks face similar risks and similar costs of funding. Hence they will lend Euros at approximately the same rate. But LIBOR and Euribor may have some slight differences due to the composition of the panels used.

The overnight indexed swap (OIS) rate is an increasingly important interest rate. The index rate is the rate for overnight unsecured lending between banks, for example, the Federal funds rate for US dollars, Eonia for Euros or Sonia for sterling. The LIBOR-OIS spread remained historically around 10 bps in the United States, but in 2007 and in 2011 during the GFC, the spread widened to 40−50 bps which led to a reconsideration of the LIBOR rate as a risk-free rate. Chapter 24 discusses the recent debate about the choice of LIBOR versus OIS as a discount rate. Because of lower counterparty risk, the fixed rate of OIS is considered less risky than the corresponding interbank rate (LIBOR).[9]

Eonia is the daily average of overnight rates for unsecured interbank lending in the euro-zone; in other words, the equivalent of the federal funds rate in the United States. The banks contributing to Eonia are the same as the Panel Banks contributing to Euribor. Important LIBOR maturities are overnight, 1 week, 1, 2, 3, 6, and 12 months. A plot of LIBOR rates against their maturities is called the LIBOR curve. Derivatives written on LIBOR are called LIBOR instruments. Using these derivatives and the underlying Euromarket loans, banks create LIBOR exposure. SHIBOR (Shanghai), TIBOR (Tokyo), and HIBOR (Hong Kong) are examples of other benchmarks that are used for the same purpose. When we use the term "interest rates" in this chapter, we often mean LIBOR rates or OIS rates.

3.5 FIXED INCOME MARKET CONVENTIONS

In the previous chapter, we saw that bond markets quote prices rather than yields. To avoid confusion in financial engineering, it is crucial to be aware of the conventions underlying fixed income trades and the instruments. In this section, we briefly review some of these market conventions.

---

[9]OIS is an interest rate swap where the periodic floating rate of the swap is equal to the geometric average of an overnight rate (or overnight index rate) over every day of the payment period.

3.5.1 HOW TO QUOTE YIELDS

Markets use three different ways to quote yields. These are, respectively, the money market yield, the bond equivalent yield, and the discount rate.[10] We will discuss these using *default-free pure discount bonds* with maturity T as an example. Let the time-t price of this bond be denoted by $B(t, T)$. The bond is default free and pays 100 at time T. Now, suppose R represents the time-t yield of this bond.

It is clear that $B(t, T)$ will be equal to the present value of 100, discounted using R, but how should this present value be *expressed?* For example, assuming that $(T - t)$ is measured in days and that this period is less than 1 year, we can use the following definition:

$$B(t,T) = \frac{100}{(1+R)^{((T-t)/365)}} \tag{3.2}$$

where $((T - t)/365)$ is the remaining life of the bond as a fraction of year, which here is "defined" as 365 days.

But we can also think of discounting using the alternative formula:

$$B(t,T) = \frac{100}{(1 + R((T - t)/365)} \tag{3.3}$$

Again, suppose we use neither formula but instead set

$$B(t,T) = 100 - R\left(\frac{T-t}{365}\right)100 \tag{3.4}$$

Some readers may think that given these formulas, Eq. (3.2) is the right one to use. But this is not correct! In fact, they may *all* be correct, given the proper *convention.*

The best way to see this is to consider a simple example. Suppose a market quotes prices $B(t, T)$ instead of the yields R.[11] Also suppose the observed market price is

$$B(t,T) = 95.00 \tag{3.5}$$

with $(T - t) = 180$ days and the year defined as 365 days. We can then ask the following question: Which one of the formulas in Eqs. (3.2)−(3.4) will be more correct to use? It turns out that these formulas can *all* yield the same price, 95.00, *if* we allow for the use of different Rs.

In fact, with $R_1 = 10.9613\%$ the first formula is "correct," since

$$B(t,T) = \frac{100}{(1+.109613)^{(180/365)}} \tag{3.6}$$

$$= 95.00 \tag{3.7}$$

On the other hand, with $R_2 = 10.6725\%$ the second formula is "correct," since

$$B(t,T) = \frac{100}{(1 + .106725(180/365))} \tag{3.8}$$

$$= 95.00 \tag{3.9}$$

---

[10]This latter term is different from the special interest rate used by the US Federal Reserve System, which carries the same name. Here the discount rate is used as a general category of yields.
[11]Emerging market bonds are in general quotes in terms of yields. In treasury markets, the quotes are in terms of prices. This may make some difference from the point of view of both market psychology, pricing, and risk management decisions.

Finally, if we let $R_3 = 10.1389\%$, the third formula will be "correct":

$$B(t, T) = 100 - .101389 \left(\frac{180}{365} \right) 100 \tag{3.10}$$

$$= 95.00 \tag{3.11}$$

Thus, for (slightly) different values of R_i, all formulas give the *same* price. But which one of these is the "right" formula to use?

That is exactly where the notion of *convention* comes in. A market can adopt a convention to quote yields in terms of formula (3.2), (3.3), or (3.4). Suppose formula (3.2) is adopted. Then, once traders see a quoted yield in this market, they would "know" that the yield is defined in terms of formula (3.2) and *not* by (3.3) or (3.4). This convention, which is only an implicit understanding during the execution of trades, will be expressed precisely in the actual contract and will be known by all traders. A newcomer to a market can make serious errors if he or she does not pay enough attention to such market conventions.

EXAMPLE

In the United States, bond markets quote the yields in terms of formula (3.2). Such values of R are called bond equivalent yields.

Money markets that deal with interbank deposits and loans use the *money market yield* convention and utilize formula (3.3) in pricing and risk management.

Finally, the Commercial Paper and Treasury Bills yields are quoted in terms of formula (3.4). Such yields are called *discount rates*.

Finally, the continuous discounting and the *continuously compounded* yield r is defined by the formula

$$B(t, T) = 100 e^{-r(T-t)} \tag{3.12}$$

where the e^x is the exponential function. It turns out that markets do not like to quote continuously compounded yields. One exception is toward retail customers. Some retail bank accounts quote the continuously compounded savings rate. On the other hand, the continuously compounded rate is often used in some theoretical models and was, until lately, the preferred concept for academics.

One final convention needs to be added at this point. Markets have an *interest payments convention* as well. For example, the offer side interest rate on major Euroloans, the LIBOR, is paid at the conclusion of the term of the loan as a *single* payment. We say that LIBOR is paid *in-arrears*. On the other hand, many bonds make periodic *coupon payments* that occur on dates earlier than the maturity of the relevant instrument. Most OISs have one payment if shorter than 1 year and a 12-month period for longer swaps.[12]

---

[12]For further details about fixed income market conventions see the "Interest Rate Instruments and Market Conventions Guide," December 2013 by OpenGamma available at http://developers.opengamma.com/quantitative-research/Interest-Rate-Instruments-and-Market-Conventions.pdf

3.5.2 DAY-COUNT CONVENTIONS

The previous discussion suggests that ignoring quotation conventions can lead to costly numerical errors in pricing and risk management. A similar comment can be made about *day-count* conventions. A financial engineer will always check the relevant day-count rules in the products that he or she is working on. The reason is simple. The definition of a "year" or of a "month" may change from one market to another and the quotes that one observes will depend on this convention. The major day-count conventions are as follows:

1. *The 30/360 basis.* Every month has 30 days regardless of the actual number of days in that particular month, and a year always has 360 days. For example, an instrument following this convention and purchased on May 1 and sold on July 13 would earn interest on

$$30 + 30 + 12 = 72 \tag{3.13}$$

   days, while the actual calendar would give 73 days.

   More interestingly, this instrument purchased on February 28, 2003, and sold the next day, on March 1, 2003, would earn interest for 3 days. Yet, a money market instrument such as an interbank deposit would have earned interest on only 1 day (using the actual/360 basis mentioned below).

2. *The 30E/360 basis.* This is similar to 30/360 except for a small difference at the end of the month, and it is used mainly in the Eurobond markets. The difference between 30/360 and 30E/360 is illustrated by the following table, which shows the number of days interest is earned starting from March 1 according to the two conventions:

Convention	March 1–March 30	March 1–March 31	March 1–April 1
30E/360	29 days	29 days	30 days
30/360	29 days	30 days	30 days

   According to this, a Eurobond purchased on March 1 and sold on March 31 gives an extra day of interest in the case of 30/360, whereas in the case of 30E/360, one needs to hold it until the beginning of the next month to get that extra interest.

3. *The actual/360 basis.* If an instrument is purchased on May 1 and sold on July 13, then it is held 73 days under this convention. This convention is used by most money markets. Actual/360 is also sometimes written as ACT/360, where ACT is an abbreviation for actual.

4. *The actual/365 basis.* This is the case for Eurosterling money markets, for example.

5. *Actual/actual.* Many bond markets use this convention.

An example will show why these day-count conventions are relevant for pricing and risk management. Suppose you are involved in an interest rate swap. You pay LIBOR and receive fixed. The market quotes the LIBOR at 5.01, and quotes the swap rate at 6.23/6.27. Since you are receiving fixed, the relevant cash flows will come from paying 5.01 and receiving 6.23 at regular intervals. But these numbers are somewhat misleading. It turns out that LIBOR is quoted on an ACT/360 basis. That is to say, the number 5.01 assumes that there are 360 days in a year.

However, the swap rates may be quoted on an ACT/365 basis, and all calculations may be based on a 365-day year.[13] Also the swap rate may be *annual* or *semiannual*. Thus, the two interest rates where one pays 5.01 and receives 6.23 are *not* directly comparable.

EXAMPLE

Swap markets are the largest among all financial markets, and the swap curve has become the central pricing and risk management tool in finance. Hence, it is worth discussing swap market conventions briefly.

- USD swaps are liquid against 3m-LIBOR and 6m-LIBOR. The day-count basis is annual, ACT/360.
- Japanese yen (JPY) swaps are liquid against 6m-LIBOR. The day-count basis is semi-annual, ACT/365.
- British pound (GBP) swaps are *semiannual*, ACT/365 versus 6m-LIBOR.
- Finally, Euro (EUR) swaps are liquid against 6m-LIBOR *and* against 6m-Euribor. The day-count basis is *annual* 30/360.

Table 3.1 summarizes the day-count and yield/discount conventions for some important markets around the world. A few comments are in order. First note that the table is a summary of three types of conventions. The first is the day-count, and this is often ACT/360. However, when the 30/360 convention is used, the 30E/360 version is more common. Second, the table tells us about the yield quotation convention. Third, we also have a list of coupon payment conventions concerning long-term bonds. Often, these involve semiannual coupon payments.[14]

Finally, note that the table also provides a list of the major instruments used in financial markets. The exact definitions of these will be given gradually in the following chapters.

3.5.2.1 Holiday conventions

Financial trading occurs across borders. But holidays adopted by various countries are always somewhat different. There are special independence days, special religious holidays. Often during Christmas time, different countries adopt different holiday schedules. In writing financial contracts, this simple point should also be taken into account, since we may not receive the cash we were counting on if our counterparty's markets were closed due to a special holiday in that country.

Hence, all financial contracts stipulate the particular holiday schedule to be used (London, New York, and so on), and then specify the date of the cash settlement if it falls on a holiday. This could be *the next business day* or the *previous business day*, or other arrangements could be made.

---

[13]Swaps are sometimes quoted on a 30/360 basis and at other times on an ACT/365 basis. One needs to check the confirmation ticket.

[14]To be more precise, day-count is a convention in measuring time. Properties like semiannual, quarterly, and so on are compounding frequency and would be part of yield quote convention.

Table 3.1 Day-Count and Yield/Discount Conventions

	Day-Count	Yield Cash
United States		
Depo/CD	ACT/360	Yield
T-Bill/CP/BA	ACT/360	Discount
Treasuries	ACT/ACT, semiannual	B-E yield
Repo	ACT/360	Yield
Euromarket		
Depo/CD/ECP	ACT/360 (ACT/365 for sterling)	Yield
Eurobond	30E/360	Yield
United Kingdom		
Depo/CD/Sterling CP	ACT/365	Yield
BA/T-bill	ACT/365	Discount
Gilt	ACT/365 (semiannual)	B-E yield
Repo	ACT/365	Yield
Germany		
Depo/CD/Sterling CP	ACT/360	Yield
Bund	30E/360 (annual)	B-E yield
Repo	ACT/360	Yield
Japan		
Depo/CD	ACT/365	Yield
Repo domestic	ACT/365	Yield
Repo international	ACT/360	Yield

3.5.3 TWO EXAMPLES

We consider how day-count conventions are used in two important cases. The first example summarizes the confirmation of short-term money market instruments, namely a Eurodollar deposit. The second example discusses the confirmation summary of a Eurobond. Eurocurrency markets and Eurobonds were introduced in Chapter 2.

EXAMPLE: *A EURODOLLAR DEPOSIT*

Amount	$100,000
Trade date	Tuesday, June 5, 2002
Settlement date	Thursday, June 7, 2002
Maturity	Friday, July 5, 2002
Days	30
Offer rate	4.789%
Interest earned	$(100,000) \times 0.04789 \times 30/360$

Note three important points. First, the depositor earns interest on the settlement date, but does not earn interest for the day the contract matures. This gives 30 days until maturity. Second, we are looking at the deal from the bank's side, where the bank sells a deposit, since the interest rate is the offer rate. Third, note that interest is calculated using the formula

$$(1 + r\delta)100,000 - 100,000 \tag{3.14}$$

and not according to

$$(1+r)^{\delta}100,000 - 100,000 \tag{3.15}$$

where

$$\delta = 30/360 \tag{3.16}$$

is the day-count adjustment.

The second example involves a Eurobond trade.

EXAMPLE: *A EUROBOND*

European Investment Bank, 5.0% (Annual Coupon)

Trade date	Tuesday, June 5, 2002
Settlement date	Monday, June 11, 2002
Maturity	December 28, 2006
Previous coupon	April 25, 2001
Next coupon	April 25, 2002
Days in coupon period	360
Accrued coupon	Calculate using money market yield

We have two comments concerning this example. The instrument is a Eurobond, and Eurobonds make coupon payments annually, rather than semiannually (as is the case of Treasuries, for example). Second, the Eurobond year is 360 days. Finally, accrued interest is calculated the same way as in money markets.

We can now define the major instruments that will be used. The first of these are the forward loans. These are not liquid, but they make a good starting point. We then move to FRAs and to Eurocurrency futures.

3.6 A CONTRACTUAL EQUATION

Once an instrument is replicated with a portfolio of other (liquid) assets, we can write a *contractual equation* and create new synthetics. In this section, we will obtain a contractual equation. In the next section, we will show several applications. This section provides a basic approach to constructing static replicating portfolios and hence is central to what will follow.

3.6.1 FORWARD LOAN

A *forward loan* is engineered like any forward contract, except that what is being bought or sold is not a currency or commodity, but instead, a *loan*. At time t_0, we write a contract that will settle at a future date t_1. At settlement, the trader receives (delivers) a loan that matures at t_2, $t_1 < t_2$. The contract will specify the interest rate that will apply to this loan. This interest rate is called the *forward rate* and will be denoted by $F(t_0, t_1, t_2)$. The forward rate is determined at t_0. t_1 is the start date of the future loan, and t_2 is the date at which the loan matures.

The situation is depicted in Figure 3.10. We write a contract at t_0 such that at a future date, t_1, USD100 are received; the principal and interest are paid at t_2. The interest is $F_{t_0}\delta$, where δ is the day-count adjustment, ACT/360:

$$\delta = \frac{t_2 - t_1}{360} \tag{3.17}$$

To simplify the notation, we abbreviate the $F(t_0, t_1, t_2)$ as F_{t_0}. The day-count convention needs to be adjusted if a year is defined as having 365 days.

Forward loans permit a great deal of flexibility in balance sheet, tax, and risk management. The need for forward loans arises under the following conditions:

- A business would like to *lock in* the "current" *low* borrowing rates from money markets.
- A bank would like to *lock in* the "current" *high* lending rates.
- A business may face a floating-rate liability at time t_1. The business may want to *hedge* this liability by securing a future loan with a known cost.

It is straightforward to see how forward loans help to accomplish these goals. With the forward loan of Figure 3.10, the party has agreed to receive 100 dollars at t_1 and to pay them back at t_2 with interest. The key point is that the interest rate on this forward loan is fixed at time t_0. The forward rate $F(t_0, t_1, t_2)$ "locks in" an unknown future variable at time t_0 and thus eliminates the risk associated with the unknown rate. The L_{t_1} is the LIBOR interest rate for a $(t_2 - t_1)$ period loan and can be observed only at the future date t_1. Fixing $F(t_0, t_1, t_2)$ will eliminate the risk associated with L_{t_1}.

The chapter discusses several examples involving the use of forward loans and their more recent counterparts, FRAs.

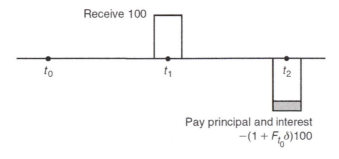

FIGURE 3.10

A forward loan.

3.6.2 REPLICATION OF A FORWARD LOAN

In this section, we apply the financial engineering outlined in Chapter 1 to forward loans and thereby obtain synthetics for this instrument. More than the synthetic itself, we are concerned with the methodology used in creating it. Although forward loans are not liquid and rarely traded in the markets, the synthetic will generate a contractual equation that will be useful for developing contractual equations for FRAs, and the latter *are* liquid instruments.

We begin the engineering of a synthetic forward loan by following the same strategy described in Chapter 1. We first decompose the forward loan cash flows into separate diagrams and then try to convert these into known liquid instruments by adding and subtracting appropriate new cash flows. This is done so that, when added together, the extra cash flows cancel each other out and the original instrument is recovered. Figure 3.11 displays the following steps:

1. We begin with the cash flow diagram for the forward loan shown in Figure 3.11a. We *detach* the two cash flows into separate diagrams. Note that at this stage, these cash flows cannot

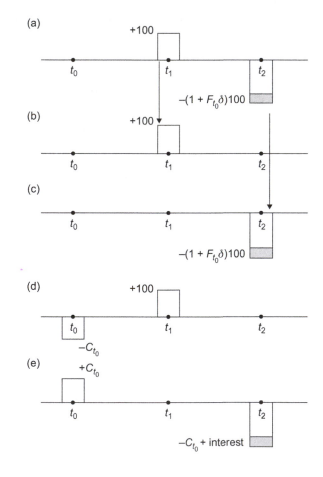

FIGURE 3.11

Replication of a forward loan using bonds.

form tradeable contracts. Nobody would want to buy 3.11c, and everybody would want to have 3.11b.

2. We need to transform these cash flows into tradeable contracts by adding compensating cash flows in each case. In Figure 3.11b, we add a *negative* cash flow, preferably at time t_0.[15] This is shown in Figure 3.11d. Denote the size of the cash flow by $-C_{t_0}$.

3. In Figure 3.11c, add a *positive* cash flow at time t_0, to obtain Figure 3.11e. The cash flow has size $+C_{t_0}$.

4. Make sure that the vertical addition of Figure 3.11d and e will replicate what we started with in Figure 3.11a. For this to be the case, the two newly added cash flows have to be identical in absolute value but different in sign. A vertical addition of Figure 3.11d and e will cancel any cash exchange at time t_0, and this is exactly what is needed to duplicate Figure 3.11a.[16]

At this point, the cash flows of Figure 3.11d and e need to be interpreted as specific financial contracts so that the components of the synthetic can be identified. There are many ways to do this. Depending on the interpretation, the synthetic will be constructed using different assets.

3.6.2.1 Bond market replication

As usual, we assume credit risk away. A first synthetic can be obtained using bond and T-bill markets. Although this is not the way preferred by practitioners, we will see that the logic is fundamental to financial engineering. Suppose default-free pure discount bonds of specific maturities denoted by $\{B(t_0, t_i), i = 1, \ldots, n\}$ trade actively.[17] They have par value of 100.

Then, within the context of a pure discount bond market, we can interpret the cash flows in Figure 3.11d as a *long* position in the t_1-maturity discount bond. The trader is paying C_{t_0} at time t_0 and receiving 100 at t_1. This means that

$$B(t_0, t_1) = C_{t_0} \tag{3.18}$$

Hence, the value of C_{t_0} can be determined if the bond price is known.

The synthetic for the forward loan will be fully described once we put a label on the cash flows in Figure 3.11e. What do these cash flows represent? These cash flows look like an appropriate *short* position in a t_2-maturity discount bond.

Does this mean we need to short *one* unit of the $B(t_0, t_2)$? The answer is no, since the time t_0 cash flow in Figure 3.11e has to equal C_{t_0}.[18] However, we know that a t_2-maturity bond will necessarily be cheaper than a t_1-maturity discount bond.

$$B(t_0, t_1) < B(t_0, t_1) = C_{t_0} \tag{3.19}$$

Thus, shorting *one* t_2-maturity discount bond will not generate sufficient time t_0 funding for the position in Figure 3.11d. The problem can easily be resolved, however, by shorting not one but λ bonds such that

$$\lambda B(t_0, t_2) = C_{t_0} \tag{3.20}$$

---

[15]Otherwise, if we add it at any other time, we get another forward loan.

[16]That is why both cash flows have size C_{t_0} and are of opposite sign.

[17]The $B(t_0, t_i)$ are also called default-free discount factors.

[18]Otherwise, time t_0 cash flows will not cancel out as we add the cash flows in Figure 3.10d and e vertically.

But we already know that $B(t_0, t_1) = C_{t_0}$. So the λ can be determined easily:

$$\lambda = \frac{B(t_0, t_1)}{B(t_0, t_2)} \tag{3.21}$$

According to Eq. (3.3) λ will be greater than one. This particular short position *will* generate enough cash for the long position in the t_1-maturity bond. Thus, we finalized the first synthetic for the forward loan:

$$\left\{ \text{Buy one } t_1\text{-discount bond, short } \frac{B(t_0, t_1)}{B(t_0, t_2)} \text{ units of the } t_2\text{-discount bond} \right\} \tag{3.22}$$

To double-check this result, we add Figure 3.11d and e vertically and recover the original cash flow for the forward loan in Figure 3.11a.

3.6.2.2 Pricing

If markets are liquid and there are no other transaction costs, arbitrage activity will make sure that the cash flows from the forward loan and from the replicating portfolio (synthetic) are the same. In other words, the sizes of the time t_2 cash flows in Figure 3.11a and e should be equal. This implies that

$$1 + F(t_0, t_1, t_2)\delta = \frac{B(t_0, t_1)}{B(t_0, t_2)} \tag{3.23}$$

where the δ is, as usual, the day-count adjustment.

This arbitrage relationship is of fundamental importance in financial engineering. Given liquid bond prices $\{B(t_0, t_1), B(t_0, t_1)\}$, we can price the forward loan *off* the bond markets using this equation. More important, equality (3.23) shows that there is a crucial relationship between forward rates at different maturities and discount bond prices. But discount bond prices are *discounts* which can be used in obtaining the present values of future cash flows. This means that forward rates are of primary importance in pricing and risk managing financial securities.

Before we consider a second synthetic for the forward loan, we prefer to discuss how all this relates to the notion of arbitrage.

3.6.2.3 Arbitrage

What happens when the equality in formula (3.23) breaks down? We analyze two cases assuming that there are no bid-ask spreads. First, suppose market quotes at time t_0 are such that

$$(1 + F_{t_0}\delta) > \frac{B(t_0, t_1)}{B(t_0, t_2)} \tag{3.24}$$

where the forward rate $F(t_0, t_1, t_2)$ is again abbreviated as F_{t_0}. Under these conditions, a market participant can secure a synthetic forward loan in bond markets at a cost below the return that could be obtained from lending in forward loan markets. This will guarantee positive arbitrage gains. This is the case since the "synthetic" *funding cost*, denoted by $F_{t_0}^*$,

$$F_{t_0}^* = \frac{B(t_0, t_1)}{\delta B(t_0, t_2)} - \frac{1}{\delta} \tag{3.25}$$

will be less than the forward rate, F_{t_0}. The position will be riskless if it is held until maturity date t_2.

These arbitrage gains can be secured by (i) shorting $B(t_0, t_1)/B(t_0, t_2)$ units of the t_2-bond, which generates $B(t_0, t_1)$ dollars at time t_0, then (ii) using these funds buying one t_1-maturity bond, and (iii) at time t_1 lending, at rate F_{t_0}, the 100 received from the maturing bond. As a result of these operations, at time t_2, the trader would owe $(B(t_0, t_1)/B(t_0, t_2))100$ and would receive $(1 + F_{t_0}\delta)100$.

The latter amount is greater, given the condition (3.24).

Now consider the second case. Suppose time t_0 markets quote:

$$(1 + F_{t_0}\delta) < \frac{B(t_0, t_1)}{B(t_0, t_2)} \qquad (3.26)$$

Then, one can take the reverse position. Buy $B(t_0, t_1)/B(t_0, t_2)$ units of the t_2-bond at time t_0. To fund this, short a $B(t_0, t_1)$ bond and borrow 100 forward. When time t_2 arrives, receive the $(B(t_0, t_1)/B(t_0, t_2))100$ and pay off the forward loan. This strategy can yield arbitrage profits since the funding cost during $[t_1, t_2]$ is lower than the return.

3.6.2.4 Money market replication

Now assume that all maturities of deposits up to 1 year are quoted actively in the interbank money market. Also assume there are no arbitrage opportunities. Figure 3.12 shows how an alternative synthetic can be created. The cash flows of a forward loan are replicated in Figure 3.12a. Figure 3.12c shows a Euromarket loan. C_{t_0} is borrowed at the interbank rate $L_{t_0}^2$.[19] The time t_2 cash flow in Figure 3.12c needs to be discounted using this rate. This gives

$$C_{t_0} = \frac{100(1 + F_{t_0}\delta)}{(1 + L_{t_0}^2 \delta^2)} \qquad (3.27)$$

where $\delta^2 = (t_2 - t_0)/360$.

Then, C_{t_0} is immediately redeposited at the rate $L_{t_0}^1$ at the shorter maturity. To obtain

$$C_{t_0}(1 + L_{t_0}^1 \delta^1) = 100 \qquad (3.28)$$

with $\delta^1 = (t_1 - t_0)/360$. This is shown in Figure 3.12b.

Adding Figure 3.12b and c vertically, we again recover the cash flows of the forward loan. Thus, the two Eurodeposits form a second synthetic for the forward loan.

3.6.2.5 Pricing

We can obtain another pricing equation using the money market replication. In Figure 3.12, if the credit risks are the same, the cash flows at time t_2 would be equal, as implied by Eq. (3.27). This can be written as

$$(1 + F_{t_0}\delta)100 = C_{t_0}(1 + L_{t_0}^2 \delta^2) \qquad (3.29)$$

where $\delta = (t_2 - t_1)/360$. We can substitute further from formula (3.28) to get the final pricing formula:

$$(1 + F_{t_0}\delta)100 = \frac{100(1 + L_{t_0}^2 \delta^2)}{(1 + L_{t_0}^1 \delta^1)} \qquad (3.30)$$

---

[19]Here the $L_{t_0}^2$ means the time t_0 LIBOR rate for a "cash" loan that matures at time t_2.

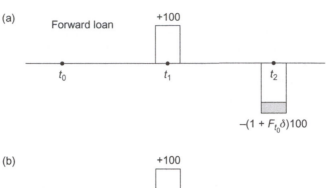

(a)

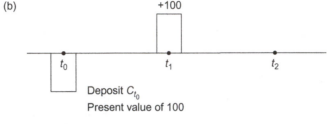

(b)

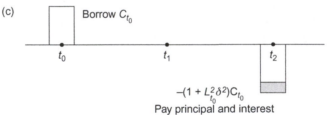

(c)

FIGURE 3.12

Money market replication of a forward loan.

Simplifying,

$$(1 + F_{t_0}\delta) = \frac{1 + L_{t_0}^2 \delta^2}{1 + L_{t_0}^1 \delta^1} \tag{3.31}$$

This formula prices the forward loan off the money markets. The formula also shows the important role played by *LIBOR* interest rates in determining the forward rates.

3.6.3 CONTRACTUAL EQUATIONS

We can turn these results into analytical contractual equations. Using the bond market replication, we obtain

$$
\boxed{\begin{array}{l}\text{Forward loan}\\\text{that begins at } t_1\\\text{and ends at } t_2\end{array}} = \boxed{\begin{array}{l}\text{Short } B(t_0, t_1)/\\B(t_0, t_2) \text{ units}\\\text{of } t_2 \text{ maturity}\\\text{bond}\end{array}} + \boxed{\begin{array}{l}\text{Long } t_1\text{-maturity}\\\text{bond}\end{array}} \tag{3.32}
$$

The expression shown in formula (3.32) is a *contractual equation*. The left-hand-side contract leads to the same cash flows generated *jointly* by the contracts on the right-hand side. This does not necessarily mean that the monetary *value* of the two sides is always the same. In fact, one or more of the contracts shown on the right-hand side may not even *exist* in that particular economy and the markets may not even have the opportunity to put a price on them.

Essentially the equation says that the risk and cash flow attributes of the two sides are the same. If there is no credit risk, no transaction costs, and if the markets in *all* the involved instruments are liquid, we expect that *arbitrage* will make the values of the two sides of the contractual equation equal.

If we use the money markets to construct the synthetic, the contractual equation above becomes

$$
\boxed{\begin{array}{c} \text{Forward loan} \\ \text{that begins at } t_1 \\ \text{and ends at } t_2 \end{array}} = \boxed{\begin{array}{c} \text{Loan with} \\ \text{maturity } t_2 \end{array}} + \boxed{\begin{array}{c} \text{Deposit with} \\ \text{maturity } t_1 \end{array}} \qquad (3.33)
$$

These contractual equations can be exploited for finding solutions to some routine problems encountered in financial markets although they do have drawbacks. Ignoring these for the time being we give some examples.

3.6.4 APPLICATIONS

Once a contractual equation for a forward loan is obtained, it can be algebraically manipulated to create further synthetics. We discuss two such applications in this section.

3.6.4.1 Application 1: creating a synthetic bond

Suppose a trader would like to buy a t_1-maturity bond at time t_0. The trader also wants this bond to be *liquid*. Unfortunately, he discovers that the only bond that is liquid is an *on-the-run* Treasury with a longer maturity of t_2. All other bonds are *off-the-run*.[20] How can the trader create the liquid short-term bond synthetically assuming that all bonds are of discount type and that, contrary to reality, forward loans are liquid?

Rearranging Eq. (3.32), we get

$$
\boxed{\begin{array}{c} \text{Long } t_1\text{-maturity} \\ \text{bond} \end{array}} = \boxed{\begin{array}{c} \text{Forward loan} \\ \text{from } t_1 \text{ to } t_2 \end{array}} - \boxed{\begin{array}{c} \text{Short } B(t_0, t_1)/ \\ B(t_0, t_2) \text{ units} \\ \text{of } t_2\text{-maturity bond} \end{array}} \qquad (3.34)
$$

The minus sign in front of a contract implies that we need to *reverse* the position. Doing this, we see that a t_1-maturity bond can be constructed synthetically by arranging a forward loan from t_1

---

[20]An on-the-run bond is a liquid bond that is used by traders for a given maturity. It is the latest issue at that maturity. An off-the-run bond has already ceased to have this function and is not liquid. It is kept in investors' portfolios.

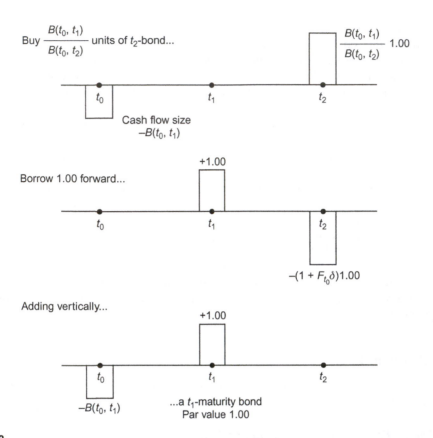

FIGURE 3.13

Constructing a bond synthetically.

to t_2 and then by going *long* $B(t_0, t_1)/B(t_0, t_2)$ units of the bond with maturity t_2. The resulting cash flows would be identical to those of a short bond. More important, if the forward loan and the long bond are liquid, then the synthetic will be more liquid than any existing off-the-run bonds with maturity t_1. This construction is shown in Figure 3.13.

3.6.4.2 Application 2: covering a mismatch

Consider a bank that has a *maturity mismatch* at time t_0. The bank has borrowed t_1-maturity funds from Euromarkets and lent them at maturity t_2. Clearly, the bank has to roll over the short-term loan that becomes due at time t_1 with a new loan covering the period $[t_1, t_2]$. This new loan carries an (unknown) interest rate L_{t_1} and creates a mismatch risk. The contractual equation in formula (3.33) can be used to determine a *hedge* for this mismatch, by creating a synthetic forward loan, and, in this fashion, locking in time t_1 funding costs.

In fact, we know from the contractual equation in formula (3.33) that there is a relationship between short and long maturity loans:

$$
\boxed{\begin{array}{c} t_2\text{-maturity} \\ \text{loan} \end{array}} = \boxed{\begin{array}{c} \text{Forward loan} \\ \text{from } t_1 \text{ to } t_2 \end{array}} - \boxed{\begin{array}{c} t_1\text{-maturity} \\ \text{deposit} \end{array}} \qquad (3.35)
$$

Changing signs, this becomes

$$
\boxed{\begin{array}{c} t_2\text{-maturity} \\ \text{loan} \end{array}} = \boxed{\begin{array}{c} \text{Forward loan} \\ \text{from } t_1 \text{ to } t_2 \end{array}} + \boxed{\begin{array}{c} t_1\text{-maturity} \\ \text{loan} \end{array}} \qquad (3.36)
$$

According to this, the forward loan converts the short loan into a longer maturity loan and in this way eliminates the mismatch.

3.7 FORWARD RATE AGREEMENTS

An interest rate FRA is an interest rate forward contract, an over-the-counter contract, between parties that specifies the rate of interest to be paid or received on an obligation beginning at a future start date. The next section develops a contractual agreement for an interest rate FRA. Currency FRAs are discussed in Chapter 6.

A forward loan contract implies not one but *two* obligations. First, 100 units of currency will have to be received at time t_1, and second, interest F_{t_0} has to be paid. One can see several drawbacks to such a contract:

1. The forward borrower may not necessarily want to receive cash at time t_1. In most hedging and arbitraging activities, the players are trying to *lock in* an unknown interest rate and are not necessarily in need of "cash." A case in point is the convergence play described in Section 3.2, where practitioners were receiving (future) Italian rates and paying (future) Spanish rates. In these strategies, the objective of the players was to *take a position* on Spanish and Italian interest rates. None of the parties involved had any wish to end up with a loan in 1 or 2 years.
2. A second drawback is that forward loan contracts involve *credit risk*. It is not a good idea to put a credit risk on a balance sheet if one wanted to lock in an interest rate.[21]
3. These attributes may make speculators and arbitrageurs stay away from any potential forward loan markets, and the contract may be *illiquid*.

These drawbacks make the forward loan contract a less-than-perfect financial engineering instrument. A good instrument would *separate* the credit risk and the interest rate commitment that coexist in the forward loan. It turns out that there is a nice way this can be done.

---

[21]Note that the forward loan in Figure 3.10 assumes the credit risk away.

3.7.1 ELIMINATING THE CREDIT RISK

First, note that a player using the forward loan *only* as a tool to lock in the future LIBOR rate L_{t_1} will immediately have to relend the USD100 received at time t_1 at the going market rate L_{t_1}. Figure 3.14a displays a forward loan committed at time t_0. Figure 3.14b shows the corresponding *spot* deposit. The practitioner waits until time t_1 and then makes a deposit at the rate L_{t_1}, which will be known at that time. This "swap" *cancels* an obligation to receive 100 and ends up with only the fixed rate F_{t_0} commitment.

Thus, the joint use of a forward loan, and a spot deposit *to be* made in the future, is sufficient to reach the desired objective—namely, to eliminate the risk associated with the unknown LIBOR rate L_{t_1}. These steps will lock in F_{t_0}. We consider the result of this strategy in Figure 3.14c. Add vertically the cash flows of the forward loan (3.14a) and the spot deposit (3.14b). Time t_1 cash flows cancel out since they are in the same currency. Time t_2 payment and receipt of the

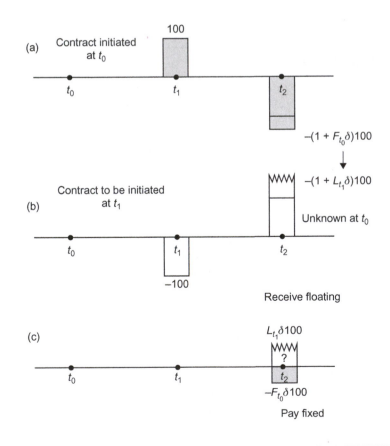

FIGURE 3.14

A synthetic FRA.

principal will also cancel. What is left is the respective interest payments. This means that the portfolio consisting of

$$\{\text{A forward loan for } t_1 \text{ initiated at } t_0, \text{ a spot deposit at } t_1\} \tag{3.37}$$

will lead, according to Figure 3.14c, to the following (net) cash flows:

	Cash Paid	Cash Received	Total
Time t_1	-100	$+100$	0
Time t_2	$-100(1 + F_{t_0}\delta)$	$100(1 + L_{t_1}\delta)$	$100(L_{t_1} - F_{t_0})\delta$

Thus, letting the principal of the forward loan be denoted by the parameter N, we see that the portfolio in expression (3.37) results in a time t_2 net cash flow equaling

$$N(L_{t_1} - F_{t_0})\delta \tag{3.38}$$

where δ is the day's adjustment to interest, as usual.

3.7.2 DEFINITION OF THE FRA

This is exactly where the FRA contract comes in. If a client has the objective of locking in the future borrowing or lending *costs* using the portfolio in Eq. (3.37), why not offer this to him or her in a *single* contract? This contract will involve *only* the exchange of two interest payments shown in Figure 3.14c.

In other words, we write a contract that specifies a notional amount, N, the dates t_1 and t_2, and the "price" F_{t_0}, with payoff $N(L_{t_1} - F_{t_0})\delta$.[22] This instrument is a *paid-in-arrears* FRA.[23] In an FRA contract, the *purchaser* accepts the receipt of the following sum at time t_2:

$$(L_{t_1} - F_{t_0})\delta N \tag{3.39}$$

if $L_{t_1} > F_{t_0}$ at date t_1. On the other hand, the purchaser pays

$$(F_{t_0} - L_{t_1})\delta N \tag{3.40}$$

if $L_{t_1} < F_{t_0}$ at date t_1. Thus, the buyer of the FRA will pay *fixed* and *receive* floating.

In the case of market-traded FRA contracts, there is one additional complication. The settlement is *not* done in-arrears at time t_2. Instead, FRAs are settled at time t_1, and the transaction will involve the following discounted cash flows. The

$$\frac{(L_{t_1} - F_{t_0})\delta N}{1 + L_{t_1}\delta} \tag{3.41}$$

will be received at time t_1, if $L_{t_1} > F_{t_0}$ at date t_1. On the other hand,

$$\frac{(F_{t_0} - L_{t_1})\delta N}{1 + L_{t_1}\delta} \tag{3.42}$$

---

[22] The N represents a *notional* principal since the principal amount will never be exchanged. However, it needs to be specified in order to determine the amount of interest to be exchanged.
[23] It is paid-in-arrears because the unknown interest, L_{t_1}, will be known at time t_1, the interest payments are exchanged at time t_2, when the forward (fictitious) loan is due.

will be paid at time t_1, if $L_{t_1} < F_{t_0}$. Settling at t_1 instead of t_2 has one subtle advantage for the FRA seller, which is often a bank. If during $[t_0, t_1]$ the interest rate has moved in favor of the bank, time t_1 settlement will reduce the marginal credit risk associated with the payoff. The bank can then operate with a lower credit line.

3.7.2.1 An interpretation

Note one important interpretation. An FRA contract can be visualized as an *exchange* of two interest payments. The purchaser of the FRA will be paying the known interest $F_{t_0} \delta N$ and is accepting the (unknown) amount $L_{t_1} \delta N$. Depending on which one is greater, the settlement will be a receipt or a payment. The sum $F_{t_0} \delta N$ can be considered, as of time t_0, as the fair payment market participants are willing to make against the random and unknown $L_{t_1} \delta N$. It can be regarded as the time to "market value" of $L_{t_1} \delta N$.

3.7.3 FRA CONTRACTUAL EQUATION

We can immediately obtain a synthetic FRA using the ideas displayed in Figure 3.14. Figure 3.14 displays a swap of a fixed-rate loan of size N, against a floating-rate loan of the same size. Thus, we can write the contractual equation

$$
\boxed{\text{Buying a FRA}} = \boxed{\begin{array}{c}\text{Fixed rate loan}\\ \text{starting } t_1 \\ \text{ending } t_2\end{array}} + \boxed{\begin{array}{c}\text{Floating rate deposit}\\ \text{starting } t_1 \text{ ending } t_2\end{array}} \qquad (3.43)
$$

It is clear from the construction in Figure 3.14 that the FRA contract eliminates the credit risk associated with the principals—since the two N's will cancel out—but leaves behind the exchange of interest rate risk. In fact, we can push this construction further by "plugging in" the contractual equation for the fixed rate forward loan obtained in formula (3.33) and get

$$
\boxed{\text{Buying a FRA}} = \boxed{\begin{array}{c}\text{Loan with}\\ \text{maturity } t_2\end{array}} + \boxed{\begin{array}{c}\text{Deposit with}\\ \text{maturity } t_1\end{array}} + \boxed{\begin{array}{c}\text{Floating rate (spot)}\\ \text{deposit starting}\\ t_1 \text{ ending } t_2\end{array}} \qquad (3.44)
$$

This contractual equation can then be exploited to create new synthetics. One example is the use of FRA strips.

3.7.3.1 Application: FRA strips

Practitioners use portfolios of FRA contracts to form *FRA strips*. These in turn can be used to construct synthetic loans and deposits and help to hedge swap positions. The best way to understand FRA strips is with an example that is based on the contractual equation for FRAs obtained earlier.

Suppose a market practitioner wants to replicate a 9-month fixed-rate borrowing synthetically. Then the preceding contractual equation implies that the practitioner should take a cash loan at time t_0, pay the LIBOR rate L_{t_0}, *and* buy an *FRA strip* made of *two* sequential FRA contracts, a (3×6) FRA and a (6×9) FRA. This will give a synthetic 9-month fixed-rate loan. Here the symbol (3×6) means t_2 is 6 months and t_1 is 3 months.

Before we discuss interest rate futures and compare them to interest rate forwards we introduce some basic risk measures for fixed income markets. This discussion will help us to appreciate some of the difference between interest rate forwards and futures.

3.8 FIXED INCOME RISK MEASURES: DURATION, CONVEXITY AND VALUE-AT-RISK

As we saw in the introductory chapter, volatility trading is made possible in the presence of two instruments, one whose value moves linearly as the underlying risk changes, while the other's value moves according to a convex curve. As we will see later in this section there are also convexity differences between interest rate forwards and interest rate futures. This is due to the fact that the pricing equation for Eurocurrency futures is linear whereas the market-traded FRAs have a pricing equation that is nonlinear in the corresponding LIBOR rate.

We will encounter several other convex instruments, including options, in later chapters but one of the simplest *convex* instruments is a default-free bond. A bond's price is a nonlinear function of its yield. This is shown in Figure 3.15.[24] The solid line represents the price—yield relationship for a

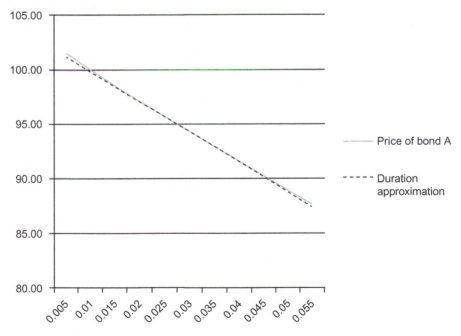

FIGURE 3.15

Price—yield curve and duration approximation.

[24]Curves with shapes such as that of the price—yield relationship are said to be *convex*.

coupon bond with 3 years to maturity, semiannual coupons and a face value of $100. The pricing function for the bond is

$$B(0, T) = \frac{C}{(1 + y/2)} + \frac{C}{(1+y/2)^2} + \ldots + \frac{C + F}{(1+y/2)^6}.$$

(3.45)

In this example, the coupon rate is 1% per year and the bond makes semiannual coupon payments of 50 cents. The above figure captures that an increase (decrease) in the bond yield decreases (increases) the price of the bond.

Risky bonds are exposed to various risk factors including credit risk, but a default-free bond is only exposed to interest rate risk. In this section, we review some of the most important measures used to calculate and hedge risk in fixed income markets. These are DV01, *duration* and *convexity*. As we will see duration is a linear approximation to the nonlinear bond price−yield relationship, while convexity measures the curvature of the relationship. One of the underlying simplifying assumptions here will be that only one risk factor is driving changes in the term structure of interest rates. More complex models exist but will not be discussed here.

3.8.1 DV01 AND PV01

The solid line in Figure 3.15 shows that the price of Bond A is $94.30 when the yield to maturity is 3%. If the yield was to increase from 3% to 3.10%, the bond price would fall from $94.30 to $94.03. This implies a slope of the curve equal to $((93.03 - 94.30)/(3.10\% - 3.00\%)) = -274.687$ or -2.74 cents per basis point. The price sensitivity itself is not constant as yields change. For example, if yields were to fall from 2% to 1.9%, the bond price would rise from 97.10 to 97.38. This would imply a slope of $((97.38 - 97.10)/(1.9\% - 2\%)) = -285.286$ or 2.85 cents per basis point. We have thus calculated by how much the bond price changes when the yield changes by 1 basis point. It turns out that this sensitivity is the definition of the *dollar duration* or DV01 measure.

3.8.1.1 Dollar duration DV01

DV01 is short for "*dollar value of a 01*", that is 1 basis point. More formally, if we denote the change in the bond price by ΔB and the change in the yield by Δy then DV01 is defined as follows

$$DV01 = -\frac{\Delta B}{10,000 \times \Delta y}$$

(3.46)

Note that dollar duration or DV01 is the change in price in *dollars*, not in *percentage*. The negative sign is there by convention to assume that DV01 is positive for most fixed income securities if which prices fall as yields rise. As we will see in later chapters, DV01 is the equivalent of the *delta*, one of the *Greeks* in option pricing, which measures the sensitivity of an option to changes in the underlying, for example. One of the advantages of DV01 over other measures of interest rate risk (such as modified duration) is that it is useful for fixed income instruments with zero present value at inception such as interest rate swaps.

In the above example, we have chosen several points on the yield axis to calculate the sensitivity of bond prices to yields. We saw that the sensitivity changes with yields, but if we were to move the different points on the yield curve together we would obtain the tangent to the price−yield curve at the given yield level. The slope of the tangent line at a given yield level is called the derivative of the price yield function and is denoted by dB/dy. In the case of the simple coupon bond that we examined above the derivative of the price−yield curve is available in closed form, that is, as a simple mathematical formula:

$$DV01 = -\frac{dB}{10,000 \times dy} \tag{3.47}$$

For more complex instruments the derivative needs to be approximated numerically by means of simulations. Many market participants take DV01 to refer to the yield-based DV01, but there are other more general versions of DV01 which measure how a specific risk factor changes by one basis point and how this affects different parts of the term structure of interest rates.

3.8.1.2 PV01

There are some standard bond market terms that are often used in swap markets. The *present value of a basis point* or PV01 is the present value of an annuity of 0.0001 paid periodically at times t_i, calculated using the proper LIBOR rates, or the corresponding forward rates:

$$PV01 = \sum_{i=1}^{n} \frac{(0.01)\delta}{\sum_{j=1}^{i}(1+L_{t_j}\delta)} \tag{3.48}$$

In order to get the sensitivity to one basis point, the number obtained from this formula needs to be divided by 100. The concepts, PV01 and DV01, are routinely used by market practitioners for the pricing of swap-related instruments and other fixed income products.[25]

3.8.2 DURATION

There are several measures of risk in fixed income markets. As we saw above, DV01 measures the dollar change in the value of an asset in response to a basis point change in the yield. A related measure of interest rate sensitivity focuses on the percentage change in the value of an asset in response to a unit change in yields. This measure is called *modified* duration or D and is defined as follows:

$$D = -\frac{1}{P}\frac{\Delta P}{\Delta y} \tag{3.49}$$

Since for a coupon bond $B(t,T)$ an explicit formula is available for the bond−yield relationship, we can calculate duration D by differentiating with respect to y. For simplicity, consider a bond with annual coupon payments C. The price of such bond, $B(0,T)$, at time $t=0$ with a maturity T years and a face value of F is given by

$$B(0,T) = \frac{C}{(1+y)} + \frac{C}{(1+y)^2} + \ldots + \frac{C+F}{(1+y)^T} \tag{3.50}$$

---

[25]In the preceding formulas, the "bond price" always refers to dirty price of a bond. This price equals the true market value, which is denoted by "clean price" plus any accrued interest.

Modified duration D_{modified} can therefore be expressed as

$$D_{\text{Modified}} = -\frac{1}{P}\frac{dB}{dy} = \frac{1}{1+y} \times \frac{1}{P}\left[\frac{C}{(1+y)^1} \times 1 + \frac{C}{(1+y)^2} \times 2 + \ldots + \frac{C+F}{(1+y)^T} \times T\right] \qquad (3.51)$$

This can be rewritten as

$$D_{\text{Modified}} = \frac{1}{1+y} \times \frac{1}{\text{Sum of PV of CF}}\sum_{t=1}^{T} t \times \text{PV}(\text{CF}_t) = \frac{1}{1+y} \times \frac{1}{B}\sum_{t=1}^{T} t \times \text{PV}(\text{CF}_t) \qquad (3.52)$$

The above equation nicely illustrates that duration is linked to the average time to maturity of the cash flows (CF_t) embedded in the bond. Specifically, it is related to the weighted average where each time to maturity is weighted by the present value of a given cash flow relative to the bond price (which itself is the sum of the present value of all cash flows, of course). We can apply the above duration formula to the 3-year bond example in Figure 3.15. What is the duration of the bond if the yield to maturity is 3%? If we apply Eq. (3.52) to this bond, we obtain a modified duration of 2.9178. This means that a one basis point increase in the yield from 3% to 3.01% is going to lead to an approximate decrease in the bond price of 2.9178×0.0001 or 0.029178%. Since the price of the bond at 3% is $94.30, this percentage change corresponds to a dollar change of $0.029178\% \times \$94.30$ or $2.74 per basis point, which, not surprisingly, is the DV01 that we calculated at the yield level of 3%. The dashed line in Figure 3.15 shows the linear *duration approximation* at the yield level of 3%. As we can see, for larger changes in yields, duration is a poor approximation to the convex bond—yield relationships.

EXAMPLE

Figure 3.16 presents a real-world example of risk measures for a fixed income security. It shows a screenshot from Bloomberg on June 16, 2014 for the T $2\frac{1}{2}$ 05/15/24, that is the 10-year Treasury note with a 2.5% fixed coupon, a $1m face value and a maturity of May 5, 2024. The DV01 at the current yield level of 2.59% is 865 and the modified duration is 8.701.

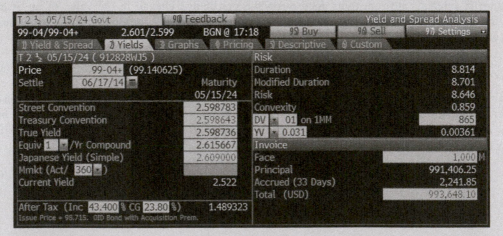

FIGURE 3.16

Ten-year Treasury note with 2.5% fixed coupon.

3.8.3 CONVEXITY

As we saw in the context of the DV01 measure, the sensitivity of bond prices to yields is not constant but changes with the level of yields. This sensitivity is measured by *convexity* and captures the curvature of the bond−yield relationship in Figure 3.15. Duration is based on the *first* derivative and convexity is based on the second derivative of the price−yield curve. Convexity is defined as

$$Convexity = \frac{1}{B}\frac{d^2B}{dy^2} \qquad (3.53)$$

In the case of a coupon bond whose price is given in Eq. (3.50) above, the formula for convexity is

$$Convexity = \frac{1}{B \times (1+y)^2} \sum_{t=1}^{T} \left[\frac{CF_t}{(1+y)^t}(t^2 + t) \right] \qquad (3.54)$$

where CF_t is the cash flow paid to the bondholder at date t. Convexity is the rate of change of the slope of the price−yield curve. The Treasury security in Figure 3.16 has a convexity of 0.859. Convexity is an attractive feature. This applies for buy-and-hold investors as well as volatility traders. Bonds with greater curvature gain more in price when yields fall than they lose when yields rise.

For noncallable bonds, convexity is positive. In other words, the slope increases and becomes less negative as yields increase. Convexity is however not always positive for fixed income securities. As we will see in Chapter 17, for bonds with embedded options, such as callable bonds, for example, there is a *range* of yields for which convexity is negative and the curvature of the price−yield relationship increases as yields rise. CF_t with T years to maturity and annual coupon bonds. One of the advantages of convexity is that it can be used to improve on the duration approximation when measuring the sensitivity of the bond−yield relationship. Another important use is that it can help to measure the potential for volatility trading and convexity gains when forming portfolios of bonds and hedging them. The exercises at the end of this chapter contain an exercise on the duration and convexity of a barbell strategy that illustrates one of the uses of convexity measures for bonds.

3.8.4 IMMUNIZATION

Hedging fixed income portfolios against interest rate risk fluctuations is called *immunization*. Immunization strategies can be based on duration. The duration of a portfolio is the weighted average of the duration of the portfolio constituents. Thus, a portfolio of two bonds A and B has a duration D_P:

$$D_P = w \times D_A + (1 - w)D_B \qquad (3.55)$$

The above equation can also be used to calculate the optimal weights for an immunized portfolio. Asset A could be a bond and asset B could be another bond or a bond futures. To immunize the portfolio we would want to calculate the weights w and $(1 - w)$ that set the portfolio duration D_P to zero. This implies $w = (D_B/(D_B - D_A))$. Another application is the context of

Asset–Liability Management, as we will see in the next chapter. Pension funds have liabilities whose present value is typically calculated by dividing by a bond market rate. The assets of a pension fund can include fixed income instruments as well as other assets. One typical risk management objective of a pension fund is to immunize the balance sheet against interest rate risk. This can be achieved by making sure that the duration of the assets is equal to the duration of the liabilities. As interest rates go up, for example, the present value of pension liabilities tends to fall. If the asset side is dominated by fixed income securities, the value of the assets would also fall at the same time. As we will see in the next chapter, swaps can also be used to immunize a portfolio or balance sheet.

3.8.5 VALUE-AT-RISK, EXPECTED SHORTFALL, BASEL CAPITAL REQUIREMENTS AND FUNDING COSTS

One of the distinguishing features of our approach in this book is to assume the perspective of the market maker when considering financial engineering applications. All risk takers in banks including market makers need to *fund* their activities. Funding costs are driven by capital requirements that banks in different countries are subject to and these are guided by the Basel Committee on Banking Supervision. Funding costs for hedge funds are also affected by regulation, either directly or indirectly through the prime brokerage relationship. The ability of a prime brokerage desk to finance hedge fund trades is controlled by the bank's internal capital requirements. Banking regulation distinguishes banks' lending and trading activities by separating them into *banking* and *trading books.* One of the most widely used measures of risk for trading books since 1990s was *value-at-risk* (VaR).[26] VaR is a risk measure of the risk of loss on a given portfolio of financial assets. VaR is defined as the maximum loss not exceeded with a given probability defined as the *confidence level*, over a given period of time. For example, if a portfolio of bonds has a 1-day 5% VaR of $100 million, then there is a 0.05 probability that the portfolio will fall in value by more than $100 million over a 1-day period. On October 2013, the Basel Committee on Banking Supervision, in its review of trading book rules proposed scrapping VaR as the basis for modeling market risk capital requirements. The committee recommended replacing VaR with *expected shortfall*, which measures the expected value of losses above a given confidence internal. The reputation of VaR was badly damaged during the GFC, since it became clear to regulators, banks and other market participants that stressed market conditions can lead to losses far in excess of the maximum amounts forecast by standard VaR applications. One of the theoretical advantages of expected shortfall is that, unlike VaR, expected shortfall is a coherent risk measure. This means that VaR might theoretically discourage diversification. Although one of the stated intentions of the Basel committee is to better capture *tail risk*, it is important to note that the failure of VaR during the crisis is not so much due to its mathematical properties, but rather the results of how it was applied in practice, which often relied on using short historical time periods and distributional assumptions that did not allow for

---

[26]In addition to the determination of regulatory capital, VaR is also used in risk management, financial control and financial reporting. See McNeil et al. (2005) for further details about risk measures including VaR and expected shortfall and their uses.

scenarios with extreme volatilities and asset correlations. Expected shortfall applications have their own practical challenges such as data requirements and backtesting issues. By its nature, expected shortfall requires orders of magnitude more data than VaR since it incorporates less likely scenarios. Regulation goes through waves and whichever risk measure dominates capital requirement calculations in the future, financial engineers will need to be familiar with them since these rules determine funding costs and, therefore, which trades and derivative products are profitable in practice and which are not.

3.9 FUTURES: EUROCURRENCY CONTRACTS

Forward loans do not trade in the OTC market because FRAs are much more cost-effective. Eurocurrency futures are another attractive alternative. In this section, we discuss Eurocurrency futures using the Eurodollar (ED) futures as an example and then compare it with FRA contracts. This comparison illustrates some interesting aspects of successful contract design in finance.

FRA contracts involve exchanges of interest payments associated with a floating and a fixed-rate loan. The Eurodollar futures contracts trade future loans *indirectly*. The settlement will be in cash and the futures contract will again result only in an exchange of interest rate payments. However, there are some differences with the FRA contracts.

Eurocurrency futures trade the forward loans (deposits) shown in Figure 3.10 as homogenized contracts. These contracts deal with loans and deposits in *Euromarkets*, as suggested by their name. The *buyer* of the Eurodollar futures contract is a potential *depositor* of 3-month Eurodollars and will lock in a future deposit rate.

Eurocurrency futures contracts do not deliver the deposit itself. At expiration date t_1, the contract is cash settled. Suppose we denote the price of the futures contract quoted in the market by Q_{t_0}. Then the *buyer* of a 3-month Eurodollar contract "promises" to deposit $100(1 - \tilde{F}_{t_0}\frac{1}{4})$ dollars at expiration date t_1 and receive 100 in 3 months. The *implied* annual interest rate on this loan is then calculated by the formula

$$\tilde{F}_{t_0} = \frac{100.00 - Q_{t_0}}{100} \tag{3.56}$$

This means that the price quotations are related to forward rates through the formula

$$Q_{t_0} = 100.00(1 - \tilde{F}_{t_0}) \tag{3.57}$$

However, there are important differences with forward loans. The interest rate convention used for forward loans is equivalent to a *money market yield*. For example, to calculate the time t_1 present value at time t_0 we let

$$PV(t_0, t_1, t_2) = \frac{100}{(1 + F_{t_0}\delta)} \tag{3.58}$$

Futures contracts, on the other hand, use a convention similar to *discount rates* to calculate the time t_1 value of the forward loan

$$\tilde{PV}(t_0, t_1, t_2) = 100(1 - \tilde{F}_{t_0}\delta) \tag{3.59}$$

If we want the amount traded to be the same:

$$PV(t_0, t_1, t_2) = P\tilde{V}(t_0, t_1, t_2) \tag{3.60}$$

the two forward rates on the right-hand side of formulas (3.58) and (3.59) *cannot* be identical. Of course, there are many other reasons for the right-hand side and left-hand side in formula (3.60) not to be the same. Futures markets have mark-to-market; FRA markets, in general, do not. With mark-to-market, gains and losses occur daily, and these daily cash flows may be *correlated* with the overnight funding rate. Thus, the forward rates obtained from FRA markets need to be adjusted to get the forward rate in the Eurodollar futures, and vice versa.

EXAMPLE

Suppose at time t_0, futures markets quote a price

$$Q_{t_0} = 95.76 \tag{3.61}$$

for a Eurodollar contract that expires on the third Wednesday of December 2020. This would mean two things. First, the implied forward rate for that period is given by:

$$F_{t_0} = \frac{100.00 - 95.76}{100} = 0.0424 \tag{3.62}$$

Second, the contract involves a position on the delivery of

$$100\left(1 - 0.04240\frac{1}{4}\right) = 98.94 \tag{3.63}$$

dollars on the third Wednesday of December 2020.

At expiry, these funds will never be deposited explicitly. Instead, the contract will be cash settled. For example, if on expiration the exchange has set the delivery settlement price at $Q_{t_1} = 96.0$, this would imply a forward rate

$$F_{t_1} = \frac{100 - 96.0}{100} = 0.04 \tag{3.64}$$

and a settlement

$$100\left(1 - 0.04\frac{1}{4}\right) = 99.00 \tag{3.65}$$

Thus, the buyer of the original contract will be compensated as if he or she is making a deposit of 98.94 and receiving a loan of 99.00. The net gain is

$$99.00 - 98.94 = 0.06 \text{ per } 100 \text{ dollars} \tag{3.66}$$

This gain can be explained as follows. When the original position was taken, the (forward) rate for the future 3-month deposit was 4.24%. Then at settlement this rate declined to 4.0%.

Actually, the above example is a simplification of reality as the gains would never be received as a lump sum at the expiry due to marking-to-market. The mark-to-market adjustments would lead to a gradual accumulation of this sum in the buyer's account. The gains will earn some interest as well. This creates another complication. Mark-to-market gains losses may be correlated with daily interest rate movements applied to these gains (losses).[27]

3.9.1 OTHER PARAMETERS

There are some other important parameters of futures contracts. Instead of discussing these in detail, we prefer to report contract descriptions directly. The following table describes this for the CME Eurodollar contract.[28]

Delivery months	: March, June, September, December (for 10 years)
Delivery (Expiry) day	: Third Wednesday of delivery month
Last trading day	: 11.00 Two business days before expiration
Minimum tick	: 0.0025 (for spot-month contract)
"Tick value"	: USD6.25 for nearest month, otherwise USD12.50
Settlement rule	: ICE LIBOR on the settlement date

The design and the conventions adopted in the Eurodollar contract may seem a bit odd to the reader, but the contract is a successful one. First of all, quoting Q_{t_0} instead of the forward rate $\tilde{F}_{t_0}$ makes the contract similar to buying and selling a futures contract on T-bills. This simplifies related hedging and arbitrage strategies. Second, as mentioned earlier, the contract is settled in cash. This way, the functions of securing a loan and locking in an interest rate are successfully separated.

Third, the convention of using a *linear* formula to represent the relationship between Q_{t_0} and $\tilde{F}_{t_0}$ is also a point to note. Suppose the underlying time t_1 deposit is defined by the following equation

$$D(t_0, t_1, t_2) = 100(1 - \tilde{F}_{t_0}\delta) \tag{3.67}$$

A small variation of the forward rate $\tilde{F}_{t_0}$ will result in a *constant* variation in $D(t_0, t_1, t_2)$:

$$\frac{\partial D(t_0, t_1, t_2)}{\partial \tilde{F}_{t_0}} = -\delta 100 = -25 \tag{3.68}$$

Thus, the *sensitivity* of the position with respect to the underlying interest rate risk is constant, and the product is truly *linear* with respect to $\tilde{F}_{t_0}$.

---

[27]An additional difference is due to the interest payments on margin/collateral. No interest is paid on the variation margin of futures.

[28]See www.cmegroup.com/trading/interest-rates/stir/eurodollar.html for recent quotes and contract specifications.

3.9.1.1 The "TED spread"

The difference between the interest rates on 3-month Treasury Bills (T-Bills) and 3-month Eurodollar (ED) futures is called the *TED spread*.[29] T-bill (T-Note) rates provide a measure of the US government's short-term (medium-term) borrowing costs. Eurodollar futures relate to short-term private sector borrowing costs while T-bills are considered risk free (for a country like the United States, but not all countries as the GFC has demonstrated). Thus the "TED spread" has credit risk elements in it.[30]

The TED spread average from 1986 to 2014 is around 62 bps.[31] On October 10, 2008, the TED spread reached 458 bps as liquidity in the interbank lending market dried up and lenders' perception of the default on interbank loans (counterparty risk) spiked. This reading eclipsed the level of 278 bps reached on October 28, 1987. Figure 3.17 plots the TED spread against the 3-month USD LIBOR and T-BILL rates from January 2, 1986 until April 17, 2014.

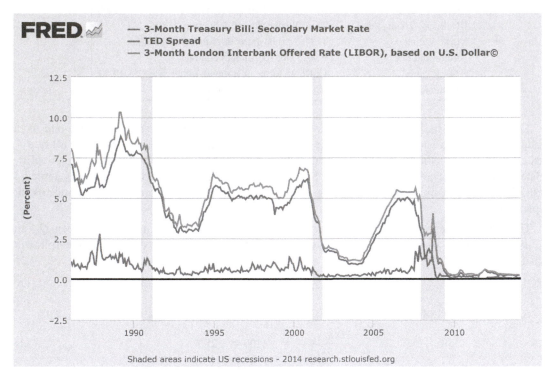

FIGURE 3.17

TED spread versus 3-month USD LIBOR and T-Bill.

---

[29]TED is an acronym resulting from the combination of *T-Bill* and *ED*, the ticker symbol for the Eurodollar futures contract.

[30]During the credit crisis of 2007−2008, TED spread was often used as a measure of banking sector credit risk.

[31]The TED spread time series is available at http://research.stlouisfed.org/fred2/series/TEDRATE#.

Traders form strips of Eurodollar futures and trade them against T-Bills of similar maturity. A similar spread can be put together using Treasury Notes (T-notes) and Eurodollars as well. Given the different ways of quoting yields, calculation of the spread involves some technical adjustments. T-Notes use bond equivalent yields whereas Eurodollars are quoted similar to discount rate basis. The calculation of the TED spread requires putting together strips of futures while adjusting for these differences. There are several technical points that arise along the way.

Once the TED spread is calculated, traders put on trades to benefit from changes in the yield curve slope and in private sector credit risk. For example, traders would *long* the TED spread if they expected the yield spread to *widen*. In the opposite case, they would *short* the TED spread and would thus benefit from the *narrowing* of the yield spread. Experienced fixed income relative value traders are wary of being short the TED spread, however, due to the occasionally extreme downside.

3.9.2 COMPARING FRAs AND EURODOLLAR FUTURES

A brief comparison of FRAs with Eurocurrency futures may be useful. (i) Being OTC contracts, FRAs are more flexible instruments, since Eurodollar futures trade in terms of preset *homogeneous* contracts. (ii) FRAs have the advantage of *confidentiality*. There is no requirement that the FRA terms be announced. The terms of a Eurocurrency contract are known. (iii) As discussed in Chapter 2, until recently there were no margin requirements for FRAs and the *mark-to-market requirements* are less strict. With FRAs, money changes hands only at the settlement date. Eurocurrency futures come with margin requirements as well as with mark-to-market requirements. (iv) FRAs have *counterparty risk*, whereas the credit risk of Eurocurrency futures contracts is insignificant. (v) FRAs are quoted on an interest rate basis while Eurodollar futures are quoted on a price basis. Thus a trader who sells an FRA will hedge this position by selling a Eurodollar contract as well. (vi) Finally, an interesting difference occurs with respect to *fungibility*. Eurocurrency contracts are fungible, in the sense that contracts with the same expiration can be netted against each other even if they are entered into at different times and for different purposes. FRA contracts cannot be netted against each other even with respect to the same counterparty, unless the two sides have a specific agreement.

3.9.2.1 Convexity differences

Besides these structural differences, FRAs and Eurocurrency futures have different convexities. The pricing equation for Eurocurrency futures is linear in $\tilde{F}_{t_0}$, whereas the market-traded FRAs have a pricing equation that is nonlinear in the corresponding LIBOR rate. We will see that this requires *convexity adjustments*, which is one reason why we used different symbols to denote the two forward rates.

3.9.3 HEDGING FRAs WITH EUROCURRENCY FUTURES

For short-dated contracts, convexity and other differences may be negligible, and we may ask the following question. Putting convexity differences aside, can we hedge an FRA position with futures, and vice versa?

It is best to answer this question using an example. The example also illustrates some real-world complications associated with this hedge.

EXAMPLE

Suppose we are given the following Eurodollar futures prices on June 17, 2002:

September price (delivery date: September 16) 96.500 (implied rate = 3.500)
December price (delivery date: December 16) 96.250 (implied rate = 3.750)
March price (delivery date: March 17) 96.000 (implied rate = 4.000)

A trader would like to *sell* a (3×6) FRA on June 17, with a notional amount of USD100,000,000. How can the deal be hedged using these futures contracts?

Note first that according to the value and settlement date conventions, the FRA will run for the period September 19–December 19 and will encompass 92 days. It will settle against the LIBOR fixed on September 17. The September futures contract, on the other hand, will settle against the LIBOR fixed on September 16 and is quoted on a 30/360 basis. Thus, the implied forward rates will not be identical for this reason as well.

Let f be the FRA rate and ϵ be the differences between this rate and the forward rate implied by the futures contract. Using formula (3.41), the FRA settlement, with notional value of 100 million USD, may be written as

$$\frac{100\,m((0.035 + \epsilon) - \text{Libor})\frac{92}{360}}{\left(1 + \text{Libor}\frac{92}{360}\right)} \tag{3.69}$$

Note that this settlement is discounted to September 19 and will be received once the relevant LIBOR rate becomes known. Ignoring mark-to-market and other effects, a futures contract covering similar periods will settle at

$$\alpha\left(1\,m(0.0350 - \text{Libor})\frac{90}{360}\right) \tag{3.70}$$

Note at least two differences. First, the contract has a nominal value of USD1 million. Second, 1 month is, by convention, taken as 30 days, while in the case of FRA it was the actual number of days. The α is the number of contracts that has to be chosen so that the FRA position is correctly hedged.

The trader has to choose α such that the two settlement amounts are as close as possible. This way, by taking opposite positions in these contracts, the trader will hedge his or her risks.

3.9.3.1 *Some technical points*

The process of hedging is an approximation that may face several technical and practical difficulties. To illustrate them we look at the preceding example once again.

1. Suppose we tried to hedge (or price) a *strip* of FRAs rather than having a single FRA be adjusted to contract using a strip of available futures contracts. Then the strip of FRAs will have to deal with increasing notional amounts. Given that futures contracts have *fixed* notional amounts, contract numbers need to be adjusted instead.

2. As indicated, a 3-month period in futures markets is 90 days, whereas FRA contracts count the actual number of days in the corresponding 3-month period.
3. Given the convexity differences in the pricing formulas, the forward rates implied by the two contracts are not the same and, depending on LIBOR volatility, the difference may be large or small.
4. There may be differences of 1 or 2 days in the fixing of the LIBOR rates in the two contracts.

These technical differences relate to this particular example, but they are indicative of most hedging and pricing activity.

3.10 REAL-WORLD COMPLICATIONS

Up to this point, the discussion ignored some real-life complications. We made the following simplifications. (i) We ignored bid-ask spreads. (ii) Credit risk was assumed away. (iii) We ignored the fact that the fixing date in an FRA is, in general, different from the settlement date. In fact, this is another date involved in the FRA contract. Let us now discuss these issues.

3.10.1 BID-ASK SPREADS

We begin with bid-ask spreads. The issue will be illustrated using a bond market construction. When we replicate a forward loan via the bond market, we buy a $B(t_0, t_1)$ bond and short-sell a $B(t_0, t_2)$ bond. Thus, we have to use ask prices for $B(t_0, t_1)$ and bid prices for $B(t_0, t_2)$. This means that the asking price for a forward interest rate will be

$$1 + F_{t_0}^{ask}\delta = \frac{B(t_0, t_1)^{ask}}{B(t_0, t_2)^{bid}} \tag{3.71}$$

Similarly, when the client sells an FRA, he or she has to use the bid price of the dealers and brokers. Again, going through the bond markets we can get

$$1 + F_{t_0}^{bid}\delta = \frac{B(t_0, \ t_1)^{bid}}{B(t_0, \ t_2)^{ask}} \tag{3.72}$$

This means that

$$F_{t_0}^{bid} < F_{t_0}^{ask} \tag{3.73}$$

The same bid-ask spread can also be created from the money market synthetic using the bid-ask spreads in the money markets

$$1 + F_{t_0}^{ask}\delta = \frac{1 + L_{t_0}^{1\ bid}\delta^1}{1 + L_{t_0}^{2\ ask}\delta^2} \tag{3.74}$$

Clearly, we again have

$$F_{t_0}^{bid} < F_{t_0}^{ask} \tag{3.75}$$

Thus, pricing will normally yield two-way prices.

In market practice, FRA bid-ask spreads are not obtained in the manner shown here. The bid-ask quotes on the FRA rate are calculated by first obtaining a rate from the corresponding LIBORs and then adding a spread to both sides of it. Many practitioners also use the more liquid Eurocurrency futures to "make" markets.

3.10.2 AN ASYMMETRY

There is another aspect to using FRAs for hedging purposes. The net return and net cost from an interest rate position will be asymmetric since, whether you buy (pay fixed) or sell (receive fixed), an FRA *always* settles against LIBOR. But LIBOR is an offer (asking) rate, and this introduces an asymmetry.

We begin with a hedging of floating borrowing costs. When a company hedges a floating *borrowing* cost, both interest rates from the cash and the hedge will be LIBOR based. This means that:

- The company pays LIBOR + margin to the bank that it borrows funds from.
- The company pays the fixed FRA rate to the FRA counterparty for hedging this floating cost against which the company receives LIBOR from the FRA counterparty.

Adding all receipts and payments, the net borrowing cost becomes *FRA rate + margin*.

Now consider what happens when a company hedges, say, a 3-month floating *receipt*. The relevant rate for the cash position is *Libid*, the bid rate for placing funds with the Euromarkets.

But an FRA always settles always against LIBOR. So the picture will change to:

- Company receives Libid, assuming a zero margin.
- Company receives FRA rate.
- Company pays LIBOR.

Thus, the net return to the company will become FRA-(LIBOR-Libid).

3.11 FORWARD RATES AND TERM STRUCTURE

A detailed framework for fixed income engineering will be discussed in Chapter 14. However, some preliminary modeling of the term structure is in order. This will clarify the notation and some of the essential concepts.

3.11.1 BOND PRICES

Let $\{B(t_0, t_i), i = 1, 2..., n\}$ represent the *bond price family*, where each $B(t_0, t_i)$ is the price of a default-free zero-coupon bond that matures and pays $1 at time t_i. These $\{B(t_0, t_i)\}$ can also be viewed as a vector of *discounts* that can be used to value default-free cash flows.

For example, given a complicated default-free asset, A_{t_0}, that pays deterministic cash flows $\{C_{t_i}\}$ occurring at arbitrary times, t_i, $i = 1, ..., k$, we can obtain the value of the asset easily if we assume the following bond price process:

$$A_{t_0} = \sum_i C_{t_i} B(t_0, t_i) \tag{3.76}$$

That is to say, we just multiply the t_ith cash flow with the current value of one unit of currency that belongs to t_i, and then sum over i.

This idea has an immediate application in the pricing of a coupon bond. Given a coupon bond with a nominal value of $1 that pays a coupon rate of $c\%$ at times t_i, the value of the bond can easily be obtained using the preceding formula, where the last cash flow will include the principal as well.

3.11.2 WHAT FORWARD RATES IMPLY

In this chapter, we obtained the important arbitrage equality

$$1 + F(t_0, t_1, t_2)\delta = \frac{B(t_0, t_1)}{B(t_0, t_2)} \tag{3.77}$$

where $F(t_0, t_1, t_2)$ is written in the expanded form to avoid potential confusion.[32] It implies a forward rate that applies to a loan starting at t_1 and ending at t_2. Writing this arbitrage relationship for *all* the bonds in the family $\{B(t_0, t_i)\}$, we see that

$$1 + F(t_0, t_0, t_1)\delta = \frac{B(t_0, t_0)}{B(t_0, t_1)} \tag{3.78}$$

$$1 + F(t_0, t_1, t_2)\delta = \frac{B(t_0, t_1)}{B(t_0, t_2)} \tag{3.79}$$

$$\cdots\cdots \tag{3.80}$$

$$1 + F(t_0, t_{n-1}, t_n)\delta = \frac{B(t_0, t_{n-1})}{B(t_0, t_n)} \tag{3.81}$$

Successively substituting the numerator on the right-hand side using the previous equality and noting that for the first bond we have $B(t_0, t_0) = 1$, we obtain

$$B(t_0, t_n) = \frac{1}{(1 + F(t_0, t_0, t_1)\delta)\ldots(1 + F(t_0, t_{n-1}, t_n)\delta)} \tag{3.82}$$

We have obtained an important result. The bond price family $\{B(t_0, t_i)\}$ can be expressed using the forward rate family,

$$\{F(t_0, t_0, t_1), \ldots, F(t_0, t_{n-1}, t_n)\} \tag{3.83}$$

Therefore, if all bond prices are given we can determine the forward rates.

3.11.2.1 Remark

Note that the "first" forward rate $F(t_0, t_0, t_1)$ is contracted at time t_0 and applies to a loan that starts at time t_0. Hence, it is also the t_0 spot rate:

$$(1 + F(t_0, t_0, t_1)\delta) = (1 + L_{t_0}\delta) = \frac{1}{B(t_0, t_1)} \tag{3.84}$$

---

[32]Here the δ has no i subscript. This means that the periods $t_i - t_{i-1}$ are constant across i and are given by $(t_i - t_{i-1})/360$.

We can write this as

$$B(t_0, t_1) = \frac{1}{(1 + L_{t_0}\delta)} \qquad (3.85)$$

The bond price family $B(t_0, t_i)$ is the relevant *discounts* factors that market practitioners use in obtaining the present values of default-free cash flows. We see that modeling F_{t_0}'s will be quite helpful in describing the modeling of the yield curve or, for that matter, the discount curve.

3.12 CONVENTIONS

FRAs are quoted as two-way prices in bid-ask format, similar to Eurodeposit rates. A typical market contributor will quote a 3-month and a 6-month series.

EXAMPLE

The 3-month series will look like this:

1×4	4.87	4.91
2×5	4.89	4.94
3×6	4.90	4.95
etc.		

The first row implies that the interest rates are for a 3-month period that will start in 1 month. The second row gives the forward rate for a loan that begins in 2 months for a period of 3 months and so on.

The 6-month series will look like this:

1×7	4.87	4.91
2×8	4.89	4.94
3×9	4.90	4.95
etc.		

According to this table, if a client would like to lock in a fixed payer rate in 3 months for a period of 6 months and for a notional amount of USD1 million, he or she would buy the *3s against 9s* and pay the 4.95% rate. For 6 months, the actual net payment of the FRA will be

$$\frac{1,000,000\left(\frac{L_{t_3}}{100} - 0.0495\right)\frac{1}{2}}{\left(1 + \frac{1}{2}\frac{L_{t_3}}{100}\right)} \qquad (3.86)$$

where L_{t_3} is the 6-month LIBOR rate that will be observed in 3 months.

Another convention is the use of *LIBOR* rate as a *reference rate* for both the sellers and the buyers of the FRA. *LIBOR* being an asking rate, one might think that a client who sells an FRA may receive a lower rate than *LIBOR*. But this is not true, as the reference rate does not change.

3.13 A DIGRESSION: STRIPS

Before finishing this chapter we discuss an instrument that is the closest real-life equivalent to the default-free pure discount bonds $B(t_0, t_i)$. This instrument is called *strips*. US strips have been available since 1985 and UK strips since 1997.

Consider a long-term *straight* Treasury bond, a German bund, or a British gilt and suppose there are no implicit options. These bonds make *coupon* payments during their life at regular intervals. Their day-count and coupon payment intervals are somewhat different, but in essence they are standard long-term debt obligations. In particular, they are not the zero-coupon bonds that we have been discussing in this chapter.

Strips are obtained from coupon bonds. The market practitioner buys a long-term coupon bond and then "strips" each coupon interest payment and the principal and trades them *separately*. Such bonds will be equivalent to zero-coupon bonds except that, if needed, one can put them back together and reconstruct the original coupon bond.

The institution overseeing the government bond market, the Bank of England in the United Kingdom or the Treasury in the United States, arranges the necessary infrastructure to make stripping possible and also designates the strippable securities.[33] Note that only some particular dealers are usually allowed to strip and to reconstruct the underlying bonds. These dealers put in a request to strip a bond that they already have in their account and then they sell the pieces separately.[34] As an example, a 10-year gilt is strippable into 20 coupons plus the principal. There will be 21 zero-coupon bonds with maturities 6, 12, 18, 24 (and so on) months.

3.14 CONCLUSIONS

This chapter has developed two main ideas. First, we considered the engineering aspects of simple interest rate derivatives such as forward loans and FRAs. Second, we developed our first contractual equation. This equation was manipulated to obtain synthetic loans, synthetic deposits, and synthetic FRAs. A careful use of such contractual equations may provide useful techniques that are normally learned only after working in financial markets. We introduced the term structure of

---

[33]Stripping a Gilt costs less than $2 and is done in a matter of minutes at the touch of a button. Although it changes depending on the market environment, about 40% of a bond issue is stripped in the United States and in the United Kingdom.
[34]The reason for designing some bonds as strippable is because (i) large bond issues need to be designated and (ii) the coupon payment dates need to be such that they fall on the same date, so that when one strips a 2- and a 4-year bond, the coupon strips for the first 2 years become interchangeable. This will increase the liquidity of the strips and also make their maturity more homogeneous.

interest rates and introduced the forward rate (LIBOR) processes. The chapter continued to build on the simple graphical financial engineering methods that are based on cash flow manipulations.

Before concluding, we would like to emphasize some characteristics of forward contracts that can be found in other swap-type derivatives as well. It is these characteristics that make these contracts very useful instruments for market practitioners.

First, at the time of initiation, the forward (future) contract did not require any initial *cash payments*. This is a convenient property if our business is trading contracts continuously during the day. We basically don't have to worry about "funding" issues.

Second, because forward (future) contracts have zero initial value, the position taker does not have anything to put on the balance sheet. The trader did not "*buy*" or "*sell*" something tangible. With a forward (futures) contract, the trader has simply taken a position. So these instruments are *off-balance sheet* items.

Third, forward contracts involve an exchange at a *future date*. This means that if one of the counterparties "defaults" before that date, the damage will be limited, since no principal amount was extended. What is at risk is simply any capital gains that may have been earned.

SUGGESTED READING

Futures and forward markets have now been established for a wide range of financial contracts, commodities, and services. This chapter dealt only with basic engineering aspects of interest rate forwards and futures contracts, and a comprehensive discussion was avoided. It may be best to go over a survey of existing futures and forward contracts. We recommend two good introductory sources. The first is the Foreign Exchange and Money Markets, an introductory survey prepared by **Reuters** and published by Wiley. **Hull** (2014), **Das** (1994), and **Wilmott** (2006) are among the best sources for a detailed analysis of forward and futures contracts. There are many more fixed income instruments involving more complicated parameters than those discussed here. Some of these will certainly be examined in later chapters. But reading some market-oriented books that deal with technical aspects of these instruments may be helpful at this point. Two such books are **Serrat and Tuckman** (2011) and **Questa** (1999). **McNeil et al.** (2005) provide a review of quantitative risk management and VaR.

APPENDIX—CALCULATING THE YIELD CURVE

How do we calculate the (government) yield curve?[35] The traditional way of calculating yield curves starts with liquid bond prices and then obtains the discounts and related yields. Thus, the method first calculates the implied zero-coupon prices, and then the corresponding yields and forward rates from observed coupon bond prices.

We will briefly review this approach to yield curve calculation. It may still be useful in markets where liquid interest rate derivatives do not trade. First, we need to summarize the concepts.

---

[35]We will discuss the swap yield curve, its construction and its relative advantages over the government yield curve in Chapter 4.

PAR YIELD CURVE

Consider a straight coupon bond with coupon rate c exactly equaling the yield at time t for that maturity. The current price of this "par" bond will be exactly 100, the par value. Such a bond will have a yield to maturity, the *par yield*. The current price of these bonds is equal to 100 and their coupon would be indicative of the correct yield for that maturity and credit at that particular time.

We can write the present value of a three-period par bond, paying interest annually, as

$$100 = \frac{100c}{(1+y)} + \frac{100c}{(1+y)^2} + \frac{100(1+c)}{(1+y)^2} \tag{3.87}$$

where the par yield implies that $c = y$.

This property is desirable because with coupon bonds, the maturity does not give the correct timing for the average cash receipt, and if we consider bonds with coupons different than the par yield, the durations of the bonds would be different and the implied yields would also differ.

ZERO-COUPON YIELD CURVE

We can also calculate a yield curve using zero-coupon bonds with par value 100 by exploiting the equality,

$$B(t, T) = \frac{100}{(1+y_t^T)^{T-t}} \tag{3.88}$$

where $B(t, T)$ is time-t price of default-free zero-coupon bond that pays 100 at time T. The y_t^T will correspond to the $(T - t)$-maturity zero-coupon yield.

It turns out that the par yield curve and the zero-coupon yield curve are different in general. We now show the calculations in an example.

EXAMPLE

We would like to show the relationship between par yields and zero-coupon yields. Suppose we are given the following zero-coupon bond prices:

$$
\begin{aligned}
B(0, 1) &= 96.00 \\
B(0, 2) &= 91.00 \\
B(0, 3) &= 87.00
\end{aligned} \tag{3.89}
$$

1. What are the zero-coupon yields?
2. What are the par yields for the same maturities?

To calculate the zero-coupon yields, we use the following formula:

$$B(t, T) = \frac{100}{(1+y_t^T)^{T-t}} \tag{3.90}$$

and obtain:

$$96.00 = \frac{100}{(1+y_0^1)^1}$$

$$91.00 = \frac{100}{(1+y_0^2)^2}$$

$$87.00 = \frac{100}{(1+y_0^3)^3} \tag{3.91}$$

Solving for the unknown zero-coupon yields:

$$y_0^1 = 0.04167, \quad y_0^2 = 0.04828, \quad y_0^3 = 0.04752 \tag{3.92}$$

We now calculate par yields using the relationship with $t_n = T$

$$P(t_0, T) = \sum_{i=0}^{n} \tilde{y}B(t_0, t_i) + B(t_0, T) = 100 \tag{3.93}$$

The $\tilde{y}$ that satisfies this equation will be the par yield for maturity T. The idea here is that, when discounted by the correct discount rate, the sum of the cash flows generated by a par bond should equal 100; i.e., we must have $P(t_0, T) = 100$. *Only one $\tilde{y}$ will make this possible for every T.*

Calculating the par yields, we obtain

$$\tilde{y}_0^1 = y_0^1 = 0.04167, \quad \tilde{y}_0^2 = 0.04813, \quad \tilde{y}_0^3 = 0.04745 \tag{3.94}$$

As these numbers show, the par yields and the zero-coupon yields are slightly different.

ZERO-COUPON CURVE FROM COUPON BONDS

Traditional methods of calculating the yield curve involve obtaining a zero-coupon yield curve from arbitrary coupon bond prices. This procedure is somewhat outdated now, but it may still be useful in economies with newly developing financial markets. Also, the method is a good illustration of how synthetic asset creation can be used in yield curve construction. It is important to remember that all these calculations refer to default-free bonds.

Consider a 2-year coupon bond. The *default-free* bond carries an annual coupon of c percent and has a current price of $P(t, t+2)$. The value at maturity is 100. Suppose we know the level of the current annual interest rate r_t.[36] Then the portfolio

---

[36]Alternatively, we can assume that the price of the one-period coupon bond $P(t, t+1)$ is known.

$$\left\{ \frac{100c}{(1+r_t)} \text{ units of time } t \text{ borrowing, and buying two-period coupon bond, } P(t, t+2) \right\} \qquad (3.95)$$

will yield the cash flow equivalent to $1 + c$ units of a two-period discount bond. Thus, we have

$$P(t, t+2) - \frac{100c}{1+r_t} = (100(1+c))/(1+y_t^2)^2 \qquad (3.96)$$

If the coupon bond price $P(t, t+2)$ and the 1-year interest rate r_t are known, then the 2-year zero-coupon yield y_t^2 can be calculated from this expression. Zero-coupon yields for other maturities can be calculated by forming similar synthetics for longer maturity zero-coupon bonds, recursively.

EXERCISES

1. You have a 4-year coupon bond that pays semiannual interest. The coupon rate is 8% and the par value is 100.
   a. Can you construct a synthetic equivalent of this bond? Be explicit and show your cash flows.
   b. Price this coupon bond assuming the following term structure (represented by bid/ask prices at annual frequency) and by using a linear interpolation to discount semi-annual cashflows:
   $$B_1 = 0.90/0.91, \quad B_2 = 0.87/0.88, \quad B_3 = 0.82/0.83, \quad B_4 = 0.80/0.81$$
   c. What is the 1×2 FRA rate?
2. You have purchased 1 Eurodollar contract at a price of $Q_0 = 94.13$, with an initial margin of 5%. You keep the contract for 5 days and then sell it by taking the opposite position. In the meantime, you observe the following settlement prices:
   $$\{Q_1 = 94.23, Q_2 = 94.03, Q_3 = 93.93, Q_4 = 93.43, Q_5 = 93.53\} \qquad (3.97)$$
   a. Calculate the string of mark-to-market losses or gains.
   b. Suppose the spot interest rate during this 5-day period was unchanged at 6.9%. What is the total interest gained or paid on the clearing firm account?
   c. What are the total gains and losses at settlement?
3. The treasurer of a small bank has borrowed funds for 3 months at an interest rate of 6.73% and has lent funds for 6 months at 7.87%. The total amount is USD38 million.
   To cover his exposure created by the mismatch of maturities, the dealer needs to borrow another USD38 million for months, in 3 months' time, and hedge the position now with an FRA.
   The market has the following quotes from three dealers:

BANK A	3×6	6.92–83
BANK B	3×6	6.87–78
BANK C	3×6	6.89–80

   a. What is (are) the exposure(s) of this treasurer? Represent the result on cash flow diagrams.
   b. Calculate this treasurer's break-even forward rate of interest, assuming no other costs.
   c. What is the best FRA rate offered to this treasurer?

   d. Calculate the settlement amount that would be received (paid) by the treasurer if, on the settlement date, the LIBOR fixing was 6.09%.

4. A corporation will receive USD7 million in 3 months' time for a period of 3 months. The current 3-month interest rate quotes are 5.67−5.61. The Eurodollar futures price is 94.90.

   Suppose in 3 months the interest rate becomes 5.25% for 3-month Eurodeposits and the Eurodollar futures price is 94.56.

   a. How many ticks has the futures price moved?

   b. How many futures contracts should this investor buy or sell if she wants to lock in the current rates?

   c. What is the profit (loss) for an investor who bought the contract at 94.90?

5. Suppose practitioners learn that the ICE Benchmark Administration (IBA) will change the panel of banks used to calculate the yen LIBOR. One or more of the "weaker" banks will be replaced by "stronger" banks at a future date. IBA replaced the British Banker's Association (BBA) in 1 February 2014.

   The issue here is not whether yen LIBOR will go down, as a result of the panel now being "stronger." In fact, due to market movements, even with stronger banks in the panel, the yen LIBOR may in the end go up significantly. Rather, what is being anticipated is that the yen LIBOR should decrease in London *relative* to other yen fixings, such as Tibor. Thus, to benefit from such a BBA move, the market practitioner must form a position where the risks originating from market movements are eliminated and the "only" relevant variable remains the decision by the BBA.

   a. How would a trader benefit from such a change without taking on too much risk?

   b. Using cash flow diagrams, show how this can be done.

   c. In fact, show which *spread* FRA position can be taken. Make sure that the position is (mostly) neutral toward market movements and can be created, the only significant variable being the decision by the IBA.

   (From Thomson Reuters IFR, issue 1267) Traders lost money last week following the British Bankers' Association (BBA) decision to remove one Japanese bank net from the yen LIBOR fixing panel. The market had been pricing in no significant changes to the panel just the day before the changes were announced.

   Prior to the review, a number of dealers were reported to have been short the LIBOR/ Tibor spread by around 17 bp, through a twos into fives forward rate agreement (FRA) spread contract. This was in essence a bet that the Japanese presence on the LIBOR fixing panel would be maintained.

   When the results of the review were announced on Wednesday January 20, the spread moved out by around 5 bp to around 22 bp—leaving the dealers with mark-to-market losses. Some were also caught out by a sharp movement in the one-year yen/dollar LIBOR basis swap, which moved in from minus 26 bp to minus 14 bp.

   The problems for the dealers were caused by BBA's decision to alter the nature of the fixing panel, which essentially resulted in one Japanese bank being removed to be replaced by a foreign bank. Bank of China, Citibank, Tokai Bank and Sakura were taken out, while Deutsche Bank, Norinchukin Bank, Rabobank and WestLB were added.

   The move immediately increased the overall credit quality of the grouping of banks responsible for the fixing rate. This caused the yen LIBOR fix—the average cost of panel banks raising funds in the yen money market—to fall by 8 bp in a single day. Dealers said that one Japanese bank was equivalent to a 5 bp lower yen LIBOR rate and that the removal of the Bank of China was equivalent to a 1 bp or 2 bp reduction.

Away from the immediate trading losses, market reaction to the panel change was mixed. The move was welcomed by some, who claimed that the previous panel was unrepresentative of the yen cash business being done.

"Most of the cash is traded in London by foreign banks. It doesn't make sense to have half Japanese banks on the panel," said one yen swaps dealer. He added that because of the presence of a number of Japanese banks on the panel, yen LIBOR rates were being pushed above where most of the active yen cash participants could fund themselves in the market.

Others, however, disagreed. "It's a domestic [Japanese] market at the end of the day. The BBA could now lose credibility in Japan," said one US bank money markets trader.

BBA officials said the selections were made by the BBA's FX and Money Markets Advisory Panel, following private nominations and discussions with the BBA LIBOR Steering Group. They said the aim of the advisory panel was to ensure that the contributor panels broadly reflected the "balance of activity in the interbank deposit market."

6. You are given the following information:

3-m LIBOR	3.2%	92 days
3 × 6 FRA	3.3%–3.4%	90 days
6 × 9 FRA	3.6%–3.7%	90 days
9 × 12 FRA	3.8%–3.9%	90 days

   a. Show how to construct a synthetic 9-month loan with fixed rate beginning with a 3-month loan. Plot the cash flow diagram.

   b. What is the fixed 9-month borrowing cost?

7. Consider the following instruments and the corresponding quotes.

   a. Rank these instruments in increasing order of their yields.

Instrument	Quote
30-day US T-bill	5.5
30-day UK T-bill	5.4
30-day ECP	5.2
30-day interbank deposit USD	5.5
30-day US CP	5.6

   b. You purchase an ECP (Euro) with the following characteristics

Value date	July 29, 2002
Maturity	September 29, 2002
Yield	3.2%
Amount	10,000,000 USD

   What payment do you have to make?

8. Determine the dollar settlement of the 3 × 6 FRA for the amount $300,000 assuming (a) the settlement occurs on the date of loan initiation and (b) if settlement occurs on the date of loan repayment. Use the following data to carry out the calculations. Interest rates are expressed as stated annual interest rates.

$$3 \times 6 \text{ FRA rate } F_{t_0} = 4.38\% \text{ and 3-m LIBOR rate } L_{t_3} = 4.02\%$$

9. Assume the Eurodollar futures price at time t_0 is 93.83 and the contract expires in 3 months time
  a. Calculate the 3-month forward rate implied by this price.
  b. Calculate the repayment amount for bonds with maturities of 3, 6, 9 and 12 months if the investor bought $5 million future contracts.

10. Using the following bond price data calculate the 3 × 6, 6 × 9, 3 × 9 and 6 × 12 FRA rates.

$$B(t_0, t_1) = 98.79, \quad B(t_0, t_2) = 97.21, \quad B(t_0, t_3) = 95.84, \quad B(t_0, t_4) = 93.82$$

11. Barbell example.

On June 16, 2014, a fund manager is considering the purchase of $1 million face amount of US Treasury $1\frac{1}{2}$s with a maturity date of May 31, 2019 at a cost of 990,468.75. The manager examines the bond further and, using his Bloomberg terminal, notices that the bond has a yield of 1.701% and a modified duration of 4.748. Before deciding to buy the bond the manager considers the information in the following table about two other bonds:

Data on Three US Treasury Bonds as of June 16, 2014					
Coupon	Maturity	Price	Yield	Duration	Convexity
$0\frac{3}{8}$	May 31, 2016	99.8398438	0.457%	1.943	0.048
$1\frac{1}{2}$	May 31, 2019	99.046875	1.701%	4.748	0.254
$2\frac{1}{2}$	May 15, 2024	99.140625	2.598%	8.701	0.859

The three bonds in the table have maturities of approximately 2, 5, and 10 years. The fund manager could purchase the 1½ bond or alternatively a barbell portfolio consisting of the 0⅜ and the 2½ bonds. In a barbell portfolio, part of the portfolio is made up of long-term bonds and the other part comprises very short-term bonds. The term "barbell" is based on the notion that the strategy resembles a barbell, heavily weighted at both ends and with nothing in between. Construct such a barbell portfolio and evaluate the alternative barbell strategy compared to the 1½ bond investment. Such barbell portfolio should be constructed to cost the same and have the same duration as the bullet investment. Your evaluation should consider the yield and the convexity of the 1½ bond relative to the barbell strategy.

INTRODUCTION TO INTEREST-RATE SWAP ENGINEERING

4.1 THE SWAP LOGIC

Swaps are the first basic tool that we introduced in Chapter 1. It should be clear by now that swaps are essentially the generalization of what was discussed in Chapters 1 and 3. We start this chapter by providing a general logic for swaps.

It is important to realize that essentially all swaps can be combined under one single logic. Consider *any* asset. Suppose we add to this asset another contract and form a basket. But, suppose we choose this asset so that the market risk, or the volatility associated with it, is exactly zero. Then the volatility (or the risk) of the basket is identical to the volatility (or the risk) of the original asset. Yet, the addition of this "zero" can change other characteristics of the asset and make the whole portfolio much more liquid, practical, and useful for hedging, pricing, and administrative reasons. This is what happens when we move from original "cash" securities to swaps. We take a security and augment it with a "zero-volatility" asset. This is the swap strategy.

4.1.1 THE EQUIVALENT OF ZERO IN FINANCE

First we would like to develop the equivalent of zero in finance as was done in Chapter 1. Why? Because, in the case of standard algebra, we can add (subtract) zero to a number and its value does not change. Similarly if we have the equivalent of a "zero" as a security, then we could add this security to other securities and this addition would not change the original *risk characteristics* of the original security. But in the mean time, the cash flow characteristics, regulatory requirements, tax exposure, and balance sheet exposure of the portfolio may change in a desirable way.

What is a candidate for such a "zero"? Consider the interbank money market loan as shown in Figure 4.1. The loan principal is 100 and is paid at time t_1. Interest and principal is received at t_2. Hence this is a default-free loan to be made in the future. The associated interest rate is the LIBOR rate L_{t_1} to be observed at time t_1.

We write a forward contract on this loan. According to this, 100 is borrowed at t_1 and for this the prevailing interest rate is paid at that time. What is, then, the value of this forward loan contract for all $t \in [t_0, t_1]$?

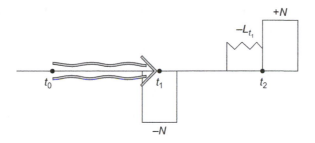

FIGURE 4.1

An interbank money market loan.

It turns out that one can, in fact, calculate this value exactly at time t_0 even though the future LIBOR rate L_{t_1} is not known then. Consider the following argument.

The t_2 cash flows are

$$+100 + 100L_{t_1}\delta \tag{4.1}$$

Discounting this value to time t_1, we get

$$\frac{+(1 + L_{t_1}\delta)100}{(1 + L_{t_1}\delta)} = +100 \tag{4.2}$$

Adding this to the initial 100 that was lent, we see that the total value of the cash flows generated by the forward loan contract is exactly zero for all times t during the interval $[t_0, t_1]$, *no matter what* the market thinks about the future level of L_{t_1}.[1]

Denoting the value of this forward contract by V_t, we can immediately see that

$$\text{Volatility}(V_t) \equiv 0 \quad \text{for all } t \in [t_0, t_1] \tag{4.3}$$

Hence adding this contract to any portfolio would not change the risk (volatility) characteristics of that portfolio. This is important and is a special property of such LIBOR contracts.[2] Thus let V_t

---

[1] Another way of saying this is to substitute the forward rate F_{t_0} for L_{t_1}. As Δ amount of time passes, this forward rate would change to $F_{t_0} + \Delta$. But the value of the loan would not change, because

$$\frac{-(1 + F_{t_0}\delta)100}{(1 + F_{t_0}\delta)} = \frac{-(1 + F_{t_0+\Delta}\delta)100}{(1 + F_{t_0+\Delta}\delta)} = -100$$

[2] For example, if the forward contract specified a forward rate F_{t_0} at time t_0, the value of the contract would not stay the same, since starting from time t_0 as Δ amount of time passes, a forward contract that specifies a F_{t_0} will have the value

$$\frac{-(1 + F_{t_0}\delta)100}{(1 + F_{t_0+\Delta}\delta)} \neq \frac{-(1 + F_{t_0}\delta)100}{-(1 + F_{t_0}\delta)} = -100$$

This is the case since, normally,

$$F_{t_0} \neq F_{t_0+\Delta}$$

denote the value of a security with a sequence of cash flows so that the security has a value equal to zero identically for all $t \in [t_0, t_1]$,

$$V_1 = 0 \tag{4.4}$$

Let S_t be the value of any other security, with

$$0 < \text{Volatility}(S_t) \quad t \in [t_0, t_1] \tag{4.5}$$

Suppose both assets are *default-free*.[3] Then, because the loan contract has a value identically equal to zero for *all* $t \in [t_0, t_1]$, we can write

$$S_t + V_t = S_t \tag{4.6}$$

in the sense that

$$\text{Volatility}(S_t + V_t) = \text{Volatility}(S_t) \tag{4.7}$$

Hence the portfolio consisting of an S_t *and* a V_t asset has the identical volatility and correlation characteristics as the original asset S_t. It is in this sense that the asset V_t is equivalent to zero. By adding it to any portfolio, we do not change the market risk characteristics of this portfolio.

Still, the addition of V_t may change the original asset in important ways. In fact, with the addition of V_t:

1. The asset may move the S_t off-balance sheet. Essentially, nothing is purchased for cash.
2. Registration properties may change. Again no basic security is purchased.[4]
3. Regulatory and tax treatment of the asset may change.
4. No up-front cash will be needed to take the position. This will make the modified asset much more liquid.[5]

We will show these using three important applications of the swap logic: but first some advantages of the swaps. Swaps have the following important advantages among others.

Remark 1: When you buy a US Treasury bond or a stock issued by a US company, you can only do this in the United States. But, when you work with the swap, $S_t + V_t$, you can do it anywhere, since you are *not* buying/selling a cash bond or a "cash" stock. It will consist of only swapping cash.

Remark 2: The swap operation is a natural extension of a market practitioner's daily work. When a trader buys an asset, the trader needs to *fund* this trade. "Funding" an asset with a LIBOR loan amounts to the same scheme as adding V_t to the S_t. In fact, the addition of the zero asset eliminates the initial cash payments.

---

[3]The GFC during which LIBOR rose substantially highlighted the counterparty risk associated with interbank lending and LIBOR. This has led to a reassessment of LIBOR as the appropriate proxy for the riskless rate. The OIS (Overnight Index Swap) rate is nowadays a commonly used discount rate for the pricing of certain derivatives. See Chapter 24 for further details.

[4]What is purchased is its derivative.

[5]There are collateral requirements associated with OTC swaps. New regulation requires the posting of initial and variation margin for centrally cleared and noncentrally cleared swaps (with the exception of FX swaps).

Remark 3: The new portfolio will have no *default* risk.[6] In fact with a swap, no loan is extended by any party.

Remark 4: Finally the accounting, tax, and regulatory treatment of the new basket may be much more advantageous.

4.1.2 A GENERALIZATION

We can generalize this notion of "zero." Consider Figure 4.2. This figure adds *vertically n* such deposits, all having the same maturity but starting at different times, t_i, $i = 1, 2, \ldots$. The resulting cash flows can be interpreted in two ways. First, the cash flows can be regarded as coming from a floating rate note (FRN) that is purchased at time t_i with maturity at $t_n = T$. The note pays LIBOR flat. The value of the FRN at time t_i will be given by

$$\text{Value}_t[\text{FRN}] = V_t^1 + V_t^2 + \cdots + V_t^n \quad t \in [t_0, t_1] \tag{4.8}$$
$$= 0$$

where V_t^i is the time t value of the period deposit starting at time t_1.

The second interpretation is that the cash flows shown in Figure 4.2 are those of a *sequence* of money market loans that are rolled over at periods $t_1, t_2, \ldots, t_{n-1}$.

4.2 APPLICATIONS

In order to see how powerful such a logic can be, we apply the procedure to different types of assets as was done in Chapter 1. In this chapter, we consider an equity portfolio and add the zero-volatility asset to it. This way we obtain an *equity swap*. A similar approach can be applied to currencies and we discuss *cross-currency* and *FX swaps* in Chapter 6. A *commodity swap* can be obtained similarly and we explain exactly how in Chapter 7.

In Chapter 19, we do the same with a *defaultable bond*. The operation will lead to a credit default swap (CDS). These swaps lead to some of the most liquid and largest markets in the world. They are all obtained from a single swap logic.

4.2.1 EQUITY SWAP

Consider a portfolio of stocks whose *fair* market value at time t_0 is denoted by S_{t_0}. Let $t_n = T$, $t_0 < \cdots < t_n$, where T is a date that defines the expiration of an equity-swap contract. For simplicity, think of $t_n - t_0$ as a 1-year period. We divide this period into equally spaced intervals of length δ, with $t_1, t_2, t_3, \ldots, t_n = T$ being the settlement dates.

Let $\delta = \frac{1}{4}$ so that the t_i's are 3 months apart. During a 1-year interval with $n = 4$, the portfolio's value will be changed by

$$S_{t_4} - S_{t_0} = [(S_{t_1} - S_{t_0}) + (S_{t_2} - S_{t_1}) + (S_{t_3} - S_{t_2}) + (S_{t_4} - S_{t_3})] \tag{4.9}$$

---

[6]Although there will be a counterparty risk.

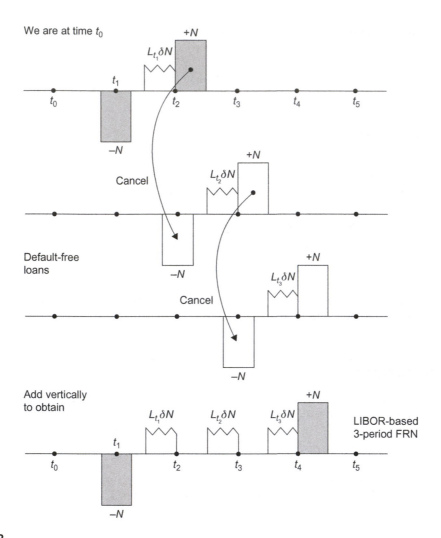

FIGURE 4.2

A sequence of FRN purchases.

This can be rewritten as

$$S_{t_4} - S_{t_0} = \Delta S_{t_0} + \Delta S_{t_1} + \Delta S_{t_2} + \Delta S_{t_3} \tag{4.10}$$

We consider buying and marking this portfolio to market in the following manner:

1. $N = 100$ is invested at time t_1.
2. At every t_1, $i = 1, 2, 3, 4$ total dividends amounting to d are collected.[7]

---

[7]Note that we are assuming constant and known dividend payments throughout the contract period.

3. At the settlement dates, we collect (pay) the cash due to the appreciation (depreciation) of the portfolio value.

4. At time $t_n = T$, collect the original USD100 invested.

This is exactly what an equity investor would do. The investor would take the initial investment (principal), buy the stocks, collect dividends, and then sell the stocks. The final capital gains or losses will be $S_{t_n} - S_{t_0}$. In our case, this is monetized at each settlement date. The cash flows generated by this process can be shown in Figure 4.3.

Now we follow the swap logic discussed above and add to the stock portfolio the contract V_t which denotes the time t value of the cash flows implied by a forward LIBOR deposit. Let g_{t_i} be the percentage decline or increase in portfolio value at each date, and let the initial investment be denoted by the notional amount N:

$$S_{t_0} = N \tag{4.11}$$

Then,

1. The value of the stock portfolio has not changed any time between t_0 and t_1, since the forward FRN has value identically equal to zero at any time $t \in [t, t_0]$.

2. But the initial and final N's cancel.

3. The outcome is an exchange of

$$L_{t_{i-1}} \delta N \tag{4.12}$$

against

$$(\Delta S_{t_i} + d)\delta N \tag{4.13}$$

at each t_i.

4. Then we can express the cash flows of an equity swap as the exchange of

$$(L_{t_{i-1}} - d_i)\delta N \tag{4.14}$$

against

$$\Delta S_{t_i} \delta N \tag{4.15}$$

at each t_i. The d_i being an unknown percentage dividend yield, the market will trade this as a spread. The market maker will quote the "expected value of" d_i and any incremental supply−demand imbalances as the *equity-swap spread*. The buyer of the equity swap will need to pay LIBOR $L_{t_{i-1}} \delta N$ plus a spread while receiving the total return on the equity index, i.e., $(\Delta S_{t_i} + d)\delta N$ at each t_i.

5. The swap will involve no up-front payment.

This construction proves that the market *expects* the portfolio S_{t_i} to change by $L_{t_{i-1}} - d_{t_i}$ each period, in other words, we have

$$E_t^{\tilde{P}}[\Delta S_{t_i}] = L_{t_{i-1}} - d_{t_i} \tag{4.16}$$

This result is proved normally by using the fundamental theorem of asset pricing and the implied risk-neutral probability.

Now we move on to interest-rate swaps.

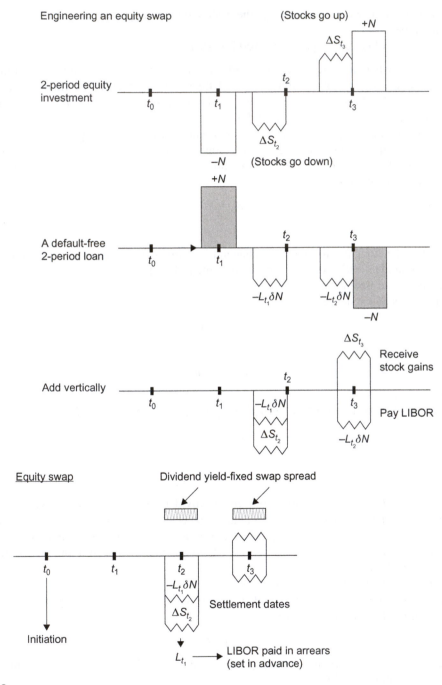

FIGURE 4.3

Construction of an equity swap.

4.3 THE INSTRUMENT: SWAPS

Imagine *any* two sequences of cash flows with different characteristics. These cash flows could be generated by any process—a financial instrument, a productive activity, a natural phenomenon. They will also depend on different risk factors. Then one can, in principle, devise a contract where these two cash flow sequences are exchanged. This contract will be called a *swap*. To design a swap, we use the following principles:

1. A swap is arranged as a pure exchange of cash flows and hence should not require any additional net cash payments at initiation.[8] In other words, the *initial* value of the swap contract should be *zero*.
2. The contract specifies a *swap spread*. This variable is adjusted to make the two counterparties willing to exchange the cash flows.

A generic exchange is shown in Figure 4.4. In this figure, the first sequence of cash flows starts at time t_1 and continues periodically at $t_2, t_3, \ldots, t_k$. There are k *floating* cash flows of differing sizes denoted by

$$\{C(s_{t_0}, x_{t_1}), C(s_{t_0}, x_{t_2}), \ldots, C(s_{t_0}, x_{t_k})\} \tag{4.17}$$

These cash flows depend on a vector of market or credit risk factors denoted by x_{t_i}. The cash flows depend also on the s_{t_0}, a swap *spread* or an appropriate *swap rate*. By selecting the value of s_{t_0}, the initial value of the swap can be made zero.

Figure 4.4b represents another strip of cash flows:

$$\{B(y_{t_0}), B(y_{t_1}), B(y_{t_2}), \ldots, B(y_{t_k})\} \tag{4.18}$$

which depend potentially on some other risk factors denoted by y_{t_i}.

The swap consists of exchanging the $\{C(s_{t_0}, x_{t_i})\}$ against $\{B(y_{t_i})\}$ *at settlement dates* $\{t_i\}$. The parameter s_{t_0} is selected at time t_0 so that the two parties are willing to go through with this exchange without any initial cash payment. This is shown in Figure 4.4c. One will *pay* the $C(\cdot)$'s and *receive* the $B(\cdot)$'s. The counterparty will be the "other side" of the deal and will do the reverse.[9] Clearly, if the cash flows are in the *same* currency, there is no need to make two different payments in each period t_i. One party can simply pay the other the *net* amount. Then actual wire transfers will look more like the cash flows as shown in Figure 4.5. Of course, what one party receives is equal to what the counterparty pays.

Now, if two parties who are willing to exchange the two sequences of cash flows without any up-front payment, the market value of these cash flows must be the same no matter how different they are in terms of implicit risks. Otherwise one of the parties will require an up-front net payment. Yet, as time passes, a swap agreement may end up having a positive or negative net value, since the variables x_{t_i} and y_{t_i} will change, and this will make one cash flow more "valuable" than the other.

---

[8]Here we abstract away from any real-world complications such as the exchange of collateral according to ISDA CSAs.
[9]Here we use the term "cash flows," but it could be that what is exchanged is physical goods.

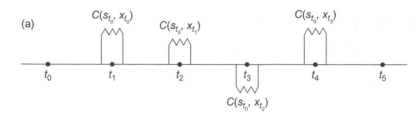

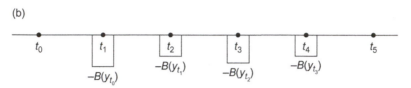

Adding vertically, we get a swap.

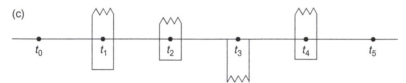

Note that time-t_0 value is zero...

FIGURE 4.4

Payoffs of a generic swap.

If cash flows are in the same currency,
then the counterparty will receive the net amounts...

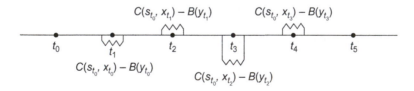

FIGURE 4.5

Exchange of net amounts in an interest-rate swap.

EXAMPLE

Suppose you signed a swap contract that entitles you to a 7% return in dollars, in return for a 6% return in euros. The exchanges will be made every 3 months at a *predetermined* exchange rate e_{t_0}. At initiation time t_0, the net value of the commitment should be zero, given the correct *swap spread*. This means that at time t_0, the market value of the receipts and payments are the same. Yet, after the contract is initiated, USD interest rates may fall relative to European rates. This would make the receipt of 7% USD funds relatively more valuable than the payments in euro.

As a result, from the point of view of the USD-receiving party, the value of the swap will move from zero to positive, while for the counterparty, the swap will have a negative value.

Of course, actual exchanges of cash flows at times $t_1, t_2, \ldots, t_n$ may be a more complicated process than the simple transactions as shown in Figure 4.5. What exactly is paid or received? Based on which price? Observed when? What are the penalties if deliveries are not made on time? What happens if a t_i falls on a holiday? A typical swap contract needs to clarify many such parameters. These and other issues are specified in the *documentation* set by the International Swaps and Derivatives Association (ISDA).

4.4 TYPES OF SWAPS

Swaps are a very broad instrument category. Practically, every cash flow sequence can be used to generate a swap. It is impossible to discuss all the relevant material in this book. So, instead of spreading the discussion thinly, we adopt a strategy where a number of critical swap structures are selected and the discussion is centered on these. We hope that the extension of the implied swap engineering to other swap categories will be straightforward.

4.4.1 NONINTEREST-RATE SWAPS

Most swaps are interest rate related given the LIBOR and yield curve exposures on corporate and bank balance sheets. But swaps form a broader category of instruments and to emphasize this point, we start the discussion with one particular type of noninterest-rate swap, the equity swap. One of the most important recent innovations is the CDS, but we will examine it in detail in a separate chapter. Similarly we will discuss currency- and commodity-related swaps in separate chapters on financial engineering applications in foreign exchange and commodity markets respectively.

4.4.1.1 Equity swaps

Equity-swaps exchange equity-based returns against LIBOR as seen earlier. These swaps are sometimes also labeled as total return swaps.

In equity swaps, the parties will exchange two sequences of cash flows. One of the cash flow sequences will be generated by dividends and capital gains (losses), while the other will depend on a money market instrument, in general LIBOR. Once clearly defined, each cash flow can be valued separately. Then, adding or subtracting a *spread* to the corresponding LIBOR rate would make the two parties willing to exchange these cash flows with no initial payment. The contract that makes this exchange legally binding is called an equity swap.

Thus, a typical equity swap consists of the following. Initiation time will be t_0. An equity index I_{t_i} and a money market rate, say LIBOR L_{t_i}, are selected. At times $\{t_1, t_2, \ldots, t_n\}$, the parties will exchange cash flows based on the percentage change in I_{t_i}, written as

$$N_{t_{i-1}} \left(\frac{I_{t_i} - I_{t_{i-1}}}{I_{t_{i-1}}} \right) \tag{4.19}$$

against LIBOR-based cash flows, $N_{t_{i-1}} L_{t_{i-1}} \delta$ plus or minus a spread. The N_{t_i} is the notional amount, which is not exchanged.

Note that the notional amount is allowed to be reset at every $t_0, t_1, \ldots, t_{n-1}$, allowing the parties to adjust their position in the particular equity index periodically. In equity swaps, this notional principal can also be selected as a constant, N.

EXAMPLE

In Figure 4.6, we have a 4-year sequence of capital gains (losses) plus dividends generated by a certain equity index. They are exchanged every 90 days, against a sequence of cash

(a) Total return from equity index

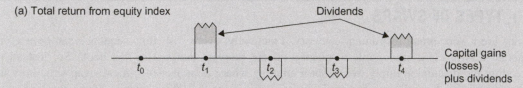

(b) LIBOR-based cash flows

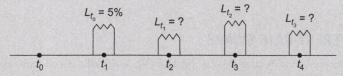

(c) Cash flows from equity swap

Receive capital gains (losses) and dividends

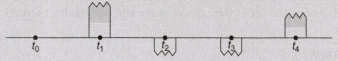

Pay LIBOR-based cash flows plus a negative or positive spread...

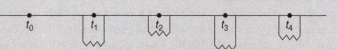

FIGURE 4.6

Components of an equity swap.

flows based on 3-month LIBOR—20 bp. Figure 4.6a and b depicts the cash flows from the total return—associated with the equity index and the LIBOR payments, respectively. Figure 4.6c shows the payoffs of the equity swap.

The *notional principal* is USD 1 million. At time t_0, the elements of these cash flows will be unknown.

At time t_1, the respective payments can be calculated once the index performance is observed. Suppose we have the following data:

$$I_{t_0} = 800 \tag{4.20}$$

$$I_{t_1} = 850 \quad L_{t_0} = 5\% \quad \text{spread} = 0.20 \tag{4.21}$$

Then the time t_1 equity-linked cash flow is

$$1\,m\left(\frac{I_{t_1} - I_{t_0}}{I_{t_0}}\right) = 1,000,000(0.0625) = 62,500 \tag{4.22}$$

The LIBOR-linked cash flows will be

$$1\,m(L_{t_0} - s_{t_0})\frac{90}{360} = 1,000,000(0.05 - 0.002)\frac{1}{4} = 12,000 \tag{4.23}$$

The remaining unknown cash flows will become known as time passes, dividends are paid, and prices move. The spread is subtracted from the interest rate.

Some equity swaps are between *two* equity indices. The following example illustrates the idea.

EXAMPLE

In an equity swap, the holder of the instrument pays the total return of the S&P 500 and receives the return on another index, say the Nikkei. Its advantage for the holder lies in the fact that, as a swap, it does not involve paying any up-front premium.

Of course, the same trade could also be created by selling S&P futures and buying futures on another equity index. But the equity swap has the benefit that it simplifies tracking the indices.

Later in this chapter, we will discuss several uses of equity swaps.

4.4.2 INTEREST-RATE SWAPS

This is the largest swap market. It involves exchanging cash flows generated by different interest rates. The most common case is when a *fixed* swap rate is paid (received) against a floating LIBOR rate in the *same* currency. Interest-rate swaps have become a fundamental instrument in world financial markets. The following reading illustrates this for the case of plain vanilla interest-rate swaps.

EXAMPLE

The swap curve is being widely touted as the best alternative to a dwindling Treasury market for benchmarking US corporate bonds.... This has prompted renewed predictions that the swap curve will be adopted as a primary benchmark for corporate bonds and asset-backed securities.

... Investors in corporate bonds say there are definite benefits from the increasing attention being paid to swap spreads for valuing bonds. One is that the mortgage-backed securities market has already to a large degree made the shift to use of LIBOR-based valuation of positions, and that comparability of corporate bonds with mortgage holdings is desirable.

... Swap dealers also point out that while the agency debt market is being adroitly positioned by Fannie Mae and Freddie Mac as an alternative to the Treasury market for benchmarking purposes, agency spreads are still effectively bound to move in line with swap spreads.

... Bankers and investors agree that hedging of corporate bond positions in the future will effectively mean making the best use of whatever tools are available. So even if swaps and agency bonds have limitations, and credit costs edge up, they will still be increasingly widely used for hedging purposes

(Thomson Reuters IFR, Issue 6321).

This reading illustrates the crucial position held by the swap market in the world of finance. The "swap curve" obtained from *interest-rate swaps* is considered by many as a benchmark for the term structure of interest rates, and this means that most assets could eventually be priced off the interest-rate swaps, in one way or another. Also, the reading correctly points out some major sectors in markets. In particular, (1) the mortgage-backed securities (MBS) market, (2) the market for "agencies," which means securities issued by Fannie Mae or Freddie Mac, etc., and (3) the corporate bond market have their own complications, yet, swaps play a major role in all of them. At this point, it is best to define formally the interest-rate swap and then look at an example.

A *plain vanilla interest-rate swap* initiated at time t_0 is a commitment to exchange interest payments associated with a notional amount N, settled at clearly identified settlement dates, $\{t_1, t_2,\ldots, t_n\}$. The *buyer* of the swap will make fixed payments of size $s_{t_0}N\delta$ each and receive floating payments of size $L_{t_i}N\delta$. The LIBOR rate L_{t_i} is determined at *set dates* $\{t_0, t_1,\ldots, t_{n-1}\}$. The maturity of swap is m years.[10] The s_{t_0} is the swap rate.

EXAMPLE

An interest-rate swap has a notional amount N of USD 1 million, a 7% fixed rate for 2 years in semiannual (s.a.) payments against a cash flow generated by 6-month LIBOR. This is shown in Figure 4.7a. There are two sequences of cash flows. One involves four payments of USD35,000 each. They are known at t_0 and paid at the *end* of each 6-month period.

---

[10]Here, $m = n\delta$. The δ is the days count parameter.

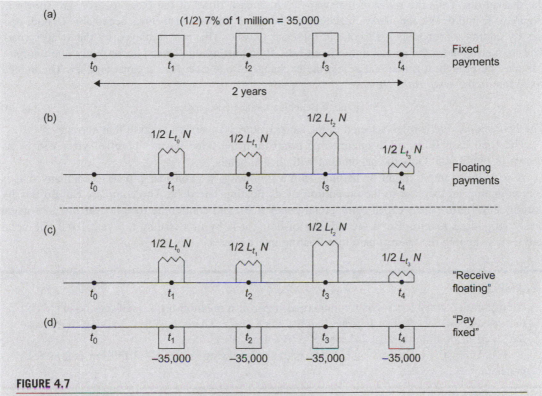

FIGURE 4.7

Cash flows of fixed payer and fixed receiver in an interest-rate swap.

The second is shown in Figure 4.7b. These cash flows are determined by the value of 6-month USD LIBOR to be observed at set dates. Four separate LIBOR rates will be observed during this period. The L_{t_0} is known at the initial point t_0. The remaining LIBOR rates, L_{t_1}, L_{t_2}, and L_{t_3}, will be observed gradually as time passes but are unknown initially.

In Figure 4.7, the floating cash flows depending on L_{t_i} are observed at time t_i but are *paid-in-arrears* at times t_{i+1}. Swaps that have this characteristic are known as paid-in-arrears swaps.

Clearly, we have two sets of cash flows with different market risk characteristics. The market will price them separately. Once this is done, market participants can trade them. A *fixed payer* will pay the cash flows in Figure 4.7a and receive the one in Figure 4.7b. This institution is the *buyer* of the interest-rate swap.

The market participant on the other side of the deal will be doing the reverse—receiving cash flows based on a fixed interest rate at time t_0, while paying cash flows that become gradually known as time passes and the LIBOR rates L_{t_i} are revealed. This party is the *fixed receiver*, whom the market also calls the *seller* of the swap.[11] These cash flows of the fixed receiver are shown in Figure 4.7c and d. We can always make the exchange of the two cash flows acceptable to both parties by adding a proper spread to one of the cash flows.[12] This role is played by the swap spread. The market includes the spread in the fixed rate. By adjusting this spread accordingly, the two parties may be brought together and accept the exchange of one cash flow against another. The agreed fixed rate is the *swap rate*. We have

$$\text{Swap rate} = \text{Benchmark rate} + \text{Swap spread} \tag{4.24}$$

The *benchmark rate* is often selected as the same maturity sovereign bond in that currency.

The final cash flows of an interest-rate swap from the fixed payer's point of view will be as shown in Figure 4.5. Only the net amount will change hands.

A real-life example might be helpful. In the following, we consider a private company that is contemplating an increase in the proportion of its floating rate debt. The company can do this by issuing short-term paper, called commercial paper (CP), and continuing to roll over the debt when these obligations mature. But a second way of doing it is by first issuing a 5-year fixed-rate bond and then swapping the interest paid into floating interest rates.

EXAMPLE

A corporation considers issuing commercial paper or a medium-term fixed-rate bond (MTN) that it can convert to a floating-rate liability via a swap. The company is looking to increase the share of floating-rate liabilities to 50–55% from 30%.

The alternative to tapping the MTN market is drawing on its $700 million commercial paper facility.

This reading shows one role played by swaps in daily decisions faced by corporate treasuries. The existence of swaps makes the rates observed in the important CP sector more closely related to the interest rates in the MTN sector.

4.4.2.1 Basis swaps

Basis swaps are similar to currency swaps except that often there is only one currency involved. A basis swap involves exchanging cash flows in one *floating* rate, against cash flows in another *floating* rate, in the same currency. One of the involved interest rates is often a non-LIBOR-based rate, and the other is LIBOR.

The following reading gives an idea about the basis swap. Fannie Mae, a US government *agency*, borrows from international money markets in USD LIBOR and then lends these funds to mortgage banks. Fannie Mae faces a *basis risk* while doing this. There is a small difference between the interest rate that it eventually pays, which is USD LIBOR, and the interest rate it

---

[11]Similarly to the FRA terminology, those who pay a fixed rate are in general players who are looking to lock in a certain interest rate and reduce risks associated with floating rates. These are clients who need "protection." Hence, it is said that they are buying the swap.

[12]After all, apples and oranges are rarely traded one to one.

eventually receives, the USD discount rate. To hedge its position, Fannie Mae needs to convert one floating rate to the other. This is the topic of the reading that follows:

EXAMPLE

Merrill Lynch has been using Fannie Mae benchmark bonds to price and hedge its billion dollar discount/ basis swap business. "We have used the benchmark bonds as a pricing tool for our discount/LIBOR basis swaps since the day they were issued. We continue to use them to price the swaps and hedge our exposure," said the head of interest-rate derivatives trading.

He added that hedging activity was centered on the 5- and 10-year bonds—the typical discount/LIBOR basis swap tenors. Discount/LIBOR swaps and notes are employed extensively by US agencies, such as Fannie Mae, to hedge their basis risk. They lend at the US discount rate but fund themselves at the LIBOR rate and as a result are exposed to the LIBOR/discount rate spread. Under the basis swap, the agency/ municipality receives LIBOR and in return pays the discount rate.

Major US derivative providers began offering discount/LIBOR basis swaps several years ago and now run billion dollar books.

(Thomson Reuters IFR, Issue 1229).

This reading illustrates two things. Fannie Mae needs to swap one floating rate to another in order to allow the receipts and payments to be based on the same risk. But at the same time, *because* Fannie Mae is hedging using basis swaps and because there is a large amount of such Fannie Mae bonds, some market practitioners may think that these *agency bonds* make good pricing tools for basis swaps themselves.

4.4.2.2 *What is an asset swap?*

The term *asset swap* can, in principle, be used for any type of swap. After all, sequences of cash flows considered thus far are generated by some assets, indices, or reference rates. Also, swaps linked to equity indices or reference rates such as LIBOR can easily be visualized as FRNs, corporate bond portfolios, or portfolios of stocks. Exchanging these cash flows is equivalent to exchanging the underlying asset.

Yet, the term asset swap is often used with a more precise meaning. Consider a defaultable *par* bond that pays annual coupon c_{t_0}. Suppose the payments are semiannual. Then we can imagine a swap where coupon payments are exchanged against 6-month LIBOR L_{t_i} plus a spread s_{t_0}, every 6 months. The coupon payments are fixed and known at t_0. The floating payments will be random, although the spread component, s_{t_0}, is known at time t_0 as well. This structure is often labeled an asset swap.[13] The reader can easily put together the cash flows implied by this instrument, if the issue of default is ignored. Such a cash flow diagram would follow the exchanges as shown in Figure 4.4. One sequence of cash flows would represent coupons, the other LIBOR plus a spread.

Asset swaps interpreted this way offer a useful alternative to investors. An investor can always buy a bond and receive the coupon c_{t_0}. But by using an asset swap, the investor can also swap out of the coupon payments and receive only floating LIBOR plus the spread s_{t_0}. This way the

---

[13]In an asset, swap credit risk remains with the bond holder.

exposure to the *issuer* is kept and the exposure to *fixed interest rates* is eliminated. In fact, treasury bonds or fixed receiver interest swaps may be better choices if one desires exposure to fixed interest rates. Given the use of LIBOR in this structure, the s_{t_0} is calculated as the spread to the corresponding fixed swap rates.

4.4.2.3 More complex swaps

The swaps discussed thus far are liquid and are traded actively. One can imagine many other swaps. Some of these are also liquid, others are not. Amortizing swaps, bullion swaps, MBS swaps, quanto (differential) swaps, inflation swaps, longevity swaps[14] are some that come to mind. We will not elaborate on them at this point; some of these swaps will be introduced as examples or exercises in later chapters.

An interesting special case is constant maturity swaps (CMS), which will be discussed in detail in Chapter 14. The CMS swaps have an interesting convexity dimension that requires taking into account volatilities and correlations across various forward rates along a yield curve. A related swap category is constant maturity treasury (CMT) swaps.

4.4.3 SWAP CONVENTIONS

Interest-rate swap markets have their own conventions. In some economies, the market quotes the swap *spread*. This is the case for USD interest-rate swaps. USD interest-rates swaps are quoted as a spread to Treasuries. In Australia, the market also quotes swap spreads. But the spreads are to bond futures.

In other economies, the market quotes the swap *rate*. This is the case for euro interest-rate swaps.

Next, there is the issue of how to quote swaps. This is done in terms of two-way interest-rate quotes. But sometimes the quoted swap rate is on an annual basis, and sometimes it is on a semi-annual basis. Also, the day-count conventions change from one market to another. In USD swaps, the day-count is in general ACT/360. In EUR swaps day-count is 30/360.

According to market conventions, a fixed payer, called *the payer*, is *long* the swap, and has *bought* a swap. On the other hand, a fixed receiver, called *the receiver*, is *short* the swap, and has *sold* a swap.

4.5 ENGINEERING INTEREST-RATE SWAPS

We now study the financial engineering of swaps. We focus on plain vanilla *interest-rate* swaps. Engineering of other swaps is similar in many ways and is left to the reader. For simplicity, we deal with a case of only three settlement dates. Figure 4.8 shows a *fixed-payer*, three-period

---

[14]A longevity swap can help hedge the risk that pension scheme members live longer than expected. From an asset—liability management perspective, the main risks that pension funds face are interest rate, inflation, and and longevity (or mortality) risks. In a longevity swap, the pension fund would make fixed payments based on the agreed mortality assumptions to a counterparty (such as an insurer or investment bank) and receive floating payments based on the scheme's future realized mortality rates.

(a) An interest rate swap (with annual settlements)...

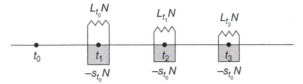

(b) Add and subtract $N = 1.00$ at the start and end dates...

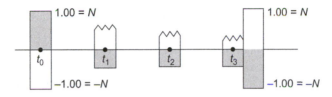

(c) Then detach the two cash flows...

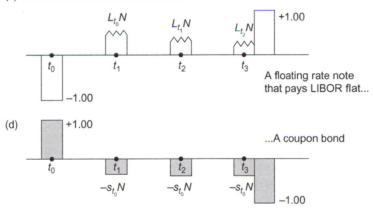

A floating rate note that pays LIBOR flat...

(d)

+1.00

...A coupon bond

FIGURE 4.8

Decomposition of an interest-rate swap.

interest-rate swap, with start date t_0. The swap is initiated at date t_0. The party will make three fixed payments and receive three floating payments at dates t_1, t_2, and t_3 in the same currency. The dates t_1, t_2, and t_3 are the settlement dates. The t_0, t_1, and t_2 are also the *reset* dates, dates on which the relevant LIBOR rate is determined.

We select the notional amount N as unity and let $\delta = 1$, assuming that the floating rate is 12-month LIBOR:[15]

$$N = \$1 \qquad\qquad (4.25)$$

---

[15]This is a simplification. In reality, the floating rate is either 3- or 6-month LIBOR.

Under these simplified conditions, the fixed payments equal s_{t_0}, and the LIBOR-linked payments equal L_{t_0}, L_{t_1}, and L_{t_2}, respectively. The *swap spread* will be the difference between s_{t_0} and the treasury rate on the bond with the same maturity, denoted by y_{t_0}.[16] Thus, we have

$$\text{Swap spread} = s_{t_0} - y_{t_0} \qquad (4.26)$$

We will study the engineering of this interest-rate swap. More precisely, we will discuss the way we can replicate this swap. More than the exact synthetic, what is of interest is the way(s) one can approach this problem.

A swap can be reverse-engineered in at least two ways:

1. We can first decompose the swap *horizontally*, into two streams of cash flows, one representing a floating stream of payments (receipts), the other a fixed stream. If this is done, then each stream can be interpreted as being linked to a certain type of bond.
2. Second, we can decompose the swap *vertically*, slicing it into n cash exchanges during n time periods. If this is done, then each cash exchange can be interpreted similarly (but not identically) to an FRA paid-in-arrears, with the property that the fixed rate is constant across various settlement dates.

We now study each method in detail.

4.5.1 A HORIZONTAL DECOMPOSITION

First we simplify the notation and the parameters used in this section. To concentrate on the engineering aspects only, we prefer to eliminate some variables from the discussion. For example, we assume that the swap will make payments every year so that the day-count parameter is $\delta = 1$, unless assumed otherwise. Next, we discuss a *forward* swap that is signed at time t_0, but starts at time t_1, with $t_0 < t_1$. During this discussion, we may occasionally omit the use of the term "forward" and refer to the forward swap simply as a swap.[17]

The traditional way to decompose an interest-rate swap is to do this horizontally. The original swap cash flows are shown in Figure 4.8a. Before we start, we need to use a trick. We add and subtract the same notional amount N at the start, and end dates, for both sequences of cash flows. Since these involve identical currencies and identical amounts, they cancel out and we recover the standard exchanges of floating versus fixed-rate payments. With the addition and subtraction of the initial principals, the swap will look as shown in Figure 4.8b.

Next, "detach" the cash flows in Figure 4.8b horizontally, so as to obtain two separate cash flows as shown in Figure 4.8c and d. Note that each sequence of cash flows is already in the form of a meaningful financial contract.

In fact, Figure 4.8c can immediately be recognized as representing a long forward position in an FRN that pays LIBOR flat. At time t_1, the initial amount N is paid and L_{t_1} is set. At t_2, the first

---

[16]This could be any interest rate accepted as a benchmark by the market.
[17]Remember also that there is no credit or counterparty risk, and that time is discrete. Finally, there are no bid-ask spreads.

interest payment is received, and this will continue until time t_4 where the last interest is received along with the principal.

Figure 4.8d can be recognized as a short forward position on a par coupon bond that pays a coupon equal to s_{t_0}. We (short) sell the bond to receive N. At every payment date, the fixed coupon is paid and then, at t_4, we pay the last coupon and the principal N.

Thus, the immediate decomposition suggests the following synthetic:

$$\text{Interest rate swap} = \{\text{Long FRN with LIBOR coupon, short par coupon bond}\} \tag{4.27}$$

Here the bond in question needs to have the same credit risk as in a flat LIBOR-based loan. Using this representation, it is straightforward to write the contractual equation:

$$
\boxed{\text{Long swap}} = \boxed{\begin{array}{l}\text{Short par}\\ \text{bond with}\\ \text{coupon } s_{t_0}\end{array}} + \boxed{\begin{array}{l}\text{Long}\\ \text{FRN paying}\\ \text{LIBOR flat}\end{array}} \tag{4.28}
$$

Using this relationship, one can follow the methodology introduced earlier and immediately generate some interesting synthetics.

4.5.1.1 A synthetic coupon bond

Suppose a AAA-rated entity with negligible default risk issues only 3-year FRNs that pay LIBOR—10 bp every 12 months.[18] A client would like to buy a coupon bond from this entity, but it turns out that no such bonds are issued. We can help our client by synthetically creating the bond. To do this, we manipulate the contractual equation so that we have a long coupon bond on the right-hand side:

$$
\boxed{\begin{array}{l}\text{Long}\\ \text{par bond with}\\ \text{coupon}\\ s_{t_0}-10\text{ bp}\end{array}} = \boxed{\begin{array}{l}\text{Sell a swap}\\ \text{with rate } s_{t_0}\end{array}} + \boxed{\begin{array}{l}\text{Long}\\ \text{FRN paying}\\ \text{LIBOR}-10\text{ bp}\end{array}} \tag{4.29}
$$

The geometry of this engineering is shown in Figure 4.9. The synthetic results in a coupon bond issued by the same entity and paying a coupon of $s_{t_0} - 10$ bp. The 10 bp included in the coupon accounts for the fact that the security is issued by a AAA-rated entity.

4.5.1.2 Pricing

The contractual equation obtained in Eq. (4.35) permits pricing the swap off the *debt markets*, using observed prices of fixed and floating coupon bonds.[19] To see this, we write the present value of the cash streams generated by the fixed and floating rate bonds using appropriate discount factors. Throughout this section, we will work with a special case of a 3-year spot swap that makes fixed

---

[18]Before the credit crisis of 2007−2008, AAA entities did have "negligible" credit risk, and spread to LIBOR was negative.

[19]As we discussed above, in practice, debt markets use swap rates as benchmarks and discount rates.

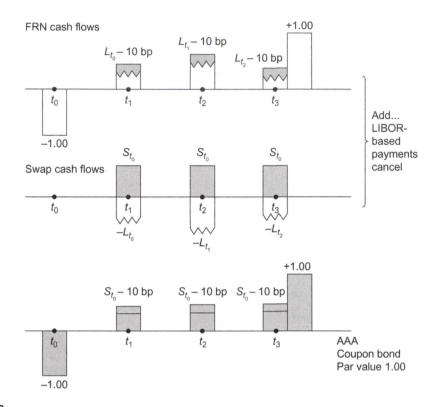

FIGURE 4.9

Creation of a synthetic bond.

payments against 12-month LIBOR. This simplifies the discussion. It is also straightforward to generalize the ensuing formulas to an n-year swap.

Suppose the swap makes three annual coupon payments, each equaling $s_{t_0}N$. We also have three floating rate payments, each with the value $L_{t_{i-1}}N$, where the relevant LIBOR $L_{t_{i-1}}$ is set at $t_i - 1$ but is paid at t_i.

4.5.1.3 Valuing fixed cash flows

To obtain the present value of the fixed cash flows, we discount them by the relevant floating rates. Note that, *if* we knew the floating rates $\{L_{t_0}, L_{t_1}, L_{t_2}\}$, we could write

$$\text{PV-fixed} = \frac{s_{t_0}N}{(1 + L_{t_0})} + \frac{s_{t_0}N}{(1 + L_{t_0})(1 + L_{t_1})} + \frac{s_{t_0}N + N}{(1 + L_{t_0})(1 + L_{t_1})(1 + L_{t_2})} \tag{4.30}$$

where we added N to date t_3 cash flows. But at $t = 0$, the LIBOR rates L_{t_i}, $i = 1, 2$, are unknown. Yet, we know that against each L_{t_i}, $i = 1, 2$, the market is willing to pay the known forward, (FRA) rate, $F(t_0, t_i)$. Thus, using the FRA rates *as if* they are the time t_0 market values of the unknown LIBOR rates, we get

$$\text{PV-fixed as of } t_0 = \frac{s_{t_0} N}{(1 + F(t_0, t_0))} + \frac{s_{t_0} N}{(1 + F(t_0, t_0))(1 + F(t_0, t_1))} \tag{4.31}$$

$$+ \frac{s_{t_0} N + N}{(1 + F(t_0, t_0))(1 + F(t_0, t_1))(1 + F(t_0, t_2))} \tag{4.32}$$

All the right-hand side quantities are known, and the present value can be calculated exactly, given the s_{t_0}.

4.5.1.4 Valuing floating cash flows

For the floating rate cash flows, we have[20]

$$\text{PV-floating as of } t_0 = \frac{L_{t_0} N}{(1 + L_{t_0})} + \frac{L_{t_1} N}{(1 + L_{t_0})(1 + L_{t_1})} + \frac{L_{t_2} N + N}{(1 + L_{t_0})(1 + L_{t_1})(1 + L_{t_2})} \tag{4.33}$$

Here, to get a numerical answer, we don't even need to use the forward rates. This present value can be written in a much simpler fashion, once we realize the following transformation:

$$\frac{L_{t_2} N + N}{(1 + L_{t_0})(1 + L_{t_1})(1 + L_{t_2})} = \frac{(1 + L_{t_2})N}{(1 + L_{t_0})(1 + L_{t_1})(1 + L_{t_2})} \tag{4.34}$$

$$= \frac{N}{(1 + L_{t_0})(1 + L_{t_1})} \tag{4.35}$$

Then, add this to the second term on the right-hand side of the present value in Eq. (4.33) and use the same simplification,

$$\frac{L_{t_1} N}{(1 + L_{t_0})(1 + L_{t_1})} + \frac{N}{(1 + L_{t_0})(1 + L_{t_1})} = \frac{N}{(1 + L_{t_0})} \tag{4.36}$$

Finally, apply the same trick to the first term on the right-hand side of Eq. (4.33) and obtain, somewhat surprisingly, the expression

$$\text{PV of floating payments as of } t_0 = N \tag{4.37}$$

According to this, the present value of an FRN equals the par value N at *every settlement date*. Such recursive simplifications can be applied to present values of floating rate payments at reset dates.[21] We can now combine these by letting

$$\text{PV of fixed payments as of } t_0 = \text{PV of floating payments as of } t_0 \tag{4.38}$$

---

[20]We remind the reader that the day's adjustment factor was selected as $\delta = 1$ to simplify the exposition. Otherwise, all LIBOR rates and forward rates in the formulas will have to be multiplied by the δ.

[21] Of course, this result does not hold *between* reset dates since the numerator and the denominator terms will, in general, be different. The payments will be made according to $(1 + L_{t_i})$, but the present values will use the observed LIBOR *since the last reset date*:

$$(1 + L_{t_0 + u} \delta)(1 + L_{t_1})$$

where u is the time elapsed since the last reset date. The $L_{t_0 + u}$ is the rate observed at time $t_0 + u$. The value of the cash flow will be a bit greater or smaller, depending on the value of $L_{t_0 + u}$.

This gives an equation where s_{t_0} can be considered as an unknown:

$$\frac{s_{t_0}N}{(1 + F(t_0, t_0))} + \frac{s_{t_0}N}{(1 + F(t_0, t_0))(1 + F(t_0, t_1))}$$
$$+ \frac{s_{t_0}N + N}{(1 + F(t_0, t_0))(1 + F(t_0, t_1))(1 + F(t_0, t_2))} = N \tag{4.39}$$

Canceling the N and rearranging, we can obtain the numerical value of s_{t_0} given $F(t_0, t_0)$, $F(t_0, t_1)$, and $F(t_0, t_2)$. This would value the swap off the FRA markets.

4.5.1.5 An important remark

Note a very convenient, but *very delicate* operation that was used in the preceding derivation. Using the liquid FRA markets, we "replaced" the unknown L_{t_i} by the known $F(t_0, t_i)$ in the appropriate formulas. Yet, these formulas were nonlinear in L_{t_i} and even if the forward rate is an unbiased forecast of the appropriate LIBOR,

$$F(t_0, t_i) = E_{t_0}^{P*}[L_{t_i}] \tag{4.40}$$

under some appropriate probability $P*$, it is not clear whether the substitution is justifiable. For example, it is known that the conditional expectation operator at time t_0, represented by $E_{t_0}^{P*}[.]$, cannot be moved *inside* a nonlinear formula due to Jensen's inequality:

$$E_{t_0}^{P*}\left[\frac{1}{1 + L_{t_i}\delta}\right] > \frac{1}{1 + E_{t_0}^{P*}[L_{t_i}]\delta} \tag{4.41}$$

So, it is not clear how L_{t_i} can be replaced by the corresponding $F(t_0, t_i)$, even when the relation in Eq. (4.48) is true. These questions will have to be discussed after the introduction of risk-neutral and forward measures in Chapters 12 and 13. Such "substitutions" are delicate and depend on many conditions. In our case, we are allowed to make the substitution, because the forward rate is what the market "pays" for the corresponding LIBOR rates at time t_0.

4.5.2 A VERTICAL DECOMPOSITION

We now study the second way of decomposing the swap. We already know what FRA contracts are. Consider an annual FRA where the $\delta = 1$. Also, let the FRA be paid-in-arrear. Then, at some time $t_i + \delta$, the FRA parties will exchange the cash flow:

$$(L_{t_i} - F(t_0, t_i))\delta N \tag{4.42}$$

where N is the notional amount, $\delta = 1$, and the $F(t_0, t_i)$ is the FRA rate determined at time t_0. We also know that the FRA amounts to exchanging the fixed payment $F(t_0, t_i)\delta N$ against the floating payment $L_{t_i}\delta N$.

Is it possible to decompose a swap into n FRAs, each with an FRA rate $F(t_0, t_i)$, $i = 1,\ldots, n$? The situation is shown in Figure 4.10 for the case $n = 3$. The swap cash flows are split by slicing the swap vertically at each payment date. Figure 4.10b–d represents each swap cash flow separately. A *fixed* payment is made against an unknown floating LIBOR rate, in each case.

Are the cash exchanges shown in Figure 4.10–d tradeable contracts? In particular, are they valid FRA contracts, so that the swap becomes a portfolio of three FRAs? At first glance, the cash

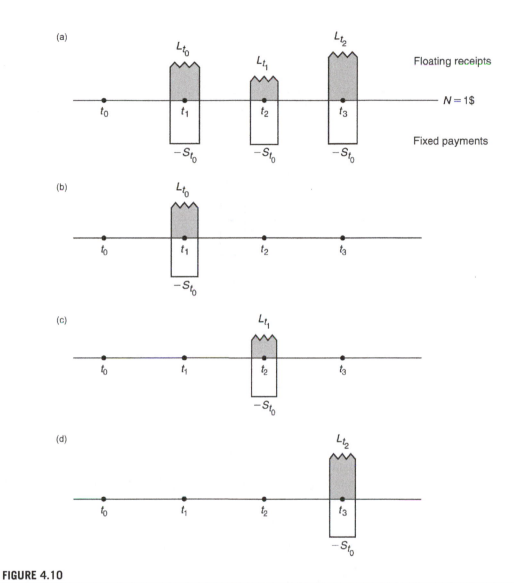

FIGURE 4.10

Decompose a swap into FRAs.

flows indeed look like FRAs. But when we analyze this claim more closely, we see that these cash flows are not valid contracts individually.

To understand this, consider the time t_2 settlement in Figure 4.10b together with the FRA cash flows for the same settlement date, as shown in Figure 4.11. This figure displays an important phenomenon concerning cash flow analysis. Consider the FRA cash flows initiated at time t_0 and settled in arrears at time t_2 and compare these with the corresponding swap settlement. The two

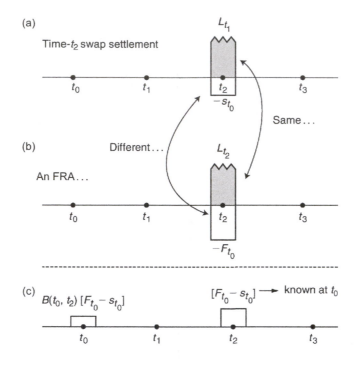

FIGURE 4.11

Settlement differences between FRAs and swaps.

cash flows look similar. A fixed rate is exchanged in the same fashion against LIBOR rate L_{t_1} observed at time t_1. But there is still an important difference.

First of all, note that the FRA rate $F(t_0, t_i)$ is determined by supply and demand or by pricing through money markets. Thus, in general

$$F(t_0, t_i) \neq s_{t_0} \tag{4.43}$$

This means that if we buy the cash flow in Figure 4.11a and sell the cash flow in Figure 4.11b, LIBOR-based cash flows will cancel out, but the fixed payments won't. As a result, the portfolio will have a *known* negative or positive net cash flow at time t_2, as shown in Figure 4.11c. Since this cash flow is known exactly, the present value of this portfolio *cannot* be zero. But the present value of the FRA cash flow *is* zero, since (newly initiated) FRA contracts do not have any initial cash payments. All these imply that the time t_2 cash flow shown in Figure 4.11c will have a known present value,

$$B(t_0, t_2)[F(t_0, t_1) - s_{t_0}]\delta N \tag{4.44}$$

where the $B(t_0, t_2)$ is the time t_0 value of the default-free zero coupon bond that matures at time t_2, with par value USD 1. This present value will be positive or negative depending on whether $F(t_0, t_1) < s_{t_0}$ or not.

Hence, slicing the swap contract vertically into separate FRA-like cash exchanges does *not* result in tradeable financial contracts. In fact, the time t_2 exchange shown in Figure 4.11c has a missing time t_0 cash flow of size $B(t_0, t_2)[F(t_0, t_1) - s_{t_0}]\delta N$. Only by adding this, does the exchange become a tradeable contract. The s_{t_0} is a fair exchange for the risks associated with L_{t_0}, L_{t_1}, and L_{t_2} *on the average.* As a result, the time t_2 cash exchange that is part of the swap contract ceases to have a zero present value when considered individually.

4.5.2.1 Pricing

We have seen that it is not possible to interpret the individual swap settlements as FRA contracts directly. The two exchanges have a nonzero present value, while the (newly initiated) FRA contracts have a price of zero. But the previous analysis is still useful for pricing the swap since it gives us an important relationship.

In fact, we just showed that the time t_2 element of the swap has the present value $B(t_0, t_2)[F(t_0, t_1) - s_{t_0}]\delta N$, which is not, normally, zero. This must be true for all swap settlements individually. Yet, the swap cash flows *altogether* do have zero present value. This leads to the following important pricing relation:

$$B(t_0, t_1)[F(t_0, t_0) - s_{t_0}]\delta N + B(t_0, t_2)[F(t_0, t_1) - s_{t_0}]\delta N + B(t_0, t_3)[F(t_0, t_2) - s_{t_0}]\delta N = 0 \tag{4.45}$$

Rearranging provides a formula that ties the swap rate to FRA rates:

$$s_{t_0} = \frac{B(t_0, t_1)[F(t_0, t_0) + B(t_0, t_2)F(t_0, t_1) + B(t_0, t_3)F(t_0, t_2)}{B(t_0, t_1) + B(t_0, t_2) + B(t_0, t_3)} \tag{4.46}$$

This means that we can price swaps off the FRA market as well. The general formula, where n is the *maturity* of the swap,

$$s_{t_0} = \frac{\sum_{i=1}^{n} B(t_0, t_i)F(t_0, t_{i-1})}{\sum_{i=1}^{n} B(t_0, t_i)} \tag{4.47}$$

will be used routinely in this book. It is an important arbitrage relationship between swap rates and FRA rates.

4.6 USES OF SWAPS

The general idea involving the use of swaps in financial engineering is easy to summarize. A swap involves exchanges of cash flows. But cash flows are generated by assets, liabilities, and other commitments. This means that swaps are simply a standardized, liquid, and cost-effective alternative to trading cash assets. Instead of trading the cash asset or liability, we are simply trading the cash flows generated by it. Because swaps, in general, have zero value at the time of initiation, and are very liquid, this will indeed be a cost-effective alternative—hence their use in position taking, hedging, and risk management. What are these uses of swaps? We begin the discussion by looking at the uses of interest-rate swaps in this chapter. The uses of commodity, credit default, and FX swaps are somewhat different and we discuss their uses in detail in separate chapters on financial engineering applications to those asset classes. We will see that these swaps have convenient balance sheet implications. Regulatory and tax treatment of equity swaps are also relevant.

4.6.1 USES OF EQUITY SWAPS

Equity swaps illustrate the versatility of swap instruments.

4.6.1.1 Fund management

There is a huge industry of fund management where the fund manager tries to track some equity *index*. One way to do this is by buying the underlying portfolio of stocks that replicates the index and constantly readjusting it as the market moves, or as new funds are received, or paid by the fund. This is a fairly complex operation. Of course, one can use the S&P 500 futures to accomplish this. But futures contracts need to be rolled over and they require mark-to-market adjustments. Using equity swaps is a cost-effective alternative.

The fund manager could enter into an equity swap using the S&P 500 in which the fund will pay, quarterly, a LIBOR-related rate *and* a (positive or negative) spread and receive the return on the S&P 500 index for a period of *n* years.[22]

The example below is similar to the one seen earlier in this chapter. Investors were looking for cost-effective ways to diversify their portfolios.

> **EXAMPLE**
>
> In one equity swap, the holder of the instrument pays the total return on the S&P 500 and receives the return on the FTSE 100. Its advantage for the investor is the fact that, as a swap, it does not involve paying any up-front premium.
>
> The same position cannot be replicated by selling S&P stocks and buying FTSE 100 stocks.

The second paragraph emphasizes one convenience of the equity swap. Because it is based directly on an index, equity-swap payoffs do not have any "tracking error." On the other hand, the attempt to replicate an index using underlying stocks is bound to contain some replication error.

4.6.1.2 Tax advantages

Equity swaps are not only "cheaper" and more efficient ways of taking a position on indices, but may have some *tax* and *ownership* advantages as well. For example, if an investor wants to sell a stock that has appreciated significantly, then doing this through an outright sale will be subject to capital gains taxes. Instead, the investor can keep the stock, but, get into an equity swap where he or she pays the capital gains (losses) and dividends acquired from the stock, and receives some LIBOR-related return and a spread. This is equivalent to selling the stock and placing the received funds in an interbank deposit.

---

[22]The return on the S&P 500 index will be made of capital gains (losses) as well as dividends.

4.6.1.3 Regulations

Finally, equity swaps help in executing some strategies that otherwise may not be possible to implement due to regulatory considerations. In the following example, with the use of equity-swaps investment in an emerging market becomes feasible.

EXAMPLE

Since the Kospi 200 futures were introduced foreign securities houses and investors have been frustrated by the foreign investment limits placed on the instrument. They can only execute trades if they secure an allotment of foreign open interest first, and any allotments secured are lost when the contract expires. Positions cannot be rolled over. Foreigners can only hold 15% of the 3-month daily average of open interest, while individual investors with "Korean Investor IDs" are limited to 3%. Recognizing the bottleneck, Korean securities houses such as Hyundai Securities have responded by offering foreign participants equity swaps which are not limited by the restrictions.

The structure is quite simple. A master swap agreement is established between the foreign client and an offshore subsidiary or a special purpose vehicle of the Korean securities company. Under the master agreement, foreigners execute equity swaps with the offshore entity which replicate the futures contract. Because the swap transactions involve two nonresident parties and are booked overseas, the foreign investment limits cannot be applied.

The Korean securities houses hedge the swaps in the futures market and book the trades in their proprietary book. Obviously, as a resident entity, the foreign investment limits are not applied to the hedging trade. Once the master swap agreement is established, the foreign client can contact the Korean securities company directly in Seoul, execute any number of trades, and have them booked and compiled against the master swap agreement

(Thomson Reuters IFR, January 27, 1996).

The reading shows how restrictions on (1) ownership, (2) trading, and (3) rolling positions over can be handled using an equity swap. The reading also displays the structure of the equity swap that is put in place and some technical details associated with it.

One of the strongest growth areas in the asset management industry has been the area of absolute return UCITS[23] in Europe and '40 Act funds in the United States as hedge fund techniques area applied in more traditional fund vehicles.[24] However, some of the techniques used to implement hedge fund techniques in these retail structures are designed to circumvent regulatory restrictions and have generated controversy as the following example shows.

EXAMPLE

In recent years some fund management companies have launched UCITS funds that use alternative investment strategies typically found in hedge funds. Some funds have succeeded in circumventing UCITS restrictions on eligible assets by using total return swaps (TRS). Typically UCITS funds cannot hold

---

[23]UCITS (Undertakings for Collective Investment in Transferable Securities) funds are subject to EU regulation but are of global interest since they can be marketed not just in the EU but also in countries such as Switzerland, Chile, and Taiwan.
[24]Pooled investment vehicles offered by a registered investment company as defined in the 1940 Investment Copanies Act are refereed to as '40 Act funds.

commodity futures, for example, but some funds have used TRSs with banks in which the fund receives the returns on a commodity futures in return for a swap rate paid to the bank counterparty. Some of the most active users of TRS were Commodity Trading Advisors (CTAs) that sometimes used TRS to create the majority of the funds return. CTAs are alternative investment funds that use long and short positions in liquid futures markets across all asset classes to express their investment views. Other examples of TRS use included long/short equity funds that use TRS structures to create short exposures since physical shorting is not permitted in UCITS funds. Such total return swaps can be used not just to hold assets that are not normally eligible in UCITS structures but also to raise leverage beyond levels prescribed by UCITS regulation as the following example illustrates: 'The Dunn WMA Ucits Fund, for example, holds a total return swap on a proprietary index that can be up to 35 times leveraged [. . .] Industry figures described this figure as "horrid" [Source: Financial Times 13/11/2011]' Such UCITS funds are the exception but leverage levels of this magnitude are clearly not in the spirit of UCITS regulation.

Although the majority of UCITS funds do not use such TRS structures, there is a fear that poor performance by a UCITS fund that uses such structures could tarnish the brand. This is of particular concern as a result of European attempts to sell UCITS structures in different Asian jurisdictions. Asian regulators can be expected to take a dim view of UCITS funds if they do not protect retail investors that they are typically marked to.

(Source: Financial Times, 13/11/2011).[25]

One can distinguish two different types of total returns swaps. The first type is a so-called *funded TRS*, in which a fund, such as a UCITS fund, pays the swap counterparty, such as a bank, to finance the swap. The fund receives collateral from the (bank) counterparty and if the collateral is less than the value of the swap this exposes the fund to counterparty risk. Such a structure can be used by UCITS funds for example that try to replicate a CTA strategy or index. The second type is an unfunded swap, in which the assets are held by custodian. The custodian manages the payments between the counterparties. One counterparty can be a UCITS fund, for example, and the other a swap provider. The fund pays a fee or financing spread to compensate the swap provider for having to borrow to finance the swap. However, TRS are a rather imperfect way of replicating certain alternative investment strategies such as CTA or managed futures strategies as the following reading illustrates:

Much of the expected tracking error in managed futures funds using a TRS is down to the additional costs imposed by the swaps, underlying hidden charges and [. . .] administering a bespoke index

(Source: Financial Times, 13/11/2011).[26]

These issues raise the question whether TRS are a cost effective tool for long-term investors to achieve their investment objectives.

4.6.1.4 Creating synthetic positions

The following example is a good illustration of how equity swaps can be used to create synthetic positions.

---

[25]"Swap tactic threatens UCITS brand" by Steve Johnson.
[26]"Necessary evil' but concerns are growing" by Steve Johnson.

Equity derivative bankers have devised equity-swap trades to (handle) the regulations that prevent them from shorting shares in Taiwan, South Korea, and possibly Malaysia. The technique is not new but has reemerged as convertible bond (CB) issuance has picked up in the region, and especially in these three countries.

Bankers have been selling equity swaps to CB arbitrageurs, who need to short the underlying shares but have been prohibited from doing so by local market regulations.

It is more common for a CB trader to take a short equity-swap position with a natural holder of the stock. The stockholder will swap his long stock position for a long equity-swap position. This provides the CB trader with more flexibility to trade the physical shares. When the swap matures—usually 1 year later— the shares are returned to the institution and the swap is settled for cash.

(Thomson Reuters IFR, December 5, 2001).

In this example, a CB trader needs to short a security by an amount that changes continuously.[27] A convenient way to handle such operations is for the CB trader to write an equity swap that pays the equity returns to an investor and gets the investors' physical shares to do the hedging.

4.6.1.5 Stripping credit risk

Suppose we would like to strip the credit risk implicit in a defaultable coupon bond. Note that the main problem is that the bond yield will depend on *two* risks. First is the credit spread and second is the interest-rate risk. An asset swap, where we pay a fixed swap rate and receive LIBOR, will then eliminate the interest-rate risk in the bond. The result is called asset swap spread.

4.6.2 USING INTEREST-RATE AND OTHER SWAPS

Interest-rate swaps play a much more fundamental role than equity swaps. In fact, all swaps can be used in *balance sheet management*. Balance sheets contain several cash flows; using the swap, one can switch these cash flow characteristics. Swaps are used in *hedging*. They have *zero* value at time of initiation and hence don't require any funding. A market practitioner can easily cover his positions in equity, commodities, and fixed income by quickly arranging proper swaps and then unwind these positions when there is no need for the hedge.

Finally, swaps are also *trading* instruments. In fact, one can construct *spread trades* most conveniently by using various types of swaps. Some possible spread trades are given by the following:

- Pay n-period fixed rate s_{t_0} and receive floating LIBOR with notional amount N.
- Pay L_{t_i} and receive r_{t_i} both floating rates, in the same currency. This is a basis swap.
- Pay and receive two floating rates in different currencies. This will be a currency swap.

As these examples show, swaps can pretty much turn every interesting instrument into some sort of "spread product." This will reduce the underlying credit risks, make the value of the swap zero at initiation, and, if properly designed, make the position relatively easy to value.

[27]This is required for the hedging of the implicit option, as will be seen in later chapters.

4.6.3 TWO USES OF INTEREST-RATE SWAPS

We now consider two examples of the use of interest-rate swaps.

4.6.3.1 Changing portfolio duration

As we saw in the previous chapter, duration is the "average" maturity of a fixed-income portfolio. It turns out that in general the largest fixed-income liabilities are managed by governments, due to the existence of government debt. Depending on market conditions, governments may want to adjust the average maturity of their debt. Swaps can be very useful here. The following example illustrates this point.

EXAMPLE

France and Germany are preparing to join Italy in using interest-rate swaps to manage their debt. Swaps can be used to adjust debt duration and reduce interest-rate costs, but government trading of over-the-counter derivatives could distort spreads and tempt banks to front-run sovereign positions. The United States and United Kingdom say they have no plans to use swaps to manage domestic debt.

As much as E150 billion of swap use by France is possible over the next couple of years, though the actual figure could be much less, according to an official at the French debt management agency. That is the amount that would be needed if we were to rely on only swaps to bring about "a [significant] shortening of the average duration of our debt," an official said. France has E644.8 billion of debt outstanding, with an average maturity of 6 years and 73 days.

The official said decisions would be made in September about how to handle actual swap transactions. "If E150 billion was suddenly spread in the market, it could produce an awful mess," he said

(Thomson Reuters IFR, Issue 1392).

Using swap instruments, similar adjustments to the duration of liabilities can routinely be made by corporations as well.

How can measures of interest-rate risk such as DV01 and duration discussed in the previous chapter be applied to swaps? In principle, the DV01 of receiving fixed in a fixed-for-floating interest-rate swap is equal to the DV01 of its fixed leg minus the DV01 of its floating leg. Thus the duration of a swap can be calculated like that of a fixed and a floating rate bond. However, for practical purposes, this is not very useful since the different legs of the swap are exposed to movements in different parts of the term structure of interest rates. The floating leg is more exposed to movements in long-term interest rates while the short leg is exposed to movements in the short leg. Therefore, market participants in practice use different approaches to hedge the short and the long legs of the swap. These more advanced methodological approaches involve multifactor risk metrics and address the weakness resulting from our assumption in the previous chapter that movements in the entire term structure can be described by one interest-rate factor.

4.6.3.2 Using swaps to reduce a country's debt level

Although governments oftentimes fret about the use of swaps by companies, sometimes governments use swaps to achieve their own objectives. Italy and Greece, for example, were two countries that used swaps to mask and reduce their debt levels in the run-up to their membership of the European Monetary Union.

EXAMPLE

The Goldman Sachs transaction swapped debt issued by Greece in dollars and yen for euros using an historical exchange rate, a mechanism that implied a reduction in debt, Sardelis said. It also used an off-market interest-rate swap to repay the loan. Those swaps allow counterparties to exchange two forms of interest payment, such as fixed or floating rates, referenced to a notional amount of debt.

The trading costs on the swap rose because the deal had a notional value of more than 15 billion euros, more than the amount of the loan itself, said a former Greek official with knowledge of the transaction who asked not to be identified because the pricing was private. The size and complexity of the deal meant that Goldman Sachs charged proportionately higher trading fees than for deals of a more standard size and structure, he said.

"It looks like an extremely profitable transaction for Goldman," said Saul Haydon Rowe, a partner in Devon Capital LLP, a London-based firm that advises global investors on derivatives disputes.

[. . .]

Cross-currency swaps are contracts borrowers use to convert foreign currency debt into a domestic-currency obligation using the market exchange rate. As first reported in 2003, Goldman Sachs used a fictitious, historical exchange rate in the swaps to make about 2% of Greece's debt disappear from its national accounts. To repay the 2.8 billion euros it borrowed from the bank, Greece entered into a separate swap contract tied to interest-rate swings.

Falling bond yields caused that bet to sour, and tweaks to the deal failed to prevent the debt from almost doubling in size by the time the swap was restructured in August 2005.

Greece, which last month secured a second, 130 billion-euro bailout, is sitting on debt equal to about 160% of its GDP as of last year.

(Source: Bloomberg, 6 March 2012).[28]

4.6.3.3 Technical uses

Swaps have technical uses. The following example shows that they can be utilized in designing new bond futures contracts where the delivery is tied not to bonds, but to swaps.

EXAMPLE

LIFFE is to launch its swap-based LIBOR Financed Bond on October 18. Both contracts are designed to avoid the severe squeeze that has afflicted the Deutsche Terminboerse Bund future in recent weeks.

LIFFE's new contract differs from the traditional bond future in that it is swap-referenced rather than bond-referenced. Instead of being settled by delivery of cash bonds chosen from a delivery basket, the LIBOR Financed Bond is linked to the International Swap and Derivatives Association benchmark swap rate. At expiry, the contract is cash-settled with reference to this swap rate.

Being cash-settled, the Liffe contract avoids the possibility of a short squeeze—where the price of the cheapest to deliver bond is driven up as the settlement day approaches. And being referenced to a swap curve rather than a bond basket, the contract eliminates any convexity and duration risk. The LIBOR Financed Bond replicates the convexity of a comparable swap position and therefore reduces the basis risk resulting from hedging with cash bonds or bond futures.

---

[28]"Goldman Secret Greece Loan Shows Two Sinners as Client Unravels" by Nicolas Dunbar and Elisa Martinuzzi.

> *An exchange-based contender for benchmark status, the DTB Bund, has drawn fire in recent weeks following a short squeeze in the September expiry. In the week before, the gross basis between the cheapest to deliver Bund and the Bund future was driven down to −3.5.*
>
> *The squeeze had been driven by a flight to quality on the back of the emerging market crisis. Open interest in the Bund future is above 600,000 contracts or DM15 billion. In contrast, the total deliverable basket for German government bonds is roughly DM74 billion and the cheapest to deliver account to DM30 billion.*
>
> *Officials from the DTB have always contended that there will be no lack of deliverable Bunds. They claim actual delivery has only been made in about 4% of open positions in the past*
>
> **(Thomson Reuters IFR Issue 1327).**

In fact, several new cash-settled futures contracts were recently introduced by LIFFE and CME on swaps. CME launched EUR-denominated deliverable interest-rate swap futures (Euro DSF) on April 14, 2014.[29] Swaps are used as the underlying instrument. Without the existence of liquid swap markets, a swap futures contract would have no such reference point and would have to be referenced to a bond basket.

4.7 MECHANICS OF SWAPPING NEW ISSUES

The swap engineering introduced in this chapter has ignored several minor technical points that need to be taken into account in practical applications. Most of these are minor and are due to differences in market conventions in bonds, money markets, and swap markets. In this section, we provide a discussion of some of these technical issues concerning interest-rate swaps. In other swap markets, such as in commodity swaps, further technicalities may need to be taken into account. A more or less comprehensive list is as follows:

1. Real-world applications of swaps deal with *new* bond issues, and new bond issues imply fees, commissions, and other expenses that have to be taken into account in calculating the true cost of the funds. This leads to the notion of *all-in-cost*, which is different to the "interest rate" that will be paid by the issuer.

   We will show in detail how all-in-cost is calculated, and how it is handled in swapping new issues.

2. Interest-rate swaps deal with fixed and floating rates simultaneously. The corresponding LIBOR is often taken as the floating rate, while the swap rate, or the relevant swap spread, is taken as the fixed rate. Another real-world complication appears at this point.

   Conventions for quoting money market rates, bond rates, and swap rates usually differ. This requires converting rates defined in one basis, into another.

   In particular, money market rates such as LIBOR are quoted on an *ACT*/360 basis, while *some* bonds are quoted on an annual or semiannual 30/360 basis. In swap engineering, these cash flows are exchanged at regular times, and hence appropriate adjustments need to be made.

---

[29]See "CME Group to launch euro-denominated deliverable interest rate swap futures," hedgeweek, November, 2, 2014, for further details.

Table 4.1 Details of the New Issue

Shinhan Bank	
Amount	USD200 million
Maturity	3 years (due July 2009)
Coupon	4%
Reoffer price	99.659
Spread at reoffer	168.8 bp over the 2-year US Treasury
Launch date	July 23
Payment	July 29
Fees	20 bp
Listing	London
Governing law	London
Negative pledge	Yes
Cross-default	Yes
Sales restrictions	US, UK, South Korea
Joint lead managers	ABN AMRO, BNP Paribas, UBS Warburg

Source: *Thomson Reuters IFR issue 1444.*

3. In this chapter we mostly ignored credit risk. This greatly simplified the exposition because swap rates and corporate rates of similar maturities became equal. In financial markets, they usually are not. Issuers have different credit ratings and bonds sold by them and carry credit spreads that are different from the swap spread. This gives rise to new complications in matching cash flows of coupon bonds and interest-rate swaps. We need to look at some examples of this as well.

4. Finally, the mechanics of how new issues are swapped into fixed or floating rates and how this may lead to *sub-LIBOR* financing is an interesting topic by itself.

The discussion will be conducted with a real-life, *new issue*, explained next. First we report the "market reaction" to the bond, and second we have the details of the new issue (Table 4.1).

EXAMPLE

South Korea's Shinhan Bank, rated Baa1/BBB by Moody's and S&P, priced its USD200 million 3-year bond early last week (...). The deal came with a 4% coupon and offered a spread of 168.8 bp over the 2-year US Treasury, equivalent to 63 bp over LIBOR.

This was some 6 bp wide of the Korea Development Bank (KDB) curve, although it was the borrower's intention to price flat to it. Despite failing to reach this target, the borrower still managed to secure a coupon that is the lowest on an Asian bond deal since the regional crisis, thanks to falling US Treasury yields which have shrunk on a renewed flight to safe haven assets

(Thomson Reuters IFR, issue 1444).

Consider now the basic steps of swapping this new issue into floating USD funds.[30] The issuer has to enter into a 3-year interest-rate swap agreement. How should this be done, and what are the relevant parameters? Suppose at the time of the issue, the market makers were quoting the swap spreads as given in Table 4.2. First we consider the calculation of all-in-cost for the preceding deal.[31]

4.7.1 ALL-IN-COST

The information given in the details of the new issue implies that the coupon is 4%. But, this is not the true costs of funds from the point of view of the issuer. There are at least three additional factors that need to be taken into account. (1) The reoffer price is not 100, but 99.659. This means that for each bond, the issuer will receive less cash than the par. (2) Fees have to be paid. (3) Although not mentioned in the information in Table 4.1, the issuer has legal and documentation expenses. We assume that these were USD75,000.

To calculate the fixed all-in-cost (30/360 basis), we have to calculate the *proceeds* first. Proceeds is the net cash received by the issuer after the sale of the bonds. In our case, using the terminology of Table 4.1,

$$\text{Proceeds} = \text{Amount} \times \left(\frac{\text{Price}}{100} - \text{Fees}\right) - \text{Expenses} \tag{4.48}$$

Plugging in the relevant amounts,

$$\text{Proceeds} = 200,000,000(0.99659 - 0.0020) - 75,000 \tag{4.49}$$

$$= 198,843,000 \tag{4.50}$$

Table 4.2 USD Swap Index Versus 12 month LIBOR, Semi, 30/360			
Maturity	Bid Spread	Ask Spread	The Bid–Ask Swap Rate
2 years	42	46	2.706–2.750
3 years	65	69	3.341–3.384
4 years	70	74	3.796–3.838
5 years	65	69	4.147–4.187
7 years	75	79	4.653–4.694
10 years	61	65	5.115–5.159
12 years	82	86	5.325–5.369
15 years	104	108	5.545–5.589
20 years	126	130	5.765–5.809
30 years	50	54	5.834–5.885

---

[30]The actual process may differ slightly from our simplified discussion here.
[31]Liquid swaps are against 3- or 6-month LIBOR. Here we use 12-month LIBOR for notational simplicity.

Next, we see that the bond will make three coupon payments of 8 million each. Finally, the principal is returned in 3 years. The cash flows associated with this issue are summarized in Figure 4.12. What is the internal rate of return of this cash flow? This is given by the formula

$$198,843,000 = \frac{8,000,000}{(1+y)} + \frac{8,000,000}{(1+y)^2} + \frac{8,000,000 + 200,000,000}{(1+y)^3} \tag{4.51}$$

The y that solves this equation is the internal rate of return. It can be interpreted as the true cost of the deal, and it is the fixed all-in-cost under the (30/360) day-count basis. The calculation gives

$$y = 0.04209 \tag{4.52}$$

This is the fixed all-in-cost.

The next step is to swap this issue into floating and obtain the floating all-in-cost. Suppose we have the same notional amount of $200 million and consider a fixed to floating 3-year interest-rate swap. Table 4.2 gives the 3-year *receiver* swap rate as 3.341%. This is, by definition, a 30/360, semiannual rate.

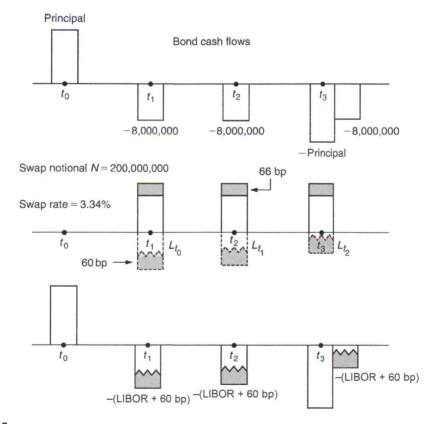

FIGURE 4.12

New issue and interest-rate swap cash flows.

This requires converting the semiannual swap rate into an annual 30/360 rate, denoted by r. This is done as follows:

$$(1 + r) = \left(1 + 0.03341\frac{1}{2}\right)^2 \tag{4.53}$$

which gives

$$r = 3.369\% \tag{4.54}$$

With a \$200 million notional, this is translated into three fixed receipts of

$$200,000,000(0.03369) = 6,738,000 \tag{4.55}$$

each. The cash flows are shown in Figure 4.12.

Clearly, the fixed swap receipts are not equal to the fixed annual coupon payments, which are \$8 million each. Apparently, the issuer pays a higher rate than the swap rate due to higher credit risk. To make these two equal, we need to increase the fixed receipts by

$$8,000,000 - 6,738,000 = 1,262,000 \tag{4.56}$$

This can be accomplished by increasing both the floating rate paid and the fixed rate received by equivalent amounts. This can be accomplished if the issuer accepts paying LIBOR *plus* a spread equivalent to the 66 bp. Yet, here the 66 bp is p.a. 30/360, whereas the LIBOR convention is p.a. ACT/360. So the basis point difference of 66 bp may need to be adjusted further.[32] The final figure will be the floating all-in-cost and will be around 60 bp.

4.8 SOME CONVENTIONS

If you have a coupon bond and the payment date falls on a nonworking day, then the payment will be made on the first following working day. But the amount does not change. In swaps, this convention is slightly different. The payment is again delayed to the next working day.[33] But the payment amount will be adjusted according to the actual number of days. This means that the payment dates and the amounts may not coincide exactly in case swaps are used as hedges for fixed-income portfolios. Most swaps involve terminology and documentation that have been standardized by the International Swap and Derivatives Association (ISDA) and involve a *Master ISDA* agreement in place before dealing.

4.8.1 QUOTES

Suppose we see quotes on interest-rate swaps or some other liquid swap. Does this mean we can deal on them? Not necessarily. Observed swap rates may be available as such only to a bank's best customers; others may have to pay more. In practice, the bid−ask spreads on liquid instruments are very tight, and a few large institutions dominate the market.

---

[32] Our calculations provided a slightly different number than the 63 basis points mentioned in the market reaction mainly because we used a swap against 12-month LIBOR for simplicity.

[33] If this next day is in the following calendar month, then the payment is made during the *previous* working day.

4.9 ADDITIONAL TERMINOLOGY

We would like to introduce some additional terms and instruments before moving on.

A *par swap* is the formal name of the interest-rate swaps that we have been using in this chapter. It is basically a swap structure calculated over an initial and final (nominal) exchange of a principal equal to 100. This way, there will be no additional cash payments at the time of initiation.

An *accrual swap* is an interest-rate swap in which one party pays a standard floating reference rate, such as LIBOR, and receives LIBOR plus a spread. But the interest payments to the counterparty will accrue only for days on which LIBOR stays within preset upper and lower boundaries.

An *extendible swap* is a swap in which one party has the right to extend a swap beyond its original term.

A *power LIBOR swap* is a swap that pays LIBOR squared or cubed (and so on), less a fixed amount/rate, in exchange for a floating rate.

A *LIBOR-in-arrears swap* is a swap in which the interest paid on a particular date is determined by that date's interest rate rather than the interest rate of the previous payment date.

4.10 CONCLUSIONS

Why buy and sell securities when you can swap the corresponding returns and achieve the same objective efficiently, and at minimal cost?

In fact, selling or buying a security may not be practical in many cases. First, these operations generate cash which needs to be taken care of. Second, the security may not be very liquid and selling it may not be easy. Third, once a security is sold, search costs arise when, for some reason, we need it back. Can we find it? For how much? What are the commissions? Swapping the corresponding returns may cost less.

Due to their eliminating the need to use cash in buying and selling transactions, combining these two operations into one, and eliminating potential credit risks, swaps have become a major tool for financial engineers.

SUGGESTED READING

Swaps are vanilla products, and there are several recommended books that deal with them. This chapter has provided a nontechnical introduction to swaps, hence we will list references at the same level. For a good introduction to swap markets, we recommend **McDougall** (1999). **Flavell** (2009) and **Das** (1994) give details of swaps and discuss many examples. **Serrat and Tuckman** (2011) describe risk management and hedging involving swaps in greater detail than this chapter allowed for space reasons.

EXERCISES

1. On March 3, 2000, the Financial Accounting Standards Board, a crucial player in financial engineering problems, published a series of important new proposals concerning the accounting of certain derivatives.[34] It is known as Statement 133 and affects the daily lives of risk managers and financial engineers significantly. One of the treasurers who is affected by the new rules had the following comment on these new rules:

    Statement 133 in and of itself will make it a problem from an accounting point of view to do swaps. The amendment does not allow for a distinction to be made between users of aggressive swap hedges and those involved in more typical swaps. According to Thomson Reuters IFR, this treasurer has used synthetic swaps *to get around [the FAS 133].*[35]

    a. Ignoring the details of swaps as an instrument, what is the main point in FAS 133 that disturbs this market participant?
    b. How does the treasurer expect to get around this problem by constructing synthetics?

2. Read the following episode carefully.

    Italian Asset Swap Volumes Soar on Buyback Plans,
    Volumes in the basis-swap spread market doubled last week as traders entered swaps in response to the Italian treasury's announcement that it "does not rule out buybacks." Traders said the increase in volume was exceptional given that so many investors are on holiday at this time of year.
    Traders and investors were entering trades designed to profit if the treasury initiates a buyback program and the bonds increase in value as they become scarcer and outperform the swaps curve. A trader said in a typical trade the investor owns the 30-year Italian government bond and enters a swap in which it pays the 6% coupon and receives 10.5 basis points over 6-month Euribor. "Since traders started entering the position last Monday, the spread has narrowed to 8 bps over Euribor," he added. The trader thinks the spread could narrow to 6.5 bps over Euribor within the next month if conditions in the equity and emerging markets improve. A trader at a major European bank predicts this could go to Euribor flat over the next 6 months. The typical notional size of the trades is E50 million (USD43.65 million) and the maturity is 30 years (Thomson Reuters IFR, Issue 1217).

    a. Suppose there is an Italian swap curve along with a yield curve obtained from Italian government bonds (sovereign curve). Suppose this latter is upward sloping. Discuss how the two curves might shift relative to each other if the Italian government buys back some bonds.
    b. Is it important which bonds are bought back? Discuss.
    c. Show the cash flows of a 5-year Italian government coupon bond (paying 6%) and the cash flows of a fixed-payer interest-rate swap.
    d. What is the reason behind the existence of the 10.15 bp spread?
    e. What happens to this spread when government buys back bonds? Show your conclusions using cash flow diagrams.

---

[34]Updates on FASB statement 133 are available on the FASB webpage http://www.fasb.org/derivatives/issuindex.shtml.
[35]IFR, Issue 1325.

3.   You are a swap dealer and you have the following deals on your book:

Long

- 2-year receiver vanilla interest-rate swap, at 6.75% p.a. 30/360. USD $N = 50$ million.
- 3-year receiver vanilla interest-rate swap, at 7.00% p.a. 30/360. USD $N = 10$ million.

Short

- 5-year receiver vanilla interest-rate swap, at 7.55% p.a. 30/360. USD $N = 10$ million.

a. Show the cash flows of each swap.

b. What is your net position in terms of cash flows? Show this on a graph.

c. Calculate the present values of each swap using the swap curve:

Maturity	Bid–Ask
2	6.75–6.80
3	6.88–6.92
4	7.02–7.08
5	7.45–7.50
6	8.00–8.05

d. What is your net position in terms of present value?

e. How would you hedge this with a 4-year swap? Which position would you take, and what should the notional amount be?

f. Where would you go to get this hedge?

g. Can you suggest another hedge?

REPO MARKET STRATEGIES IN FINANCIAL ENGINEERING

5.1 INTRODUCTION

This is a nontechnical chapter which deals with repos, an important operation. The idea behind a repo is simple and easy to understand since it is similar to a secured loan such as the purchase of a house using a mortgage. There are some differences between repos and secured loans, but the basic idea is similar since an asset is used as collateral for a loan. The chapter briefly reviews *repo* markets and some uses of repo. Repo is short for *repurchase agreement*, which is a transaction that involves the sale of securities together with an agreement for the seller to buy back the securities at a later date.

Figure 5.1 illustrates a simplified example of a repo. It involves two parties, a buyer and a seller. The repo has a start and a maturity date and the repo's term is 7 days. On the start date, the seller delivers collateral, such as bonds, to the buyer, in exchange for an amount of cash (€1,000,000).[1] On the maturity date, the buyer returns the collateral to the seller who repays the loan with interest. The interest rate is called the *repo rate* and, in the example, it is assumed to be 2 percent (per year) and calculated on a money market basis, i.e., actual/360. Therefore, the payment on the maturity date is

$$€1,000,389 = €1,000,000 \times \left(1 + \text{repo rate} \times \frac{\text{actual}}{360}\right) = €1,000,000 \times \left(1 + 2\% \times \frac{\text{actual}}{360}\right)$$

Below we will explore the details and different types of repos further. In the above example, we made the simplifying assumption that the amount of cash exchanged at the start date was equal to

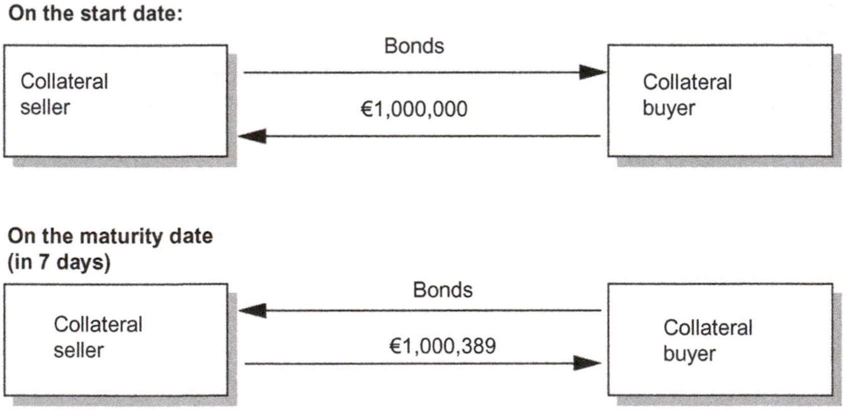

On the start date:

Collateral seller	Bonds →	Collateral buyer
	€1,000,000 ←	

On the maturity date (in 7 days)

Collateral seller	← Bonds	Collateral buyer
	€1,000,389 →	

FIGURE 5.1

A simplified repo example.

---

[1]The amount of cash is equal to the market value of the collateral which includes any accrued interest, which was introduced in Chapter 4.

the market value of the collateral. But what if the collateral value was to suddenly drop? This would lead to losses for the collateral buyer. One way to protect against such potential losses would be for the collateral buyer to pay less than the market value of collateral to the seller. The difference between the market value of the collateral and the amount paid by the buyer to the seller is called the *haircut*. It is a way of imposing a margin on the collateral seller similar to that of initial margin in futures markets. If in the above example we assumed a haircut of 1% which would mean that the buyer would only pay €990,000 to the seller and the repo interest is calculated based on €990,000 and not €1,000,000. Note that if a client faces a 1% haircut when he or she borrows cash in the repo market, the repo dealer can repo the same security with zero haircut and benefit from this transaction.

Many financial engineering strategies require the use of the repo market. The repo market is both a complement and an alternative to swap markets. During a swap transaction, the market practitioner conducts a simultaneous "sale" and "purchase" of two sequences of cash flows generated by two different securities. In an equity swap, for example, returns of an equity instrument are swapped for a floating rate LIBOR. With no exchange of cash, flexible maturities, and liquid markets, swaps become a fundamental tool. Using swaps, a complex sequence of operations can be accomplished efficiently, quickly, and with little risk.

Repo transactions provide similar efficiencies, with two major differences. In swaps, the use of cash is minimized and the ownership of the underlying instruments does not change. In a repo transaction, both cash and (temporary) ownership change hands. Suppose a practitioner *does* need cash or needs to own a security. Yet, he or she does not want to give up or assume the ownership of the security permanently. Swaps are of no help, but a repo is.

Repo is a tool that can provide us cash without requiring the sale or giving up eventual ownership of the involved assets. Alternatively, we may need a security, but we may not want to own this security permanently. Then we must use a tool that secures ownership, without really requiring the purchase of the security. In each case, these operations require either a temporary use of cash or a temporary ownership of securities. Repo markets provide tools for such operations. With repo transactions, we can "buy" without really buying, and we can "sell" without really selling. This is similar to swaps in a sense, but most repo transactions involve exchanges of cash or securities, and this is the main difference with swap instruments.

In each case, the purpose behind these operations is not "long term." Rather, the objective is to conduct daily operations rather smoothly, take directional positions, or hedge a position more efficiently.

5.2 REPO DETAILS

We begin with the standard definition. As Figure 5.1 showed, a repo is a *repurchase agreement* where a *repo dealer* sells a security to a counterparty and *simultaneously* agrees to buy it back at a predetermined price and at a predetermined date. Thus, it is a sale and a repurchase written on the same ticket. In other words, a repo is legally recognized as a single transaction and not as a disposal and a repurchase for tax purposes.[2]

---

[2]This is a valuable feature in the event of counterparty insolvency.

5.2.1 REPO TERMINOLOGY

To avoid any confusion, it is important to learn the naming conventions used in repo markets. In repo markets, most of the terminology is set from the point of view of the repo *dealer*. Also, words such as "borrowing" and "lending" are used as if the item that changes hands is not cash, but a security such as a bond or equity. In particular, the terms "lender" or "borrower" are determined by the lending and borrowing of a security and not of "cash"—although in the actual exchange, cash is changing hands. How can we apply these concepts to Figure 5.1? In Figure 5.1, the collateral seller was the lender and the collateral buyer was the borrower (of the security).

Accordingly, in a repo transaction where the security is first delivered to a client and cash is received, the repo dealer is the "lender"—he or she lends the security and gets cash. This way, the repo dealer has raised cash. If, on the other hand, the same operation was initiated by a client and the counterparty was a repo dealer, the deal becomes a reverse repo, or is simply referred to as *reverse*. The dealer is borrowing the security, the reverse of what happens in a repo operation.

At first glance, the repo operation looks like a fairly simple transaction that would not contribute to the methodology of financial engineering. This is not true. In fact, in terms of practical applications of financial engineering repo may be as common as swaps.

Consider the following example. Suppose an investor wants to buy a security using short-term funding. If he borrows these funds from a bank and then goes to another dealer to buy the bond, the original loan will be *nonsecured*. This implies higher interest costs. Now, if the investor uses repo by buying first, and then repoing the security, he can get the needed funds cheaper because there will be collateral behind the "loan." As a result, both the transaction costs and the interest rate will be lower. In addition, transactions are grouped and written on a single ticket. Given the lower risks, higher flexibility, and other conveniences, repo transactions are very liquid and practical.

With a repo, the sequence of transactions changes. In a typical outright purchase a market professional would

$$\text{Secure funds} \rightarrow \text{Pay for the security} \rightarrow \text{Receive the security} \tag{5.1}$$

When repo markets are used for buying a security, the sequence of transactions becomes:

$$\text{Buy the bond} \rightarrow \text{Immediately repo it out} \rightarrow \text{Secure the funds} \rightarrow \text{Pay for the bond} \tag{5.2}$$

In this case, the repo market is used for finding cheap funding for the purchases the practitioner needs to make. The bond is used as collateral. If this is a default-free security, borrowed funds will come with a relatively low *repo rate*.

Similarly, shorting securities also become possible. The market participant will use the repo market and go through the following steps:

$$\text{Deliver the cash and borrow the bond} \rightarrow \text{Return the bond and receive cash plus interest} \tag{5.3}$$

The market practitioner will earn the repo rate while borrowing the bond. This is equivalent to the market practitioner holding a short-term bond position. The bond is not purchased, but it is "leased."

5.2.2 SPECIAL VERSUS GENERAL COLLATERAL

Repo transactions can be classified into two categories. Sometimes, specific securities receive special attention from markets. For example, some bonds become *cheapest-to-deliver* (CTD). The "shorts" who promised delivery in the bond futures markets are interested in a particular bond and not in others that are similar. This particular bond becomes very much in demand and *goes special* in the repo markets. A repo transaction that specifies the particular security in detail is called a *special repo*. The security remains *special* as long as the relative scarcity persists in the market.

Otherwise, in a repo deal, the party that lends the securities can lend any security of a similar risk class. This type of security is called *general collateral*. One party lends US government bonds against cash, and the counterparty does not care about the particular bonds this basket contains. Then the collateral could be any Treasury bond.

The *special* security will have a higher price than its peers, as long as it remains special. This means that to borrow this security, the client gives up his or her cash at a lower interest rate. After all, the client really needs this particular bond and will therefore have to pay a "price"—agreeing to a lower repo rate.

The interest rate for general collateral is called the *repo rate*. Specials command a repo rate that is significantly lower. In this case, the cash can be relent at a higher rate via a general repo and the original owner of the "special" benefits.

EXAMPLE

Suppose repo rate quotes are 4.5−4.6%. You own a bond worth USD100, which by chance goes special the next day. You can lend your bond for, say, USD100, and get cash for 1 week and pay only 2.5%. This is good, since you can immediately repo this sum against general collateral and earn an annual rate of 4.5% on the 100. You have earned an enhanced return on your bond because you just happened to hold something special.

When using bond market data in research, it is important to take into account the existence of specials in repo transactions. If "repo specials" are mixed with transactions dealing with general collateral, the data may exhibit strange variations and may be quite misleading. This point is quite relevant since about 20% of repo transactions involve specials.

5.2.2.1 Why do bonds go special?

There are at least two reasons why some securities go special systematically. For one, some bonds are cheapest-to-deliver (CTD) in bond futures trading (see the case study at the end of this chapter). The second reason is that *on-the-run* issues are more liquid and are therefore more in demand by traders in order to support hedging and position-taking activities. Such "benchmark" bonds often go special. This is somewhat paradoxical, as the more liquid bonds become the more expensive they are to obtain relative to others.[3]

---

[3]An *on-the-run* issue is the latest issue for a particular maturity, in a particular risk class. For example, an on-the-run 10-year treasury will be the last 10-year bond sold in a treasury auction. Other 10-year bonds will be *off-the-run*.

As an example, consider the so-called *butterfly trades* in the fixed-income sector. Nonparallel shifts that involve the *belly* of the yield curve are sometimes called *butterfly shifts*. These shifts may have severe implications for balance sheets and fixed-income portfolios. Traders use 2−5−10-year on-the-run bonds to put together hedging trades, to guard, or speculate against such yield curve movements. These trades are called butterfly trades. The on-the-run bonds used in such strategies may become "benchmarks" and may go special.

5.2.3 SUMMARY

We can now summarize the discussion. What are the advantages of repo transactions?

1. A repo provides *double security* when lending cash. These are the (high) credit rating of a repo dealer and the collateral.
2. A "special" repo is a unique and convenient way to enhance returns.
3. By using repo markets, traders can short the market and raise funding efficiently. This improves general market efficiency and trading.
4. Financial strategies and product structuring will benefit due to lower transaction costs, more efficient use of time, and lower funding costs.

We now consider various types of repo or repo-type transactions.

5.3 TYPES OF REPO

The term "repo" is used for selling and then simultaneously repurchasing the same instrument. But in practice, this operation can be done in different ways, and these lead to slightly different repo categories.

5.3.1 CLASSIC REPO

A *classic repo* is also called a US style repo. This is the operation that we just discussed. A repo dealer owns a security that he or she sells at a price, 100. This security he or she immediately promises to repurchase at 100, say in 1 month. At that time, the repo dealer returns the original cash received, *plus* the repo interest due on the sum.

EXAMPLE

An investor with a fixed-income portfolio wants to raise cash for a period of 1 week only. This will be done through lending a bond on the portfolio. Suppose the trade date is Monday morning. The parameters of the deal are as follows:

Value date: Deal date + 2 days
Start proceeds: 50 million euro
Collateral: 6 3/4% 4/2003 Bund (the NOMINAL value equals 47.607 m)
Term: 7 days

Repo rate: 4.05%

End proceeds: Start proceeds + (start proceeds × repo rate × term)

This gives

EUR 50m + (EUR 50m × 0.0405 × 7/360) = EUR 50,039,375

Repo interest: 39,375

Thus, by lending 47,607,000 of nominal bonds (DBRs), the investor borrows EUR 50 million (Figure 5.2).

The difference between the nominal and 50m is due to the existence of accrued interest. Accrued interest needs to be added to the nominal. That is to say, the calculations are done using bond's dirty price.[4]

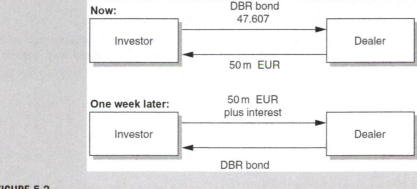

FIGURE 5.2

A classic repo.

Before we look at further real-life examples, we need to consider other repo types.

5.3.2 SELL AND BUY-BACK

A second type of repo is called *sell and buy-back*. The end result of a sell and buy-back is no different from the classic repo. But, the legal foundations differ, which means that credit risks may also be different. In fact, sell and buy-backs exist in two different ways. Some are undocumented. Two parties write two separate contracts at the same time t_0. One contract involves a spot sale of a security, while the other involves a forward repurchase of the same security at a future date. Everything else being the same, the two prices should incorporate the same interest component as in the classic repo. In the documented sell and buy-back, there is a single contract, but the two operations are again treated as separate.

---

[4]As explained in Chapter 3, the dirty price of a coupon bond is the price that includes the present value of all future cash flows, including accrued interest.

EXAMPLE

We use the same parameters as in the previous example, but the way we look at the operation is different although the interest earned is the same:

Nominal: EUR 47.607 million Bund 6 3/4% 4/2003
Start price: 101.971
Plus accrued interest: 3.05625
Total price: 105.02725
Start proceeds: EUR 50,000,322.91
End price: 101.922459
Plus accrued interest: 3.1875
Total price 105.109959
End proceeds: 50,039,698.16
Repo interest earned: EUR $50000322.91 \times 0.0405 \times 7/360 = 39,375$.

In this case, the investor's interest cost is the difference between the purchase price and selling price. The interest earned is exactly the same as in classic repo, but the way interest rate is characterized is different. We show the deal in Figure 5.3.

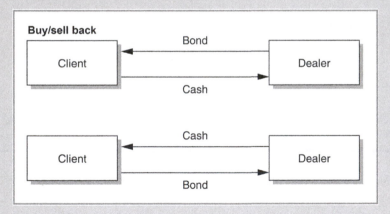

FIGURE 5.3

Sell and buy-back example.

The major difference between the two repo types lies not in the mechanics, or in interest earned, but in the legal and risk management aspects. First of all, sell and buy-backs have no mark-to-market. So they are "easier" to book. Second, in case of undocumented sell and buy-backs, no documentation means lower legal expenses and lower administrative costs. Yet, associated credit risks may be higher. In particular, with sell and buy-backs there is no specific right to offset during default.

5.3.3 SECURITIES LENDING

Securities lending is older than repo as a transaction. It is also somewhat less practical than repo. However, the mechanics of the operation are similar. The main difference is that one of the parties

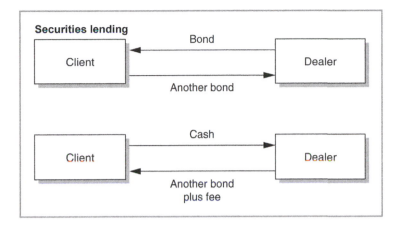

FIGURE 5.4

Securities lending example.

to the transaction may not really need the cash that a repo would generate. But this party may still want to earn a return, hence, the party simply lends out the security for a fee. Any cash received may be deposited as collateral with another entity.

Clearing firms such as Euroclear and Clearstream do securities lending. Suppose a bond dealer is a member of Euroclear. The dealer sold a bond that he or she did not own, and could not find in the markets for an on-time delivery. This may result in a *failure to deliver*. Euroclear can automatically lend this dealer a security by borrowing (at random) from another member.

Note that here securities can be lent not only against cash but against other securities as well. The reason is simple: the lender of the security does not need cash, but rather needs collateral. The collateral can even be a letter of credit or any other acceptable form.

One difference between securities lending and repo is in their quotation. In securities lending, a fee is quoted instead of a repo rate.

EXAMPLE

Nominal: GBP 10 million 8.5% 12/07/05 is lent for 2 weeks
Collateral: GBP 10.62 million 8% 10/07/06
Fee: 50 bp
Total fee: 50 bp $\times$ (14 days/360) $\times$ GBP 10 million.

Obviously, the market value of the collateral will be at least equal to the value of borrowed securities. All other terms of the deal will be negotiated depending on the credit of the borrowing counterparty and the term. This transaction is shown in Figure 5.4.

5.3.4 REPO AND CUSTODY TYPES

There are two main types of repos: *bilateral* and *tri-party* repos. At the beginning of the repo market, there was only the bilateral repo. The bilateral repo market allows for the exchange of cash

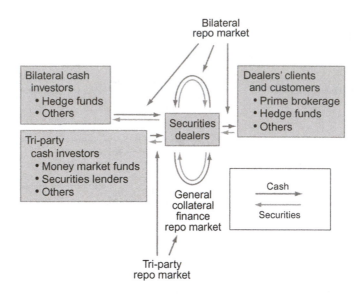

FIGURE 5.5

US repo market.

and securities directly between collateral and cash providers. Moreover, we can distinguish two different ways of holding the collateral. A classic type is *delivery repo*. Here the security is delivered to the counterparty. It is done either by physical delivery or as a transfer of a book entry. A second category is called *hold-in-custody repo*, where the "seller" (lender) keeps the security on behalf of the buyer during the *term* of the repo. This is either because it is impossible to make the transfer or because it is not worth due to time or other limitations.

The third type of *custody* handling is through a *tri-party repo*, where a third party holds the collateral on behalf of the "buyer" (borrower). Often the two parties have accounts with the same custodian, in which case the tri-party repo involves simply a transfer of securities from one account to another. This will be cheaper since fewer fees or commissions are paid. In this case, the custodian also handles the technical details of the repo transaction such as (i) ensuring that delivery versus payment (DVP) is made and (ii) ensuring marking to market of the collateral. The third party is responsible for settlement of the repo on its books and for collateral management. As a result the cash lender does not require the back-office capability to take possession of the collateral.

Based on all of this, a good clearing, custody, and settlement infrastructure is an essential prerequisite for a well-functioning repo market. The above discussion is largely based on the US repo market. Figure 5.5 illustrates the structure of the US repo market including bilateral and tri-party repos.[5] Securities dealers are active in both borrowing and lending. In a repo one institution with cash to invest, the collateral buyer, purchases securities from another institution, the collateral seller. In practice, there are many different types of collateral buyers such as money market funds, asset managers, securities lending agents, and investors who require specific securities as collateral

---

[5]For further details about the US repo market and the source of the above figure, see Copeland et al. (2012).

in order to hedge or speculate in certain markets. In a triparty repo, a third party, called a clearing bank, acts as an intermediary. For simplicity, the figure does not show a clearing bank. In the United States, Bank of New York Mellon and JP Morgan Chase are the only two clearing banks that handle triparty repos.

5.3.4.1 What is a matched-book repo dealer?

Repo dealers are in the business of writing repo contracts. At any time, they post bid and ask repo rates for general, as well as special, collateral. In a typical repo contributor page of Reuters or Bloomberg, the specials will be clearly indicated and will command special prices (i.e., special repo rates). At any time, the repo dealer is ready to borrow and lend securities, whether they are special or general collateral. This way, books are "matched." But this does not mean that dealers don't take one-way positions in the repo book. Their profit comes from bid-ask spreads and from taking market exposure when they think it is appropriate to do so.

5.3.5 ASPECTS OF THE REPO DEAL

We briefly summarize some further aspects of repo transactions.

1. A repo is a temporary exchange of securities against cash. But it is important to realize that the party who borrows the security *has* temporary ownership of the security. The underlying security can be sold. Thus, repo can be used for short-selling.
2. Because securities borrowed through repo can, in general, be sold, the securities returned in the second leg of the repo do not have to be identical. They can be "equivalent," unless specified otherwise in the repo deal.
3. In a repo deal, the lender of the security transfers the title for a short period of time. But the original owner *keeps* the risk and the return associated with the security. Thus, coupon payments due during the term of the repo are passed on to the original owner of the security.

   The risk remains with the original owner also because of the marking to market of the borrowed securities. For example, during the term of the repo, markets might crash and the value of the collateral may decrease. The borrower of the security then has the right to demand additional collateral. If the value of the securities increases, some of the collateral has to be returned.
4. Coupon or dividend payments during the term of the repo are passed on to the original owner. This is called *manufactured dividend* and can occur at the end of the repo deal or some time during the term of the repo.[6]
5. In the United States and the United Kingdom, repo documentation is standardized. A standard repo contract is known as SIFMA/ICMA global master repurchase agreement (the "GMRA").[7]

---

[6]Manufactured dividend is due on the same date as the date of the coupon. But for sell and buy-back this changes. The coupon is paid at the second leg.

[7]The first version of the GMRA was published in 1992. The latest version is the 2011 version published by the International Capital Market Association (ICMA) in conjuction with the Securities Industry and Financial Markets Association (SIFMA). The GMRA's basic structure consists of provisions for all repurchase transactions between the seller and the buyer and various annexes. CMA obtains and annually updates opinions from numerous jurisdictions worldwide on the GMRA 1995, 2000, and 2011 versions. See http://www.icmagroup.org/ for further details.

6. In the standard repo contract, it is possible to substitute other securities for the original collateral, if the lender desires so.
7. As mentioned earlier, the legal title of the repo passes on to the borrower (in a classic repo), so that in case of default, the security automatically belongs to the borrower (buyer). There will be no need to establish ownership.

Finally, we should mention that settlement in a classic repo is DVP. For international securities, the parties will in general use Euroclear and Clearstream.

There are three possible ways to settle repo transactions: (i) *cash settlement*, which involves the same-day receipt of "cash"; (ii) *regular settlement*, where the cash is received on the first business date following a trade date; and (iii) *skip settlement*, when cash is received 2 business days after the trade date.

5.3.6 TYPES OF COLLATERAL

The best-known repo collateral is, of course, government bonds, such as US Treasuries. Every economy with a liquid government bond market will also have a liquid repo market if it is permissible legally. Yet, there are many types of collateral other than sovereign bonds. One of the most common collateral types is Mortgage-Backed Securities (MBS) or Asset-Backed Securities (ABS). Many hedge funds carry such securities with repo funding. Other collateral types are emerging market repo and equity repo, discussed below.

5.3.7 REPO AND CREDIT RISK

During the unfolding of the credit crisis of 2007, repo strategies played a significant role. Several "vehicles" established by banks had repoed structured assets to secure funding. Among these were senior tranches of CDOs that carried a AAA rating. However, repo is senior to senior tranches. During a margin call, the repo dealer has the first right to the collateralized assets, if additional margin is not posted. In this sense, repo funding does introduce additional credit risk.

5.4 EQUITY REPOS

If we can repo bonds out, can we do the same with equities? This would indeed be very useful. Equity repos are becoming more popular, but, from a financial engineering perspective, there are potential technical difficulties:

1. Equities pay dividends and make rights issues. There are mergers and acquisitions. How would we take these events into consideration in a repo deal? It is easy to account for coupons because these are homogeneous payouts. But mergers, acquisitions, and rights issues imply much more complicated changes in the underlying equity.
2. It is relatively easy to find 100 million USD of a single bond to repo out; how do we proceed with equities? To repo equities worth 100 million USD, a portfolio needs to be put together. This complicates the instrument and makes it harder to design a liquid contract.

3. The nonexistence of a standard equity repo agreement also hampers liquidity. In the United Kingdom, this business is conducted with an equity annex to the standard repo agreement.

4. Finally, we should remember that equity has higher volatility which implies more frequent marking to market.

We should also point out that some investment houses carry old-fashioned equity swaps and equity loans, and then label them as equity repos. The following article highlights some of the recent issues and risks related to equity repos.

EXAMPLE

Repurchase agreements, known as repos, backed by equities rose 40 percent during the year ended Jan. 10, according to Federal Reserve data. Rising equity-collateral usage combined with a slide in repos backed by government securities pushed equities share to 9.6 percent of the $1.55 trillion tri-party repo market in January, up from 5.7 percent a year earlier, [. . .]

 "One reason for the equity-backed repo increase is because of the higher yield available on them, that is, from the lender's perspective," Robert Grossman, managing director of macro credit research at Fitch in New York said in a telephone interview yesterday. "All other things being equal, in a distressed market the nongovernment collateral can have more liquidity issues. [. . .]

 Rates on short-term equity-backed repo transactions were 35 basis points, compared with about 11 basis points on agency repos, according to a separate Fitch study using August 2013 data.

 Fed policy makers and researchers have discussed during the past year that the repo market still requires changes to reduce risks related to fire-sales triggered by a dealer default or lenders' perceptions that it may default. [. . .] The Fed has been seeking to strengthen the tri-party repurchase agreement market, which almost collapsed in 2008 amid the demise of Bear Stearns Cos. and bankruptcy of Lehman Brothers Holdings Inc. [. . .]

 Even as U.S. share prices have surged since 2011, reaching record levels last month, "the degree of leverage achievable through repo borrowing for equity collateral is unchanged," with median discounts holding stable at about 8 percent, according to the Fitch report. The discount refers to the percent difference between the value of a security used as collateral in a repo transaction and the amount of cash a lender is willing to exchange for it.

 "This is something to watch," Grossman said, noting that during the financial crisis equity discounts weren't raised until after the market started to decline.

                 (Bloomberg, February 12, 2014)[8]

The above article illustrates the issue of financing costs and haircuts. A small discount makes the borrowing cheaper from the borrower's perspective, but it provides less of a safety cushion from the perspective of the securities lender in case the value of the collateral was to decrease suddenly as a result of equity market turmoil for example.

5.5 REPO MARKET STRATEGIES

The previous sections dealt with repo mechanics and terminology. In this section, we start using repo instruments to devise financial engineering strategies.

---

[8]"Fire Sale Risk Climbs With Equity-Backed Repo Rise, Fitch Says" by Liz Capo McCormick, Bloomberg, February 12, 2014.

5.5.1 FUNDING A BOND POSITION

The most classic use of repo is in funding fixed-income portfolios. A dealer thinks that it is the right time to buy a bond. But, as is the case for market professionals, the dealer does not have cash in hand, but he can use the repo market. A bond is bought and repoed out at the same time to secure the funds needed to pay for it. The dealer earns the bond return; his cost will be the repo rate.

The same procedure may be used to fund a fixed-income portfolio and to benefit from any opportunities in the market, as the following reading shows.

EXAMPLE

Foreign fund managers have recently been putting on bond versus swap spread plays in the Singapore dollar-denominated market to take advantage of an expected widening in the spread between the term repo rate and swap spreads. "It's one of the oldest trades in the book," said [a trader] noting that it has only recently become feasible in the local market...

In a typical trade, an investor buys 10-year fixed-rate Singapore government bonds yielding 3.58%, and then raises cash on these bonds via the repo market and pays an annualized funding rate of 2.05%... At the same time the investor enters a 10-year interest-rate swap in which it pays 3.715% fixed and receives the floating swap offer rate, currently 2.31%. While the investor is paying out 13.5 basis points on the difference between the bond yield it receives and the fixed rate it pays in the swap, the position makes 26 bp on the spread between the floating rate the investor receives in the swap and the term repo funding rate. The absolute levels of the repo and swap offer rate may change, but the spread between them is most likely to widen, increasing the profitability of the transaction.

One of the most significant factors that has driven liquidity in the repo is that in the last few months the Monetary Authority of Singapore has started using the repo market for monetary authority intervention, rather than the foreign exchange market which it had traditionally used...

(Based on an article in Derivatives Week).

We will analyze this episode in detail using the financial engineering tools developed in earlier chapters. For simplicity, we assume that the underlying are par bonds and that the swap has a 3-year maturity with the numerical values given in the example above.[9] The bond position of the trader is shown in Figure 5.6a. A price of 100 is paid at t_0 to receive the coupons, denoted by C_{t_0}, and the principal. Figure 5.6b shows the swap position. The swap "hedges" the fixed coupon payments and "converts" the fixed coupon receipts from the bond into floating interest receipts. The equivalent of LIBOR in Singapore is SIBOR. After the swap, the trader receives SIBOR-13.5 bp. This is shown in Figure 5.6c which adds the first two cash flows vertically. At this point, we see another characteristic of the position. The trader receives the floating payments, but still has to make the initial payment of 100. This means the trader has to get these funds from somewhere.

One possibility is to borrow them from the market. A better way to obtain them is the repo. By lending the bond as collateral, the player can get the needed funds, 100—assuming zero haircut. This situation is shown in Figure 5.7. We consider, artificially, a 1-year repo contract and assume that the repo can be rolled over at unknown repo rates R_{t_1} and R_{t_2} in future periods. According to the reading, the current repo rate is known:

$$R_{t_0} = 2.05\% \tag{5.4}$$

---

[9]It is straightforward for the reader to extend the graphs given here to 10-year cash flows.

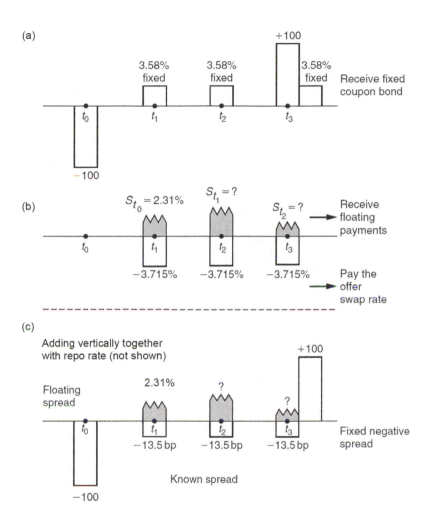

FIGURE 5.6

Bond and swap cash flows.

Adding the first two positions in Figure 5.7 vertically, we obtain the final exposure of the market participant.

The market participant has a 12.5 bp net gain for 1 year. But, more important, the final position has the following characteristic: the market participant is long a forward floating rate bond, which pays the floating SIBOR rates S_{t_1} and S_{t_2}, minus the spread, with the following expectation:

$$S_{t_1} > R_{t_1} + 13.5 \text{ bp} \tag{5.5}$$

$$S_{t_2} > R_{t_2} + 13.5 \text{ bp} \tag{5.6}$$

That is to say, if the spread between future repo rates and SIBOR tightens below 13.5 bp, the position will be losing money. This is one of the risks implied by the overall position. The lower part

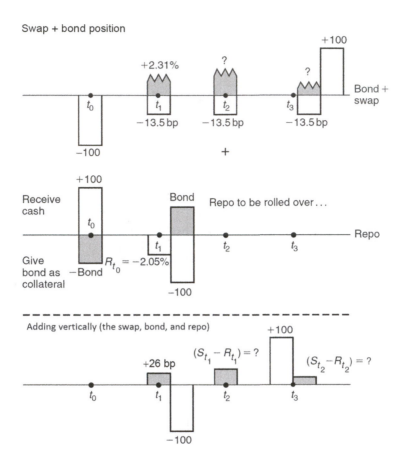

FIGURE 5.7

Bond, swap, and repo cash flows.

of Figure 5.7 shows how this exposure can be hedged. To hedge the position, we would need to go short the same bond forward.

5.5.1.1 A subtle risk

There is another, more subtle risk in this "classical" position. The investor is short the bond and is paying a fixed swap rate. It is true that if the rates move in a similar way, the par bond and the par swap gains or losses would cancel each other.

Yet, the swap spread, $S_{t_0} - C_{t_0}$ can also change. For example, suppose S_t remains the same but C_t increases significantly, implying a lower swap spread. Then, the value of the swap would remain the same, but the value of the bond would decline. Overall, the bond plus swap position would lose money.

More important, the repo dealer would ask for more collateral since the original collateral is now less than the funds lent.

5.5.1.2 The asset swap

There is another way we can describe this position. The investor is buying the bond using repo and *asset swapping* it. This terminology is more current.

5.5.1.3 Risks and pricing aspects

The position studied in the previous section is quite common in financial markets. Practitioners call these *arbitrage plays* or just *arb*. But it is clear from the cash flow diagrams that this is not the arbitrage that an academic would refer to. In the preceding example, there was no initial investment. The immediate net gain was positive, but the practitioner had an open position which was risky. The position was paying net 12.5 bp today, however, the trader was taking the risk that the future spreads between repo rates and SIBOR could tighten below 13.5 basis points. It is true that a 6-month SIBOR has a longer tenor than, say, a 1-month repo rate and, assuming a positively sloped yield curve, the spread will be positive; but this cannot be guaranteed.

Second, the player is assuming different credit risks. He or she is paying a low 2.05% on the repo financing because it is backed by Singapore government bonds. On the other hand, the 2.31% received from the SIBOR side is on a loan made to a high-quality *private* sector credit. Thus, the question remains: Is the net return of 12.5 bp worth the risks taken?

5.5.1.4 An arbitrage approach

There *is* a way to evaluate the appropriateness of the 12.5 bp return mentioned in the example. In fact, the market practitioner's final position is equivalent to owning a *basis swap* between the repo rate and the floating swap reference rate (assuming swap spreads do not change). After all, the position taker is receiving the floating rate in the swap and paying the repo rate.

Suppose the repo and swap have identical settlement dates t_i. The final position is one where, at each settlement date, the position taker will receive

$$(L_{t_{i-1}} + 12.5 \text{ bp} - R_{t_{i-1}})\delta N \tag{5.7}$$

Clearly, this is similar to the settlement of a basis swap with a 12.5 bp spread and notional amount N. If such basis swaps traded actively in the Singapore market, one could evaluate the strategy by comparing the net return of 12.5 bp with the basis swap spread observed in the market. If they are equal, then the same position can be taken directly in the basis swap market. Otherwise, if the basis swap rate is different than 12.5 bp, then a true arbitrage position may be put in place by buying the cheaper one and simultaneously selling the more expensive position.

5.5.2 FUTURES ARBITRAGE

Repo plays a special role in bond and T-bill futures markets. Consider a futures position with expiration $t_0 + 30$ days. In 30 days, we will take possession of a default-free zero coupon bond with maturity T at the predetermined futures price P_{t_0}. Hence, at settlement, P_{t_0} dollars will be paid and the 1-year bond will be received. Of course, at $t_0 + 30$ the market value of the bond will be given by $B(t_0 + 30, T)$ and will, in general, be different than the contracted P_{t_0}. The repo market can be used to hedge this position. This leads us to the important notion of *implied repo rate*.

How can we use repo to hedge a short bond futures position? We dealt with this idea earlier in the discussion of cash-and-carry trades: secure funding, and buy a $T + 30$-day maturity bond at t_0. When time $t_0 + 30$ arrives, the maturity left on this bond will be T, and thus the cash and carry will result in the same position as the futures. The practitioner borrows USD at t_0, buys the $B(t_0, T + 30)$ bond, and keeps this bond until time $t_0 + 30$.

The novelty here is that we can collapse the two steps into one by buying the bond, and then immediately repoing it to secure financing. The result should be a futures position with an equivalent price.

This means that the following equation must be satisfied:

$$P_{t_0} = B(t_0, T + 30)\left(1 + R_{t_0}\frac{30}{360}\right) \tag{5.8}$$

In other words, once the carry cost of buying the $T + 30$-day maturity bond is included, the total amount paid should equal P_{t_0}, the futures price of the bond.

Given the market quotes on the $P_{t_0}, B(t_0, T + 30)$, market practitioners solve for the unknown R_{t_0} and call this the *implied repo rate*:

$$R_{t_0} = \left[\frac{P_{t_0}}{B(t_0, T + 30)} - 1\right]\frac{360}{30} \tag{5.9}$$

The implied repo rate is a pure arbitrage concept and shows the carry cost for fixed-income dealers.

5.5.3 HEDGING A SWAP

Repo can also be used to hedge swap positions. Suppose a dealer transacts a 100 million 2-year swap with a client. The dealer will *pay* the fixed 2-year treasury plus 30 bp, which brings the bid swap rate to, say, 5.95%. As usual, LIBOR will be received. The dealer hedges the position by buying a 2-year treasury.

In doing this, the dealer expects to transact another 2-year swap "soon" with another client, and *receives* the fixed rate. Given that the asking rate is higher than the bid swap rate, the dealer will capture the bid-ask spread. Suppose the ask side swap spread is 33 bp. Where does the repo market come in? The dealer has hedged the swap with a 2-year treasury, but how is this treasury funded? The answer is the repo market. The dealer buys the treasury and then immediately repos it out overnight. The repo rate is 5.61%. The dealer expects to find a matching order in a few days. During this time, the trader has exposure to (i) changes in the swap spread and (ii) changes in the repo rate.

5.5.4 TAX STRATEGIES

Consider the following situation:

- Domestic bond holders pay a withholding tax, while foreign owners don't. Foreign investors receive the gross coupons.

The following operation can be used. The domestic bond holder repos out the bond just before the coupon payment date to a foreign dealer (i.e., a tax-exempt counterparty). Then, the lender receives a manufactured dividend, which is a gross coupon.[10]

This is legal in some economies. In others, the bond holder would be taxed on the theoretical coupon he or she would have received if the bond had not been repoed out. Repoing out the bond to avoid taxation is called *coupon washing*.

EXAMPLE

Demand for Thai bonds for both secondary trading and investment has partly been spurred by the emergence of more domestic mutual funds, which have been launching fixed-income funds. However, foreign participation in the Thai bond market is limited because of withholding taxes.

"Nobody's figured out an effective way to wash the coupon to avoid paying withholding taxes," said one investment banker in Hong Kong. Coupon washing typically involves an offshore investor selling a bond just before the coupon payment date to a domestic counterparty. Offshore entities resident in a country having a tax treaty with the country of the bond's origin can also serve to wash coupons.

In return, the entity washing the coupon pays the offshore investor the accrued interest earned for the period before it was sold—less a small margin. Coupon washing for Thai issues is apparently widespread but is becoming more difficult, according to some sources.

(Thomson Reuters IFR, Issue 1129).

Another example of this important repo application is from Indonesia.

EXAMPLE

A new directive from Indonesia's Ministry of Finance has put a temporary stop to coupon washing activities undertaken by domestic institutions on behalf of offshore players. The new directive, among other things, requires that tax be withheld on the accrued interest investors earn from their bond holdings...

Before the directive was issued a fortnight ago, taxes were withheld only from institutions that held the bond on coupon payment date. Offshore holders of Indonesian bonds got around paying the withholding tax by having the coupons washed.

Typically, coupon washing involves an offshore institution selling and buying its bonds—just before and after the coupon payment dates—to tax-exempt institutions in Indonesia. As such, few bond holders— domestic or offshore—paid withholding taxes on bond holdings. Because the new directive requires that accrued interest on bonds be withheld, many domestic institutions have stopped coupon washing for international firms.

(Thomson Reuters IFR, Issue 1168).

The relevance of repo to taxation issues is much higher than what these readings indicate. The following example shows another use of repo.

---

[10]Note that one of the critical points is "when" a manufactured dividend is paid. If this is paid at the expiration date, coupons can be transferred into the next tax year.

> **EXAMPLE**
>
> *In Japan there is a transaction tax on buying/selling bonds—the transfer tax. To (cut costs), repo dealers lend and borrow Japanese Government Bonds (JGBs) and mark them to market every day.*
>
> *The traders don't trade the bond but trade the* name registration forms *(NRF). NRF are "memos" sent to Central Bank asking for ownership change. They are delivered to local custodians. The bond remains in the hands of the original owner, which will be the issuer of the NRF.*
>
> *JGB trading also has a* no-fail *rule, that is to say failure to deliver carries a very high cost and is considered taboo.*
>
> **(Thomson Reuters IFR, Issue 942).**

Many of the standard transactions in finance have their roots in taxation strategies as these examples illustrate.

5.6 SYNTHETICS USING REPOS

We will now analyze repo strategies by using contractual equations that we introduced in previous chapters. We show several examples. The first example deals with using repos in *cash-and-carry arbitrage*, we then manipulate the resulting contractual equations to get further synthetics.

5.6.1 A CONTRACTUAL EQUATION

Let F_t be the forward price observed at time t, for a Treasury bond to be delivered at a future date T, with $t < T$. Suppose the bond to be delivered at time T needs to have a maturity of U years. Then, at time t, we can (i) buy a $(T - t) + U$ year Treasury bond, (ii) repo it out to get the necessary cash to pay for it, and (iii) hold this repo position until T. At time T, cash plus the repo interest has to be returned to the repo dealer and the bond is received. The bond will have a maturity of U years. As seen above, these steps will result in exactly the same outcome as a bond forward. We express these steps using a contractual equation. This equation provides a synthetic forward.

$$
\boxed{\begin{array}{l}\text{Forward purchase}\\ \text{a } U - \text{year}\\ \text{bond to be delivered}\\ \text{at } T\end{array}} = \boxed{\begin{array}{l}\text{Buy a } T + U{-}t\\ \text{year bond at } t\end{array}} + \boxed{\begin{array}{l}\text{Repo the bond with}\\ \text{term } T - t\end{array}} \qquad (5.10)
$$

According to this, futures positions can be fully hedged by transactions shown on the right-hand side of the equation. This contractual equation can be used in several interesting applications of repo transactions. We discuss two examples.

5.6.2 A SYNTHETIC REPO

Now rearrange the preceding contractual equation so that repo is on the left-hand side:

$$
\boxed{\begin{array}{l}\text{Bond repo with}\\\text{term } T-t\end{array}} = \boxed{\begin{array}{l}\text{Forward purchase}\\\text{of } U-\text{year}\\\text{bond to be}\\\text{delivered at } T\end{array}} - \boxed{\begin{array}{l}\text{Buy a } T+U-t\\\text{year bond at } t\end{array}} \qquad (5.11)
$$

Thus, we can easily create a synthetic repo transaction by using a spot *sale* along with a forward purchase of the underlying asset.[11]

5.6.3 A SYNTHETIC OUTRIGHT PURCHASE

Suppose for some reason we don't want to buy the underlying asset directly. We can use the contractual equation to create a spot purchase synthetically. Moving the spot operation to the left-hand side,

$$
\boxed{\begin{array}{l}\text{Outright purchase}\\\text{of } T+U-t\\\text{year bond at } t\end{array}} = - \boxed{\begin{array}{l}\text{Repo with term}\\T-t\end{array}} + \boxed{\begin{array}{l}\text{Forward purchase}\\\text{of a } U-\text{year bond}\\\text{at } T\end{array}} \qquad (5.12)
$$

The right-hand side operations are equivalent to the outright purchase of the security.

5.6.4 SWAPS VERSUS REPO

There may be some interesting connections between strips, swaps, and repo market strategies. For example, if strips are purchased by investors who hold them until maturity, there will be fewer whole-coupon bonds. This by itself raises the probability that these bonds will trade as "special." As a result, the repo rate will on the average be lower, since the trader who is *short* the instrument will have to accept a "low" repo rate to get the security that is "special" to him or her.[12]

According to some traders, this may lead to an increase in the average swap spread because the availability of cheap funding makes *paying* fixed relatively more attractive than *receiving* fixed.

---

[11]Remember that a minus sign before a contract means that the transaction is reversed. Hence, the spot purchase becomes a spot sale.
[12]But according to other traders, there is little relation between strips and US repo rates, because what is mostly stripped are off-the-run bonds. And off-the-run issues do not, in general, go "special."

5.7 DIFFERENCES BETWEEN REPO MARKETS AND THE IMPACT OF THE GFC

The institutional structure of the repo market had an important effect on the financial system and financial markets in different countries during the GFC. In the United States, both the bilateral and tri-party repo markets experienced runs. The demise of Lehman Brothers was accelerated when creditors in the tri-party repo market refused to extend repo financing to the bank, since they were concerned about the creditworthiness and counterparty risk of Lehman Brothers. The bilateral repo market for different asset classes experienced a different kind of run during the GFC as collateral buyers increased haircuts, thus forcing some collateral sellers to liquidate assets.[13] Financial engineers and traders need to take such potential runs into account when designing products and risk managing their positions.

The GFC did not just have a differential impact on different types of repo markets in the United States, but there were also differences between continents due to different institutional arrangements. The majority of euro repo transactions are conducted in the interbank market which is due to the fact that banks play a dominate role in the provision of capital in Europe.[14] Moreover, more than 60% of the interbank repo transactions in the euro area are conducted via central counter parties (CCP). Repos involving standard securities such as government bonds or other relatively safe securities are typically executed via CCP. The difference between CCP and non-CCP repos is that non-CCP repos typically involve less standard securities as collateral and more bespoke contract terms. Trading in the CCP-based segment of the euro interbank repo market was resilient during the GFC. Repo volume, risk premiums, and lengths of average term repos were not adversely affected for repos that were conducted via CCPs. Two of the reasons for the greater stability of the CCP-based euro interbank repo market were probably the anonymous electronic trading and the reliance on safe collateral. Whereas the CCP-based repo market fared well, the *non-CCP-based* euro interbank repo market experienced similar difficulties similar to those of the US repo market during the GFC. This episode highlights the importance of repo markets and their institutional structure for financial markets and financial engineering applications.

5.8 CONCLUSIONS

The basic concepts underlying repo markets are simple, but due to the fact that repos are dominated by institutions, knowledge about repo markets is not widespread compared to other aspects of financial engineering. Yet, repo markets are crucial for the smooth operation of financial systems. Many financial strategies would be difficult to implement if it weren't for the repo. This chapter has shown that repos can be analyzed with the same techniques discussed in earlier chapters.

---

[13]See Copeland et al. (2010), Krishnamurthy et al. (2014), and Adrian et al. (2013) for recent studies of the repo market.
[14]Mancini et al. (2014) analyze the euro interbank repo market and contrast it with the US repo market.

SUGGESTED READING

Relatively few sources are available on repos, but the ones that exist are good. One good text is **Steiner** (2012). Risk, Euromoney, and similar publications have periodic supplements that deal with repo. These supplements contain interesting examples in terms of recent repo market strategies. Many of the examples in this chapter are taken from such past supplements. Repo markets differ between countries. For an introduction to Equity Repos, see **Choudhry** (2010).

EXERCISES

1. Suppose you are a newly employed investment banker and you find that your bank got stuck in a deal where it has to deliver a bond to close a short future position. The bond has just gone "special" and you do not own it yet. What strategy do you suggest to address this situation and avoid a heavy penalty in case of failure to deliver?

2. A dealer needs to borrow EUR 30 million. He uses a Bund as collateral. The Bund has the following characteristics:
- Collateral 4.3% Bund, June 12, 2004
- Price: 100.50
- Start date: September 10
- Term: 7 days
- Repo rate: 2.7%
- Haircut: 0%
- **a.** How much collateral does the dealer need?
- **b.** Two days after the start of the repo, the value of the Bund increases to 101. How much of the securities will be transferred to whom?
- **c.** What repo interest will be paid?

3. A dealer repos $10 million T-bills. The haircut is 5%. The parameters of the deal are as follows:
- T-bill yield: 2%
- Maturity of T-bills: 90 days
- Repo rate: 2.5% term: 1 week
- **a.** How much cash does the dealer receive?
- **b.** How much interest will be paid at the end of the repo deal?

4. A dealer wants to borrow $10 million using a bond repo but due to certain market restriction he cannot repo out the bond for 2 months.
- **a.** Suggest a strategy so that he can borrow the cash for the period without losing the permanent ownership of the bond.
- **b.** Calculate the forward price of bond trading at $43.59 if the current repo rate is 4% p.a.

5. Answer the questions related to the following case study:

CASE STUDY: CTD AND REPO ARBITRAGE

Two readings follow. Please read them carefully and answer the questions that follow. You may have to first review three basic concepts: (i) special repo versus general collateral, (ii) the notion of CTD bonds, and (iii) failure to deliver. You must understand these well, otherwise the following strategies will not make sense.

Readings

DB Bank is believed to have pocketed over EUR100 million (USD89.4 million) after reportedly squeezing repo traders in a massive interest-rate futures position. The bank was able to take advantage of illiquidity in the cheapest-to-deliver bond that would have been used to settle a long futures position it entered, in a move that drew sharp criticism from some City rivals.

In the trade, the Bank entered a calendar spread in which it went long the Eurex-listed BOBL March'01 future on German medium-term government bonds and sold the June'01 contract to offset the long position, said traders familiar with the transaction. One trader estimated the Bank had bought 145,000 March'01 contracts and sold the same number of June'01 futures. At the same time the Bank built up a massive long position via the repo market in the cheapest-to-deliver bond to settle the March future, in this case a 10-year Bund maturing in October 2005.

Since the size of the '05 Bund issue is a paltry EUR10.2 billion, players short the March future would have needed to round up 82% of the outstanding bonds to deliver against their futures obligations. "It is almost inconceivable that this many of the Bunds can be delivered," said a director-derivatives strategy in London. "Typically traders would be able to rustle up no more than 25% of a cheapest-to-deliver bond issue," he added.

At the same time it was building the futures position, the Bank borrowed the cheapest-to-deliver bonds in size via the repo market. Several traders claim the Bank failed to return the bonds to repo players by the agreed term, forcing players short the March future to deliver more expensive bonds or else buy back the now more expensive future.

The Bank was able to do this because penalties for failure to deliver in the repo market are less onerous than those governing failure to deliver on a future for physical delivery. Under Eurex rules, traders that fail to deliver on a future must pay 40 basis points of the face value of the bond per day. After a week the exchange is entitled to buy any eligible bond on behalf of the party with the long futures position and send the bill to the player with the short futures position, according to traders. Conversely, the equivalent penalty for failure to deliver in the repo market is 1.33 bp per day.

(Thomson Reuters IFR, March 2001).

Eurex Reforms Bobl Future

Eurex is introducing position limits for its September contracts in its two-, five-, and 10-year German government bond futures. "If we want, we will do it in December as well," said a spokesman for the exchange.

The move is aimed at supporting the early transfer of open positions to the next trading cycle and is a reaction to the successful squeeze of its Bundesobligation (Bobl) or five-year German government bond futures contract in March.

"The new trading rules limit the long positions held by market participants, covering proprietary and customer trading positions," said Eurex's spokesman. Position limits will be set in relation to the issue size of the cheapest-to-deliver bond and will be published six exchange trading days before the rollover period begins.

(Thomson Reuters IFR, June 9, 2001).

Part A. First Reading

1. What is a calendar spread? Show DB's position using cash flow diagrams.
2. Put this together with DB's position in the repo market.
3. What is DB's position aiming for?
4. What is the importance of the size of '05 Bund issue? How do traders "rustle up" such bonds to be delivered?

5. Why are penalties for failure to deliver relevant?
6. Would an asset swap (e.g., swapping LIBOR against the relevant bond mentioned in the paper) have helped the *shorts*? Explain.
7. Could taking a carefully chosen position in the relevant maturity FRA, offset the losses that *shorts* have suffered? Explain carefully.
8. Explain how CTD bonds are determined. For needed information go to Web sites of futures exchanges.

Part B. Second Reading

1. Eurex has made some changes in the Bund futures trading rules. What are these?
2. Suppose these rules had been in effect during March, would they have prevented DB's arbitrage position?
3. Would there be ways DB can still take such a position? What are they?

CASH FLOW ENGINEERING IN FOREIGN EXCHANGE MARKETS

CHAPTER OUTLINE

6.1 INTRODUCTION

Forwards and futures contracts are ideal for creating synthetic instruments for many reasons. Forwards and futures are, in general, linear permitting static replication. They are often very liquid and, in the case of currency forwards, have homogeneous underlying. Many technical complications are automatically eliminated by the homogeneity of a currency. Consider the following interpretation.

Financial instruments are denominated in different currencies. A market practitioner who needs to perform a required transaction in US dollars (USD) normally uses instruments denoted in USD. In the case of the dollar, this works out fine since there exists a broad range of liquid markets. Market professionals can offer all types of services to their customers using these. On the other hand, there is a relatively small number of, say, *liquid* Swiss Franc (CHF) denoted instruments. Would the Swiss market professionals be deprived of providing the same services to their clients? It turns out that liquid foreign exchange (FX) forward contracts in USD/CHF can, in principle, make USD-denominated instruments available to CHF-based clients as well.

Instead of performing an operation in CHF, one can first buy and sell USD at t_0 and then use a USD-denominated instrument to perform any required operation. Liquid FX forwards permit *future* USD cash flows to be reconverted into CHF as of time t_0. Thus, entry into and exit from a different currency is fixed at the initiation of a contract. As long as liquid forward contracts exist, market professionals can use USD-denominated instruments in order to perform operations in any other currency without taking FX risk.

As an illustration, we provide the following example where a synthetic zero coupon bond is created using an FX forward and the bond markets of another country.

EXAMPLE

Suppose we want to buy, at time t_0, a USD-denominated default-free discount bond, with maturity at t_1 and current price $B(t_0, t_1)$. We can do this synthetically using bonds denominated in any other currency, as long as an FX forward exists and the relevant credit risks are the same.

First, we buy an appropriate number of, say, euro-denominated bonds with the same maturity, default risk, and the price $B(t_0, t_1)^{\text{EUR}}$. This requires buying euros against dollars in the spot market at an exchange rate e_{t_0}. Then, using a forward contract on euro, we sell forward the euros that will be received on December 31, 2015, when the bond matures. The forward exchange rate is F_{t_0}.

The final outcome is that we pay USD now and receive a known amount of USD at maturity. This should generate the same cash flows as a USD-denominated bond under no-arbitrage conditions. This operation is shown in Figure 6.1.

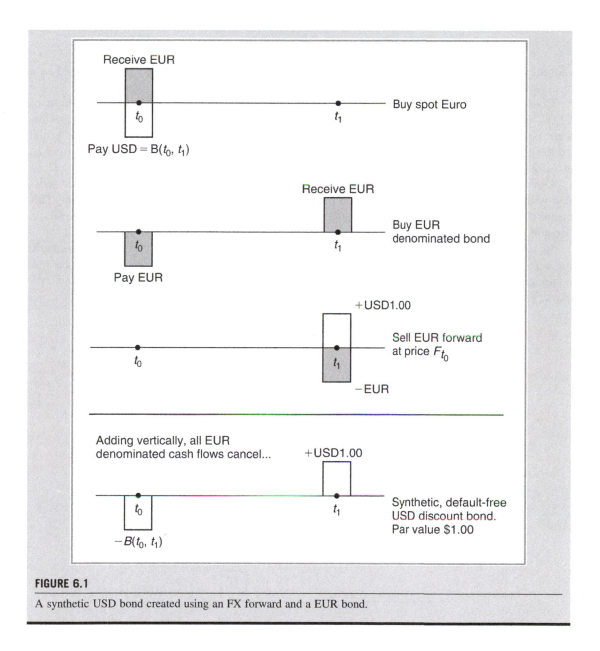

FIGURE 6.1

A synthetic USD bond created using an FX forward and a EUR bond.

In principle, such steps can be duplicated for any (linear) underlying asset, and the ability to execute forward purchases or sales plays a crucial role here.

Next we discuss the engineering of one of the simplest and most liquid contracts; namely, the currency *forward*.

6.2 CURRENCY FORWARDS

Currency forwards are very liquid instruments. Although they are elementary, they are used in a broad spectrum of financial engineering problems.

Consider the EUR/USD exchange rate.[1] The cash flows implied by a forward purchase of USD100 against euros are represented in Figure 6.2a. At time t_0, a contract is written for the forward purchase (sale) of USD100 against $100/F_{t_0}$ euros. The settlement—that is to say, the actual exchange of currencies—will take place at time t_1. The forward exchange rate is F_{t_0}. At time t_0, nothing changes hands.

Obviously, the forward exchange rate F_{t_0} should be chosen at t_0 so that the two parties are satisfied with the future settlement and thus do not ask for any immediate compensating payment. This means that the time t_0 value of a forward contract *concluded* at time t_0 is *zero*. It may, however, become positive or negative as time passes and markets move.

In this section, we discuss the structure of this instrument. How do we create a synthetic for an instrument such as this one? How do we decompose a forward contract? Once this is understood, we consider applications of our methodology to hedging, pricing, and risk management.

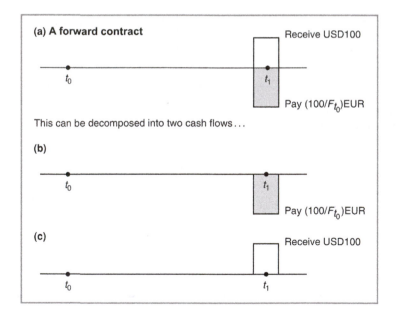

FIGURE 6.2

(a–c) A currency forward. (d–f) A money market synthetic.

[1]Written as EUR/USD in this quote, the *base currency* is the euro.

A general method of engineering a (currency) forward—or, for that matter, any linear instrument—is as follows:

1. Begin with the cash flow diagram as shown in Figure 6.2a.
2. Detach and carry the (two) rectangles representing the cash flows into Figure 6.2b and c.
3. Then, add *and* subtract new cash flows at carefully chosen dates so as to *convert* the detached cash flows into meaningful financial contracts that players will be willing to buy and sell.
4. As you do this, make sure that when the diagrams are added *vertically*, the newly added cash flows cancel out and the original cash flows are *recovered*.

This procedure will become clearer as it is applied to progressively more complicated instruments. Now we consider the details.

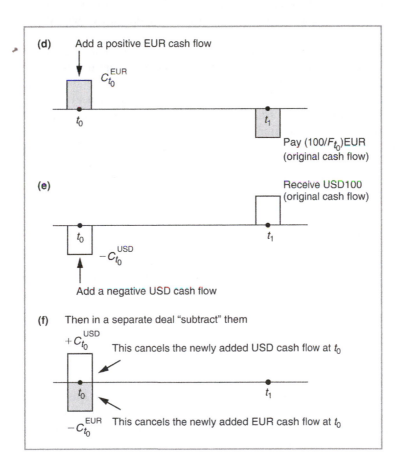

FIGURE 6.2

(Continued)

6.2.1 ENGINEERING THE CURRENCY FORWARD

We apply this methodology to engineering a currency forward. Our objective is to obtain a contractual equation at the end and, in this way, express the original contract as a *sum* of two or more elementary contracts. The steps are discussed in detail.

Begin with cash flows as shown in Figure 6.2a. If we detach the two cash flows, we get Figure 6.2b and c. At this point, nobody would like to buy cash flows in Figure 6.2b, whereas nobody would sell the cash flows in Figure 6.2c. Indeed, why pay something without receiving anything in return? So at this point, Figure 6.2b and c cannot represent tradeable financial instruments.

However, we can *convert* them into tradeable contracts by inserting new cash flows, as in step 3 of the methodology. In Figure 6.2b, we *add* a corresponding cash *in*flow. In Figure 6.2c, we add a cash *out*flow. By adjusting the size and the timing of these new cash flows, we can turn the transactions in Figure 6.2b and c into meaningful financial contracts.

We keep this as simple as possible. For Figure 6.2b, add a positive cash flow, preferably at time t_0.[2] This is shown in Figure 6.2d. Note that we denote the size of the newly added cash flow by $C_{t_0}^{EUR}$.

In Figure 6.2c, add a negative cash flow at time t_0, to obtain Figure 6.2e. Let this cash flow be denoted by $C_{t_0}^{USD}$. The size of $C_{t_0}^{USD}$ is not known at this point, except that it has to be in USD.

The *vertical addition* of Figure 6.2d and e should replicate what we started with in Figure 6.2a. At this point, this *will* not be the case, since $C_{t_0}^{USD}$ and $C_{t_0}^{EUR}$ do not cancel out at time t_0 as they are denominated in different currencies. But, there is an easy solution to this. The "extra" time t_0 cash flows can be eliminated by considering a third component for the synthetic. Consider Figure 6.2f where one exchanges $C_{t_0}^{USD}$ against $C_{t_0}^{EUR}$ at time t_0. After the addition of this component, a vertical sum of the cash flows in Figure 6.2d−f gives a cash flow pattern identical to the ones shown in Figure 6.2a. If the credit risks are the same, we have succeeded in replicating the forward contract with a synthetic.

6.2.2 WHICH SYNTHETIC?

Yet, it is still not clear what the synthetic in Figure 6.2d−f consists of. True, by adding the cash flows in these figures, we recover the original instrument shown in Figure 6.2a, but what kind of contracts do these figures represent? The answer depends on how the synthetic instruments shown in Figure 6.2d−f are interpreted.

This can be done in many different ways. We consider two major cases. The first is a deposit−loan interpretation. The second involves Treasury bills.

6.2.2.1 A money market synthetic

The first synthetic is obtained using money market instruments. To do this, we need a brief review of money market instruments. The following lists some important money market instruments, along with the corresponding quote, registration, settlement, and other conventions that will have cash flow patterns similar to Figure 6.2d and e. The list is not comprehensive.

---

[2]We *could* add it at another time, but it would yield a more complicated synthetic. The resulting synthetic will be less liquid and, in general, more expensive.

EXAMPLE

Deposits/loans. These mature in less than 1 year. They are denominated in domestic and Eurocurrency units. Settlement is on the same day for domestic deposits and in 2 business days for Eurocurrency deposits. There is no registration process involved and they are not negotiable.

Certificates of deposit (CD). Generally these mature in up to 1 year. They pay a coupon and are sometimes sold in discount form. They are quoted on a yield basis, and exist both in domestic and Eurocurrency forms. Settlement is on the same day for domestic deposits and in 2 working days for Eurocurrency deposits. They are usually bearer securities and are negotiable.

Treasury bills. These are issued at 13-, 26-, and 52-week maturities. In France, they can also mature in 4–7 weeks; in the United Kingdom, also in 13 weeks. They are sold on a discount basis (United States, United Kingdom). In other countries, they are quoted on a yield basis. Issued in domestic currency, they are bearer securities and are negotiable.

Commercial paper (CP). Their maturities are 1–270 days. They are very short-term securities, issued on a discount basis. The settlement is on the same day; they are bearer securities and are negotiable.

Euro-CP. The maturities range from 2 to 365 days but most have 30- or 180-day maturities. Issued on a discount basis, they are quoted on a yield basis. They can be issued in any Eurocurrency, but in general they are in Eurodollars. Settlement is in 2 business days, and they are negotiable.

How can we use these money market instruments to interpret the synthetic for the FX forward as shown in Figure 6.2?

One money market interpretation is as follows. The cash flow shown in Figure 6.2e involves making a payment of $C_{t_0}^{USD}$ at time t_0, to receive USD100 at a later date, t_1. Clearly, an interbank *deposit* will generate exactly this cash flow pattern. Then, the $C_{t_0}^{USD}$ will be the present value of USD100, where the discount factor can be obtained through the relevant Eurodeposit rate.

$$C_{t_0}^{USD} = \frac{100}{1 + L_{t_0}^{USD}((t_1 - t_0)/360)} \tag{6.1}$$

Note that we are using an ACT/360-day basis for the deposit rate $L_{t_0}^{USD}$, since the cash flow is in Eurodollars. Also, we are using money market conventions for the interest rate.[3] Given the observed value of $L_{t_0}^{USD}$, we can numerically determine the $C_{t_0}^{USD}$ by using this equation.

---

[3]We remind the reader that, as discussed in Chapter 3, if this was a domestic UK or eurosterling deposit, for example, the day basis would be 365. This is another warning that in financial engineering, conventions matter.

How about the cash flows shown in Figure 6.2d? This can be interpreted as a *loan* obtained in interbank markets. One receives $C_{t_0}^{EUR}$ at time t_0 and makes a euro-denominated payment of $100/F_{t_0}$ at the later date t_1. The value of this cash flow will be given by

$$C_{t_0}^{EUR} = \frac{100/F_{t_0}}{1 + L_{t_0}^{EUR}((t_1 - t_0)/360)} \tag{6.2}$$

where $C_{t_0}^{EUR}$ is the relevant interest rate in euros.

Finally, we need to interpret the last diagram shown in Figure 6.2f. These cash flows represent an exchange of $C_{t_0}^{USD}$ against $C_{t_0}^{EUR}$ at time t_0. Thus, what we have here is a *spot purchase* of dollars at the rate e_{t_0}.

The synthetic is now fully described:

- Take an interbank loan in euros (Figure 6.2d).
- Using these euro funds, buy spot dollars (Figure 6.2f).
- Deposit these dollars in the interbank market (Figure 6.2f).

This portfolio would exactly replicate the currency forward, since by adding the cash flows in Figure 6.2d−f, we recover exactly the cash flows generated by a currency forward as shown in Figure 6.2a.

6.2.2.2 A synthetic with T-bills

We can also create a synthetic currency forward using Treasury-bill markets. In fact, let $B(t_0, t_1)^{USD}$ be the time t_0 price of a default-free discount bond that pays USD100 at time t_1. Similarly, let $B(t_0, t_1)^{EUR}$ be the time t_0 price of a default-free discount bond that pays EUR100 at time t_1. Then the cash flows shown in Figure 6.2d−f can be reinterpreted so as to represent the following transactions:[4]

- Figure 6.2d is a *short* position in $B(t_0, t_1)^{EUR}$ where $1/F_{t_0}$ units of this security is borrowed and sold at the going market price to generate $B(t_0, t_1)^{EUR}/F_{t_0}$ euros.
- In Figure 6.2f, these euros are exchanged into dollars at the going exchange rate.
- In Figure 6.2e, the dollars are used to buy one dollar-denominated bond $B(t_0, t_1)^{USD}$.

At time t_1 these operations would amount to exchanging EUR $100/F_{t_0}$ against USD100, given that the corresponding bonds mature at par.

6.2.2.3 Which synthetic should one use?

If synthetics for an instrument can be created in many ways, which one should a financial engineer use in hedging, risk management, and pricing? We briefly comment on this important question.

As a rule, a market practitioner would select the synthetic instrument that is most desirable according to the following attributes: (1) The one that *costs* the least. (2) The one that is most *liquid*, which, *ceteris paribus*, will, in general, be the one that costs the least. (3) The one that is most convenient for *regulatory* purposes. (4) The one that is most appropriate given *balance sheet* considerations. Of course, the final decision will have to be a compromise and will depend on the particular needs of the market practitioner.

---

[4]Disregard for the time being whether such liquid discount bonds exist in the desired maturities.

6.2.2.4 Credit risk

Section 6.2.2.1 displays a list of instruments that have similar cash flow patterns to loans and T-bills. The assumption of no-credit risk is a major reason why we could alternate between loans and T-bills in Sections 6.2.2.1 and 6.2.2.2. If credit risk were taken into account, the cash flows would be significantly different. In particular, for loans we would have to consider a diagram as shown in Figure 6.1, whereas T-bills would have no default risks.[5]

6.3 SYNTHETICS AND PRICING

A major use of synthetic assets is in pricing. Everything else being the same, a replicating portfolio must have the same price as the original instrument. Thus, adding up the values of the constituent assets, we can get the *cost* of forming a replicating portfolio. This will give the price of the original instrument once the market practitioner adds a proper margin.

In the present context, *pricing* means obtaining the unknowns in the currency forward, which is the forward exchange rate, F_{t_0}, introduced earlier. We would like to determine a set of pricing equations which result in *closed-form* pricing formulas. Let us see how this can be done.

Begin with Figure 6.2f. This figure implies that the time t_0 market values of $C_{t_0}^{USD}$ and $C_{t_0}^{EUR}$ should be the same. Otherwise, one party will not be willing to go through with the deal. This implies

$$C_{t_0}^{USD} = C_{t_0}^{EUR} e_{t_0} \tag{6.3}$$

where e_{t_0} is the *spot* EUR/USD exchange rate. Replacing from Eqs. (6.1) and (6.2):

$$F_{t_0} \left[\frac{100}{1 + L_{t_0}^{USD}((t_1 - t_0)/360)} \right] = \left[\frac{100}{1 + L_{t_0}^{EUR}((t_1 - t_0)/360)} \right] e_{t_0} \tag{6.4}$$

Solving for the forward exchange rate F_{t0},

$$F_{t_0} = e_{t_0} \left[\frac{1 + L_{t_0}^{USD}((t_1 - t_0)/360)}{1 + L_{t_0}^{EUR}((t_1 - t_0)/360)} \right] \tag{6.5}$$

This is the well-known *covered interest rate parity* (CIRP) equation. Note that it expresses the "unknown" F_{t_0} as a function of variables that can be observed at time t_0. Hence, using the market quotes, F_{t_0} can be numerically calculated at time t_0 and does not require *any* forecasting effort.[6]

The second synthetic using T-bills gives an alternative pricing equation. Since the values evaluated at the current exchange rate, e_t, of the two bond positions needs to be the same, we have

$$F_{t_0} B(t_0, t_1)^{USD} = e_{t_0} B(t_0, t_1)^{EUR} \tag{6.6}$$

---

[5]As the GFC has reminded us, this is a simplifying assumption since in reality short- and long-term government bonds *are* risky.
[6]In fact, bringing in a forecasting model to determine the F_{t_0} will lead to the wrong market price and may create arbitrage opportunities.

Hence, the F_{t_0} priced *off the T-bill markets* will be given by

$$F_{t_0} = e_{t_0} \frac{B(t_0, t_1)^{\text{EUR}}}{B(t_0, t_1)^{\text{USD}}} \tag{6.7}$$

If the bond markets in the two currencies are as liquid as the corresponding deposits and loans, and if there is no credit risk, the F_{t_0} obtained from this synthetic will be very close to the F_{t_0} obtained from deposits.[7]

6.4 A CONTRACTUAL EQUATION

In this section, we will obtain a contractual equation for a currency forward. In the next section, we will show several applications. We have just created a synthetic for a currency forward. The basic idea was that a portfolio consisting of the following instruments:

{Loan in EUR, Deposit of USD, spot purchase of USD against EUR}

would generate the same cash flows, at the same *time* periods, with the same *credit risk* as the currency forward. This means that under the (unrealistic) assumptions of

1. No transaction costs
2. No bid—ask spreads
3. No credit risk
4. No counterparty risk
5. Liquid markets

we can write the equivalence between the related synthetic and the original instrument as a contractual equation that can conveniently be exploited in practice. In fact, the synthetic using the money market involved three contractual deals that can be summarized by the following contractual equation:

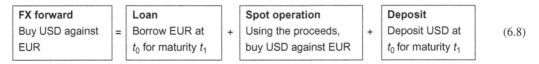

$$\tag{6.8}$$

This operation can be applied to any two currencies to yield the corresponding FX forward.

   Under the simplifying assumptions listed above, we expect that *arbitrage* will make the values of the two sides of the contractual equation equal. Due to arbitrage activity by market participants, CIRP is one of the financial engineering concepts that holds very closely in *normal* market conditions. In an interesting recent paper, Mancini-Griffoli and Ranaldo (2013) document deviations from CIRP for several months after the Lehman bankruptcy due to *limits of arbitrage*. Limits of arbitrage refers to a theory that, due to restrictions that are placed on funds that would normally be used by rational traders to exploit arbitrage opportunities, prices may remain in a nonequilibrium state for long periods of time.[8]

---

[7]Remember the important point that, in practice, both the liquidity and the credit risks associated with the synthetics could be significantly different. Then the calculated F_{t_0} would diverge.

[8]See also Shleifer and Vishny (1997) and Gromb and Vayanos (2010), who provide a recent review of the limits of arbitrage theory.

6.5 APPLICATIONS

The contractual equation derived earlier and the manipulation of cash flows that led to it may initially be thought of as theoretical constructs with limited practical application. This could not be further from the truth. We now discuss four examples that illustrate how the equation can be skillfully exploited to find solutions to practical, common problems faced by market participants.

6.5.1 APPLICATION 1: A WITHHOLDING TAX PROBLEM

We begin with a practical problem of withholding taxes on interest income. Our purpose is not to comment on the taxation aspects but to use this example to motivate uses of synthetic instruments.

The basic idea here is easy to state. If a government imposes withholding taxes on gains from a particular instrument, say a *bond*, and if it is possible to synthetically replicate this instrument, then the synthetic may not be subject to withholding taxes. If one learns how to do this, then the net returns offered to clients will be significantly higher—with, essentially, the same risk.

EXAMPLE

Suppose an economy has imposed a withholding tax on interest income from government bonds. Let this withholding tax rate be 20%. The bonds under question have zero default probability and make no coupon payments. They mature at time T and their time t price is denoted by $B(t, T)$. This means that if

$$B(t, T) = 92 \tag{6.9}$$

one pays 92 dollars at time t to receive a bond with face value 100. The bond matures at time T, with the maturity value

$$B(T, T) = 100 \tag{6.10}$$

Clearly, the interest the bondholder has *earned* will be given by

$$100 - B(t, T) = 8 \tag{6.11}$$

But because of the withholding tax, the interest *received* will only be 6.4:

$$\text{Interest received} = 8 - 0.2 \cdot (8) = 6.4 \tag{6.12}$$

Thus the bondholder receives significantly less than what he or she earns.

We can immediately use the ideas put forward to form a synthetic for any discount bond using the contractual equation in formula (6.8). We discuss this case using two arbitrary currencies called Z and X. Suppose T-bills in both currencies trade actively in their respective markets. The contractual equation written in terms of T-bills gives

$$
\boxed{\begin{array}{l}\textbf{FX forward}\\[2pt]\text{Sell currency } Z\\[2pt]\text{against currency } X\end{array}}=\boxed{\begin{array}{l}\textbf{Short } Z\text{-}\\[2pt]\textbf{denominated}\\[2pt]\textbf{T-bill}\end{array}}+\boxed{\begin{array}{l}\textbf{Spot operation}\\[2pt]\text{Buy currency } X\\[2pt]\text{with currency } Z\end{array}}+\boxed{\begin{array}{l}\textbf{Buy } X\text{-}\\[2pt]\textbf{denominated}\\[2pt]\textbf{T-bill}\end{array}} \tag{6.13}
$$

Manipulating this as an algebraic equation, we can transfer the Z-denominated T-bill to the left-hand side and group all other instruments on the right-hand side.

$$
-\boxed{\begin{array}{l}\textbf{Short } Z\text{-}\\[2pt]\textbf{denominated}\\[2pt]\textbf{T-bill}\end{array}}=-\boxed{\begin{array}{l}\textbf{FX forward}\\[2pt]\text{Sell } Z \text{ against } X\end{array}}+\boxed{\begin{array}{l}\textbf{Spot operation}\\[2pt]\text{Buy currency } X\\[2pt]\text{with } Z\end{array}}+\boxed{\begin{array}{l}\textbf{Buy } X\text{-}\\[2pt]\textbf{denominated}\\[2pt]\textbf{T-bill}\end{array}} \tag{6.14}
$$

Now, we change the negative signs to positive, which reverses the cash flows, and obtain a synthetic Z-denominated T-bill:

$$
\boxed{\begin{array}{l}\textbf{Long } Z\text{-}\\[2pt]\textbf{denominated}\\[2pt]\textbf{T-bill}\end{array}}=\boxed{\begin{array}{l}\textbf{FX forward}\\[2pt]\text{Buy } Z \text{ against } X\end{array}}+\boxed{\begin{array}{l}\textbf{Spot operation}\\[2pt]\text{Buy currency } X\\[2pt]\text{with } Z\end{array}}+\boxed{\begin{array}{l}\textbf{Buy } X\text{-}\\[2pt]\textbf{denominated}\\[2pt]\textbf{T-bill}\end{array}} \tag{6.15}
$$

Thus, in order to construct a synthetic for Z-denominated discount bonds, we first need to use money or T-bill markets of another economy where there is no withholding tax. Let the currency of this country be denoted by the symbol X. According to Eq. (6.15), we exchange Z's into currency X with a *spot* operation at an exchange rate e_{t_0}. Using the X's obtained this way, we buy the relevant X-denominated T-bill. At the same time, we *forward purchase* Z's for time t_1. The geometry of these operations is shown in Figure 6.3. We see that by adding the cash flows generated by the right-hand-side operations, we can get exactly the cash flows of a T-bill in Z.

There is a simple logic behind these operations. Investors are taxed on Z-denominated bonds. So they use *another* country's markets where there is no withholding tax. They do this in a way that ensures the recovery of the needed Z's at time t_1 by buying Z forward. In a nutshell, this is a strategy of carrying funds over time using another currency as a *vehicle* while making sure that the *entry* and *exits* of the position are pinned down at time t_0.[9]

---

[9]For more information on recent issues related to TRS and taxes, see http://www.rkmc.com/~/media/PDFs/Financial%20Litigation%20Insights%20Fall%202013.pdf and "United States: Final and Proposed Regulations Address U.S. Withholding Tax on U.S. Equity Derivatives' (article published on www.mondaq.com on 24 December, 2014).

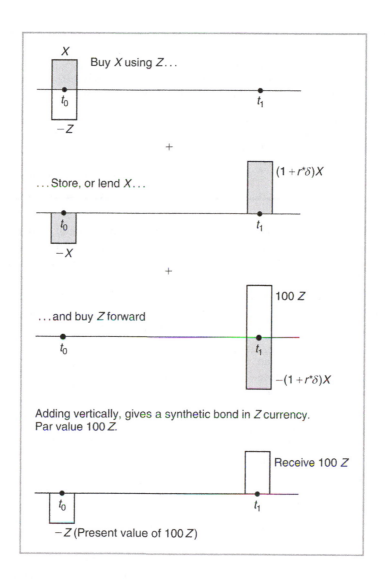

FIGURE 6.3

A synthetic Z-denominated bond.

6.5.2 APPLICATION 2: CREATING SYNTHETIC LOANS

The second application of the contractual equation has already been briefly discussed in Chapter 1. Consider the following market event from the year 1997. [10]

---

[10]Note that a basis point (bp) is one-hundredth of 1%. In other words, 1% equals 100 bs.

EXAMPLE

Following the collapse of Hokkaido Takushoku Bank, the "Japanese premium," the extra cost to Japanese banks of raising money in the Eurodollar market increased last week in dramatic style. Japanese banks in the dollar deposit market were said to be paying around 40 bp over their comparable US credits, against less than 30 bp only a week ago.

Faced with higher dollar funding costs, Japanese banks looked for an alternative source of dollar finance. Borrowing in yen and selling yen against the dollar in the spot market, they bought yen against dollars in the forward market, which in turn caused the USD/yen forward rate to richen dramatically.

Thomson Reuters IFR (November 22, 1997).

Readers with no market experience may consider this episode difficult to understand. Yet, the contractual equation in formula (6.8) can be used skillfully, to explain the strategy of Japanese banks mentioned in the example. In fact, what Japanese banks were trying to do was to create synthetic USD loans. The USD loans were either too expensive or altogether unavailable due to lack of *credit lines*. As such, the excerpt provides an excellent example of a use for synthetics.

We now consider this case in more detail. We begin with the contractual equation in formula (6.8) again, but this time we write it for the USD/JPY exchange rate:

FX forward		Loan		Spot operation		Deposit	
Sell USD against JPY for time t_1	=	Borrow USD with maturity t_1	+	Buy JPY pay USD at t_0	+	Deposit JPY for maturity t_1	(6.16)

Again, we manipulate this like an algebraic equation. Note that on the right-hand side, there is a loan contract. This is a genuine USD loan, and it can be isolated on the left-hand side by rearranging the right-hand-side contracts. The loan would then be expressed in terms of a replicating portfolio.

Loan		FX forward		Spot operation		Deposit	
Borrow USD with maturity t_1	=	Sell USD against JPY for time t_1	−	Buy JPY pay USD at t_0	−	Deposit JPY for maturity t_1	(6.17)

Note that because we moved the deposit and the spot operation to the other side of the equality, signs changed. In this context, a deposit with a minus sign would mean reversing the cash flow diagrams and hence it becomes a loan. A spot operation with a minus sign would simply switch the currencies exchanged. Hence, the contractual equation can finally be written as

USD loan		FX forward		Spot operation		A loan	
	=	Sell USD against JPY for time t_1	+	Buy USD against JPY at t_0	+	Borrow JPY for maturity t_1	(6.18)

This contractual equation can be used to understand the previous excerpt. According to the quote, Japanese banks that were hindered in their effort to borrow Eurodollars in the interbank

(Euro) market instead borrowed Japanese yen in the domestic market, which they used to buy (cash) dollars. But, at the same time, they sold dollars forward against yen in order to hedge their future currency exposure. Briefly, they created exactly the synthetic that the contractual equation implies on the right-hand side.

6.5.3 APPLICATION 3: CAPITAL CONTROLS

Several countries have, at different times, imposed restrictions on capital movements. These are known as *capital controls.* Suppose we assume that a spot purchase of USD against the local currency X is *prohibited* in some country.

A financial engineer can construct a *synthetic spot operation* using the contractual relationship, since such spot operations were one of the constituents of the contractual equation as shown in formula (6.8). Rearranging formula (6.8), we can write

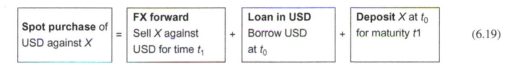

$$\boxed{\begin{array}{c}\text{Spot purchase of}\\\text{USD against } X\end{array}} = \boxed{\begin{array}{l}\textbf{FX forward}\\\text{Sell } X \text{ against}\\\text{USD for time } t_1\end{array}} + \boxed{\begin{array}{l}\textbf{Loan in USD}\\\text{Borrow USD}\\\text{at } t_0\end{array}} + \boxed{\begin{array}{l}\textbf{Deposit } X \text{ at } t_0\\\text{for maturity } t1\end{array}} \qquad (6.19)$$

The right-hand side will be equivalent to a spot purchase of USD even when there are capital controls. Precursors of such operations were called *parallel loans* and were extensively used by businesses, especially in Brazil and some other emerging markets.[11] The geometry of this situation is shown in Figure 6.4.[12]

6.5.4 APPLICATION 4: "CROSS" CURRENCIES

Our final example does not make use of the contractual equation in formula (6.8) directly. However, it is an interesting application of the notion of contractual equations, and it is appropriate to consider it at this point.

One of the simplest synthetics is the "cross rates" traded in FX markets. A cross currency exchange rate is a price that does not involve USD. The major "crosses" are EUR/JPY, EUR/CHF, and GBP/EUR. Other "crosses" are relatively minor. In fact, if a trader wants to purchase Swiss francs in, for example, Taiwan, the trader would carry out two transactions instead of a single spot transaction. He or she would buy USD with Taiwan dollars and then sell the USD against the Swiss franc. At the end, Swiss francs are paid by Taiwan dollars. Why would one go through two transactions instead of a direct purchase of Swiss francs in Taiwan? It is cheaper to do so because of lower transaction costs and higher liquidity of the USD/CHF and USD/TWD exchange rates.

---

[11]One may ask the following question: If it is not possible to buy foreign currency in an economy, how can one borrow in it? The answer to this is simple. The USD borrowing is done with a foreign counterparty.
[12]For a recent example of the effect of capital controls on interest rate parity relationships, see the following BIS working paper that studies the effect of capital controls on the onshore and offshore renminbi interest rates and uncovered interest rate parity: http://www.bis.org/publ/work233.pdf.

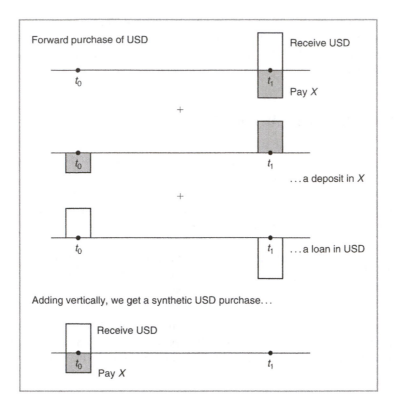

FIGURE 6.4

A synthetic spot operation.

We can formulate this as a contractual equation:

Spot purchase of CHF using Taiwan dollars	=	Buy USD against Taiwan dollar	+	Sell USD against Swiss francs

(6.20)

It is easy to see why this contractual equation holds. Consider Figure 6.5. The addition of the cash flows in the top two graphs results in the elimination of the USD element, and one creates a synthetic "contract" of spot purchase of CHF against Taiwan dollars.

This is an interesting example because it shows that the price differences between the synthetic and the actual contract cannot always be exploited due to transaction costs, liquidity, and other rigidities such as the legal and organizational framework. It is also interesting in this particular case, that it is the synthetic instrument which turns out to be *cheaper*. Thus, before buying and selling an instrument, a trader should always try to see if there is a cheaper synthetic that can do the same job.

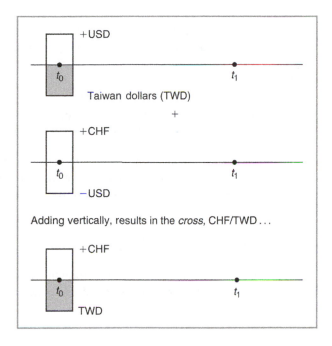

FIGURE 6.5

A synthetic cross rate.

6.6 CONVENTIONS FOR FX FORWARD AND FUTURES

6.6.1 QUOTING CONVENTIONS FOR FX FORWARD

Forwards in foreign currencies have special quotation conventions. Markets do not quote outright forward rates, but the so-called *forward points*. The related terminology and conventions are illustrated in the following example.

EXAMPLE

Suppose outright forward EUR/USD quotes are given by

Bid	Ask
1.0210	1.0220

and that the spot exchange rate quotes are as given by

Bid	Ask
1.0202	1.0205

Then, instead of the outright forward quotes, traders prefer to quote the *forward points* obtained by subtracting the corresponding spot rate from the outright forward:

Bid	Ask
8	15

In reality, forward points are determined directly from Eq. (6.5) or (6.7).

Market conventions sometimes yield interesting information concerning trading activity and the forward FX quotes is a case in point. Let F_{t_0} and e_{t_0} be time t_1 forward and time t_0 spot exchange rates respectively as given by Eq. (6.5). Using the expression in Eq. (6.5) and ignoring the bid–ask spreads, we can write approximately

$$F_{t_0} - e_{t_0} \cong \left(r_{t_0}^d - r_{t_0}^f\right)\left(\frac{t_1 - t_0}{360}\right)e_{t_0} \tag{6.21}$$

where $r_{t_0}^d, r_{t_0}^f$ are the relevant interest rates in domestic and foreign currencies, respectively.[13] *Forward points* are the difference between the forward rate found using the pricing equation in formula (6.21) and the spot exchange rate:

$$F_{t_0} - e_{t_0} \tag{6.22}$$

They are also called "pips" and written as bid/ask. We give an example for the way forward points are quoted and used.

EXAMPLE

Suppose the spot and forward rate quotes are as follows:

EUR/USD	Bid	Ask
Spot	0.8567	0.8572
1 Year	−28.3	−27.3
2 Year	44.00	54.00

From this table, we can calculate the outright forward exchange rate F_{t_0}.

For year 1, subtract the negative pips in order to get the outright forward rates:

$$\left(0.8567 - \frac{28.3}{10,000}\right) \Big/ \left(0.8572 - \frac{27.3}{10,000}\right) \tag{6.23}$$

---

[13]This assumes a day count basis of 360 days. If one or both of the interest rates have a 365-day convention, this expression needs to be adjusted accordingly.

For year 2, the quoted pips are *positive*. Thus, we add the positive points to get the outright forward rates:

$$\left(0.8567 + \frac{44}{10,000}\right) \bigg/ \left(0.8572 + \frac{54}{10,000}\right) \tag{6.24}$$

Forward points give the amount needed to adjust the spot rate in order to obtain the outright forward exchange rate. Depending on the market, they are either added to or subtracted from the spot exchange rate. We should discuss briefly some related conventions.

There are two cases of interest. First, suppose we are given the following forward point quotes (second row) and spot rate quotes (first row) for EUR/USD:

Bid	Ask
1.0110	1.0120
12	16

Next note that the forward point listed in the "bid" column is lower than the forward point listed in the "ask" column. If forward point quotes are presented this way, then the points will be *added* to the last two digits of the corresponding spot rate.

Thus, we will obtain

$$\text{Bid forward outright} = 1.0110 + 0.0012 = 1.0122 \tag{6.25}$$

$$\text{Ask forward outright} = 1.0120 + 0.0016 = 1.0136 \tag{6.26}$$

Note that the bid−ask spread on the forward outright will be greater than the bid−ask spread on the spot.

Second, suppose, we have the following quotes:

Bid	Ask
1.0110	1.0120
23	18

Here the situation is reversed. The forward point listed in the "bid" column is greater than the forward point listed in the "ask" column. Under these conditions, the forward points will be *subtracted* from the last two digits of the corresponding spot rate. Thus, we will obtain

$$\text{Bid forward outright} = 1.0110 - 0.0023 = 1.0087 \tag{6.27}$$

$$\text{Ask forward outright} = 1.0120 - 0.0018 = 1.0102 \tag{6.28}$$

Note that the bid−ask spread on the forward outright will again be greater than the bid−ask spread on the spot. This second case is due to the fact that sometimes the minus sign is ignored in quotations of forward points.

6.6.2 FX FORWARD VERSUS FX FUTURES

The OTC FX market turnover has historically dwarfed trading volume of standardized FX products on exchanges. According to the BIS, in 2013, the OTC FX market was about 17 times larger than the exchange-traded FX derivatives market. Although market participants typically use forward instead of futures for FX trading, it is instructive to review contract specifications of FX futures.

FX futures contracts exist on many currency pairs. The CME, for example, offers the following currency contract: EUR/USD, JPY/USD, GBP, USD, CHF/USD, CDN/USD, AUD/USD, MXN/USD, NZD/USD, RUB/USD, ZAR/USD, BRL/USD, and most recently RMB/USD and KRW/USD. Major *cross-rate contracts* include EUR/GBP, EUR/JPY, EUR/CHF, GBP/CHF, and others.

Exchange-traded currency futures have historically differed from OTC FX transactions in terms of their standardization and flexibility or customization inherent in working with a dealer. However, derivatives exchanges are introducing greater degrees of flexibility in their trading practices. FX futures are predominantly traded electronically. The typical contract calls for delivery of a specified currency, or a cash settlement, during the months of March, June, September, and December.

6.7 SWAP ENGINEERING IN FX MARKETS

So far we have discussed simple currency derivatives in the form of FX forward. Now we will apply the more advanced financial engineering concepts related to interest rate swaps that we discussed in Chapter 4 and illustrate swap engineering in FX markets. We will encounter two different types of swaps linked to FX markets. First we will discuss *FX swaps*, then we will turn to *currency swaps*. As we will see there are important differences in the replication and uses of FX and currency swaps respectively. In practice, in terms of turnover, the FX swap market is an order of magnitude larger than the currency swap market.

In the previous sections, we created two synthetics for forward FX contracts. We can now ask the next question: Is there an optimal way of creating a synthetic? Or, more practically, can a trader buy a synthetic cheaply, and sell it to clients after adding a margin, and still post the smallest bid−ask spreads?

6.7.1 FX SWAPS

We can use the so-called *FX swaps* and pay a single bid−ask spread instead of going through two separate bid−ask spreads, as is done in contractual equation (6.8). The construction of an FX swap is shown in Figure 6.6.

According to this figure, there are at least two ways of looking at a FX swap. The FX swap is made of a money market deposit and a money market loan in different currencies written on the same "ticket." The second interpretation is that we can look at a FX swap as if the two counterparties *spot* purchase and *forward* sell two currencies against each other, again on the same deal ticket.

When combined with a spot operation, FX swaps duplicate forward currency contracts easily, as shown in Figure 6.7. Because they are swaps of a deposit against a loan, interest rate differentials

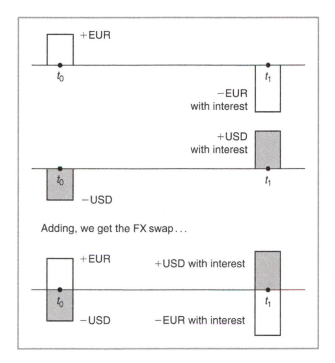

FIGURE 6.6

A simplified FX swap.

will play an important role in FX swaps. After all, one of the parties will be giving away a currency that can earn a higher rate of interest and, as a result, will demand compensation for this "loss." This compensation will be returned to him or her as a proportionately higher payment at time t_1. The parties must exchange *different* amounts at time t_1, as compared to the original exchange at t_0.

As the above figure shows, an FX swap is a contract in which one party borrows one currency from, and at the same time lends another to, the second party. Each party uses the repayment obligation to its counterparty as collateral and the amount of repayment is fixed at the FX forward rate at the start of the contract. Hence, FX swaps can be interpreted as collateralized borrowing/lending, although the collateral does not necessarily cover the entire counterparty risk.

6.7.1.1 Advantages

Why would a bank prefer to deal in FX swaps instead of outright forward? This is an important question from the point of view of financial engineering. It illustrates the advantages of spread products.

FX swaps have several advantages over the synthetic seen earlier. First of all, FX swaps are interbank instruments and, normally, are not available to clients. Banks deal with each other every day, and thus, compared to other counterparties, one could argue that counterparty risk is relatively lower in writing such contracts. In liquid markets, the implied bid–ask spread for synthetics

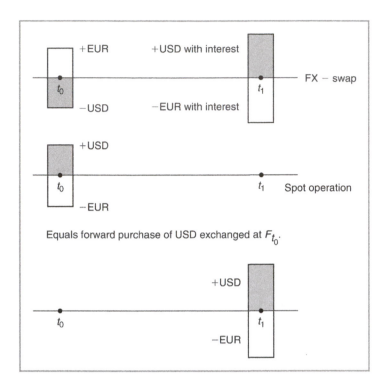

FIGURE 6.7

FX swap combined with spot operation leads to forward.

constructed using FX swaps will be smaller than the synthetic constructed from deposits and loans, or T-bills for that matter.

The second issue is liquidity. How can a market participant borrow and lend in both currencies without moving prices? An FX swap is again a preferable way of doing this. With an FX swap, traders are not *buying* or *selling* deposits, rather they are exchanging them.

The final advantage of FX swaps resides in their balance sheet effects, or the lack thereof. The synthetic developed in Figure 6.2 leads to increased assets and liabilities. One borrows new funds and lends them. Such transactions may lead to new credit risks and new capital requirements. FX swaps are off-balance sheet items, and the synthetic shown in Figure 6.7 will have minor balance sheet effects. FX swaps can affect a bank's capital adequacy ratio (CAR) since they change the capital requirement for market risk and credit market risk.[14]

6.7.1.2 *Uses of FX swaps*

Financial institutions and their clients, such as exporters and importers as well as investors, use FX swaps to hedge their positions. FX swaps can also be used for speculation, typically by

---
[14]See Barkbu and Ong (2010) for further details.

combining two offsetting FX swap positions with different maturity dates. Financial institutions with a need for foreign currency can either (a) borrow directly in the uncollateralized money market for the foreign currency or (b) borrow in another (normally the domestic) currency's uncollateralized money market, and then convert the proceeds into a foreign currency obligation through an FX swap. When a bank, for example, raises dollars by means of an FX swap using the euro as a funding currency, it exchanges euros for dollars at the FX spot rate. The borrowing cost in dollars implied by the EUR/USD FX swap is called the FX swap-implied dollar rate $r_{t_0}^d$. It is defined as

$$r_{t_0}^d = \frac{F_{t_0}}{e_{t_0}} \times \left(1 + r_{t_0}^f\right)$$

where $r_{t_0}^d$ and $r_{t_0}^f$ are the relevant interest rates in domestic and foreign currencies, respectively, and F_{t_0} and e_{t_0} are the forward and spot rates between the euro and the dollar. The choice of whether the bank should raise dollars using (a) or (b) above should be a function of the relative cost. If the costs associated with (a) and (b) are the same, i.e., the FX swap-implied dollar rate is equal to the dollar LIBOR rate in our example, then CIRP holds. Deviations from CIRP occurred during the GFC and other turbulent financial market episodes in the past, but arbitrageurs must be careful when interpreting any test of CIRP deviations since these must carefully take into account real-life complications such as bid−ask spreads, differences between the costs of secured and unsecured borrowing as well as whether rates such as LIBOR reflect actual interbank borrowing costs in periods of stress.

6.7.1.3 *Quotation conventions*

There is an important advantage to quoting swap points over the outright forward quotes. This indicates a subtle aspect of market activity. A quote in terms of forward points will essentially be independent of spot exchange rate movements and will depend only on interest rate differentials. An outright forward quote, on the other hand, will depend on the spot exchange rate movements as well. Thus, by quoting forward points, market professionals are essentially *separating* the risks associated with interest rate differentials and spot exchange rate movements respectively. The exchange rate risk will be left to the spot trader. The forward FX trader will be trading the risk associated with interest rate differentials only.

To see this better, we now look at the details of the argument.

Taking partial derivatives of Eq. (6.21) gives

$$\partial\left(F_{t_0} - e_{t_0}\right) \cong \left(r_{t_0}^d - r_{t_0}^f\right)\left(\frac{t_1 - t_0}{360}\right)\partial e_{t_0}$$

$$\cong 0$$

(6.29)

If the *daily* movement of the spot rate e_{t_0} is small, the right-hand side will be negligible. In other words, the forward swap quotes would not change for normal daily exchange rate movements, if interest rates remain the same and *as long as* exchange rates are quoted to four decimal places. The following example illustrates what this means.

EXAMPLE

Suppose the relevant domestic and foreign interest rates are given by

$$r_{t_0}^d = 0.03440 \tag{6.30}$$

$$r_{t_0}^f = 0.02110 \tag{6.31}$$

where the domestic currency is euro and the foreign currency is USD. If the EUR/USD exchange rate has a daily volatility of, say, *0.01%* a day, which is a rather significant move, then, for FX swaps with 3-month maturity, we have the following change in forward points:

$$\partial\left(F_{t_0} - e_{t_0}\right) = 0.01330\left(\frac{90}{360}\right)0.0100$$

$$= 0.00003325 \tag{6.32}$$

which, in a market that quotes only four decimal points, will be negligible.

Hence, forward points depend essentially on the interest rate differentials. This "separates" exchange rate and interest rate risk and simplifies the work of the trader. It also shows that forward FX contracts can be interpreted as if they are "hidden" interest rate contracts.

6.8 CURRENCY SWAPS VERSUS FX SWAPS

Although they have similar names, currency swaps and FX swaps are different instruments. First we introduce currency swaps and apply the swap engineering methodology from Chapter 4 to their replication. Then we will highlight differences between currency swaps and FX swaps.

6.8.1 CURRENCY SWAPS

Currency swaps[15] are similar to interest rate swaps, but there are some differences. First, the exchanged cash flows are in *different currencies*. This means that two different yield curves are involved in swap pricing instead of just one. Second, in the large majority of cases, a floating rate is exchanged against another floating rate. A third difference lies in the exchange of principals at initiation and a re-exchange at maturity. In the case of interest rate swaps, this question does not arise since the notional amounts are in the same currency. Yet, currency swaps can be engineered almost the same way as interest rate swaps.

Formally, a currency swap will have the following components. There will be two currencies, say USD ($) and euro (€). The swap is initiated at time t_0 and involves (1) an exchange of a principal amount $N^\$$ against the principal $M^€$ and (2) a series of floating interest payments associated with the

---

[15]Currency swaps are also referred to as *cross-currency swaps* or *cross-currency basis swaps*. There are variations of currency swaps depending on whether fixed or floating interest rates are exchanged, for example.

principals $N^\$$ and $M^\euro$, respectively. They are settled at settlement dates, $\{t_1, t_2, \ldots, t_n\}$. One party will pay the floating payments $L^\$_{t_i} N^\$ \delta$ and receive floating payments of size $L_{t_i} M \delta$. The *two* LIBOR rates $L^\$_{t_i}$ and L_{t_i} will be determined at *set dates* $\{t_0, t_1, \ldots, t_{n-1}\}$. The maturity of swap will be m years.

A small *spread* s_{t_0} can be added to one of the interest rates to make both parties willing to exchange the cash flows. The market maker will quote bid/ask rates for this spread.

EXAMPLE

Figure 6.8 shows a currency swap. The USD notional amount is 1 million. The current USD/EUR exchange rate is at 0.95. The agreed spread is 6 bp. The initial 3-month LIBOR rates are

$$L^\$_{t_i} = 3\% \tag{6.33}$$

$$L^\$_{t_i} = 3.5\% \tag{6.34}$$

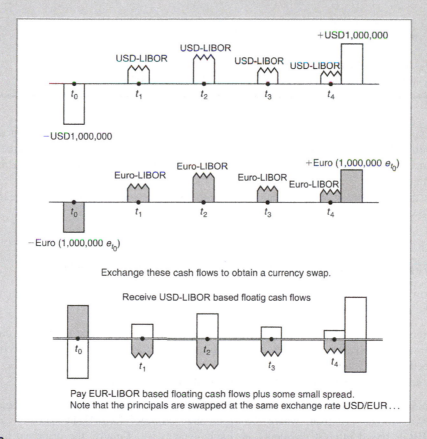

FIGURE 6.8

Three-period currency swap.

This means that at the first settlement date

$$(1,000,000)(0.03 + 0.0006)\frac{1}{4} = \$7650 \tag{6.35}$$

will be exchanged against

$$0.95(1,000,000)(0.035)\frac{1}{4} = 8312.5 \tag{6.36}$$

All other interest payments would be unknown. Note that the euro principal amount is related to the USD principal amount according to

$$e_{t_0} N^{\$} = M \tag{6.37}$$

where e_{t_0} is the spot exchange rate at t_0.
Also, note that we added the swap spread to the USD LIBOR.

Pricing currency swaps will follow the same principles as interest rate swaps. A currency swap involves well-defined cash flows and consequently we can calculate an arbitrage-free value for each sequence of cash flows. Then these cash flows are traded. An appropriate *spread* is added to either floating rate.

By adjusting this spread, a swap dealer can again make the two parties willing to exchange the two cash flows.

6.8.2 DIFFERENCES BETWEEN CURRENCY SWAPS AND FX SWAPS

We will now compare currency swaps with FX swaps introduced earlier. To recap, a currency swap has the following characteristics:

1. Two principals in different currencies and of equal value are exchanged at the start date t_0.
2. At settlement dates, interest will be paid and received in different currencies and according to the agreed interest rates.
3. At the end date, the principals are re-exchanged at the *same* exchange rate.

A simple example is the following. 100,000,000 euros are received and against these $100,000,000 e_{t_0}$ dollars are paid, where the e_{t_0} is the "current" EUR/USD exchange rate. Then, 6-month LIBOR-based interest payments are exchanged twice. Finally, the principal amounts are exchanged at the end date *at the same* exchange rate e_{t_0}, even though the actual exchange rate e_{t_2} at time t_2 may indeed be different than e_{t_0}. See Figure 6.9.

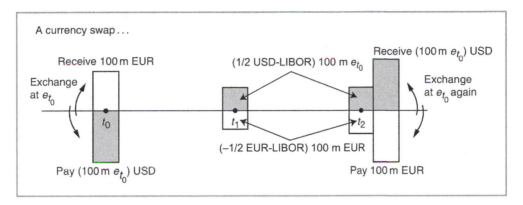

FIGURE 6.9

Simple currency swap.

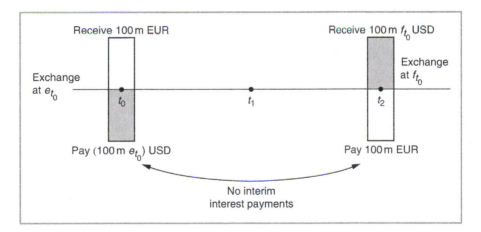

FIGURE 6.10

FX swap.

The FX swap for the same period is shown in Figure 6.10. Here, we have no interim interest payments, but instead the principals are re-exchanged at a *different* exchange rate equal to

$$f_{t_0} = e_{t_0} \frac{1 + L_{t_0}^{USA}\delta}{1 + L_{t_0}^{EUR}\delta} \tag{6.38}$$

Why this difference? Why would the same exchange rate be used to exchange the principals at start and end dates of a currency swap while different exchange rates are used for an FX swap?

We can look at this question from the following angle. The two parties are exchanging currencies for a period of 1 year. At the end of the year, they are getting back their original currency.

But during the year, the interest rates in the two currencies would normally be different. This difference is explicitly paid out in the case of currency swaps during the life of the swap as interim interest payments. As a result, the counterparties are ready to receive the exact original amounts back. The interim interest payments would compensate them for any interest rate differentials.

In the case of FX swaps, there are no interim interest payments. Hence, the compensation must take place at the end date. Thus, the interest payments are bundled together with the exchange of principals at the end date.

6.8.3 ANOTHER DIFFERENCE

Looked at from a financial engineering perspective, the currency swap is like an exchange of two FRNs with different currencies and no credit risk. The FX swap, on the other hand, is like an exchange of two zero-coupon bonds in different currencies.

Because the LIBOR rates at time t_1 are unknown as of time t_0, the currency swap is subject to slightly different risks than FX swaps of the same maturity. Note that FX and currency swaps are in principle subject to significantly more exposure to *counterparty risk* than are interest rate swaps, due to the exchange of notional amounts in FX and currency swaps.

6.8.4 USES OF CURRENCY SWAPS

Although there are structural differences between FX and currency swaps, the former basically serve the same economic purpose as the latter, except for the exchange of (floating) interest rates during the contract term. Financial institutions and their clients, including multinational corporations involved in foreign direct investment, use currency swaps to fund foreign currency investments. Other uses involve converting currencies of liabilities, particularly by issuers of bonds denominated in foreign currencies, as we will see in an example at the end of the chapter.

6.8.5 RELATIVE SIZE AND LIQUIDITY OF FX SWAP AND CURRENCY SWAP MARKETS

We discussed similarities and differences between FX and currency swaps above. Which instrument is more widely used and more important in practice? FX swaps account for a much larger proportion of FX market turnover than currency swaps. In 2013, the FX swap market was about *40 times larger* than the currency swap market. According to the BIS Triennial Central Bank Survey 2013, FX swaps remained the most actively traded FX instrument in 2013 and their daily volume of £2.2 trillion accounted for 42% of all FX-related transactions. Spot FX transactions and outright forward transactions made up 37% and 12%, respectively, of FX market turnover. In comparison, currency swaps contribute only a small share of FX turnover. Currency swap turnover stood at $54 billion per day, or around 1% of the total. FX swaps remain most liquid at maturities below 1 year, but FX swap turnover has recently been growing for transactions with longer maturities. In contrast to FX swaps, most currency swaps are long term with maturities ranging from 1 to 30 years. The reason for this is that the currency swap maturity reflects the maturity of the transactions that they are funding.

6.8.6 THE EFFECT OF THE GFC ON THE FX MARKET, MARGINS, AND CLEARING

As we saw above, the advantages of FX forward and swaps are linked to their relatively high liquidity and low cost. During the GFC, there were significant deviations from CIRP, i.e., the FX swap-implied USD rates and the LIBOR rates for major currencies such as the euro and pound sterling diverged. This has been partly attributed to European financial institutions that required USD but faced increasing concerns about their creditworthiness and counterparty risk in dollar interbank markets. As a result, these institutions used the FX swap market to raise dollars using both euro and pound sterling as funding currencies, causing a shift toward one-sided order flow in the FX swap market.

The GFC took a toll on volumes in FX derivatives, but the regulatory overhaul of the financial system that followed the crisis has left the FX derivatives market largely unchanged due to exemptions that were granted. On November 16, 2012, the US Department of the Treasury issued its final determination that exempts FX swaps and forwards from mandatory derivative requirements, including central clearing and exchange trading. Following the footsteps of the BIS and the US Treasury, on April 14, 2014, European supervisory authorities proposed to exempt uncleared FX swaps and forwards from initial margin requirements. The EU proposals envisage that users of the FX swaps and forward market will only have to post variation margin on these trades, rather than putting up hefty initial margins as well, leaving these instruments less capital-intensive than other asset classes and boding well for liquidity and financial engineering applications in this market.

EXAMPLE

While participants to an interest rate swap might be paying variation margin on a trade for one or two decades, trade duration in FX is typically much shorter. Of the current gross notional amount of FX swaps and forward in the EU, €5.7 trillion has a maturity of less than 6 months, and only €1.2 trillion stretches out over 6 months. The consultation paper estimates that only 18% of all OTC FX trades will be processed through central counterparties. The figure for rates trades is expected to be 55%, and for credit 62%. Forex market-makers have already begun offering replication products to clients, which mimic the effect of rates or credit hedges but at a fraction of the cost.

FX week (April 16, 2014).

As the above reading illustrates the cost advantages implied by lower margin requirements for FX forwards and swaps have already had an impact on financial engineering practice. They have led market makers at major dealers to explore cross-hedges in the form of FX products. Clients, such as investors or banks, that have an incentive to hedge interest rate or credit risk may face higher capital charges and costs on such hedges. If FX products can be devised by dealers that mimic interest rate or credit hedges due to higher correlations with these markets, this can be a more cost-efficient hedge solution for clients and a new revenue stream for dealers. The example also illustrates the continuing trends toward financial engineering solutions involving multiple asset classes and asset class correlations.

6.9 MECHANICS OF SWAPPING NEW ISSUES

6.9.1 INTEREST RATE AND CURRENCY SWAP EXAMPLE

In the context of interest rate swap engineering, we noted that the basic cash flow engineering for swaps ignores several minor technical points that need to be considered in practice. We provided a real-life example of interest rate swap usage and highlighted quoting conventions and costs. Similarly, in this chapter, our discussion of swap engineering in currency markets has abstracted away from important real-world considerations. Therefore, here we provide another real-world application to new bond issues.

Suppose there is an A-rated British entity that would like to borrow 100 million sterling (GBP) for a period of 3 years. The entity has no preference toward either floating or fixed-rate funding, and intends to issue in Euromarkets. Market research indicates that if the entity went ahead with its plans, it could obtain fixed-rate funds at 6.5% annually.[16] But the bank recommends the following approach.

It appears that there are nice *opportunities* in USD—GBP currency swaps, and it makes more sense to issue a floating rate Eurobond in the USD sector with fixed coupons. The swap market quotes funding at LIBOR + 95 bp in GBP against USD rates for this entity. Then the proceeds can be swapped into sterling for a lower all-in-cost. How would this operation work? And what are the risks?

We begin with the data concerning the new issue. The parameters of the newly issued bond are provided in Table 6.1.[17] Now, the issuer would like to swap these proceeds to floating rate GBP

Table 6.1 The New Issue	
Amount	USD100 million
Maturity	2 years
Coupon	6% p.a.
Issue price	$100\frac{3}{4}$
Options	None
Listing	Luxembourg
Commissions	$1\frac{1}{4}$
Expenses	USD75000
Governing law	English
Negative pledge	Yes
Pari passu	Yes
Cross default	Yes

---

[16]With a day-count basis of 30/360.

[17]Some definitions: *Negative pledge* implies that the investors will not be put in a worse position at a later date by the issuer's decision to improve the risks of other bonds. *Pari passu* means that no investor who invested in these bonds will have an advantageous position. *Cross default* means the bond will be considered in default even if there is time to maturity and if the issuer defaults on another bond.

Table 6.2 Swap Market Quotes	
Spot exchange rate GBP–USD	1.6701/1.6708
GBP 2-year interest rate swap	5.46/51
USD–GBP currency swap	+4/−1

funds. In doing this, the issuer faces the market conditions as provided in Table 6.2. We first work out the original and the swapped cash flows and then calculate the all-in-cost, which is the real cost of funds to the issuer after the proceeds are swapped into GBP.

The first step is to obtain the amount of cash the issuer will receive at time t_0 and then determine how much will be paid out at t_1, t_2. To do this, we again need to calculate the *proceeds* from the issue.

- The issue price is 100.75 and the commissions are 1.25%. This means that the amount received by the issuer before expenses is

$$\frac{(100.75 - 1.25)}{100} 100,000,000 = 99,500,000 \tag{6.39}$$

We see that the issue is sold at a premium which increases the proceeds, but once commissions are deducted, the amount received falls below 100 million. Thus, expenses must be deducted

$$\text{Proceeds} = 99,500,000 - 75,000 = 99,425,000 \tag{6.40}$$

Given the proceeds, we can calculate the effective cost of fixed-rate USD funds for this issuer. The issuer makes two coupon payments of 6% (out of the 100 million) and then pays back 100 million at maturity. At t_0, the issuer receives only 99,425,000. This cash flow is shown in Figure 6.11. Note that unlike the theoretical examples, the principal paid is not the same as the principal received. This is mainly due to commissions and expenses.

- From this cash flow, we can calculate the internal rate of return y_{t_0} by solving the equation

$$99,425,000 = \frac{60,000}{\left(1 + y_{t_0}\right)} + \frac{60,000}{\left(1 + y_{t_0}\right)^2} + \frac{100,000,000}{\left(1 + y_{t_0}\right)^2} \tag{6.41}$$

The solution is

$$y_{t_0} = 6.3150\% \tag{6.42}$$

Hence, the true fixed cost of USD funds is greater than 6%.

The issuer will first convert this into floating rate USD funds. For this purpose, the issuer will *sell* a swap. That is to say, the issuer will receive fixed 5.46% and pay floating LIBOR flat. This is equivalent to paying approximately USD LIBOR + 54 bp. Finally, the issuer will convert these USD floating rate funds into GBP floating rate funds by paying floating GBP and receiving floating USD.

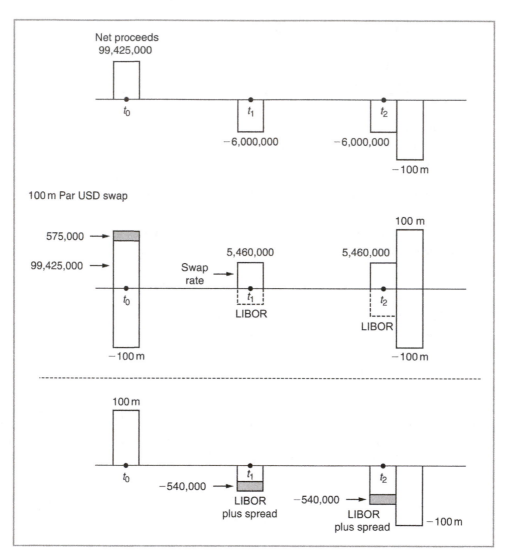

FIGURE 6.11

New bond issue and currency swap.

6.10 CONCLUSIONS

This chapter has developed two main ideas. First we considered the engineering aspects of FX forward contracts and developed a contractual equation for them. We discussed two methodologies to FX forward replication based on money market and T-bill synthetics. Second, we applied the swap engineering methodology to FX swaps and currency swaps and highlighted differences between them. Real-life complications such as quoting conventions, clearing, and margin requirements are important for financial engineering in practice. We discussed the relative size of FX and currency swap markets and highlighted the effect of recent regulatory exemptions of FX forward and swaps on financial engineering practice in FX markets.

SUGGESTED READING

Weithers (2006) provides a good practical introduction to FX markets. **Baba et al.** (2008) examine the spillover of money market turbulence in 2007 to FX swap and currency swap markets. **Cloyle** (2000) provides an introduction to the basics of currency swaps.

EXERCISES

1. Today is March 1, 2004. The day-count basis is actual/365. You have the following contracts on your FX book.

   CONTRACT A: On March 15, 2004, you will sell 1,000,000 EUR at a price F_t^1 dollars per EUR.

   CONTRACT B: On April 30, 2004, you will buy 1,000,000 EUR at a price F_t^2 dollars per EUR.

   a. Construct one synthetic equivalent of each contract.

   b. Suppose the spot EUR/USD is 1.1500/1.1505. The USD interest rates for loans under 1 year equal 2.25/2.27, and the German equivalents equal 2.35/2.36. Calculate the F_t^i numerically.

   c. Suppose the forward points for F_t^1 that we observe in the markets is equal to 10/20. How can an arbitrage portfolio be formed?

2. You are hired by a financial company in New Zealand and you have instant access to markets. You would like to lock in a 3-month borrowing cost in NZ$ for your client. You consider a NZ$ 1×4 FRA. But you find that it is overpriced as the market is thin.

   So you turn to Aussie. A$ FRAs are very liquid. It turns out that the A$ and NZ$ forward are also easily available.

   In particular, you obtain the following data from Reuters:

A$/NZ$	Spot the:	1.17/18
	1-month forward:	1.18/22
	3-month forward:	1.19/23
	4-month forward:	1.28/32
A$ FRA's	1×4	8.97

   a. Show how you can create a 1×4 NZ$ from these data.

   b. Show the cash flows.

   c. What are the risks of your position (if any) compared to a direct 1×4 NZ$ FRA?

   d. To summarize the lessons learned from this exercise (if any), do you think there must be arbitrage relationships between the FRA markets and currency forward? Explain. Or better, provide the relevant formulas.

3. Suppose at time $t = 0$, we are given prices for four zero-coupon bonds (B_1, B_2, B_3, B_4) that mature at times $t = 1, 2, 3$, and 4. This forms the term structure of interest rates.

We also have the *one-period* forward rates (f_0, f_1, f_2, f_3), where each f_i is the rate contracted at time $t = 0$ on a loan that begins at time $t = i$ and ends at time $t = i + 1$. In other words, if a borrower borrows N GBP at time i, he or she will pay back $N(1 + f_i)$ GBP at time $t = i + 1$.

The spot rates are denoted by r_i. By definition, we have

$$r = f_0 \tag{6.43}$$

The $\{B_i\}$ and all forward loans are default-free, so that there is no credit risk. You are given the following *live* quotes:

$$B_1 = 0.92/0.94, B_2 = 0.85/0.88, B_3 = 0.82/0.85 \tag{6.44}$$

and

$$f_0 = 8.10/8.12, f_1 = 9.01/9.03, f_2 = 10.12/10.16, f_3 = 18.04/18.10 \tag{6.45}$$

a. Given the data on forward rates, obtain arbitrage-free prices for the zero-coupon bonds, B_1 and B_2.

b. What is the three-period swap rate under these conditions?

4. Going back to the previous exercise, suppose you are given, in addition, data on FRAs both for USD and for EUR. Also suppose you are looking for arbitrage opportunities. Would these additional data be relevant for you? Discuss briefly.

5. Foreigners buying Australian dollar instruments issued in Australia have to pay withholding taxes on interest earnings. This withholding tax can be exploited in tax-arbitrage portfolios using swaps and bonds. First let us consider an episode from the markets related to this issue.

Under Australia's withholding tax regime, resident issuers have been relegated to second cousin status compared with nonresident issuers in both the domestic and international markets. Something has to change.

In the domestic market, bond offerings from resident issuers incur the 10% withholding tax. Domestic offerings from nonresident issuers, commonly known as Kangaroo bonds, do not incur withholding tax because the income is sourced from overseas. This raises the spectre of international issuers crowding out local issuers from their own markets.

In the international arena, punitive tax rules restricting coupon washing have reduced foreign investor interest in Commonwealth government securities and semigovernment bonds. This has facilitated the growth of global Australian dollar offerings by Triple A-rated issuers such as Fannie Mae, which offer foreign investors an attractive tax-free alternative.

The impact of the tax regime is aptly demonstrated in the secondary market. Exchangeable issues in the international markets from both Queensland Treasury Corporation and Treasury Corporation of NSW are presently trading through comparable domestic issues. These exchangeable issues are exempt from withholding tax.

If Australia wishes to develop into an international financial center, domestic borrowers must have unfettered access to the international capital markets—which means compliance costs and uncertainty over tax treatment must be minimized. Moreover, for the Australian domestic debt markets to continue to develop, the inequitable tax treatment between domestic and foreign issues must be corrected (Thomson Reuters IFR, Issue 1206).

We now consider a series of questions dealing with this problem. First, take a 4-year straight coupon bond issued by a local government that pays interest annually. We let the coupon rate be denoted by $c\%$.

Next, consider an Aussie\$ Eurobond issued at the same time by a Spanish company. The Eurobond has a coupon rate $d\%$. The Spanish company will use the funds domestically *in Spain.*

Finally, you know that interest rate swaps or FRAs in Aussie\$ are not subject to any tax.

a. Would a foreign investor have to pay the withholding tax on the Eurobond? Why or why not?

b. Suppose the Aussie\$ IRSs are trading at a swap rate of $d + 10$ bp. Design a 4-year interest rate swap that will benefit from tax arbitrage. Display the relevant cash flows.

c. If the swap notional is denoted by N, how much would the tax-arbitrage yield?

d. Can you benefit from the same tax arbitrage using a strip of FRAs in Aussie\$?

e. Which arbitrage portfolio would you prefer, swaps or FRAs? For what reasons?

f. Where do you think it is more profitable for the Spanish company to issue bonds under these conditions, in Australian domestic markets or in Euromarkets? Explain.

6. Consider a 2-year currency swap between USD and EUR involving floating rates only. The EUR benchmark is selected as 6-month Euro LIBOR, the dollar benchmark is 6-month BBA LIBOR. You also have the following information:

$$\text{Notional amount} = \text{USD}10{,}000{,}000$$

$$\text{Exchange rate EUR/USD} = 0.84$$

a. Show the cash flow diagrams of this currency swap. Make sure to quantify every cash flow exactly (i.e., use a graph as well as the corresponding number).

b. Show that this currency swap is equivalent to two floating rate loans.

c. Suppose a company is trying to borrow USD10,000,000 from money markets.

The company has the following information concerning available rates on 6-month loans:

EUR LIBOR = 5.7%, USD LIBOR = 6.7%

EUR–USD currency swap spread: 1 year—75, 2 years—90.

Should this company borrow USD directly? Would the company benefit if it borrowed EUR first and then swapped them into USD?

7. A treasurer in Europe would like to borrow USD for 3 months. But instead of an outright loan, the treasurer decides to use the repo market. The company has holdings of EUR40 million bonds. The treasurer uses a *cross-currency* repo. The details of the transaction are as follows:

- Clean price of the bonds: 97.00
- Term: September 1 to December 1
- Last coupon date on the bond: August 12
- Bond coupon 4% item EUR/USD exchange rate: 1.1150

- 3-month USD repo rate: 3%
- Haircut: 3%
- **a.** What is the invoice price *(dirty price)* of the bond in question?
- **b.** Should the repo be done on the dirty price or the clean price?
- **c.** How much in dollars is received on September 1?
- **d.** How much repo interest is paid on December 1?

8. Suppose the quoted swap rate is 5.06/5.10. Calculate the amount of fixed payments for a fixed payer swap for the currencies below in a 100 million swap.
   - USD.
   - EUR.

   Now calculate the amount of fixed payments for a fixed receiver swap for the currencies below in a 100 million swap.
   - JPY.
   - GBP.

CASH FLOW ENGINEERING AND ALTERNATIVE CLASSES (COMMODITIES AND HEDGE FUNDS)

CHAPTER OUTLINE

7.1 INTRODUCTION

We have so far applied our cash flow engineering methodology to fixed income and currency instruments. Unlike these asset classes, commodities incur significant storage costs and this real-world complication needs to be taken into account in commodities cash flow engineering. For example, in contrast to currencies which are a homogeneous asset, commodities differ in their quality or *grade*. We provide more examples of the replication of forwards, futures, and swaps, and examine cash flow engineering in commodities. In contrast to interest rates and equities, commodities are tangible assets. This makes them useful for illustrating some of the abstract derivatives market-related concepts such as the term structure of futures prices as well as open interest and volume. Three practical reasons motivate a careful look at financial engineering applications in commodity markets. As interest in commodities investing and hedging has increased, this has also led to what has been described as the financialization of commodities. Institutional investors are increasingly investing in commodities through commodity futures and swaps. ETFs linked to various commodity indices and commodities have been launched, thus allowing retail investors to invest in commodities as well.

7.2 PARAMETERS OF A FUTURES CONTRACT

To broaden the examination of futures and forwards, in this section, we concentrate on *commodities* that are generally traded via futures contracts on organized exchanges. Thus, in contrast to currency markets, exchange traded commodity futures are more liquid than OTC commodity forwards.

Crude oil is one of the economically most important commodities and crude oil futures are among the most liquidly traded commodity futures. Advanced financial engineering methodologies have been extended to crude oil. These include simple swaps and calendar swaps, discussed in this chapter, as well as options, implied volatility indices, and crude oil volatility index futures, discussed in future chapters. Crude oil is also one of the largest constituents in commodity indices such as the GSCI on which many commodity-related structured products are written. Therefore, we use a crude oil futures contract to introduce key concepts about futures markets. CME and ICE are the exchanges with the most liquid crude oil futures contracts.

It may come initially as a surprise that crude oil is *not* a very homogeneous commodity. This creates various trading opportunities but may also be a limitation to arbitrage strategies and require careful consideration in financial engineering applications. There are different *grades* of crude oil depending on its chemical composition and its sulfur content in particular. The New York Mercantile Exchange (NYMEX) designates crude oil with less than 0.42% sulfur as *sweet*. Crude oil containing higher levels of sulfur is called *sour* crude oil and is considered lower quality. Light sweet crude oil is of higher quality since it can be more easily refined into gasoline, kerosene, and high-quality diesel (gas oil) and typically trades at a premium to sour crude oil on futures exchanges. In addition to differences in grade, the geographical location also has to be taken account. West Texas Intermediate (WTI) and the North Sea Brent crude are both light sweet grades and used as benchmarks in oil pricing.

The following example shows the contract specification of the WTI Crude Oil Futures Contract on CME NYMEX/Globex. The terminologies introduced here apply to all futures contracts, not just commodity futures.

EXAMPLE: *WTI CRUDE OIL CONTRACT ON CME NYMEX/GLOBEX*

1. **Contract size:** 1000 barrels (bbl) ($\sim$42,000 US gallons/159,000 l).
2. **Units of trading:** Any multiples of 1000 bbl.
3. **Minimum fluctuation:** $0.01 per bbl.
4. **Price quote:** US dollars and cents per bbl.
5. **Contract months:** Crude oil futures are some of the most flexible commodity contracts due to the large range of contract months. Crude oil futures are listed 9 years forward using the following listing schedule: consecutive months are listed for the current year and the next 5 years; in addition, the June and December contract months are listed beyond the sixth year. On June 23, 2014, the *most distant* contract month available was December 2022.
6. **Delivery:** Delivery shall be made free-on-board ("FOB")[1] at any pipeline or storage facility in Cushing, Oklahoma, with pipeline access to Enterprise, Cushing storage or Enbridge, Cushing storage.
7. **Delivery grades:** Because of quality differences, there must be some standardization so that the buyer knows what she is receiving. Various domestic crudes such as WTI, New Mexican Sweet, Oklahoma Sweet and Foreign Crudes (at a discount of 15–55 cents per bbl) such as UK Brent Blend and Nigeria Bonny Light, for example. For details, see the CME/NYMEX rulebook.[2]
8. **Last trading day:** The business day prior to the 15th calendar day of the contract month.
9. **Last delivery day:** Delivery shall take place no earlier than the first calendar day of the delivery month and no later than the last calendar day of the delivery month.
10. **Trading hours:** Open outcry, 9:00 a.m. to 2:30 p.m. Chicago time, Monday through Friday. Electronic, 5 p.m. to 4:15 p.m. Chicago time, Sunday through Friday.
11. **Daily price limit:** $10.00 per bbl above or below the previous day's settlement price for such contract month.

Table 7.1 provides prices, volumes, and open interest for CME WTI crude oil contracts with maturity up to 1 year as of June 20, 2014. Volume represents the total amount of trading activity or WTI futures contracts that were traded on that day. The August 2014 contract, for example, had a volume of 213,130 which corresponds to approximately 213.1 million bbl with an estimated value of $24.5 billion (at an assumed oil price of $115/bbl). Open interest is defined as the total number

---

[1]According to the Incoterms 2010 standard, published by the International Chamber of Commerce, FOB, means that the seller pays for transportation of the goods to the port of shipment and loading costs. The buyer is responsible for paying the cost of marine freight transport, insurance, unloading, and transportation from the arrival port to the final destination.
[2]http://www.cmegroup.com/rulebook/NYMEX/2/200.pdf.

| Table 7.1 WTI Crude Oil Contract Prices, Volume, and Open Interest ||||
Month	(Settlement) Price	Volume	Open Interest
July 14	107.26	31,672	24,798
August 14	106.83	213,130	312,344
September 14	105.97	72,503	186,919
October 14	104.9	36,431	126,676
November 14	103.85	20,751	66,259
December 14	102.86	51,854	229,847
January 15	101.88	6,744	68,163
February 15	100.93	1,548	35,424
March 15	100.04	4,301	64,348
April 15	99.15	630	24,723
May 15	98.31	890	23,460
June 15	97.55	11,556	100,963

Source: *CME GLOBEX on June 20, 2014.*

of outstanding contracts that are held by market participants at the end of each day. Note that for each seller of a futures contract, there is a buyer of that contract. A seller and a buyer combine to create only one contract. Hence, to calculate the total open interest, we need only to know the totals from one side or the other, buyers or sellers, not the sum of both. The total open interest across all contracts on that day is not given in the table, but was 1,709,934 with an estimated market value of almost $200 billion (at a price of $115/bbl).

The contract month with an expiration date closest to the current date is called the *front month* or *prompt month*. On June 23, 2014, the front month is July 2014, i.e., the July 2014 contract is the shortest duration crude oil contract that could be purchased in the futures market. The front month contract is also referred to as the *nearby* or *nearest* contract.

Contracts that are a month or more behind the front month contracts are referred to as *back month* contracts. The liquidity, as approximated by volume, for front month contracts given in the table is higher than for back month contracts. This is generally the case and applies not just to this particular day and this commodity but also to other commodities. Therefore, a trader that seeks the most liquid contract to minimize transaction costs would typically choose a nearby contract. Considerations other than liquidity determine however which contract month is chosen by a particular trader and we discuss this in further detail below. Consider an oil producer that hedges her production using futures. If the hedger liquidates the futures contract and the asset before contract maturity, the hedger bears basis risk, because the futures price and the spot price may not move in lockstep before the delivery date. In energy commodity markets, there are in fact *three* primary types of basis risk: locational basis risk, product/quality basis risk, and calendar (spread) basis risk. Assume that the hedger in our example produces oil in Saudi Arabia, for example. If the hedge is designed to hedge Saudi oil that will be produced in September of a given year, for example, and a NYMEX WTI futures with December maturity in the same year is used as a hedge, then all three risks are relevant. First, the delivery location specified in the futures contract will be different from that of the physical location in Saudi Arabia where the oil will be produced. Second, the sulfur content will be different since Saudi oil is heavier, thus not qualifying as a delivery grade for the NYMEX WTI contract. Third, there is a difference of 3 months between the delivery date specified in the futures contract and the expected physical delivery of the oil.

7.3 THE TERM STRUCTURE OF COMMODITY FUTURES PRICES

Table 7.1 provides the futures prices corresponding to different contract months. This is the term structure of futures prices. If we plot the future prices against all contract months, we obtain a curve. The nearby contracts prices are on the left and the furthest contract months are on the right.

7.3.1 BACKWARDATION AND CONTANGO

Traders use different terms to describe the shape of the curve. If the resulting curve is downward sloping, we say that it is *backwardation*. In other words, backwardation describes a situation where the price for forward delivery is below the price for spot delivery. If the curve is upward sloping, we say that it is in *contango*. Figure 7.1a and b plots examples of curves in backwardation and in contango. In Chapter 2, we introduced the concept of the basis, which was defined as the difference between the spot price of an underlying asset and its futures price. We can apply this terminology to the term structure of futures prices. A commodity whose futures curve term structure is in backwardation will have a positive basis, whereas a commodity whose futures curve term structure is in contango will have a negative basis. Below, we will discuss, in more detail, what economic mechanism can explain the shape of the commodity futures curves. Contango and backwardation also exist in the pricing of traditional financial products, but, as we will see below, the role of the underlying physical markets in commodities is much more important, especially when demand exceeds supply.

The term structure of futures prices changes over time by commodity and it differs across commodities at the same point in time. Some commodities have a tendency to exhibit contango; others exhibit backwardation most of the time. As Figure 7.1 illustrates, the crude oil futures curve often tends to be in backwardation while the gold futures curve tends to be in contango. The term structure of crude oil futures prices in Table 7.1 implies that on June 20, 2014, crude oil futures were in backwardation. To illustrate how much the term structure can change and that it can flip from backwardation to contango, Figure 7.2 shows the WTI crude oil futures price term structure on four different dates. The orange line represents the curve on June 16, 2014, which is the baseline. The green line plots the term structure 1 month earlier, on May 16, 2014, the blue line plots the curve 1 year earlier, on June 14, 2013, whereas the red line shows the term structure of futures prices on June 16, 2009. As we see the curves in 2013−2014 were in backwardation, but in November 2009, the crude oil futures curve was in contango. Below we will discuss the trading opportunities that this relatively unusual situation gave rise to. First we will turn to potential explanations for the shape of the term structure of futures prices. To do so we will build on the cash flow engineering methodology developed in earlier chapters and the relationship between commodity futures prices and the underlying replication portfolio.

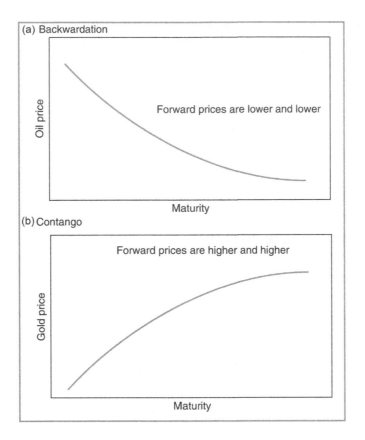

FIGURE 7.1

Backwardation and contango.

7.3.2 CONTRACTUAL EQUATION AND SYNTHETIC COMMODITIES

Consider the example of an oil refiner that contacts a bank asking for the price for the purchase of oil in, say, 12 months. The price quoted by the bank is not a forecast of where it thinks the price of oil will trade at the time of delivery. The bank will quote a price which depends on the cost of hedging the bank's own exposure. This is another example of the application of cash flow engineering to derivatives products. The cost of the product will be a function of the cost of the hedge. To avoid the risk of a rise in the price of oil, the bank would enter a series of transactions on the trade date to hedge its risk. Since the bank is agreeing to sell a fixed amount of oil in the future, it must first fund the purchase of oil in the spot market now and store the oil until delivery. Oil is traded in US dollars. The spot oil purchase could be financed by borrowing in US dollars at LIBOR from another bank. The bank buys the agreed amount of oil in the spot market from another institution, say another investment bank or an oil producer. The bank also incurs storage and insurance costs to store the oil until delivery. The above example is framed in the context of a bilateral OTC forward contract, but the same principle also applies to an oil futures contract.

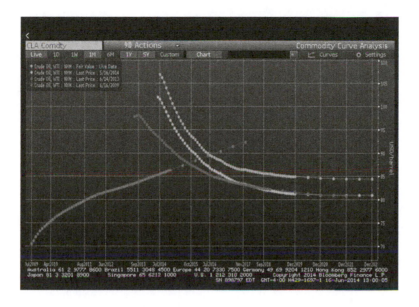

FIGURE 7.2

Historical crude oil term structure of futures prices.

We can apply the contractual equation developed earlier to create *synthetic commodities*. For example, suppose S_t represents spot oil, which is the underlying asset for a futures contract with price F_t and expiration date T, $t_0 < T$. How can we create a synthetic for this contract? The answer is quite similar to the case of currencies. Using the same logic, we can write a contractual equation:

$$
\boxed{\begin{array}{c} \text{Long oil futures} \\ \text{expiration } T \end{array}} = \boxed{\begin{array}{l} \text{A loan:} \\ \text{borrow USD at} \\ t_0 \text{ for maturity } T \end{array}} + \boxed{\begin{array}{l} \text{Spot operation} \\ \text{buy 1 unit of spot} \\ \text{oil for } S_{t_0} \end{array}} + \boxed{\begin{array}{l} \text{Store the oil at a} \\ \text{cost } q_{t_0} \text{ a day} \\ \text{until } T \end{array}} \quad (7.1)
$$

The contractual equation above incorporates one of the distinguishing features of commodities, which is storage costs. Most commodities can be stored at a cost. Note, however, that for some commodities, storage is either not possible (e.g., due to seasons) or prohibitive (e.g., wholesale electricity). Figure 7.3 shows the cash flows associated with a long futures and a synthetic long futures position (ignoring margin payments).

We can use Eq. (7.1) to obtain two results. First, by rearranging the contracts, we create a synthetic spot:

$$
\boxed{\begin{array}{l} \text{Spot operation} \\ \text{buy one unit of} \\ \text{spot oil for } S_{t_0} \end{array}} = - \boxed{\begin{array}{l} \text{A loan:} \\ \text{borrow USD at } t_0 \\ \text{for maturity } T \end{array}} + \boxed{\begin{array}{l} \text{Long oil futures} \\ \text{Expiration } T \end{array}} - \boxed{\begin{array}{l} \text{Store the oil at a} \\ \text{cost } q_{t_0} \text{ a day} \\ \text{until } T \end{array}} \quad (7.2)
$$

In other words, after changing signs, we need to *borrow* one unit of oil, make a deposit of S_{t_0} dollars, and go long an oil futures contract. This will yield a synthetic spot.

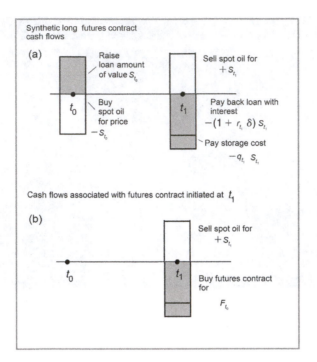

FIGURE 7.3

Synthetic long crude oil futures position.

Second, the contractual equation can be used in pricing. In fact, the contractual equation gives the *carry cost* of a position. To see this, first note that according to Eq. (7.1), the value of the synthetic is the same as the value of the original contract. Then, we must have

$$F_{t_0} = \left(1 + r_{t_0}\delta\right)S_{t_0} + q_{t_0}(T - t_0)$$ (7.3)

where δ is the factor of days' adjustment to the interest rate denoted by the symbol r_{t_0} and represents storage costs $q_{t_0}(T - t_0)$. The interest rate r_{t_0} could be the LIBOR rate, for example. Storage costs include warehousing and insurance costs. The *carry cost* is the interest plus storage costs here.

If storage costs are expressed as a percentage of the price, at an annual rate, just like the interest rates, this formula becomes

$$F_{t_0} = \left(1 + r_{t_0}\delta + q_{t_0}\delta\right)S_{t_0}$$ (7.4)

This replication approach allows us to identify all the cash flows associated with the position and, thus, quote a *fair value* or theoretical price that will ensure no loss at the point of delivery, independent of the actual future spot price. Equation (7.4) is also sometimes called the *spot–futures parity* relationship.

7.3.3 CONVENIENCE YIELD

Note that according to Eq. (7.4), the more distant the expiration of the contracts is, the higher its price. This means that future term structures would normally be upward sloping, or in contango, as shown in Figure 7.1b. However, as we discussed above, many commodity markets such as oil and certain base metals tend to exhibit a futures price term structure in backwardation. To be able to explain a backwardated term structure, we can extend Eq. (7.3) and incorporate an additional component called the *convenience yield (cy)*. This leads to the following equation:

$$F_{t_0} = (1 + r_{t_0}\delta)S_{t_0} + q_{t_0}(T - t_0) - cy \tag{7.5}$$

We can interpret the convenience yield as the premium that a consumer is willing to pay for being able to consume the commodity now rather than at some point in the future. One advantage of having a secure supply of raw materials is that it reduces the risk of *stockouts*, for example. However, the convenience yield remains a somewhat abstract concept that is intangible since it is really a mathematical placeholder that is used to reconcile the relationship between futures and spot prices in the above equation. The convenience yield cy changes in practice in response to the physical supply and demand imbalance for the underlying commodity. The convenience yield will be high if there are supply shortages and it will be lower when demand and supply are in balance. The convenience yield may be quite high for crude oil which counts oil refiners as its main consumers. First, it is expensive to store oil. Second, if there were supply shortages, the cost of closing a refinery for several months is high. For these two reasons, oil refiners may be willing to pay a premium for spot delivery. This could rationalize but not really economically (in a fundamental sense) explain a futures price that is below the spot oil price (backwardation).

7.3.4 CASH AND CARRY ARBITRAGE

We will now present a real-world arbitrage example that helps to develop the intuition for spot and futures markets dynamics as a function of supply and demand as well as storage costs. Note that Eq. (7.5) can also be used to identify potential arbitrage opportunities. If the current forward price F_{t_0} was greater than the right-hand side—which is more likely if the curve was in contango—one would sell the forward contract, buy the commodity in the spot market (financed by a loan), pay the cost of storage, benefit from holding the physical commodity over the interval $(t_0; T)$, and realize at maturity T a *cash and carry arbitrage*. The following reading gives an example of real-world storage and shipping costs, applies our contractual equation (7.4) and illustrates the cash and carry arbitrage in the heating oil market:

EXAMPLE

JPMorgan Hires Supertanker for Storage, Brokers Say
   —JPMorgan Chase & Co., the second-largest US bank by deposits, hired a newly built supertanker to store heating oil off Malta, shipbrokers reported, in the company's first such booking in at least 5 years.
   The bank hired the Front Queen for 9 months, according to daily reports from Oslo-based SeaLeague A/S and Athens-based Optima Shipbrokers Ltd. David Wells, a spokesman for JPMorgan in London, declined to comment.

> *JPMorgan, which has never hired an oil tanker based on data compiled by Bloomberg going back 5 years, follows companies including Citigroup Inc.'s Phibro LLC unit and BP Plc in hiring ships to store crude or oil products at sea. The firms are seeking to take advantage of higher prices later in the year.*
>
> *"It's opportunity-driven," Sverre Bjorn Svenning, an analyst at Fearnley Consultants AS in Oslo, said by phone. "I doubt it's going to be a permanent or new sort of trade."*
>
> *Heating oil for immediate delivery costs $553 a metric ton in northwest Europe and supplies for August are at $580, according to data compiled by Bloomberg.[...]*
>
> *Front Queen can hold about 273,000 tons of heating oil, more than three times the amount held by a more conventional Long Range 2 tanker, according to Riverlake Shipping SA, Switzerland's largest shipbroker.*
>
> *JPMorgan hired the ship at $35,000–41,000 a day, according to the broker reports. The bank is also paying $1.6 million for the ship to sail from Singapore to Europe without a cargo, the brokers said. Long Range 2 tankers cost about $25,000 a day for storage, according to Riverlake Shipping.*
>
> *Based on a full cargo, the monthly cost for the Front Queen works out at $3.85–4.50 a ton, excluding what JPMorgan paid to get the ship from Asia to Europe. Benchmark supertanker rates plunged 87% from their peak in July as oil producers curbed supply and ship owners expanded the global fleet. The Paris-based International Energy Agency expects the first back-to-back drop in global annual oil demand since 1983.*
>
> *Traders were already using smaller tankers to store record volumes of jet fuel and heating oil in Europe as onshore tanks filled up, D/S Torm A/S, Europe's biggest oil products shipping line, said April 3. Inventories of heating oil at Amsterdam, Rotterdam, and Antwerp stand at 2.7 million tons, the highest for at least 2 years, according to data from PJK International BV.[...]*
>
> **Bloomberg (extract from article "JPMorgan Hires Supertanker for Storage, Broker Say" by Alaric**
>
> **Nightingale – June 3, 2009)**

What is the profit that JPMorgan could expect to make on the transaction? If we make a few simplifying assumptions we can calculate the price of the synthetic futures based on the contractual equation in Eq. (7.1) and compare it with the actual futures price available in the market. This way, we can approximate the expected profit. In other words, we assume the bank buys spot heating oil for $553, pays storage and other costs and sells a futures for $580 in the hope of making a profit. To draw a representative futures curve, we would require futures prices for more contract months, but based on these two price points, the heating oil futures price term structure is in contango. This is the same shape as the June 2009 crude oil term structure shown in Figure 7.2. Let's assume that JPMorgan bought 273,000 tons heating oil in the spot market in June 2009 and entered an appropriately sized short heating oil future. At an assumed spot price of $553 per metric ton, the physical heating oil purchase in June 2009 has a value of $150.969 million. If we make a conservative assumption and take borrowing costs to be, say, 1% for 2 months, total borrowing costs would be $1.50969 million over the period. Insurance costs for oil tankers can vary from 0.25% to more than 3% of hull value depending on the geopolitical and macroeconomic situation. If we assume an insurance cost of 0.25% of the value of the spot purchase for 2 months, then the total insurance cost is $377,423. The cost of shipping the tanker from Singapore to Europe is $1.6 million according to the reading. Storage costs are $3.85 per ton amount or 1.051 million for the whole cargo. Table 7.2 provides the breakdown of expected costs and revenues. Total costs are $155.507163 million. If we assume a futures price of $580, this corresponds to revenue of $158.34 million. The expected profit is then $2.833 million.

Table 7.2 Expected Profit Calculations for Heating Oil Example	
Cost of buying spot heating oil	$150.969 million
Cost of borrowing (at 1% for 2 months)	$1.50969 million
Insurance costs (at 0.25% of hull value)	$0.377423 million
Costs of shipping the tanker from Singapore to Europe	$1.6 million
Costs of storage (at $3.85 per ton)	$1.05105 million
Total costs	$155.507163 million
Total revenue from futures sale	$158.34 million
Total profit	$2.832838 million

The above example not only illustrates the construction of a synthetic futures position, but also the exploitation of a potential arbitrage opportunity. We can ask what the effect of such transactions involving spot heating oil purchases and futures markets sales is likely to be. The extract from the Bloomberg article above suggests that many market participants engaged in such trades. This can be expected to lead to an increase in spot prices and a decrease in futures prices, thus reducing the contango and potentially moving it to the backwardation shape that energy commodities typically exhibit. The above is just an example, but it illustrates drivers of the term structure that have been found to be important in more comprehensive studies of the term structure of futures prices such as those listed at the end of this chapter.

7.3.5 ANOTHER INTERPRETATION OF SPOT–FUTURES PARITY

There are no upfront payments but buying futures or forward contracts is not costless. Ignoring any guarantees or *margins* that may be required for taking futures positions, taking forward or futures positions *does* involve a cost. Suppose we consider a storable commodity with spot price P_{t_0}. Let the forward price be denoted by $P_{t_0}^{\mathrm{f}}$. Finally, suppose storage costs and all such effects are zero. Then the futures price is given by

$$P_{t_0}^{\mathrm{f}} = \left(1 + r_{t_0}\delta\right)P_{t_0} \tag{7.6}$$

where r_{t_0} is the appropriate interest rate that applies for the trader and δ is the time to expiration as a proportion of a year. Now, suppose the spot price remains the *same* during the life of the contract. This means that the difference

$$P_{t_0}^{\mathrm{f}} - P_{t_0} = r_{t_0}\delta P_{t_0} \tag{7.7}$$

is the cost of taking this position. Note that *this is as if we had borrowed P_{t_0} dollars for a "period"* δ in order to carry a long position. Yet there has been no exchange of principals. In the case of a default, no principal will be lost.

7.4 SWAP ENGINEERING FOR COMMODITIES

The overall structure of commodity swaps is similar to equity swaps, which we mentioned in Chapter 4 and discuss in greater detail in Chapter 18 on cash flow engineering applications in equity markets. As with equity swaps, there are two major types of commodity swaps. Parties to the swap can either (1) exchange fixed for floating payments based on a commodity index or (2) exchange payments when one payment is based on an index and the other on a money market reference rate.

The vast majority of commodity swaps involve oil. Consider a refinery, for example. Refineries buy crude oil and sell refined products. They may find it useful to lock in a fixed price for crude oil. This way, they can plan future operations better. Hence, using a swap, a refiner may want to receive a floating price of oil and pay a fixed price per bbl.

Such *commodity* or *oil swaps* can be arranged for all sorts of commodities, metals, precious metals, and energy prices.

EXAMPLE

Japanese oil companies and trading houses are naturally short in crude oil and long in oil products. They use the short-term swap market to cover this exposure and to speculate, through the use of floating/fixed-priced swaps. Due to an overcapacity of heavy-oil refineries in the country, the Japanese are long in heavy-oil products and short in light-oil products. This has produced a swap market of Singapore light-oil products against Japanese heavy-oil products.

There is also a "paper balance" market, which is mainly based in Singapore but developing in Tokyo. This is an oil instrument, which is settled in cash rather than through physical delivery of oil.

Thomson Reuters IFR (Issue 946)

7.4.1 SWAP CASH FLOW ENGINEERING IN COMMODITY MARKETS

Let's examine how we can engineer a commodity swap. Here we assume that the counterparties exchange payments where one payment is based on an index and the other on a money market reference rate. Suppose the S_t represents a commodity. It could be oil for example. Then, the analysis would be identical to engineering an equity swap.

One could invest $N = 100$ and "buy" Q units of the commodity in question. The price S_t would move over time. One can think of the investment paying (receiving) any capital gains (losses) to the investor at regular intervals, $t_0, t_1, \ldots, t_n$. At the maturity of the investment, the N is returned to the investor. All this is identical to the case of stocks covered in Chapter 4. This is illustrated in Figure 7.4a.

One can put together a *commodity swap* by adding the n-period FRN to this investment, as shown in Figure 7.4b. The initial and final payments of the N would cancel and the swap would consist of paying any capital gains and receiving the capital losses and the LIBOR $+ d_t$, where d_t is the swap spread (see Figure 7.4c).

Note that the swap spread may deviate from zero due to any *convenience yield* the commodity may offer, or due to supply−demand imbalances during short periods of time. The convenience yield here would be the equivalent of the dividends paid by the stock.

7.4.2 PRICING AND VALUATION OF COMMODITY SWAPS

Consider a commodity swap in which the counterparties exchange fixed for floating payments based on a commodity index instead of the swap type shown in Figure 7.4 in which there was an

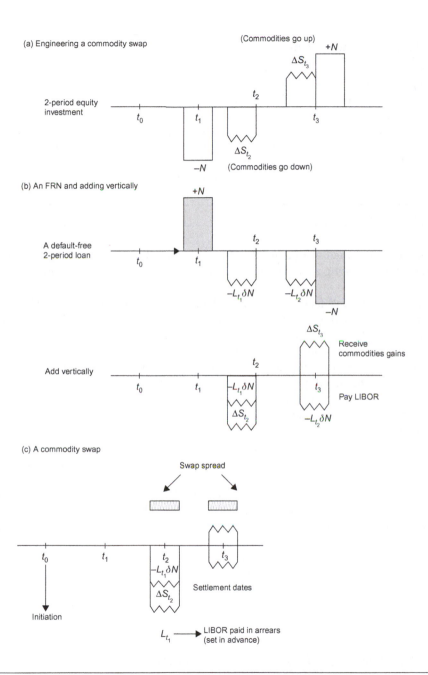

FIGURE 7.4

Commodity swap replication.

Table 7.3 Sample Swap Contract	
Fixed-price payer:	Swap dealer X
Floating-price payer:	Crude oil producer
Commodity reference price:	NYMEX WTI futures contract
Fixed-price payment:	$95.00 bbl × volume
Floating-price payment:	The average over the calendar month of the daily settlement prices for the NYMEX prompt contract times the monthly volume
Monthly volume:	100,000 bbl/month
Term:	January 2013–December 2013
Payments:	14 business days following each settlement period
Source: *risk.net/energy-risk, Blanco and Piere (2013)*.	

exchange of payments when one payment is based on an index and the other on a money market reference rate. The following example involves an oil producer that sells a swap to hedge revenues over a certain period.

Here we make the assumption that the oil producer is attempting to hedge its revenues against an unexpected decline of crude oil prices in 2013. According to Table 7.3, the oil producer and the swap dealer X would agree to enter into a calendar year 2013 swap with a fixed price of $95/bbl and a monthly volume of 100,000 bbl per month. The net payment of each month is the difference between the fixed-price payment of $95/bbl and the realized average price, multiplied by the monthly volume. As a result of these terms and conditions, the seller of the swap (the producer) will receive a payment from the swap dealer if the average oil price is lower than $95/bbl. The opposite will happen if the average oil price is higher than $95/bbl. Assume, for example, that the floating leg of the swap settles at $85/bbl at the end of January 2013. In this case, there would be a payment from the swap dealer to the oil producer of $100,000 \text{ bbl} \times (\$95 - \$85) = \$1,000,000$. Figures 7.5a and 7.5b illustrate the payments received by the swap seller (i.e. the oil producer) and the swap buyer (i.e., the swap dealer). For the purpose of illustration, we assume in Figure 7.5 that the floating leg of the swap settles at $100/bbl and $105/bbl in February 2013 and March 2013. Figure 7.5c shows the net payments to the swap seller, i.e., the oil producer. This example shows that the swap fulfills its purpose of hedging the oil producer against unexpected decreases in oil prices. Table 7.4 explains the key contract terms used in Table 7.3. It is common for energy commodity sales and purchase contracts to have *ratable* delivery provisions such that the final invoice price is based on the average price during the pricing window. This provides hedgers such as producers or consumers with an incentive to use average price swaps or a strip of futures contracts to match the pricing provisions of the physical deals.

7.4.2.1 Calculating swap prices from the futures curve

As we saw in the context of interest rate swaps, the swap rate at inception is set so that the present value of future fixed and floating payments is equal. The mark-to-market value of the interest rate swap is calculated by discounting the future cash flows using forward rates, for example. Similarly, in the case of the oil swap, we can calculate the mark-to-market gains and losses by valuing the

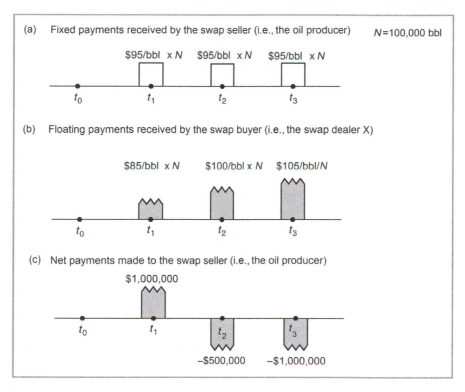

FIGURE 7.5

Cash flow example of oil commodity.

Table 7.4 Key Contract Terms of an Over-the-Counter Swap	
Parameter	**Explanation**
Commodity reference price	The commodity price index that will be used to determine the floating leg price. Common choices for the underlying are the prompt futures contract settlement, spot prices, month-ahead indices, and deferred (second-prompt prices)
Floating price	The formula that will be used to determine the floating leg payment. It is often the average of the daily settlements of the commodity reference price during the pricing or fixing window
Fixed price	Negotiated price for the fixed leg of the contract
Fixing or pricing period(s)	A schedule of dates in which the underlying price will be observed to set the value of the floating leg for each settlement period
Notional or volume	Notional amount of commodity used to determine the swap cash flows
Source: *risk.net/energy-risk, Blanco and Piere (2013).*	

swap and constructing swap prices. Commodity swap prices also require a forward curve. The price quotes for the forward curve, which are the equivalent of the forward curve implied LIBOR rates in an interest rate swap, could come from futures, forward, or swap prices or a mixture of the three. Let's assume that we use the futures curve to calculate the swap curve. This is illustrated in Table 7.5. Here we further assume that the swap curve is based on the *expected* average of the futures prompt contract during a given month which involves combining two futures contract maturities for each swap price. In constructing the curve day count conventions, contract expiry dates and holidays have to be carefully taken into account. Consider, for example, the swap curve price for January 2013 in Table 7.5. It is based on the following calculation:

$$P_{swap,Jan} = P_{future,Feb} \times w_1 + P_{future,March} \times w_2$$
$$= \$88.40 \times 52\% + 88.92 \times 48\% = \$88.65$$

Table 7.5 provides the swap curve construction for months January 2013 until December 2013.

Similar to an interest swap, the information in Table 7.5 can be used to calculate the mark-to-market value of the swap and its "fair value." For this, we require information about the discount rate and yield curve. This is left as an exercise at the end of this chapter. The exercise also provides information about the discount rate to use to discount the cash flows.

Table 7.5 Building the Calendar Year 2013 Prompt Month Swap Curve from the Futures Curve

Futures Curve			Swap Curve Construction					
Month	Expiration	Price ($/bbl)	Date	Days Until Roll Date	Days After Roll Date	Weight Contract #1 (%)	Weight Contract #2 (%)	Swap Curve (Prompt) ($)
January 13			January 13	12	11	52	48	88.65
February 13	16/02/2013	88.40	February 13	9	10	47	53	89.19
March 13	13/02/2013	88.92	March 13	10	11	48	52	89.68
April 13	14/03/2013	89.43	April 13	11	10	52	48	90.07
May 13	15/04/2013	89.90	May 13	12	10	55	45	90.35
June 13	16/05/2013	90.25	June 13	9	10	47	53	90.52
July 13	13/06/2013	90.47	July 13	12	10	55	45	90.60
August 13	16/07/2013	90.57	August 13	11	10	52	48	90.63
September 13	15/08/2013	90.63	September 13	10	10	50	50	90.63
October 13	13/09/2013	90.63	October 13	12	11	52	48	90.62
November 13	16/10/2013	90.62	November 13	10	11	48	52	90.56
December 13	14/11/2013	90.61	December 13	9	13	41	59	90.46
January 14	12/12/2013	90.52						
February 14	11/01/2013	90.41						

Source: *risk.net/energy-risk, Blanco and Piere (2013).*

7.4.3 REAL-WORLD COMPLICATIONS

OTC commodity swaps such as the one discussed above can be customized. However, commodity swap contract specifications and liquidity differ even in the exchange traded market. On October 19, 2014, for example, Dalian Commodity Exchange (DCE) launched an iron ore commodity swap. The contract differs from the iron ore contract offered by the Singapore Exchange (SGX) since 2009 because there are currency and settlement differences. The DCE contract is in RMB, whereas the SGX contract is in US dollars. In contrast to the Dalian contract, which is physically settled, the SGX contract is financially settled. In 2013, SGX cleared around 95% of the world's iron ore swaps—the remainder accounted for by trades on the Chicago Mercantile Exchange and ICE. The following example illustrates potential arbitrage opportunities and limitations in commodity swap markets as well as real-world complications such as seasonality and time-varying liquidity.

EXAMPLE

In line with a number of Chinese commodity contracts, there are only three liquid months on DCE, for contracts that settle in January, May, and September. With liquidity along larger parts of the curve in Singapore, this leaves traders with China operations able to conduct spread trades based on the arbitrage between DCE and SGX [...]

The reason SGX volumes grew after the DCE launch is the significant difference between the two products: currency (RMB versus US$) and physical settlement in Dalian versus financial settlement on the SGX. This has led to the creation of different, and complementary, dynamics between the exchanges.

"The Dalian contract and the Singapore swap are fundamentally different contracts," says Leong Chean Wai, Singapore-based head of commodity derivatives at DBS Bank. "Dalian is a closed market—it isn't accessible to international players, unlike Singapore. It has, however, introduced the spread trade, which is a new element to the iron ore market, even though it is something that is commonly executed in other base metal markets. This encourages more players to enter the market and means the DCE contract has led to an increase in Singapore liquidity, rather than a fall." [...]

A lot of iron ore traders in DCE trade coke, coking coal, and steel. They frequently adjust their positions among different contracts for arbitrage. This means it is difficult to predict which contract will be the most active one. [...]

"China's futures market is still taking baby steps. A well-developed futures market usually has market-makers who can guarantee that every listed contract has enough liquidity. But there are no market-makers in China's futures market. That's part of the reason why some contracts have great liquidity, but some do not."

But it's not so much a lack of market-makers that is holding back liquidity along large parts of the DCE curve; the seasonal nature of demand for commodities is also significant, according to Lin Ling, Shanghai-based analyst at Citic Futures.

May and September are peak seasons for consuming iron ore. The consumption in January is quite low, but it's immediately before Chinese New Year and steel mills need to have some winter storage for the next year. This rule is applicable to both coke and coking coal as well. [...] The real test comes when the May contract—the first liquid month on DCE—becomes due. "It's not possible to predict how that will turn out until the actual day. A lot will depend on where the physical is trading compared with the exchange. China has a large active physical market and traders will wait—if the physical is cheaper, they will buy that, and sell out their futures positions and if the price is lower on the DCE than the physical delivery, they will do that."

There is, essentially, a three-way arbitrage on the China iron ore swaps market: the DCE futures price versus the physical onshore versus the international swap market. While the former is a straightforward arbitrage of the futures price versus physical, the onshore/offshore arbitrage is a function of the different curves in SGX and DCE.

Backward look

> *Chinese commodities markets are typically characterized by contango—where the forward price is higher than the spot—due to factors such as tax, storage costs, and the limited access to onshore markets from international players, and iron ore was initially no exception. However, it has now moved into backwardation (where the futures price is below the expected future spot price for a particular commodity).*
>
> *East says: "What we saw when the DCE launched its iron ore futures contract was that it was in contango, whereas the SGX market was in backwardation. What happened was that all the traders came in and bought SGX backwardation and sold DCE contango." Arbitrage is inherently risky and East points to the dangers posed by a steepening of either the backwardation or contango, which can lead to mark-to-market losses—and of course, having access to the China physical market is critical to arbitraging the DCE and SGX markets. [...] These markets move—there is an arbitrage between the two, but also it causes problems with regard to margining. "If the price goes up, the trader will be making a profit in one and a loss in the other, but because these positions are in different exchanges he can't offset the margin," he says.*
>
> **Risknet/Asia Risk (April 2, 2014)**

The reading above provides example of an application of the contractual equation in Eq. (7.1) to the iron ore market. Traders can use the spot—futures parity relationship to identify arbitrage opportunity between the onshore physical and the futures market. Since iron ore contracts are traded at the DCE and SGX but exhibited different term structures, traders bet on a convergence of the curves and essentially bought low and sold high. They sold the DCE contango (where initially nearby contract prices were below back month contracts price) and bought SGX backwardation (nearby contract prices above back month ones).

This example not only illustrates the use of commodity swaps for spread arbitrage trades, but also the continuing spread of financial engineering applications around the world. Currently the US dollar is still the currency in which most derivatives contracts are denominated, but it is very likely that over the next decades, currencies such as the RMB and other Asian currencies will see growth in derivatives and financial engineering applications.

7.4.4 OTHER COMMODITY SWAPS

A *commodity-linked interest rate swap* is a hybrid swap in which LIBOR is exchanged for a fixed rate, linked to a commodity price. A buyer of crude oil may wish to tie costs to the cost of debt. The buyer could elect to receive LIBOR and pay a crude-oil-linked rate such that, as the price of crude oil rises, the fixed rate the buyer pays declines.

A *crack spread swap* is a swap used by oil refiners. They pay the floating price of the refined product and receive the floating price of crude oil plus a fixed margin, the crack spread. This way, refiners can hedge a narrowing of the spread between crude oil prices and the price of their refined products.

7.5 THE HEDGE FUND INDUSTRY

Throughout the book we refer to the application of financial engineering techniques to trading opportunities and arbitrage as well as derivatives product design. One of the most prominent types of arbitrageur in financial markets is hedge funds. Their strategies are very diverse, but many funds use financial engineering techniques to design their strategies and they trade derivatives. Similar to commodities, hedge funds are often referred to as an *alternative asset class*. Therefore, in this section, we review the hedge fund industry and its role in financial markets.

7.5.1 ALTERNATIVE INVESTMENT FUNDS

Hedge funds are alternative investment funds (AIFs) that use a diverse set of techniques to improve performance and reduce volatility. The hedge fund industry has significantly changed over recent years and grown from what in retrospective appears to have been a loosely regulated multibillion dollar cottage industry dominated by high net worth individuals as its main investors in the 1990s to a tightly regulated and institutionalized sector of the asset management industry that absorbed many activities previously predominantly found in banks.

Hedge funds differ from traditional investment funds such as mutual funds in three main ways: *regulation*, *investment mandate*, and *legal structure*.

7.5.2 HEDGE FUND REGULATION

First, to understand the effect of hedge fund regulation, it is important to distinguish the alternative investment fund *manager* (AIFM) from the hedge *funds* that it manages.[3] The object of recent regulation such as the 2010 Dodd-Frank Walter Street Reform and Consumer Protection Act (Dodd-Frank), the JOBS Act and AIFMD/EMIR is the AIFM. AIFMs such as Bridgewater Associates and the Man Group, for example, are headquartered in Connecticut and London, respectively, but their products include onshore and offshore funds. Most hedge funds are located in lightly regulated, low-tax, offshore centers like the Cayman Islands or Bermuda. The Cayman Islands are the single most popular *location* for hedge funds, with almost one in two registered there. However, even before the global financial crisis (GFC), there have been many onshore hedge funds in the United States or Europe where they were registered with regulators.

Dodd-Frank changed the regulation of hedge funds in the United States by requiring that all hedge funds with £100 million or more in assets be registered with the Securities and Exchange Commission (SEC). It also instructed the SEC to collect information, on a confidential basis, from private fund advisers regarding the risk-profiles of their funds. The 2012 JOBS (Jumpstart Our Business Startups) act directed the SEC to remove the decades-old ban on general solicitation that applied to companies or funds that make private securities offerings. As a result, hedge fund managers in the United States are now free to communicate freely with and advertise to the public.

The European Union (EU) Directive on AIFMs passed in 2010 requires hedge funds to register with national regulators and increased disclosure requirements for AIFMs operating in the EU. The directive also increases capital requirements for hedge funds and places further restrictions on leverage utilized by hedge funds. Non-EU AIFMs marketing funds in the EU are subject to reporting requirements under an enhanced national private placement regime until they are eligible to or required to market under an EU passport in the future.[4]

---

[3]The AIFM is also known as the *management company* or *hedge fund adviser*.
[4]See https://www.managedfunds.org/issues-policy/issues/globalhedge-fund-regulation/.

The GFC and the Maddof scandal affected both the supply of and demand for investor capital for hedge funds. The scandal negatively affected the perception of hedge funds by high net worth individuals who often invested in hedge funds via funds of funds.[5] Arbitrages in the real world are not riskless and they require a stable sources of funding. Therefore, sudden redemptions by funds of funds and high net worth individuals during the GFC led many hedge funds to set out and seek more *sticky* sources of capital such as institutional investors. As a result, the investor base of hedge funds has changed. Since 2010, institutional investors such as pension funds, SWFs, endowments, and foundations account for the majority of hedge fund investor assets, while HNWI and family offices are in the minority.

7.5.3 HEDGE FUND INVESTMENT MANDATE AND TECHNIQUES

The second distinguishing feature of hedge funds has to do with their investment mandate and techniques. Historically, most registered mutual funds had essentially two choices: They could either go long in the stated asset class or stay in cash. Often there is a limit on the latter. Thus, mutual funds have no room for maneuver during bear markets. Hedge funds on the other hand were less restricted and could *short* the markets. This was perceived at that time to be the "hedge" for a down-market, hence the term "hedge fund" emerged. In addition to shorting, hedge funds could also use derivatives to speculate or hedge and reduce risk. In recent years, a *convergence* of the alternative investment industry and the traditional investment industry has been witnessed as *40 Act* funds in the United States and absolute return UCITS funds in the EU increasingly employ hedge fund techniques while being marketed to retail investors.[6]

Regulation explains why hedge funds are allowed to use shorting and derivative, but less well-known differences in investment mandate explain why they choose to. While traditional mutual funds normally only offer *relative* returns, hedge funds aim to offer *absolute* returns on an investment. For example, traditional funds set for themselves stock or bond market *benchmarks* and then measure their performance relative to these benchmarks. The fund may be down, but if the benchmark is down even more, the fund is said to *outperform*.[7] This is because a typical mutual fund has to be long the underlying securities. When the price of the security goes down, they have few means to make positive returns. Hedge funds, on the other hand, can realize profits in down-markets as well. Unlike most traditional fund managers, hedge funds can (1) sell short, (2) use derivatives, and (3) leverage to make bigger investments.

7.5.4 HEDGE FUNDS' LEGAL STRUCTURE

The use of leverage is explained by the third main feature of hedge funds, which is its legal structure. Hedge funds' legal structure grants them contractual flexibility, such as lockups for investors,

---

[5]Most institutional investors are not allowed by regulators to directly invest in hedge funds. However, they can invest in *funds of hedge funds.*

[6]Joenväärä and Kosowski (2013) provide a comprehensive review of the techniques used by hedge funds and the convergence between mainstream and alternative asset management.

[7]For example, a manager is said to have outperformed his or her benchmark even if the fund loses 2%, but the benchmark is down 5%.

whose legal rights are those of a limited partner. These characteristics facilitate leverage and allow the prime broker to grant special funding conditions to hedge funds. Leverage arises not just as a result of (a) borrowing, but also (b) the use of financial instruments with intrinsic leverage such as futures or (c) construction leverage resulting from short positions that fund long positions.[8] Nevertheless, hedge funds have had on average lower leverage than banks, which had leverage ratios of 20−30, in the run up to the GFC.[9] In February 2014, average hedge fund leverage was 2.04, but it varied from 1.02 for Distressed Securities to 3.44 for Global Macro funds according to Citi Prime Finance.[10] Because some hedge funds *leverage* their capital significantly and some funds trade frequently, hedge funds are much more influential than their *equity capital* indicates.

Assets under management in hedge funds were around $2.7 trillion in 2014 compared to less than $200 billion in 1994.[11] However, most of the hedge fund AuM are concentrated in the largest funds. Although only 9% of funds in 2012 have AuM above $1 billion, these funds manage more than 70% of the industry AuM. The majority of hedge fund assets is therefore managed by large institutionalized AIFMs.[12]

Hedge funds are important market participants. According to Reuters, hedge fund trading activity accounted for up to half the daily turnover on the New York Stock Exchange and the London Stock Exchange in 2005. More importantly, hedge funds make up almost 60% of US credit derivatives trading, and about half of emerging market bond trading. According to the same sources, about one-third of equity market activity is due to hedge funds. It is estimated that they are responsible for more than 50% of trading in commodities.

7.5.5 INCENTIVES

Historically, hedge fund managers often started as prop desk traders at major banks. They then left the bank to set up their own businesses. The *Volcker Rule*, part of the 2010 Dodd-Frank financial overhaul, came into effect on April 1, 2014, and restricts the ability of banking institutions and certain nonbank financial companies to engage in proprietary trading. In anticipation of the implementation of the rule, many investment banks closed proprietary trading desks with many traders moving to less regulated areas of financial markets such as hedge funds or privately held proprietary trading firms. Client-flow trading desks, of course, remain in investment banks and provide a training ground for young traders and a source of talent for hedge funds.

Successful traders in banks are attracted by the hedge funds' potentially generous remuneration. Annual *fees* at traditional mutual funds are normally between 30 and 50 bp. Hedge funds charge an annual 1−2% management fee. However, they also receive up to 20% of any outperformed amount. Funds of hedge funds charge an additional annual 1−1.5% *management fee* and an average 10% *performance fee*.

---

[8]See "An Overview of Leverage" AIMA Canada Strategy Paper Series: Companion Document, October 2006, Number 4.
[9]See http://www.worldbank.org/financialcrisis/pdf/levrage-ratio-web.pdf.
[10]See Citi Prime Finance Hedge Fund Industry Snapshot. February 2014. The report uses the following measure of leverage: Gross Market Value Leverage = (Long Market Value + Absolute Value of Short Market Value)/Net Equity).
[11]HFR Global Hedge Fund Industry Report for Q1 2014.
[12]See "The Effect of Investment Constraints on Hedge Fund Investor Returns" by Joenväärä, Kosowski and Tolonen (2014).

Many hedge funds have *high water marks*. If the value of the portfolio they are managing falls below, say, last year's value, the fund does not receive performance fees until this level is exceeded again. This means that after a sharp fall in the portfolio, such funds will be dissolved and reestablished under a different entity.

7.5.6 STRATEGIES

Hedge funds are classified according to the strategies they employ. The market a hedge fund uses is normally the basis for its strategy classification.

Global macro funds bet on trends in financial markets based on macroeconomic factors. Positions are in general levered using derivatives. As we saw in our discussion of futures and margin requirements, such derivatives cost a fraction of the outright purchase.

Managed futures/Commodity Trading Advisors (CTAs). These strategies speculate on market trends using futures markets. Systematic CTAs differ from discretionary CTAs in that they use computer programs that use technical analysis tools like relative strength and momentum indicators to make investment decisions.

Long/short strategists buy stocks they think are cheap and short those they think are expensive. The overall position can be net long or net short. This means that they are not necessarily market neutral. Within this category, the class of *short bias* funds can take long equity positions, but their overall position must be short.

Emerging market funds use equities or fixed income. Managers using this strategy tend to buy securities and sell only those they own. Note that this is in contrast to short-selling which is a technique that requires borrowing a security before selling it. Many emerging market countries do not allow short selling and derivatives markets in these countries are normally not developed.

Event driven. There are two groups here. *Merger or risk arbitrage* trades the shares of firms in takeover battles, normally with a view that the bid will have to be raised to win over shareholders in the target company and will cost the bidder a lot more money. A higher buyout price will usually weigh on the bidding company's share price as it could deplete cash reserves or force it to issue bonds to pay the extra money.

Distressed debt funds trade the bonds of a company in financial distress, where prices have collapsed, but where the chances of repayment are seen as high or there is a possibility that debt could be converted into equity. Distressed debt normally trades at a deep discount to its nominal value and could be bought against the company's investment grade bonds, which, because of collateral agreements, may not have crashed to the same extent.

Relative value. These strategies generally buy stocks or bonds managers think are cheap and sell those they think are expensive.

Equity neutral or hedged strategies should be cash neutral, which means the dollar value of stocks bought should equal the dollar value of stocks they have shorted. They can also be market neutral, which means the correlation between a portfolio and the overall market should be zero.

Fixed income strategies look at interest rates, sovereign bonds, corporate bonds, and mortgage and asset-backed securities. Managers can trade corporate against government debt, cash versus futures, or a yield curve—short maturity bonds against long maturity bonds. Bonds of different governments are often traded against each other where interest rate cycles are seen to be out of sync.

Convertible arbitrage funds trade convertible bonds. These are the implicit equity, bond, credit, and derivatives such as options.

Credit derivatives are a key tool for hedge funds. Within this class, *capital structure arbitrage*, put simply, involves a hedge fund manager trading corporate bonds against the company's stock on the basis that one is cheap and the other expensive.[13]

7.5.7 PRIME BROKERS

Prime brokers offer settlement, custody, and securities lending services to hedge funds. Prime brokers earn their money from commissions and by charging a premium over money market lending rates for loans.

Prime brokers also provide trade execution, stock lending, leveraged finance, and other essential services to hedge funds. In fact, without prime brokerages, hedge fund activity would be very different than where it is at the moment.

One factor is the level of leverage prime brokers are offering. Such leverage can multiply a hedge fund's activities by a factor of 3 or more, but increased capital requirements for banks mean that leverage levels are low in hedge funds, especially compared to banks where they are typically 10 times higher than for hedge funds. The second major help provided by prime brokers is in execution of trades. For example, short selling an asset requires borrowing it. How would the hedge fund find a place to borrow such an asset? Prime brokers' securities lending divisions have better information on this.

An important prime brokerage function is risk management and position keeping. Prime brokers keep the positions of the hedge funds[14] and have developed elaborate risk management systems that a small hedge fund may find too costly to own.

Global prime brokerage revenues have fallen from $15 billion in 2008 to $12 billion in 2012.[15] This trend is due to pressure on profit margins and the multiprime trend in the prime brokerage industry. The collapse of several prime brokerage firms during the GFC, which accounted for a large share of the prime brokerage business, highlighted the need for hedge funds to diversify their prime broker exposure in order to reduce their counterparty risk. Counterparty risk has become a much more important consideration in financial engineering since the GFC and we will discuss it in a later chapter in detail.

7.6 CONCLUSIONS

In this chapter, we extended the application of cash flow engineering techniques to commodity markets. As we saw storage costs and quality and grade differences in commodities make the financial engineering for commodities more complex than for homogeneous assets such as currencies, for example. We used the contractual equation for a commodity futures to derive a synthetic commodity futures position and applied it to a real-world cash and carry arbitrage example in the heating oil market. Commodity swap engineering is similar to that for other assets such as equities and we provided

---

[13]Capital structure arbitage strategies are discussed in detail in Chapter 19 in the context of structural models of default.
[14]This also helps watching the health of the fund closely.
[15]See EY 2012/2013 prime brokerage survey report.

examples of different types of commodity swaps. Just as the commodity market has undergone a financialization with the inflow of investor capital and new investment products such as commodity ETNs and ETFs, the hedge fund industry has significantly changed in recent years and undergone an institutionalization. We discussed how recent regulation and the GFC have significantly transformed the hedge fund industry.

SUGGESTED READING

Our discussion of commodities in this chapter was focused on highlighting cash flow engineering applications in commodities. For reasons of space, we could not provide a comprehensive overview of commodity markets and derivatives. **Schofield** (2007) is an excellent practical overview of different commodity markets, their economics and dynamics as well as derivatives pricing. For a recent commodity futures guide, see **Kleinman** (2013). Energy commodity markets are some of the economically most important markets and **Edwards** (2010) provides a good introduction to them. The commodity swap valuation example in this chapter is taken from **Blanco and Pierce** (2013). **Cheng and Xiong** (2013) discuss the effects of the financialization of commodity markets. **Gorton et al.** (2012) show how commodity term structure dynamics are related to inventories. Finally, the **CME Commodity Trading Manual** provides more detail on one of the most important commodity futures markets in the world. **Lhabitant** (2009) and **Stefanini** (2010) are comprehensive guides on hedge funds and their strategies.

EXERCISES

1. In this question we consider a gold miner's hedging activities.
    a. What is the natural position of a gold miner? Describe using payoff diagrams.
    b. How would a gold miner hedge her position if gold prices are expected to drop steadily over the years? Show using payoff diagrams.
    c. Would this hedge ever lead to losses?

2. Consider the JPMorgan heating oil example in the text of this chapter. We calculated the expected profit from the spot oil purchases and forward sale after making assumptions about borrowing and storage costs. Now assume that storage costs are toward the upper end of the range provided in the Bloomberg article, i.e., $4.5 instead of $3.85 per ton and that insurance costs are 1.5% and not 0.25% of hull value. What is the expected profit in this case?

3. Consider the oil calendar swap example provided in the text. Table 7.5 provided the swap curve derived from the futures curve. To calculate the mark-to-market value of the swap and its "fair price," we just require information about the discount factor or yield curve. Assume the following discount factor below, taken from Blanco and Pierce (2013) and calculate the floating payments, fixed payments, net payments, NPV of the net payment and swap mark-to-market and fair price for the swap. Assume a valuation date of December 18, 2012.

Settlements	Floating Payment ($)	Fixed Payment ($)	Net Payment ($)	Discount Factor	NPV Net Payment ($)
31/01/2013				0.99973	
28/02/2013				0.99947	
31/03/2013				0.99915	
30/04/2013				0.99886	
28/06/2013				0.99859	
31/07/2013				0.99834	
30/08/2013				0.99804	
30/09/2013				0.99776	
31/10/2013				0.99747	
27/11/2013				0.99717	
31/12/2013				0.99656	
Swap mark-to-market					?
Fair fixed price					?

4. Answer the questions related to the following case study.

CASE STUDY: HKMA AND THE HEDGE FUNDS, 1998

The Hong Kong Monetary Authority (HKMA) has been in the news because of you and your friends, hedge fund managers. In 1998, you are convinced of the following:

1. The HK$ is overvalued by about 20% against the US$.
2. Hong Kong's economy is based on the real estate industry.
3. High interest rates cannot be tolerated by property developers (who incidentally are among Hong Kong's biggest businesses) and by the financial institutions.
4. Hong Kong's economy has entered a recession.

You decide to speculate on Hong Kong's economy with a "double play" that is made possible by the mechanics of the currency board system. You will face the HKMA as an adversary during this "play."

You are provided some background readings. You can also have the descriptions of various futures contracts that you may need for your activities as a hedge fund manager. Any additional data that you need should be searched for in the Internet. Answer the following questions:

1. What is the rationale of your double-play strategy?
2. In particular, how are HIBOR, HSI, and HSI futures related to each other?
3. Display your position explicitly using precise futures contract data.
4. How much will your position cost during 1 year?
5. How do you plan to roll your position over?
6. Looking back, did Hong Kong drop the peg?

Hedge Funds Still Bet the Currency's Peg Goes

HONG KONG—The stock market continued to rally last week in the belief the government is buying stocks to drive currency speculators out of the financial markets, though shares ended lower on Friday on profit-taking.

Despite the earlier rally, Hong Kong's economy still is worsening; the stock market hit a 5-year low 2 weeks ago, and betting against the Hong Kong dollar is a cheap and easy wager for speculators.

The government maintains that big hedge funds that wager huge sums in global markets had been scooping up big profits by attacking both the Hong Kong dollar and the stock market.

Under this city's pegged-currency system, when speculators attack the Hong Kong dollar by selling it, that automatically boosts interest rates. Higher rates lure more investors to park their money in Hong Kong, boosting the currency. But they also slam the stock market because rising rates hurt companies' abilities to borrow and expand.

Speculators make money in a falling stock market by short-selling shares—selling borrowed shares in expectation that their price will fall and that the shares can be replaced more cheaply. The difference is the short-seller's profit.

"A lot of hedge funds which operate independently happen to believe that the Hong Kong dollar is overvalued" relative to the weak economy and to other Asian currencies, said Bill Kaye, managing director of hedge fund outfit Pacific Group Ltd. Mr. Kaye points to Singapore where, because of the Singapore dollar's depreciation in the past year, office rents are now 30% cheaper than they are in Hong Kong, increasing the pressure on Hong Kong to let its currency fall so it can remain competitive.

Hedge funds, meanwhile, "are willing to take the risk they could lose money for some period," he said, while they bet Hong Kong will drop its 15-year-old policy of pegging the local currency at 7.80 Hong Kong dollars to the US dollar.

These funds believe they can wager hundreds of millions of US dollars with relatively little risk. Here's why: If a hedge fund bets the Hong Kong dollar will be toppled from its peg, it's a one-way bet, according to managers of such funds. That's because if the local dollar is dislodged from its peg, it is likely only to fall. And the only risk to hedge funds is that the peg remains, in which case they would lose only their initial cost of entering the trade to sell Hong Kong dollars in the future through forward contracts.

That cost can be low, permitting a hedge fund to eat a loss and make the same bet all over again. When a hedge fund enters a contract to sell Hong Kong dollars in, say, a year's time, it is committed to buying Hong Kong dollars to exchange for US dollars in 12 months. If the currency peg holds, the cost of replacing the Hong Kong dollars it has sold is essentially the difference in 12-month interest rates between the United States and Hong Kong.

On Thursday, that difference in interbank interest rates was about 6.3 percentage points. So a fund manager making a US$1 million bet Thursday against the Hong Kong dollar would have paid 6.3%, or US$63,000.

Whether a fund manager wanted to make that trade depends on the odds he assigned to the likelihood of the Hong Kong dollar being knocked off its peg and how much he expected it then to depreciate.

If he believed the peg would depreciate about 30%, as a number of hedge fund managers do, then it would have made sense to enter the trade if he thought there was a one-in-four chance of the peg going in a year. That's because the cost of making the trade—US$63,000—is less than one-fourth of the potential profit of a 30% depreciation, or US$300,000. For those who believe the peg might go, "it's a pretty good trade," said Mr. Kaye, the hedge fund manager. He said that in recent months, he hasn't shorted Hong Kong stocks or the currency (Wall Street Journal, August 24, 1998).

DYNAMIC REPLICATION METHODS AND SYNTHETICS ENGINEERING

8

CHAPTER OUTLINE

8.1 INTRODUCTION

The previous chapters have dealt with *static* replication of cash flows. The synthetic constructions we discussed were static in the sense that the replicating portfolio did not need any *adjustments* until the target instrument matured or expired. As time passed, the fair value of the synthetic and the value of the target instrument moved in an identical fashion.

However, static replication is not always possible in financial engineering, and replicating portfolios may need constant adjustment (rebalancing) to maintain their equivalence with the targeted instrument. This is the case for many different reasons. First of all, the implementation of static replication methods depends on the existence of other assets that permit the use of what we called *contractual equations.* To replicate the targeted security, we need a minimum number of "right-hand side" instruments in the contractual equation. If markets in the component instruments do not exist, contractual equations cannot be used directly and the synthetics cannot be created this way.

Second, the instruments themselves may exist, but they may not be *liquid.* If the components of a theoretical synthetic do not trade actively, the synthetic may not really replicate the original asset satisfactorily, even though sensitivity factors with respect to the underlying risk factors are the same. For example, if constituent assets are illiquid, the price of the original asset cannot be obtained by "adding" the prices of the instruments that constitute the synthetic. These prices cannot be readily obtained from markets. Replication and marking-to-market can only be done using assets that are liquid and "similar" but *not* identical to the components of the synthetic. Such replicating portfolios may need periodic adjustments.

Third, the asset to be replicated can be highly *nonlinear.* Using linear instruments to replicate nonlinear assets will involve various approximations. At a minimum, the replicating portfolios need to be rebalanced periodically. This would be the case with assets containing optionality. As the next two chapters will show, options are convex instruments, and their replication requires dynamic hedging and constant rebalancing.

Finally, the parameters that play a role in the valuation of an asset may change, and this may require rebalancing of the replicating portfolio.

In this chapter, we will see that creating synthetics by *dynamic* replication methods follows the same general principles as those used in static replication, except for the need to rebalance periodically. In this sense, dynamic replication may be regarded as merely a generalization of the static replication methods discussed earlier. In fact, we could have started the book with principles of dynamic replication and then shown that, under some special conditions, we would end up with static replication. Yet, most "bread-and-butter" market techniques are based on the static replication of basic instruments. Static replication is easier to understand, since it is less complex. Hence, we dealt with static replication methods first. This chapter extends them now to dynamic replication.

8.2 AN EXAMPLE

Dynamic replication is traditionally discussed within a theoretical framework. It works "exactly" only in continuous time, where continuous, infinitesimal rebalancing of the replicating portfolio is possible. This exactness in replication may quickly disappear with transaction costs, jumps in asset prices, and other complications. In discrete time, dynamic replication can be regarded as an approximation. Yet, even when it does not lead to the exact replication of assets, dynamic replication is an essential tool for the financial engineer.

In spite of the many practical problems, discrete time *dynamic hedging* forms the basis of pricing and hedging of many important instruments in practice. The following reading shows how dynamic replication methods are spreading to areas quite far from their original use in financial engineering—namely, for pricing and hedging plain vanilla options.

EXAMPLE

A San Francisco-based institutional asset manager is selling an investment strategy that uses synthetic bond options to supply a guaranteed minimum return to investors...

Though not a new concept—option replication has been around since the late 1980s... the bond option replication portfolio... replicates call options in that it allows investors to participate in unlimited upside while not participating in the downside.

The replicating portfolio mimics the price behaviour of the option every day until expiration. Each day the model provides a hedge ratio or delta, which shows how much the option price will change as the underlying asset changes.

"They are definitely taking a dealer's approach, rather than an asset manager's approach in that they are not buying options from the Street; they are creating them themselves," [a dealer] said.

(Thomson Reuters IFR, February 28, 1998)

This reading illustrates *one* use of dynamic replication methods. It shows that market participants may replicate nonlinear assets in a cheaper way than buying the same security from the dealers. In the example, dynamic replication is combined with *principal preservation* to obtain a product that investors may find more attractive. Hence, dynamic replication is used to create synthetic options that are more expensive in the marketplace.

8.3 A REVIEW OF STATIC REPLICATION

The following briefly reviews the steps taken in static replication.

1. First, we write down the cash flows generated by the asset to be replicated. Figure 8.1 repeats the example of replicating a deposit. The figure represents the cash flows of a T-maturity *Eurodeposit*. The instrument involves two cash flows at two different times, t and T, in a given currency, US dollars (USD).

2. Next, we decompose these cash flows in order to recreate some (liquid) assets such that a vertical addition of the new cash flows match those of the targeted asset. This is shown in the top part of Figure 8.1. A forward currency contract written against a currency X, a foreign deposit in currency X, and a spot FX operation have cash flows that duplicate the cash flows of the Eurodeposit when added vertically.

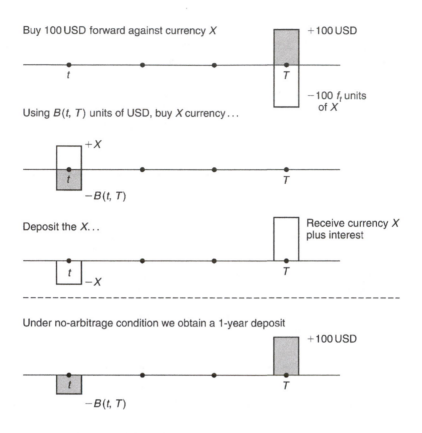

Buy 100 USD forward against currency X

$+100$ USD

t T

$-100\, f_t$ units of X

Using $B(t, T)$ units of USD, buy X currency...

$+X$

t T

$-B(t, T)$

Deposit the X...

Receive currency X plus interest

t T

$-X$

Under no-arbitrage condition we obtain a 1-year deposit

$+100$ USD

t T

$-B(t, T)$

FIGURE 8.1

Eurodeposit example.

3. Finally, we have to make sure that the (credit) risks of the targeted asset and the proposed synthetic are indeed the same. The constituents of the synthetic asset form what we call the *replicating portfolio*.

We have seen several examples for creating such synthetic assets. It is useful to summarize two important characteristics of these synthetics.

First of all, a synthetic is created at time t by taking positions on three *other* instruments. But, and this is the point that we would like to emphasize, once these positions are taken we *never* again have to modify or readjust the *quantity* of the instruments purchased or sold until the expiration of the targeted instrument. This is in spite of the fact that market risks would certainly change during the interval (t, T). The decision concerning the weights of the replicating portfolio is made at time t, and it is kept until time T. As a result, the synthetic does not require further *cash injections* or *cash withdrawals*, and it matches all the cash flows generated by the original instrument.

Second, the goal is to match the expiration cash flows of the target instrument. Because the replication does not require any cash injections or withdrawals during the interval $[t, T]$, the time t value of the target instrument will then match the value of the synthetic.

8.3.1 **THE FRAMEWORK**

Let us show how nonexistence or illiquidity of markets and the convexity of some instruments change the methodology of static synthetic asset creation. We first need to illustrate the difficulties of using static methods under these circumstances. Second, we need to motivate *dynamic* synthetic asset creation.

The treatment will naturally be more technical than the simple approach adopted prior to this chapter. It is clear that as soon as we move into the realm of portfolio rebalancing and dynamic replication, we will need a more analytical underlying framework. In particular, we need to be more careful about the timing of adjustments, and especially how they can be made *without* any cash injections or withdrawals.

We adopt a simple environment of dynamic synthetic asset creation using a basic example—we use discount bonds and assume that risk-free borrowing and lending is the only other asset that exists. We assume that there are no markets in FX, interest rate forwards, and Eurodeposit accounts beyond the very short maturity. We will try to create synthetics for discount bonds in this simple environment. Later in the chapter, we move into equity instruments and options and show how the same techniques can be implemented there.

We consider a sequence of intervals of length δ:

$$t_0 < \cdots < t_i < \cdots < T \qquad (8.1)$$

with

$$t_{i+1} - t_i = \delta \qquad (8.2)$$

Suppose the market practitioner faces only *two* liquid markets. The first is the market for one-period lending/borrowing, denoted by the symbol B_t.[1] The B_t is the time t value of \$1 invested at time t_0. Growing at the annual floating interest rate L_{t_i} with tenor δ, the value of B_t at time t_n can be expressed as

$$B_{t_n} = \left(1 + L_{t_0}\delta\right)\left(1 + L_{t_1}\delta\right)\cdots\left(1 + L_{t_{n-1}}\delta\right) \qquad (8.3)$$

The second liquid market is for a default-free pure discount bond whose time t price is denoted by $B(t, T)$. The bond pays 100 at time T and sells for the price $B(t, T)$ at time t. The practitioner can use only these two liquid instruments, $\{B_t, B(t, T)\}$, to construct synthetics. No other liquid instrument is available for this purpose.

It is clear that these are not very realistic assumptions except maybe for some emerging markets where there is a liquid overnight borrowing−lending facility and one other liquid, on-the-run discount bond. In mature markets, not only is there a whole set of maturities for borrowing and lending and for the discount bond, but rich interest rate and FX derivative markets also exist. These facilitate the construction of complex synthetics as seen in earlier chapters. However, for discussing dynamic synthetic asset creation, the simple framework selected here will be very useful. Once the methodology is understood, it will be straightforward to add new markets and instruments to the picture.

---

[1]Some texts call this instrument a *savings account.*

8.3.2 SYNTHETICS WITH A MISSING ASSET

Consider a practitioner operating in the environment just described. Suppose this practitioner would like to *buy*, at time t_0, a two-period default-free pure discount bond denoted by $B(t_0, T_2)$ with maturity date $T_2 = t_2$. It turns out that the only bond that is liquid is a *three-period* bond with price $B(t_0, T_3)$ and maturity $T_3 = t_3$. The $B(t_0, T_2)$ either does not exist or is illiquid. Its current fair price is unknown. So the market practitioner decides to create the $B(t_0, T_2)$ synthetically.

One immediate consideration is that a *static* replication would *not* work in this setting. To see this, consider Figures 8.2 and 8.3. Figure 8.2 shows the cash flow diagrams for B_t, the one-period borrowing/lending, combined with the cash flows of a two-period bond. Figure 8.3 shows the cash flows of the three-period bond. The top portion of Figure 8.2 shows that $B(t_0, T_2)$ is paid at time t_0

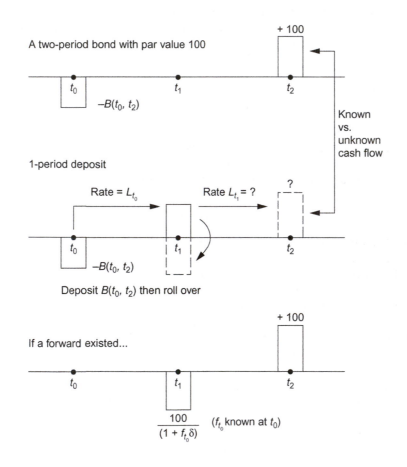

FIGURE 8.2

Cash flows of a two-period bond, a one-period deposit, and a forward.

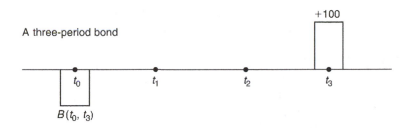

FIGURE 8.3

Cash flows of a three-period bond.

to buy the bond that yields 100 at maturity T_2. These simple cash flows cannot, unfortunately, be reconstructed using one-period borrowing/lending B_t *only*, as can be seen in the second part of Figure 8.2. The two-period bond consists of two known cash flows at times t_0 and T_2. It is impossible to duplicate, *at time* t_0, the cash flow of 100 at T_2 using B_t, *without* making any cash injections and withdrawals, as the next section will show.

8.3.2.1 A synthetic that uses B_t only

Suppose we adopt a rollover strategy: (i) lend money at time t_0 for one period, at the known rate L_{t_0}, (ii) collect the proceeds from this at t_1, and (iii) lend it again at time t_1 at a rate L_{t_1}, so as to achieve a net cash inflow of 100 at time t_2. There are *two* problems with this approach. First, the rate L_{t_1} is *not known* at time t_0, and hence we cannot decide, at t_0, how much to lend in order to duplicate the time t_2 cash flow. The amount

$$\frac{100}{\left(1 + L_{t_0}\delta\right)\left(1 + L_{t_1}\delta\right)} \tag{8.4}$$

that needs to be invested to recover the USD100 needed at time t_2 is not known. This is in spite of the fact that L_{t_0} is known.

Of course, we could guess how much to invest and then make any necessary additional cash injections into the portfolio when time t_1 comes. We can invest B_{t_0} at t_0, and then once L_{t_1} is observed at t_1, we add or subtract an amount ΔB of *cash* to make sure that

$$\left[B_{t_0}\left(1 + L_{t_0}\delta\right) + \Delta B\right]\left(1 + L_{t_1}\delta\right) = 100 \tag{8.5}$$

But, and this is the second problem, this strategy requires *injections* or *withdrawals* ΔB of an unknown amount at t_1. This makes our strategy useless for hedging, as the portfolio is not self-financing and the need for additional funds is not eliminated.

Pricing will be imperfect with this method. Potential injections or withdrawals of cash imply that the *true* cost of the synthetic at time t_0 is not known.[2] Hence, the one-period borrowing/lending cannot be used by itself to obtain a static synthetic for $B(t_0, T_2)$. As of time t_0, the creation of the

---

[2]If there are injections, we cannot use the synthetic for pricing because the cost of the synthetic is not only what we pay at time t_0. We may end up paying more or less than this amount. This means that the true cost of the strategy is not known at time t_0.

synthetic is not complete, and we need to make an additional decision at date t_1 to make sure that the underlying cash flows match those of the targeted instrument.

8.3.2.2 Synthetics that use B_t and $B(t, T_3)$

Bringing in the liquid longer-term bond $B(t, T_3)$ will not help in the creation of a *static* synthetic either. Figure 8.4 shows that no matter what we do at time t_0, the three-period bond will have an *extra* and nonrandom cash flow of \$100 at maturity date T_3. This cash flow, being "extra" (an exact duplication of the cash flows generated by $B(t, T_2)$ as of time t_0), is not realized.

Up to this point, we did not mention the use of interest rate forward contracts. It is clear that a straightforward synthetic for $B(t_0, T_2)$ could be created if a market for forward loans or forward rate agreements (FRAs) existed along with the "long" bond $B(t_0, T_3)$. In our particular case, a 2×3

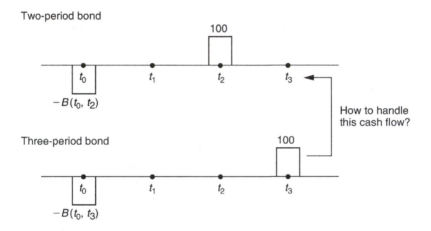

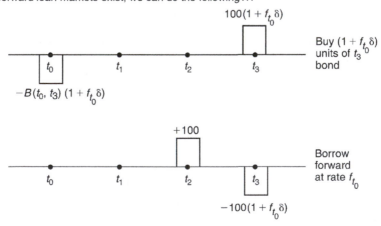

FIGURE 8.4

Replicating two-period loan with FRA and three-period bond.

FRA would be convenient as shown in Figure 8.4. The synthetic consists of buying $\left(1 + f_{t_0}\delta\right)$ units of the $B(t_0, T_3)$ and, at the same time, taking out a one-period forward loan at the forward rate f_{t_0}. This way, we would successfully recreate the two-period bond in a *static* setting. But this approach assumes that the forward markets exist and that they are liquid. If these markets do not exist, dynamic replication is our only recourse.

8.4 *"AD HOC"* SYNTHETICS

Then how can we replicate the two-period bond? There are several answers to this question, depending on the level of accuracy a financial engineer expects from the "synthetic." An accurate synthetic requires dynamic replication which will be discussed later in this chapter. But, there are also less accurate, *ad hoc*, solutions. As an example, we consider a simple, yet quite popular way of creating synthetic instruments in the fixed income sector, referred to as the *immunization* strategy.

In this section, we will temporarily deviate from the notation used in the previous section and let, for simplicity, $\delta = 1$; so that t_i represents years. We assume that there are three instruments. They depend on the *same* risk factors, yet they have different *sensitivities* due to strong nonlinearities in their respective valuation formulas. We adopt a slightly more abstract framework compared to the previous section and let the three assets $\{S_{1t}, S_{2t}, S_{3t}\}$ be defined by the pricing functions:

$$S_{1t} = f(x_t) \tag{8.6}$$

$$S_{2t} = g(x_t) \tag{8.7}$$

$$S_{3t} = h(x_t) \tag{8.8}$$

where the functions $h(.)$, $f(.)$, and $g(.)$ are nonlinear. x_t is the common risk factor to all prices. S_{1t} will play the role of targeted instrument and $\{S_{2t}, S_{3t}\}$ will be used to form the synthetic.

We again begin with *static* strategies. It is clear that as the sensitivities are different, a static methodology such as the one used in Chapters 3–7 cannot be implemented. As time passes, x_t will change randomly, and the response of S_{it}, $i = 1, 2, 3$, to changes in x_t will be different. However, one *ad hoc* way of creating a synthetic for S_{1t} by using S_{2t} and S_{3t} is the following.

At time t, we form a portfolio with a value equal to S_{1t} and with weights θ^2 and θ^3 such that the sensitivities of the portfolio

$$\theta^2 S_{2t} + \theta^3 S_{3t} \tag{8.9}$$

with respect to the risk factor x_t are as close as possible to the corresponding sensitivities of S_{1t}. Using the first-order sensitivities, we obtain two equations in two unknowns, $\{\theta^2, \theta^3\}$:

$$S_1 = \theta^2 S_2 + \theta^3 S_3 \tag{8.10}$$

$$\frac{\partial S_1}{\partial x} = \theta^2 \frac{\partial S_2}{\partial x} + \theta^3 \frac{\partial S_3}{\partial x} \tag{8.11}$$

A strategy using such a system may have some important shortcomings. It will in general require cash injections or withdrawals over time, and this violates one of the requirements of a synthetic instrument. Yet, under some circumstances, it may provide a practical solution to problems faced by the financial engineer. The following section presents an example.

8.4.1 IMMUNIZATION

Suppose that, at time t_0, a bank is considering the purchase of the previously mentioned two-period discount bond at a price $B(t_0, T_2)$, $T_2 = t_0 + 2$. The bank can fund this transaction either by using 6-month floating funds or by selling short a three-period discount bond $B(t_0, T_3)$, $T_3 = t_0 + 3$ or a combination of both. How should the bank proceed?

The issue is similar to the one that we pursued earlier in this chapter—namely, how to construct a synthetic for $B(t_0, T_2)$. The best way of doing this is, of course, to determine an exact synthetic that is liquid and least expensive—using the 6-month funds and the three-period bond—and then, if a hedge is desired, *sell* the synthetic. This will also provide the necessary funds for buying $B(t_0, T_2)$. The result will be a fully hedged position where the bank realizes the bid−ask spread. We will learn later in the chapter how to implement this "exact" approach using dynamic strategies.

An approximate way of proceeding is to *match the sensitivities* as described earlier. In particular, we would try to match the *first-order* sensitivities of the targeted instrument. The following strategy is an example for the *immunization* of a fixed-income portfolio. As we saw in Chapter 3, the first-order sensitivities are called duration. In order to work with a simple risk factor, we assume that the yield curve displays parallel shifts only. This assumption rarely holds, but it is still used quite frequently by some market participants as a first-order approximation. In our case, we use it to simplify the exposition.

EXAMPLE

Suppose the zero-coupon yield curve is flat at $y = 8\%$ and that the shifts are parallel. Then, the values of the 2-year, 3-year, and 6-month bonds in terms of the corresponding yield y will be given by

$$B(t_0, T_2) = \frac{100}{(1+y)^2} = 85.73 \tag{8.12}$$

$$B(t_0, T_3) = \frac{100}{(1+y)^3} = 79.38 \tag{8.13}$$

$$B(t_0, T_{0.5}) = \frac{100}{(1+y)^{0.5}} = 96.23 \tag{8.14}$$

Using the "long" bond $B(t_0, T_3)$ and the "short" $B(t_0, T_{0.5})$, we need to form a portfolio with initial cost 85.73. This will equal the time t_0 value of the target instrument, $B(t_0, T_2)$. We also want the sensitivities of this portfolio with respect to y to be the same as the sensitivity of the original instrument. We therefore need to solve the equations

$$\theta^1 B(t_0, T_3) + \theta^2 B(t_0, T_{0.5}) = 85.73 \tag{8.15}$$

$$\theta^1 \frac{\partial B(t_0, T_3)}{\partial y} + \theta^2 \frac{\partial B(t_0, T_{0.5})}{\partial y} = \frac{\partial B(t_0, T_2)}{\partial y} \tag{8.16}$$

We can calculate the "current" values of the partials:

$$\frac{\partial B(t_0, T_{0.5})}{\partial y} = \frac{-50}{(1+y)^{1.5}} = -44.55 \tag{8.17}$$

$$\frac{\partial B(t_0, T_2)}{\partial y} = -158.77 \tag{8.18}$$

$$\frac{\partial B(t_0, T_3)}{\partial y} = -220.51 \tag{8.19}$$

Replacing these in Eqs. (8.15) and (8.16), we get

$$\theta^1 79.38 + \theta^2 96.23 = 85.73 \tag{8.20}$$

$$\theta^1(220.51) + \theta^2(44.55) = 158.77 \tag{8.21}$$

Solving

$$\theta^1 = 0.65, \theta^2 = 0.36 \tag{8.22}$$

Hence, we need to short 0.36 units of the 6-month bond and short 0.65 units of the 3-year bond to create an approximate synthetic that will fund the 2-year bond. This will generate the needed cash and has the same first-order sensitivities with respect to changes in y at time t_0. This is a simple example of immunizing a fixed-income portfolio.

According to this, the asset being held, $B(t_0, T_2)$, is "funded" by a portfolio of other assets, in a way to make the response of the total position insensitive to first-order changes in y. In this sense, the position is "immunized."

The preceding example shows an approximate way of obtaining "synthetics" using dynamic methods. In our case, portfolio weights were selected so that the response to a small change in the yield, dy, was the same. But, note the following important point.

- The second- and higher-order sensitivities were not matched. Thus, the funding portfolio was not really an exact synthetic for the original bond $B(t_0, T_2)$. In fact, the second partials of the "synthetic" and the target instrument would respond differently to dy. Therefore, the portfolio weights θ^i, $i = 1, 2$ need to be recalculated as time passes and new values of y are observed.

It is important to realize in what sense(s) the method is approximate. Even though we can adjust the weights θ^i as time passes, these adjustments would normally *require cash injections or withdrawals*. This means that the portfolio is not self-financing.

In addition, the shifts in the yield curve are rarely parallel, and the yields for the three instruments may change by different amounts, destroying the equivalence of the first-order sensitivities as well.

8.5 PRINCIPLES OF DYNAMIC REPLICATION

We now go back to the issue of creating a satisfactory synthetic for a "short" bond $B(t_0, T_2)$ using the savings account B_t and a "long" bond $B(t_0, T_3)$. The best strategy for constructing a synthetic for $B(t_0, T_2)$ consists of a "clever" position taken in B_t and $B(t_0, T_3)$ such that, at time t_1, the extra cash *generated* by B_t adjustment is sufficient for adjusting $B(t_0, T_3)$.

In other words, we give up *static* replication, and we decide to adjust the time t_0 positions at time t_1, in order to match the time T_2 cash payoff of the two-period bond. *However*, we adjust the positions in a way that *no net cash injections or withdrawals take place*. Whatever cash is needed at time t_1 for the adjustment of one instrument will be provided by the adjustment of the *other* instrument. If this is done while at the same time it is ensured that the time T_2 value of this adjusted portfolio is 100, replication will be complete. It will not be static; it will require adjustments, but, importantly, we *would* know, at time t_0, how much cash to put down in order to receive \$100 at T_2.

Such a strategy works because both B_{t_1} and $B(t_0, T_3)$ depend on the same L_{t_1}, the interest rate that is unknown at time t_0, and both have known valuation formulas. By cleverly taking offsetting positions in the two assets, we may be able to eliminate the effects of the unknown L_{t_1} as of time t_0.

The strategy will combine imperfect instruments that are correlated with each other to get a synthetic at time t_0. However, this synthetic will need constant rebalancing due to the dependence of the portfolio weights on random variables unknown as of time t_0. Yet, if these random variables were correlated in a certain fashion, these correlations can be used *against each other* to eliminate the need for cash injections or withdrawals. The cost of forming the portfolio at t_0 would then equal the arbitrage-free value of the original asset.

What are the general *principles* of dynamic replication according to the discussion thus far?

1. We need to make sure that during the life of the security there are no dividends or other payouts. The replicating portfolio must match the *final* cash flows exactly.
2. During the replication process, there should be no net cash injections or withdrawals. The cash deposited at the initial period should equal the *true cost* of the strategy.
3. The credit risks of the proposed synthetic and the target instrument should be the same.

As long as these principles are satisfied, any replicating portfolio whose weights change during $[t, T]$ can be used as a synthetic of the original asset. In the rest of the chapter, we apply these principles to a particular setting and learn the mechanics of dynamic replication.

8.5.1 DYNAMIC REPLICATION OF OPTIONS

For replicating options, we use the same logic as in the case of the two-period bond discussed in the previous section. We will explore options in the next chapter. However, for completeness we repeat a brief definition. A European call option entitles the holder to buy an underlying asset, S_t, at a strike price K, at an expiration date T. Thus, at time T, $t < T$, the call option payoff is given by the broken line shown in Figure 8.5. If price at time T is lower than K, there is no payoff. If S_T exceeds K, the option is worth $(S_T - K)$. The value of the option *before* expiration involves an additional component called the *time value* and is given by the curve shown in Figure 8.5.

Let the underlying asset be a stock whose price is S_t. Then, when the stock price rises, the option price also rises, everything else being the same. Hence the stock is highly *correlated* with the option.

This means that we can form at time t_0 a porfolio using B_{t_0} and S_{t_0} such that as time passes, the gains from adjusting one asset compensate the losses from adjusting the other. Constant rebalancing can be done without cash injections and withdrawals, and the final value of the portfolio would equal the expiration value of the option. If this can be done with reasonably close approximation,

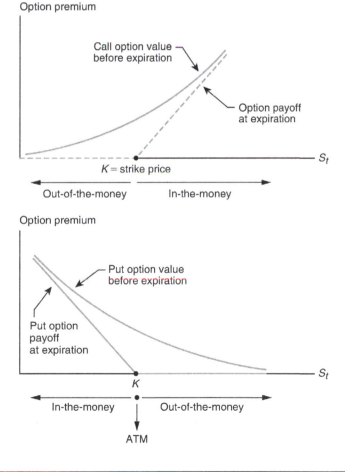

FIGURE 8.5

Payoffs of call and put option.

the cost of forming the portfolio would equal the arbitrage-free value of the option. We will discuss this case in full detail later in this chapter, and will see an example when interest rates are assumed to be constant.

8.5.2 DYNAMIC REPLICATION IN DISCRETE TIME

In practice, dynamic replication cannot be implemented in continuous time. We do need some time to adjust the portfolio weights, and this implies that dynamic strategies need to be analyzed in discrete time. We prefer to start with bonds again, and then move to options. Suppose we want to replicate the two-period default-free discount bond $B(t_0, T_2)$, $T_2 = t_2$, using B_t, $B(t_0, T_3)$ with $T_2 < T_3$, similar to the special case discussed earlier. How do we go about doing this in practice?

8.5.2.1 The method

The replication period is $[t_0, T_2]$, and rebalancing is done in discrete intervals during this period. First, we select an interval of length Δ, and divide the period $[t_0, T_2]$ into n such finite intervals:

$$n\Delta = T_2 - t_0 \tag{8.23}$$

At each $t_i = t_{i-1} + \Delta$, we select new portfolio weights θ_{t_i} such that

1. At T_2, the dynamically created synthetic has exactly the same value as the T_2-maturity bond.
2. At each step, the adjustment of the replicating portfolio requires no net cash injections or withdrawals.

To implement such a replication strategy, we need to deviate from static replication methods and make some *new* assumptions. In particular, we just saw that correlations between the underlying assets play a crucial role in dynamic replication. Hence, we need a *model* for the way B_t, $B(t, T_2)$, and $B(t, T_3)$ move *jointly* over time.

This is a delicate process, and there are at least three approaches that can be used to model these dynamics: (i) binomial-tree or trinomial-tree methods; (ii) partial differential equation (PDE) methods, which are similar to trinomial-tree models but are more general; and (iii) direct modeling of the risk factors using stochastic differential equations and Monte Carlo simulation. In this section, we select the *simplest* binomial tree methods to illustrate important aspects of creating synthetic assets dynamically.

8.5.3 BINOMIAL TREES

We simplify the notation significantly. We let $j = 0, 1, 2, \ldots$ denote the "time period" for the binomial tree. We choose Δ so that $n = 3$. The tree will consist of three periods, $j = 0, 1$, and 2. At each *node* there are two possible states only. This implies that at $j = 1$ there will be two possible states, and at $j = 2$ there will be four altogether.[3]

In fact, by adjusting Δ and selecting the number of possible states at each node as two, three, or more, we obtain more and more complicated trees. With two possible states at every node, the tree is called binomial; with three possible states, the tree is called trinomial. The implied binomial

---

[3]In general, for nonrecombining trees at $j = n$, there are 2^n possible states.

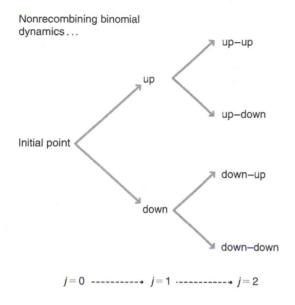

Nonrecombining binomial dynamics . . .

FIGURE 8.6

A binomial tree.

tree is in Figure 8.6. Here, possible states at every node are denoted, as usual, by *up* or *down*. These terms do not mean that a variable necessarily goes up or down. They are just shortcut names used to represent what traders may regard as "bullish" and "bearish" movements.

8.5.4 THE REPLICATION PROCESS

In this section, we let $\Delta = 1$, for notational convenience. Consider the two binomial trees shown in Figure 8.7 that give the joint dynamics of B_t and $B(t, T)$ over time. The top portion of the figure represents a binomial tree that describes an investment of \$1 at $j = 0$. This investment, called the *savings account*, is rolled over at the going spot interest rate. The bottom part of the figure describes the price of the "long bond" over time. The initial point $j = 0$ is equivalent to t_0, and $j = 3$ is equivalent to t_3 when the *long bond* $B(t_0, T_3)$ matures. The tree is *nonrecombining*, implying that a fall in interest rates following an increase would not give the same value as an increase that follows a drop. Thus, the *path* along which we get to a time node is important.[4] We now consider the *dynamics* implied by these binomial trees.

8.5.4.1 The B_t, $B(t, T_3)$ dynamics

First consider a tree for B_t, the savings account or risk-free borrowing and lending. The practitioner starts at time t_0 with one dollar. The observed interest rate at $j = 0$ is 10%, and the dollar invested

---

[4]See Jarrow (2002) for an excellent introductory treatment of such trees and their applications to the arbitrage-free pricing of interest-sensitive securities.

initially yields 1.10 regardless of which state of the world is realized at time $j = 1$.[5] There are two possibilities at $j = 1$. The *up* state is an environment where interest rates have *fallen* and bond prices, in general, have increased. Figure 8.7 shows a new spot rate of 8% for the *up* state in period $j = 1$. For the *down* state, it displays a spot rate that has increased to 15%.

Thus, looking at the tree from the initial point t_0, we can see four possible paths for the spot rate until maturity time t_2 of the bond under consideration. Starting from the top, the spot interest rate paths are

$$\{10\%, 8\%, 6\%\} \tag{8.24}$$

$$\{10\%, 8\%, 9\%\} \tag{8.25}$$

$$\{10\%, 15\%, 12\%\} \tag{8.26}$$

$$\{10\%, 15\%, 18\%\} \tag{8.27}$$

These imply four possible paths for the value of the savings account B_t:

$$\{1, 1.10, 1.188, 1.26\} \tag{8.28}$$

$$\{1, 1.10, 1.188, 1.29\} \tag{8.29}$$

$$\{1, 1.10, 1.26, 1.42\} \tag{8.30}$$

$$\{1, 1.10, 1.26, 1.49\} \tag{8.31}$$

It is clear that as Δ becomes smaller, and n gets larger, the number of possible paths will increase.

The tree for the *"long"* bond is shown in the bottom part of Figure 8.7. Here the value of the bond is $100 at $j = 3$, since the bond matures at that point. Because there is no default risk, the maturity value of the bond is the same in any state of the world. This means that one period *before* maturity the bond will mimic a one-period risk-free investment. In fact, no matter which one of the next two states occurs, in going from a node at time $j = 2$ to a relevant node at time $j = 3$, we always invest a constant amount and receive 100. For example, at point A, we pay

$$B(2, 3)^{\text{down}} = 91.7 \tag{8.32}$$

for the bond and receive 100, regardless of the spot rate move. This will change, however, as we move toward the origin. For example, at point B, we have either a "good" return:

$$R^{\text{up}} = \frac{94.3}{85.0} \tag{8.33}$$

or a "bad" return:

$$R^{\text{down}} = \frac{91.7}{85.0} \tag{8.34}$$

Hence, Figure 8.7 shows the dynamics of two different default-free fixed-income instruments: the savings account B_t, which can also be interpreted as a shorter maturity bond, and a three-period long bond $B(t, T_3)$. The question is how to combine these two instruments so as to form a synthetic medium-term bond $B(t, T_2)$.

---

[5]Hence the term, "risk-free investment."

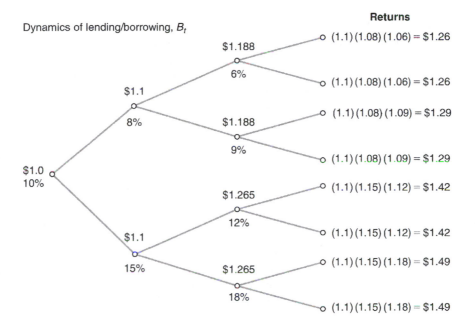

Dynamics of lending/borrowing, B_t

Returns

- $(1.1)(1.08)(1.06) = \$1.26$
- $(1.1)(1.08)(1.06) = \$1.26$
- $(1.1)(1.08)(1.09) = \$1.29$
- $(1.1)(1.08)(1.09) = \$1.29$
- $(1.1)(1.15)(1.12) = \$1.42$
- $(1.1)(1.15)(1.12) = \$1.42$
- $(1.1)(1.15)(1.18) = \$1.49$
- $(1.1)(1.15)(1.18) = \$1.49$

Time 0 --------→ 1 ------------→ 2 ------------→ 3

Dynamics of the 3-period bond price

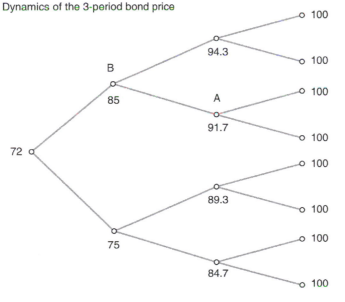

At $j = 3$, bond value is known, and is constant at 100

FIGURE 8.7

Binomial tree with lending/borrowing and three-period bond dynamics.

8.5.4.2 Mechanics of replication

We will now discuss the mechanics of replication. Consider Figure 8.8, which represents a binomial tree for the price of a two-period bond, $B(t, T_2)$. This tree is assumed to describe exactly the same states of the world as the ones shown in Figure 8.7. The periods beyond $j = 2$ are not displayed, given that $B(t, T_2)$ matures then. According to this tree, we know the value of the two-period bond only at $j = 2$. This value is 100, since the bond matures. Earlier values of the bond are not known and hence are left blank. The most important unknown is, of course, the time $j = 0$ value $B(t_0, T_2)$. This is the "current" price of the two-period bond. The problem we deal with in this section is how to "fill in" this tree.

The idea is to use the information given in Figure 8.7 to form a portfolio with (time-varying) weights θ_t^{lend} and θ_t^{bond} for B_t and $B(t, T_3)$. The portfolio should mimic the value of the medium-term bond $B(t, T_2)$ at *all* nodes at $j = 0, 1, 2$. The first condition on this portfolio is that, at T_2, its value must equal 100.

The second important condition to be satisfied by the portfolio weights is that the $j = 0, 1$ adjustments do not require any cash injections or withdrawals. This means that, as the portfolio weights are adjusted or *rebalanced*, any cash needed for increasing the weight of *one* asset should come from adjustment of the *other* asset. This way, cash flows will consist of a payment at time t_0, and a receipt of \$100 at time T_2, with no interim net payments or receipts in between—which is exactly the cash flows of a two-period discount bond.

Then, by arbitrage arguments the value of this portfolio should track the value of the $B(t, T_2)$ at all relevant times. This means that the θ_t^{lend} and θ_t^{bond} will also satisfy

$$\theta_t^{\text{lend}} B_t + \theta_t^{\text{bond}} B(t, T_3) = B(t, T_2) \tag{8.35}$$

for all t, or j.

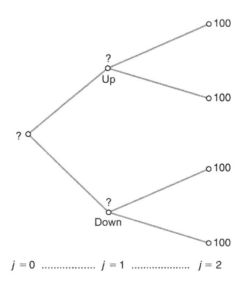

FIGURE 8.8

Binomial tree for two-period bond.

8.5.4.3 Guaranteeing self-financing

How can we guarantee that the adjustments of the weights θ_j^{lend} and θ_j^{bond}, observed along the tree paths $j = 0, 1, 2$ will not lead to any cash injections or withdrawals? The following additional conditions at $j = 0, 1$, will be sufficient to do this:

$$\theta_j^{lend} B_{j+1}^{up} + \theta_j^{bond} B(j+1,3)^{up} = \theta_{j+1}^{lend} B_{j+1}^{up} + \theta_{j+1}^{bond} B(j+1,3)^{up} \tag{8.36}$$

$$\theta_j^{lend} B_{j+1}^{down} + \theta_j^{bond} B(j+1,3)^{down} = \theta_{j+1}^{lend} B_{j+1}^{down} + \theta_{j+1}^{bond} B(j+1,3)^{down} \tag{8.37}$$

Let us see what these conditions mean. On the left-hand side, the portfolio weights have the subscript j, while the asset prices are measured as of time $j + 1$. This means that the left-hand side is the value of a portfolio *chosen* at time j, and valued at a new *up* or *down* state at time $j + 1$. The left-hand side is, thus, a function of "new" asset prices, but "old" portfolio weights.

On the right-hand side of these equations, we have "new" portfolio weights, θ_{j+1}^{lend} and θ_{j+1}^{bond} multiplied by the time $j + 1$ prices. Thus, the right-hand side represents the cost of a new portfolio chosen at time $j + 1$, either in the *up* or *down* state. Putting these two together, the equations imply that, regardless of which state occurs, the previously chosen portfolio generates just enough cash to put together a new replicating portfolio.

If θ_{j+1}^{lend} and θ_{j+1}^{bond} are chosen so as to satisfy Eqs. (8.36) and (8.37), there will be no need to inject or withdraw any cash during portfolio rebalancing. The replicating portfolio will be *self-financing*. This is what we mean by dynamic replication. By following these steps, we can form a portfolio at time $j = 0$ and rebalance at *zero cost* until the final cash flow of $\$100$ is reached at time $j = 2$. Given that there is no credit risk, and all the final cash flows are equal, the initial cost of the replicating portfolio must equal the value of the two-period bond at $j = 0$:

$$\theta_0^{lend} B_0 + \theta_0^{bond} B(0,3) = B(0,2) \tag{8.38}$$

Hence, dynamic replication would create a true synthetic for the two-period bond.

Finally, consider rewriting Eq. (8.37) after a slight manipulation:

$$(\theta_j^{lend} - \theta_{j+1}^{lend}) B_{j+1}^{down} = -(\theta_j^{bond} - \theta_{j+1}^{bond}) B(j+1,3)^{down} \tag{8.39}$$

This shows that the cash obtained from adjusting one weight will be just sufficient for the cash needed for the adjustment of the second weight. Hence, there will be no need for extra cash injections or withdrawals. Note that this "works" even though B_{j+1}^i and $B(j+1, 3)^i$ are random. The trees in Figure 8.7 implicitly assume that these random variables are *perfectly correlated* with each other.

8.5.5 TWO EXAMPLES

We apply these ideas to two examples. In the first, we determine the current value of the two-period default-free pure discount bond using the dynamically adjusted replicating portfolio from Figure 8.7. The second example deals with replication of options.

8.5.5.1 Replicating the bond

The top part of Figure 8.7 shows the behavior of savings account B_t. The bottom part displays a tree for the two-period discount bond $B(t, T_3)$. Both of these trees are considered as given exogenously, and their arbitrage-free characteristic is not questioned at this point. The objective is to fill in the future and current values in Figure 8.8 and price the two-period bond $B(t, T_2)$ under these circumstances.

EXAMPLE

To determine $\{B(j, 2), j = 0, 1, 2\}$, we need to begin with period $j = 2$ in Figure 8.8. This is the maturity date for the two-period bond, and there is no default possibility by assumption. Thus, the possible values of the two-period bond at $j = 2$, denoted by $B(2, 2)^i$, can immediately be determined:

$$B(2,2)^{\text{up}-\text{up}} = B(2,2)^{\text{down}-\text{up}} = B(2,2)^{\text{up}-\text{down}} = B(2,2)^{\text{down}-\text{down}} = 100 \tag{8.40}$$

Once these are placed at the $j = 2$ nodes in Figure 8.8, we take one step back and obtain the values of $\{B(1, 2)^i, i = \text{up}, \text{down}\}$. Here, the principles that we developed earlier will be used. As "time" goes from $j = 1$ to $j = 2$, the value of the portfolio put together at $j = 1$ using B_1 and $B(1, 3)^i$ should match the possible values of $B(2, 2)$ at all nodes. Consider first the top node, $B(1, 2)^{\text{up}}$. The following equations need to be satisfied:

$$\theta_1^{\text{lend},\text{up}} B_2^{\text{up}-\text{up}} + \theta_1^{\text{bond},\text{up}} B(2,3)^{\text{up}-\text{up}} = B(2,2)^{\text{up}-\text{up}} \tag{8.41}$$

$$\theta_1^{\text{lend},\text{up}} B_2^{\text{up}-\text{down}} + \theta_1^{\text{bond},\text{up}} B(2,3)^{\text{up}-\text{down}} = sB(2,2)^{\text{up}-\text{down}} \tag{8.42}$$

Here, θs have $j = 1$ subscript, hence the left-hand side is the value of the replicating portfolio put together at time $j = 1$, but valued as of $j = 2$. In these equations, all variables are known except portfolio weights $\theta_1^{\text{lend},\text{up}}$ and $\theta_1^{\text{bond},\text{up}}$. Replacing from Figure 8.7

$$\theta_1^{\text{lend},\text{up}} 1.188 + \theta_1^{\text{bond},\text{up}} 94.3 = 100 \tag{8.43}$$

$$\theta_1^{\text{lend},\text{up}} 1.188 + \theta_1^{\text{bond},\text{up}} 91.7 = 100 \tag{8.44}$$

Solving these two equations for the two unknowns, we get the replicating portfolio weights for $j = 1$, $i = \text{up}$. These are in units of securities, not in dollars.

$$\theta_1^{\text{lend},\text{up}} = 84.18 \tag{8.45}$$

$$\theta_1^{\text{bond},\text{up}} = 0 \tag{8.46}$$

Thus, if the market moves to $i = \text{up}$, 84.18 units of B_1 will be sufficient to replicate the future values of the bond at time $j = 2$. In fact, this position will have the $j = 2$ value of

$$84.18(1.188) = 100 \tag{8.47}$$

Note that the weight for the long bond is zero.[6] The cost of the portfolio at time $j = 1$ can be obtained using the just calculated $\{\theta_1^{\text{lend},\text{up}}, \theta_1^{\text{bond},\text{up}}\}$; this cost should equal $B(1, 2)^{\text{up}}$:

---

[6]This is to be expected. Because the bond is similar to a risk-free investment right before the maturity, the replicating portfolio puts a nonzero weight on the B_t only. This is the case, since the three-period bond will be a risky investment. Also, if $\theta_1^{\text{bond, up}}$ was nonzero, the two equations would be inconsistent.

$$\theta_1^{\text{lend,up}}(1.1) + \theta_1^{\text{bond,up}}(85.0) = 92.6 \tag{8.48}$$

Similarly, for the state $j = 1$, $i = \text{down}$, we have the two equations:

$$\theta_1^{\text{lend,down}} 1.265 + \theta_1^{\text{bond,down}} 89.3 = 100 \tag{8.49}$$

$$\theta_1^{\text{lend,down}} 1.265 + \theta_1^{\text{bond,down}} 84.7 = 100 \tag{8.50}$$

Solving, we get the relevant portfolio weights:

$$\theta_1^{\text{lend,down}} = 79.05 \tag{8.51}$$

$$\theta_2^{\text{bond,down}} = 0 \tag{8.52}$$

We obtain the cost of the portfolio for this state:

$$\theta_1^{\text{lend,down}}(1.1) + \theta_1^{\text{bond,down}}(75) = 86.9 \tag{8.53}$$

This should equal the value of $B(1,2)^{\text{down}}$. Finally, we move to the initial period to determine the value $B(0,2)$. The idea is again the same. At time $j = 0$ choose the portfolio weights θ_0^{lend} and θ_0^{bond} such that, as time passes, the value of the portfolio equals the possible future values of $B(1,2)$:

$$\theta_0^{\text{lend}} 1.1 + \theta_0^{\text{bond}} 85.00 = 92.6 \tag{8.54}$$

$$\theta_0^{\text{lend}} 1.1 + \theta_0^{\text{bond}} 75.00 = 86.9 \tag{8.55}$$

Here, the left-hand side is the value of the portfolio put together at time $j = 0$ such that its value equals those of the two-period bond at $j = 1$. Solving for the unknowns,

$$\theta_0^{\text{lend}} = 40.1 \tag{8.56}$$

$$\theta_0^{\text{bond}} = 0.57 \tag{8.57}$$

Thus, at time $j = 0$ we need to make a deposit of 40.1 dollars and buy 0.57 units of the three-period bond with price $B(0,3)$. This will replicate the two possible values $\{B(1,2)^i$, $i = \text{up, down}\}$. The cost of this portfolio must equal the current fair value of $B(0,2)$, if the trees for B_t and $B(j,3)$ are arbitrage-free. This cost is given by

$$B(0,2) = 40.1 + 0.57(72) = 81.14 \tag{8.58}$$

This is the fair value of the two-period bond at $j = 0$.

The arbitrage-free market value of the two-period bond is obtained *by calculating all the current and future weights* for a dynamic self-financing portfolio that duplicates the final cash flows of a two-period bond. At every step, the portfolio weights are adjusted so that the rebalanced portfolio keeps matching the values of $B(j, 2)$, $j = 0, 1, 2$. The fact that there were only two possible moves from every node gave a system of two equations, in two unknowns.

Note the (important) analogy to static replication strategies. By following this dynamic strategy and adjusting the portfolio weights, we guarantee to match the final cash flows generated by the two-period bond, while never really making any cash injections or withdrawals. Each time a future node is reached, the previously determined portfolio will always yield just enough cash to do necessary adjustments.[7]

8.5.6 APPLICATION TO OPTIONS

We can apply the replication technique to options and create appropriate synthetics. Thus, consider the same risk-free lending and borrowing B_t dynamics shown in Figure 8.7. This time, we would like to replicate a call option C_t written on a stock S_t. The call has the following *plain vanilla* properties. It expires at time t_2 and has a strike price $K = 100$. The option is European and cannot be exercised before the expiration date. The underlying stock S_t does not pay any dividends. Finally, there are no transaction costs such as commissions and fees in trading S_t or C_t.

Suppose the stock price S_t follows the tree shown in Figure 8.9. Note that unlike a bond, the stock never "matures" and future values of S_t are always random. There is no terminal time period where we know the future value of S_t, as was the case for the bond that expired at time T_3.

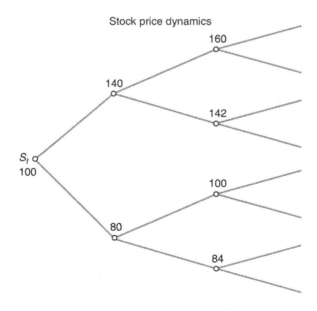

FIGURE 8.9

Binomial tree for stock price.

---

[7]Readers should also remember the assumption that the asset to be replicated makes no interim payouts.

However, the corresponding binomial tree for the call option still has *known* values at expiration date $j = 2$. This is the case since, at expiration, we know the possible values that the option may assume due to the formula:

$$C_2 = \max[S_2 - 100, 0] \tag{8.59}$$

Given the values of S_2, we can determine the possible values of C_2. But, the values of the call at *earlier* time periods still need to be determined.

How can this be done? The logic is essentially the same as the one utilized in the case of a two-period default-free bond. We need to determine the current value of the call option, denoted by C_0, using a dynamically adjusted portfolio that consists of the savings account and of the stock S_t.

EXAMPLE

Start with the expiration period and use the boundary condition:

$$C_2^i = \max[S_2^i - 100, 0] \tag{8.60}$$

where the i superscript represents gain in the states of the world {up–up, up–down, down–up, down–down}. Using these, we determine the four possible values of C_2^i at expiration:

$$C_2^{up-up} = 60,\ C_2^{up-down} = 42,\ C_2^{down-up} = 0,\ C_2^{down-down} = 0 \tag{8.61}$$

Next, we take one step back and consider the value C_1^{up}. We need to replicate this with a portfolio using B_1, S_1, such that as "time" passes, the value of this portfolio stays identical to the value of the option C_2^i. Thus, we need

$$\theta_1^{lend,up} B_2^{up-up} + \theta_1^{stock,up} S_2^{up-up} = C_2^{up-up} \tag{8.62}$$

$$\theta_1^{lend,up} B_2^{up-down} + \theta_1^{stock,up} S_2^{up-down} = C_2^{up-down} \tag{8.63}$$

Replacing the known values from Figures 8.7 and 8.9, we have two equations and two unknowns:

$$\theta_1^{lend,up}(1.188) + \theta_1^{stock,up}(160) = 60 \tag{8.64}$$

$$\theta_1^{lend,up}(1.188) + \theta_1^{stock,up}(142) = 42 \tag{8.65}$$

Solving for the portfolio weights $\theta_1^{lend,up}$ and $\theta_1^{stock,up}$, we get

$$\theta_1^{lend,up} = -84.18 \tag{8.66}$$

$$\theta_1^{stock,up} = 1 \tag{8.67}$$

Thus, at time $j = 1$, $i = $ up, we need to sell 84.18 units of B_t and buy one stock. The behavior of this portfolio in the immediate future will be equal to the future values of $\{C_2^i\}$ where i denotes the four possible states at $j = 2$. The cost of this portfolio is C_1^{up}:

$$C_1^{\text{up}} = -84.18(1.1) + 140 \tag{8.68}$$

$$= 47.40 \tag{8.69}$$

Similarly, in order to determine C_1^{down}, we first form a replicating portfolio by solving the equations

$$\theta_1^{\text{lend,down}}(1.26) + \theta_1^{\text{stock,down}}(100) = 0 \tag{8.70}$$

$$\theta_1^{\text{lend,down}}(1.26) + \theta_1^{\text{stock,down}}(84) = 0 \tag{8.71}$$

which gives

$$\theta_1^{\text{lend,down}} = 0 \tag{8.72}$$

$$\theta_1^{\text{stock,down}} = 0 \tag{8.73}$$

The cost of this portfolio is zero and hence the option is worthless if we are at $j = 1$, $i = \text{down}$:

$$C_1^{\text{down}} = 0 \tag{8.74}$$

Finally, the fair value C_0 of the option can be determined by finding the initial portfolio weights from

$$\theta_0^{\text{lend}}(1.1) + \theta_0^{\text{stock}}(140) = 47.40 \tag{8.75}$$

$$\theta_0^{\text{lend}}(1.1) + \theta_0^{\text{stock}}(80) = 0 \tag{8.76}$$

We obtain

$$\theta_0^{\text{lend}} = -57.5 \tag{8.77}$$

$$\theta_0^{\text{stock}} = 0.79 \tag{8.78}$$

Thus, we need to borrow 57.5 dollars and then buy 0.79 units of stock at $j = 0$. The cost of this will be the current value of the option:

$$C_0 = -57.5 + 0.79(100) \tag{8.79}$$

$$= 21.3 \tag{8.80}$$

This will be the fair value of the option if the exogenously given trees are arbitrage-free.

Note again the important characteristics of this dynamic strategy.

1. To determine the current value of the option, we started from the expiration date and used the boundary condition.
2. We kept adjusting the portfolio weights so that the replicating portfolio eventually matched the final cash flows generated by the option.
3. Finally, there were no cash injections or cash withdrawals, so that the initial amount invested in the strategy could be taken as the cost of the synthetic.

8.6 SOME IMPORTANT CONDITIONS

In order for these methods to work, some important assumptions are needed. These are discussed in detail here.

8.6.1 ARBITRAGE-FREE INITIAL CONDITIONS

The methods discussed in this chapter will work only if we start from dynamics that originally exclude any arbitrage opportunities. Otherwise, the procedures shown will give "wrong" results. For example, some bond prices $B(j, T_2)^i$, $j = 0, 1$ or the option price may turn out to be *negative*.

There are many ways the arbitrage-free nature of the original dynamics can be discussed. One obvious condition concerns the returns associated with the savings account and the other constituent asset. It is clear that, at all nodes of the binomial trees in Figure 8.7, the following condition needs to be satisfied:

$$R_j^{\text{down}} < L_j < R_j^{\text{up}} \tag{8.81}$$

where L_j is the one-period spot rate that is observed at that node and the R_j^{down} and R_j^{up} are two possible returns associated with the bond at the same node.

According to this condition, the risk-free rate should be between the two possible returns that one can obtain from holding the "risky" asset, $B(t, T)$. For the case of bonds, before expiration we must also have, due to arbitrage,

$$R_j^{\text{down}} = L_j = R_j^{\text{up}} \tag{8.82}$$

Otherwise, we could buy or sell the bond, and use the proceeds in the risk-free investment to make unlimited gains.

Yet, the arbitrage-free characteristic of binomial trees normally requires more than this simple condition. As Chapter 11 will show, the underlying dynamics should be conformable with proper *Martingale* dynamics in order to make the trees arbitrage-free.

8.6.2 ROLE OF BINOMIAL STRUCTURE

There is also a very strong assumption behind the binomial tree structure that was used during the discussion. This assumption does not change the logic of the dynamic replication strategy, but can make it numerically more complicated if it is not satisfied.

Consider Figure 8.7. In these trees, it was assumed that when the short rate dropped, the long rate always dropped along with it. Conversely, when the short rate increased, the long rate increased with it. That is to say, the long bond return and the short rate were *perfectly* correlated. It is thanks to this assumption that we were able to associate a future value of B_t with another future value of $B(t, T_3)$. These "associations" were never random. A similar assumption was made concerning the binomial trees for S_t and C_t. The movements of these two assets were perfectly correlated.

This is a rather strong assumption and is due to the fact that we are using the so-called *one-factor* model. It is assumed that there is a single random variable that determines the future value of the assets under consideration at every node. In reality, given a possible movement in the short rate L_t, we may not know whether a bond price $B(t, T)$ will go up or down in the immediate future, since *other* random factors may be at play. Under such conditions, it would be impossible to obtain the same equations, since the *up* or *down* values of the two assets would not be associated with certainty.

Yet, introducing further random factors will only increase the numerical complexity of the tree models. We can, for example, move from binomial to trinomial or more complicated trees. The general logic of the dynamic replication does not change. However, we may need further base assets to form a proper synthetic.

8.7 REAL-LIFE COMPLICATIONS

Real-life complications make dynamic replication a much more fragile exercise than static replication. The problems that are encountered in static replication are well known. There are operational problems, counterparty risk, and the theoretically exact synthetics may not be identical to the original asset. There are also liquidity problems and other transaction costs. But, all these are relatively minor and in the end, static replicating portfolios used in practice generally provide good synthetics.

With dynamic replication, these problems are magnified because the underlying positions need to be readjusted many times. For example, the effect of transaction costs is much more serious if dynamic adjustments are required frequently. Similarly, the implications of liquidity problems will also be more serious. But more important, the real-life use of dynamic replication methods brings forth *new* problems that would not exist with static synthetics. We study these briefly.

8.7.1 BID—ASK SPREADS AND LIQUIDITY

Consider the simple case of bid—ask spreads. In static replication, the portfolio that constitutes the synthetic is put together at time t and is never altered until expiration T. In such an environment, the existence of bid—ask spreads may be non-negligible, but this is hardly a major aspect of the problem. After all, any bid—ask spread will end up increasing (or lowering) the cost of the associated synthetic, and in the unlikely case that these are prohibitive, then the synthetic will not be put together.

Yet, with dynamic replication, the practitioner is constantly adjusting the replicating portfolio. Such a process is much more vulnerable to widening bid−ask spreads or the underlying liquidity changes. At the time dynamic replication is initiated, the future movements of bid−ask spreads or of liquidity will not be known exactly and cannot be factored into the initial cost of the synthetic. Such movements will constitute additional risks and increase the costs even when the synthetic is held until maturity.

8.7.2 MODELS AND JUMPS

Dynamic replication is never perfect in real life. It is done using *models* in discrete time. But models imply assumptions and discrete time means approximation. This leads to a model risk. Many factors and the possibility of having jumps in the underlying risks may have serious consequences if not taken into account properly during the dynamic replication process.

8.7.3 MAINTENANCE AND OPERATIONAL COSTS

It is easy to obtain a dynamic replication strategy theoretically. But in practice, this strategy needs to be implemented using appropriate position-keeping and risk-management tools. The necessary software and human skills required for these tasks may lead to significant new costs.

8.7.4 CHANGES IN VOLATILITY

Often, dynamic replication is needed because the underlying instruments are nonlinear. It turns out that, in dealing with nonlinear instruments, we will have additional exposures to new and less transparent risks such as movements in the *volatility* of the associated risk factors. Because risk-managing volatility exposures are much more delicate (and difficult) than the management of interest rate or exchange rate risks, dynamic replication often requires additional skills.

In the exercises at the end of this chapter, we briefly come back to this point and provide a reading (and some questions) concerning the role of volatility changes during the dynamic hedging process.

8.8 CONCLUSIONS

We finish the chapter with an important observation. Static replication was best done using cash flow diagrams and resulted in contractual equations with *constant* weights.

Creating synthetics dynamically requires constant adjustments and careful selection of portfolio weights θ_t^i, in order to make the synthetic *self-financing*. Thus, we again use contractual equations, but this time, the weights placed on each contract change as time passes. This requires the use of algebraic equations and is done with computers.

Finally, the *dynamic* synthetic is nothing but the sequence of weights $\{\theta_1^i, \theta_2^i, \ldots, \theta_n^i\}$ that the financial engineer will determine at time t_0.

SUGGESTED READING

Several books deal with dynamic replication. Often these are intermediate-level textbooks on derivatives and financial markets. We have two preferred sources that the reader can consult for further examples. The first is **Jarrow** (2002). This book deals with fixed-income examples only. The second is **Jarrow and Turnbull** (1999), where dynamic replication methods are discussed in much more detail with a broad range of applications. The reader can also consult the original **Cox and Ross** (1976a) article. It remains a very good summary of the procedure.

EXERCISES

1. Consider the immunization example given in Section 8.4. Assume that the zero-coupon yield curve is flat at $y = 5\%$ and not 8% as in the example. The shifts in the yield curve are parallel. What positions and how many units of the 6-month bond and the 3-year bond are needed to create an approximate synthetic that will fund the 2-year bond?
2. Suppose you are given the following data:
   - The risk-free interest rate is 6%.
   - The stock price follows:

$$dS_t = \mu S_t \, dt + \sigma S_t \, dW_t \tag{8.83}$$

   - Volatility is 12% a year.
   - The stock pays no dividends and the current stock price is 100.
     Using these data, you are asked to approximate the current value of a European call option on the stock. The option has a strike price of 100 and a maturity of 200 days.
   a. Determine an appropriate time interval Δ, such that an implied binomial tree has five steps.
   b. What is the implied *up* probability? *Hint: To obtain the probability we need to discretize the stochastic differential equation.*
   c. Determine the tree for the stock price S_t.
   d. Determine the tree for the call premium C_t.
3. Suppose the stock discussed in Exercise 1 pays dividends. Assume all parameters are the same. Consider three forms of dividends paid by the firm:
   a. The stock pays a continuous, known stream of dividends at a rate of 4% per time.
   b. The stock pays 5% of the value of the stock at the third node. No other dividends are paid.
   c. The stock pays a $5 dividend at the third node.
     In each case, determine the tree for the ex-dividend stock price. For the first two cases, determine the premium of the call. In what way(s) does the third type of dividend payment complicate the binomial tree?
4. You are going to use binomial trees to value American-style options on the British pound. Assume that the British pound is currently worth $1.40. Volatility is 10%. The current British risk-free rate is 5%, and the US risk-free rate is 2%. The put option has a strike price of $1.50. It expires in 200 days. American-style options can be exercised before expiration.

a. The first issue to be settled is the role of US and British interest rates. This option is being purchased in the United States, so the relevant risk-free rate is 2%. But British pounds can be used to earn British risk-free rates. So this variable can be treated as a continuous rate of dividends.

Taking this into account, determine a Δ such that the binomial tree has five periods.

b. Determine the relevant probabilities. Use a similar method to the one used to value the European stock call option above.

c. Determine the tree for the exchange rate.

d. Determine the tree for a European put with the same characteristics.

e. Determine the price of an American style put with these properties.

5. Consider the reading that follows which deals with the effects of straightforward *delta hedging*. Read the events described and then answer the questions that follow.

Dynamic Hedging

US equity option market participants were of one voice last week in refuting the notion that [dynamic hedging due to] equity options trading had exacerbated the stock market correction of late October, which saw the Dow Jones Industrial Average fall 554.26 points, or some 7%.

Dynamic hedging is a strategy in which investors buy and sell stocks to create a payout, which is the same as going long and short options. Thus, if the market takes a big drop, a writer of puts sells stock to cut their losses. Dynamic hedgers buy and sell stock to achieve the position they desire to equalize their exposure to volatility.

The purpose of dynamic hedging, also known as delta hedging, is to remain market-neutral. The hedger's objective is to have no directional exposure to the market. For example, the hedgers will buy puts, giving them the right to sell stock. They are thus essentially short the market. To offset this short position, the hedger will purchase the underlying stock. The investor is now long the put and long the stock, and thereby market-neutral.

If the market falls, the investor's put goes in-the-money, increasing the short exposure to the market. To offset this, the investor will sell the underlying...

"It is my humble opinion that few investors use dynamic hedging. If somebody is selling options, i.e. selling volatility, they will have an offsetting position where they are long volatility. People don't take big one-sided bets," said a senior official at another U.S. derivatives exchange. (Thomson Reuters IFR, issue 832)

a. Suppose there are a lot of put writers. How would these traders hedge their position? Show using appropriate payoff diagrams.

b. What would these traders do when markets start falling? Show on payoff diagrams.

c. Now suppose an option's trader is short volatility as the last paragraph implies. Describe how this trader can be long volatility "somewhere else."

d. Is it possible that the overall market is a bit short volatility, yet that this amount is still very substantial for the underlying (cash) markets?

There are many special terms in this reading, but at this point we would like to emphasize one important aspect of dynamic hedging that was left unmentioned in the chapter. As mentioned in the reading, in order to dynamically hedge a nonlinear asset, we need a *delta.*

Delta is the sensitiveness of the option to underlying price changes. Now if this asset is indeed nonlinear, then the delta will depend on the volatility of underlying risks. If this volatility is itself dependent on many factors, such as the strike price, then there will be a *volatility smile* and delta hedging may be inaccurate.

To this effect, suppose you have a long options' position on FTSE-100. How would you delta hedge this position? More important, how would this delta hedge be affected by the observations in the last paragraph of the reading?

6. How could you determine whether the trees in Figure 8.7 are arbitrage-free or not?

7. Consider the replication of a European call option. Write a VBA program to show the dynamic hedging strategy using only stocks and a saving account to replicate a short European option. Calculate the position in both the stock and the savings account for all the intermediate time points and all possible states. Use the binomial model with the following data:

- $S(0) = 100$; $K = 110$; $T = 3$; $\sigma = 30\%$; $M = 3$
- $L(0) = 1$; $\sigma = 5\%$
- Use the value of $u = e^{(\sigma\sqrt{\Delta t}+(r-\sigma^2/2)\Delta t)}$ and $d = e^{(-\sigma\sqrt{\Delta t}+(r-\sigma^2/2)\Delta t)}$.

MECHANICS OF OPTIONS

CHAPTER OUTLINE

9.1 INTRODUCTION

This chapter is an introduction to methods used in dealing with *optionality* in financial instruments. Compared to most existing textbooks, the present text adopts a different way of looking at options. We discuss options from the point of view of an options market maker. In our setting, options are *not* presented as instruments to bet on or hedge against the *direction* of an underlying risk. Instead, options are motivated as instruments of *volatility*.

In the traditional textbook approach, options are introduced as *directional* instruments. This is not how market professionals think of options. In most textbooks, a call option becomes in-the-money and hence profitable if the underlying price increases, indirectly associating it with a bullish view. The treatment of put options is similar. Puts are seen as appropriate for an investor who thinks the price of the underlying asset is going to decrease. For an end investor or retail client, such *directional motivation* for options may be natural. But, looking at options this way is misleading if we are concerned with the interbank or interdealer market. In fact, motivating options as directional tools will disguise the fundamental aspect of these instruments, namely that options are tools for trading *volatility*. The intuition behind these two views of options is quite different, and we would like the reader to think like an option trader or market maker.

This chapter intends to show that an option exposure, when fully put in place, is an impure position on the way volatility is expected to change. A market maker with a net *long* position in options is someone who is "expecting" the volatility to *increase*. A market maker who is *short* the option is someone who thinks that the volatility of the underlying is going to decrease. Sometimes such positions are taken as funding vehicles.

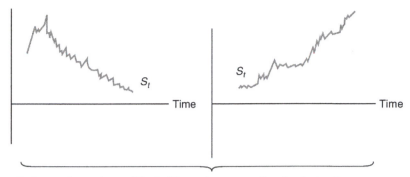

Option market makers will be indifferent between selling (buying) calls or puts

An option market maker anticipating the following future movements will . . .

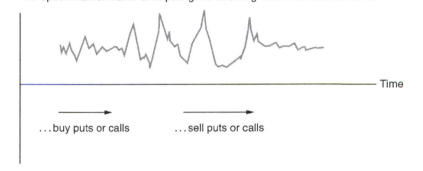

. . . buy puts or calls   . . . sell puts or calls

FIGURE 9.1

Two possible stock price trajectories.

In this sense, a trader's way of looking at puts and calls is in complete contrast to the directional view of options. For example, market makers look at European calls and puts as if they were *identical* objects. As we will see in this chapter, from an option *market maker's* point of view, there is really no difference between buying a call or buying a put. Both of these transactions, in the end, result in the same payoff. Consider Figure 9.1, where we show two possible intraday trajectories of an underlying price, S_t. In one case prices are falling rapidly, while in the other, prices are rising. An option trader will sell puts or calls with the same ease. As we will see, the trader may be concerned with whether he should *sell* or *buy* any options, rather than *which* type of option to sell.

In this chapter and the next, we intend to clarify the connection between volatility and option prices. However, we first review some basics.

9.2 WHAT IS AN OPTION?

From a market practitioner's point of view, options are instruments of volatility. A *retail* investor who owns a call on an asset, S_t, may feel that a persistent upward movement in the price of this asset is "good" for him or her. But, a market maker who may be long in the same call may prefer that the underlying price S_t *oscillate* as much as possible, as *often* as possible. The more frequently and violently prices oscillate, the more *long* (short) positions in option books will gain (lose), regardless of whether calls or puts are owned.

The following reading is a good example as to how option traders look at options.

EXAMPLE

Wall Street firms are gearing up to recommend long single-stock vol positions on companies about to report earnings. While earnings seasons often offer opportunities for going long vol via buying Calls or Puts, this season should present plenty of opportunities to benefit from long vol positions given overall negative investor sentiment. Worse-than-expected earnings releases from one company can send shockwaves through the entire market.

The big potential profit from these trades is from gamma, in other words, large moves in the underlying, rather than changes in implied vol. One promising name... announced in mid-February that manufacturing process and control issues have led to reduced sales of certain products in the United States, which it expected to influence its first quarter and full-year sales and earnings. On Friday, options maturing in August had a mid-market implied vol of around 43%, which implies a 2.75% move in the stock per trading day. Over the last month, the stock has been moving on average 3% a day, which means that by buying options on the company, you're getting vol cheap.

(Derivatives Week, now part of GlobalCapital, April 1, 2001)

This reading illustrates several important characteristics of options. First, we clearly see that puts and calls are considered as similar instruments by market practitioners. The issue is not to buy puts or calls, but whether or not to buy them.

Second, and this is related to the first point, note that market participants are concerned with volatilities and not with the direction of prices—referring to volatility simply as vol. Market professionals are interested in the difference between *actual* daily volatilities of stock prices and the volatilities *implied* by the options. *Implied volatility* is calculated by taking the observed option price in the market and a pricing formula such as the Black–Scholes formula that will be introduced below and backing out the volatility that is consistent with the option price given other input parameters such as the strike price of the option, for example. The last sentence in the reading is a good (but potentially misleading) example of this. The reading suggests that options *imply* a daily volatility of 2.75%, while the *actual* daily volatility of the stock price is 3%. According to this, options are considered "cheap," since the actual underlying moves more than what the option price implies on a given day.[1] This distinction between implied volatility and "actual volatility" should be kept in mind. Actual volatility is also sometimes referred to as realized volatility.

Finally, the reading seems to refer to two different types of gains from volatility. One, from "large movements in the underlying price," leads to *gamma gains*, and the other, from implied volatility, leads to *vega gains*. During this particular episode, market professionals were expecting implied volatility to remain the same, while the underlying assets exhibited sizable fluctuations. It is difficult, at the outset, to understand this difference. The present chapter will clarify these notions

---

[1]This analysis should be interpreted carefully. In the option literature, there are many different measures of volatility. As this chapter will show, it is perfectly reasonable that the two values be different, and this may not necessarily imply an arbitrage possibility.

and reconcile the market professional's view of options with the directional approach the reader may have been exposed to earlier.[2]

9.3 OPTIONS: DEFINITION AND NOTATION

Option contracts are generally divided into the categories of *plain vanilla* and *exotic options*, although many of the options that used to be known as exotic are vanilla instruments today. In discussing options, it is good practice to start with a simple benchmark model, understand the basics of options, and then extend the approach to more complicated instruments. This simple benchmark will be a plain vanilla option treated within the framework of the *Black–Scholes* model.

The buyer of an option does not buy the underlying instrument; he or she buys a *right*. If this right can be exercised only at the expiration date, then the option is *European*. If it can be exercised anytime during the specified period, the option is said to be *American*. A *Bermudan* option is "in between," given that it can be exercised at more than one of the dates during the life of the option.

In the case of a European plain vanilla call, the option holder has purchased the right to "buy" the underlying instrument at a certain price, called the *strike* or exercise price, at a specific date, called the *expiration* date. In the case of the European plain vanilla put, the option holder has again purchased the right to an action. The action in this case is to "sell" the underlying instrument at the strike price and at the expiration date.

American-style options can be exercised anytime until expiration and hence may be more expensive. They may carry an early exercise premium. At the expiration date, options cease to exist. In this chapter, we discuss basic properties of options using mostly plain vanilla calls. Obviously, the treatment of puts would be similar.

9.3.1 NOTATION

We denote the strike prices by the symbol K, and the expiration date by T. The price or value of the underlying instrument will be denoted by S_t if it is a cash product, and by F_t if the underlying is a forward or futures price. The fair price of the call at time t will be denoted by $C(t)$, and the price of the put by $P(t)$.[3] These prices depend on the variables and parameters underlying the contract. We use S_t as the underlying, and write the corresponding call option *pricing function* as

$$C(t) = C(S_t, t | r, K, \sigma, T) \tag{9.1}$$

---

[2]The previous example also illustrates a technical point concerning volatility calculations in practice. Consider the way *daily* volatility was calculated once *annualized* percentage volatility was given. Suppose there are 246 trading days in a year. Then, note that an annual percentage volatility of 43% is not divided by 246. Instead, it is divided by the *square root* of 246 to obtain the "daily" 2.75% volatility. This is known as *the square root rule* and has to do with the role played by Wiener processes in modeling stock price dynamics. Wiener process increments have a *variance* that is proportional to the time that has elapsed. Hence, the standard deviation or volatility will be proportional to the square root of the elapsed time.

[3]The way we characterize and handle the time index is somewhat different from the treatment up to this chapter. Option prices are not written as C_t and P_t, as the notation of previous chapters may suggest. Instead, we use the notation $C(t)$ and $P(t)$. The former notation will be reserved for the partial derivative of an option's price with respect to time t.

Here, σ is the volatility of S_t and r is the spot interest rate, assumed to be constant. In more compact form, this formula can be expressed as

$$C(t) = C(S_t, t) \tag{9.2}$$

This function is assumed to have the following partial derivatives:

$$\frac{\partial C(S_t, t)}{\partial S_t} = C_s \tag{9.3}$$

$$\frac{\partial^2 C(S_t, t)}{\partial S_t^2} = C_{ss} \tag{9.4}$$

$$\frac{\partial C(S_t, t)}{\partial t} = C_t \tag{9.5}$$

More is known on the properties of these partials. Everything else being the same, if S_t increases, the call option price, $C(t)$, also increases. If S_t declines, the price declines. But the changes in $C(t)$ will never exceed those in the underlying asset, S_t. Hence, we should have

$$0 < C_s < 1 \tag{9.6}$$

At the same time, everything else being the same, as t increases, the life of the option gets shorter and the time value declines,

$$C_t < 0 \tag{9.7}$$

Finally, the expiration payoff of the call (put) option is a convex function, and we expect the $C(S_t, t)$ to be convex as well. This means that

$$0 < C_{ss} \tag{9.8}$$

This information about the partial derivatives is assumed to be known even when the exact form of $C(S_t, t)$ itself is not known.

The notation in Eq. (9.1) suggests that the partials themselves are functions of S_t, r, K, t, T, and σ. Hence, one may envisage some further higher-order partials. The traditional Black–Scholes *vanilla* option pricing environment uses the partials, $\{C_s, C_{ss}, C_t\}$ only. Further partial derivatives are brought into the picture as the Black–Scholes assumptions are relaxed gradually.

Figure 9.2 shows the expiration date payoffs of plain vanilla put and call options. In the same figure we have the time t, $t < T$ value of the calls and puts. These values trace a smooth convex curve obtained from the Black–Scholes formula.

We now consider a real-life application of these concepts. The following example looks at *Microsoft options* traded at the Chicago Board of Options Exchange and discusses various parameters within this context.

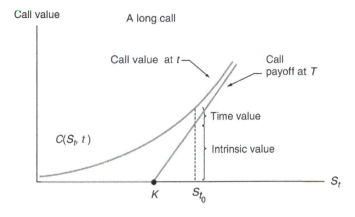

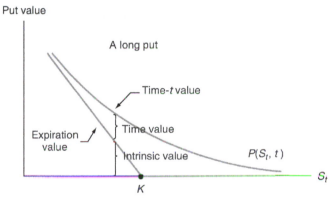

FIGURE 9.2

Expiration date payoffs of plain vanilla put and call options.

EXAMPLE

Suppose Microsoft (MSFT) is "currently" trading at 61.15 at Nasdaq. Further, the overnight rate is 2.7%. We have the following quotes from the Chicago Board of Options Exchange (CBOE).

In the table, the first column gives the expiration date and the strike level of the option. The exact time of expiration is the third Friday of every month. These equity options in CBOE are of American style. The bid price is the price at which the market maker is willing to buy this option from the client, whereas the ask price is the price at which he or she is willing to sell it to the client.

Calls	Bid	Ask	Volume
Nov 55.00	7.1	7.4	78
Nov 60.00	3.4	3.7	6291
Nov 65.00	1.2	1.3	1456
Nov 70.00	0.3	0.4	98
Dec 55.00	8.4	8.7	0
Dec 60.00	5	5.3	29
Dec 65.00	2.65	2.75	83
Dec 70.00	1.2	1.25	284

Puts	Bid	Ask	Volume
Nov 55.00	0.9	1.05	202
Nov 60.00	2.3	2.55	5984
Nov 65.00	5	5.3	64
Nov 70.00	9	9.3	20
Dec 55.00	2.05	2.35	10
Dec 60.00	3.8	4.1	76
Dec 65.00	6.3	6.6	10
Dec 70.00	9.8	10.1	25

Note: *October 24, 2002, 11:02 A.M. data from CBOE.*

CBOE option prices are multiplied by $100 and then invoiced. Of course, there are some additional costs to buying and selling options due to commissions and possibly other expenses. The last column of the table indicates the trading volume of the relevant contract.

For example, consider the November 55 put. This option will be in-the-money, if the Microsoft stock is below 55.00. If it stays so until the third Friday of November 2001, the option will have a positive payoff at expiration.

100 such puts will cost

$$1.05 \times 100 \times 100 = \$10,500 \tag{9.9}$$

plus commissions to buy, and can be sold at

$$0.90 \times 100 \times 100 = \$9000 \tag{9.10}$$

if sold at the bid price. Note that the bid–ask spread for one "lot" had a value of $1500 that day.

We now study option mechanics more closely and introduce further terminology.

9.3.2 ON RETAIL USE OF OPTIONS

Consider a *retail client* and an *option market maker* as the two sides of the transaction. Suppose a business uses the commodity S_t as a production input and would like to "cap" the price S_T at a future date T. For this insurance, the business takes a *long* position using call options on S_t. The call option premium is denoted by $C(t)$. By buying the call, the client makes sure that he or she can buy one unit of the underlying at a *maximum* price K, at expiration date T. If at time T, S_T is lower than K, the client will not exercise the option. There is no need to pay K dollars for something that is selling for less in the marketplace. The option will be exercised only if S_T equals or exceeds K at time T.

Looked at this way, options are somewhat similar to standard *insurance* against potential increases in commodities prices. In such a framework, options can be motivated as *directional* instruments. One has the impression that an increase in S_t is harmful for the client, and that the call "protects" against this risk. The situation for puts is symmetrical. Puts appear to provide protection against the risk of undesirable "declines" in S_t. In both cases, a certain direction in the change of the underlying price S_t is associated with the call or put, and these appear to be fundamentally different instruments.

Figure 9.3 illustrates these ideas graphically. The upper part shows the payoff diagram for a call option. Initially, at time t_0, the underlying price is at S_{t_0}. Note that $S_{t_0} < K$, and the option is *out-of-the-money*. Obviously, this does not mean that the right to buy the asset at time T for K dollars has no value. In fact, from a *client's point of view*, S_t may move *up* during interval $t \in [t_0, T]$ and end up exceeding K by time T. This will make the option *in-the-money*. It would then be profitable to exercise the option and buy the underlying at a price K. The option payoff will be the difference $S_T - K$, if S_T exceeds K. This payoff can be shown either on the horizontal axis or, more explicitly, on the vertical axis.[4] Thus, looked at from the *retail* client's point of view, even at the price level S_{t_0}, the out-of-the money option is valuable, since it *may* become in-the-money later. Often, the directional motivation of options is based on these kinds of arguments.

If the option expires at $S_T = K$, the option will be *at-the-money* (ATM) and the option holder may or may not choose to receive the underlying. However, as the costs associated with delivery of the call underlying are, in general, *less* than the transaction costs of buying the underlying in the open market, some holders of ATM options prefer to exercise.

Hence, we get the typical price diagram for a plain vanilla European call option. The option price for $t \in [t_0, T]$ is shown in Figure 9.3 as a smooth convex curve that converges to the piecewise linear option payoff as expiration time T approaches. The vertical distance between the payoff line and the horizontal axis is called *intrinsic value*. The vertical distance between the option price curve and the expiration payoff is called the *time value* of the option. Note that for a fixed t, the time value appears to be at a maximum when the option is ATM—that is to say, when $S_t = K$.

9.3.3 SOME INTRIGUING PROPERTIES OF THE DIAGRAM

Consider point A in the top part of Figure 9.3. Here, at time t, the option is *deep* out-of-the money. S_t is close to the origin and the time value is close to zero. The tangent at point A has a positive

---

[4]As usual, the upward-sloping line in Figure 9.3 has slope $+1$, and thus "reflects" the profit, $S_T - K$ on the horizontal axis, toward the vertical axis.

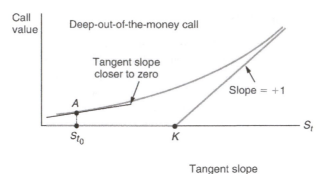

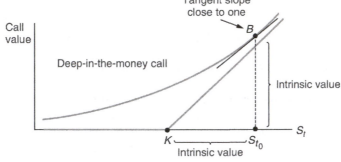

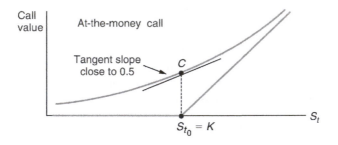

FIGURE 9.3

Value of in-the-money, out-of-the-money, and ATM call.

slope that is little different from zero. The curve is almost "linear" and the *second* derivative is also close to zero. This means that for small changes in S_t, the slope of the tangent will not vary much.

Now, consider the case represented by point B in Figure 9.3. Here, at time t, the option is deep in-the-money. S_t is significantly higher than the strike price. However, the time value is *again* close to zero. The curve approaches the payoff line and hence has a slope close to $+1$. Yet, the second derivative of the curve is once again very close to zero. This again means that for small changes in S_t, the slope of the tangent will not vary much.[5]

---

[5]That is, it will stay close to 1.

The third case is shown as point C in the lower part of Figure 9.3. Suppose the option was ATM at time t, as shown by point C. The value of the option is entirely made of time value. Also, the slope of the tangent is close to 0.5. Finally, it is interesting that the *curvature* of the option is highest at point C and that if S_t changes a little, the slope of the tangent will change *significantly*.

This brings us to an interesting point. The more convex the curve is at a point, the higher the associated time value seems to be. In the two extreme cases where the slope of the curve is diametrically different, namely at points A and B, the option has a *small* time value. At both points, the *second* derivative of the curve is small. When the curvature reaches its maximum, the time value is greatest. The question, of course, is whether or not this is a coincidence.

Pursuing this connection between *time value* and *curvature* further will lead us to valuing the underlying volatility. Suppose, by holding an option, a market maker can somehow generate "cash" earnings as S_t oscillates. Could it be that, everything else being the same, the greater the curvature of $C(t)$, the greater the cash earnings are? Our task in the next section is to show that this is indeed the case.

9.4 OPTIONS AS VOLATILITY INSTRUMENTS

In this section we see how convexity is translated into cash earnings, as S_t oscillates and creates time value.[6] The discussion is conducted in a highly *simplified* environment to facilitate understanding of the relationship between volatility and cash gains (losses) of long (short) option positions.

Consider a *market maker* who quotes two-way prices for a European vanilla call option $C(t)$, with strike K, and expiration T, written on a nondividend paying asset, denoted by S_t.[7] Let the risk-free interest rate r be constant. For simplicity, consider an ATM option, $K = S_t$. In the following, we first show the initial steps taken by the market maker who buys an option. Then, we show how the market maker hedges this position dynamically and earns some cash due to S_t oscillations.

9.4.1 INITIAL POSITION AND THE HEDGE

Suppose this market maker *buys* a call option from a client.[8] The initial position of the market maker is shown in the top portion of Figure 9.4. It is a standard *long call* position. The market maker is not an investor or speculator, and this option is bought with the purpose of keeping it on the books and then selling it to another client. Hence, some mechanical procedures should be followed. First, the market maker needs to *fund* this position. Second, he or she should *hedge* the associated risks.

---

[6]It is important to emphasize that this way of considering options is from an interbank point of view. For end investors, options can still be interpreted as directional investments, but the pricing and hedging of options can only be understood when looked at from the dealer's point of view. The next chapter will present applications related to classical uses of options.

[7]Remember that market makers have the obligation to buy and sell at the prices they are quoting.

[8]This means that the client has "hit" the bid price quoted by the market maker.

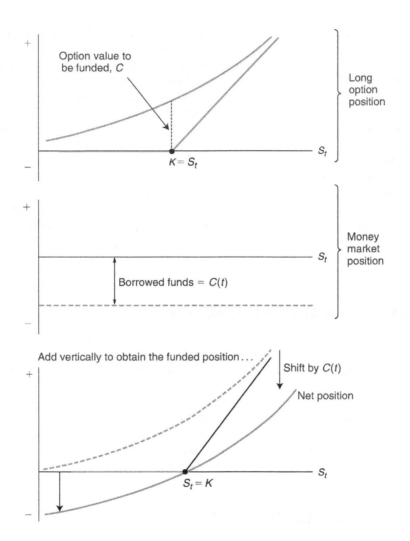

FIGURE 9.4

Funding a long call position with borrowing.

We start with the first requirement. Unlike the end investor, market makers never have "money" of their own. The trade needs to be *funded.* There are at least two ways of doing this. One is to *short* an appropriate asset in order to generate the needed funds, while the other is to borrow these funds directly from the money market desk.[9] Suppose the second possibility is selected and the market maker borrows $C(t)$ dollars from the money market desk at an interest rate $r_t = r$. The *net* position that puts together the option and the borrowed funds is shown in the bottom part of Figure 9.4.

[9]The market maker may also wait for some other client to show up and buy the option back. Market makers have position limits and can operate for short periods without closing open positions.

Now, consider the risks of the position. It is clear from Figure 9.4 that the long call position funded by a money market loan is similar to going long the S_t. If S_t decreases, the position's value will decrease, and a market maker who takes such positions many times on a given day cannot afford this. The market maker must hedge this risk by taking another position that will offset these possible gains or losses. When S_t declines, a *short* position in S_t gains. As S_t changes by ΔS_t, a short position will change by $-\Delta S_t$. Thus, we might think of using this short position as a hedge.

But there is a potential problem. The long call position is described by a *curve*, whereas the short position in S_t is represented by a *line*. This means that the responses of $C(t)$ and S_t, to a change in S_t, are not going to be identical. Everything else being the same, if the underlying changes by ΔS_t, the change in the option price will be *approximately*[10]

$$\Delta C(t) \cong C_s \Delta S_t \tag{9.11}$$

The change in the short position on the other hand will equal $-\Delta S_t$. In fact, the net response of the portfolio

$$V_t = \{\text{long } C(t), \text{short } S_t\} \tag{9.12}$$

to a small change in S_t will be given by the *first-order* approximation,

$$\Delta V_t \cong C_s \Delta S_t - \Delta S_t \tag{9.13}$$

$$= (C_s - 1)\Delta S_t < 0 \tag{9.14}$$

due to the condition $0 < C_s < 1$. This position is shown in Figure 9.5. It is *still* a risky position and, interestingly, the risks are reversed. The market maker will now lose money if S_t increases. In fact, this position amounts to a long put financed by a money market loan.

How can the risks associated with the movements in S_t be eliminated? In fact, consider Figure 9.5. We can approximate the option value by using the tangent at point $S_t = K$. This would also be a line. We can then adjust the short position accordingly. According to Eq. (9.14), short selling *one* unit of S_t overdid the hedge. Figure 9.5 suggests that the market maker should short h_t units of S_t, selecting the h_t according to

$$h_t = \frac{\partial C(S_t, t)}{\partial S_t} = C_s \tag{9.15}$$

To see why this might work, consider the new portfolio, V_t:

$$V_t = \{\text{long } 1 \text{ unit of } C(t), \text{borrow } C(t) \text{ dollars}, \text{short } C_s \text{ units of } S_t\} \tag{9.16}$$

If S_t changes by ΔS_t, everything else being the same, the change in this portfolio's value will be approximately

$$\Delta V_t \cong [C(S_t + \Delta S_t, t) - C(S_t, t)] - C_s \Delta S_t \tag{9.17}$$

---

[10]Due to the assumption of everything else being the same, ΔS_t and $\Delta C(t)$ should be interpreted within the context of partial differentiation.

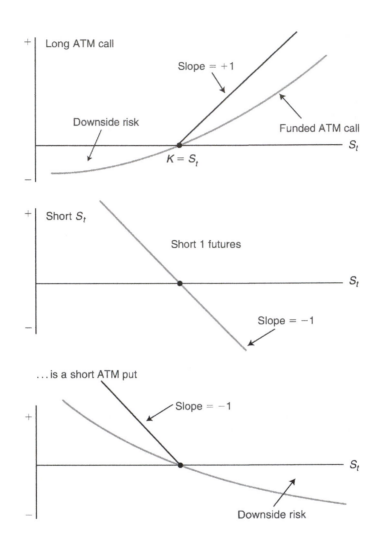

FIGURE 9.5

Long call and short futures position.

We can use a first-order Taylor series approximation of $C(S_t + \Delta S_t, t)$, around point S_t, to simplify this relationship:[11]

---

[11] Let $f(x)$ be a continuous and infinitely differentiable function of x. The kth-order Taylor series approximation of $f(x)$, at point x_0, is given by

$$f(x) = f(x_0) + f'(x_0)(x - x_0) + \frac{1}{2}f''(x_0)(x - x_0)^2 + \cdots + \frac{1}{k!}f^k(x_0)(x - x_0)^k$$

where $f^k(x_0)$ is the kth derivative of $f(.)$ evaluated at $x = x_0$.

$$C(S_t + \Delta S_t, t) = C(S_t, t) + \frac{\partial C(S_t, t)}{\partial S_t} \Delta S_t + R \tag{9.18}$$

Here, R is the *remainder*. The right-hand side of this formula can be substituted in Eq. (9.17) to obtain

$$\Delta V_t \cong \left[\frac{\partial C(S_t, t)}{\partial S_t} \Delta S_t + R \right] - C_s \Delta S_t \tag{9.19}$$

After using the definition

$$\frac{\partial C(S_t, t)}{\partial S_t} = C_s \tag{9.20}$$

and simplifying, this becomes

$$\Delta V_t \cong R \tag{9.21}$$

That is to say, this portfolio's sensitivity toward changes in S_t will be the remainder term, R. It is related to Ito's Lemma, shown in Appendix 9.2. The biggest term in the remainder is given by

$$\frac{1}{2} \frac{\partial^2 C(S_t, t)}{\partial S_t^2} (\Delta S_t)^2 \tag{9.22}$$

Since the second partial derivative of $C(t)$ is always positive, the portfolio's value will always be *positively* affected by small changes in S_t. This is shown in the bottom part of Figure 9.6. A portfolio such as this one is said to be *delta neutral*. That is to say, the *delta exposure*, represented by the first-order sensitivity of the position to changes in S_t, is zero. Note that during this discussion the time variable, t, was treated as a constant.

This way of constructing a hedge for options is called *delta hedging* and h_t is called the hedge ratio. It is important to realize that the procedure will need constant updating of the hedge ratio, h_t, as time passes and S_t changes. After all, the idea depends on a first-order Taylor series approximation of a nonlinear instrument using a linear instrument. Yet, Taylor series approximations are *local* and they are satisfactory only for a reasonable neighborhood around the initial S_t. As S_t changes, the approximation needs to be adjusted. Consider Figure 9.7. When S_t moves from point A to point B, the approximation at A deteriorates and a new approximation is needed. This new approximation will be the tangent at point B.

9.4.2 ADJUSTING THE HEDGE OVER TIME

We now consider what happens to the *delta*-hedged position as S_t oscillates. According to our discussion in the previous chapter, as time passes, the replicating portfolio needs to be rebalanced. This rebalancing will generate cash gains.

We discuss these portfolio adjustments in a highly simplified environment. Considering a sequence of simple oscillations in S_t *around* an initial point $S_{t_0} = S^0$, let

$$t_0 < t_1 < \cdots < t_n \tag{9.23}$$

with

$$t_i - t_{i-1} = \Delta \tag{9.24}$$

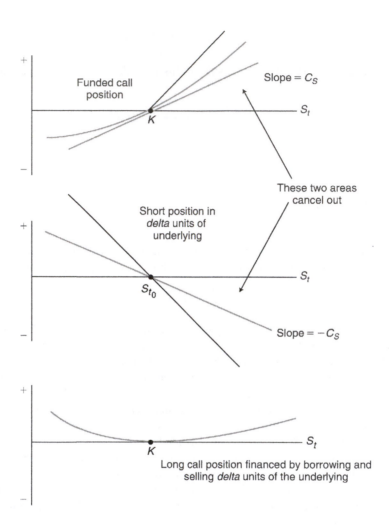

FIGURE 9.6

Delta neutral portfolio example.

denote successive time periods that are apart Δ units of time. We assume that S_t oscillates at an annual percentage rate of one standard deviation, σ, *around* the initial point $S_{t_0} = S^0$. For example, one possible round turn may be

$$S^0 \rightarrow (S^0 + \Delta S) \rightarrow S^0 \tag{9.25}$$

With $\Delta S = \sigma S^0 \sqrt{\Delta}$, the percentage oscillations will be proportional to $\sqrt{\Delta}$. The mechanics of maintaining the *delta*-hedged long call position will be discussed in this simplified setting.

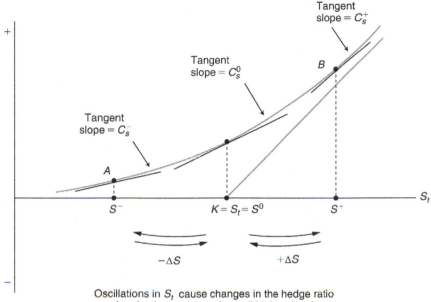

Oscillations in S_t cause changes in the hedge ratio
and make the market maker sell high... buy low

FIGURE 9.7

Oscillations in underlying and changes in hedge ratio.

Since S_{t_i} moves between three possible values only, we simplify the notation and denote the possible values of S_t by S^-, S^0, and S^+, where[12]

$$S^+ = S^0 + \Delta S \tag{9.26}$$

$$S^- = S^0 - \Delta S \tag{9.27}$$

We now show how these oscillations generate cash gains. According to Figure 9.7, as S_t fluctuates, the slope, C_s, of the $C(S_t, t)$ also changes. Ignoring the effect of time, the slope will change, say, between C_s^+, C_s^0, and C_s^-, as shown in Figure 9.7.[13] We note that

$$C_s^- < C_s^0 < C_s^+ \tag{9.28}$$

---

[12] We can represent this trajectory by a three-state *Markov chain* that has the following probabilities:

$$P(S^0 | S^+) = 1 \quad P(S^- | S^0) = \frac{1}{2} \quad P(S^+ | S^0) = \frac{1}{2} \quad P(S^0 | S^-) = 1$$

where S^0 is the sorting value. If prices are at S^+ or S^- they *always* go back to S_0. From S_0, they can either go up or down.

[13] It is important to realize that these slopes also depend on time t, although, to simplify the notation, we are omitting the time index here.

for all t_i. This means that as S_t moves, h_t, the hedge ratio will change in a particular way. In order to keep the portfolio *delta*-hedged, the market maker needs to *adjust* the number of the underlying S_t that was shorted.

Second, and unexpectedly, the hedge adjustments have a "nice" effect. When S_t moves from S^+ to S^0 or from S^0 to S^-, the market maker has to *decrease* the size of the short position in S_t. To do this, the market maker needs to "buy" back a portion of the underlying asset that was originally shorted at a higher price S^0 or S^+.

Accordingly, the market maker sells short when prices are high and covers part of the position when prices decline. This leads to cash gains.

Consider now what happens when the move is from S^0 to S^+. The new slope, C_s^+, is steeper than the old, C_s^0. This means that the market maker needs to short *more* of the S_t asset at the new price. When S_t moves back to S^0, these shorts are covered at S^0, which is lower than S^+.

Thus, as S_t oscillates around S^0, the portfolio is adjusted accordingly, and the market maker would *automatically* sell high and buy low. At every *round turn*, say, $\{S^0, S^+, S^0\}$, which takes *two* periods, the hedge adjustments will generate a cash gain equal to

$$(C_s^+ - C_s^0)[(S^0 + \Delta S) - S^0] = (C_s^+ - C_s^0)\Delta S \tag{9.29}$$

Here, $(C_s^+ - C_s^0)$ represents the number of S_t assets that were shorted after the price moved from S^0 to S^+. Once the price goes *back* to S^0, the same securities are purchased at a lower price. It is interesting to look at these trading gains as the time interval, Δ, becomes smaller and smaller.

9.4.2.1 Limiting form

As $\Delta S \to 0$, we can show an important approximation to the trading (hedging) gains

$$(C_s^+ - C_s^0)\Delta S \tag{9.30}$$

The term $(C_s^+ - C_s^0)$ is the *change* in the *first* partial derivative of $C(S_t, t)$, as S_t moves from S_{t_0} to a new level denoted by $S_{t_0} + \Delta S$. We can convert $(C_s^+ - C_s^0)$ into a *rate* of change after multiplying and dividing by ΔS:

$$(C_s^+ - C_s^0)\Delta S = \frac{C_s^+ - C_s^0}{\Delta S}(\Delta S)^2 \tag{9.31}$$

As we let ΔS go to zero, we obtain the approximation

$$\frac{C_s^+ - C_s^0}{\Delta S} \cong \frac{\partial^2 C(S_t, t)}{\partial S_t^2} \tag{9.32}$$

Thus, the *round-turn* gains from *delta*-hedge adjustments shown in Eq. (9.29) can be approximated as

$$(C_s^+ - C_s^0)\Delta S \cong \frac{\partial^2 C(S_t, t)}{\partial S_t^2}(\Delta S)^2 \tag{9.33}$$

Per time unit gains are then half of this,

$$\frac{1}{2}\frac{\partial^2 C(S_t, t)}{\partial S_t^2}(\Delta S)^2 \tag{9.34}$$

These gains are only part of the potential cash inflows and outflows faced by the market maker. The position has further potential cash flows that need to be described. This is done in the next two sections.

9.4.3 OTHER CASH FLOWS

We just showed that oscillations in S_t generate positive cash flows if the market maker *delta* hedges his or her long option position. Does this imply an arbitrage opportunity? After all, the market maker did not advance any cash yet seems to receive cash spontaneously as long as S_t oscillates. The answer is no. There are *costs* to this strategy, and the *delta*-hedged option position is *not* riskless.

1. The market maker funded his or her position with borrowed money. This means, that, as time passes, an *interest cost* is incurred. For a period of length Δ, this cost will equal

$$rC\Delta \tag{9.35}$$

   under the constant spot rate assumption. (We write $C(t)$, as C.)
2. The option has *time value*, and as time passes, everything else being the same, the value of the option will decline at the *rate*

$$C_t = \frac{\partial C(S_t, t)}{\partial t} \tag{9.36}$$

   The option value will go down by

$$\frac{\partial C(S_t, t)}{\partial t} \Delta \tag{9.37}$$

   dollars, for each Δ that passes.
3. Finally, the cash received from the short position generates $rS_t C_s \Delta$ dollars interest every time period Δ.

The trading gains and the costs can be put together to obtain an important *partial differential equation* (PDE), which plays a central role in financial engineering.

9.4.4 OPTION GAINS AND LOSSES AS A PDE

We now add all gains and costs per unit of time Δ. The options' gains per time unit from hedging adjustments are

$$\frac{1}{2} \frac{\partial^2 C(S_t, t)}{\partial S_t^2} (\Delta S)^2 \tag{9.38}$$

In case the process S_t is *geometric*, the annual percentage variance will be constant and this can be written as (see Appendix 9.2)

$$\frac{1}{2} C_{ss} \sigma^2 S_t^2 \Delta \tag{9.39}$$

The rest of the argument will continue with the assumption of a constant σ.

Interest is paid daily on the funds borrowed to purchase the call. For every period of length Δ, a long call holder will pay

$$rC\Delta \tag{9.40}$$

Another item is the interest earned from cash generated by shorting C_s units of S_t:[14]

$$rC_sS_t\Delta \tag{9.41}$$

Adding these, we obtain the net cash gains (losses) from the hedged long call position during Δ:

$$\frac{1}{2}C_{ss}\sigma^2S_t^2\Delta + rC_sS_t\Delta - rC\Delta \tag{9.42}$$

Now, in order for there to be no arbitrage opportunity, this must be equal to the daily loss of time value:

$$\frac{1}{2}C_{ss}\sigma^2S_t^2\Delta + rC_sS_t\Delta - rC\Delta = -C_t\Delta \tag{9.43}$$

We can eliminate the common Δ terms and obtain a very important relationship that some readers will recognize as the *Black–Scholes PDE*:

$$\frac{1}{2}C_{ss}\sigma^2S_t^2 + rC_sS_t - rC + C_t = 0 \tag{9.44}$$

Every PDE comes with some boundary conditions and this is no exception. The call option will expire at time T, and the expiration $C(S_T, T)$ is given by

$$C(S_T, T) = \max[S_T - K, 0] \tag{9.45}$$

Solving this PDE gives the Black–Scholes equation. In most finance texts, the PDE derived here is obtained from some mathematical derivation. In this section, we obtained the same PDE heuristically from practical trading and arbitrage arguments.

9.4.5 CASH FLOWS AT EXPIRATION

The cash flows at expiration date have three components: (i) the market maker has to pay the original loan if it is not paid off slowly over the life of the option, (ii) there is the final option settlement, and (iii) there is the final payoff from the short S_t position.

Now, at an infinitesimally short time period, dt, before expiration, the price of the underlying will be very close to S_T. Call it S_T^-. The price curve $C(S_t, t)$ will be very near the piecewise linear option payoff. Thus, the hedge ratio $h_T^- = C_s$ will be very close to either 0 or 1:

$$h_T^- \cong \begin{cases} 1 & S_T^- > K \\ 0 & S_T^- < K \end{cases} \tag{9.46}$$

This means that, at time T, any potential gains from the long call option position will be equal to losses on the short S_t position.

---

[14]If the underlying asset is not "cash" but a futures contract, then this item may drop.

The interesting question is, how does the market maker manage to pay back the original loan under these conditions? There is only one way. The only cash that is available is the accumulation of (net) trading gains from hedge adjustments during $[t, T]$. As long as Eq. (9.44) is satisfied for every t_i, the hedged long option position will generate enough cash to pay back the loan. The option price, $C(t)$, regarded this way is the discounted sum of all gains and losses from a *delta*-hedged option position the trader will incur based on *expected* S_t volatility.

We will now consider a numerical example to our highly simplified discussion of how realized volatility is converted into cash via an option position.

9.4.6 AN EXAMPLE

Consider a stock, S_t, trading at a price of 100. The stock pays no dividends and is known to have a Black–Scholes volatility of $\sigma = 45\%$ per annum. The risk-free interest rate is 4% and S_t is known to follow a geometric process, so that the Black–Scholes assumptions are satisfied.

A market maker buys 100 plain vanilla, ATM calls that expire in 5 days. The premium for one call is 2.13 dollars. This is the price found by plugging the above data into the Black–Scholes formula. Hence, the total cash outlay is $213. There are no other fees or commissions. The market maker borrows the $213, buys the call options, and immediately hedges the long position by short selling an appropriate number of the underlying stock.

EXAMPLE

Suppose that during these 5 days the underlying stock follows the path:

$$\{\text{Day 1} = 100, \text{Day 2} = 105, \text{Day 3} = 100, \text{Day 4} = 105, \text{Day 5} = 100\} \qquad (5.47)$$

What are the cash flows, gains, and losses generated by this call option that remain on the market maker's books?

1. **Day 1: The purchase date**
   Current Delta: 51 (Found by differentiating the Black–Scholes formula with respect to S_t, plugging in the data and then multiplying by 100)
   Cash paid for the call options: $213
   Amount borrowed to pay for the calls: $213
   Amount generated by short selling 51 units of the stock: $5100. This amount is deposited at a rate of 4%.

2. **Day 2: Price goes to 105**
   Current Delta: 89 (Evaluated at $S_t = 105$, 3 days to expiration)
   Interest on amount borrowed: $213(0.04)\left(\frac{1}{360}\right) = \$.02$
   Interest earned from deposit: $5100(0.04)\left(\frac{1}{360}\right) = \$.57$ (Assuming no bid–ask difference in interest rates)
   Short selling 38 units of additional stock to reach delta neutrality which generate: $38(105) = \$3990$.

---

3. **Day 3: Price goes back to 100**
   Current Delta: 51
   Interest on amount borrowed: $213(0.04) \left(\frac{1}{360}\right) = \$0.02$
   Interest earned from deposits: $(5100 + 3990)(0.04) \left(\frac{1}{360}\right) = \$1$
   Short covering 38 units of additional stock at 100 each, to reach delta neutrality generates a cash flow of: $38(5) = \$190$. Interest on these profits is ignored to the first order of approximation.

4. **Day 4: Price goes to 105**
   Current Delta: 98
   Interest on amount borrowed: $213(0.04) \left(\frac{1}{360}\right) = \$0.02$
   Interest earned from deposits: $5100(0.04) \left(\frac{1}{360}\right) = \$0.57$
   Shorting 47 units of additional stock at 105 each, to reach delta neutrality generates: $47(105) = \$4935$.

5. **Day 5: Expiration with $S_T = 100$**
   Net cash generated from covering the short position: $47(5) = \$235$ (There were 98 shorts, covered at \$100 each; 47 shorts were sold at \$105, 51 shorts at \$100)
   Interest on amount borrowed: $213(0.04) \left(\frac{1}{360}\right) = \$0.02$
   Interest earned from deposits: $(5100 + 4935)(0.04) \left(\frac{1}{360}\right) = \$1.1$. The option expires ATM and generates no extra cash.

6. **Totals**
   Total interest paid: $4(0.02) = \$0.08$
   Total interest earned: $2(0.57) + 1 + 1.1 = \$3.24$
   Total cash earned from hedging adjustments: $\$235 + \$190$
   Cash needed to repay the loan: \$213
   Total net profit ignoring interest on interest $= \$215.16$
   A more exact calculation would take into account interest on interest earned and the interest earned on \$190 for 2 days.

We can explain why total profit is positive. The path followed by S_t in this example implies a daily actual volatility of 5%. Yet, the option was sold at an annual implied volatility of 45%, which corresponds to a "daily" percentage implied volatility of:

$$0.45 \sqrt{\frac{1}{365}} = 2.36\% \tag{9.48}$$

Hence, during the life of the option, S_t fluctuated more than what the implied volatility suggested. As a result, the long convexity position had a net profit.

This example is, of course, highly simplified. It keeps implied volatility constant and the oscillations occur around a fixed point. If these assumptions are relaxed, the calculations will change.

9.4.6.1 Some caveats

Three assumptions simplified notation and discussion in this section.

- First, we considered oscillations around a *fixed* S^0. In real life, oscillations will clearly occur around points that themselves move. As this happens, the partial derivatives, C_s and C_{ss}, will change in more complicated ways.
- Second, C_s and C_{ss} are also functions of time t, and as time passes, this will be another source of change.
- The third point is more important. During the discussion, oscillations were kept constant at ΔS. In real life, volatility may change over time and be random as well. This would not invalidate the essence of our argument concerning gains from hedge adjustments, but it will clearly introduce *another risk* that the market maker may have to hedge against. This risk is known as *vega* risk.
- Finally, it should be remembered that the underlying asset did not make any payouts during the life of the option. If dividends or coupons are paid, the calculation of cash gains and losses needs to be adjusted accordingly.

These assumptions were made to emphasize the role of options as volatility instruments. Forthcoming chapters will deal with how to relax them.

9.5 TOOLS FOR OPTIONS

The Black–Scholes PDE can be exploited to obtain the major tools available to an option trader or market maker. First of these is the *Black–Scholes formula*, which gives the arbitrage-free price of a plain vanilla call (put) option under specific assumptions.

The second set of tools is made up of the "Greeks." These measure the sensitivity of an option's price with respect to changes in various market parameters. The Greeks are essential in hedging and risk managing options books. They are also used in pricing and in options strategies.

The third set of tools are *ad hoc* modifications of these theoretical constructs by market practitioners. These modifications adapt the theoretical tools to the real world, making them more "realistic."

9.5.1 SOLVING THE FUNDAMENTAL PDE

The convexity of option payoffs implies an arbitrage argument, namely that the expected net gains (losses) from S_t oscillations are equal to time decay during the same period. This leads to the Black–Scholes PDE:

$$\frac{1}{2}C_{ss}\sigma^2 S_t^2 + rC_s S_t - rC + C_t = 0 \tag{9.49}$$

with the boundary condition

$$C(T) = \max[S_T - K, 0] \tag{9.50}$$

Now, under some conditions PDEs can be solved analytically and a *closed-form formula* can be obtained. See Duffie (2001). In our case, with specific assumptions concerning the dynamics of S_t, this PDE has such a closed-form solution. This solution is the market benchmark known as the Black–Scholes formula.

9.5.2 BLACK–SCHOLES FORMULA

An introduction to the Black–Scholes formula first requires a good understanding of the underlying assumptions. Suppose we consider a plain vanilla call option written on a stock at time t. The option expires at time $T > t$ and has strike price K. It is of European style and can be exercised only at expiration date T. Further, the underlying asset price and the related market environment denoted by S_t have the following characteristics:

1. The risk-free interest rate is constant at r.
2. The underlying stock price dynamics are described in continuous time by the stochastic differential equation (SDE):[15]

$$dS_t = \mu(S_t)S_t\, dt + \sigma S_t\, dW_t \quad t \in [0, \infty) \tag{9.51}$$

where W_t represents a Wiener process with respect to real-world probability P.[16]

To emphasize an important aspect of the previous SDE, the dynamics of S_t are assumed to have a constant *percentage* variance during infinitesimally short intervals. Yet, the drift component, $\mu(S_t)S_t$, can be general and need *not* be specified further. Arbitrage arguments are used to eliminate $\mu(S_t)$ and replace it with the risk-free instantaneous spot rate r in the previous equation.

3. The stock pays no dividends, and there are no stock splits or other corporate actions during the period $[t, T]$.
4. Finally, there are no transaction costs and no bid–ask spreads.

Under these assumptions, we can solve the PDE in Eqs. (9.49) and (9.51) and obtain the Black–Scholes formula:

$$C(t) = S_t N(d_1) - Ke^{-r(T-t)}N(d_2) \tag{9.52}$$

where d_1, d_2 are

$$d_1 = \frac{\log(S_t/K) + (r + (\sigma^2/2))(T - t)}{\sigma\sqrt{T - t}} \tag{9.53}$$

$$d_2 = \frac{\log(S_t/K) + (r - (\sigma^2/2))(T - t)}{\sigma\sqrt{T - t}} \tag{9.54}$$

---

[15] Appendix 9.2 discusses SDEs further.

[16] The assumption of a Wiener process implies heuristically that

$$E_t[dW_t] = 0$$

and that

$$E_t[dW_t]^2 = d_t$$

These increments are the continuous time equivalents of sequences of normally distributed variables. For a discussion of stochastic differential equations and the Wiener process, see, for example, Øksendal (2010). Hirsa and Neftci (2013) provides the heuristics.

$N(x)$ denotes the cumulative standard normal probability:

$$N(x) = \int_{-\infty}^{x} \frac{1}{\sqrt{2\pi}} e^{-(1/2)u^2} du \tag{9.55}$$

In this formula, r, σ, T, and K are considered *parameters*, since the formula holds in this version, only when these components are kept constant.[17] The *variables* are S_t and t. The latter is allowed to change during the life of the option.

Given this formula, we can take the partial derivatives of

$$C(t) = C(S_t, t \mid r, \sigma, T, K) \tag{9.56}$$

with respect to the variables S_t and t *and* with respect to the parameters r, σ, T, and K. These partials are the *Greeks*. They represent the sensitivities of the option price with respect to a small variation in the parameters and variables. Implied volatility (σ_{IV}) is calculated by backing out the σ that gives the observed option price $C(t)$ if one was to plug it into the pricing formula (9.52) together with the other input parameters such as r, T, K, S_t, and t. Since the implied volatility is implied by the observed market price of the option, its calculation is similar to that of the yield to maturity for bonds which is calculated based on the market price of the bonds, a pricing formula and other input parameters such as maturity and frequency of cash flow payments.

9.5.2.1 Black's formula

The Black–Scholes formula in Eq. (9.52) is the solution to the fundamental PDE when *delta* hedging is done with the "cash" underlying. As discussed earlier, trading gains and funding costs lead to the PDE:

$$rC_s S_t - rC + \frac{1}{2}C_{ss}\sigma^2 S_t^2 = -C_t \tag{9.57}$$

with the boundary condition:

$$C(S_T, T) = \max[S_T - K, 0] \tag{9.58}$$

When the underlying becomes a *forward* contract, S_t will become the corresponding forward price denoted by F_t and the Black–Scholes PDE will change slightly.

Unlike a cash underlying, buying and selling a forward contract does not involve funding. Long and short forward positions are *commitments* to buy and sell at a future date T, rather than outright purchases of the underlying asset. Thus, the only cash movements will be interest expense for funding the call, and cash gains from hedge adjustments. This means that the corresponding PDE will look like

$$-rC + \frac{1}{2}C_{ss}\sigma^2 F_t^2 = -Ct \tag{9.59}$$

with the same boundary condition:

$$C(F_T, T) = \max[F_T - K, 0] \tag{9.60}$$

where F_t is now the forward price of the underlying.

---

[17]The volatility of the underlying needs to be constant during the life of the option. Otherwise, the formula will not hold, even though the logic behind the derivation would.

The solution to this PDE is given by the so-called *Black's formula* in the case where the options are of European style.

$$C(F_t, t)^{\text{Black}} = e^{-r(T-t)}[F_t N(d_1) - K N(d_2)] \tag{9.61}$$

with

$$d_1^{\text{Black}} = \frac{\log(F_t/K) + (1/2)\sigma^2(T-t)}{\sigma\sqrt{(T-t)}} \tag{9.62}$$

$$d_2^{\text{Black}} = d_1^{\text{Black}} - \sigma\sqrt{(T-t)} \tag{9.63}$$

Black's formula is useful in many practical circumstances where the Black–Scholes formula cannot be applied directly. Interest rate derivatives such as caps and floors, for example, are options written on LIBOR rates that will be observed at future dates. Such settings lend themselves better to the use of Black's formula. The underlying risk is a *forward* interest rate such as *forward LIBOR*, and the related option prices are given by Black's formula. However, the reader should remember that in the preceding version of Black's formula the spot rate is taken as constant. In Chapter 16, this assumption will be relaxed.

9.5.3 OTHER FORMULAS

The Black–Scholes type PDEs can be solved for a closed-form formula under somewhat different conditions as well. These operations result in expressions that are similar but contain further parameters and variables. We consider two cases of interest. Our first example is a *chooser* option.

9.5.3.1 Chooser options

Consider a vanilla put, $P(t)$ and a vanilla call, $C(t)$ written on S_t with strike K, and expiration T. A chooser option then is an option that gives the right to choose between $C(t)$ and $P(t)$ at some later date T_0. Its *payoff* at time T_0, with $T_0 < T$ is

$$C^h(T_0) = \max\left[C(S_{T_0}, T_0), P(S_{T_0}, T_0)\right] \tag{9.64}$$

Arbitrage arguments lead to the equality

$$P(S_{T_0}, T_0) = -(S_{T_0} - Ke^{-r(T-T_0)}) + C(S_{T_0}, T_0) \tag{9.65}$$

Using this, Equation (9.64) can be written as

$$C^h(T_0) = \max\left[C(S_{T_0}, T_0), -(S_{T_0} - Ke^{-r(T-T_0)}) + C(S_{T_0}, T_0)\right] \tag{9.66}$$

or, taking the common term out,

$$C^h(T_0) = C(S_{T_0}, T_0) + \max\left[-(S_{T_0} - Ke^{-r(T-T_0)}), 0\right] \tag{9.67}$$

In other words, the chooser option payoff is either equal to the value of the call at time T_0, or it is that plus a positive increment, in the case that

$$(S_{T_0} - Ke^{-r(T-T_0)}) < 0 \tag{9.68}$$

But, this is equal to the payoff of a put with strike price $Ke^{-r(T-T_0)}$ and exercise date T_0. Thus, the pricing formula for the chooser option is given by

$$C^h(t) = [S_t N(d_1) - Ke^{-r(T-t)} N(d_2)] + [-S_t N(-\bar{d}_1) + Ke^{-r(T-T_0)} e^{-r(T_0-t)} N(-\bar{d}_2)] \tag{9.69}$$

Simplifying:

$$C^h(t) = [S_t(N(d_1) - N(-\bar{d}_1))] + Ke^{-r(T-t)}(N(-\bar{d}_2) - N(d_2)) \tag{9.70}$$

with

$$d_{1,2} = \frac{\ln(S_t/K) + (r \pm (1/2)\sigma^2)(T-t)}{\sigma\sqrt{T-t}} \tag{9.71}$$

$$\bar{d}_{1,2} = \frac{\ln(S_t/K) + (r(T-t) \pm (1/2)\sigma^2(T_0-t))}{\sigma\sqrt{(T_0-t)}} \tag{9.72}$$

A more interesting example from our point of view is the application of the Black–Scholes approach to barrier options, which we consider next.

9.5.3.2 Barrier options

Barrier options will be treated in detail in the next chapter. Here we just define these instruments and explain the closed-form formula that is associated with them under some simplifying assumptions. This will close the discussion of the application spectrum of Black–Scholes type formulas.

Consider a European vanilla call, written on S_t, with strike K and expiration T, $t < T$. Assume that S_t satisfies all Black–Scholes assumptions. Consider a *barrier H*, and assume that $H < S_t < K$ as of time t. Suppose we write a contract stipulating that if, during the life of the contract, $[t, T]$, S_t falls below the level H, the option disappears and the option writer will have no further obligation. In other words, as long as $H < S_u$, $u \in [t, T]$, the vanilla option is in effect, but as soon as S_u falls below H, the option dies. This is a *barrier option*—specifically a *down-and-out* barrier. Two examples are shown in Figure 9.8a.

The pricing formula for the down-and-out call is given by

$$\begin{aligned} C^b(t) &= C(t) - J(t) \quad \text{for } H \le S_t \\ C^b(t) &= 0 \quad \text{for } S_t < H \end{aligned} \tag{9.73}$$

Here $C(t)$ is the value of the vanilla call, which is given by the standard Black–Scholes formula, and where $J(t)$ is the discount that needs to be applied because the option may die if S_t falls below H during $[t, T]$. See Figure 9.8b. The formula for $J(t)$ is

$$J(t) = S_t \left(\frac{H}{S_t}\right)^{(2(r-(1/2)\sigma^2)/\sigma^2)+2} N(c_1) - Ke^{-r(T-t)}\left(\frac{H}{S_t}\right)^{(2(r-\frac{1}{2}\sigma^2))/\sigma^2} N(c_2) \tag{9.74}$$

where

$$c_{1,2} = \frac{\log(H^2/S_t K) + (r \pm (1/2)\sigma^2)(T-t)}{\sigma\sqrt{T-t}} \tag{9.75}$$

It is interesting to note that when S_t touches the barrier

$$S_t = H \tag{9.76}$$

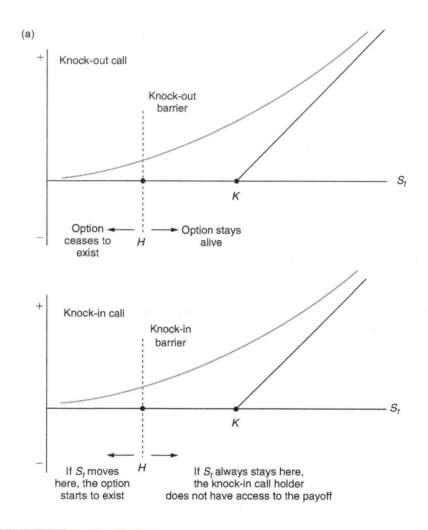

(a)

FIGURE 9.8

(a) Barrier option example. (b) Vanilla call and knock-in call.

the formula for $J(t)$ becomes

$$J(t) = S_t N(d_1) - Ke^{-r(T-t)}N(d_2) \tag{9.77}$$

That is to say, the value of $C^b(t)$ is zero:

$$C^b(t) = C(t) - C(t) \tag{9.78}$$

This characterization of a barrier option as a standard option plus or minus a discount term is very useful from a financial engineering angle. In the next chapter, we will obtain some simple contractual equations for barriers, and the use of discounts will then be useful for obtaining Black–Scholes type formulas for other types of barriers.

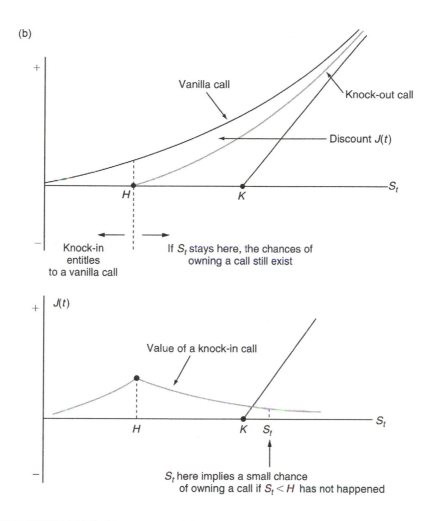

(b)

Vanilla call

Knock-out call

Discount $J(t)$

H K S_t

Knock-in entitles to a vanilla call

If S_t stays here, the chances of owning a call still exist

$J(t)$

Value of a knock-in call

H K S_t S_t

S_t here implies a small chance of owning a call if $S_t < H$ has not happened

FIGURE 9.8

(Continued)

9.5.4 USES OF BLACK—SCHOLES TYPE FORMULAS

Obviously, the assumptions underlying the derivation of the Black—Scholes formula are quite restrictive. This becomes especially clear from the way we introduced options in this book. In particular, if options are used to bet on the direction of volatility, then how can the assumption of constant percentage volatility possibly be satisfied? This issue will be discussed further in later chapters where the way market professionals use the Black—Scholes formula while trading volatility is clarified.

When the underlying asset is an interest rate instrument or a foreign currency, some of the Black–Scholes assumptions become untenable.[18] Yet, when these assumptions are relaxed, the logic used in deriving the Black–Scholes formula may not result in a PDE that can be solved for a closed-form formula.

Hence, a market practitioner may want to use the Black–Scholes formula or variants of it, and then adjust the formula in some *ad hoc*, yet practical, ways. This may be preferable to trying to derive new complicated formulas that may accommodate more realistic assumptions. Also, even though the Black–Scholes formula does not hold when the underlying assumptions change, acting as if the assumptions hold yields results that are surprisingly robust.[19] We will see that this is exactly what happens when traders adjust the volatility parameter depending on the "moneyness" of the option under consideration.

This completes our brief discussion of the first set of tools that are essential for option analysis, namely Black–Scholes type closed-form formulas that give the arbitrage-free price of an option under some stringent conditions. Next, we discuss the second set of tools that traders and market makers routinely use: various sensitivity factors called the "Greeks."

9.6 THE GREEKS AND THEIR USES

The Black–Scholes formula gives the value of a vanilla call (put) option under some specific assumptions. Obviously, this is useful for calculating the arbitrage-free value of an option. But a financial engineer needs methods for determining how the option premium, $C(t)$, *changes* as the variables or the parameters in the formula change within the market environment. This is important since the assumptions used in deriving the Black–Scholes formula *are* unrealistic. Traders, market makers, or risk managers must constantly monitor the sensitivity of their option books with respect to changes in S_t, r, t, or σ. The role of Greeks should be well understood.

> **EXAMPLE**
>
> A change in σ is a good example. We motivated option positions essentially (but not fully) as positions taken on volatility. It is clear that volatility is not constant as assumed in the Black–Scholes world. Once an option is bought and delta-hedged, the hedge ratio C_s and the C_{ss} both depend on the movements in the volatility parameter σ.
>
> Hence, the "hedged" option position will still be risky in many ways. For example, depending on the way changes in σ and S_t affect the C_{ss}, a market maker may be correct in his or her forecast of how much S_t will fluctuate, yet may still lose money on a long option position.

A further difficulty is that option sensitivities may not be uniform across the strike price K or expiration T. For options written on the *same* underlying, differences in K and T lead to what are called *smile effects* and *term structure effects*, respectively, and should be taken into account carefully.

---

[18]For example, a foreign currency pays foreign interest. This is like an underlying stock paying dividends.
[19]See for example, El-Karoui et al. (1998).

Option sensitivity parameters are called the "Greeks" in the options literature. We discuss them next and provide several practical examples.

9.6.1 DELTA

Consider the Black–Scholes formula $C(S_t, t \mid r, \sigma, T, K)$. How much would this theoretical price change if the underlying asset price, S_t, moved by an infinitesimal amount?

One theoretical answer to this question can be given by using the partial derivative of the function with respect to S_t. This is by definition the *delta* at time t:

$$\text{delta} = \frac{\partial C(S_t, t \mid r, \sigma, T, K)}{\partial S_t} \tag{9.79}$$

This partial derivative was denoted by C_s earlier. Note that *delta* is the local sensitivity of the option price to an infinitesimal change in S_t only, which incidentally is the reason behind using partial derivative notation.

To get some intuition on this, remember that the price curve for a long call has an upward slope in the standard $C(t)$, S_t space. Being the slope of the tangent to this curve, the *delta* of a long call (put) is always positive (negative). The situation is represented in Figure 9.9. Here, we consider three outcomes for the underlying asset price represented by S_A, S_B, and S_C and hence obtain three points, A, B, and C, on the option pricing curve. At each point, we can draw a tangent. The slope of this tangent corresponds to the *delta* at the respective price.

- At point C, the slope, and hence, the *delta* is close to 0, since the curve is approaching the horizontal axis as S_t falls.
- At point B, the *delta* is close to 1, since the curve is approaching a line with slope $+1$.
- At point A, the *delta* is in the "middle," and the slope of the tangent is between 0 and 1.

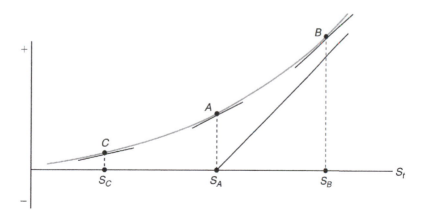

FIGURE 9.9

Payoff of a long call position and slope.

Thus, we always have $0 < delta < 1$ in case of a long call position. As mentioned earlier, when the option is ATM, the *delta* is close to 0.5.

9.6.1.1 Convention

Market professionals do not like to use decimal points. The convention in option markets is to think about trading, not one, but 100 options, so that the *delta* of option positions can be referred to in whole numbers, between 0 and 100. According to this convention, the *delta* of an ATM option is around 50. A 25-*delta* option would be out-of-the-money and a 75-*delta* option in-the-money. Especially in FX markets, traders use this terminology to trade options.

Under these conditions, an options trader may evaluate his or her exposure using *delta* points. A trader may be *long delta*, which means that the position gains if the underlying increases, and loses if the underlying decreases. A *short delta* position implies the opposite.

9.6.1.2 The exact expression

The partial derivative in Eq. (9.79) can be taken in case the call option is European and the price is given by the Black–Scholes formula. Doing so, we obtain the *delta* of this important special case:

$$\frac{\partial C(S_t, t \mid r, \sigma, T, K)}{\partial S_t} = \int_{-\infty}^{\frac{(T-t)(r+(1/2)\sigma^2)+\log(S_t/K)}{\sqrt{(T-t)\sigma}}} \frac{1}{\sqrt{2\pi}} e^{-(1/2)x^2} dx \tag{9.80}$$
$$= N(d_1)$$

This derivation is summarized in Appendix 9.1. It is shown that the *delta* is itself a function that depends on the "variables" S_t, K, r, σ, and on the remaining life of the option, $T - t$. This function is in the form of a *probability*. The *delta* is between 0 and 1, and the function will have the familiar S-shape of a continuous cumulative distribution function (CDF). This, incidentally, means that the derivative of the *delta* with respect to S_t, which is called *gamma*, will have the shape of a probability *density* function (PDF).[20] A typical *delta* will thus look like the S-shaped curve shown in Figure 9.10.

We can also see from this formula how various movements in market variables will affect this particular option sensitivity. The formula shows that whatever increases the ratio

$$\frac{\log(S_t/K) + (r + (1/2)\sigma^2)(T - t)}{\sigma\sqrt{T - t}} \tag{9.81}$$

will increase the *delta;* whatever decreases this ratio, will decrease the *delta.*

For example, it is clear that as r increases, the *delta* will increase. On the other hand, a decrease in the moneyness of the call option, defined as the ratio

$$\frac{S_t}{K} \tag{9.82}$$

decreases the *delta.* The effect of volatility changes is more ambiguous and depends on the moneyness of the option.

---

[20]Some traders use the *delta* of a particular option as if it is the probability of being in-the-money. This could be misleading.

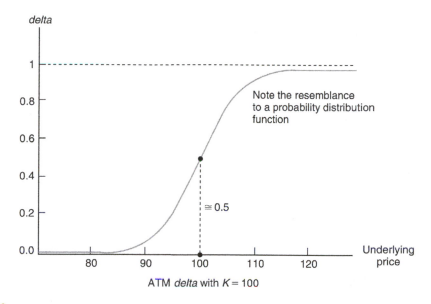

FIGURE 9.10

Delta of a long call position based on Black–Scholes formula.

EXAMPLE

We calculate the delta for some specific options. We first assume the Black–Scholes world, even though the relevant market we are operating in may violate many of the Black–Scholes assumptions. This assumes, for example, that the dividend yield of the underlying is zero and this assumption may not be satisfied in real-life cases. Second, we differentiate the function $C(t)$

$$C(t) = S_t N(d_1) - K e^{-r(T-t)} N(d_2) \qquad (9.83)$$

where d_1 and d_2 are as given in Eqs. (9.53) and (9.54), with respect to S_t. Then, we substitute values observed for $S_t, K, r, \sigma, (T-t)$.

Suppose the Microsoft December calls and puts shown in the table from our first example in this chapter satisfy these assumptions. The deltas can be calculated based on the following parameter values:

$$S_t = 61.15, r = 0.025, \sigma = 30.7\%, T - t = 58/365 \qquad (9.84)$$

Here, σ is the implied volatility obtained by solving the equation for $K = 60$,

$$C(61.115, 60, 0.025, 58/365, \sigma) = \text{Observed price} \qquad (9.85)$$

Plugging the observed data into the formula for delta yields the following values:

Calls	Delta
Dec 55.00	0.82
Dec 60.00	0.59
Dec 65.00	0.34
Dec 70.00	0.16

Puts	Delta
Dec 55.00	−0.17
Dec 60.00	−0.40
Dec 65.00	−0.65
Dec 70.00	−0.84

We can make some interesting observations:

1. The ATM calls and puts have the same price.
2. Their deltas, however, are different.
3. The calls and puts that are equally far from the ATM have slightly different deltas in absolute value.

According to the last point, if we consider 25-delta calls and puts, they will not be exactly the same.[21]

We now point out to some *questionable* assumptions used in our example. First, in calculating the *deltas* for various strikes, we always used the *same* volatility parameter σ. This is not a trivial point. Options that are identical in every other aspect, except for their strike K, *may* have different implied volatilities. There may be a *volatility smile*. Using the ATM implied volatility in calculating the *delta* of *all* options may not be the correct procedure. Second, we assumed a zero dividend yield, which is not realistic either. Normally, stocks have positive expected dividend yields and some correction for this should be made when option prices and the relevant Greeks are calculated. A rough way of doing it is to calculate an annual expected percentage dividend yield and subtract it from the risk-free rate r. Third, should we use S_t or a futures market equivalent, in case this latter exists, the *delta* evaluated in the futures or forward price may be more desirable.

9.6.2 GAMMA

Gamma represents the rate of change of the delta as the underlying risk S_t changes. Changes in delta were seen to play a fundamental role in determining the price of a vanilla option. Hence,

---

[21]We ignore the fact that these CBOE equity options are American.

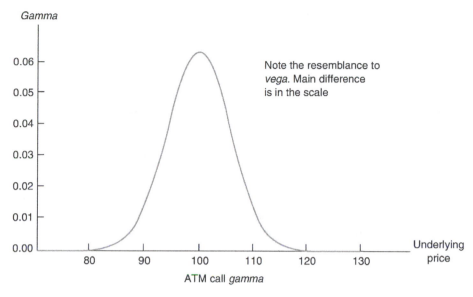

FIGURE 9.11

Gamma of a long call position based on Black−Scholes formula.

gamma is another important Greek. It is given by the second partial derivative of $C(S_t,\ t)$ with respect to S_t:

$$\text{gamma} = \frac{\partial^2 C(S_t, t \mid r, \sigma, T, K)}{\partial S_t^2} \tag{9.86}$$

We can easily obtain the exact expression for *gamma* in the case of a European call. The derivation in Appendix 9.1 gives

$$\frac{\partial^2 C(S_t, t \mid r, \sigma, T, K)}{\partial S_t^2} = \frac{1}{S_t \sigma \sqrt{T-t}} \left[\frac{1}{\sqrt{2\pi}} e^{-\frac{1}{2}\left(\frac{\log(S/k)+r(T-t)+\frac{1}{2}\sigma^2(T-t)}{\sigma\sqrt{T-t}} \right)^2} du \right] \tag{9.87}$$

Gamma shows how much the *delta* hedge should be adjusted as S_t changes. Figure 9.11 illustrates the *gamma* for the Black−Scholes formula. We see the already-mentioned property. *Gamma* is highest if the option is ATM, and approaches zero as the option becomes deep in-the-money *or* out-of-the-money.

We can gain some intuition on the shape of the *gamma* curve. First, remember that *gamma* is, in fact, the derivative of *delta* with respect to S_t. Second, remember that *delta* itself had the shape of a *cumulative* normal distribution. This means that the shape of *gamma* will be similar to that of a continuous, bell-shaped PDF, as expression (9.87) indicates.

Consider now a numerical example dealing with *gamma* calculations. We use the same data utilized earlier in the chapter.

EXAMPLE

To calculate the gamma, we use the same table as in the first example in the chapter. We take the partial derivative of the delta with respect to S_t. This gives a new function S_t, K, r, σ, $(T-1)$, which measures the sensitivity of delta to the underlying S_t. We then substitute the observed values for S_t, K, r, σ, $(T-t)$ to obtain gamma at that particular point.

For the Microsoft December calls and puts shown in the table, gammas are calculated based on the parameter values

$$S_t = 60.0, r = 0.025, \sigma = 31\%, T - t = 58/365, k = 60 \tag{9.88}$$

where σ is the implicit volatility.

Again we are using the implicit volatility that corresponds to the ATM option in calculating the delta of all options, in-the-money or out.

Plugging the observed data into the formula for gamma yields the following values:

Calls	Gamma
Dec 55.00	0.034
Dec 60.00	0.053
Dec 65.00	0.050
Dec 70.00	0.032

Puts	Gamma
Dec 55.00	0.034
Dec 60.00	0.053
Dec 65.00	0.050
Dec 70.00	0.032

The following observations can be made:

1. The puts and calls with different distance to the ATM strike have gammas that are alike but not exactly symmetric.
2. Gamma is positive if the market maker is long the option; otherwise it is negative.

It is also clear from this table that gamma is highest when we are dealing with an ATM option.

Finally, we should mention that as time passes, the second-order curvature of ATM options will increase as the *gamma* function becomes more peaked and its tails go toward zero.

9.6.2.1 Market use

We must comment on the role played by *gamma* in option trading. We have seen that long *delta* exposures can be hedged by going short using the underlying asset. But, how are *gamma* exposures hedged? Traders sometimes find this quite difficult. Especially in very short-dated, deep out-of-the-money options, *gamma* can suddenly go from zero to very high values and may cause significant losses (or gains).

EXAMPLE

The forex option market was caught short gamma in GBP/EUR last week. The spot rate surged from GBP0.6742 to GBP0.6973 late the previous week, 1-month volatilities went up from about 9.6% to roughly 13.3%. This move forced players to cover their gamma.

(A typical market quote)

This example shows one way *delta* and *gamma* are used by market professionals. Especially in the foreign exchange markets, options of varying moneyness characteristics are labeled according to their *delta*. For example, consider 25-*delta* Sterling puts. Given that an ATM put has a *delta* of around 50, these puts are out-of-the-money. Market makers had sold such options and, after hedging their *delta* exposure, were holding short *gamma* positions. This meant that as the Sterling–Euro exchange rate fluctuated, hedge adjustments led to higher than expected cash outflows.

9.6.3 VEGA

A critical Greek is the *vega*. How much will the value of an option change if the volatility parameter, σ, moves by an infinitesimal amount? This question relates to an option's sensitivity with respect to implied volatility movements. *Vega* is obtained by taking the partial derivative of the function with respect to σ:

$$\text{vega} = \frac{\partial C(S_t, t \mid r, \sigma, T, K)}{\partial \sigma} \tag{9.89}$$

An example of *vega* is shown in Figure 9.12 for a call option. Note the resemblance to the *gamma* displayed earlier in Figure 9.11. According to this figure, the *vega* is greatest when the option is ATM. This implies that if we use the ATM option as a vehicle to benefit from oscillations in S_t, we will also have maximum exposure to movements in the implied volatility. We consider some examples of *vega* calculations using actual data.

EXAMPLE

Vega is the sensitivity with respect to the percentage volatility parameter, σ, of the option. According to the convention, this is calculated using the Black–Scholes formula. We differentiate the formula with respect to the volatility parameter σ.

Doing this and then substituting

$$C(61.15, 0.025, 60, 58/365, \sigma) = \text{Observed price} \tag{9.90}$$

we get a measure of how this option's prices will react to small changes in σ.

For the table above, we get the following results:

Calls	Vega($)
Dec 55.00	6.02
Dec 60.00	9.4
Dec 65.00	8.9
Dec 70.00	5.6

Puts	Vega($)
Dec 55.00	6.02
Dec 60.00	9.4
Dec 65.00	8.9
Dec 70.00	5.6

We can make the following comments:

1. ATM options have the largest values of vega.
2. As implied volatility increases, the ATM vega changes marginally, whereas the out-of-the-money and in-the-money option vegas do change, and in the same direction.

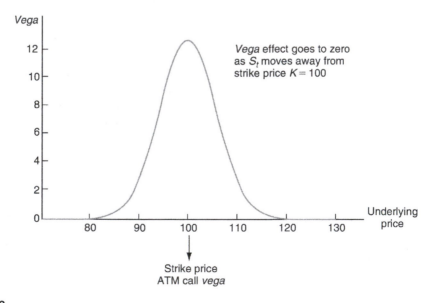

FIGURE 9.12

Vega of a long call position based on Black–Scholes formula.

Option traders can use the *vega* in calculating the "new" option price in case implied volatilities change by some projected amount. For example, in the preceding example, if the implied volatility increases by 2 percentage points, then the value of the Dec 60-put will increase approximately by 0.19, everything else being the same.

9.6.3.1 *Market use*

Vega is an important Greek because it permits market professionals to keep track of their exposure to changes in implied volatility. This is important, since the Black–Scholes formula is derived in a framework where volatility is assumed to be constant, yet used in an environment where the volatility parameter, σ, changes. Market makers often quote the σ directly, instead of quoting the Black–Scholes value of the option. Under these conditions, *vega* can be used to track exposure of option books to changes in σ. This can be followed by *vega hedging*.

The following reading is one example of the use of *vega* by the traders.

EXAMPLE

Players dumped USD/JPY vol last week in a quiet spot market, causing volatilities to go down further. One player was selling USD1 billion in six-month dollar/yen options in the market. These trades were entered to hedge vega exposure. The drop in the vols forced market makers to hedge exotic trades they had previously sold.

According to this reading, some practitioners were *long* volatility. They had bought options when the dollar–yen exchange rate volatility was higher. They faced *vega risk*. If implied volatility declined, their position would lose value at a rate depending on the position's *vega*. To cover these risky positions, they sold volatility and caused further declines in this latter. The size of *vega* is useful in determining such risks faced by such long or short volatility positions.

9.6.3.2 Vega *hedging*

Vega is the response of the option value to a change in implied volatility. In a liquid market, option traders quote implied volatility and this latter continuously fluctuates. This means that the value of an existing option position also changes as implied volatility changes. Traders who would like to eliminate this exposure use *vega hedging* in making their portfolio *vega*-neutral. *Vega* hedging in practice involves buying and selling options, since only these instruments have convexity and hence, have *vega*.

9.6.4 THETA

Next, we ask how much the theoretical price of an option would change if a small amount of time, dt, passes. We use the partial derivative of the function with respect to time parameter t, which is called *theta*:

$$\text{theta} = \frac{\partial C(S_t, t \mid r, \sigma, T, K)}{\partial t} \tag{9.91}$$

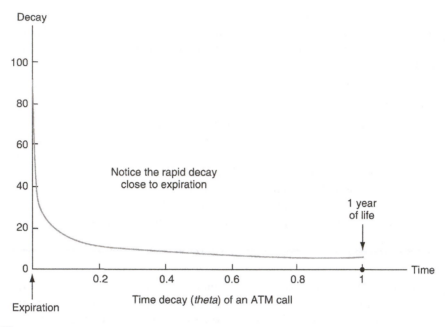

FIGURE 9.13

Theta of a long call position based on Black–Scholes formula.

According to this, *theta* measures the decay in the time value of the option. The intuition behind *theta* is simple. As time passes, one has less time to gain from future S_t oscillations. Option's time value decreases. Thus, we must have *theta* < 0.

If the Black–Scholes assumptions are correct, we can calculate this derivative analytically and plot it. The derivative is represented in Figure 9.13. We see that, all else being the same, a plain vanilla option's time value will decrease at a faster *rate* as expiration approaches.

9.6.5 OMEGA

This Greek relates to American options only and is an approximate measure developed by market professionals to measure the expected life of an American-style option.

9.6.6 HIGHER-ORDER DERIVATIVES

The Greeks seen thus far are not the only sensitivities of interest. One can imagine many other sensitivities that are important to market professionals and investors. In fact, we can calculate the sensitivity of the previously mentioned Greeks *themselves* with respect to S_t, σ, t, and r. These are higher-order cross partial derivatives and under some circumstances will be quite relevant to the trader.

Two examples are as follows. Consider the *gamma* of an option. This Greek determines how much cash can be earned as the underlying S_t oscillates. But the value of the *gamma* depends on the S_t and σ as well. Thus, a *gamma trader* may be quite interested in the following sensitivities:

$$\frac{\partial \text{ gamma}}{\partial S_t} \qquad \frac{\partial \text{ gamma}}{\partial \sigma} \tag{9.92}$$

These two Greeks are sometimes referred to as the *speed* and *zomma*, respectively.[22] It is obvious that the magnitude of these partials will be useful in determining the risks and gains of *gamma* positions. Exotic option *deltas* and *gammas* may have discontinuities, and such high-order moments may be very relevant.

Another interesting Greek is the derivative of *vega* with respect to S_t:

$$\frac{\partial \text{ vega}}{\partial S_t}$$

$$\frac{\partial \text{ vega}}{\partial \sigma} = \text{volga} \tag{9.93}$$

This derivative is of interest to a *vega trader*. In a sense, this is volatility *gamma*, hence the name. Similarly, the partial derivative of all important Greeks with respect to a small change in time parameter may provide information about the way the Greeks move over time.

9.6.7 GREEKS AND PDEs

The fundamental Black–Scholes PDE that we derived in this chapter can be reinterpreted using the Greeks just defined. In fact, we can plug the Greeks into the Black–Scholes PDE

$$\frac{1}{2} C_{ss}\sigma^2 S_t^2 + rC_s S_t - rC + C_t = 0 \tag{9.94}$$

and recast it as

$$\frac{1}{2} \text{gamma } \sigma^2 S_t^2 + r \text{ delta } S_t - rC + \text{theta} = 0 \tag{9.95}$$

In this interpretation, being long in options means, "earning" *gamma* and "paying" *theta*.

It is also worth noting that the higher-order Greeks mentioned in Eqs. (9.92) and (9.93) are *not* present in Eq. (9.95). This is because they are second-order Greeks. The first-order Greeks are related to changes in the underlying risk ΔS_t, $\Delta\sigma$, or time Δ, whereas the higher-order Greeks would relate to changes that will have sizes given by the products $(\Delta S_t \Delta\sigma)$ or $(\Delta\sigma\Delta)$. In fact, when ΔS_t, $\Delta\sigma$, Δ are "small" but non-negligible, products of two small numbers such as $(\Delta S_t \Delta\sigma)$ are even smaller and negligible, *depending* on the sizes of incremental changes in S_t, or volatility.[23]

In some real-life applications, when volatility "spikes," higher-order Greeks may become relevant. Yet, in theoretical models with standard assumptions, where $\Delta \rightarrow 0$, they fall from the overall picture, and do not contribute to the PDE in Eq. (9.94).

---

[22]Zomma will help the trader of a gamma-hedged portfolio anticipate changes to the effectiveness of the hedge as volatility changes.

[23]The Wiener process has variance dt over infinitesimal intervals, hence *gamma* relates to first-order changes.

9.6.7.1 Gamma trading

The Black–Scholes PDE can be used to explain what a *gamma trader* intends to accomplish. Assume that the real-life *gamma* is correctly calculated by choosing a formula for $C(S_t, t \mid r, K, \sigma, T)$ and then taking the derivative:

$$\text{gamma} = \frac{\partial^2 C(S_t, t \mid r, K, \sigma, T}{\partial S_t^2} \tag{9.96}$$

Following the logic that led to the Black–Scholes PDE in Eq. (9.94), a *gamma* trader would, first, form a *subjective view* on the size of expected changes in the underlying using some subjective probability P^*, as of time $t_0 < t$. The gains can be written as,[24]

$$E_{t_0}^{P*}\left[\frac{1}{2}\text{gamma}(\Delta S_t)^2\right] \tag{9.97}$$

This term would be greater, the greater the oscillations in S_t. Then these gains will be compared with interest expenses and the loss of time value. If the expected *gamma* gains are greater than these costs, then the *gamma* trader will go *long gamma*. If, in contrast, the costs are greater, the *gamma* trader will prefer to be *short gamma*.

There are at least two important comments that need to be made about trading *gammas*.

9.6.7.2 Gamma *trading versus* Vega

First of all, the *gamma* of an option position depends on the implied volatility parameter σ. This parameter represents implied volatility. It need *not* have the same value as the (percentage) oscillations anticipated by a *gamma* trader. In fact, a *gamma* trader's subjective (expected) gains, due to S_t oscillations, are given by

$$E_{t_0}^{p*}\left[\frac{1}{2}\text{gamma}(\Delta S_t)^2\right] \tag{9.98}$$

There is no guarantee that the implied volatility parameter will satisfy the equality

$$\sigma^2 S_t^2 \Delta E_{t_0}[\text{gamma}] = E_{t_0}^{P*}[\text{gamma}(\Delta S_t)^2] \tag{9.99}$$

This is *even if* the trader is correct in his or her anticipation. The right-hand side of this expression represents the anticipated (percentage) oscillations in the underlying asset that depend on a subjective probability distribution, whereas the left-hand side is the volatility value that is plugged into the Black–Scholes formula to get the option's fair price.

Thus, a *gamma* trader's gains and losses also depend on the implied volatility movements, and the option's *vega* will be a factor here. For example, a *gamma* trader may be right about increased real-world oscillations, but, may still lose money if *implied volatility*, σ, falls simultaneously. This will lower the value of the position if

$$\frac{\partial C_{ss}}{\partial \sigma} < 0 \tag{9.100}$$

---

[24]The *gamma* itself depends on S_t, so it needs to be kept inside the expectation operator.

The following reading illustrates the approaches a trader or risk manager may adopt with respect to *vega* and *gamma* risks.

EXAMPLE

The VOLX contracts, (one) the new futures based on the price volatility of three reference markets measured by the closing levels of the benchmark cash index. The three are the German (DAX), UK (FT-SE), and Swedish (OMX) markets.

The designers argue that VOLX products, by creating a term structure of volatility that is arbitrageable, offer numerous hedging and trading possibilities. This covers both vega and gamma exposures and also takes in the long-dated options positions that are traditionally very difficult to hedge with short options.

Simply put, option managers who have net short positions and therefore are exposed to increases in volatility, can hedge those positions by being long the VOLX contract. The reverse is equally true. As a pure form of vega, the contracts offer particular benefits for vega hedging. Their vega profile is constant for any level of spot ahead of the rate setting period, and then diminishes linearly once the RSP has begun.

The gamma of VOLX futures, in contrast, is very different from those of traditional options. Although a risk manager would traditionally hedge an option position by using a product with a similar gamma profile, hedging the gamma of a complex book with diversified strikes can become unwieldy. VOLX gamma, regardless of time and the level of the underlying spot, is evenly distributed. VOLX will be particularly useful for the traditionally hard to hedge out-of-the-money wings of an option portfolio.

(Thomson Reuters IFR, November 23, 1996)

9.6.7.3 Which expectation?

We characterized trading gains expected from S_t oscillations using the expression:

$$E_{t_0}^{P*}\left[\frac{1}{2}\text{gamma}(\Delta S_t)^2\right] \tag{9.101}$$

Here the expectation $E_{t_0}^{P*}[(\Delta S_t)^2]$ is taken with respect to subjective probability distribution P^*. The behavior of *gamma* traders depends on their subjective probability, but the market-determined arbitrage-free price will be objective and the corresponding expectation has to be arbitrage-free. The corresponding pricing formulas will depend on objective *risk-adjusted probabilities*.

9.7 REAL-LIFE COMPLICATIONS

In actual markets, the issues discussed here should be applied with care, because there will be significant deviations from the theoretical Black—Scholes world. By *convention*, traders consider the Black—Scholes world as the benchmark to use, although its shortcomings are well known.

Every assumption in the Black—Scholes world can be violated. Sometimes these deviations are harmless or can easily be accommodated by modifying the formula. Some such modifications of the formula would be minor, and others more significant, but in the end they take care of the problem at a reasonable effort.

Yet, there are two cases that require substantial modifications. The first concerns the behavior of volatility. In financial markets, not only is volatility *not* constant, but it also has some unexpected characteristics. One of these anomalies is the *smile effects*.[25] Volatility has, also, a *term structure*.

The second case is when interest rates are stochastic, and the underlying asset is an interest rate-related instrument. Here, the deviation from the Black—Scholes world, again, leads to significant changes.

9.7.1 DEALING WITH OPTION BOOKS

This chapter discussed *gamma*, *delta*, and *vega* risks for single option positions. Yet, market makers do not deal with single options. They have option books and they try to manage the *delta*, *gamma*, and *vega* risks of *portfolios* of options. This complicates the hedging and risk management significantly. The existence of exotic options compounds these difficulties.

First of all, option books consist of options on different, possibly correlated, assets. Second, implied volatility may be different across strikes and expiration dates, and a straightforward application of *delta*, *gamma*, and *vega* concepts to the portfolio may become impossible. Third, while for single options *delta*, *vega*, and *gamma* have known shapes and dynamics, for portfolios of options, the shapes of *delta*, *gamma*, and *vega* are more complex and their movement over time may be more difficult to track.

9.7.2 FUTURES AS UNDERLYING

This chapter has discussed options written on cash instruments. How would we analyze options that are written on a futures or forward contract? There are two steps in designing option contracts. First, a futures or a forward contract is introduced on the cash instrument, and second, an option is written on the futures. The holder of the option has the right to buy one or more futures contracts.

Why would anyone write an option on futures (forwards), instead of writing it on the cash instrument directly?

In fact, the advantages of such contracts are many, and the fact that option contracts written on futures and forwards are the most liquid is not a coincidence. First of all, if one were to buy and sell the underlying in order to hedge the option positions, the futures contracts are more convenient. They are more liquid, and they do not require upfront cash payments. Second, hedging with cash instruments could imply, for example, selling or buying thousands of barrels of oil. Where would a trader put so much oil, and where would he get it? Worse, dynamic hedging requires adjusting such positions continuously. It would be very inconvenient to buy and sell a cash underlying. Long and short positions in futures do not result in delivery until the expiration date. Hence, the trader can constantly adjust his or her position without having to store barrels of oil at each rebalancing of the hedge. Futures are also more liquid and the associated transactions costs and counterparty risks are much smaller.

Thus, the choice of futures and forwards as the underlying instead of cash instruments is, in fact, clever contract design. But we must remember that futures come with daily marking to market. Forward contracts, on the other hand, may not require any marking to market until the expiration date.

---

[25]Smile is the change in implied volatility as strike price changes. It will be dealt with in Chapter 16.

9.7.2.1 *Delivery mismatch*

Note the possibility of a mismatch. The option may result in the delivery of a futures contract at time T, but the futures contract may not expire at that same time. Instead, it may expire at a time $T + \Delta$ and may result in the delivery of the cash commodity. Such timing mismatches introduce new risks.

9.8 CONCLUSION: WHAT IS AN OPTION?

This chapter has shown that an option is essentially a volatility instrument. The critical parameter is how much the underlying risk oscillates within a given interval. We also saw that there are many other risks to manage. The implied volatility parameter, σ, may change, interest rates may fluctuate, and option sensitivities may behave unexpectedly. These risks are not "costs" of maintaining the position perhaps, but they affect pricing and play an important role in option trading.

SUGGESTED READING

Most textbooks approach options as directional instruments. There are, however, some nontechnical sources that treat options as volatility instruments directly. The first to come to mind is **Natenberg** (1994). A reader who prefers a technical approach has to consider more abstract treatments such as **Musiela** and **Rutkowski** (1998). Several texts discuss Black−Scholes theory. The one that we recommend is **Duffie** (2001). Readers should look at **Wilmott** (2006) for the technical details. For the useful combination of options analysis with Mathematica, the reader can consult **Stojanovic** (2003). Risk publications have several books that collect articles that have the same approach used in this chapter. There, the reader will find a comprehensive discussion of the Black−Scholes formula. Examples on Greeks were based on the terminology used in Derivatives Week (now part of GlobalCapital).

APPENDIX 9.1

In this appendix, we derive formulas for *delta* and *gamma*. The relatively lengthy derivation is for *delta*.

DERIVATION OF DELTA

The Black−Scholes formula for a plain vanilla European call expiration T, strike, K, is given by

$$C(S_t, t) = S_t \int_{-\infty}^{\frac{\log(S_t/K)+(r+(1/2)\sigma^2)(T-t)}{\sigma\sqrt{T-t}}} \frac{1}{\sqrt{2\pi}} e^{-(1/2)u^2}\,\mathrm{d}u - e^{-r(T-t)}K \int_{-\infty}^{\frac{\log(S_t/K)+(r-(1/2)\sigma^2)(T-t)}{\sigma\sqrt{T-t}}} \frac{1}{\sqrt{2\pi}} e^{-(1/2)u^2}\,\mathrm{d}u \qquad (9.102)$$

Rearrange and let $x_t = (S_t/Ke^{-r(T-t)})$, to get

$$C(x_t, t) = Ke^{-r(T-t)} \left[x_t \int_{-\infty}^{\frac{\log x_t + (1/2)\sigma^2(T-t)}{\sigma\sqrt{T-t}}} \frac{1}{\sqrt{2\pi}} e^{-(1/2)u^2} du - \int_{-\infty}^{\frac{\log x_t - (1/2)\sigma^2(T-t)}{\sigma\sqrt{T-t}}} \right] \tag{9.103}$$

$$\frac{1}{\sqrt{2\pi}} e^{-(1/2)u^2} du \tag{9.104}$$

Now differentiate with respect to x_t:

$$\frac{dC(x_t, t)}{dx_t} = Ke^{-r(T-t)} \left[\int_{-\infty}^{\frac{\log x_t + (1/2)\sigma^2(T-t)}{\sigma\sqrt{T-t}}} \frac{1}{\sqrt{2\pi}} e^{-(1/2)u^2} du \right] + \frac{1}{\sigma\sqrt{T-t}} \tag{9.105}$$

$$\left[\frac{1}{\sqrt{2\pi}} e^{-\frac{1}{2}\left(\frac{\log x_t + (1/2)\sigma^2(T-t)}{\sigma\sqrt{T-t}} \right)^2} \right] \tag{9.106}$$

$$- \left[\frac{1}{x_t \sigma\sqrt{T-t}} \frac{1}{\sqrt{2\pi}} e^{-\frac{1}{2}\left(\frac{\log x_t - (1/2)\sigma^2(T-t)}{\sigma\sqrt{T-t}} \right)^2} \right] \tag{9.107}$$

Now we show that the last two terms in this expression sum to zero and that

$$\frac{1}{\sigma\sqrt{T-t}} \left[\frac{1}{\sqrt{2\pi}} e^{-\frac{1}{2}\left(\frac{\log x_t + (1/2)\sigma^2(T-t)}{\sigma\sqrt{T-t}} \right)^2} \right] = \frac{1}{x_t \sigma\sqrt{T-t}} \frac{1}{\sqrt{2\pi}} e^{-\frac{1}{2}\left(\frac{\log x_t - (1/2)\sigma^2(T-t)}{\sigma\sqrt{T-t}} \right)^2} \tag{9.108}$$

To see this, on the right-hand side, use the substitution:

$$\frac{1}{x_t} = e^{-\log x_t} \tag{9.109}$$

and then rearrange the exponent in the exponential function.
   Thus, we are left with

$$\frac{\partial C(x_t, t)}{\partial x_t} = Ke^{-r(T-t)} \left[\int_{-\infty}^{\frac{\log x_t + (1/2)\sigma^2(T-t)}{\sigma\sqrt{T-t}}} \frac{1}{\sqrt{2\pi}} e^{-(1/2)u^2} du \right] \tag{9.110}$$

Now use the chain rule and obtain

$$\frac{\partial C(S_t, t)}{\partial S_t} = \left[\int_{-\infty}^{\frac{\log x_t + (1/2)\sigma^2(T-t)}{\sigma\sqrt{T-t}}} \frac{1}{\sqrt{2\pi}} e^{-(1/2)u^2} du \right] \tag{9.111}$$

$$= N(d_1) \tag{9.112}$$

DERIVATION OF GAMMA

Once *delta* of a European call is obtained, the *gamma* will be the derivative of the *delta*. This gives

$$\frac{\partial^2 C(S_t, t)}{\partial S_t^2} = \frac{1}{S_t \sigma \sqrt{T-t}} \frac{1}{\sqrt{2\pi}} e^{-\frac{1}{2}\left(\frac{\log x_t + (1/2)\sigma^2(T-t)}{\sigma\sqrt{T-t}}\right)^2} \tag{9.113}$$

with $x_t = (S_t / K e^{-r(T-t)})$

APPENDIX 9.2

In this appendix, we review some basic concepts from stochastic calculus. This brief review can be used as a reference point for some of the concepts utilized in later chapters. Øksendal (2010) is a good source that provides an introductory discussion on stochastic calculus. Heuristics can be found in Hirsa and Neftci (2013).

STOCHASTIC DIFFERENTIAL EQUATIONS

An SDE, driven by a Wiener process W_t, is written as

$$dS_t = a(S_t, t)dt + b(S_t, t)dW_t \quad t \in [0, \infty) \tag{9.114}$$

This equation describes the dynamics of S_t over time. The *Wiener process* W_t has increments ΔW_t that are normally distributed with mean zero and variance Δ, where the Δ is a small time interval. These increments are uncorrelated over time. As a result, the future increments of a Wiener process are unpredictable given the information at time t, the I_t.

The $a(S_t, t)$ and the $b(S_t, t)$ are known as the *drift* and the *diffusion* parameters. The drift parameter models *expected* changes in S_t. The diffusion component models the corresponding *volatility*. When unpredictable movements occur as jumps, this will be referred as a *jump component*.

A jump component would require adding terms such as $\lambda(S_t, t)dJ_t$ to the right-hand side of the SDE shown above. Otherwise S_t will be known as a *diffusion process*. With a jump component it becomes a *jump-diffusion process*.

Examples

The simplest SDE is the one where the drift and diffusion coefficients are independent of the information received over time:

$$dS_t = \mu dt + \sigma \, dW_t \quad t \in [0, \infty) \tag{9.115}$$

Here, W_t is a standard Wiener process with variance t. In this SDE, the coefficients μ and σ do not have time subscripts t, as time passes, they do not change.

The standard SDE used to model underlying asset prices is the *geometric process*. It is the model assumed in the Black and Scholes world:

$$dS_t = \mu S_t dt + \sigma S_t dW_t \quad t \in [0, \infty) \tag{9.116}$$

This model implies that drift and the diffusion parameters change proportionally with S_t.

An SDE that has been found useful in modeling interest rates is the *mean reverting* model:

$$dS_t = \lambda(\mu - S_t)dt + \sigma S_t dW_t \quad t \in [0, \infty) \tag{9.117}$$

According to this, as S_t falls below a "long-run mean" μ, the term $(\mu - S_t)$ will become positive, which makes dS_t more likely to be positive, hence, S_t will revert back to the mean μ.

ITO'S LEMMA

Suppose $f(S_t)$ is a function of a *random* process S_t having the dynamics:

$$dS_t = a(S_t, t)dt + b(S_t, t)dW_t \quad t \in [0, \infty) \tag{9.118}$$

We want to expand $f(S_t)$ around a known value of S_t, say S_0 using Taylor series expansions. The expansion will yield:

$$f(S_t) = f(S_0) + f_s(S_0)[S_t - S_0] + \frac{1}{2}f_{ss}(S_0)[S_t - S_0]^2 + R(S_t, S_0) \tag{9.119}$$

where $R(S_t, S_0)$ represents all the remaining terms of the Taylor series expansion.

First note that $f(S_t)$ can be rewritten as, $f(S_0 + \Delta S_t)$, if we define ΔS_t as:

$$\Delta S_t = S_t - S_0 \tag{9.120}$$

Then, the Taylor series approximation will have the form:

$$f(S_0 + \Delta S_t) - f(S_0) \cong f_s \Delta S_t + \frac{1}{2}f_{ss}\Delta S_t^2 \tag{9.121}$$

The ΔS_t is a "small" change in the random variable S_t. In approximating the right-hand side, we *keep* the term $f_s \Delta S_t$.

Consider the second term $(1/2)f_{ss}(\Delta S_t)^2$. If S_t is deterministic, one can say that the term $(\Delta S_t)^2$ is small. This could be justified by keeping the size of ΔS_t non-negligible, yet small enough that its square $(\Delta S_t)^2$ *is* negligible. However, here, changes in S_t will be random. Suppose these changes have zero mean. Then the variance is

$$0 < E[\Delta S_t]^2 \cong b(S_t, t)^2 \Delta \tag{9.122}$$

This equality means that as long as S_t is random, the right-hand side of Eq. (9.121) must keep the second-order term in any type of Taylor series approximation.

Moving to infinitesimal time dt, this gives Ito's Lemma, which is the stochastic version of the Chain rule,

$$df(S_t) = f_s dS_t + \frac{1}{2}f_{ss}b(S_t, t)^2 dt \tag{9.123}$$

This equation can be regarded as the dynamics of the process $f(S_t)$, which is driven by S_t. The dS_t term in the above equation can be substituted out using the S_t dynamics.

GIRSANOV THEOREM

Girsanov Theorem provides the general framework for transforming one probability measure into another "equivalent" measure. It is an abstract result that plays a very important role in pricing.

In heuristic terms, this theorem says the following. If we are given a Wiener process W_t, then, we can multiply the probability distribution of this process by a special function ξ_t that depends on time t, and on the information available at time t, the I_t. This way we can obtain a *new* Wiener process $\tilde{W}_t$ with probability distribution $\tilde{P}$. The two processes will relate to each other through the relation:

$$d\tilde{W}_t = dW_t - X_t dt \qquad (9.124)$$

That is to say, $\tilde{W}_t$ is obtained by subtracting an I_t-dependent term X_t, from W_t.

Girsanov Theorem is often used in the following way: (i) we have an expectation to calculate, (ii) we transform the original probability measure, such that expectation becomes easier to calculate, and (iii) we calculate the expectation under the new probability.

EXERCISES

1. Consider the following comment dealing with options written on the euro—dollar exchange rate:

    Some traders, thinking that implied volatility was too high entered new trades. One example was to sell one-year in-the-money euro Puts with strikes around USD1.10 and buy one-year at-the-money euro Puts. If the euro is above USD1.10 at maturity, the trader makes the difference in the premiums. The trades were put on across the curve. (Based on an article in Derivatives Week (now part of GlobalCapital)).

    a. Draw the profit/loss diagrams of this position at expiration for each option separately.
    b. What would be the gross payoff at expiry?
    c. What would be the net payoff at expiry?
    d. Why would the traders buy "volatility" given that they buy and sell options? Don't these two cancel each other in terms of volatility exposure?

2. (Delta Hedge Portfolio)

    Write a VBA program to show the delta-hedged portfolio adjustments and cash flows for a long call from the dealers' perspective with the following data:
    - Total number of long calls $N = 100$
    - Number of adjustments $M = 15$
    - $S(0) = 100$; $K = 100$; $T = 1$; $r = 8\%$;
    - $\sigma = 30\%$; and realized volatility $= 50\%$.

3. Consider the following quote:

    Implied U.S. dollar/New Zealand dollar volatility fell to 10.1%/11.1% on Tuesday. Traders bought at-the-money options at the beginning of the week, ahead of the Federal Reserve interest-rate cut. They anticipated a rate cut which would increase short-term volatility. They wanted to be long gamma. Trades were typically for one-week maturities, in average notionals of USD10—20 million. (Based on an article in Derivatives Week (now part of GlobalCapital)).

    a. Explain why traders wanted to be long *gamma* when the volatility was expected to increase.
    b. Show your argument using numerical values for Greeks and the data given in the reading.

   c. How much money would the trader lose under these circumstances? Calculate approximately, using the data supplied in the reading. Assume that the position was originally for USD30 million.

4. Consider the following episode:

EUR/USD one-month implied volatility sank by 2.7% to 10% Wednesday as traders hedged this euro exposure against the greenback, as the euro plunged to historic lows on the spot market. After the European Central Bank raised interest rates by 25 basis points, the euro fell against leading to a strong demand for euro Puts. The euro touched a low of USD0.931 Wednesday. (Based on an article in Derivatives Week.)

   a. In the euro/dollar market, traders rushed to stock up on *gamma* by buying short-dated euro puts struck below USD0.88 to hedge against the possibility that the interest rates rise. Under normal circumstances, what would happen to the currency?

   b. When the euro failed to respond and fell against major currencies, why would the traders then rush to buy euro puts? Explain using payoff diagrams.

   c. Would a trader "stock up" *gamma* if euro-triggered barrier options?

5. You are given the following table concerning the price of a put option satisfying all Black–Scholes assumptions. The strike is 20 and the volatility is 30%. The risk-free rate is 2.5%.

Option Price	Underlying Asset Price
10	10
5	15
1.3	20
0.25	25
0.14	30

The option expires in 100 days. Assume (for convenience), that, for every month the option loses approximately one-third of its value.

   a. How can you approximate the option *delta?* Calculate *three* approximations for the *delta* in the previous case.

   b. Suppose you bought the option when the underlying was at 20 using borrowed funds. You have hedged this position in a standard fashion. How much do you gain or lose in four equal time periods if you observe the following price sequence in that order:

$$10, 25, 25, 30 \tag{9.125}$$

   c. Suppose now that the underlying price follows the new trajectory given by

$$10, 30, 10, 30 \tag{9.126}$$

How much do you gain or lose until expiration?

   d. Explain the difference between gains and losses.

6. Search the Internet for the following questions.

   a. Which sensitivities do the Greeks, volga and Vanna represent?

   b. Why are they relevant for *vega* hedging?

CASE STUDY: SWISS CENTRAL BANK, 2012

"'Exploiting the franc peg'
   [...]
   The Swiss currency is no longer rallying the way it did during market distress on Eurozone debt concerns. It all changed when the Swiss National Bank (SNB) announcement pegging its currency against the euro at the EUR/CHF rate of 1.20, aimed at preventing excessive franc acceleration against the debt-ridden euro. As credit rating agencies rushed to downgrade the sovereign debt of Southern Europe in late 2009, investors rushed their savings out of the single currency and into safe-haven francs. The exodus took the form of cash flight, property sales and bank transfers as "default" became a recurring theme in Greece, Spain, Portugal, Ireland and Italy.
   Consequently, the franc soared 35% and 40% against the euro and the USD respectively from 2009 to September 2011.
   The SNB began massive interventions in March 2009 to sell its currency for euros to stem the tide of the soaring franc. But the surge of franc-bound capital caused the SNB to lose more than CHF 20 billion from early 2009 to end of 2010. When the central bank's losses became a matter of national urgency, it ultimately went with the "nuclear option."
   On Sept. 6, 2011 the SNB announced it "will no longer tolerate a EUR/CHF exchange rate below the minimum rate of CHF 1.20. The SNB will enforce this minimum rate with the utmost determination and is prepared to buy foreign currency in unlimited quantities."
   SNB had pegged its currency once before against the Deutsche mark in 1978 after francs soared close to 100%, near doubling in the prior six years as a result of U.S. stagflation following the oil crisis. The franc/mark peg lasted two years and dragged down the franc's effective trade index by about 14%.
   The question then becomes whether the SNB remains successful in stemming further CHF strength. Will it make sense to sell the franc into 2012? As of this writing, the Swiss franc lost nearly one third of its value against the euro and U.S. dollar since the euro peg began in September.
   [...]
   Our favorite pick against the Swiss franc would be the Canadian dollar because of ongoing volatility in oil prices stemming from Mideast uncertainty. Aside from the energy argument sustaining the loonie, any surprise bounce in U.S. growth is likely to help loonie sentiment. CAD/CHF is seen extending gains towards 0.96 from the current 0.90, with support cropping up at 0.88. An alternative but similar trade would be to combine longs in USD/CHF with shorts in USD/CAD." (source: www.futuresmag.com, Exploiting the franc peg, by Ashraf Laidi, January 1, 2012)

7. Answer the questions related to the case study Swiss Central Bank, 2012:
    a. What is the rationale for the proposed futures strategy
    b. What options strategy could you consider that would also benefit from the peg?
8. (MATLAB exercise on European Option)
        Write a MATLAB program to determine the initial price of European Call and European Put option using the BSM formula with the following data:
        $S(0) = 100$; $K = 105$; $T = 1$; $r = 8\%$; $\sigma = 30\%$
        Now plot the initial price of both call and put options by varying the following parameters at a time while keeping the other parameters fixed.
    a. $S(0)$
    b. K
    c. r
    d. σ
        Also plot the initial call and put price juxtaposed by varying two parameters at a time and keeping other parameters fixed.

e. $S(0)$ & K

f. K & σ

g. $S(0)$ & σ

9. (MATLAB exercise on Chooser Option)

Write a MATLAB program to determine the initial price of chooser option using the BSM formula with the following data:

$S(0) = 100$; $K = 105$; $T_0 = 0.5$; $T = 1$; $r = 8\%$; $\sigma = 30\%$

Now plot the initial price of option by varying the following parameters at a time while keeping the other parameters fixed.

a. $S(0)$

b. K

c. T_0

Also plot the variation of option price with T_0 & T together and keeping other parameters fixed.

10. (MATLAB exercise on Barrier Option)

Write a MATLAB program to determine the initial price of Barrier Down and In Call option and Barrier Down and Out Call option using the BSM formula with the following data:

$S(0) = 100$; $K = 105$; $T = 1$; $r = 8\%$; $\sigma = 30\%$; $H = 95$

Now plot the initial price of both call and put options by varying the following parameters at a time while keeping the other parameters fixed.

a. $S(0)$

b. K

c. H

11. MATLAB exercise on a Delta-Hedged Portfolio

Write a MATLAB program to delta hedge the portfolio consisting only of the stock and the risk-free asset to cover the long call option. Observe the number of shorted shares and the cash flow during the adjustment of the portfolio to make it delta neutral at each time point. Report the final cash position at the expiration after settling all the open positions.

Gradually increase the frequency of adjustment and observe the net cash position and also plot the performance vs frequency of this delta hedging that how correctly it covers the long call position.

Use the following data:

$S(0) = 100$; $K = 105$; $T = 1$; $r = 8\%$; $\sigma = 30\%$ and realized volatility $= 50\%$

Frequency: [6, 12, 50, 100, 300]

ENGINEERING CONVEXITY POSITIONS

CHAPTER OUTLINE

10.1 INTRODUCTION

How can anyone trade volatility? Stocks, yes. Bonds, yes. But volatility is not even an *asset*. Several difficulties are associated with defining precisely what volatility is. For example, from a technical point of view, should we define volatility in terms of the estimate of the conditional *standard deviation* of an asset price S_t?

$$\sqrt{E_t[S_t - E_t[S_t]]^2} \qquad (10.1)$$

Or should we define it as the average *absolute* deviation?

$$E_t[|S_t - E_t[S_t]|] \qquad (10.2)$$

There is no clear answer, and these two definitions of statistical volatility will yield *different* numerical values. Leaving statistical definitions of volatility aside, there are many instances where traders quote, directly, the volatility instead of the dollar value of an instrument. For example, interest rate derivatives markets quote *cap-floor* and *swaption* volatilities.[1] Equity options provide *implied volatility*. Implied volatility is calculated by taking the market price of an option and backing out the implied volatility that results in the market price given a particular option pricing model and other input parameters. Traders and market makers trade the quoted volatility. Hence, there must be some way of isolating and pricing what these traders call volatility in their respective markets.

We started seeing how this can be done in Chapter 9. Options became *more valuable* when "volatility" increased, everything else being the same. Chapter 9 showed how these strategies can quantify and measure the "volatility" of an asset in *monetary* terms. This was done by forming *delta*-neutral portfolios, using assets with different degrees of convexity. In this chapter, we develop this idea further, apply it to instruments other than options, and obtain some generalizations. The plan for this chapter is as follows.

First, we show how convexity of a long bond relates to *yield volatility*. The higher the volatility of the associated yield, the higher the benefit from holding the bond. We will discuss the mechanics of valuing this convexity. Then, we compare these mechanics with option-related convexity trades. We see some close similarities and some differences. At the end, we generalize the results to any instrument with different convexity characteristics. The discussion associated with *volatility trading* itself has to wait until Chapter 15, since it requires an elementary treatment of arbitrage pricing theory.

10.2 A PUZZLE

Here is a puzzle. Consider the yield curve shown in Figure 10.1. The 10-year zero-coupon bond has a yield to maturity that equals 5.2%. The 30-year zero, however, has a yield to maturity of just 4.94%. In other words, if we buy and hold the latter bond 20 *more* years, we would receive a *lower* yield during its lifetime.

---

[1]Caps, floors, and swaptions are discussed in detail in a future chapter. Caps and floors are sequences of call and put options on interest rates, respectively. A swaption is an option on a swap.

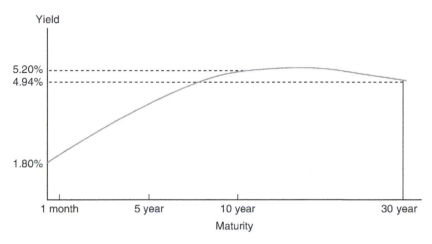

FIGURE 10.1

Yield curve.

It seems a bit strange that the longer maturity is compensated with a lower yield. There are several economic or institutional explanations for this phenomenon. For example, expectations for inflation 20 years down the line may be less than the inflationary expectations for the next 10 years only. Or, the relative demands for these maturities may be determined by institutional factors and, because players don't like to move out of their "preferred" maturity, the yield curve may exhibit such inconsistencies. Insurance companies, for example, need to hedge their positions on long-term retirement contracts and this preference may lower the yield and raise the price of long bonds.

But these explanations can hardly fully account for the observed anomaly. Institutional reasons such as *preferred habitat* and treasury debt retirement policies that reduce the supply of 30-year treasuries may account for some of the difference in yield, but it is hard to believe that an additional 20-year duration is compensated so little. Can there be another explanation?

In fact, the yield to maturity may not show all the gains that can be realized from holding a long bond. This may be hard to believe, as yield to maturity is *by definition* how much the bond will yield per annum if kept until maturity.

Yet, there can be additional gains to holding a long bond, due to the convexity properties of the instrument, depending on what else is available to trade "against" it, and depending on the underlying volatility. These could explain the "puzzle" shown in Figure 10.1. The 4.94% paid by the 30-year treasury, *plus* some additional gains, could exceed the total return from the 10-year bond. This is conceivable since the yield to maturity and the *total return* of a bond are, in fact, quite different ways of measuring financial returns on fixed-income instruments.

10.3 BOND CONVEXITY TRADES

We have already seen convexity trades within the context of vanilla options. Straightforward discount bonds, especially those with long maturities, can be analyzed in a similar fashion and have exposure to interest rate volatility. In fact, a "long" bond and a vanilla option are *both* convex instruments and they both coexist with instruments that are either linear or have less convexity.[2] Hence, a *delta*-neutral portfolio can be put together for long maturity bonds to benefit from volatility shifts. The overall logic will be similar to the options discussed in the previous chapter.

Consider a long maturity default-free discount bond with price $B(t, T)$, with $t < T$. This bond's price at time t can be expressed using the corresponding time t yield, y_t^T:

$$B(t, T) = \frac{1}{(1 + y_t^T)^T} \qquad (10.3)$$

For $t = 0$, and $T = 30$, this function is plotted against various values of the 30-year zero-coupon yield, in Figure 10.2. It is obvious that the price is a *convex* function of the yield.

A short bond, on the other hand, can be represented in a similar space with an almost linear curve. For example, Figure 10.3 plots a 1-year bond price $B(0, 1)$ against a 1-year yield y_0^1. We see that the relationship is essentially linear.[3]

The main point here is that, under some conditions, using these two bonds we can put together a portfolio that will isolate bond convexity gains similar to the convexity gains that the dynamic hedging of options has generated. Thus, suppose movements in the two yields y_t^1 and y_t^{30} are perfectly correlated over time t.[4] Next, consider a trader who tries to duplicate the strategy of the option market maker discussed in the previous chapter. The trader buys the long bond with borrowed funds and *delta* hedges the first-order yield exposure by shorting an appropriate amount of the shorter maturity bond.

This trader will have to borrow $B(0, 30)$ dollars to buy and fund the long bond position. The payoff of the portfolio

$$\{\text{Long bond, loan of } B(0, 30) \text{ dollars}\} \qquad (10.4)$$

is as shown in Figure 10.2b as curve BB'. Now compare this with Figure 10.2c. Here we show the profit/loss position of a market maker who buys an at-the-money "put option" on the yield y_t^{30}. At expiration time T, the option will pay

$$P(T) = \max[y_0^{30} - y_T^{30}, 0] \qquad (10.5)$$

This option is financed by a money market loan so that the overall position is shown as the downward sloping curve BB'.[5] We see a great deal of resemblance between the two positions.

---

[2] The short maturity bonds are almost linear. In the case of vanilla options, positions on underlying assets such as stocks are also linear.

[3] In fact, a first-order Taylor series expansion around zero yields

$$B(0, 1) \quad = \frac{1}{(1 + y_0^1)}$$

$$\cong (1 - y_0^1)$$

if the y_0^1 is "small."

[4] This simplifying assumption implies that all bonds are affected by the *same* unpredictable random shock, albeit to a varying degree. It is referred to as the 1 factor model.

[5] The option price is the curve BB'. The curve shifts down by the money market loan amount P0, which makes the position one of zero cost.

(a)

Bond price

Current
price
$B(0, 30)$

y_0^{30}

y_t^T

(b)

Net position

B

Net bond position after borrowing $B(0, 30)$,
and buying the 30-year bond

y_0^{30}

y_t^T

B'

(c)

Option value

B

Put option with strike $K = y_0^{30}$
financed by a money market loan

Current
premium

K

y_t^T

B'

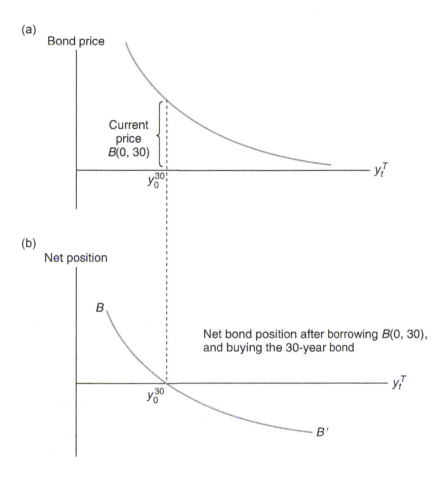

FIGURE 10.2

Bond and put option as examples of convex instruments.

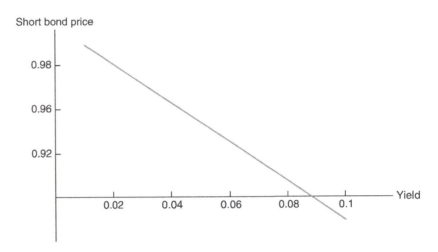

FIGURE 10.3

Price—yield relationship for a short bond.

Given this similarity between bonds and options, we should be able to isolate convexity or *gamma* trading gains in the case of bonds as well. In fact, once this is done, using an arbitrage argument, we should be able to obtain a partial differential equation (PDE) that default-free discount bond prices will satisfy. This PDE will have close similarities to the Black—Scholes PDE derived in Chapter 8.

The discussion below proceeds under some simplifying and unrealistic assumptions. We use the so-called one-factor model. Our purpose is to *understand* the mechanics of volatility trading in the case of bonds and this assumption simplifies the exposition significantly. Our context is different than in real life, where fixed-income instruments are affected by more than a single common random factor. Thus, we make two initial assumptions:

1. There is a *short* and a *long* default-free discount bond with maturities T^s and T, respectively. Both bonds are *liquid* and can be traded without any transaction costs.
2. The two bond prices depend on the same risk factor denoted by r_t. This can be interpreted as a spot interest rate that captures all the randomness at time t and is the single factor mentioned earlier.

The second assumption means that the two bond prices are a function of the short rate r_t. These functions can be written as

$$B(t, T^S) = S(r_t, t, T^S) \tag{10.6}$$

and

$$B(t, T) = B(r_t, t, T) \tag{10.7}$$

where $B(t, T^s)$ is the time t price of the short bond and $B(t, T)$ is the time t price of the long bond. We postulate that the maturity T^s is such that the short bond price $B(t, T^s)$ is (almost) a *linear* function of r_t, meaning that the second derivative of $B(t, T^s)$ with respect to r_t is negligible.

Thus, we will proceed as if there was a single underlying risk that causes price fluctuations in a convex and a quasi-linear instrument, respectively. We will discuss the cash gains generated by the dynamically hedged bond portfolio in this environment.

10.3.1 DELTA-HEDGED BOND PORTFOLIOS

The trader buys the long bond with borrowed funds and then hedges the downside risk implied by the curve AA' in Figure 10.4. The hedge for the downside risk will be a position that makes money when r_t increases and loses money when r_t declines. This can be accomplished by shorting an appropriate number of the short bond.

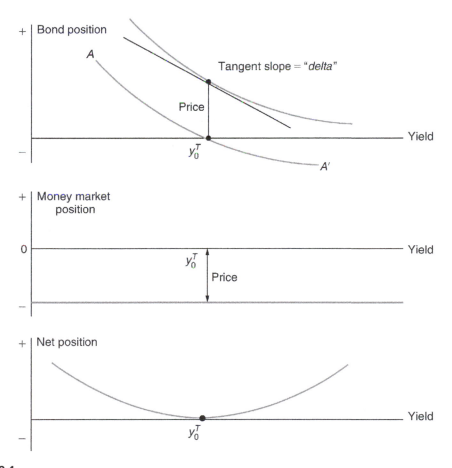

FIGURE 10.4

Delta-hedged bond portfolio.

In fact, the trick to form a *delta*-neutral portfolio is the same as in Chapter 9. Take the partial derivative of the functions $S(r_t, t, T^s)$ and $B(r_t, t, T)$ with respect to r_t, evaluate them at point r_{t_0}, and use these to form a hedge ratio, h_t:

$$h_t = \frac{(\partial B(r_t, t, T)/\partial r_t)}{(\partial S(r_t, t, T^s)/\partial r_t)} \tag{10.8}$$

$$= \frac{B_r}{S_r} \tag{10.9}$$

S_r is assumed to be a constant, given the quasi-linearity of the short bond price with respect to r_t. h_t is a function of r_t, since B_r is not constant due to the long bond's convexity. Given the value of r_{t_0}, h_t can be numerically calculated, and h_{t_0} units of the short maturity bond would be *sold* short at t_0.

The change in the value of this portfolio due to a small change in the spot rate Δr_t only is given by

$$\Delta[B(r_t, t, T) - h_t S(r_t, t, T^s)] = B_r \Delta r_t - \frac{B_r}{S_r} S_r \Delta r_t + R \tag{10.10}$$

$$= R \tag{10.11}$$

since the S_r terms cancel out. R is the remainder term of the implied Taylor series approximation, or Ito's Lemma in this case, which depends essentially on the second derivative, B_{rr}, and on r_t volatility. S_r is approximately constant. This means that the net position,

$$\{\text{Borrow } B(t, T) \text{ dollars, Buy one } B(t, T), \text{Short } h_t \text{ units of } B(t, T^s)\} \tag{10.12}$$

will have the familiar volatility position shown in the bottom part of Figure 10.4. As r_t fluctuates, this position is adjusted by buying and (short) selling an appropriate number of the nonconvex asset. The new value of partial derivative, h_t, is used at each readjustment. Again, just as in Chapter 8, this will make the practitioner "sell high" and "buy low" (or vice versa). As a result of these hedge adjustments, the counterparty who *owns* the long bond will earn *gamma* profits. These trading gains will be greater as volatility increases. Hence, we reach the result:

- Everything else being the same, the greater the volatility of r_t, the more "valuable" the long bond.

This means that as volatility increases, *ceteris paribus*, the yield of the convex instruments should decline, since more market participants will try to put this trade in place and drive its price higher.

EXAMPLE

Suppose that initially the yield curve is flat at 5%. The value of a 30-year default-free discount bond is given by

$$B(0, 30) = \frac{1}{(1 + 0.05)^{30}} \tag{10.13}$$

$$= 0.23 \tag{10.14}$$

The original delta of the bond, D_{t_0} at $r_{t_0} = 0.05$ will be:

$$D_{t_0} = -\frac{30}{\left(1+r_{t_0}\right)^{31}} \tag{10.15}$$

$$= -6.61 \tag{10.16}$$

A 1-year short bond is assumed to have an approximately linear pricing formula

$$B(t_0, T^s) = \left(1 - r_{t_0}\right) \tag{10.17}$$

$$= 0.95 \tag{10.18}$$

The market maker will borrow 0.23 dollars, buy one long bond, and then hedge this position by shorting

$$\frac{-6.61}{-1.0} \tag{10.19}$$

units of the short bond. (Given linearity approximation the short bond has unit interest sensitivity.)

A small time, Δ, passes. All rates change, r_t moves to 6%. The portfolio value will move

$$\Delta B(t, T) - h_t \Delta B(t, T^s) = \left[\frac{1}{(1+0.06)^{30}} - \frac{1}{(1+0.05)^{30}}\right] \\ -6.61[(1-0.6) - (1-0.05)] \tag{10.20}$$

$$= 0.009 \tag{10.21}$$

Note that in calculating this number, we are assuming that Δ is small. Only the effect of changing r_t is taken into account. In a sense, we are using a framework similar to partial derivatives.

The new delta is calculated as -4.9. The adjusted portfolio should be short 4.9 units of the short bond. Thus,

$$(6.6 - 4.9) = 1.7 \tag{10.22}$$

units need to be covered at a price of 0.94 each to bring the position to the desired delta-neutral state.

This leaves a trading profit equal to

$$1.7(0.95 - 0.94) = \$0.017 \tag{10.23}$$

Another period passes, with r_t going back to $r_{t_2} = 0.05$. The cycle repeats itself. The delta will change again, the portfolio will be readjusted, and trading profits will continue to accumulate.

This example is approximate, since not all costs of the position are taken into account. The example started with the assumption of a flat yield curve, which was later relaxed and the yields became volatile. However, we never mentioned what causes this change. It turns out that volatility leads to additional gains for long bond holders and this increases the demand for them. As a result, *ceteris paribus*, long bond yields would *decline* relative to short bond yields. Hence, the introduction of yield volatility changed the structure of the initial yield curve.

10.3.2 COSTS

What are the costs (and other gains) of putting together such a long volatility position using default-free discount bonds? First, there is the funding cost. To buy the long bond, $B(t, T)$ funds were borrowed at r_t percent per annum. As long as the position is kept open, interest expense will be incurred. Second, as time passes, the pricing function of the bond becomes less and less convex, and hence the portfolio's trading gains will respond less to volatility changes. Finally, as time passes, the value of the bonds will increase automatically even if the rates don't come down.

10.3.3 A BOND PDE

A PDE consisting of the gains from convexity of long bonds and costs of maintaining the volatility position can be put together. Under some conditions, this PDE has an analytical solution, and an analytical formula can be obtained the way the Black–Scholes formula was obtained.

First we discuss the PDE informally. We start with the trading gains due to convexity. These gains are given by the continuous adjustment of the hedge ratio h_t, which essentially depends on B_r, except for a constant of proportionality, since the hedging instrument is quasi-linear in r_t. As r_t changes, the partial B_r changes, and this will be captured by the second derivative. Then, convexity gains during a small time interval Δ is a function, as in Chapter 9, of

$$\frac{1}{2}\frac{\partial^2 B(t, T)}{\partial r_t^2}\left(\sigma(r_t, t)r_t\sqrt{\Delta}\right)^2 \tag{10.24}$$

This is quite similar to the case of vanilla options, except that here the $\sigma(r_t, t)$ is the percentage short rate volatility. Short bond interest sensitivity will cancel out.

If we model the *risk-neutral dynamics* of the short rate r_t as

$$dr_t = \mu(r_t, t)dt + \sigma/r_t\, dW_t \quad t\in[0, T] \tag{10.25}$$

where percentage volatility σ is constant, these *gamma* gains simplify to

$$\frac{1}{2}B_{rr}\sigma^2 r_t^2\Delta \tag{10.26}$$

during a small period Δ.[6]

---

[6] Note that we are using the notation

$$\frac{\partial^2 B(t, T)}{\partial r_t^2} = B_{rr}$$

To these, we need to add (subtract) other costs and gains that the position holder is subject to. The interest paid during the period Δ on borrowed funds will be

$$r_t B(t, T)\Delta \tag{10.27}$$

The other gain (loss) is the direct effect of passing time

$$\frac{\partial B(t, T)}{\partial t}\Delta = B_t \Delta \tag{10.28}$$

As time passes, bonds earn accrued interest, and convexity declines due to "roll-down" on the yield curve. The interest earned due to shorting the linear instrument will cancel out the cost of this short position.

The final component of the gains and losses that the position is subject to during Δ is more complex than the case of a vanilla call or put. In the case of the option, the underlying stock, S_t, provided a very good *delta* hedging tool. The market maker sold $(\partial/\partial S_t)C(S_t, t)$ units of the underlying S_t in order to hedge a long call position. In the present case, the underlying risk is not the stock price S_t or some futures contract. The underlying risk is the spot rate r_t, and this is *not* an asset. That is to say, the "hedge" is not r_t itself, but instead it is an asset *indirectly influenced* by r_t. Also, randomness of interest rates requires projecting future interest gains and costs. All these complicate the cash flow analysis.

These complications can be handled by positing that the *drift* term $\mu(r_t, t)$ in the dynamics,[7]

$$dr_t = \mu(r_t, t)dt + \sigma r_t dW_t, \quad t \in [0, T] \tag{10.29}$$

represents the *risk-free expected change* in the spot rate over an infinitesimal interval dt.[8] Using this drift, we can write the last piece of gains and losses over a small interval Δ as (Vasicek, 1977):

$$\mu(r_t, t)B_r\Delta \tag{10.30}$$

Adding all gains and losses during the interval Δ, we obtain the *net* gains from the convexity position:

$$\frac{1}{2}B_{rr}\sigma^2 r_t^2\Delta + \mu(r_t, t)B_r\Delta - r_t B\Delta + B_t\Delta \tag{10.31}$$

In order to preclude arbitrage opportunities, this sum must equal zero. Canceling the common Δ terms, we get the PDE for the bond:

$$\frac{1}{2}B_{rr}\sigma^2 r_t^2 + \mu(r_t, t)B_r - r_t B + B_t = 0 \tag{10.32}$$

The boundary condition is simpler than in the case of vanilla options and is given by

$$B(T, T) = 1 \tag{10.33}$$

the par value of the default-free bond at maturity date T.

---

[7]See Appendix 9.2 in Chapter 9 for a definition of this SDE.
[8]Chapter 12 will go into the details of this argument that uses risk-neutral probabilities.

10.3.4 PDEs AND CONDITIONAL EXPECTATIONS

In this PDE, the *unknown* is again a *function B(t, T)*. This function will depend on the random process r_t, t, as well as other parameters of the model. The most important of these is the short rate volatility, σ. If r_t is the continuously compounded short rate, the solution is given by the conditional expectation

$$B(t, T) = E_t^{\tilde{P}}\left[e^{-\int_t^T r_u du} \right] \tag{10.34}$$

where $\tilde{P}$ is an appropriate probability. In other words, taking appropriate partial derivatives of the right-hand side of this expression, and then plugging these in the PDE would make the sum on the left-hand side of Eq. (10.32) equal to zero.[9]

It is interesting to look at the parallel with options. The pricing function for $B(t, T)$ was based on a particular *conditional expectation* and solved the bond PDE. In the case of vanilla options written on a stock S_t, and satisfying all Black–Scholes assumptions, the call price $C(S_t, t)$ is given by a similar conditional expectation,

$$C(S_t, t) = E_t^{\tilde{P}}[e^{-r(T-t)}C(S_T, T)] \tag{10.35}$$

where T is the expiration date and $\tilde{P}$ is the appropriate probability. If this expectation is differentiated with respect to S_t and t, the resulting partial derivatives will satisfy the Black–Scholes PDE with the corresponding boundary condition. The main difference is that the Black–Scholes assumptions take the short rate r_t to be constant, whereas in the case of bonds, it is a stochastic process.

These comments reconcile the two views of options that were mentioned in Chapter 9. If we interpret options as directional instruments, then Eq. (10.35) will give the expected gains of the optional at expiration, under an appropriate probability. The argument above shows that this expectation solves the associated PDE which was approached as an arbitrage relationship tying *gamma* gains to other costs incurred during periodic rebalancing. In fact, we see that the two interpretations of options are equivalent.

10.3.5 FROM BLACK–SCHOLES TO BOND PDE

Comparing the results of trading bond convexity with those obtained in trading vanilla options provides good insights into the general characteristics of PDE methods that are commonly used in finance.

In Chapter 9, we derived a PDE for a plain vanilla call, $C(t)$ using the argument of convexity trading. In this chapter, we discussed a PDE that is satisfied by a default-free pure discount bond $B(t, T)$. The results were as follows.

1. The price of a plain vanilla call, written on a nondividend-paying stock S_t, strike K, expiration T, was shown to satisfy the following "arbitrage" equality:

$$\frac{1}{2}C_{SS}(\sigma(S_t, t)S_t)^2 \Delta = (rC - rC_s S_t)\Delta - C_t \Delta \tag{10.36}$$

---

[9]The major condition to be satisfied for this is the Markovness of r_t.

where $\sigma(S_t, t)$ is the percentage volatility of S_t during 1 year. The way it is written here, this percentage volatility could very well depend on time t and S_t.

According to this equation, in order to preclude any arbitrage opportunities, trading gains obtained from dynamic hedging during a period of length Δ should equal the net funding cost, plus loss of time value. Canceling common terms and introducing the boundary condition yielded the Black–Scholes PDE for a vanilla call:

$$\tfrac{1}{2}C_{SS}(\sigma(S_t, t)S_t)^2 + rC_S S_t - rC + C_t = 0 \tag{10.37}$$

$$C(T) = \max[S_T - K, 0] \tag{10.38}$$

Under the additional assumption that $\sigma(S_t, t)S_t$ is proportional to S_t with a constant factor of proportionality σ,

$$\sigma(S_t, t)S_t = \sigma S_t \tag{10.39}$$

this PDE could be solved analytically, and a closed-form formula could be obtained for the $C(t)$. This formula is the Black–Scholes equation:

$$C(t) = S_t N(d_1) - Ke^{-r(T-t)}N(d_2) \tag{10.40}$$

$$d_{1,2} = \frac{\log(S_t/K) + r(T - t) \pm (1/2)\sigma^2(T - t)}{\sigma\sqrt{T - t}} \tag{10.41}$$

The partial derivatives of this $C(t)$ would satisfy the preceding PDE.

2. The procedure for a default-free pure discount bond $B(t, T)$ followed similar steps, with some noteworthy differences. Assuming that the continuously compounded spot interest rate, r_t, is the only factor in determining bond prices, the convexity gains due to oscillations in r_t and to dynamic hedging can be isolated, and a similar "arbitrage relation" can be obtained:

$$\frac{1}{2}B_{rr}(\sigma(r_t, t)r)^2\Delta = (r_t B - \mu(r_t, t)B_r)\Delta - B_t\Delta \tag{10.42}$$

Here, $\sigma(r_t, t)$ is the percentage volatility of the short rate r_t during 1 year.

Canceling common terms, and adding the boundary condition, we obtain the bond PDE:

$$\tfrac{1}{2}B_{rr}\sigma^2 r_t^2 + \mu(r_t, t)B_r - r_t B + B_t = 0 \tag{10.43}$$

$$B(T, T) = 1 \tag{10.44}$$

Under some special assumptions on the dynamic behavior of r_t, this bond PDE can be solved analytically, and a closed-form formula can be obtained.

We now summarize some important differences between these parallel procedures. First, note that the PDE for the vanilla option is obtained in an environment where the only risk comes from the *asset price* S_t, whereas for bonds the only risk is the interest rate r_t, which is *not* an asset *per se*. Second, the previously mentioned difference accounts for the emergence of the term $\mu(r_t, t)$ in the bond PDE, while no such nontransparent term existed in the call option PDE. The $\mu(r_t, t)$ represents the expected change in the spot rate during dt once the effect of interest rate risk is taken out. Third, the $\mu(r_t, t)$ may itself depend on other parameters that affect interest rate dynamics. It is obvious that under these conditions, the closed-form solution for $B(t, T)$ would depend on the same parameters. Note that in the case of the vanilla option, there was no such issue and the only relevant parameter was σ. This point is important since it could make the bond price formula depend on *all* the parameters of the underlying random process, whereas in the case of vanilla options, the Black–Scholes formula depended on the characteristics of the *volatility* parameter only.

Before we close this section, a final parallel between the vanilla option and bond prices should be discussed. The PDE for a call option led to the closed-form Black–Scholes formula under some assumptions concerning the volatility of S_t. Are there similar closed-form solutions to the bond PDE? The answer is yes.

10.3.6 CLOSED-FORM BOND PRICING FORMULAS

Under different assumptions concerning short rate dynamics, we can indeed solve the bond PDE and obtain closed-form formulas. We consider three cases of increasing complexity. The cases are differentiated by the assumed short rate dynamics.

10.3.6.1 Case 1

The first case is simple. Suppose r_t is constant at r. This gives the trivial dynamics,

$$dr_t = 0 \tag{10.45}$$

where σ and $\mu(r_t, t)$ are both zero. The bond PDE in Eq. (10.43) then reduces to

$$-rB + B_t = 0 \tag{10.46}$$

$$B(T, T) = 1 \tag{10.47}$$

This is a simple, ordinary differential equation. The solution $B(t, T)$ is given by

$$B(t, T) = e^{-r(T-t)} \tag{10.48}$$

10.3.6.2 Case 2

The second case is known as the Vasicek model.[10] Suppose the *risk-adjusted* dynamics of the spot rate follows the *mean-reverting* process given by[11]

$$dr_t = \alpha(\kappa - r_t)dt + \sigma dW_t \quad t \in [0, T] \tag{10.49}$$

where W_t is a Wiener process defined for a *risk-adjusted* probability.[12]

Note that the volatility structure is restricted to constant *absolute* volatility denoted by σ. Suppose, further, that the parameters α, κ, σ, are known exactly. The fundamental PDE for a typical $B(t, T)$ will then reduce to

$$B_r\alpha(\kappa - r_t) + B_t + \frac{1}{2}B_{rr}\sigma^2 - r_tB = 0 \tag{10.50}$$

---

[10]See Vasicek (1977).
[11]The fact that this dynamic is *risk-adjusted* is not trivial. Such dynamics are calculated under risk-neutral probabilities and may differ significantly from *real-world* dynamics. These issues will be discussed in Chapter 12.
[12]The adjustments for risk and the associated probabilities will be discussed in Chapter 12.

Using the boundary condition $B(T, T) = 1$, this PDE can be solved analytically, to provide a closed-form formula for $B(t, T)$. The closed-form solution is given by the expression

$$B(t, T) = A(t, T)e^{-C}(t, T)r_t \tag{10.51}$$

where

$$C(t, T) = \frac{(1 - e^{-\alpha(T-t)})}{\alpha} \tag{10.52}$$

$$A(t, T) = e^{\frac{(C(t,T) - (T-t))(\alpha^2\kappa - (1/2)\sigma^2)}{\alpha^2} - \frac{\sigma^2 C(t,T)^2}{4\alpha}} \tag{10.53}$$

Here, r_t is the "current" observation of the spot rate.

10.3.6.3 *Case 3*

The third well-known case, where the bond PDE in Eq. (10.43) can be solved for a closed form, is the Cox−Ingersoll−Ross (CIR) model. In the CIR model, the spot rate r_t is assumed to obey the slightly different mean-reverting stochastic differential equation

$$dr_t = \alpha(\kappa - r_t)dt + \sigma\sqrt{r_t}dW_t \quad t \in [0, T] \tag{10.54}$$

which is known as the square-root specification of interest rate volatility. Here W_t is a Wiener process under the *risk-neutral* probability.

The closed-form bond pricing equation here is somewhat more complex than in the Vasicek model. It is given by

$$B(t, T) = A(t, T)e^{-C(t,T)r_t} \tag{10.55}$$

where the functions $A(t, T)$ and $C(t, T)$ are given by

$$A(t, T) = \left(2\frac{\gamma e^{1/2(\alpha+\gamma)(T-t)}}{(\alpha+\gamma)(e^{\gamma(T-t)} - 1) + 2\gamma} \right)^{2\frac{\alpha\kappa}{\sigma^2}} \tag{10.56}$$

$$C(t, T) = 2\frac{e^{\gamma(T-t)} - 1}{(\alpha + \gamma)(e^{\gamma(T-t)} - 1) + 2\gamma} \tag{10.57}$$

and where γ is defined as

$$\gamma = \sqrt{(\alpha)^2 + 2\sigma^2} \tag{10.58}$$

The bond volatility σ determines the *risk premia* in expected discount bond returns.

10.3.7 A GENERALIZATION

The previous sections showed that whenever two instruments depending on the same risk factor display different degrees of convexity, we can, in principle, put together a *delta* hedging strategy similar to the *delta* hedging of options discussed in Chapter 9. Whether this is worthwhile depends, of course, on the level of volatility relative to transaction costs and bid−ask spreads.

When a market practitioner buys a convex instrument and short sells an appropriate number of a linear (or less convex) instrument, he or she will benefit from *higher* volatility. We then say that the position is *long* volatility or *long gamma*. This trader has purchased *gamma*. If, in contrast, the convex instrument is shorted and the linear instrument is purchased at proper ratios, the position will benefit when the volatility of the underlying decreases.

As the case of long bonds shows, the idea that volatility can be isolated (to some degree), and then traded is very general, and can be implemented when instruments of different convexities are available on the *same* risk. Of course, volatility can be such that transaction costs and bid−ask spreads make trading it unfeasible, but that is a different point. More important, if the yield curve slope changes due to the existence of a second factor, the approach presented in the previous sections will not guarantee convexity gains.

10.4 SOURCES OF CONVEXITY

There is more than one reason for the convexity of pricing functions. We discuss some simple cases briefly, using a broad definition of convexity.

10.4.1 MARK TO MARKET

We start with a minor case due to daily *marking-to-market* requirements. Let f_t denote the daily *futures* settlement price written on an underlying asset S_t, let F_t be the corresponding *forward* price, and let r_t be the overnight interest rate.

Marking to market means that the futures position makes or loses money every day depending on how much the futures settlement price has changed,

$$\Delta f_t = f_t - f_{t-1} \qquad (10.59)$$

where the time index t is measured in days and hence is discrete.

Suppose the overnight interest rate r_t is stochastic. Then if the trader receives (pays) mark-to-market gains daily, these can be deposited or borrowed at higher or lower overnight interest rates. If Δf_t were *uncorrelated* with interest rate changes,

$$\Delta r_t = r_t - r_{t-1} \qquad (10.60)$$

marking to market would not make a difference.

But, when S_t is itself an interest rate product or an asset price related to interest rates, the random variables Δf_t and Δr_t will, in general, be correlated. For illustration, suppose the correlation between Δf_t and Δr_t is positive. Then, when f_t increases, r_t is likely to increase also, which means that the mark-to-market gains can now be invested at a higher overnight interest rate. If the correlation between Δf_t and Δr_t is negative, the reverse will be true. Forward contracts do not, normally, require such daily marking-to-market. The contract settles only at the expiration date. This means that daily paper gains or losses on forward contracts cannot be reinvested or borrowed at higher or lower rates.

Thus, a futures contract written on an asset S_t whose price is *negatively* correlated with r_t will be cheaper than the corresponding forward contract. If the correlation between S_t and r_t is positive, then the futures contract will be more expensive. If S_t and r_t are uncorrelated, then futures and forward contracts will have the same price, everything else being the same.

EXAMPLE

Consider any Eurocurrency future. We saw in Chapter 3 that the price of a 1-year Eurodollar future, settling at time $t + 1$, is given by the linear function

$$V_t = 100(1 - f_t) \tag{10.61}$$

Normally, we expect overnight interest rate r_t to be positively correlated with the futures rate f_t. Hence, the price V_t, which is not a convex function, would be negatively correlated with r_t. This means that the Eurodollar futures will be somewhat cheaper than corresponding forward contracts, which in turn means that futures interest rates are somewhat higher than the forward rates.

Mark-to-market is one reason why futures and forward rates may be different.

10.4.2 CONVEXITY BY DESIGN

Some products have convexity by design. The contract specifies payoffs and underlying risks, and this specification may make the contract price a nonlinear function of the underlying risks. Among the most important classes of instruments that permit such convexity gains are, of course, options.

We also discussed convexity gains from bonds. Long maturity default-free discount bond prices, when expressed as a function of yield to maturity y_t, are simple nonlinear functions, such as

$$B(t, T) = \frac{100}{(1 + y_t)^T} \tag{10.62}$$

Coupon bond prices can be expressed using similar discrete time yield to maturity. The price of a coupon bond with coupon rate c, and maturity T, can be written as

$$P(t, T) = \left(\sum_{i=1}^{T} \frac{100c}{(1 + y_t)^i} \right) + \frac{100}{(1 + y_t)^T} \tag{10.63}$$

It can be shown that default-free pure discount bonds, or strips, have more convexity than coupon bonds with the same maturity.

10.4.2.1 Swaps

Consider a plain vanilla, fixed payer interest rate swap with immediate start date at $t = t_0$ and end date, $t_n = T$. Following market convention, the floating rate set at time t_i is paid at time t_{i+1}. For simplicity, suppose the floating rate is 12-month USD LIBOR. This means that $\delta = 1$. Let the time $t = t_0$ swap rate be denoted by s and the notional amount, N, be 1.

Then, the time t_0 value of the swap is given by

$$V_{t_0} = E_{t_0}^{\tilde{P}} \left[\frac{L_{t_0} - s}{(1 + L_{t_0})} + \frac{L_{t_1} - s}{(1 + L_{t_0})(1 + L_{t_1})} + \cdots + \frac{L_{t_{n-1}} - s}{\prod_{i=0}^{n-1}(1 + L_{t_i})} \right] \tag{10.64}$$

where $\{L_{t_0}, \ldots, L_{t_{n-1}}\}$ are random LIBOR rates to be observed at times $t_0, \ldots, t_{n-1}$, respectively, and $\tilde{P}$ is an appropriate probability measure. With a proper choice of measure, we can act as if we can substitute a *forward* LIBOR rate, $F(t_0, t_i)$, for the future spot LIBOR L_{t_i} for all t_i.[13] If liquid markets exist where such forward LIBOR rates can be observed, then after this substitution we can write the previous pricing formula as

$$V_{t_0} = \frac{L_{t_0} - s}{(1 + L_{t_0})} + \frac{F(t_0, t_1) - s}{(1 + L_{t_0})(1 + F(t_0, t_1))}$$
$$+ \frac{F(t_0, t_2) - s}{(1 + L_{t_0})(1 + F(t_0, t_1))(1 + F(t_0, t_2))} + \cdots + \frac{F(t_0, t_{n-1}) - s}{\prod_{i=0}^{n-1}(1 + F(t_0, t_i))} \qquad (10.65)$$

where $F(t_0, t_0) = L_{t_0}$, by definition. Clearly, this formula is nonlinear in each $F(t_0, t_i)$. As the forward rates change, V_{t_0} changes in a nonlinear way.

This can be seen better if we assume that the yield curve is flat and that all yield curve shifts are parallel. Under such unrealistic conditions, we have

$$L_{t_0} = F(t_0, t_0) = F(t_0, t_1) = \cdots = F(t_0, t_{n-1}) = F_{t_0} \qquad (10.66)$$

The swap formula then becomes

$$V_{t_0} = \frac{F_{t_0} - s}{(1 + F_{t_0})} + \frac{F_{t_0} - s}{(1 + F_{t_0})^2} + \cdots + \frac{F_{t_0} - s}{(1 + F_{t_0})^T} \qquad (10.67)$$

which simplifies to[14]

$$V_{t_0} = (F_{t_0} - s)\frac{\left((1 + F_{t_0})^T - 1\right)}{F_{t_0}(1 + F_{t_0})^T} \qquad (10.68)$$

The second derivative of this expression with respect to F_{t_0} will be negative, for all $F_{t_0} > 0$.

As this special case indicates, the fixed payer swap is a nonlinear instrument in the underlying forward rates. Its second derivative is negative, and the function is *concave* with respect to a "typical" forward rate. This is not surprising since a fixed payer swap has risks similar to the *issuing* of a 30-year bond. This means that a *fixed-receiver* swap will have a convex pricing formula and will have a similar profile as a long position in a 30-year coupon bond.

EXAMPLE

Figure 10.5 plots the value of a fixed payer swap under the restrictive assumption that the yield curve is flat and that it shifts only parallel to itself. The parameters are as follows:

$$t = 0 \qquad (10.69)$$

$$s = 7\% \qquad (10.70)$$

---

[13]This substitution is delicate and depends on many conditions, among them the fact that the LIBOR decided at reset date i is settled at date $i + 1$.

[14] Factor out the numerator and use the geometric series sum:

$$1 + a + a^2 + \cdots + a^T = \frac{1 - a^{T+1}}{1 - a}$$

$$T = 30 \qquad (10.71)$$

$$F_{t_0} = 6\% \qquad (10.72)$$

We see that the function is nonlinear and concave.

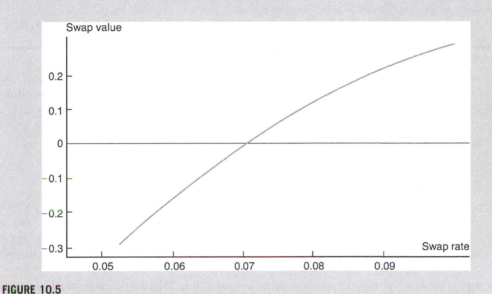

FIGURE 10.5

Value of a fixed payer swap.

In Chapter 20, we will consider a different type of swap called constant maturity swap. The convexity of constant maturity swaps is due to their structure. This convexity will, in general, be more pronounced and at the same time more difficult to correctly account for.

Taking convexity characteristics of financial instruments into account is important. This is best illustrated by the Chicago Board of Trade's (CBOT) attempt to launch a new contract with proper convexity characteristics.

EXAMPLE

The Chicago Board of Trade's board of directors last week approved a plan to launch 5- and 10-year US dollar denominated interest rate swap futures and options contracts. Compared with the over-the-counter market, trading of swaps futures will reduce administrative cost and eliminate counterparty risk, the exchange said.

The CBOT's move marks the second attempt by the exchange to launch a successful swap futures contract. Treasuries were the undisputed benchmark a decade ago. They are not treated as a benchmark for valuation anymore. People price off the swap curve instead, said a senior economist at the CBOT.

The main difference between the new contract and the contract that the CBOT de-listed in the mid-1990s is that the new one is convex in form rather than linear. It's one less thing for end users to worry about, the economist said, noting that swap positions are marked to market on a convex basis. Another critical flaw in the old contract was that it launched in the 3- and 5-year, rather than the 5- and 10-year maturities, which is where most business takes place.

The new swaps contracts will offer institutional investors such as bank treasurers, mortgage passthrough traders, originators, service managers, portfolio managers, and other OTC market participants a vehicle for hedging credit and interest rate exposure, the exchange said.

(Thomson Reuters IFR Issue 1393, July 21, 2001)

This is an excellent example that shows the importance of convexity in contract design. Futures contracts are used for hedging by traders. If the convexity of the hedging instrument is different than the convexity of the risks to be hedged, then the hedge may deteriorate as volatility changes. In fact, as volatility increases, the more convex instrument may yield higher *gamma* gains and this will influence its price.

10.4.2.2 Convexity of FRAs

Now consider the case of forward rate agreements (FRAs). As discussed in Chapter 3, FRAs are instruments that can be used to fix, at time t_0, the risk associated with a LIBOR rate L_{t_i}, that will be observed at time t_i, and that has a tenor of δ expressed in *days per year*.[15] The question is *when* would this FRA be settled. This can be done in different ways, leading to slightly different instruments. We can envisage three types of FRAs.

One way is to set L_{t_i} at time t_i, but then, settle at time $t_i + \delta$. This would correspond to the "natural" way interest is paid in financial markets. Hence, at time $t = t_0$, the value of the FRA will be zero and at time $t_i + \delta$ the FRA buyer will receive or pay

$$\left[L_{t_i} - F_{t_0}\right] N\delta \tag{10.73}$$

depending on the sign of the difference. The FRA seller will have the opposite cash flow.

The second type of FRA trades much more frequently in financial markets. The description of these is the same, except that the FRA is settled at time t_i, instead of at $t_i + \delta$. At time t_i, when the LIBOR rate L_{t_i} is observed, the buyer of the FRA will make (receive) the payment

$$\frac{\left[L_{t_i} - F_{t_0}\right]\delta}{1 + L_{t_i}\delta} N \tag{10.74}$$

This is the previous settlement amount *discounted* from time $t_i + \delta$ to time t_i, using the time t_i LIBOR rate. Figure 10.6 shows an example to the payoff of a 12-month FRA.

Of even more interest for our purpose is a third type of FRA contract, a LIBOR-in-arrears FRA, where the LIBOR observed at time t_i is used to settle the contract at time t_i, according to

$$\left[L_{t_i} - f_{t_0}\right]\delta N \tag{10.75}$$

---

[15]The year is assumed to be 360 days.

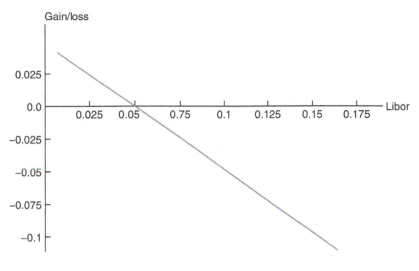

FIGURE 10.6

Payoff of a 12-month FRA.

Here, f_{t_0} is the FRA rate that applies to this particular type of FRA.[16] Note that we are using a symbol different than F_{t_0}, because the two FRA rates are, in general, different from each other due to convexity differences in the two contracts.

The question to ask here is under what conditions would the rates F_{t_0} and f_{t_0} differ from each other? The answer depends indeed on the convexity characteristics of the underlying contracts. In fact, market practitioners approximate these differences using *convexity adjustment* factors.

10.4.3 PREPAYMENT OPTIONS

A major class of instruments that have convexity by design is the broad array of securities associated with *mortgages.* A mortgage is a loan secured by the purchaser of a residential or commercial property. Most fixed-rate mortgages have a critical property. They contain the right to prepay the loan. The mortgage receiver has the right to pay the remaining balance of the loan at any time, and incur only a small transaction cost. This is called a prepayment option and introduces negative convexity in mortgage-related securities. In fact, the prepayment option is equivalent to an American-style put option written on the mortgage rate R_t. If the mortgage rate R_t falls below a limit R^K, the mortgage receiver will pay back the original amount denoted by N, by refinancing at the new rate R_t. Instead of making a stream of fixed annual interest payments $R_{t_0}N$, the mortgage receiver has the option (but not the obligation) to pay the annual interest $R_{t_i}N$ at some time t_i. The mortgage receiver may exercise this option if $R_{t_i} < R_{t_0}$. The situation is reversed for the mortgage issuer.

---

[16]Similarly, we can have LIBOR-in-arrear swaps on the generalization of this type of FRA contract.

The existence of such prepayment options creates negative convexity for mortgage-backed securities (MBSs) and other related asset classes. Since the prepayment option involves an exchange of one fixed stream of payments against another fixed stream, it is clear that interest rate swaps play a critical role in hedging and risk-managing these options dynamically. We will deal with this important topic in Chapter 17.

10.5 A SPECIAL INSTRUMENT: QUANTOS

Quanto-type financial products form a major class of instruments where price depends on *correlations*. At the end of this chapter, we will look at these in detail and study the financial engineering of quantos by discussing their characteristics and other issues. This can be regarded as another example to the methods introduced in Chapter 9. We will consider pricing of quantos in Chapter 13.

10.5.1 A SIMPLE EXAMPLE

Consider the standard currency swap in Figure 10.7. There are two cash flows, in two currencies, USD and EUR. The principal amounts are exchanged at the start date and reexchanged at the end date. During the life of the swap, floating payments based on USD LIBOR are exchanged for floating payments based on EUR LIBOR. There will be a small known spread involved in these exchanges as well.

The standard currency swap of Figure 10.7 will now be modified in an interesting way. We keep the two floating LIBOR rates the same, but *force* all payments to be made in one currency *only*, say USD. In other words, the calculated EUR LIBOR indexed cash flows will be paid (received) in USD. This instrument is called a *quanto swap* or *differential swap*. In such a swap, the principal amounts would be in the same currency, and there would be no need to exchange them. Only net interest rate cash flows will be exchanged.

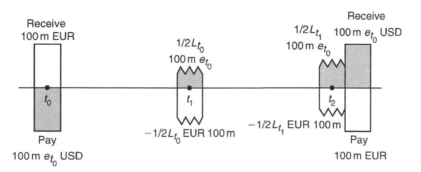

FIGURE 10.7

Standard currency swap.

EXAMPLE

Suppose the notional principal is USD30 million. Quotes on LIBOR are as follows:

TENOR	YEN LIBOR	DOLLAR LIBOR
3-month	0.055	1.71
6-month	0.185	1.64
12-month	0.065	1.73

In a quanto swap, one party would like to receive 6-month USD LIBOR and pay 6-month JPY LIBOR for 1 year. However, all payments are made in USD. For example, if the first settlement is according to the quotes given in the table, in 6 months this party will receive:

$$30,000,000(0.0164)\left(\frac{1}{2}\right) - 30,000,000(0.00185)\left(\frac{1}{2}\right) - 30,000,000\left(\frac{1}{2}\right)c \qquad (10.76)$$

where c is a constant spread that needs to be determined in the pricing of this quanto swap. Note that the JPY interest rate is applied to a USD denominated principal.

In this type of swap, the two parties are exposed to the risk of interest rate differentials. However, at least one of them is not exposed to currency risk.

Why would anyone be interested in quanto swaps? Note that even after the spread c is included, the interest cost paid in *dollars*,

$$\text{JPY LIBOR} + c \qquad (10.77)$$

may be significantly less than USD LIBOR rates. This way, the party that receives USD LIBOR and pays JPY LIBOR (in USD) may be lowering funding costs substantially. Accordingly, the market would see interest in such quanto swaps when the short ends of the yield curves in two major currencies are significantly different. Banks could then propose these instruments to their clients as a way of "reducing" funding costs. Of course, from the clients' point of view, quanto swaps still involve an interest rate risk and, possibly, an exchange rate risk. If the underlying yield curves shift in unexpected ways, losses may be incurred.

The following example illustrates these from the point of view of British pound and Swiss franc interest rates.

EXAMPLE

With European economies at a very different point in the trade cycle, corporates are looking to switch their debts into markets offering the cheapest funding. But whereas most would previously have been dissuaded by foreign exchange risk, the emergence of quanto products has allowed them to get the best of both worlds.

With quanto swaps, interest is paid in a different currency to that of the reference index, the exchange rate being fixed at the outset of the swap. As a result, the product can provide exposure to a nondomestic yield curve without the accompanying exchange rate risks.

> *In recent weeks, this type of product has proved increasingly appealing to UK corporates that have entered into a swap in which the paying side is referenced to Swiss LIBOR but the returns are paid in sterling. Swiss franc LIBOR is still low relative to sterling LIBOR and although the corporate ends up paying Swiss LIBOR plus a spread, funding costs are often still considerably cheaper than normal sterling funding. Deals have also been referenced to German or Japanese LIBOR.*
>
> *However, derivatives officials were also keen to point out that quanto products are far from being risk-free. "Given that the holder of the swap ends up paying Swiss LIBOR plus a spread, the curves do not have to converge much to render the trade uneconomic," said one.*
>
> **(Thomson Reuters IFR Issue 1190, July 5 1997)**

10.5.1.1 *Quantos in equity*

The notion of a quanto instrument can be applied in other financial markets. For example, a foreign investor may want to have exposure to Japanese equity markets without having to incur currency risk. Then, a quanto contract can be designed such that the gains and losses of an index in Japanese equities are paid annually in the foreign investors' domestic currency instead of in yen.

10.5.2 PRICING

The pricing of quanto contracts raises interesting financial engineering issues.[17] We discuss a simple case to illustrate quantos. First, fix the underlying. Assume that we are dealing with a particular *foreign currency* denominated stock S_t^*. Without loss of generality, suppose the *domestic* currency is USD, the foreign currency is euro, and the stock is European.

A dollar-based investor would like to buy the stock, and benefit from potential upside in European markets, but dislikes currency exposure to euro. The investor desires exposure to underlying equity risk only. To accommodate his wish, the bank proposes purchasing the stock via a *quanto forward*. An expiration date T is chosen, and the current exchange rate EUR/USD, e_t is used to calculate the time T settlement. The forward contract has USD price F_t and settles according to

$$V_T = (e_t S_T^* - F_t) \tag{10.78}$$

Here, the V_T is the time T value of the contract. It is measured in the domestic currency and will be positive if the stock appreciates sufficiently; otherwise, it will be negative. F_t is the forward price of the quanto contract on S_T^* and has to be determined by a proper pricing strategy.

10.5.3 THE MECHANICS OF PRICING

Suppose the current time is t and a forward quanto contract on S_T^* is written with settlement date $T = t + \Delta$. Suppose also that at time T there are only three possible states of the world $\{\omega^1, \omega^2, \omega^3\}$. The following table gives the possible values of four instruments, the foreign stock, a foreign deposit, a domestic deposit, and a forward FX contract on the exchange rate e_t.

---

[17]This is an example of the measure switching techniques to be discussed in Chapter 14.

Time t Price	Value in ω^1	Value in ω^2	Value in ω^3
S_T^*	$S_{t+\Delta}^{*1}$	$S_{t+\Delta}^{*2}$	$S_{t+\Delta}^{*3}$
1 USD	$(1+r\Delta)$	$(1+r\Delta)$	$(1+r\Delta)$
1 e_t	$e_{t+\Delta}^1(1+r^*\Delta)$	$e_{t+\Delta}^2(1+r^*\Delta)$	$e_{t+\Delta}^3(1+r^*\Delta)$
0	$f_t - e_{t+\Delta}^1$	$f_t - e_{t+\Delta}^2$	$f_t - e_{t+\Delta}^3$

In this table, the first row gives the value of the foreign stock in the three future states of the world. These are measured in the foreign currency. The second row represents what happens to 1 dollar invested in a domestic savings account. The third row shows what happens when 1 unit of foreign currency is purchased at e_t dollars and invested at the foreign rate r^*.

The forward exchange rate f_t is priced as

$$f_t = e_t \frac{1+r\Delta}{1+r^*\Delta} \tag{10.79}$$

where e_t is the current exchange rate. In this example, we are assuming that the domestic and foreign interest rates are constant at r and r^*, respectively. Now consider the quanto forward contract with current price F_t mentioned earlier. F_t will be determined at time t, and the contract will settle at time $T = t + \Delta$. Depending on which state occurs, the settlement amount will be one of the following:

$$\{(S_{t+\Delta}^{*1}e_t - F_t), (S_{t+\Delta}^{*2}e_t - F_t), (S_{t+\Delta}^{*3}e_t - F_t)\} \tag{10.80}$$

These amounts are all in USD. What is the arbitrage-free value of F_t?

We can use three of the four instruments listed to form a portfolio with weights λ_i, $i = 1, 2, 3$ that replicate the possible values of $e_t S_{t+\Delta}^*$ at each state exactly. This will be similar to the cases discussed in Chapter 8. For example, using the first three instruments, for each state we can write

$$\text{State } \omega^1 \quad \lambda_1 S_{t+\Delta}^{*1} e_{t+\Delta}^1 + \lambda_2(1+r\Delta) + \lambda_3 e_{t+\Delta}^1(1+r^*\Delta) = S_{t+\Delta}^{*1} e t \tag{10.81}$$

$$\text{State } \omega^2 \quad \lambda_1 S_{t+\Delta}^{*2} e_{t+\Delta}^2 + \lambda_2(1+r\Delta) + \lambda_3 e_{t+\Delta}^1(1+r^*\Delta) = S_{t+\Delta}^{*2} e t \tag{10.82}$$

$$\text{State } \omega^3 \quad \lambda_1 S_{t+\Delta}^{*3} e_{t+\Delta}^3 + \lambda_2(1+r\Delta) + \lambda_3 e_{t+\Delta}^1(1+r^*\Delta) = S_{t+\Delta}^{*3} e t \tag{10.83}$$

In these equations, the right-hand side is the future value of the foreign stock measured at the current exchange rate. The left-hand side is the value of the replicating portfolio in that state.

These form three equations in three unknowns, and, in general, can be solved for the unknown λ_i. Once these portfolio weights are known, the current cost of putting the portfolio together leads to the price of the quanto:

$$\lambda_1 S_t^* e_t + \lambda_2 + \lambda_3 e_t \tag{10.84}$$

This USD amount needs to be carried to time T, since the contract settles at T. This gives

$$F_t = [\lambda_1 S_t^* e_t + \lambda_2 + \lambda_3 e_t](1+r\Delta) \tag{10.85}$$

EXAMPLE

Suppose we have the following data on the first three rows of the previous table:

Time t Price	Value in ω^1	Value in ω^2	Value in ω^3
100	115	100	90
1 USD	$(1+0.05\Delta)$	$(1+0.05\Delta)$	$(1+0.05\Delta)$
1 EUR $\times$ 0.98	$(1+0.03\Delta)1.05$	$(1+0.03\Delta)0.98$	$(1+0.03\Delta)0.90$

What is the price of the quanto forward?
We set up the three equations

$$\lambda_1(1.05)115 + \lambda_2(1+0.05\Delta) + \lambda_3 1.05(1+0.03\Delta) = 0.98(115) \tag{10.86}$$

$$\lambda_1(0.98)100 + \lambda_2(1+0.05\Delta) + \lambda_3 0.98(1+0.03\Delta) = 0.98(100) \tag{10.87}$$

$$\lambda_1(0.90)90 + \lambda_2(1+0.05\Delta) + \lambda_3 0.90(1+0.03\Delta) = 0.98(90) \tag{10.88}$$

We select the expiration $\Delta = 1$, for simplicity, and obtain

$$\lambda_1 = 0.78 \tag{10.89}$$

$$\lambda_2 = 60.67 \tag{10.90}$$

$$\lambda_3 = -41.53 \tag{10.91}$$

Borrowing 42 units of foreign currency, lending 61 units of domestic currency, and buying 0.78 units of the foreign stock would replicate the value of the quanto contract at time $t+1$. The price of this portfolio at t will be

$$100\lambda_1 0.98 + \lambda_2 + 0.98\lambda_3 = 96.41 \tag{10.92}$$

If this is to be paid at time $t + \Delta$, then it will be equal to the arbitrage-free value of F_t:

$$F_t = (1.05)96.41 = 101.23 \tag{10.93}$$

This example shows that the value of the quanto feature is related to the correlation between the movements of the exchange rate and the foreign stock. If this correlation is zero, then the quanto will have the same value as a standard forward. If the correlation is positive (negative), then the quanto forward will be less (more) valuable than the standard forward. In the example above, the exchange rates and foreign stock were positively correlated and the quantoed instrument cost less than the original value of the foreign stock.

10.5.4 WHERE DOES CONVEXITY COME IN?

The discussion of the previous section has shown that, in a simple *one period* setting with three possible states of the world, we can form a replicating portfolio for the quantoed asset payoffs at a future date. As the number of states increases and time becomes continuous, this type of replicating portfolio needs readjustment. The portfolio adjustments would, in turn, lead to negative or positive trading gains depending on the sign of the correlation, similar to the case of options. This is where volatilities become relevant. In the case of quanto assets there are, at least, *two* risks involved, namely, exchange rate and foreign equity or interest rates. The covariance between these affects pricing as well.

The quanto feature will have a positive or negative value at time t_0 due to the trading gains realized during rebalancing. Thus, quantos form another class of assets where the non-negligibility of second-order sensitivities leads to dependence of the asset price on variances and covariances.

10.5.5 PRACTICAL CONSIDERATIONS

At first glance, quanto assets may appear very attractive to investors and portfolio managers. After all, a contract on foreign assets is purchased and all currency risk is eliminated. Does this mean we should always *quanto?*

Here again, some real-life complications are associated with the instrument. First of all, the purchase of a quanto may involve an upfront payment and the quanto characteristics depend on risk premia, bid–ask spreads, and on transaction costs associated with the underlying asset and the underlying foreign currency. These may be high and an approximate hedge using foreign currency forwards may be cheaper in the end.

Second, quanto assets have expiration dates. If, for some unforeseen reason, the contract is unwound before expiration, further costs may be involved. More important, if the foreign asset is held beyond the expiration date, the quanto feature would no longer be in effect.

Finally, the quanto contract depends on the *correlation* between two risk factors, and this correlation may be *unstable*. Under these conditions, the parties that are long or short the quanto have exposure to changes in this correlation parameter. This may significantly affect the mark-to-market value of the quanto contracts.

10.6 CONCLUSIONS

Pricing equations depends on one or more risk factors. When the pricing functions are nonlinear, replicating portfolios that use linear assets with periodically adjusted weights will lead to positive or negative cash flows during the hedging process. If the underlying volatilities and correlations are significant, trading gains from these may exceed the transaction costs implied by periodic rebalancing, and the underlying nonlinearity can be traded.

In this chapter, we saw two basic examples of this: one from the fixed-income sector which made convexity of bonds valuable, and the second from quanto instruments, which also brought in the covariance between risks. The example on quantos is a good illustration of what happens when term structure models depend on more than one factor. In such an environment, the covariances as well as the volatilities between the underlying risks may become important.

SUGGESTED READING

Two introductory sources discuss the convexity gains one can extract from fixed-income instruments. They are **Serrat and Tuckman** (2011) and **Jegadeesh and Tuckman** (2000). The convexity differences between futures and forwards are clearly handled in **Hull** (2014). The discussion of the quanto feature used here is from **Piros** (1998), which is in **DeRosa** (1998). **Wilmott** (2006) has a nice discussion of quantoed assets as well.

EXERCISES

1. You are given the following default-free long bond:
   Face value: 100
   Issuing price: 100
   Currency: USD
   Maturity: 30 years
   Coupon: 6%
   No implicit calls or puts.
   Further, in this market there are no bid−ask spreads and no trading commissions. Finally, the yield curve is flat and moves only parallel to itself.
   There is, however, a futures contract on the 1-year LIBOR rate. The price of the contract is determined as

$$V_t = 100(1 - f_t) \tag{10.94}$$

   where f_t is the "forward rate" on 1-year LIBOR.
   a. Show that if the yield of the 30-year bond is y_t, then at all times we have

$$y_t = f_t \tag{10.95}$$

   b. Plot the pricing functions for V_t and the bond.
   c. Suppose the current yield y_0 is at 7%. Put together a zero-cost portfolio that is delta-neutral toward movements of the yield curve.
   d. Consider the following yield movements over 1-year periods:

$$9\%, 7\%, 9\%, 7\%, 9\%, 7\% \tag{10.96}$$

   What are the convexity gains during this period?
   e. What other costs are there?
2. You are given a 30-year bond with yield y. The yield curve is flat and will have only parallel shifts. You have a liquid 3-month Eurodollar contract at your disposition. You can also borrow and lend at a rate of 5% initially.

a. Using the long bond and the Eurodollar contract, construct *a delta*-hedged portfolio that is immune to interest rate changes.

b. Now suppose you observe the following interest rate movements over a period of 1 year:

$$\{0.06, 0.04, 0.06, 0.04, 0.06, 0.04\} \tag{10.97}$$

These observations are each 2 months apart. What are your convexity gains from a long volatility position?

3. Consider the data given in the previous question.

a. Suppose an anticipated movement as in the previous question. Market participants suddenly move to an anticipated trajectory such as

$$\{0.08, 0.02, 0.08, 0.02, 0.08, 0.02\} \tag{10.98}$$

Assuming that this was the only exogenous change in the market, what do you think will happen to the yield on the 30-year bond?

4. Assuming that the yield curve is flat and has only parallel shifts, determine the spread between the paid-in-arrear FRAs and market-traded linear FRAs if the FRA rates are expected to oscillate as follows around an initial rate:

$$\{+0.02, -0.02, +0.02, -0.02, +0.02, -0.02\} \tag{10.99}$$

5. Answer the following questions related to the case study "Convexity of long bonds, convexity and arbitrage."

a. First the preliminaries. Explain what is meant by convexity of long-dated bonds.

b. What is meant by the convexity of long-dated interest rate swaps?

c. Explain the notion of convexity using a graph.

d. If bonds are convex, which fixed-income instrument is not convex?

e. Describe the cash flows of FRAs. When are FRAs settled in the market?

f. What is the convexity adjustment for FRAs?

g. What is a cap? What volatility do you buy or sell using caps?

h. Now the real issue. Explain the position taken by "knowledgeable" professionals.

i. In particular, is this a position on the direction of rates or something else? In fact, can you explain why the professionals had to hedge their position using caps or floors?

j. Do they have to hedge using caps only? Can floors do as well? Explain your answer graphically.

k. Is this a true arbitrage? Are there any risks?

MATLAB EXERCISES

6. Consider the Vasicek model

$$dr = \alpha(\mu - r)dt + \sigma \, dW\_t$$

Plot the term structure (i.e., plot yield versus time) for the following parameter sets $[\alpha, \mu, \sigma, r(0)]$:

$$[5.9, 0.3, 0.2, 0.1]$$

$$[3.9, 0.1, 0.3, 0.2]$$

$$[0.1, 0.4, 0.11, 0.1]$$

Now for the first parameter set plot the term structure by varying only one parameter at a time and report your observation.

7. Consider the "CIR (Cox-Ingersoll-Ross)" model

$$dr = \alpha(\mu - r)dt + \sigma\sqrt{r}dW\_t$$

Plot the term structure (i.e., plot yield versus time) for the following parameter sets $[\alpha, \mu, \sigma, r(0)]$:

$$[0.02, 0.7, 0.02, 0.1]$$

$$[0.7, 0.1, 0.3, 0.2]$$

$$[0.06, 0.09, 0.5, 0.02]$$

Now for the third parameter set, plot the term structure by varying only one parameter at a time and report your observation.

CASE STUDY: CONVEXITY OF LONG BONDS, SWAPS, AND ARBITRAGE

The yield of a long bond tells you how much you can earn from this bond. Correct? Wrong. You can earn more.

The reason is that long bonds and swaps have convexity. If there are two instruments, one linear and the other nonlinear, and if these are a function of the same risk factors, we can form a portfolio that is *delta*-neutral and that guarantees some positive return.

This is a complex and confusing notion and the purpose of this case study is to clarify this notion a bit.

At first, the case seems simple. Take a look at the following single reading provided on an arbitrage position taken by market professionals and answer the questions that follow.

The more sophisticated traders in the swaps market—or at least those who have been willing to work alongside their in-house quants—have until recently been playing a game of one-upmanship to the detriment of their more naive interbank counterparties. By taking into account the convexity effect on long-dated swaps, they have been able to profit from the ignorance of their counterparties who saw no reason to change their own valuation methods.

More specifically, several months ago several leading Wall Street US dollar swaps houses—reportedly JP Morgan and Goldman Sachs among them—realized that there was more value than met the eye when pricing LIBOR-in-arrears swaps. According to London traders, they began to arbitrage the difference between their own valuation models and those of "swap traders who still relied on naive, traditional methods" and transacting deals where they would receive LIBOR in arrears and pay LIBOR at the start of the period, typically for notional amounts of US$100m and over.

Depending on the length of the swap and the LIBOR reset intervals, they realized that they could extract up to an additional 8–10 bp from the transaction, irrespective of the shape of the yield curve. The counterparty, on the other hand, would see money "seep away over the life of the swap, even if it thought it was fully hedged," said a trader.

The added value is only significant on long-dated swaps—typically between five and 10 years—and in particular those based on 12-month LIBOR rather than the more traditional six-month LIBOR basis. This value is due to the convexity effect more commonly associated with the relationship between yields and the price of fixed income instruments.

It therefore pays to be long convexity, and when applied to LIBOR-in-arrears structures proved to be profitable earlier this year. The first deals were transacted in New York and were restricted to the US dollar market, but in early May several other players were alerted to what was going on in the market and decided to apply the same concept in London. One trader expressed surprise at the lack of communication between dealers at different banks, a fact which allowed the arbitrage to continue both between banks directly and through swaps brokers.

Also, "none of the US banks active in the market was involved in trying to exploit the same opportunities in other currencies," he said, adding "you could play the same game in sterling—convexity applies to all currencies."

In fact, there was one day in May when the sterling market was flooded with these transactions, and it "lasted for several days" according to a sterling swaps dealer, "until everyone moved their prices out," effectively putting a damper on further opportunities as well as making it difficult to unwind positions.

Further, successful structures depend on cap volatility as the extra value is captured by selling caps against the LIBOR-in-arrears being received, in addition to delta hedging the swap. In this way value can be extracted from yield curves irrespective of the slope.

"In some cap markets such as the yen, volatility isn't high enough to make the deal work," said one dealer. Most of the recent interbank activity has taken place in US dollars, sterling, and Australian dollars.

As banks have become aware of the arbitrage, opportunities have become rarer, at least in the interbank market. But as one dealer remarked, "the reason this [structure] works is because swap traders think they know how to value LIBOR-in-arrears swaps in the old way, and they stick to those methods."

"Paying LIBOR in arrears without taking the convexity effect into account," he added, "is like selling an option for free, but opportunities will still exist where traders stick to the old pricing method."

Many large swap players last week declined to comment, suggesting that the market is still alive, although BZW in London, which has been active in the market, did say that it saw such opportunities as a chance to pass on added value to its own customers. (Thomson Reuters IFR issue 1092, July 29, 1995)

CHAPTER OUTLINE

11.1 INTRODUCTION

This chapter discusses traditional option strategies from the financial engineering perspective and provides market-based examples. It then moves on to discuss exotic options. We are concerned with portfolios and positions that are taken with a precise *gain−loss profile* in mind. The players consciously take risks in the hope of benefiting or protecting themselves from an expected movement in a certain risk factor. Most investor behavior is of this kind. Investors buy a stock with a higher (systematic) risk, in anticipation of higher returns. A high-yield bond carries a higher default probability, which the bond holder is willing to bear. For all the different instruments, there are one or more risk factors that influence the gains and losses of the position taken. The investor weighs the risks due to potentially adverse movements in these factors against the gains that will result, if these factors behave in the way the investor expected. Some of the hedging activity can be interpreted in a similar way. This chapter deals with techniques and strategies that use options in doing this. We consider classical (vanilla) as well as exotic options tools.

According to an important theorem in modern finance, if options of all strikes exist, carefully selected option portfolios can replicate *any* desired gain−loss profile that an investor or a hedger desires. We can synthetically create any asset using a (static) portfolio of carefully selected options,[1] since financial positions are taken with a payoff in mind. Hence, we start our discussion by looking at payoff diagrams.

11.1.1 PAYOFF DIAGRAMS

Let x_t be a random variable representing the time-t value of a risk factor, and let $f(x_T)$ be a function that indicates the payoff of an *arbitrary* instrument at "maturity" date T, given the value of x_T at time $T > t$. We call $f(x_T)$ a *payoff function*. The functional form of $f(.)$ is known if the contract is well defined.[2] It is customary in textbooks to represent the pair $\{f(x_T), x_T\}$ as in Figure 11.1 or 11.2. Note that, here, we have a nonlinear upward sloping payoff function that depends on the values assumed by x_T only. The payoff diagram in Figure 11.1 is drawn in a completely arbitrary fashion, yet, it illustrates some of the general principles of financial exposures. Let us review these.

---

[1]This is a theoretical result, and it depends on options of all strikes existing. In practice this is not the case. Yet the result may still hold as an approximation.

[2]Here x_t can be visualized as a *kxl* vector of risk factors. To simplify the discussion, we will proceed as if there is a single risk factor, and we assume that x_t is a scalar random variable.

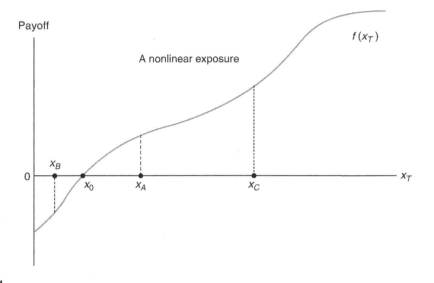

FIGURE 11.1

Payoff of nonlinear exposure.

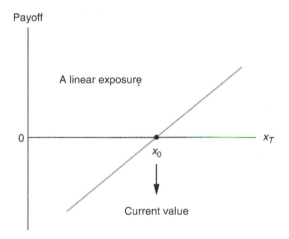

FIGURE 11.2

Payoff of a linear exposure.

First of all, for fairly priced exposures that have zero value of initiation, *net* exposures to a risk factor, x_T, must be negative for *some* values of the underlying risk. Otherwise, we would be making positive gains, and there would be no risk of losing money. This would be an arbitrage opportunity. Swap-type instruments fall into this category. If, on the other hand, the final payoffs of the contract are non-negative for all values of x_T, the exposure has a positive value at initiation, and to take the position an upfront payment will have to be made. Option positions have this characteristic.[3]

Second, exposures can be convex, concave, or linear with respect to x_T, and this has relevance for an investor or market professional. The implication of linearity is obvious: the sensitivity of the position to movements in x_T is constant. The relevance of convexity was discussed in Chapters 9 and 10. With convexity, movements in volatility need to be priced in, and again options are an important category here.

Finally, it is preferable that the payoff functions $f(x_T)$ depend *only* on the underlying risk, x_T, and do not move due to extraneous risks. We saw in Chapters 9 and 10 that volatility positions taken with options may not satisfy this requirement. The issue will be discussed in Chapter 15.

11.1.1.1 *Examples of* x_t

The discussion thus far dealt with an abstract underlying, x_t. This underlying can be almost any risk the human mind can think of. The following lists some well-known examples of x_t:

- Various *interest rates*: The best examples are LIBOR rates and swap rates. But the commercial paper (CP) rate, the federal funds rate, the index of overnight interest rates (an example of which is EONIA, Euro Over Night Index Average), and many others are also used as reference rates.
- *Exchange rates*, especially major exchange rates such as dollar–euro, dollar–yen, dollar–sterling ("cable"), and dollar–Swiss franc.
- *Equity indices*: Here also the examples are numerous. Besides the well-known US indices such as the Dow, NASDAQ, and the S&P500, there are European indices such as CAC40, DAX, and FTSE100, as well as various Asian indices such as the Nikkei 225 and emerging market indices.
- *Commodities* are also quite amenable to such positions. Futures on coffee, soybeans, and energy are other examples for x_T.
- *Bond price indices*: One example is the EMBI + prepared by JPMorgan to track emerging market bonds.

Besides these well-known risks, there are more complicated underlyings that, nevertheless, are central elements in financial market activity:

1. The underlying to the option positions discussed in this chapter can represent *volatility* or variance. If we let the percentage volatility of a price, at time t, be denoted by σ_t, then the time T value of the underlying x_T may be defined as

$$x_T = \int_t^T \sigma_u^2 S_u^2 \, du \tag{11.1}$$

where S_u may be any risk factor. In this case, x_T represents the *total variance* of S_u during the interval $[t, T]$. Volatility is the square root of x_T.
2. The *correlation* between two risk factors can be traded in a similar way.
3. The underlying, x_t, can also represent the *default probability* associated with a counterparty or instrument. This arises in the case of credit instruments.

---

[3]The market maker will borrow the needed funds and buy the option. Position will still have zero value at initiation.

4. The underlying can represent the probability of an extraordinary *event* happening. This would create a "Cat" instrument that can be used to buy insurance against various catastrophic events.

5. The underlying, x_t, can also be a *nonstorable item* such as electricity, weather, or bandwidth.

Readers who are interested in the details of such contracts or markets should consult Hull (2014). In this chapter, we limit our attention to the engineering aspects of option contracts.

11.2 OPTION STRATEGIES

We divide the engineering of option strategies into two broad categories. First, we consider the classical option-related methods. These will cover strategies used by market makers as well as retail investors. They will themselves be divided into two groups, those that can be labeled *directional* strategies, and those that relate to views on the *volatility* of some underlying instrument. The second category involves exotic options, which we consider as more efficient and sometimes cheaper alternatives to the classical option strategies. The underlying risks can be any of those mentioned in the previous section.

11.2.1 SYNTHETIC LONG AND SHORT POSITIONS

We begin with strategies that utilize options essentially as directional instruments, starting with the creation of *long* and *short positions* on an asset. Options can be used to create these positions *synthetically*.

Consider two plain vanilla options written on a *forward* price F_t of a certain asset. The first is a short put, and the second a long call, with prices $P(t)$ and $C(t)$, respectively, as shown in Figure 11.3. The options have the same strike price K, and the same expiration time T.[4] Assume that the Black–Scholes conditions hold, and that both options are of European style. Importantly, the underlying asset does not have any payouts during $[t, T]$. Also, suppose the appropriate short rate to discount future cash flows is constant at r.

Now consider the portfolio

$$\{1 \text{ Long } K\text{-Call}, 1 \text{ Short } K\text{-Put}\} \qquad (11.2)$$

At expiration, the payoff from this portfolio will be the vertical sum of the graphs in Figure 11.3 and is as shown in Figure 11.4. This looks like the payoff function of a *long forward contract* entered into at K. If the options were at-the-money (ATM) at time t, the portfolio would exactly duplicate the long forward position and hence would be an exact synthetic. But there is a close connection between this portfolio and the forward contract, even when the options are not ATM.

At expiration time T, the value of the portfolio is

$$C(T) - P(T) = F_T - K \qquad (11.3)$$

---

[4]Short calls and long puts lead to symmetric results and are not treated here.

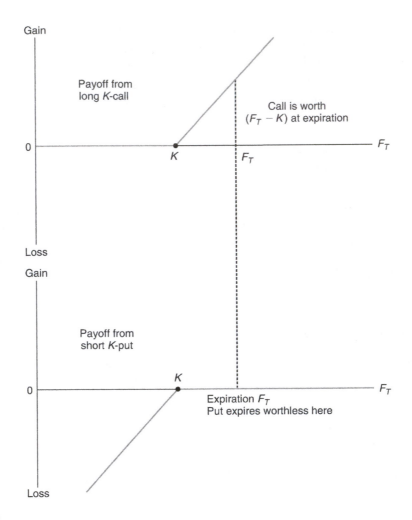

FIGURE 11.3

Payoff of a long call and a short put position.

where F_T is the time-T value of the forward price. This equation is valid because at T, only *one* of the two options can be in-the-money. Either the call option has a value of $F_T - K$ while the other is worthless, or the put is in-the-money and the call is worthless, as shown in Figure 11.3.

Subtract the time-t forward price, F_t, from both sides of this equation to obtain

$$C(T) - P(T) + (K - F_t) = F_T - F_t \qquad (11.4)$$

This expression says that the sum of the payoffs of the long call and the short put *plus* $(K - F_t)$ units of cash should equal the time-T gain or loss on a forward contract entered into at F_t, at time t.

Take the expectation of Eq. (11.4). Then the time-t value of the portfolio,

$$\{1 \text{ Long } K\text{-Call}, 1 \text{ Short } K\text{-Put}, e^{-r(T-t)}(K - F_t)\text{Dollars}\} \qquad (11.5)$$

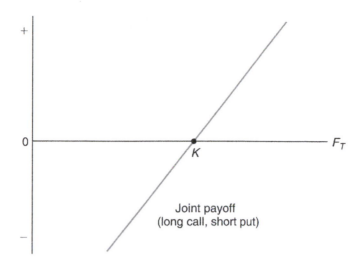

FIGURE 11.4

Joint payoff of a long call and short put.

should be zero at t, since credit risks and the cash flows generated by the forward and the replicating portfolio are the same. This implies that

$$C(t) - P(t) = e^{-r(T-t)}(F_t - K) \tag{11.6}$$

This relationship is called *put-call parity*. It holds for European options. It can be expressed in terms of the spot price, S_t, as well. Assuming zero storage costs, and no convenience yield:[5]

$$F_t = e^{r(T-t)}S_t \tag{11.7}$$

Substituting in the preceding equation gives

$$C(t) - P(t) = (S_t - e^{-r(T-t)}K) \tag{11.8}$$

Put-call parity can thus be regarded as another result of the application of contractual equations, where options and cash are used to create a synthetic for the S_t. This situation is shown in Figure 11.5.

11.2.1.1 An application

Option market makers routinely use the put-call parity in exploiting windows of arbitrage opportunities. Using options, market makers construct synthetic futures positions and then trade them

---

[5]Here r is the borrowing cost and, as discussed in Chapter 3, is a determinant of forward prices. The convenience yield is the opposite of carry cost. Some stored cash goods may provide such convenience yield and affect F_t.

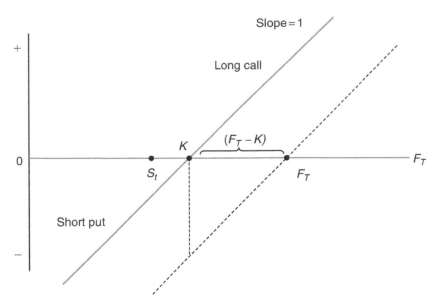

FIGURE 11.5

Put-call parity illustration and payoff of a synthetic stock position.

against futures contracts. This way, small and temporary differences between the synthetic and the true contract are converted into "riskless" profits. In this section, we discuss an example.

Suppose, without any loss of generality, that a stock is trading at

$$S_t = 100 \tag{11.9}$$

and that the market maker can buy and sell ATM options that expire in 30 days. Suppose also that the market maker faces a funding cost of 5%. The stock never pays dividends and there are no corporate actions.

Also, and this is the *real-life* part, the market maker faces a transaction cost of 20 cents per traded option and a transaction cost of 5 cents per traded stock. Finally, the market maker has calculated that to be able to continue operating, he or she needs a margin of 0.25 cent per position. Then, we can apply put-call parity and follow the *conversion* strategy displayed in Figure 11.6.

> Borrow necessary funds overnight for 30 days, and buy the stock at price S_t. At the same time, sell the S_t-call and buy the S_t-put that expires in 30 days, to obtain the position shown in Figure 11.6.

The position is *fully hedged*, as any potential gains due to movement in S_t will cover the potential losses. This means that the only factors that matter are the *transaction costs* and any *price differentials* that may exist between the call and the put. The market maker will monitor the difference between the put and call premiums and take the arbitrage position shown in Figure 11.6 if this difference is bigger than the total cost of the conversion.

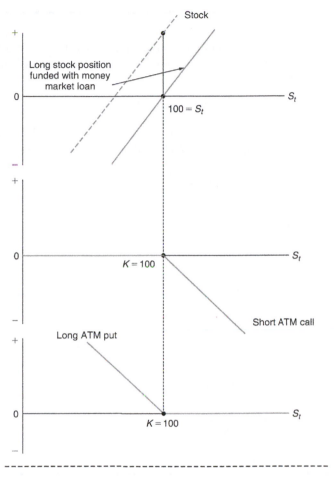

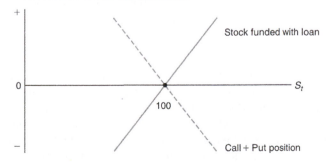

FIGURE 11.6

Put-call parity application and conversion strategy.

EXAMPLE

Suppose $S_t = 100$, and 90-day call and put options trade actively. The interest cost is 5%. A market maker has determined that the call premium, $C(t)$, exceeds the put premium, $P(t)$, by $2.10:

$$C(t) - P(t) = 2.10 \qquad (11.10)$$

The stock will be purchased using borrowed funds for 90 days, and the ATM put is purchased and held until expiration, while the ATM call is sold. This implies a funding cost of

$$100(0.05)\left(\frac{90}{360}\right) = \$1.25 \qquad (11.11)$$

Add all the costs of the conversion strategy:

Cost per Security	$
Funding cost	1.25
Stock purchase	0.05
Put purchase	0.20
Call sale	0.20
Operating costs	0.25
Total cost	1.95

The market maker incurs a total cost of $1.95. It turns out that under these conditions, the net cash position will be positive:

$$\text{Net profit} = 2.10 - 1.95 \qquad (11.12)$$

and the position is worth taking.

If, in the example just discussed, the put-call premium difference is negative, then the market maker can take the opposite position, which would be called a *reversal*.[6]

11.2.1.2 Arbitrage opportunity?

An outside observer may be surprised to hear that such "arbitrage" opportunities exist, and that they are closely monitored by market makers on the trading floor. Yet, such opportunities are available *only* to market makers on the "floor" and may not even constitute arbitrage in the usual sense.

This is because of the following: (i) Off-floor investors pay much higher transaction costs than the on-floor market makers. Total costs of taking such a position may be prohibitive for off-floor investors. (ii) Off-floor investors cannot really make a *simultaneous* decision to buy (sell) the underlying, and buy or sell the implied puts or calls to construct the strategy. By the time these strategies are

---

[6]This is somewhat different from the upcoming strategy known as risk reversals.

communicated to the floor, prices could move. (iii) Even if such opportunities are found, net gains are often too small to make it worthwhile to take such positions sporadically. It is, however, worthwhile to a market maker who specializes in these activities. (iv) Finally, there is also a serious risk associated with these positions, known as the *pin risk*. Pin risk refers to the situation when the market price of the underlying of an option contract at the time of the contract's expiration is close to the option's strike price. Traders pay attention to whether an underlying such as a stock, for example, may be close to *pinning*, since a small movement of the price of the underlying through the strike can have a large impact on a trader's net position in the underlying on the trading day after expiration.

11.2.2 A REMARK ON THE PIN RISK

It is worthwhile to discuss the *pin risk* in more detail, since similar risks arise in hedging and trading some exotic options as well. Suppose we put together a *conversion* at 100, and waited 90 days until expiration to unwind the position. The positions will expire some 90 days later during a Friday. Suppose at expiration S_t is exactly 100. This means that the stock closes exactly at the strike price. This leads to a dilemma for the market maker.

The market maker owns a stock. If he or she does *not* exercise the long put and if the short call is not assigned (i.e., if he or she does not get to sell at K exactly), then the market maker will have an open long position in the stock during the weekend. Prices may move by Monday and he or she may experience significant losses.

If, on the other hand, the market maker does exercise the long put (i.e., he or she sells the stock at K) and if the call is assigned (i.e., he or she needs to deliver a stock at K), then the market maker will have a short stock position during the weekend. These risks may not be great for an end investor who takes such positions occasionally, but they may be substantial for a professional trader who depends on these positions. There is no easy way out of this dilemma. This type of risk is known as the pin risk.

The main cause of the pin risk is the kink in the expiration payoff at $S_T = K$. A kink indicates a sudden change in the slope—for a long call, from zero to one or vice versa. This means that even with small movements in S_t, the hedge ratio can be either zero or one, and the market maker may be caught significantly off guard. If the slope of the payoff diagram changed smoothly, then the required hedge would also change smoothly. Thus, a risk similar to pin risk may arise whenever the *delta* of the instrument shows discrete jumps.

11.2.3 RISK REVERSALS

A more advanced version of the synthetic long and short futures positions is known as *risk reversals*. These are liquid synthetics especially in the foreign exchange markets, where they are traded as a commodity. Risk reversals are directional positions but differ in more than one way from synthetic long−short futures positions discussed in the previous section.

The idea is again to buy and sell calls and puts in order to replicate long and short futures positions—but this time using options with *different* strike prices. Figure 11.7 shows an example. The underlying is S_t. The strategy involves a short put struck at K_1, and a long call with strike K_2. Both options are out-of-the-money initially, and the S_t satisfies

$$K_1 < S_t < K_2 \tag{11.13}$$

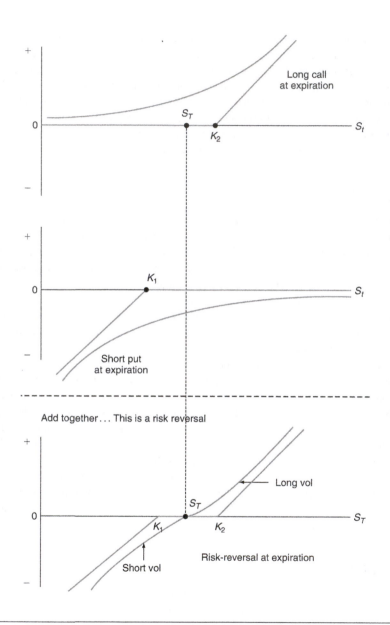

FIGURE 11.7

Payoff of a long call, short put, and a risk reversal position.

Since strikes can be chosen such that the put and call have the same premium, the risk reversal can be constructed so as to have zero initial price.

By adding vertically the option payoffs in the top portion of Figure 11.7, we obtain the expiration payoff shown at the bottom of the figure. If, at expiration, S_T is between K_1 and K_2, the strategy has zero payoff. If, at expiration, $S_T < K_1$, the risk reversal loses money, but under $K_2 < S_T$, it makes money. Clearly, what we have here is similar to a long position but the position is neutral for small movements in the underlying starting from S_t. If taken naked, such a position would imply a bullish view on S_t.

We consider an example from foreign exchange (FX) markets where risk reversals are traded as commodities.

EXAMPLE

Twenty-five delta 1-month risk reversals showed a stronger bias in favor of euro calls (dollar puts) in the last 2 weeks after the euro started to strengthen against the greenback.

Traders said market makers in EUR calls were buying risk reversals expecting further euro upside. The 1-month risk reversal jumped to 0.91 in favor of euro calls Wednesday from 0.3 3 weeks ago. Implied volatility spiked across the board. One-month volatility was 13.1% Wednesday from 11.78% 3 weeks ago as the euro appreciated to USD1.0215 from USD1.0181 in the spot market.

The 25-*delta* risk reversals mentioned in this reading are shown in Figure 11.8. The risk reversal is constructed using two options, a call and a put. Both options are out-of-the-money and have a "current" *delta* of 0.25. According to the reading, the 25-*delta* EUR call is more expensive than the 25-*delta* EUR put.

11.2.3.1 *Uses of risk reversals*

Risk reversals can be used as "cheap" hedging instruments. Here is an example.

EXAMPLE

A travel company in Paris last week entered a zero-cost risk reversal to hedge US dollar exposure to the USD. The company needs to buy dollars to pay suppliers in the United States, China, Indonesia, and South America.

The head of treasury said it bought dollar calls and sold dollar puts in the transaction to hedge 30% of its USD200–300 million dollar exposure versus the USD. The American-style options can be exercised between November and May.

The company entered a risk reversal rather than buying a dollar call outright because it was cheaper. The head of treasury said the rest of its exposure is hedged using different strategies, such as buying options outright. (Based on an article in Derivatives Week (now part of GlobalCapital).)

Here we have a corporation that has EUR receivables from tourists going abroad but needs to make payments to foreigners in dollars. Euros are received at time t, and dollars will be paid at some future date T, with $t < T$. The risk reversal is put together as a zero-cost structure, which means that the premium collected from selling the put (on the USD) is equal to the call premium on the USD. For small movements in the exchange rate, the position is neutral, but for large movements it represents a hedge similar to a futures contract.

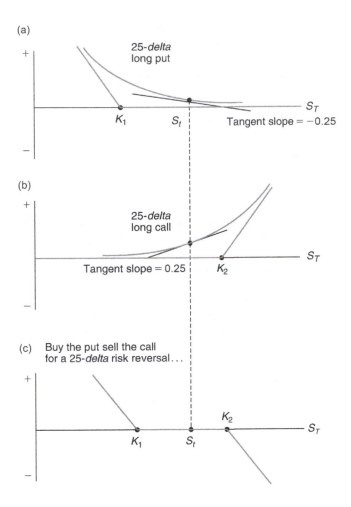

FIGURE 11.8

Payoff of a long put, short call a risk reversal position.

Of course, such a position could also be taken in the futures market. But one important advantage of the risk reversal is that it is "composed" of options, and hence involves, in general, no daily mark-to-market adjustments.

11.2.4 YIELD ENHANCEMENT STRATEGIES

The class of option strategies that we have studied thus far is intended for creating synthetic short and long futures positions. In this section, we consider option synthetics that are said to lead to *yield enhancement* for investment portfolios.

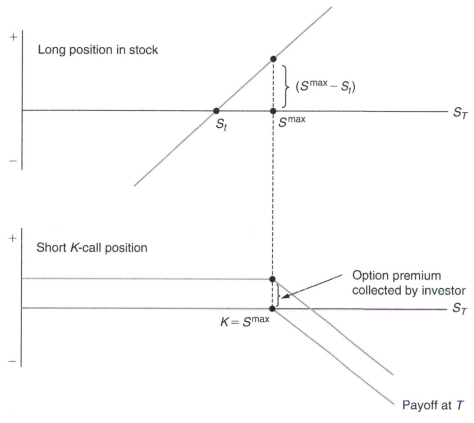

FIGURE 11.9

Payoff of call overwriting strategy components.

11.2.4.1 Call overwriting

The simplest case is the following. At time t, an investor takes a long position in a stock with current price S_t, as shown in Figure 11.9. If the stock price increases, the investor gains; if the price declines, he or she loses. The investor has, however, a *subjective* expected return, $\hat{R}_t$, for an interval of time Δ, that can be expressed as

$$\hat{R}_t = E_t^{\hat{P}}\left[\frac{S_{t+\Delta} - S_t}{S_t}\right] \tag{11.14}$$

where $\hat{P}$ is a subjective conditional probability distribution for the random variable $S_{t+\Delta}$. According to the formula, the investor is expecting a gain of $\hat{R}_t$ during period Δ. The question is whether we can provide a *yield-enhancing* alternative to this investor. The answer depends on what we mean by "yield enhancement."

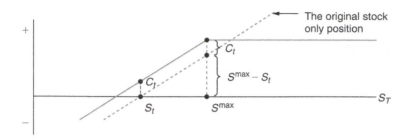

FIGURE 11.10

Payoff of call overwriting strategy.

Suppose we ask the investor the following question: "What is the maximum gain you would like to make on this stock position?" and the investor indicates S^{max} as the price he or she is willing to sell the stock and realize the "maximum" desired gain:

$$(S^{max} - S_t) \tag{11.15}$$

Next, consider a call option $C(t)^{max}$ that has the strike

$$K = S^{max} \tag{11.16}$$

and that expires at $T = t + \Delta$. This option sells for $C(t)^{max}$ at time t. We can then recommend the following portfolio to this investor:

$$\text{Yield enhanced portfolio} = \{\text{Long } S_t, \text{Short } C(t)^{max}\} \tag{11.17}$$

Assuming zero interest rates, at time $T = t + \Delta$, this portfolio has the following value, $V_{t+\Delta}$:

$$V_{t+\Delta} = \begin{cases} C(t)^{max} + S_{t+\Delta} & \text{Option not exercised} \\ C(t)^{max} + S_{t+\Delta}(S_{t+\Delta} - S^{max}) = C(t)^{max} + S^{max} & \text{Option exercised} \end{cases} \tag{11.18}$$

According to this, if at expiration, the price stays below the level S^{max}, the investor "makes" an extra $C(t)^{max}$ dollars. If $S_{t+\Delta}$ exceeds the S^{max}, the option will be exercised, and the gains will be truncated at $S^{max} + C(t)^{max}$. But, this amount is higher than the price at which this investor was willing to sell the stock according to his or her subjective preferences. As a result, the option position has enhanced the "yield" of the original investment. However, it is important to realize that what is being enhanced is not the objective risk-return characteristics, but instead, the *subjective* expected returns of the investor.

Figure 11.9 shows the situation graphically. The top portion is the long position in the stock. The bottom profile is the payoff of the short call, written at strike S^{max}. If $S_{t+\Delta}$ exceeds this strike, the option will be in-the-money and the investor will have to surrender his or her stock, worth $S_{t+\Delta}$ dollars, at a price of S^{max} dollars. But, the investor was *willing* to sell at S^{max} anyway. The sum of the two positions is illustrated in the final payoff diagram in Figure 11.10.

This strategy is called *call overwriting* and is frequently used by some investors. The following reading illustrates one example. Fund managers who face a stagnant market use call overwriting to enhance yields.

EXAMPLE

Fund manager motivation for putting on options strategies ahead of the Russell indices annual rebalance next month is shifting, say some options strategists.

"The market has had no direction since May last year," said a head of equity derivatives strategy in New York. Small cap stocks have only moved up slightly during the year, he added.

Fund managers are proving increasingly willing to test call overwriting strategies for the rebalance as they seek absolute returns, with greater competition from hedge funds pushing traditional fund managers in this direction, [a] head of equity derivatives strategy said. Employing call overwriting strategies—even though they suppress volatility levels—looks attractive, because the worst outcome is that they outperform the stock on the downside. As such, it can help managers enhance their returns.

(Thomson Reuters IFR, Issue 1433, May 11, 2002)

The situation described in this reading is slightly more complicated and would not lend itself to the simple call overwriting position discussed earlier. The reading illustrates the periodic and routine rebalancing that needs to be performed by fund managers. Many funds "track" well-known indices. But, these indices are periodically revised. New names enter, others leave, at known dates. A fund manager who is trying to track a particular index, has to rebalance his or her portfolio as indices are revised.

11.3 VOLATILITY-BASED STRATEGIES

The first set of strategies dealt with *directional* uses of options. Option portfolios combined with the underlying were used to take a *view* on the direction of the underlying risk. Now we start looking at the use of options from the point of view of volatility positioning. The strategy used in putting together volatility positions in this section is the following: First, we develop a *static* position that eliminates exposure to market direction. This can be done using *straddles* and their cheaper version, *strangles*. Second, we combine strangle and straddle portfolios to get more complicated volatility positions, and to reduce costs.

Thus, the basic building blocks of volatility positions considered in this section are straddles and strangles. The following example indicates how an option position is used to take a *view on volatility*, rather than the price of the underlying.

EXAMPLE

An Italian bank recommended the following position to a client. We will analyze what this means for the client's expectations [views] on the markets. First we read the episode.

A bank last week sold 4% out-of-the-money puts and calls on ABC stock, to generate a premium on behalf of an institutional investor. The strangle had a tenor of six weeks. . .. The strategy generated 2.5% of the equity's spot level in premium.

At the time of the trade, the stock traded at roughly USD1874.6. Volatilities were at 22% when the options were sold. ABC was the underlying, because the investor does not believe the stock will move much over the coming weeks and thus is unlikely to break the range and trigger the options.

(Based on an article in Derivatives Week (now part of GlobalCapital))

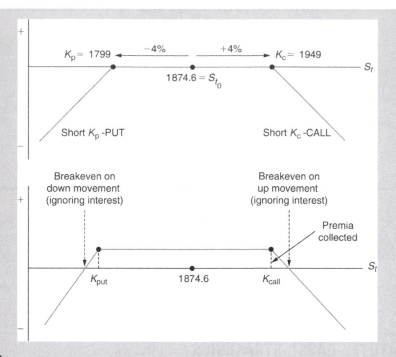

FIGURE 11.11

Payoff of a strangle position related to ABC stock.

Figure 11.11 shows the payoff diagram of these option positions at expiration. Adding the premiums received at the initial point we get the second diagram in the bottom part of the figure. This should not be confused with the anticipated payoff of the client. Note that the eventual objective of the client is to benefit from volatility realizations. The option position is only a vehicle for doing this.

We can discuss this in more detail. The second part of Figure 11.11 shows that at expiration, the down and up breakeven points for the position are 1762 and 1987, respectively. These are obtained by subtracting and adding the $37.5 received from the strangle position, to the respective strike prices.

But the reading also gives the implied volatility in the market. From here we can use the square root formula and calculate the implied volatility for the period under consideration

$$\sigma S_t \cdot \Delta = 0.22\sqrt{\frac{6}{52}}1874.6 = 140.09 \tag{11.19}$$

Note that the breakeven points are set according to 4% movements toward either side, whereas the square root formula gives 7.5% expected movements to either side.

According to this, the client who takes this position expects the realized volatility to be significantly less than the 7.5% quoted by the market. In fact, the client expects volatility to be somewhat less than 4%.

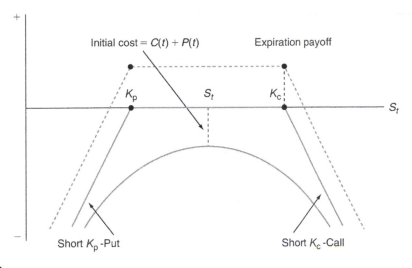

FIGURE 11.12

Typical short strangle's expiration payoff.

This brings us to a formal discussion of strangles and straddles, which form the main building blocks for classical volatility positions.

11.3.1 STRANGLES

Assume that we sell (buy) two plain vanilla, European-style options with *different strikes* on the asset S_t. The first is a put, and has strike K_p; the second is a call, and has strike K_c, with $K_p < K_c$. Suppose at the time of purchase, we have $K_p < S_{t_0} < K_c$. The expiration date is T. This position discussed in the previous example is known as a *strangle*. Because these options are sold, the seller collects a premium, at time t, of

$$C(t) + P(t) \tag{11.20}$$

The position makes money if, by expiration, S_t has moved by a "moderate" amount, otherwise the position loses money. Clearly, this way of looking at a strangle suggests that the position is *static*. A typical *short* strangle's expiration payoff is shown in Figure 11.12. The same figure indicates the value of the position at time t, when it was initially put in place.

11.3.1.1 Uses of strangles

The following is an example of the use of strangles from foreign exchange markets. First there is a switch in terminology: Instead of talking about options that are out-of-the-money by $k\%$ of the strike, the episode uses the terminology "10-*delta* options." This is the case because, as mentioned earlier, FX markets like to trade 10-*delta*, 25-*delta* options, and these will be more liquid than, say, an arbitrarily selected $k\%$ out-of-the-money option.

EXAMPLE

A bank is recommending its clients to sell 1-month 10-delta euro/dollar strangles to take advantage of low holiday volatility. The strategists said the investors should sell 1-month strangles with puts struck at USD1.3510 and calls struck at USD1.3610. This will generate a premium of 0.3875% of the notional size. Spot was trading at USD1.3562 when the trade was designed last week. The bank thinks this is a good time to put the trade on because implied volatility traditionally falls over Christmas and New Years, which means spot is likely to stay in this range.

(Based on Derivatives Week (now part of GlobalCapital))

This is a straightforward use of strangles. According to the strategist, the premium associated with the FX options implies a volatility that is higher than the expected future *realized* volatility during the holiday season due to seasonal factors. If so, the euro/dollar exchange rate is likely to be range-bound, and the options used to create the strangle will expire unexercised.[7]

11.3.2 STRADDLE

A straddle is similar to a strangle, except that the strike prices, K_p and K_c, of the constituent call and put options sold (bought), are *identical*:

$$K_p = K_c \tag{11.21}$$

Let the underlying asset be S_t, and the expiration time be T. The expiration payoff and time value of a long straddle are shown in Figure 11.13. The basic configuration is similar to a long strangle. One difference is that a straddle will *cost* more. At the time of purchase, an ATM straddle is more convex than an ATM strangle, and hence has "maximum" *gamma*.

11.3.2.1 Static or dynamic position?

It is worthwhile to emphasize that the strangle or straddle positions discussed here are *static*, in the sense that, once the positions are taken, they are not *delta*-hedged. However, it is possible to convert them into dynamic strategies. To do this, we would *delta*-hedge the position dynamically. At initiation, an ATM straddle is automatically market-neutral, and the associated *delta* is zero. When the price moves up, or down, the *delta* becomes positive, or negative. Thus, to maintain a market-neutral position, the hedge needs to be adjusted periodically.

Note a major difference between the static and dynamic approaches. Suppose we take a *static* straddle position, and S_t fluctuates by small amounts very frequently and never leaves the region $[S_1, S_2]$ shown in Figure 11.14. Then, the static position will lose money, while the dynamic *delta*-hedged position may make money, depending on the size and frequency of oscillations in S_t.

---

[7]Clearly, the issue about the seasonal movement in volatility is open to debate and is an empirically testable proposition, but it illustrates some possible seasonality left in the volatility of the data.

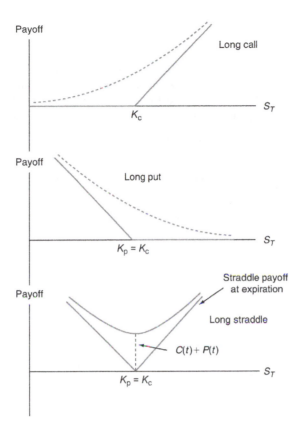

FIGURE 11.13

Expiration payoff and time value of a long straddle.

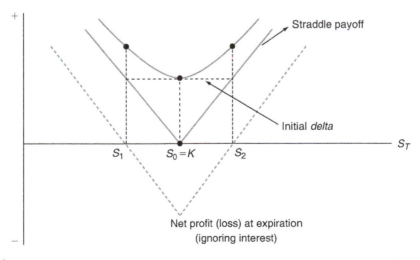

FIGURE 11.14

Payoff and net profit from straddle position.

11.3.3 BUTTERFLY

A *butterfly* is a position that is built using *combinations* of strangles and straddles. Following the same idea used throughout the book, once we develop strangle and straddle payoffs as building blocks, we can then combine them to generate further synthetic payoffs. A long butterfly position is shown in Figure 11.15. The figure implies the following contractual equation:

$$\text{Long butterfly} = \text{Long ATM straddle} + \text{Short } k\% \text{ out-of-the-money strangle} \qquad (11.22)$$

This equation immediately suggests one objective behind butterflies. By selling the strangle, the trader is, in fact, lowering the cost of buying the straddle. In the case of the short butterfly, the situation is reversed:

$$\text{Short butterfly} = \text{Short ATM straddle} + \text{Long } k\% \text{ out-of-the-money strangle} \qquad (11.23)$$

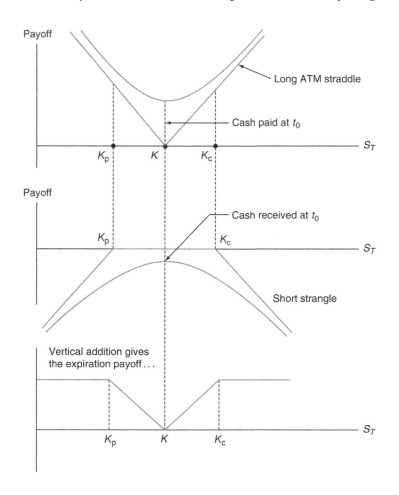

FIGURE 11.15

Long butterfly position.

A short straddle generates premiums but has an *unlimited* downside. This may not be acceptable to a risk manager. Hence, the trader buys a strangle to limit these potential losses. But this type of insurance involves costs and the net cash receipts become smaller. The following shows a practical use of the short butterfly strategy.

EXAMPLE

As the Australian dollar continues to strengthen on the back of surging commodity prices, dealers are looking to take advantage of an anticipated lull in the currency's bull run by putting in place butterfly structures. One structure is a 3-month butterfly trade. The dealer sells an ATM Aussie dollar call and put against the US dollar, while buying an Aussie call struck at AUD0.682 and buying puts struck at AUD0.6375. The structure can be put in place for a premium of 0.3% of notional, noted one trader, adding that there is value in both the puts and the calls.

(Based on an article in Derivatives Week (now part of GlobalCapital))

This structure can also be put in place by making sure that the exposure is *vega*-neutral.

11.4 EXOTICS

Up to this point, the chapter has dealt with option strategies that used only *plain vanilla* calls and puts. The more complicated volatility building blocks, namely straddles and strangles, were generated by putting together plain vanilla options with different strike prices or expiration. But the use of plain vanilla options to take a view on the direction of markets or to trade volatility may be considered by some as "outdated." There are now more practical ways of accomplishing similar objectives.

The general principle is this. Instead of combining plain vanilla options to create desired payoff diagrams, lower costs, and reach other objectives, a trader would *directly* design new option contracts that can do similar things in a "better" fashion. Of course, these new contracts imply a hedge that is, in general, made of the underlying plain vanilla options, but the new instruments themselves would sell as *exotic* options.[8] Before closing this chapter, we would like to introduce further option strategies that use exotic options as building blocks. We will look at a limited number of exotics, although there are many others that we relegate to the exercises at the end of the chapter.

11.4.1 BINARY, OR DIGITAL, OPTIONS

To understand binary options, first remember the static strangle and straddle strategies. The idea was to take a long (short) volatility position, and benefit if the underlying moved more (less) than what the implied volatility suggested. Binary options form *essential* building blocks for similar volatility strategies, which can be implemented in a cheaper and perhaps more efficient way. Also, binary options are excellent examples of option engineering. We begin with a brief description of a European-style binary option.

---

[8]The term "exotic" may be misleading. Many of the exotic options have become commoditized and trade as vanilla products.

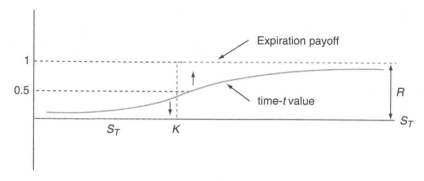

FIGURE 11.16

Intrinsic value of a binary call.

11.4.1.1 A binary call

Consider a European call option with strike K and expiration time T. S_t denotes the underlying risk. This is a standard call, except that if the option expires at or in-the-money, the payoff will be either (i) a constant cash amount or (ii) a particular asset. In this section, we consider binaries with cash payoffs only.

Figure 11.16 shows the payoff structure of this call whose time-t price is denoted by $C^{\text{bin}}(t)$. The time-T payoff can be written as

$$C^{\text{bin}}(T) = \begin{cases} R & \text{If } K \leq S_T \\ 0 & \text{Otherwise} \end{cases} \tag{11.24}$$

According to this, the binary call holder receives the cash payment R as long as S_T is not less than K at time T. Thus, the payoff has an R-or-nothing binary structure. Binary puts are defined in a similar way.

The diagram in Figure 11.16 shows the intrinsic value of the binary where $R = 1$. What would the *time value* of the binary option look like? It is, in fact, easy to obtain a closed-form formula that will price binary options. Yet, we prefer to answer this question using financial engineering. More precisely, we first create a *synthetic* for the binary option. The value of the synthetic should then equal the value of the binary.

The logic in forming the synthetic is the same as before. We have to duplicate the final payoffs of the binary using other (possibly liquid) instruments, and make sure that the implied cash flows and the underlying credit risks are the same.

11.4.1.2 Replicating the binary call

Expiration payoff of the binary is displayed by the step function shown in Figure 11.16. Now, make two additional assumptions. First, assume that the underlying S_t is the price of a futures contract traded at an exchange, and that the exchange has imposed a *minimum tick* rule such that, given S_t, the next instant's price, $S_{t+\Delta}$, can only equal

$$S_{t+\Delta} = S_t + ih \tag{11.25}$$

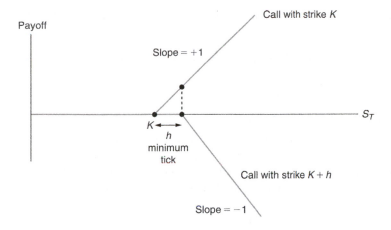

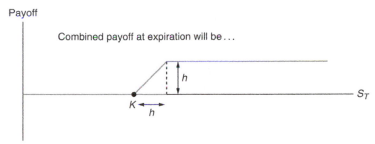

FIGURE 11.17

Expiration payoff of a binary call.

where i is an integer, and h is the *minimum tick* chosen by the exchange. Second, we assume without any loss of generality that

$$R = 1 \tag{11.26}$$

Under these conditions, the payoff of the binary is a step function that shows a jump of size 1 at $S_T = K$.

It is fairly easy to find a replicating portfolio for the binary option under these conditions. Suppose the market maker buys one vanilla European call with strike K, and, at the same time, sells one vanilla European call with strike $K + h$ on the S_t. Figure 11.17 shows the time-T payoff of this portfolio. The payoff is similar to the step function in Figure 11.16, except that the height is h, and not 1. But this is easy to fix. Instead of buying and selling 1 unit of each call, the market maker can buy and sell $\frac{1}{h}$ units. This implies the approximate contractual equation

Binary call, strike K	$\cong$	Long $\frac{1}{h}$ units of vanilla K-call	$+$	Short $\frac{1}{h}$ units of vanilla $(K+h)$-call	(11.27)

The existence of a minimum tick makes this approximation a true equality, since

$$|S_t - S_{t+\Delta} < h|$$

cannot occur due to minimum tick requirements. We can use this contractual equation and get two interesting results.

11.4.1.3 Delta and price of binaries

There is an interesting analogy between binary options and the *delta* of the constituent plain vanilla counterparts. Let the price of the vanilla K and $K + h$ calls be denoted by $C^K(t)$ and $C^{K+h}(t)$, respectively. Then, *assuming that the volatility parameter σ does not depend on K*, we can let $h \to 0$ in the previous contractual equation, and obtain the exact price of the binary, $C^{\text{bin}}(t)$, as

$$C^{\text{bin}}(t) = \lim_{h \to 0} \frac{C^K(t) - C^{K+h}(t)}{h} \tag{11.28}$$

$$= \frac{\partial C^K(t)}{\partial K} \tag{11.29}$$

assuming that the limit exists.

That is to say, at the limit the price of the binary is, in fact, the partial derivative of a vanilla call with respect to the strike price K. If all Black–Scholes assumptions hold, we can take this partial derivative analytically, and obtain[9]

$$C^{\text{bin}}(t) = \frac{\partial C^K(t)}{\partial K} = e^{-r(T-t)} N(d_2) \tag{11.30}$$

where d_2 is, as usual,

$$d_2 = \frac{\text{Log}(S_t/K) + r(T-t) - \frac{1}{2}\sigma^2(T-t)}{\sigma\sqrt{(T-t)}} \tag{11.31}$$

σ being the constant percentage volatility of S_t, and, r being the constant risk-free spot rate.

This last result shows an interesting similarity between binary option prices and vanilla option deltas. In Chapter 9 we showed that a vanilla call's *delta* is given by

$$\text{Delta} = \frac{\partial C^K(t)}{\partial S_t} = N(d_1) \tag{11.32}$$

Here we see that the price of the binary has a similar form. Also, it has a shape similar to that of a probability distribution:

$$C^{\text{bin}}(t) = e^{-r(T-t)} N(d_2) = e^{-r(T-t)} \int_{-\infty}^{\frac{\log(S_t/K) + r(T-t) - \frac{1}{2}\sigma^2(T-t)}{\sigma\sqrt{(T-t)}}} \frac{1}{\sqrt{2\pi}} e^{-\frac{1}{2}u^2} du \tag{11.33}$$

This permits us to draw a graph of the binary price, $C^{\text{bin}}(t)$. Under the Black–Scholes assumptions, it is clear that this price will be as indicated by the S-shaped curve in Figure 11.16.

---

[9]See Appendix 9.1 in Chapter 9.

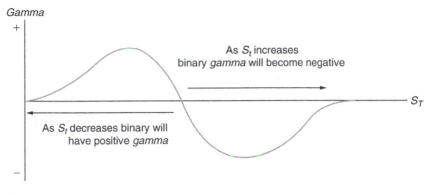

FIGURE 11.18

Gamma of a binary as a function of underlying.

11.4.1.4 Time value of binaries

We can use the previous result to obtain convexity characteristics of the binary option shown in Figure 11.16. The deep out-of-the-money binary[10] will have a positive price close to zero. This price will increase and will be around $\frac{1}{2}$ when the option becomes ATM. On the other hand, an in-the-money binary will have a price less than one, but approaching it as S_t gets larger and larger. This means that the time value of a European in-the-money binary is negative for $K < barrier$. The $C^{bin}(t)$ will never exceed 1 (or R), since a trader would never pay more than \$1 in order to get a chance of earning \$1 at T.

From this figure, we see that a market maker who buys the binary call will be *long* volatility if the binary is out-of-the-money, but will be *short* volatility, if the binary option is in-the-money. This is because, in the case of an in-the-money option, the curvature of the $C^{K+h}(t)$ will dominate the curvature of the $C^K(t)$, and the binary will have a concave pricing function. The reverse is true if the binary is out-of-the-money. An ATM binary will be neutral toward volatility.

To summarize, we see that the price of a binary is similar to the *delta* of a vanilla option. This implies that the *delta* of the *binary* looks like the *gamma* of a vanilla option. This logic tells us that the *gamma* of a binary looks like that in Figure 11.18, and is similar to the third partial with respect to S_t of the vanilla option.

11.4.1.5 Uses of the binary

A *range option* is constructed using binary puts and calls with the same payoff. This option has a payoff depending on whether the S_t remains within the range $[H^{\min}, H^{\max}]$ or not. Thus, consider the portfolio

$$\text{Range option} = \{\text{Long } H^{\min} - \text{Binary call}, \text{Short } H^{\max} - \text{Binary call}\} \qquad (11.34)$$

The time-T payoff of this range option is shown in Figure 11.19. It is clear that we can use binary options to generate other, more complicated, *range structures*.

---

[10]Remember that the payoff of the binary is still assumed to be $R = 1$.

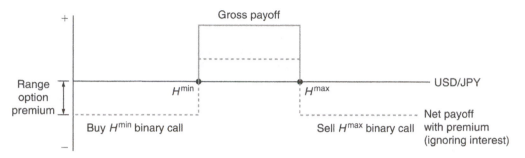

FIGURE 11.19

Payoff of a range option.

The expiration payoff denoted by $C^{\text{range}}(T)$ of such a structure will be given by

$$C^{\text{range}}(T) = \begin{cases} R & \text{if} \quad H^{\min} < S_u < H^{\max} \quad u \in [t, T] \\ 0 & \text{Otherwise} \end{cases} \qquad (11.35)$$

Thus, in this case, the option pays a constant amount R if S_u is range-bound during the *whole* life of the option, otherwise the option pays nothing. The following example illustrates the use of such binaries.

EXAMPLE

Japanese exporters last week were snapping up 1- to 3-month Japanese yen/US dollar binary options, struck within a JPY114-119 range, betting that the yen will remain bound within that range. Buyers of the options get a predetermined payout if the yen trades within the range, but forfeit a principal if it touches either barrier during the life of the option. The strategy is similar to buying a yen strangle, although the downside is capped.

(Based on an article in Derivatives Week (now part of GlobalCapital))

Figure 11.19 illustrates the long binary options mentioned in the example. Looked at from the angle of yen, the binary options have similarities to selling dollar strangles.[11]

11.4.2 BARRIER OPTIONS

To create a barrier option, we basically take a vanilla counterpart and then add some properly selected *thresholds*. If, during the life of the option, these thresholds are exceeded by the underlying, the option payoff will exhibit a *discrete* change. The option may be *knocked* out, or it may be *knocked in*, meaning that the option holder either loses the right to exercise or gains it.

Let us consider the two most common cases. We start with a European-style plain vanilla option written on the underlying, S_t, with strike K, and expiration T. Next, we consider two thresholds

---

[11]Which means buying yen strangles as suggested in the text.

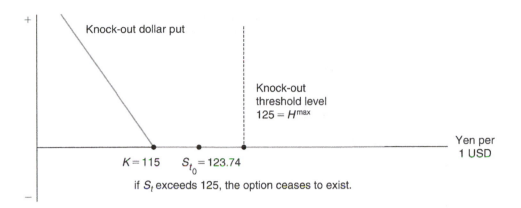

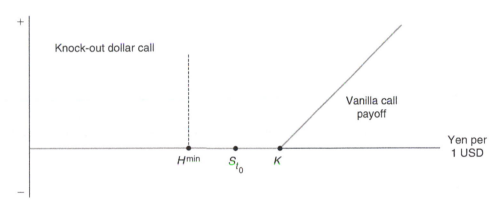

FIGURE 11.20

Payoff examples of a knock-out put and a knock-out call.

$H^{\min}$ and $H^{\max}$, with $H^{\min} < H^{\max}$. If, during the life of the option, S_t exceeds one or both of these limits in some precise ways to be defined, then the option *ceases to exist*. Such instruments are called *knock-out* options. Two examples are shown in Figure 11.20. The lower part of the diagram is a knock-out call. If, during the life of the option, we observe the event

$$S_u < H^{\min} \quad u \in [t, T] \tag{11.36}$$

then the option ceases to exist. In fact, this option is *down-and-out*. The upper part of the figure displays an up-and-out put, which ceases to exist if the event

$$H^{\max} < S_u \quad u \in [t, T] \tag{11.37}$$

is observed.

An option can also *come into existence* after some barrier is hit. We then call it a *knock-in* option. A knock-in put is shown in Figure 11.21. In this section, we will discuss an H knock-out

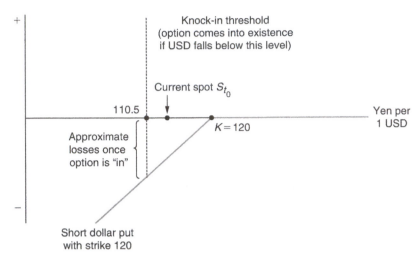

FIGURE 11.21

Payoff of a knock-in put.

call and an H knock-in call with the same strike K. These barrier options we show here have the characteristic that when they knock-in or -out, they will be out-of-the-money. Barrier options with positive intrinsic value at knock-in and -out also exist but are not dealt with (for these, see James (2003)).

11.4.2.1 A contractual equation

We can obtain a contractual equation for barrier options and the corresponding vanilla options. Consider two European-style barrier options with the same strike K. The underlying risk is S_t, and, for simplicity, suppose all Black–Scholes assumptions are satisfied. The first option, a knock-out call, whose premium is denoted by $C^O(t)$, has the standard payoff if the S_t never touches, or falls below, the barrier H. The premium of the second option, a knock-in call, is denoted by $C^I(t)$. It entitles its holder to the standard payoff of a vanilla call with strike K, only if S_t *does* fall below the barrier H. These payoffs are shown in Figure 11.22. In each case, H is such that, when the option knocks in or out, this occurs in a region with zero *intrinsic* value. Now consider the following logic that will lead to a contractual equation.

1. Start with the case where S_t is below the barrier, $S_t < H$. Here, the S_t is already below the threshold H. So, the knock-out call is already worthless, while the opposite is true for the knock-in call. The knock-in is *in*, and the option holder has already earned the right to a standard vanilla call payoff. This means that for all $S_t < H$, the knock-in call has the same value as a vanilla call. These observations mean

$$\text{For the range } S_t < H, \text{Knock-in} + \text{Knock-out} = \text{Vanilla call} = \text{Knock-in} \tag{11.38}$$

   The knock-out is worthless for this range.

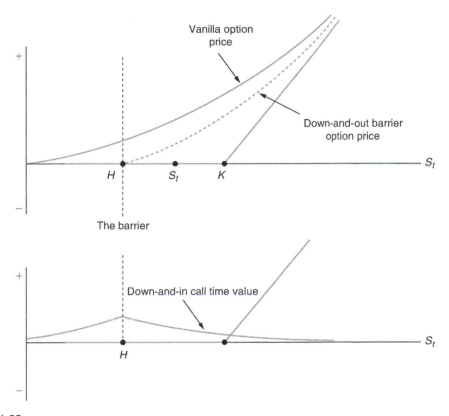

Vanilla option
price

Down-and-out barrier
option price

H

S_t

K

S_t

The barrier

Down-and-in call time value

H

S_t

FIGURE 11.22

Payoffs of a knock-out call and a knock-in call.

2. Now suppose S_t is initially above the barrier, H. There are two possibilities during the life of the barrier options: S_t either stays above H, or falls below H. One and *only* one of these events will happen during $[t, T]$. This means that, if we buy the knock-in call simultaneously with a knock-out call, we *guarantee* access to the payoff of a vanilla call. In other words,

$$\text{For the range } H < S_t, \text{Knock-in} + \text{Knock-out} = \text{Vanilla call} \tag{11.39}$$

Putting these two payoff ranges together, we obtain the contractual equation:

$$
\boxed{\begin{array}{c}\text{Vanilla call,}\\ \text{strike } K\end{array}} = \boxed{\begin{array}{c}\text{Knock-in } K\text{-Call}\\ \text{with barrier } H\end{array}} + \boxed{\begin{array}{c}\text{Knock-out } K\text{-Call}\\ \text{with barrier } H\end{array}} \tag{11.40}
$$

From here, we can obtain the pricing formulas of the knock-in and knock-out barriers. In fact, determining the pricing function of only *one* of these barriers is sufficient to determine the price of the other. In Chapter 9, we provided a pricing formula for the knock-out barrier where the underlying satisfied the Black–Scholes assumptions.[12] The formula was given by

$$C^O(t) = C(t) - J(t) \quad \text{for} \quad H \leq S_t \tag{11.41}$$

where

$$J(t) = S_t \left(\frac{H}{S_t}\right)^{\frac{2\left(r - \frac{1}{2}\sigma^2\right)}{\sigma^2} + 2} N(c_1) - Ke^{-r(T-t)} \left(\frac{H}{S_t}\right)^{\frac{2\left(r - \frac{1}{2}\sigma^2\right)}{\sigma^2}} N(c_2) \tag{11.42}$$

where

$$c_{1,2} = \frac{\ln(H^2/S_t K) + (r \pm \frac{1}{2}\sigma^2)(T - t)}{\sigma\sqrt{T - t}} \tag{11.43}$$

The $C(t)$ is the value of the vanilla call given by the standard Black–Scholes formula, and the $J(t)$ is the discount that needs to be applied because the option may disappear if S_t falls below H during $[t, T]$.

But we now know from the contractual equation that a long knock-in and a long knock-out call with the same strike K and threshold H is equivalent to a vanilla call:

$$C^O(t) + C^I(t) = C(t) \tag{11.44}$$

Using Eq. (11.41) with this gives the formula for the knock-in price as

$$C^I(t) = J(t) \tag{11.45}$$

Thus, the expressions in Eqs. (11.42)–(11.44) provide the necessary pricing formulas for barrier options that knock-out and knock-in, when they are out-of-the-money under the Black–Scholes assumptions. It is interesting to note that when S_t touches the barrier,

$$S_t = H \tag{11.46}$$

the formula for $J(t)$ reduces to the standard Black–Scholes formula:

$$J(t) = S_t N(d_1) - Ke^{-r(T-t)} N(d_2) \tag{11.47}$$

That is to say, the value of $C^O(t)$ will be zero. The knock-out call option price is shown in Figure 11.22. We see that the knock-out is cheaper than the vanilla option. The discount gets larger, the closer S_t is to the barrier, H. Also, the *delta* of the knock-out is higher everywhere and is discontinuous at H.

Finally, Figure 11.22 shows the pricing function of the knock-in. To get this graph, all we need to do is subtract $C^O(t)$ from $C(t)$, in the upper part of Figure 11.22. The reader may wonder why the knock-in call gets *cheaper* as S_t moves to the right of K. After all, doesn't the call become more in-the-money? The answer is no, because as long as $H < S_t$ the holder of the knock-in does *not* have access to the vanilla payoff yet. In other words, as S_t moves rightward, the chances that the knock-in call holder will end up with a vanilla option are going down.

---

[12]It is important to remember that these assumptions preclude a volatility smile. The smile will change the pricing.

11.4.2.2 *Some uses of barrier options*

Barrier options are quite liquid, especially in FX markets. The following examples discuss the pay-off diagrams associated with barrier options.

The next example illustrates another way knock-ins can be used in currency markets. Figures 10.20 and 10.21 illustrate these cases.

EXAMPLE

US dollar puts (yen calls) were well bid last week. Demand is coming from stop-loss trading on the back of exotic knock-in structures. At the end of December, some players were seen selling 1-month dollar puts struck at JPY119 which knock-in at JPY109.30. As the yen moved toward that level early last week, those players rushed to buy cover.

Hedge funds were not the only customers looking for cover. Demand for short-term dollar puts was widely seen. "People are still short yen," said a trader. "The risk reversal is four points in favor of the dollar put, which is as high as I have ever seen it."

(Based on an article in Derivatives Week (now part of GlobalCapital))

According to the example, as the dollar fell toward 110.6 yen, the hedge funds who had sold knock-in options were suddenly facing the possibility that these options would come into existence, and that they would lose money.[13] As a result, the funds started to cover their positions by buying out-of-the-money puts. This is a good illustration of new risks often associated with exotic structures. The changes during infinitesimal intervals in mark-to-market values of barrier options can be *discrete* instead of "gradual."

The next example concerning barrier options involves a more complex structure. The barrier may in fact relate to a *different* risk than the option's underlying. The example shows how barrier options can be used by the airline industry.

Airlines face three basic costs: labor, capital, and fuel. Labor costs can be "fixed" for long periods using wage contracts. However, both interest rate risk and fuel price risk are floating, and sudden spikes in these at any time can cause severe harm to an airline. The following example shows how airlines can hedge these two risks using a single barrier option.

EXAMPLE

Although these are slow days in the exotic option market, clients still want alternative ways to hedge cheaply, particularly if these hedges offer payouts linked to other exposures on their balance sheets. Barrier products are particularly popular. Corporates are trying to cheapen their projections by asking for knock-out options.

For example, an airline is typically exposed to both interest rate and fuel price risks. If interest rates rose above a specified level, a conventional cap would pay out, but under a barrier structure it may not if the airline is enjoying lower fuel prices. Only if both rates and fuel prices are high is the option triggered. Consequently, the cost of this type of hedge is cheaper than separate options linked to individual exposures.

(Thomson Reuters IFR, May 13, 1995)

---

[13]Assuming that they did not already cover their positions.

The use of such barriers may lower hedging costs and may be quite convenient for businesses. The exercises at the end of the chapter contain further examples of exotic options. In the next section, we discuss some of the new risks and difficulties associated with these.

11.4.3 NEW RISKS

Exotic options are often inexpensive and convenient, but they carry their own risks. Risk management of exotic options books is nontrivial because there are (i) discontinuities in the respective Greeks due to the existence of thresholds and (ii) smile effects in the implied volatility.

As the previous three chapters have shown, risk management of option books normally uses various Greeks or their modified counterparts. With threshold effects, some Greeks may *not exist* at the threshold. This introduces discontinuities and complicates risk management. We review some of these new issues next.

1. Barrier options may exhibit jumps in some Greeks. This is a new dimension in risk-managing option books. When spot is near the threshold, barrier option Greeks may change discretely even for small movements in the underlying. These extreme changes in sensitivity factors make the respective *delta, gamma*, and *vega* more complicated tools to use in measuring and managing underlying risks.
2. Barrier options are *path dependent*. For example, the threshold may be relevant at each time point until the option expires or until the barrier is hit. This makes Monte Carlo pricing and risk-managing techniques more delicate and more costly. Also, near the thresholds the spot may need further simulated trajectories and this may also be costly.
3. Barrier option hedging using vanilla and digital options may be more difficult and may be strongly influenced by *smile effects.*

We will not discuss these risk management and hedging issues related to exotic options in this book. However, smile effects will be dealt with in Chapter 16.

11.5 QUOTING CONVENTIONS

Quoting conventions in option markets may be very complicated. Given that market makers look at options as instruments of volatility, they often prefer quoting volatility directly, rather than a cash value for the option. These quotes can be very confusing at times. The best way to study them is to consider the case of risk reversals. Risk reversal quotes illustrate the role played by volatility, and show explicitly the existence of a skewness in the *volatility smile*, an important empirical observation that will be dealt with separately in Chapter 16.

One of the examples concerning risk reversals presented earlier contained the following statement:

> The 1-month risk reversal jumped to 0.81 in favor of euro calls Wednesday from 0.2 2 weeks ago.

It is not straightforward to interpret such statements. We conduct the discussion using the euro/dollar exchange rate as the underlying risk. Consider the dollar calls represented in

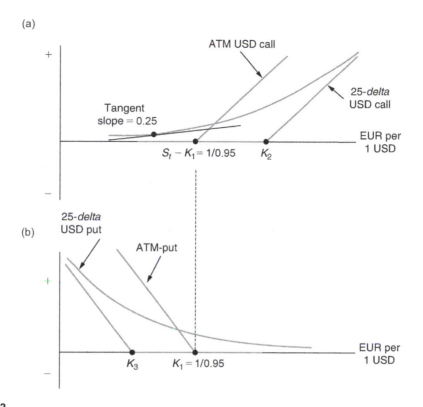

(a)

ATM USD call

+

25-*delta*
USD call

Tangent
slope = 0.25

EUR per
1 USD

$S_t - K_1 = 1/0.95$ K_2

−

(b)   25-*delta*
USD put

ATM-put

+

EUR per
1 USD

K_3 $K_1 = 1/0.95$

−

FIGURE 11.23

Payoffs of ATM and 25-*delta* calls and puts.

Figure 11.23a, where it is assumed that the spot is trading at 0.95, and that the option is ATM. In the same figure, we also show a 25-*delta* call. Similarly, Figure 11.23b shows an ATM dollar put and a 25-*delta* put, which will be out-of-the-money. All these options are supposed to be plain vanilla and European style.

Now consider the following quotes for two *different* 25-*delta* USD risk reversals:

$$\text{Example 1:"flat/0.3 USD call bid"} \tag{11.48}$$

$$\text{Example 2:"0.3/0.6 USD call bid"} \tag{11.49}$$

The interpretation of such bid−ask spreads is not straightforward. The numbers in the quotes do not relate to dollar figures, but to *volatilities*. In simple terms, the number to the right of the slash is the *volatility spread* the market maker is willing to receive for selling the risk reversal position and the number to the left is the volatility spread he is willing to pay for the position.

The numbers to the right are related to the *sale* by the market maker of the 25-*delta* USD call and simultaneously the *purchase* of a 25-*delta* USD put, which, from a *client's point of view* is the risk reversal shown in Figure 11.24a. Note that, for the client, this situation is associated with "dollar strength." If the market maker sells this risk reversal, he will be *short* this position.

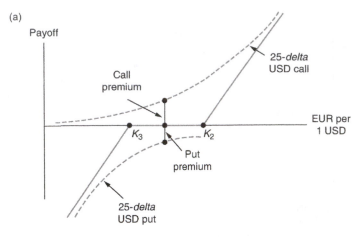

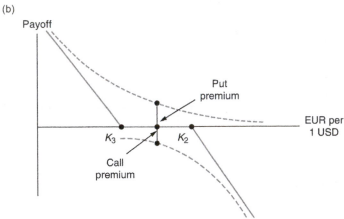

FIGURE 11.24

Payoffs of two different 25-*delta* USD risk reversals.

The numbers to the left of the slash correspond to the *purchase* of a 25-*delta* USD call and the *sale* of a 25-*delta* USD put, which is shown in Figure 11.24b. This outcome, when in demand, is associated with "dollar weakness."

11.5.1 EXAMPLE 1

Now consider the interpretation of the numerical values in the first example:

$$\text{Example 1:"flat}/0.3 \text{ USD call bid"} \tag{11.50}$$

The left side in this quote is "flat." This means that the purchase of the 25-*delta* USD call, and a simultaneous sale of the 25-*delta* USD put, would be done at the *same volatilities*. A client who sells this to the market maker pays or receives nothing extra and the deal has "zero cost." In other words, the two sides would agree on a single volatility and then plug this same number into the Black–Scholes formula to obtain the cost of the put and the cost of the call. The right-hand number in the quote shows a *bias*. It means that the market maker is willing to sell the 25-*delta* USD call, and buy the 25-*delta* USD put, only if he can earn 0.3 volatility points *net*. This implies that the volatility number used in the sale of USD call will be 0.3 points higher than the volatility used for the 25-*delta* USD put. The market maker thinks that there is a "bias" in the market in favor of dollar strength; hence, the client who purchases this risk reversal will incur a *net* cost.

11.5.2 EXAMPLE 2

The second quote given by

$$\text{Example 2:"0.3/0.6 USD call bid"} \tag{11.51}$$

is more complicated to handle, although the interpretation of the 0.6 is similar to the first example. With this number, the market maker is announcing that he or she needs to *receive* 0.6 volatility points net if a client wants to bet on the dollar strength.

However, the left-hand element of the quote is not "flat" anymore but is a positive 0.3. This implies that the bias in the market, in favor of dollar strength is so large, and so many clients demand this long position that, now the market maker is willing to *pay* net 0.3 volatility points when buying the 25-*delta* call and selling the 25-*delta* put.

Thus, in risk reversal quotes, the left-hand number is a *volatility spread that the market maker is willing to pay*, and the second number is a *volatility spread the market maker would like to earn*. In each case, to see how much the underlying options would cost, market participants have to agree on some *base volatility* and then, using it as a benchmark, bring in the volatility spreads.

11.6 REAL-WORLD COMPLICATIONS

Actual implementation of the synthetic payoff structures discussed in this chapter requires dealing with several real-world imperfections. First of all, it must be remembered that these positions are shown at expiration, and that they are piecewise linear. In real life, payoff diagrams may contain several convexities, which is an equivalent term for nonlinear payoffs. We will review these briefly.

11.6.1 THE ROLE OF THE VOLATILITY SMILE

The existence of volatility smile has especially strong effects on pricing and hedging of exotic options. If a volatility smile exists, the implied volatility becomes a function of the strike price K. For example, the expression that gave the binary option price in Eqs. (11.30)−(11.31) has to be modified to

$$C^{\text{bin}}(t) = \lim_{h \to 0} \frac{C^K(t) - C^{K+h}(t)}{h} \tag{11.52}$$

$$= \frac{\partial C^K(t)}{\partial K} + \frac{\partial C^K(t)}{\partial \sigma(K)} \frac{\partial \sigma(K)}{\partial(K)} \tag{11.53}$$

The resulting formula and the analogy to plain vanilla *delta*s will change. These types of modifications have to be applied to hedging and synthetically creating barrier options as well. Major modification will also be needed for barrier options.

11.6.2 EXISTENCE OF POSITION LIMITS

At time t before expiration, an option's value depends on many variables other than the underlying x_t. The volatility of x_t and the risk-free interest rate r_t are two random variables that affect all the positions discussed for $t < T$. This is expressed in the Black–Scholes formula for the call premium of $t < T$:

$$C_t = C(x_t, t | \sigma, r) \tag{11.54}$$

which is a function of the "parameters" r, σ. At $t = T$ this formula reduces to

$$C_T = \max[x_T - K, 0] \tag{11.55}$$

Now, if the r and σ are stochastic, then during the $t \in [0, T)$, the positions considered here will be subject to *vega* and *rho* risks as well. A player who is subject to limits on how much of these risks he or she can take, may have to *unwind* the position before T. This is especially true for positions that have *vega* risk. The existence of limits will change the setup of the problem since, until now, sensitivities with respect to the r and σ parameters, did not enter the decision to take and maintain the positions discussed.

11.7 CONCLUSIONS

In this chapter, we discussed how to synthetically create payoff diagrams for positions that take a view on the direction of markets and on the direction of volatility. These were static positions. We specifically concentrated on the payoff diagrams that were functions of a single risk factor and that were to be replicated by plain vanilla futures and options positions. The second part of the chapter discussed the engineering of similar positions using simple exotics.

SUGGESTED READING

There are several excellent books that deal with classic option strategies. **Hull** (2014) is a very good start. The reader may also consult **Natenberg** (2014) for option basics. See also the textbooks **Jarrow and Turnbull** (1999) and **Kolb** (1999). **Taleb** (1996) is a good source on exotics from a market perspective. For a technical approach, consider the chapter on exotics in **Musiela and Ruthkowski** (2007). **James** (2003) is a good source on the technicalities of option trading and option pricing formulas. It also provides a good discussion of exotics.

EXERCISES

1. Construct a payoff and profit diagram for the purchase of a 105-strike call and sale of a 95-strike call. Verify that you obtain exactly the same profit diagram for the purchase of a 105-strike put and a sale of 95-strike put. Explain the difference in the initial cost of these positions. Assume the following parameters:
   - $S(0) = 100$; $T = 1$; $r = 8\%$; $\sigma = 30\%$

2. Assume that a trader believes that during the vacation periods actual realized volatility is lower than the implied volatility. To exploit this opportunity a trader takes a short position in a strangle to cover the cost of the long straddle position. If the actual volatility is 30% less than the implied volatility, sketch out his volatility strategy. Assume the following parameters:
   - $S(0) = 1600$; $T = 3$ month; $r = 5\%$; $\sigma = 53\%$

3. Consider a *bear spread*. An investor takes a short position in a futures denoted by x_t. But he or she thinks that x_t will not fall below a level x^{min}.
   - **a.** How would you create a position that trades off gains beyond a certain level against large losses if x_t increases above what is expected?
   - **b.** How much would you pay for this position?
   - **c.** What is the maximum gain? What is the maximum loss?
   - **d.** Show your answers in an appropriate figure.

4. Consider this reading carefully and then answer the questions that follow.

   A bank suggested risk reversals to investors that want to hedge their Danish krone assets, before Denmark's Sept. 28 referendum on whether to join the Economic and Monetary Union. A currency options trader in New York said the strategy would protect customers against the Danish krone weakening should the Danes vote against joining the EMU. Danish public reports show that sentiment against joining the EMU has been picking up steam over the past few weeks, although the "Yes" vote is still slightly ahead. [He] noted that if the Danes vote for joining the EMU, the local currency would likely strengthen, but not significantly.

   Six- to 12-month risk reversals last Monday were 0.25%/0.45% in favor of euro calls. [He] said a risk-reversal strategy would be zero cost if a customer bought a euro call struck at DKK7.52 and sold a euro put at DKK7.44 last Monday when the Danish krone spot was at DKK7.45 to the euro. The options are European-style and the tenor is 6 months.

   Last Monday, 6- and 12-month euro/Danish krone volatility was at 1.55%/1.95%, up from 0.6%/0.9% for the whole year until April 10, 2000, owing to growing bias among Danes against joining the EMU. On the week of April 10, volatility spiked as a couple of banks bought 6-month and 9-month, at-the-money vol. (Based on Derivatives Week (now part of GlobalCapital), April 24, 2000.)

   - **a.** Plot the zero-cost risk-reversal strategy on a diagram. Show the DKK7.44 and DKK7.52 put and call explicitly.
   - **b.** Note that the spot rate is at DKK7.45. But, this is *not* the midpoint between the two strikes. How can this strategy have zero cost then?

    c. What would this last point suggest about the implied volatilities of the two options?

    d. What does the statement *"Last Monday, 6- and 12-month euro/Danish krone volatility was at 1.55%/1.95%,"* mean?

    e. What does at-the-money vol mean? (See the last sentence.) Is there out-of-the-money vol, then?

5. The following questions deal with *range binaries*. These are another example of exotic options. Read the following carefully and then answer the questions at the end.

> *Investors are looking to purchase range options. The product is like a straightforward range binary in that the holder pays an upfront premium to receive a fixed pay-off as long as spot maintains a certain range. In contrast to the regular range binary, however, the barriers only come into existence after a set period of time. That is, if spot breaches the range before the barriers become active, the structure is not terminated.*
>
> *This way, the buyer will have a short Vega position on high implied volatility levels. (Based on an article in Derivatives Week (now part of GlobalCapital).)*

    a. Display the payoff diagram of a range-binary option.

    b. Why would FX markets find this option especially useful?

    c. When do you think these options will be more useful?

    d. What are the risks of a *short* position in range binaries?

6. *Double no-touch options* is another name for range binaries. Read the following carefully, and then answer the questions at the end.

> *Fluctuating U.S. dollar/yen volatility is prompting option traders managing their books to capture high volatilities through range binary structures while hedging with butterfly trades. Popular trades include one-year double no touch options with barriers of JPY126 and JPY102. Should the currency pair stay within that range, traders could benefit from a USD1 million payout on premiums of 15–20%.*
>
> *On the back of those trades, there was buying of butterfly structures to hedge short vol positions. Traders were seen buying out-of-the-money dollar put/yen calls struck at JPY102 and an out-of-the-money dollar call/yen put struck at JPY126. (Based on an article in Derivatives Week (now part of GlobalCapital).)*

    a. Display the payoff diagram of the structure mentioned in the first paragraph.

    b. When do you think these options will be more useful?

    c. What is the role of butterfly structures in this case?

    d. What are the risks of a *short* position in range binaries?

    e. How much money did such a position make or lose "last Tuesday"?

7. The next question deals with a different type range option, called a *range accrual* option. Range accrual options can be used to take a view on volatility directly. When a trader is short volatility, the trader expects the actual volatility to be less than the implied volatility. Yet, within the bounds of *classical* volatility analysis, if this view is expressed using a vanilla option, it may require dynamic hedging, otherwise expensive straddles must be bought. Small shops may not be able to allocate the necessary resources for such dynamic hedging activities.

Instead, range accrual options can be used. Here, the buyer of the option receives a payout that depends on how many days the underlying price has remained within a range during the life of the option. First read the following comments then answer the questions.

The Ontario Teachers' Pension Plan Board, with CAD72 billion (USD48.42 billion) under management, is looking at ramping up its use of equity derivatives as it tests programs for range accrual options and options overwriting on equity portfolios.

The equity derivatives group is looking to step up its use now because it has recently been awarded additional staff as a result of notching up solid returns, said a portfolio manager for Canadian equity derivatives. . . . With a staff of four, two more than previously, the group has time to explore more sophisticated derivatives strategies, a trader explained. Ontario Teachers' is one of the biggest and most sophisticated end users in Canada and is seen as an industry leader among pension funds, according to market officials.

A long position in a range accrual option on a single stock would entail setting a range for the value of the stock. For every day during the life of the option that the stock trades within the range, Ontario Teachers would receive a payout. It is, hence, similar to a short vol position, but the range accrual options do not require dynamic hedging, and losses are capped at the initial premium outlay. (Based on an article in Derivatives Week (now part of GlobalCapital).)

a. How is a range accrual option similar to a strangle or straddle position?
b. Is the position taken with this option static? Is it dynamic?
c. In what sense does the range accrual option accomplish what dynamic hedging strategies accomplish?
d. How would you synthetically create a range accrual option for other "vanilla" exotics?

EXCEL EXERCISES

8. Write a VBA program to show the construction of a bull and bear spread, first using calls only and then using puts only. Also plot both the call spread and put spread in both bull and bear phases. Explain your observation for the difference in the initial cost of call and put spreads.
  Assume the following parameters:
  - $S(0) = 100$; $T = 1$; $r = 8\%$; $\sigma = 30\%$; Spread = 10%.

9. Write a VBA program to show the fabrication of the butterfly spread composed of a strangle and straddle. Assume the following parameters:
  - $S(0) = 100$; $T = 1$; $r = 8\%$; $\sigma = 30\%$; Spread = 5%
  Plot the straddle and strangle payoff along with the payoff of the butterfly spread.
  Repeat the above calculation for the short position in butterfly spread.

MATLAB EXERCISE

10. Write a MATLAB program to plot the following spreads including the time value of the spread as well as its payoff at the expiration.
  a. Bull spread

   b. Bear spread
   c. Straddle
   d. Strangle
   e. Butterfly
      Assume the following parameters:
   • $S(0) = 100$; $T = 1$; $r = 8\%$; $\sigma = 30\%$; Spread $= 5\%$.

PRICING TOOLS IN FINANCIAL ENGINEERING

CHAPTER OUTLINE

12.1 INTRODUCTION

We have thus far proceeded without a discussion of asset pricing *models* and the tools associated with them, as financial engineering has many important dimensions besides pricing. In this chapter, we will discuss models of asset pricing, albeit in a very simple context. A summary chapter on pricing tools would unify some of the previous topics and show the subtle connections between them. The discussion will approach the issue using a framework that is a natural extension of the financial engineering logic utilized thus far.

Pricing comes with at least *two* problems that seem, at first, difficult to surmount in any satisfactory way. Investors like return, but dislike risk. This means that assets associated with nondiversifiable risks will carry *risk premia*. But, how can we measure such risk premia objectively when buying assets is essentially a matter of *subjective* preferences? Modeling risk premia using *utility functions* may be feasible theoretically, but this is not attractive from a trader's point of view if hundreds of millions of dollars are involved in the process. The potential relationship between risk premia and utility functions of players in the markets is the first unpleasant aspect of practical pricing decisions.

The second problem follows from the first. One way or another, the pricing approach needs to be based on measuring the volatility of future cash flows. But volatility is associated with randomness and with some probability distribution. How can an asset pricing approach that intends to be applicable in practice obtain a reasonable set of real-world probabilities?[1]

Modern finance has found an ingenious and *practical* way of dealing with both these questions simultaneously. Instead of using a framework where risk premia are modeled explicitly, the profession *transforms* a problem with risk premia into one where there are no risk premia. Interestingly, this transformation is done in a way that the relevant probability distribution ceases to be the *real-world probability* and, instead, becomes a *market-determined probability* that can be numerically calculated at *any* point in time if there is a reasonable number of liquid instruments.[2] With this approach, the assets will be priced in an *artificial* risk-neutral environment where the risk premia are indirectly taken into account. This methodology is labeled the *Martingale approach*. It is a powerful tool in practical asset pricing and risk management.

A newcomer to financial engineering may find it hard to believe that a more or less unified theory for pricing financial assets that can be successfully applied in real-world pricing actually exists. After all, there are many different types of assets, and not all of them seem amenable to the same pricing methodology, even at a theoretical level. A market practitioner may already have heard

---

[1] Note that the subjective nature of risk premia was in the realm of pure economic theory, whereas the issue of obtaining satisfactory real-world probability distributions falls within the domain of econometric theory.

[2] That is to say, instead of using historical data, we can derive the desired probability distribution from the current quotes.

of risk-neutral pricing, but just like the newcomer to financial engineering, he or she may regard the basic theory behind it as very *abstract*. And yet, the theory is surprisingly potent. This chapter provides a discussion of this methodology from the point of view of a financial engineer. Hence, even though the topic is asset pricing, the way we approach it is based on ideas developed in previous chapters. Basically, this pricing methodology is presented as a general approach to synthetic asset creation.

Of course, like any other theory, this methodology depends on some strict assumptions. The methods used in this text will uniformly make one common assumption that needs to be pointed out at the start. Only those models that assume *complete markets* are discussed. In heuristic terms, when markets are "complete," there are "enough" liquid instruments for obtaining the working probability distribution.

This chapter progressively introduces a number of important theoretical results that are used in pricing, hedging, and risk-management application. The *main* result is called the fundamental theorem of asset pricing. Instead of a mathematical proof, we use a *financial engineering argument* to justify it, and a number of important consequences will emerge. Throughout the chapter, we will single out the results that have practical implications.

12.2 SUMMARY OF PRICING APPROACHES

In this section, we remind the reader of some important issues from earlier chapters. Suppose we want to find the *fair* market price of an instrument. First, we construct a synthetic equivalent to this instrument using liquid contracts that trade in financial markets. Clearly, this requires that such contracts are indeed available. Second, once these liquid contracts are found, an arbitrage argument is used. The cost of the *replicating portfolio* should equal the cost of the instrument we are trying to price. Third, a trader would add a proper margin to this cost and thus obtain the fair price.

In earlier chapters, we obtained synthetics for forward rate agreements (FRAs), foreign exchange (FX) forwards, and several other quasi-linear instruments. Each of these constitutes an early example of asset pricing. Obtain the synthetic and see how much it costs. By adding a profit to this cost, the fair market price is obtained. It turns out that we can *extend* this practical approach and obtain a general theory.

It should be reemphasized that *pricing* and *hedging* efforts can sometimes be regarded as two sides of the same coin. In fact, hedging a product requires finding a replicating portfolio and then using it to cover the position in the original asset. If the trader is long in the original instrument, he or she would be short in the synthetic, and vice versa. This way, exposures to risks would cancel out and the position would become "riskless." This process results in the creation of a replicating portfolio whose cost cannot be that different from the price of the original asset. Thus, a hedge will transfer unwanted risks to other parties but, at the same time, will provide a way to price the original asset.[3]

---

[3]If the hedge is not "perfect," the market maker will add another margin to the cost to account for any small deviation in the sensitivities toward the underlying risks. For example, if some exotic option cannot be perfectly hedged by the spot and the cash, then the market maker will increase or decrease the price to take into account these imperfections.

Pricing theory is also useful for the creation of "new products." A new product is basically a series of contingent cash flows. We would, first, put together a combination of financial instruments that have the same cash flows. Then, we would write a separate contract and sell these cash flows to others under a *new name*. For example, a strip of FRAs or futures can be purchased and resulting cash flows are then labeled a swap and sold to others. The new product is, in fact, a dynamically maintained portfolio of existing instruments, and its fair cost will equal the sum of the price of its constituents.

12.3 THE FRAMEWORK

The pricing framework that we use emphasizes important aspects of the theory within a real-world setting. We assume that m liquid asset prices are observed at times t_i, $i = 1, 2, \ldots$ The time t_i price of the kth asset is denoted by S_{kt_i}. The latter can represent credit, stocks, fixed-income instruments, the corresponding derivatives, or commodity prices.

In theory, a typical S_{kt_i} can assume any real value. This makes the number of possible values infinite and *uncountable*. But in practice, every price is quoted to a small number of decimal places and, hence, has a countable number of possible future values. FX rates, for example, are in general quoted to four decimal places. This brings us to the next important notion that we would like to introduce.

12.3.1 STATES OF THE WORLD

Let t_0 denote the "present," and consider the kth asset price S_{kT}, at a future date, $T = t_i$, for some $0 < i$. At time t_0, S_{kT} will be a random variable.[4] Let the symbol ω^j, with $j = 1, \ldots, n$ represent time T *states of the world* that relate to the random variable S_{kT}.[5] We assume that $n \leq m$, which amounts to saying that there are at least as many liquid assets as there are time T states of the world. For example, it is common practice in financial markets to assume a "bullish" state, a "bearish" state, and a "no-change" state. Traders expect prices in the future to be either "higher," "lower," or to "remain the same." The ω^j generalizes this characterization and makes it operational.

EXAMPLE

In this example, we construct the states of the world that relate to some asset whose time t_i price is denoted by S_{t_i}. Without any loss of generality, let

$$S_{t_0} = 100 \tag{12.1}$$

Suppose, at a future date T, with $t_n = T$, there are only $n = 4$ states of the world. We consider the task of defining these states.

---

[4]The current value of the asset S_{kt_0}, on the other hand, is known.
[5]According to this, ω^j may also need a T subscript. But we ignore it and ask the reader to remember this point.

1. Set the value of some grid parameter ΔS to assign neighboring values of S_T into a single state. For example, let

$$\Delta S = 2 \tag{12.2}$$

2. Next, pick two upper and lower bounds $[S^{\min}, S^{\max}]$ such that the probability of S_T being outside this interval is relatively small and that excursions outside this range can safely be ignored. For example, let $S^{\max} = 104$ and $S^{\min} = 96$. Accordingly, the events $104 < S_T$ and $S_T < 96$ are considered unlikely to occur, and, hence, a detailed breakdown of these states of the world is not needed. Clearly, the choice of numerical values for $[S^{\min}, S^{\max}]$ depends, among other things, on the perceived volatility of S_t during the period $[t_0, T]$.[6]

3. The states of the world can then be defined in the following fashion:

$$\omega^1 = \{S_T \text{ such that } S_T < S^{\min}\} \tag{12.3}$$

$$\omega^2 = \{S_T \text{ such that } S_T \in [S^{\min}, S^{\min} + \Delta S]\} \tag{12.4}$$

$$\omega^3 = \{S_T \text{ such that } S_T \in [S^{\min} + \Delta S, S^{\min} + 2\Delta S = S^{\max}]\} \tag{12.5}$$

$$\omega^4 = \{S_T \text{ such that } S^{\max} < S_T\} \tag{12.6}$$

This situation is shown in Figure 12.1.

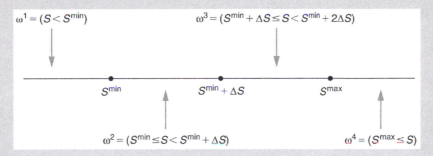

FIGURE 12.1

States of the world and asset values.

Here, the total number of states of the world depends on the size of the grid parameter ΔS, and on the choice of upper and lower bounds $[S^{\min}, S^{\max}]$. These, in turn, depend on market psychology at time t_0. For example, selecting the total number of states as $n = 4$ could be justified, if the ranges for S_T shown here were the only ones found relevant for pricing and risk-management problems faced during that particular day. If a problem under consideration requires a *finer* or *coarser* subdivision of the future, the value for n would change accordingly.

---

[6]In practice, these upper and lower bounds have to be properly *calibrated* to observed liquid, arbitrage-free prices.

12.3.2 THE PAYOFF MATRIX

The next step in obtaining the fundamental theorem of asset pricing is the definition of a payoff matrix for period T. Time T values of the assets, S_{kt}, depend on the state of the world, ω^i that will occur at time T. Given that we are working with a finite number of states of the world, *possible values* for these assets would be easy to list. Let z_k^i represent the value assumed by the kth asset in state ω^i, at time T:

$$S_{kT}^i = z_k^i \tag{12.7}$$

Then, for the first n assets, $n \leq m$, we can form the following payoff matrix for time T:

$$D = \begin{pmatrix} z_1^1 & \cdots & z_1^n \\ & \cdots & \\ z_n^1 & \cdots & z_n^n \end{pmatrix} \tag{12.8}$$

A typical row of this matrix would represent possible values of a *particular asset* in different states of the world. A typical column represents different asset prices, in a *particular state of the world*. The definition of ω^i should automatically lead to a definition of the possible values for assets under consideration, as shown in the previous example.

The fundamental theorem of asset pricing is about how "current" asset prices, S_{kt}, relate to the possible values represented by matrix D. We form a matrix equation that will play an important role in the next three chapters.

12.3.3 THE FUNDAMENTAL THEOREM

Consider the *linear* system of equations defined for a series of Q^i, indexed by the state of the world i:

$$\begin{pmatrix} S_{1t_0} \\ \cdots \\ S_{nt_0} \end{pmatrix} = \begin{pmatrix} z_1^1 & \cdots & z_1^n \\ & \cdots & \\ z_n^1 & \cdots & z_n^n \end{pmatrix} \begin{pmatrix} Q^1 \\ \cdots \\ Q^n \end{pmatrix} \tag{12.9}$$

The left-hand side shows the vector of current liquid asset prices observed at time t_0. The right-hand side has two components. The first is the matrix D of possible values for these prices at time T, and the second is a vector of constants, $\{Q^1, \ldots, Q^n\}$. The fundamental theorem of asset pricing concerns this matrix equation and the properties of the $\{Q^i\}$. The theorem can be stated heuristically as follows:

Theorem: The time t_0 prices for the $\{S_{kt_0}\}$ are arbitrage-free if and only if $\{Q^i\}$ exist and are positive.

Thus, the theorem actually works both ways. If S_{kt_0} are arbitrage-free, then Q^i exist and are all positive. If Q^i exist and are positive, then S_{kt_0} are arbitrage-free.

The fundamental theorem of asset pricing provides a *unified* pricing tool for pricing real-world assets. In the remaining part of this chapter, we derive important implications of this theorem. These can be regarded as *corollaries* that are exploited routinely in asset pricing. The first of these corollaries is the existence of *synthetic* probabilities. However, before we discuss these results we need to motivate the $\{Q^i\}$ and show why the theorem holds.

12.3.4 DEFINITION OF AN ARBITRAGE OPPORTUNITY

What is meant by arbitrage-free prices? To answer this question we need to define *arbitrage opportunity* formally. Formal definition of the framework outlined in this section provides this. Consider the asset prices $S_{1t}, \ldots, S_{kt}$. Associate the portfolio weights θ_i with asset S_{it}. Then we say that there is an arbitrage opportunity if either of the following two conditions hold.

1. A portfolio with weights θ_i can be found such that:

$$\sum_{i=1}^{k} \theta_i S_{it} = 0 \tag{12.10}$$

simultaneously with

$$0 \le \sum_{i=1}^{k} \theta_i S_{iT} \tag{12.11}$$

According to these conditions, the market practitioner advances no cash at time t to form the portfolio, but still has access to some non-zero gains at time T. This is the first type of arbitrage opportunity.

2. A portfolio with weights θ_i can be found such that:

$$\sum_{i=1}^{k} \theta_i S_{it} \le 0 \tag{12.12}$$

simultaneously with

$$\sum_{i=1}^{k} \theta_i S_{iT} = 0 \tag{12.13}$$

In this case, the market practitioner receives cash at time t while forming the portfolio, but has no liabilities at time T.

It is clear that in either case, the size of these *arbitrage portfolios* is arbitrary since no liabilities are incurred. The formal definition of arbitrage-free prices requires that such portfolios not be feasible at the "current" prices $\{S_{it}\}$.

Note that what market professionals call an *arbitrage strategy* is very different from this formal definition of arbitrage opportunity. In general, when practitioners talk about "arb" they mean positions that have a relatively small probability of losing money. Clearly this violates both of the conditions mentioned above. The methods introduced in this chapter deal with the lack of formal arbitrage opportunities and not with the market practitioners' arbitrage strategies. It should be remembered that it is the formal no-arbitrage condition that provides the important tools used in pricing and risk management.

12.3.5 INTERPRETING THE Q^i: STATE PRICES

Given the states of the world ω^i, $i = 1, \ldots, n$, we can write the preceding matrix equation for the *special case* of two important sets of instruments that are essential to understanding arbitrage-free pricing. Suppose the first asset S_{1t} is a *risk-free savings deposit* account and assume, without any loss of generality, that t_i represents years.[7] If 1 dollar is deposited at time t_0, $(1 + r_{t_0})$ can be earned at time t_1 without any risk of *default*. The r_{t_0} is the rate that is observed as of time t_0.

The second set of instruments are *elementary insurance contracts*. We denote them by C_i. These contracts are defined in the following way:

- C_1 pays \$1 at time T if ω^1 occurs. Otherwise it pays zero.
- C_n pays \$1 at time T if ω^n occurs. Otherwise it pays zero.

If a market practitioner considers state ω^i as "risky," he or she can buy a desired number of C_i's as insurance to *guarantee* any needed cash flow in that state.

Suppose now that all C_i are actively traded at time t_0. Then, according to the matrix equation, the correct arbitrage-free prices of these contracts are given by Q^i. This is the case since plugging the current prices of the savings account and the C_i at time t_0 into the matrix equation (12.9) gives[8]

$$
\begin{bmatrix} 1 \\ C_1 \\ \cdots \\ C_n \end{bmatrix} = \begin{bmatrix} (1 + r_{t_0}) & \cdots & \cdots & (1 + r_{t_0}) \\ 1 & 0 & \cdots & 0 \\ \cdots & \cdots & \cdots & \cdots \\ 0 & \cdots & 0 & 1 \end{bmatrix} \begin{bmatrix} Q^1 \\ \cdots \\ Q^n \end{bmatrix}
\tag{12.14}
$$

Following from this matrix equation, Q^i have three important properties. First, we see that they are *equal* to the prices of the elementary insurance contracts:

$$
C_i = Q^i
\tag{12.15}
$$

It is for this reason that Q^i are also called *state prices*. Second, we can show that if interest rates are positive, the sum of the time t_0 prices of C_i is *less than* one. Consider the following: a portfolio that consists of buying *one of each* insurance contract C_i at time t_0 will guarantee 1 dollar at time t_0 no matter which state, ω^i, is realized at time T. But, a guaranteed future dollar should be worth *less* than a current dollar at hand, as long as interest rates are positive. This means that the sum of Q^i paid for the elementary contracts at time t_0 should satisfy

$$
Q^1 + Q^2 + \cdots + Q^n < 1
\tag{12.16}
$$

From the first row of the matrix equation in Eq. (12.14), we can write

$$
\sum_{i=1}^{n} Q^i (1 + r_{t_0}) = 1
\tag{12.17}
$$

After rearranging,

$$
Q^1 + Q^2 + \cdots + Q^n = \frac{1}{(1 + r_{t_0})}
\tag{12.18}
$$

The *third* property is a little harder to see. As the fundamental theorem states, none of the Q^i can be *negative* or zero if C_i are indeed *arbitrage-free*. We show this with a simple counter example.

---

[7] This simplifies the notation, since the days' adjustment parameter will equal one.
[8] D in our setup is a $(n + 1) \times n$ matrix.

EXAMPLE

Suppose we have $n = 4$ and that the first elementary contract has a negative price, $C_1 = -1$. Without any loss of generality, suppose all other elementary insurance contracts have a positive price. Then the portfolio

$$\{(Q^2 + Q^3 + Q^4) \text{ unit of } C_1, 1 \text{ unit of } C_2, 1 \text{ unit of } C_3, 1 \text{ unit of } C_4\} \qquad (12.19)$$

has zero cost at time t_0 and yet will guarantee a positive return at time t_1. More precisely, the portfolio returns 1 dollar in states 2−4 and $(Q^2 + Q^3 + Q^4)$ dollars in state 1.

Hence, as long as one or more of the Q^i are negative, there will always be an arbitrage opportunity. A trader can "buy" the contract(s) with a negative price and use the cash generated to purchase the other contracts. This way, a positive return at T is *guaranteed*, while at the same time the zero-cost structure of the initial portfolio is maintained. For such arbitrage opportunities not to exist, we need $0 < Q^i$ for *all i*.

12.3.5.1 Remarks

Before going further, we ask two questions that may have already troubled the reader given the financial engineering approach we adopted in this book.

- Do insurance contracts such as C_i exist in the real world? Are they actively traded?
- Is the assumption of a small *finite* number of states of the world realistic? How can such a restrictive view of the future be useful in pricing real-world instruments?

In the next few sections, we will show that the answer to both of these questions is a qualified yes. To understand this, we need to relate the elementary insurance contracts to the concept of options. Options can be considered as ways of trading baskets of C_i's. A typical C_i pays 1 dollar if state i occurs and nothing in all other states. Thus, it is clear that option payoffs at expiration are different from those of elementary contracts. Options pay nothing if they expire out-of-the-money, but they pay, $(S_T - K)$ if they expire in-the-money. This means that depending on how we define the states ω^i, unlike the elementary contracts, options can make payments in *more* than one state. But this difference is really not that important since we can get all desired C_i from option prices, if options trade for all strikes K. In other words, the pricing framework that we are discussing here will be much more useful in practice than it seems at first.

As to the second question, it has to be said that, in practice, few strikes of an option series trade actively. This suggests that the finite state assumption may not be that unrealistic after all.

12.4 **AN APPLICATION**

The framework based on state prices and elementary insurance contracts is a surprisingly potent and *realistic* pricing tool. Before going any further and obtaining more results from the fundamental theorem of asset pricing, we prefer to provide a real-world example. The following reading deals with the S&P500 index and its associated options.

EXAMPLE

The S&P500 is an index of 500 leading stocks from the United States. It is closely monitored by market participants and traded in futures markets. One can buy and sell liquid options written on the S&P500 at the Chicago Board of Options Exchange (CBOE). These options with an expiration date of December 2001 are shown in Table 11.1 as they were quoted on August 10, 2001.

At the time these data were gathered, the index was at 1187. The three most liquid call options are

$$\{1275 - \text{Call}, 1200 - \text{Call}, 1350 - \text{Call}\} \tag{12.20}$$

The three most liquid put options, on the other hand, are

$$\{1200 - \text{Put}, 1050 - \text{Put}, 900 - \text{Put}\} \tag{12.21}$$

Not surprisingly, all the liquid options are out-of-the-money as liquid options generally are.[9]

We will now show how this information can be used to obtain (i) the states of the world ω^i, (ii) the state prices Q^i, and (iii) the corresponding synthetic probabilities associated with Q^i. We will do this in the simple setting used thus far.

12.4.1 OBTAINING THE ω^I

A financial engineer always operates in response to a particular kind of problem and the states of the world to be defined relative to the needs, *at that time*. In our present example we are working with S&P500 options, which means that the focus is on equity markets. Hence, the corresponding states of the world would relate to different states in which the US stock market might be at a future date. Also, we need to take into account that trader behavior singles out a relatively small number of *liquid* options with expirations of about 3 months. For the following example, refer to Table 12.1.

---

[9]Buying and selling in-the-money options does not make much sense for market professionals. Practitioners carry these options by borrowing the necessary funds and hedge them immediately. Hence, any intrinsic value is offset by the hedge side anyway. Yet, the convexity of in-the-money options will be the same as with those that are out-of-the-money.

Table 12.1 S&P500 Option Quotes

Calls	Last Sale	Bid	Ask	Volume	Open Interest
Dec 1175	67.1	68	70	51	1378
Dec 1200	46.5	52.8	54.8	150	8570
Dec 1225	41	40.3	42.3	1	6792
Dec 1250	28.5	29.6	31.6	0	11,873
Dec 1275	22.8	21.3	23.3	201	6979
Dec 1300	15.8	15	16.2	34	16,362
Dec 1325	9.5	10	11	0	9281
Dec 1350	6.8	6.3	7.3	125	8916
Dec 1375	4.1	4	4.7	0	2818
Dec 1400	2.5	2.5	3.2	10	17,730
Dec 1425	1.4	1.4	1.85	0	4464
Dec 1450	0.9	0.8	1.25	9	9383
Dec 1475	0.5	0.35	0.8	0	122
Puts	**Last Sale**	**Bid**	**Ask**	**Volume**	**Open Interest**
Dec 800	1.65	1.2	1.65	10	1214
Dec 900	4.3	3.4	4.1	24	11,449
Dec 950	5.4	5.3	6.3	10	8349
Dec 995	10.1	8.5	9.5	0	11,836
Dec 1025	13	11.1	12.6	11	5614
Dec 1050	13.6	14	15.5	106	19,483
Dec 1060	16.5	15.7	17.2	1	1597
Dec 1075	22.5	18	19.5	1	316
Dec 1100	26	22.7	24.7	0	17,947
Dec 1150	39	35.3	37.3	2	16,587
Dec 1175	44	44.1	46.1	14	4897
Dec 1200	53	53.9	55.9	897	26,949

EXAMPLE

We let S_T represent the value of the S&P500 at expiration and then use the strike prices K_i of the liquid options to define the future states of the world. In fact, strike prices of puts and calls discussed in the preceding example divide the S_T axis into intervals of equal length. But only a handful of these options are liquid, implying that fine subdivisions were perhaps not needed by the markets for that day and that particular expiration. Accordingly, we can use the strike prices of the three liquid out-of-the-money puts to obtain the intervals

$$\omega^1 = S_T < 900 \tag{12.22}$$

$$\omega^2 = 900 \leq S_T < 1050 \tag{12.23}$$

$$\omega^3 = 1050 \leq S_T < 1200 \tag{12.24}$$

Note that the liquid puts lead to intervals of equal length. It is interesting, but also expected, that the liquid options have this kind of regularity in their strikes. Next, we use the three out-of-the-money calls to get three intervals to define three additional states of the world as

$$\omega^4 = 1200 \leq S_T < 1275 \tag{12.25}$$

$$\omega^5 = 1275 \leq S_T < 1350 \tag{12.26}$$

$$\omega^6 = 1350 \leq S_T \tag{12.27}$$

Here, the last interval is obtained from the highest strike liquid call option. Figure 12.2 shows these options and the implied intervals. Since these intervals relate to future values of S_T, we consider them the relevant states of the world for S_T.

We pick the midpoint of the bounded intervals as an approximation to that particular state. Let these midpoints be denoted by $\{\overline{S}^i, i = 2, \ldots, 5\}$. These midpoints can then be used as a finite set of points that represent ω^i. For the first and last half-open intervals, we select the values of the two extreme points, $\overline{S}^1$, and $\overline{S}^6$, arbitrarily for the time being. We let

$$\overline{S}^1 = 750 \tag{12.28}$$

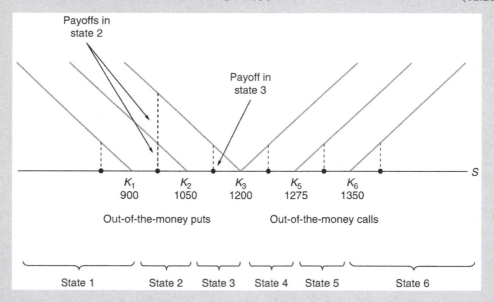

FIGURE 12.2

Option payoffs and states of the world.

$$\overline{S}^6 = 1400 \qquad\qquad (12.29)$$

so that the distance between $\overline{S}^i$ remains constant. This arbitrary selection of the "end states" is clearly unsatisfactory. In fact, by doing this we are in a sense setting the volatility of the random variables arbitrarily. We can, however, calibrate our selection. Once our educated guesses are plugged in, we can try to adjust these extreme values so that the resulting Q^i all become positive and price some other liquid asset correctly. In a sense, calibration is an attempt to see which value of the two "end states" replicates the observed prices. But for the time being, we ignore this issue and assume that the end points are selected correctly.

It is open to debate if selecting just six states of the world, as in the example, might represent future possibilities concerning S_T accurately. Traders dealing with the risk in the example must have thought so, since on that particular date trading approximately *six* liquid options was sufficient to resolve their tasks. It seems that if *a finer* subdivision of the future possibilities were more appropriate, then more liquid strikes would have been traded.

Hence, as usual in financial engineering, the specific values of ω^i that we select are based on the values of liquid instruments. In our case, the possible states of the world were chosen as dictated by liquid call and put options.

12.4.2 ELEMENTARY CONTRACTS AND OPTIONS

Elementary insurance contracts C_i do not trade directly in world financial markets. Yet, C_i are not far from a well-known instrument class—options—and they trade "indirectly." This section shows how elementary insurance contracts can be obtained from options, and vice versa. Plain vanilla options are, in fact, close relatives of elementary insurance contracts. The best way to see this is to consider a numerical example. (Generalizations are straightforward.)

EXAMPLE

Start with the first and last options selected for the previous example. Note that the 900-put is equivalent to $K_1 - \overline{S}^1$ units of C_1 because it pays approximately this many dollars if state ω_1 occurs and nothing in all other states. Similarly, the 1350-call is equivalent to $\overline{S}^6 - K_6$ units of C_6 because it pays approximately this many dollars if state 6 occurs and nothing otherwise.

The other calls and puts have payoffs in more than one state, but they also relate to elementary contracts in straightforward ways. For example, the 1050-put is equivalent to a portfolio of two elementary insurance contracts, $K_2 - \overline{S}^1$ units of C_1 and $K_2 - \overline{S}^2$ units of C_2, because it makes these payments in states 1 and 2, respectively, and nothing else in other states. In fact, pursuing this reasoning, we can obtain the following matrix equation between the payoffs of elementary contracts $C_1, \ldots, C_6$ and the option prices:

$$
\begin{bmatrix}
900 - \text{Put} \\
1050 - \text{Put} \\
1200 - \text{Put} \\
1200 - \text{Call} \\
1275 - \text{Call} \\
1350 - \text{Call}
\end{bmatrix}
=
\begin{bmatrix}
z_1^1 & 0 & 0 & 0 & 0 & 0 \\
z_2^1 & z_2^2 & 0 & 0 & 0 & 0 \\
z_3^1 & z_3^2 & z_3^3 & 0 & 0 & 0 \\
0 & 0 & 0 & z_4^4 & z_4^5 & z_4^6 \\
0 & 0 & 0 & 0 & z_5^5 & z_5^6 \\
0 & 0 & 0 & 0 & 0 & z_6^6
\end{bmatrix}
\begin{bmatrix}
C_1 \\
C_2 \\
C_3 \\
C_4 \\
C_5 \\
C_6
\end{bmatrix}
\tag{12.30}
$$

This equation holds since we have

$$
Q^i = C_i \tag{12.31}
$$

for all i.[10]

Thus, given the arbitrage-free values of traded puts and calls with different strikes but similar in every other aspect, we can easily obtain the values of the elementary insurance contracts C_i by inverting the (6×6) matrix on the right side. In fact, it is interesting to see that the matrix equation in the example contains two triangular subsystems that can be solved separately and recursively.

Hence, the existence of liquid options makes a direct application of the fundamental theorem of asset pricing possible. Given a large enough number of liquid option contracts, we can obtain the state prices, Q^i, if these exist, and, if they are all positive, use them to price other illiquid assets that depend on the same risk.[11] Obviously, when traders deal with interest rate, or exchange rate risk, or when they are interested in pricing contracts on commodities, they would use liquid options for *those* particular sectors and work with different definitions of the state of the world.

12.4.3 ELEMENTARY CONTRACTS AND REPLICATION

We now show how elementary insurance contracts and options that belong to a series can be used in replicating instruments with *arbitrary* payoffs. Consider an arbitrary financial asset, S_t, that is worth z_T^i in state of the world ω^i, $i = 1, \ldots, n$, at time T. Given n elementary insurance contracts C_i, we can immediately form a *replicating portfolio* for this asset. Assuming, without any loss of generality, that the time T payoffs of the S_t asset are denoted by $0 < z_T^i$, we can consider buying the following portfolio:

$$
\{z_T^1 \text{ units of } C_1, z_T^2 \text{ units of } C_2, \ldots, z_T^n \text{ units of } C_n\} \tag{12.32}
$$

At time T, this portfolio should be worth exactly the same as S_t, since whatever state occurs, the basket of insurance contracts will make the same time T payoff as the original asset. This provides

---

[10]If one needs to get the values of C_i from the traded puts and calls, one should start with the first put, then move to the second put, then the third. The same strategy can be repeated with the last call and so on.

[11]Here, this risk is the S&P500 index.

an immediate *synthetic* for S_t. Accordingly, if there are no-arbitrage opportunities, the value of the portfolio and the value of S_t will be identical as of time t as well.[12] We consider an example.

EXAMPLE

Take any four independent assets S_{kt}, $k = 1, \ldots, 4$ with different payoffs, z_k^i, in the states $\{\omega^i, i = 1, \ldots, 4\}$. We can express each one of these assets in terms of the elementary insurance contracts. In other words, we can find one synthetic for each S_{kt} by purchasing the portfolios:

$$\{z_k^1 \text{ unit of } C_1, z_k^2 \text{ unit of } C_2, z_k^3 \text{ unit of } C_3, z_k^4 \text{ unit of } C_4\} \tag{12.33}$$

Putting these in matrix form, we see that arbitrage-free values, S_{kt_0}, of these assets at time t_0 have to satisfy the matrix equation:

$$\begin{bmatrix} 1 \\ S_{1t_0} \\ S_{2t_0} \\ S_{3t_0} \\ S_{4t_0} \end{bmatrix} = \begin{bmatrix} 1 + r_{t_0} & 1 + r_{t_0} & 1 + r_{t_0} & 1 + r_{t_0} \\ z_1^1 & z_1^2 & z_1^3 & z_1^4 \\ z_2^1 & z_2^2 & z_2^3 & z_2^4 \\ z_3^1 & z_3^2 & z_3^3 & z_3^4 \\ z_4^1 & z_4^2 & z_4^3 & z_4^4 \end{bmatrix} \begin{bmatrix} Q^1 \\ Q^2 \\ Q^3 \\ Q^4 \end{bmatrix} \tag{12.34}$$

where the matrix on the right-hand side contains all possible values of the assets S_{kt} in states ω^i, at time T.[13]

Hence, given the prices of actively traded elementary contracts C_i, we can easily calculate the time t cost of forming the portfolio:

$$\text{Cost} = C_1 z_T^1 + C_2 z_T^2 + \cdots + C_n z_T^n \tag{12.35}$$

This can be regarded as the cost basis for the S_t asset. Adding a proper margin to it will give the fair market price S_t.

EXAMPLE

Suppose the S_t asset has the following payoffs in the states of the world $i = 1, \ldots, 4$:

$$\{z_T^1 = 10, z_T^2 = 1, z_T^2 = 14, z_T^2 = 16\} \tag{12.36}$$

Then, buying 10 units of the first insurance contract C_1 will guarantee the 10 in the first state, and so on.

---

[12]As usual, it is assumed that S_t makes no other interim payouts.
[13]For notational simplicity, we eliminated time subscripts in the matrix equation. Also remember that the time index represents years and that $C_i = Q^i$.

> Suppose we observe the following prices for the elementary insurance contracts:
>
> $$C_1 = 0.3, C_2 = 0.2, C_3 = 0.4, C_4 = 0.07 \qquad (12.37)$$
>
> Then the total cost of the insurance contracts purchased will be
>
> $$\text{Cost} = (0.3)10 + (0.2)(1) + (0.4)14 + (0.07)16 \qquad (12.38)$$
>
> $$= 9.92 \qquad (12.39)$$
>
> This should equal the current price of S_t once a proper profit margin is added.

Clearly, if such elementary insurance contracts actively traded in financial markets, the job of a financial engineer would be greatly simplified. It would be straightforward to construct synthetics for *any* asset, and then price them using the cost of the replicating portfolios as shown in the example. However, there is a close connection between C_i and options of the same series that differ only in their strikes. We saw how to obtain C_i from liquid option prices. Accordingly, if options with a broad array of strikes trade in financial markets, then traders can create *static* replicating portfolios for assets with arbitrary payoffs.[14]

12.5 IMPLICATIONS OF THE FUNDAMENTAL THEOREM

The fundamental theorem of asset pricing has a number of implications that play a critical role in financial engineering and derivatives pricing. First, using this theorem we can obtain probability distributions that can be used in asset pricing. These probability distributions will be objective and operational. Second, the theorem leads to the so-called Martingale representation of asset prices. Such a representation is useful in modeling asset price dynamics. Third, we will see that the Martingale representation can serve to objectively set expected asset returns. This property eliminates the need to model and estimate the "drift factors" in asset price dynamics. We will now study these issues in more detail.

12.5.1 RESULT 1: RISK-ADJUSTED PROBABILITIES

Q^i introduced in the previous section can be modified judiciously in order to obtain convenient probability distributions that the financial engineer can work with. These distributions do *not* provide real-world odds on the states of the world ω^i, and hence cannot be used directly in econometric prediction. Yet they *do* yield correct arbitrage-free prices. (This section shows how.) But there is more. As there are *many* such distributions, the market practitioner can also *choose* the distribution that fits his/her current needs best. How to makes this choice is discussed in the next section.

---

[14] Again, the asset with arbitrary payoffs should depend on the same underlying risk as the options.

12.5.1.1 *Risk-neutral probabilities*

Using the state prices Q^i, we first obtain the so-called *risk-neutral probability* distribution. Consider the first row of the matrix equation (12.9). Assume that it represents the savings account.[15]

$$\left(1 + r_{t_0}\right)Q^1 + \cdots + \left(1 + r_{t_0}\right)Q^n = 1 \tag{12.40}$$

Relabel, using

$$\tilde{p}_i = \left(1 + r_{t_0}\right)Q^i \tag{12.41}$$

According to Eq. (12.40), we have

$$\tilde{p}_i + \cdots + \tilde{p}_n = 1 \tag{12.42}$$

where

$$0 < \tilde{p}_i \tag{12.43}$$

since each Q^i is positive. This implies that the numbers $\tilde{p}_i$ have the properties of a discrete *probability distribution*. They are positive and they add up to one. Since they are determined by the markets, we call them "risk-adjusted" probabilities. They are, in fact, obtained as linear combinations of n current asset prices. This particular set of synthetic probabilities is referred to as the *risk-neutral* probability distribution.

To be more precise, the risk-neutral probabilities $\{\tilde{p}_i\}$ are time t_0 probabilities on the states that occur at time T. Thus, if we wanted to be more exact, they would have to carry two more subscripts, t_0 and T. Yet, these will be omitted for notational convenience and assumed to be understood by the reader.

12.5.1.2 *Other probabilities*

Several other synthetic probabilities can be generated, and these may turn out to be more useful than the risk-neutral probabilities. Given the positive Q^i, we can rescale these by any positive normalizing factor so that they can be interpreted as probabilities. There are many ways to proceed. In fact, *as long as a current asset price S_{kt_0} is nonzero and z_i^j are positive*, one can choose any kth row of the matrix equation in Eq. (12.9) to write and then define

$$1 = \sum_{i=1}^{n} \frac{z_k^i}{S_{kt_0}} Q^i \tag{12.44}$$

and then define

$$\frac{z_k^i}{S_{kt_0}} Q^i = \tilde{p}_i^k \tag{12.45}$$

$\tilde{p}_i^k$, $i = 1, \ldots, n$, can be interpreted as probabilities obtained after *normalizing* by the S_{kt_0} asset. $\tilde{p}_i^k$ will be positive and will add up to one. Hence, they will have the characteristics of a probability distribution, but again they cannot be used in prediction since they are not the actual probabilities of a particular state of the world ω^i occurring. Clearly, for *each* nonzero S_{kt_0} we can obtain a new

---

[15]Remember that $T - t_0$ is assumed to be 1 year.

probability $\tilde{p}_i^k$. These will be different across states ω^i, as long as the time T value of the asset is positive in all states.[16]

It turns out that *how* we normalize a sequence of $\{Q^i\}$ in order to convert them into some synthetic probability is important. The special case, where

$$S_{kt_0} = B(t_0, T) \qquad (12.46)$$

$B(t_0, T)$ being the current price of a T-maturity risk-free pure discount bond, is especially interesting. This yields the so-called *T-forward measure*. Because the discount bond matures at time T, the time T values of the asset are given by

$$z_k^i = 1 \qquad (12.47)$$

for all i.

Thus, we can simply divide state prices Q^i by the current price of a default-free discount bond maturing at time t_0, and obtain the T-forward measure:

$$\tilde{p}_i^T = Q^i \frac{1}{B(t_0, T)} \qquad (12.48)$$

We will see in Chapter 14 that the T-forward measure is the natural way to deal with payoffs associated with time T. Let's consider an example.

EXAMPLE

Suppose short-term risk-free rates are 5% and that there are four states of the world. We observe the following arbitrage-free bid prices for four assets at time t_0:

$$S_{1t_0} = 2.45238, S_{2t_0} = 1.72238, S_{3t_0} = 6.69429, S_{4t_0} = 3.065 \qquad (12.49)$$

It is assumed that at time $T = t_0 + 1$, measured in years, the four possible values for each asset will be given by the matrix:

$$\begin{bmatrix} 10 & 3 & 1 & 1 \\ 2 & 3 & 2 & 1 \\ 1 & 10 & 10 & 1 \\ 8 & 2 & 10 & 2 \end{bmatrix} \qquad (12.50)$$

We can form the matrix equation and then solve for the corresponding Q^i:

$$\begin{bmatrix} 1 \\ S_{1t_0} \\ S_{2t_0} \\ S_{3t_0} \\ S_{4t_0} \end{bmatrix} = \begin{bmatrix} 1+0.5 & 1+0.5 & 1+0.5 & 1+0.5 \\ 10 & 3 & 1 & 1 \\ 2 & 3 & 2 & 1 \\ 1 & 10 & 10 & 6 \\ 8 & 2 & 10 & 2 \end{bmatrix} \begin{bmatrix} Q^1 \\ Q^2 \\ Q^3 \\ Q^4 \end{bmatrix} \qquad (12.51)$$

---

[16]We emphasize that we need a positive price and positive possible time T values in order to do this.

Using the first four rows of this system, we solve for Q^i:

$$Q^1 = 0.1, Q^2 = 0.3, Q^3 = 0.07, Q^4 = 0.482 \qquad (12.52)$$

Next, we obtain the risk-neutral probabilities by using

$$\tilde{p}_i = (1 + 0.5)Q^i \qquad (12.53)$$

which gives

$$\tilde{p}_1 = 0.105, \tilde{p}_2 = 0.315, \tilde{p}_3 = 0.0735, \tilde{p}_4 = 0.5065 \qquad (12.54)$$

As a final point, note that we used the first four rows of the system shown here to determine the values of Q^i. However, the price of S_{4t_0} is also arbitrage-free:

$$\sum_{i=1}^{4} Q^i S_{4T}^i = 3.065 \qquad (12.55)$$

as required.

Interestingly, for short-term instruments, and with "normal" short-term interest rates of around $3-5\%$, the savings account normalization makes little difference. If $T - t_0$ is *small*, Q^i will be only marginally different from $\tilde{p}_i$.[17]

12.5.1.3 A remark

Can derivatives be used for normalization? For example, instead of normalizing by a savings account or by using bonds, could we normalize with a swap? The answer is no. There are probabilities called swap measures, but the normalization that applies in these cases is not a swap, but an annuity. Most derivatives are not usable in the normalization process because normalization by an S_{kt} implies, essentially, that the state prices Q^i are multiplied by factors such as

$$\frac{z_k^i}{S_{kt}} \frac{S_{kt}}{z_k^i} = 1 \qquad (12.56)$$

and then grouped according to

$$\frac{z_k^i}{S_{kt}} \frac{S_{kt}}{z_k^i} Q^i = \frac{S_{kt}}{z_k^i} \tilde{p}_i^k \qquad (12.57)$$

---

[17] For example, at a 5% short rate, a 1-month discount bond will sell for approximately

$$B(t_0, t) = \frac{1}{1 + 0.5\frac{1}{12}} = 0.9958$$

so that dividing by this scale factor will not modify Q^i very much.

But in this operation, both z_k^i, $i = 1, \ldots, n$ and S_{kt} should be nonzero. Otherwise the ratios would be undefined. This will be seen below.

12.5.1.4 Swap measure

The normalizations thus far used only *one* asset, S_{kt}, in converting Q^i into probabilities $\tilde{p}^k$. This need not be so. We can normalize using a linear combination of many assets, and sometimes this proves very useful. This is the case for the so-called swap, or annuity, measure. The swap measure is dealt with in Chapter 17.

12.5.2 RESULT 2: MARTINGALE PROPERTY

The fundamental theorem of asset pricing also provides a convenient model for pricing and risk-management purposes. All *properly normalized* asset prices have a *Martingale property* under a properly selected synthetic probability $\tilde{p}^k$. Let X_t be a stochastic process that has the following property:

$$X_t = E_t^{\tilde{p}^k}[X_T] \quad t < T \tag{12.58}$$

This essentially says that X_t have no predictable trend for all t. X_t is referred to as a Martingale. To see how this can be applied to asset pricing theory, first choose the risk-neutral probability $\tilde{P}$ as the working probability distribution.

12.5.2.1 Martingales under $\tilde{P}$

Consider any kth row in the matrix equation (12.9)

$$S_{kt_0} = (z_k^1)Q^1 + (z_k^2)Q^2 + \cdots + (z_k^n)Q^n \tag{12.59}$$

Replace Q^i with the risk-neutral probabilities $\tilde{p}_i$ using

$$Q^i = \frac{1}{1 + r_{t_0}} \tilde{p}_i \tag{12.60}$$

r_{t_0} is the interest on a risk-free 1-year deposit.

This gives

$$S_{kt_0} - \frac{1}{1 + r_{t_0}} \left[z_k^1 \tilde{p}_1 + \cdots + z_k^n \tilde{p}_n \right] \tag{12.61}$$

Here, the right-hand side is an *average* of the future values of S_{kT}, weighted by $\tilde{p}_i$. Thus, bringing back the time subscripts, the current arbitrage-free price S_{kt_0} satisfies

$$S_{kt_0} = \frac{1}{(1 + r_{t_0})} E_{t_0}^{\tilde{p}}[s_{kT}] \quad t_0 < T \tag{12.62}$$

In general terms, letting

$$X_t = \frac{\text{time-}t \text{ value of } S_{kt}}{\text{time-}t \text{ value of the savings account}} \tag{12.63}$$

we see that asset values normalized by the savings deposit have the Martingale property:

$$X_t = E_t^{\tilde{p}}[X_T] \quad t < T \tag{12.64}$$

Thus, all tools associated with Martingales immediately become available to the financial engineer for pricing and risk management.

12.5.2.2 Martingales under other probabilities

The convenience of working with Martingales is not limited to the risk-neutral measure $\tilde{P}$. A normalization with any non-zero price S_{jt} will lead to another Martingale. Consider the same kth row of the matrix equation in Eq. (12.9)

$$S_{kt_0} = (z_k^1)Q^1 + \cdots + (z_k^n)Q^n \tag{12.65}$$

This time, replace Q^i using $S_{jt_0}, j \neq k$, normalization:

$$\tilde{p}_i^j = Q^1 \frac{z_j^i}{S_{jt_0}} \tag{12.66}$$

We obtain, assuming that the denominator elements are positive:

$$S_{kt_0} = s_{jt_0} \left[z_k^1 \frac{1}{z_j^1} \tilde{p}_1^j + \cdots + z_k^n \frac{1}{z_j^n} \tilde{p}_n^j \right] \tag{12.67}$$

this means that the ratio,

$$X_t = \frac{S_{kt}}{S_{jt}} \tag{12.68}$$

is a Martingale under $\tilde{P}^j$ measure:

$$X_t - E_t^{\tilde{p}^j}[X_T] \quad t < T \tag{12.69}$$

It is obvious that the probability associated with a particular Martingale is a function of the normalization that is chosen, and that the implied Martingale property can be exploited in pricing. By choosing a Martingale, the financial engineer is also choosing the probability that he or she will be *working* with. In the remainder of this chapter and in the next, we will see several examples of how Martingale properties can be utilized.

12.5.3 RESULT 3: EXPECTED RETURNS

The next implication of the fundamental theorem is useful in modeling arbitrage-free *dynamics* for asset prices. Every synthetic probability leads to a particular *expected return* for the asset prices under consideration. These expected returns will *differ* from the true (subjective) expectations of players in the markets, but because they are agreed upon by all market participants and are associated with arbitrage-free prices, they will be even more useful than the true expectations.

We conduct the discussion in terms of the risk-neutral probability $\tilde{P}$, but our conclusions are valid for all other $\tilde{p}^k$. Consider again the Martingale property for an asset whose price is denoted by S_t, but this time reintroduce the day's adjustment parameter δ, dropping the assumption that t_i represents years. We can write, for some $0 < \delta$,

$$S_t = \frac{1}{(1 + r_t \delta)} E_t^{\tilde{P}}[S_{t+\delta}] \tag{12.70}$$

Rearrange to obtain

$$(1 + r_t \delta) = E_t^{\tilde{P}} \left[\frac{S_{t+\delta}}{S_t} \right] \tag{12.71}$$

According to this expression, under the probability $\tilde{P}$, expected net annual returns for *all* liquid assets will equal r_t, the risk-free rate observed at time t.

Similar results concerning the expected returns of the assets are obtained under other probabilities $\tilde{p}^k$. The expected returns will be different under different probabilities. Market practitioners can select the working probability so as to set the expected return of the asset to a *desired* number.[18]

In Chapter 14, we will see more complicated applications of this idea using time T forward measures. There, the expected change in the forward rates is set equal to zero by a judicious choice of probabilities.

12.5.3.1 Martingales and risk premia

Let us see how the use of Martingales "internalizes" the risk premia associated with nondiversifiable market risks. Let X_t, $t \in [t_0, T]$ be a risky asset and $\Delta > 0$ be a small time interval. The annualized *gross return* of X_t as expected by *players* at time t is defined by

$$1 + \hat{R}_t \Delta = E_t^P \left[\frac{X_{t+\Delta}}{X_t} \right] \tag{12.72}$$

where P represents the *real-world probability* used by market participants in setting up their expectation. Since this is an actual market expectation, the gross return will contain a risk premium:

$$\hat{R}_t = r_t = \mu_t \tag{12.73}$$

where r_t is the risk-free rate and μ_t is the *risk premium* commanded by the risky asset.[19] Putting these together, we have

$$(1 + r_t \Delta + \mu_t \Delta) = E_t^P \left[\frac{X_{t+\Delta}}{X_t} \right] \tag{12.74}$$

or

$$X_t = \frac{1}{(1 + r_t \Delta + \mu_t \Delta)} = E_t^P [X_{t+\Delta}] \tag{12.75}$$

This equality states that the asset price $X_{t+\Delta}$ *discounted* by the factor $(1 + r_t \Delta + \mu_t \Delta)$ is a Martingale *only* if we use the probability P. Note that in this setup there are *two* unknowns: (i) the risk premium μ_t and (ii) the real-world probability P.[20] Future cash flows accordingly need to be discounted by subjective discount factors and real-world probabilities need to be estimated. The pricing problem under these conditions is more complex. Financial engineers have to determine the value of the risk premium in addition to "projecting" future earnings or cash flows.

---

[18]Consequently, the associated risk premium need not be estimated.
[19]Under *rational* expectations, the subjective probability P is the same as the "true" distribution of X_t.
[20]Although this latter is *estimable* using econometric methods.

Now consider an alternative. Setting the (positive) risk premium equal to zero in the previous equation gives the inequality

$$X_t < \frac{1}{(1 + r_t\Delta)} E_t^P[X_{t+\Delta}] \tag{12.76}$$

But this is the same as risk-free savings account normalization. This means that by switching from P to $\tilde{P}$, we can restore the equality

$$X_t = \frac{1}{(1 + r_t\Delta)} E_t^{\tilde{P}}[X_{t+\Delta}] \tag{12.77}$$

Thus, normalization and synthetic probabilities internalize the risk premia by converting both unknowns into a known and objective probability $\tilde{P}$. Equation (12.77) can be exploited for pricing and risk management.

12.6 ARBITRAGE-FREE DYNAMICS

The last result that we derive from the fundamental theorem of asset pricing is a combination of all the corollaries discussed thus far. The synthetic probabilities and the Martingale property that we obtained earlier can be used to derive several *arbitrage-free dynamics* for an asset price. These arbitrage-free dynamics play an important role in pricing situations where an exact synthetic cannot be created, either due to differences in nonlinearities or due to a lack of liquid constituent assets. In fact, most of the pricing models will proceed along the lines of first obtaining arbitrage-free dynamics, and then either simulating paths from this or obtaining the implied binomial or trinomial trees. PDE methods also use these arbitrage-free dynamics.

12.6.1 ARBITRAGE-FREE SDEs

In this section, we briefly discuss the use of stochastic differential equations (SDEs) as a tool in financial engineering and then show how the fundamental theorem helps in specifying explicit SDEs that can be used in pricing and hedging in practice.[21] Consider an asset price S_t. Suppose we divide the time period $[t, T]$ into small intervals of equal size Δ. For each time $t + i\Delta$, $i = 1, \ldots, n$, we observe a different $S_{t+i\Delta}$. $S_{t+\Delta} - S_t$ is the *change* in asset price at time t. Choose a working probability from all available synthetic probabilities, and denote it by P^*.

Then, we can always calculate the expected value of this change under this probability. In the case of $P^* = \tilde{P}$, we obtain the risk-neutral expected net return by

$$E_t^{\tilde{P}}[S_{t+\Delta} - S_t] = r_t S_t \Delta \tag{12.78}$$

Next, note that the following statement is *always* true:

$$\text{Actual change in } S_t = \text{``expected'' change} + \text{``unexpected'' change} \tag{12.79}$$

---

[21] Appendix 9.2 in Chapter 9 provided the definition and some motivation for SDEs.

Now we can use the probability switching method and exploit the Martingale property. For example, for the risk-neutral probability we have

$$[S_{t+\Delta} - S_t]E_t^{\tilde{P}}[S_{t+\Delta} - S_t] + \epsilon_t \tag{12.80}$$

where ϵ_t represents a random variable with zero expectation under $\tilde{P}$. Replace from Eq. (12.78)

$$[S_{t+\Delta} - S_t] = r_t S_t \Delta + \epsilon_t \tag{12.81}$$

Now the error term ϵ_t can be written in the equivalent form

$$\epsilon_t = \sigma(S_t)S_t \Delta W_t \tag{12.82}$$

where ΔW_t is a Wiener process increment with variance equal to Δ.

Thus, the arbitrage-free dynamics under the $\tilde{P}$ measure can be written as

$$[S_{t+\Delta} - S_t] = r_t S_t \Delta + \sigma(S_t)S_t \Delta W_t \tag{12.83}$$

Letting $\Delta \to 0$, this equation becomes an SDE, that represents the arbitrage-free dynamics under the synthetic probability, $\tilde{P}$, during an infinitesimally short period dt. Symbolically, the SDE is written as

$$dS_t = r_t S_t dt + \sigma(S_t)S_t dW_t \tag{12.84}$$

dS_t and dW_t represent changes in the relevant variables during an *infinitesimal* time interval. Given the values for the (percentage) volatility parameter, $\sigma(S_t)$, these equations can be used to generate arbitrage-free trajectories for S_t. We deal with these in the next chapter. Note a major advantage of using the risk-neutral probability. The drift term, that is to say the first term on the right-hand side, is known. At this point, we consider a second way of obtaining arbitrage-free paths.

12.6.2 TREE MODELS

We will see another major application of the Martingale property. We develop the notion of binomial (trinomial) trees introduced in Chapter 8 and obtain an alternative way of handling arbitrage-free dynamics. Suppose the dynamics of S_t can be described by a (geometric) SDE:

$$dS_t = rS_t dt + \sigma S_t dW_t \tag{12.85}$$

where the volatility is assumed to be given by

$$\sigma(S_t) = \sigma \tag{12.86}$$

This is the constant *percentage* volatility of S_t. Also note that r_t is set to a constant. It can be shown that this SDE can be "solved" for S_t to obtain the relationship (Øksendal, 2010).

$$S_{t+\Delta} = S_t e^{r\Delta - \frac{1}{2}\sigma^2 \Delta + \sigma(W_{t+\Delta} - W_t)} \tag{12.87}$$

Our purpose is to construct an approximation to the arbitrage-free dynamics of this S_t. We will do this by considering *approximations* to possible paths that S_t can follow between t and some "expiration" date T. This approximation will be such that S_t will satisfy the Martingale property under a judiciously chosen probability. Finally, the approximation should be chosen so that as $\Delta \to 0$, the mean and the variance of the discrete approximation converge to those of the continuous time process under the relevant probability. It turns out that this can be done in *many* different ways. Each method may have its advantages and disadvantages. We discuss two different ways of building trees. As $\Delta \rightarrow 0$, the dynamics become those of continuous time.

12.6.2.1 *Case 1*

The method introduced by Cox—Ross—Rubinstein (CRR) selects the following approximation. First, the period $[t_0, T]$ is divided into N subintervals of equal length. Then, it is assumed that at each point of a path there are possible states. In the CRR case, $n = 2$ and the paths become *binomial*. An alternative trinomial tree is shown in Figure 12.3.

- At every node i of a possible path, there are only two possible states represented by the numbers $\{u_i, d_i\}$, with the (marginal) probabilities p and $(1 - p)$. The dynamics are selected as follows:

$$S_i^u = u_i S_{i-\Delta} \tag{12.88}$$

$$S_i^d = d_i S_{i-\Delta} \tag{12.89}$$

where S_i is the shortcut notation for $S_{t+i\Delta}$.

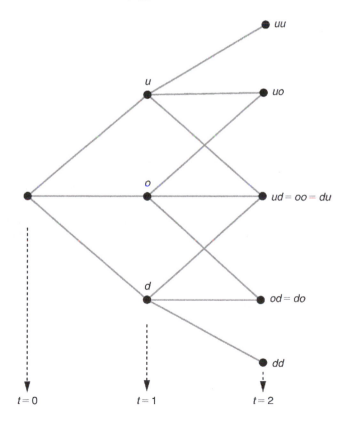

FIGURE 12.3

Trinomial tree.

- The $\{u_i, d_i\}$ are assumed to be constant at u, d.

We now show how to determine the Martingale probabilities. One approach is to find probabilities such that under p, $(1 - p)$:

$$S_i = e^{-r\Delta} E_i^{\tilde{P}}[S_{i+\Delta}] \tag{12.90}$$

or

$$S_i = e^{-r\Delta}[pS_{i+\Delta}^u + (1 - p)S_{i+\Delta}^d] \tag{12.91}$$

Using the definition of $S_{i+\Delta}^u, S_{i+\Delta}^d$, in Eqs. (12.88) and (12.89), we can write

$$S_i = e^{-r\Delta}[pS_i u + (1 - p)S_i d] \tag{12.92}$$

The mean and the variance of S_i under this probability should also be as given by the postulated dynamics of the continuous time process in the limit.[22] In other words, p should also satisfy

$$E_i^{\tilde{P}}[S_{i+\Delta}] = [pu + (1 - p)d]S_i \tag{12.93}$$

and

$$E_i^{\tilde{P}}[S_{i+\Delta}^2 - E_i^{\tilde{P}}[S_{i+\Delta}]^2] = [pu^2 + (1 - p)d^2]S_i^2 - E_i^{\tilde{P}}[S_{i+\Delta}]^2 \tag{12.94}$$

Use

$$E_i^{\tilde{P}}[S_{i+\Delta}] = S_i e^{r\Delta} \tag{12.95}$$

$$E_i^{\tilde{P}}[S_{i+\Delta}^2 - E_i^{\tilde{P}}[S_{i+\Delta}]^2] = S_i^2 e^{2r\Delta}(e^{\sigma^2\Delta} - 1) \tag{12.96}$$

and get the equations

$$e^{r\Delta} = pu + (1 - p)d \tag{12.97}$$

$$e^{2r\Delta + \sigma^2\Delta} = pu^2 + (1 - p)d^2 \tag{12.98}$$

The p, u, d that satisfy these two equations will (i) satisfy the Martingale equality for all Δ, (ii) get arbitrarily close to the mean and the variance of the continuous time process S_t as Δ goes to zero, and (iii) make the asymptotic distribution of S_i normal. However, there is one problem. Note that here we have two equations and three unknowns: u, d, and p. One more equation is needed. Choose

$$u = \frac{1}{d} \tag{12.99}$$

This makes the tree recombine and completes the system of equations. Under these conditions, the following values solve the equations

$$p = \frac{e^{r\Delta} - d}{u - d} \tag{12.100}$$

$$u = e^{\sigma\sqrt{\Delta}} \tag{12.101}$$

$$d = e^{-\sigma\sqrt{\Delta}} \tag{12.102}$$

---

[22]Here the probability $\tilde{P}$ is represented by the parameter P.

Any approximation here is in the sense that all terms containing higher orders of Δ are ignored.[23]

12.6.2.2 Case 2

The previous selection of p, u, d satisfies

$$S_i = e^{-r\Delta}\left[pS_ie^{\sigma\sqrt{\Delta}} + (1-p)S_ie^{-\sigma\sqrt{\Delta}}\right] \tag{12.103}$$

It turns out that p, u, d can be selected in other ways as well. In particular, note that during an interval Δ, S_t moves to

$$S_{t+\Delta} = S_t e^{r\Delta - \frac{1}{2}\sigma^2\Delta + \sigma[W_{t+\Delta} - W_t]} \tag{12.104}$$

Using the approximation

$$W_{t+\Delta} - W_t = \begin{cases} +\sqrt{\Delta} & \text{with probability } 0.5 \\ -\sqrt{\Delta} & \text{with probability } 0.5 \end{cases} \tag{12.105}$$

we get new values for p, u, and d:

$$u = e^{r\Delta - \frac{1}{2}\sigma^2\Delta + \sigma\sqrt{(\Delta)}} \tag{12.106}$$

$$d = e^{r\Delta - \frac{1}{2}\sigma^2\Delta - \sigma\sqrt{(\Delta)}} \tag{12.107}$$

$$p = 0.5 \tag{12.108}$$

These values will again satisfy the Martingale equality, the equality for the mean, and the variance of S_i, in the same approximate sense.

12.7 WHICH PRICING METHOD TO CHOOSE?

In general, the choice of a pricing method depends on the following factors:

- The accuracy of pricing methods does, in general, differ. Some methods are numerically more stable than others. Some methods yield coarser approximations than others. Precision is an important factor.
- The speed of pricing methods also changes from one method to another. In general, everything else being the same, the faster results are preferred.
- Some methods are easier to implement. The ease of understanding a pricing method is an important factor in its selection by practitioners.
- The parsimony associated with the model is also important. In general, we want our pricing models to depend on as few parameters as possible. This way, the model has to be calibrated to a smaller number of parameters, which means that fewer things can go wrong. Also, a trader/broker can in general compensate for a parsimonious model by adjusting the quotes based on trading experience.

---

[23]This is, in fact, a standard assumption used throughout calculus. We note that, as Δ goes to 0, the values of p will converge to $\frac{1}{2}$.

However, in the end, the method chosen depends on the circumstances and is a matter of experience. What a book like this can do is to present a brief overview of the various approaches available to the financial engineer.

12.8 CONCLUSIONS

We obtained some important results in this chapter. First, we showed that the notion of state prices can be made practical in environments with liquid option prices at different strikes. From here we showed how to obtain risk-neutral and forward measures and the corresponding arbitrage-free dynamics.

Finally, as long as liquid option prices with different strikes exist, we showed how to replicate an asset using a static portfolio of options. This is true for the following reasons:

1. Given the option prices, we can get the prices of elementary insurance contracts.
2. But we know that every asset can be synthetically created as a portfolio of elementary insurance contracts.
3. This means that every asset can be created as a portfolio of liquid options.

Hence, option markets not only provide close relatives of elementary insurance contracts, but also show us how to obtain generalized static synthetics for all assets in principle. Of course, the practical application depends on the availability of liquid options.

Finally, we must emphasize that risk management and pricing are never as straightforward in real life, since given the day, the number, and the type of liquid option, contracts change.

SUGGESTED READING

The treatment of the fundamental theorem of finance in this chapter has been heuristic and introductory, although all important aspects of the theorem have been covered. The reader can get more insight into the theorem by looking at **Duffie** (2001), which offers an excellent treatment of asset pricing. The article by **Brace et al.** (1997) is an important milestone in the use of Martingale theory, and places the right emphasis on pricing and the measure of change that fits this chapter. **Clewlow and Strickland** (1998) provide several examples.

APPENDIX 12.1: SIMPLE ECONOMICS OF THE FUNDAMENTAL THEOREM

This appendix provides a justification for the fundamental theorem from standard microeconomic theory. Consider the following setup. An investor faces a decision that involves two time periods; the time of decision, and T, the relevant future date. At T, there are only two possible states of the world ω^i, $i = 1, 2$. The investor's subjective probabilities for these are p^1 and p^2, respectively.

This investor's preferences are described by a *utility function* $U(X_t)$, where X_t is total (real) consumption at time t. Essentially, this investor is better off the higher his or her consumption:

$$0 < \frac{dU}{dX_t} \tag{12.109}$$

But, additional consumption would incrementally have less and less positive effect:

$$\frac{d^2 U}{dX_t^2} < 0 \tag{12.110}$$

This investor would like to maximize the *expected utility* associated with his or her current and future consumption:

$$E_t^P[U(X_t) + \beta U(X_T)] = U(X_t) + \beta(p^1 U(X_T^1) + p^2 U(X_T^2)) \tag{12.111}$$

where β is a constant subjective discount factor, P is the subjective personal probability, and X_T^1 and X_T^2 are the consumption levels in states 1 and 2 during period T, respectively. The maximization of this function is subject to the investor's budget constraint at time t and on the two states of the world at time T.

$$q_t X_t + S_t h_t = I$$
$$q_T^1 X_T^1 = I + h_t S_T^1 \tag{12.112}$$
$$q_T^2 X_T^2 = I + h_t S_T^2$$

S_t is a risky asset that can be purchased at time t. It has possible values S_T^1 and S_T^2 at time T. Here I is a known and constant income earned at times t and T. q_t, q_T^1, and q_T^2 are the corresponding prices of the consumption good. Note that, at time T, there are two prices, one for each state. Finally, h_t is the number of S_t purchased by the investor at time t.

According to this, we are dealing with an investor who receives a constant income that needs to be split between saving and consumption in a two-period setting. The investment can be made only by buying a desired amount of the S_t asset. The price of this asset is a random variable in the model.

The investor is risk averse and maximizes the expected utility function. There are several ways one can solve this maximization. Our intention is to show a simple example to *motivate* the fundamental theorem of finance. Hence, we are not concerned with the optimal consumption itself. Rather, we would like to obtain a relationship between "current" asset price S_t and the two possible values S_T^1 and S_T^2 at time T. The fundamental theorem of asset pricing is about these two sets of prices. Thus, we should be able to find out how the present framework can generate the state prices Q^i of the fundamental theorem.

Keeping these objectives in mind, we proceed by first substituting out X_t, X_T^1, X_T^2 from the equations in (12.112), and then differentiating the resulting expression with respect to the only remaining choice variable h_t. The substitution gives

$$U(X_t) + \beta(p^1 U(X_T^1) + p^2 U(X_T^2)) = U\left(\frac{I - S_t h_t}{q_t}\right) + \beta\left(p^1 U\left(\frac{I + h_t S_T^1}{q_T^1}\right) + p^2 U\left(\frac{I + h_t S_T^2}{q_T^2}\right)\right) \tag{12.113}$$

Differentiating the right side with respect to h_t, equating to zero, and then rearranging,

$$U'\left(\frac{I - S_t h_t}{q_t}\right)\left(\frac{S_t}{q_t}\right) = \beta\left(p^1 U'\left(\frac{I + h_t S_T^1}{q_T^1}\right)\left(\frac{S_T^1}{q_T^1}\right) + p^2 U'\left(\frac{I + h_t S_T^2}{q_T^2}\right)\left(\frac{S_T^2}{q_T^2}\right)\right) \tag{12.114}$$

where $U'(.)$ is the derivative of $U(x)$ with respect to "x."

Now comes the critical point. We can rearrange the first-order condition in Eq. (12.114) to obtain

$$S_t = \beta \left(p^1 \frac{U'((I + h_t S_T^1)/q_T^1)}{U'((I - S_t h_t)/q_t)} \frac{q_t}{q_T^1} S_T^1 + p^2 \frac{U'((I + h_t S_T^2)/q_T^2)}{U'((I - S_t h_t)/q_t)} \frac{q_t}{q_T^2} S_T^2 \right) \tag{12.115}$$

Now relabel as follows:

$$Q^1 = \beta p^1 \frac{U'((I + h_t S_T^1)/q_T^1)}{U'((I - S_t h_t)/q_t)} \frac{q_t}{q_T^1} \tag{12.116}$$

and

$$Q^2 = \beta p^2 \frac{U'((I + h_t S_T^2)/q_T^2)}{U'((I - S_t h_t)/q_t)} \frac{q_t}{q_T^2} \tag{12.117}$$

It is clear that all elements of the right-side expressions are *positive* and, as a result, Q^i, $i = 1, 2$, are positive. Substituting these Q^i back in Eq. (12.115), we get

$$S_t = S_T^1 Q^1 + S_T^2 Q^2 \tag{12.118}$$

In other words, there is a *linear* relationship between current asset price S_t and the future possible values S_T^1 and S_T^2, and $\{Q^i\}$ is the determining factor.

An interesting implication of the derivation shown here is the following. Even when the utility function $U(.)$ and the subjective probabilities p^i differ among investors, general equilibrium conditions would equate the marginal rates of substitution across these differing investors and hence the $\{Q^i\}$ would be the *same*. In other words, $\{Q^i\}$ would be unique to all consumers even when these consumers disagree on the expected future behavior of the economy.

EXERCISES

1. The following prices of the four different stocks are reported from an arbitrage-free market at time t_0

$$S_{t_0}^1 = 12.66, S_{t_0}^2 = 12.24, S_{t_0}^3 = 20.66, S_{t_0}^4 = 18.25$$

   Find out the price of the 5th asset given the following possible values of these assets at time T.

14	3	25	4
11	10	9	2
9	23	24	22
8	25	10	5
6	15	25	10

2. The current time is $t = 1$ and our framework is the LIBOR model. We consider a situation with four states of the world ω_i at time $t = 3$.

Suppose L_i is the LIBOR process with a particular tenor and $B(1, 3)$, $B(1, 4)$, and $B(1, 4)$ are zero-coupon bond prices with indicated maturities. The *possible* payoffs of these instruments in the four future states of the world are as follows:

$$L = 6\%, 6\%, 4\%, 4\% \tag{12.119}$$

$$B(1, 3) = 1, 1, 1, 1 \tag{12.120}$$

$$B(1, 4) = 0.9, 0.92, 0.95, 0.96 \tag{12.121}$$

$$B(1, 5) = 0.8, 0.84, 0.85, 0.88 \tag{12.122}$$

The current prices are, respectively,

$$1, 0.91, 0.86, 0.77 \tag{12.123}$$

Here the 1 is a dollar invested in LIBOR. It is like a savings account. Finally, current LIBOR is 5%.

a. Using Mathematica or MATLAB, determine a state price vector q_1, q_2, q_3, q_4, that corresponds to $B(1, 3)$, $B(1, 4)$, $B(1, 5)$, L as a basis.

b. Does q_i satisfy the required condition of positivity? Is there an arbitrage opportunity?

c. Let F be the 1×2 FRA rate. Can you determine its arbitrage-free value?

d. Now let C be an ATM caplet (i.e., the strike is 5%) that expires at time $t = 2$, but settled at time $t = 3$ with notional amount 1. How much is it worth?

3. Suppose you are given the following data. The risk-free interest rate is 4%. The stock price follows:

$$dS_t = \mu S_t + \sigma S_t dW_t \tag{12.124}$$

The percentage annual volatility is 18% a year. The stock pays no dividends and the current stock price is 100.

Using these data, you are asked to calculate the current value of a European call option on the stock. The option has a strike price of 100 and a maturity of 200 days.

a. Determine an appropriate time interval Δ, such that the binomial tree has five steps.

b. What would be the implied u and d?

c. What is the implied "up" probability?

d. Determine the binomial tree for the stock price S_t.

e. Determine the tree for the call premium C_t.

4. Suppose the stock discussed in the previous exercise pays dividends. Assume all parameters are the same. Consider three forms of dividends paid by the firm:

a. The stock pays a continuous, known stream of dividends at a rate of 4% per time.

b. The stock pays 5% of the value of the stock at the third node. No other dividends are paid.

c. The stock pays a $5 dividend at the third node.

   In each case, determine the tree for the ex-dividend stock price. For the first two cases, determine the premium of the call.

   In what way(s) will the third type of dividend payment complicate the binomial tree?

5. We use binomial trees to value American-style options on the British pound. Assume that the British pound is currently worth $1.40. Volatility is 20%. The current British risk-free rate is 6% and the US risk-free rate is 3%. The put option has a strike price of $1.50. It expires in 200 days.
    a. The first issue to be settled is the role of US and British interest rates. This option is being purchased in the United States, so the relevant risk-free rate is 3%. But British pounds can be used to earn the British risk-free rate. So this variable can be treated as a continuous rate of dividends. Or we can say that interest rate differentials are supposed to equal the expected appreciation of the currency.
        Taking this into account, determine a Δ such that the binomial tree has five periods.
    b. Determine the implied u and d and the relevant probabilities.
    c. Determine the tree for the exchange rate.
    d. Determine the tree for a European put with the same characteristics.
    e. Determine the price of an American-style put with the previously stated properties.

6. Barrier options belong to one of four main categories. They can be up-and-out, down-and-out, up-and-in, or down-and-in. In each case, there is a specified "barrier," and when the underlying asset price down- or up-crosses this barrier, the option either expires automatically (the "out" case) or comes into life automatically (the "in" case).
        Consider a European-style up-and-out call written on a stock with a current price of 100 and a volatility of 30%. The stock pays no dividends and follows a geometric price process. The risk-free interest rate is 6% and the option matures in 200 days. The strike price is 110. Finally, the barrier is 120. If the before-maturity stock price exceeds 120, the option automatically expires.
    a. Determine the relevant u and d and the corresponding probability.
    b. Value a call with the same characteristics but without the barrier property.
    c. Value the up-and-out call.
    d. Which option is cheaper?

EXCEL EXERCISES

7. (European Option)
        Write a VBA program to determine the initial price of a European Call and European Put option in a binomial model based on the following data:

    - $S(0) = 100$; $K = 105$; $T = 1$; $r = 8\%$; $\sigma = 30\%$; $M = 10$

        Create a function to calculate the option price using the Black–Scholes formula and compare the two results. Now gradually increase the value of M and report the subsequent option prices. Use the value of $u = e^{\sigma\sqrt{\Delta t}}$ and $d = e^{-\sigma\sqrt{\Delta t}}$

8. (European Option Nonrecombining Binomial Tree)
        Write a VBA program to determine the initial price of a European Call and European Put option in a nonrecombining binomial tree model based on the following data:

    - $S(0) = 100$; $K = 105$; $T = 1$; $r = 8\%$; $\sigma = 30\%$; $M = 10$

Create a function to calculate the option price using the Black–Scholes formula and compare the two results. Now gradually increase the value of M and report the subsequent option prices. Use the value of $u = e^{\sigma\sqrt{\Delta t}+(r-\sigma^2/2)\Delta t}$ and $d = e^{-\sigma\sqrt{\Delta t}+(r-\sigma^2/2)\Delta t}$

9. (American Option)

Write a VBA program to determine the initial price of an American Call and American Put option in a binomial model based on the following data:

- $S(0) = 100$; $K = 105$; $T = 1$; $r = 8\%$; $\sigma = 30\%$; div $= 8\%$; $M = 10$

Gradually increase the value of M and report the subsequent option prices. Use the value of $u = e^{\sigma\sqrt{\Delta t}}$ and $d = e^{-\sigma\sqrt{\Delta t}}$

10. European option on dividend paying index

Write a VBA program to determine the initial price of a European Call and European Put option in a binomial model based on the following data:

$S(0) = 100$; $K = 105$; $T = 1$; $r = 8\%$; $\sigma = 30\%$; div $= 8\%$; $M = 10$

Create function to calculate the option price using the Black–Scholes formula and compare. Now gradually increase the value of M and report the subsequent option prices. Use the value of $u = e^{\sigma\sqrt{\Delta t}}$ and $d = e^{-\sigma\sqrt{\Delta t}}$

11. European option on dividend paying index (nonrecombining binomial tree)

Write a VBA program to determine the initial price of a European Call and European Put option in nonrecombining binomial model based on the following data:

- $S(0) = 100$; $K = 105$; $T = 1$; $r = 8\%$; $\sigma = 30\%$; div $= 8\%$; $M = 10$

Create function to calculate the option price using the Black–Scholes formula and compare. Now gradually increase the value of M and report the subsequent option prices. Use the value of $u = e^{\sigma\sqrt{\Delta t}+(r-\text{div}-\sigma^2/2)\Delta t}$ and $d = e^{-\sigma\sqrt{\Delta t}+(r-\text{div}-\sigma^2/2)\Delta t}$

12. (FX American Option)

Write a VBA program to determine the initial price of an American-style options on the British pound in a binomial model based on the following data:

- $S(0) = \$1.54$; $K = \$1.54$; $T = 1$; $r = 8\%$; $r_f = 6\%$; $\sigma = 30\%$; $M = 10$

Gradually increase the value of M and report the subsequent option prices. Use the value of $u = e^{\sigma\sqrt{\Delta t}}$ and $d = e^{-\sigma\sqrt{\Delta t}}$

13. FX European Option

Write a VBA program to determine the initial price of European-style options on the British pound in the binomial model based on the following data:

- $S(0) = \$1.54$; $K = \$1.54$; $T = 1$; $r = 8\%$; $r_f = 6\%$; $\sigma = 30\%$; $M = 10$

Create a function to calculate the option price using the Black–Scholes formula and compare. Now gradually increase the value of M and report the subsequent option prices. Use the value of $u = e^{\sigma\sqrt{\Delta t}}$ and $d = e^{-\sigma\sqrt{\Delta t}}$

14. FX European Option (Nonrecombining B-Tree)

    Write a VBA program to determine the initial price of European-style options on the British pound in a nonrecombining binomial tree model based on the following data:

    - $S(0) = \$1.54$; $K = \$1.54$; $T = 1$; $r = 8\%$; $r_f = 6\%$; $\sigma = 30\%$; $M = 10$

    Create a function to calculate the option price using the Black–Scholes formula and compare. Now gradually increase the value of M and report the subsequent option prices. Use the value of $u = e^{\sigma\sqrt{\Delta t}+(r-r_f-\sigma^2/2)\Delta t}$ and $d = e^{-\sigma\sqrt{\Delta t}+(r-r_f-\sigma_2/2)\Delta t}$

CHAPTER OUTLINE

13.1 INTRODUCTION

The theorem discussed in the previous chapter establishes important no-arbitrage conditions that permit pricing and risk management using Martingale methods. According to these conditions, given unique arbitrage-free state prices, we can obtain a synthetic probability measure, $\tilde{P}$, under which all asset prices normalized by a particular Z_t become Martingales. Letting $C(S_t, t)$ represent a security whose price depends on an underlying risk S_t, we can write,

$$\frac{C(S_t, t)}{Z_t} = E_t^{\tilde{P}} \left[\frac{C(S_T, T)}{Z_T} \right] \tag{13.1}$$

As long as positive state prices exist, *many* such probabilities can be found and each will be associated with a particular normalization. The choice of the right working probability then becomes a matter of convenience and data availability.

The equality in Eq. (13.1) can be evaluated numerically using various methods. The arbitrage-free price S_t can be calculated by evaluating the expectation and then multiplying by Z_t. But to evaluate the expectation, we would need the probability $\tilde{P}$, hence, this must be obtained first. A further desirable characteristic is that the future value, Z_T, be *constant*, as it would be in the case of a default-free bond that matures at time T. Hence, T maturity bonds are good candidates for normalization.

In this chapter, we show *three* applications of the fundamental theorem. The first application is the *Monte Carlo* procedure which can be interpreted as a general method to calculate the expectation in Eq. (13.1). This method can be applied straightforwardly when instruments under consideration are of *European* type. The procedure uses the tools supplied by the fundamental theorem together with the law of large numbers.[1]

The second application of the fundamental theorem involves *calibration*. Calibration is the selection of model parameters using observed arbitrage-free prices from liquid markets. The chapter discusses simple examples of how to calibrate stochastic differential equations and tree models to market data using the fundamental theorem. This is done within the context of the Black–Derman–Toy (BDT) model.

---

[1] Let x_i, $i = 1, \ldots, N$ be independent, identically distributed observations from a random variable X with a finite first-order moment:

$$E[X] < \infty$$

Then, according to the law of large numbers, $\frac{1}{N}\sum_{i=1}^{N} x_i$ converges almost surely to $E[X]$ as N gets large.

The third application of the fundamental theorem introduced in Chapter 12 is more conceptual in nature. We use *quanto assets* to show how the theorem can be exploited in modeling. Quanto assets provide an excellent vehicle for this, since their pricing involves switches between domestic and foreign risk-neutral measures. Techniques for switching between measures are an integral part of financial engineering and will be discussed further in the next chapter. The application to quantos provides the first step.

Before we discuss these issues, a note of caution is in order. The discussion in this chapter should be regarded as an overview that presents examples for when to use the fundamental theorem, instead of being a source of how to implement such numerical techniques. Calculations using Monte Carlo or calibration are complex numerical procedures, and a straightforward application may not give satisfactory results. Interested readers can consult the sources provided at the end of the chapter.

13.2 APPLICATION 1: THE MONTE CARLO APPROACH

Consider again the expectation involving a *function* $C(S_t, t)$ of the underlying risk S_t under a working Martingale measure, $\tilde{P}$:

$$\frac{C(S_t, t)}{Z_t} = E_t^{\tilde{P}}\left[\frac{C(S_T, T)}{Z_T}\right] \tag{13.2}$$

where S_t and Z_t are two arbitrage-free asset prices at time t. Z_t is used as the normalizing asset and is instrumental in defining $\tilde{P}$. $C(S_t, t)$ may represent a European option premium or any other derivative that depends on S_t with expiration T.

This equation can be used as a vehicle to calculate a numerical value for $C(S_t, t)$, if we are given the probability measure $\tilde{P}$ and if we know Z_t. There are two ways of doing this. First, we can try to solve analytically for the expectation and obtain the resulting $C(S_t, t)$ as a *closed-form formula*. When the current value of the normalizing asset, Z_t, is known, this would amount to taking the integral:

$$C(S_t, t) = Z_t\left[\int_{-\infty}^{\infty}\int_{-\infty}^{\infty}\frac{C(S_T, T)}{Z_T}\tilde{f}(S_T, Z_T)\mathrm{d}S_T\mathrm{d}Z_T\right] \tag{13.3}$$

where $\tilde{f}(.)$ is the *joint* conditional probability density function of S_T, Z_T in terms of the $\tilde{P}$ probability.[2] Z_T on the right-hand side is considered to be random and possibly correlated with S_T. As a result, the probability $\tilde{P}$ would apply to both random variables, S_T and Z_T.

With judicious choices of Z_t, we *can* however make Z_T a constant. For example, if we choose Z_t as the default-free discount bond that matures at time T,

$$Z_T = 1 \tag{13.4}$$

It is clear that such normalization greatly simplifies the pricing exercise, since $\tilde{f}(.)$ is then a univariate conditional density.

---

[2]We assume that $\tilde{f}(S_T, Z_T)$ exists.

But, even with this there is a problem with the analytical method. Often, there are *no* closed-form solutions for the integrals, and a nice formula tying S_t to Z_t and other parameters of the distribution function $\tilde{P}$ may not exist. The value of the integral can still be calculated, although not through a closed-form formula. It has to be evaluated *numerically*.

One way of doing this is the Monte Carlo method.[3] This section briefly summarizes the procedure. We begin with a simple example. Suppose a random variable,[4] X, with a known normal distribution denoted by P, is given:[5]

$$X \sim N(\mu, \sigma) \tag{13.5}$$

Suppose we have a known function $g(X)$ of X. How would we calculate the expectation $E^P[g(X)]$, knowing that $E^P[g(X)] < \infty$? One way, of course, is by using the analytical approach mentioned earlier. Take the *integral*

$$E^P[g(X)] = \int_{-\infty}^{\infty} g(x) \left(\frac{1}{\sqrt{2\pi\sigma^2}} e^{-(1/2\sigma^2)(x-\mu)^2} \right) dx \tag{13.6}$$

if a closed-form solution exists.

But there is a second, easier way. We can invoke the *law of large numbers* and realize that given a large sample of realizations of X, denoted by x_i, the sample mean of any function of the x_i, say $g(x_i)$, will be close to the true expected value $E^P[g(X)]$. So, the task of calculating an arbitrarily good approximation of $E^P[g(X)]$ reduces to drawing a very large sample of x_i from the right distribution. Using random number generators, and the known distribution function of X, we can obtain N replicas of x_i. These would be generated independently, and the law of large numbers would apply:

$$\frac{1}{N} \sum_{i=1}^{N} g(x_i) \rightarrow E^P[g(X)] \tag{13.7}$$

The condition $E^P[g(X)] < \infty$ is sufficient for this convergence to hold. We now put this discussion in the context of asset pricing.

13.2.1 PRICING WITH MONTE CARLO

With the Monte Carlo method, an expectation is evaluated by first generating a sequence of replicas of the random variable of interest from a prespecified model, and then calculating the sample mean. The application of this method to pricing equations is immediate. In fact, the fundamental theorem provides the risk-neutral probability, $\tilde{P}$, such that for any arbitrage-free asset price S_t,

$$\frac{S_t}{B_t} = E_t^{\tilde{P}} \left[\frac{S_T}{B_T} \right] \tag{13.8}$$

---

[3]The other is the PDE approach, where we would first find the PDE that corresponds to this expectation and then solve the PDE numerically or analytically. This method will not be discussed here. Interested readers should consider Wilmott (2006) and Duffie (2001).
[4]Here the equivalent of X is S_T/B_T.
[5]In the preceding, the equivalent is $\tilde{P}$.

Here, the normalizing variable denoted earlier by Z_t is taken to be a savings account and is now denoted by B_t. This asset is defined as

$$B_t = e^{\int_0^t r_u du} \tag{13.9}$$

r_u being the continuously compounded instantaneous spot rate. It represents the time t value of an investment that was one dollar at time $t = 0$. The integral in the exponent means that r_u is not constant during $u \in [t, T]$. If r_t is a random variable, then we will need *joint* conditional distribution functions in order to select replicas of S_T and B_T. We have to postulate a *model* that describes the joint dynamics of S_T, B_T and that ties the information at time t to the random numbers generated for time T. We begin with a simple case where r_t is constant at r.

13.2.1.1 Pricing a call with constant spot rate

Consider the calculation of the price of a *European* call option with strike K and expiration T written on S_t, in a world where all Black–Scholes assumptions are satisfied. Using B_t in Eq. (13.9) as the normalizing asset, Eq. (13.8) becomes

$$\frac{C(t)}{e^{rt}} = E_t^{\tilde{P}} \left[\frac{CT}{e^{rT}} \right] \tag{13.10}$$

where $C(t)$ denotes the call premium that depends on S_t, t, K, r, and σ. After simplifying and rearranging

$$C(t) = e^{-r(T-t)} E_t^{\tilde{P}} [C(T)] \tag{13.11}$$

where

$$C(T) = \max[S_T - K, 0] \tag{13.12}$$

The Monte Carlo method can easily be applied to the right-hand side of Eq. (13.11) to obtain $C(t)$.

Using the savings account normalization, we can write down the *discretized* risk-neutral dynamics for S_t for discrete intervals of size $0 < \Delta$:

$$S_{t+\Delta} = (1 + r\Delta)S_t + \sigma S_t(\Delta W_t) \tag{13.13}$$

where it is assumed that the percentage volatility σ is constant and that the disturbance term, ΔW_t, is a normally distributed random variable with mean zero and variance Δ:

$$\Delta W_t \sim N(0, \Delta) \tag{13.14}$$

The r enters the SDE due to the use of the risk-neutral measure $\tilde{P}$. We can easily calculate replicas of S_T using these dynamics:

1. Select the size of Δ, and then use a proper pseudo-random number generator, to generate the random variable ΔW_t from a normal distribution.
2. Use the current value S_t, the parameter values r, σ, and the dynamics in Eq. (13.13) to obtain the N terminal values $S_T^j, j = 1, 2, \ldots, N$. Here j will denote a random path generated by the Monte Carlo exercise.

3. Substitute these into the payoff function,

$$C(T)^j = \max[S_T^j - K, 0] \tag{13.15}$$

and obtain N replicas of $C(T)^j$.

4. Finally, calculate the sample mean and discount it properly to get $C(t)$:

$$C(t) = e^{-r(T-t)} \frac{1}{N} \sum_{j=1}^{N} C(T)^j \tag{13.16}$$

This procedure gives the arbitrage-free price of the call option. We now consider a simple example.

EXAMPLE

Consider pricing the following European vanilla call written on S_t, the EUR/USD exchange rate, which follows the discretized (approximate) SDE:

$$S_{t_i}^j = S_{t_{i-1}}^j + (r - r^f)S_{t_{i-1}}^j \Delta + \sigma S_{t_{i-1}}^j \sqrt{\Delta} \epsilon_i^j \tag{13.17}$$

where the drift is the differential between the domestic and foreign interest rates.

We are given the following data on a call with strike $K = 1.0950$:

$$r = 2\% \quad r^f = 3\% \quad t_0 = 0, T = 5 \text{ days} = 1.09 \quad \sigma = 0.10 \tag{13.18}$$

A financial engineer decides to select $N = 3$ trajectories to price this call. The discrete interval is selected as $\Delta = 1$ day.

The software Mathematica provides the following standard normal random numbers:

$$\{0.763, 0.669, 0.477, 0.287, 1.81, -0.425\} \tag{13.19}$$

$$\{1.178, -0.109, -0.310, -2.130, -0.013, 0.421141\} \tag{13.20}$$

$$\{-0.922, 0.474, -0.556, 0.400, -0.890, -2.736\} \tag{13.21}$$

Using these in the discretized SDE,

$$S_i^j = \left(1 + (0.02 - 0.03)\frac{1}{365}\right)S_{i-1}^j + 0.10 S_{i-1}^j \sqrt{\frac{1}{365}} \epsilon_i^j \tag{13.22}$$

we get the trajectories:

Path	Day 1	Day 2	Day 3	Day 4	Day 5
1	1.0937	1.0965	1.0981	1.1085	1.1060
2	1.0893	1.0875	1.0754	1.0753	1.0776
3	1.0927	1.08946	1.0917	1.086	1.0710

For the case of a plain vanilla euro call, with strike $K = 1.095$, only the first trajectory ends in-the-money, so that

$$C(T)^1 = 0.011, \quad C(T)^2 = 0, \quad C(T)^3 = 0 \tag{13.23}$$

Using continuous compounding the call premium becomes

$$C(t) = \text{Exp}\left(-0.02\frac{5}{365}\right)\frac{1}{3}[0.011 + 0 + 0] \tag{13.24}$$

$$C(t) = 0.0037 \tag{13.25}$$

Obviously, the parameters of this model are selected to illustrate the application of the Monte Carlo procedure, and no real-life application would price securities with such a small number of trajectories. However, one important wrinkle has to be noted. The drift of this SDE was given by $(r - r^f)S_t\Delta$ and not by $rS_t\Delta$, which was the case of stock price dynamics. This modification will be dealt with below. Foreign currencies pay foreign interest rates and the risk-free interest rate differentials should be used. We discuss this in more detail in the following section.

13.2.2 PRICING BINARY FX OPTIONS

This section applies the Monte Carlo technique to pricing digital or binary options in foreign exchange markets. We consider the following elementary instrument:

If the price of a foreign currency, denoted by S_t, exceeds the level K at expiration, the option holder will receive the payoff R denoted in *domestic* currency. Otherwise the option holder receives nothing. The option is of European style and has expiration date T. The option will be sold for $C(t)$.

We would like to price this *binary* FX option using Monte Carlo. However, because the underlying is an exchange rate, some additional structure needs to be imposed on the environment and we discuss this first. This is a good example of the use of the fundamental theorem. It also provides a good occasion to introduce some elementary aspects of option pricing in FX markets.

13.2.2.1 Obtaining the risk-neutral dynamics

In the case of vanilla options written on stock prices, we assumed that the underlying stock pays no dividends and that the stock price follows a geometric continuous time process such as

$$dS_t = \mu S_t dt + \sigma S_t dW_t \tag{13.26}$$

with μ being an unknown drift coefficient representing the market's expected percentage appreciation of the stock and σ being a constant percentage volatility parameter whose value has to be obtained. W_t, finally, represents a Wiener process.

Invoking the fundamental theorem of asset pricing, we then replaced the unknown drift term μ by the risk-free interest rate r assumed to be constant. In the case of options written on *foreign exchange rates*, some of these assumptions need to be modified. We can preserve the overall geometric structure of the S_t process, but we have to change the assumption concerning *dividends*. A foreign currency is, by definition, some interbank deposit and will earn foreign (overnight) interest. According to the fundamental theorem, we can replace the real-world drift μ by the *interest rate differential*, $r_t - r_t^f$, where r_t^f is the foreign instantaneous spot rate and r_t is, as usual, the domestic rate. Thus, if spot rates are constant,

$$r_t = r, r_t^f = r^f \quad \forall t \tag{13.27}$$

This gives the arbitrage-free dynamics:[6]

$$dS_t = (r - r^f)S_t dt + \sigma S_t W_t \quad t \in [0, \infty) \tag{13.28}$$

The rationale behind using the interest rate *differential*, instead of the spot rate r, as the risk-neutral drift is a direct consequence of the fundamental theorem when the asset considered is a foreign currency. Since this chapter is devoted to applications of the fundamental theorem, we prefer to discuss this briefly.

Using the notation presented in Chapter 12, we take S_t as being the number of dollars paid for one unit of foreign currency. The fundamental theorem of asset pricing introduced in Chapter 12 implies that we can use the state prices $\{Q^i\}$ for states $i = 1, \ldots, n$, and write

$$S_t = \sum_{i=1}^{n}(1 + r^f \Delta)S_T^i Q^i \tag{13.29}$$

According to this, one unit of foreign currency will be worth S_T^i dollars in state i of time T, *and* it will also earn r^f per annum in interest during the period $\Delta = T - t$. Normalizing with the *domestic* savings account, this becomes

$$S_t = \sum_{i=1}^{n} \frac{(1 + r^f \Delta)}{(1 + r\Delta)} S_T^i (1 + r\Delta)Q^i \tag{13.30}$$

We now choose the risk-neutral probabilities as

$$\tilde{p}_i = (1 + r\Delta)Q^i \tag{13.31}$$

and rearrange Eq. (13.30) to obtain the expected gross return of S_t during Δ

$$\frac{(1 + r\Delta)}{(1 + r^f \Delta)} = E_t^{\tilde{p}}\left[\frac{S_T}{S_t}\right] \tag{13.32}$$

Here, the left-hand side can be approximated as[7]

$$(1 + r\Delta - r^f \Delta) \tag{13.33}$$

---

[6]If r_t, r_t^f were stochastic, this would require generating simultaneously random replicas of future rates as well. We would need to model interest rate dynamics.

[7]This can be done by using a first-order Taylor series approximation.

which means that S_t is expected to change at an annual rate of $(r - r^f)$ under the risk-neutral probability $\tilde{P}$. This justifies the continuous time risk-neutral drift of the dynamics:

$$dS_t = (r - r^f)S_t dt + \sigma S_t dW_t \tag{13.34}$$

Now that the dynamics are specified, the next step is selecting the Monte Carlo trajectories.

13.2.2.2 Monte Carlo process

Suppose we would like to price our digital option in such a framework. How could we do this using the Monte Carlo approach? Given that the arbitrage-free dynamics for S_t are obtained, we can simply apply the steps outlined earlier.

In particular, we need to generate random paths starting from the known current value for S_t. This can be done in *two* ways. We can first solve the SDE in Eq. (13.34) and then select random replicas from the resulting closed-form formula, if any. The second way is to discretize the dynamics in Eq. (13.34), and proceed as discussed in the previous section. Suppose we decided to proceed by first choosing a discrete interval Δ, and then *discretizing* the dynamics:[8]

$$S_{t+\Delta} = S_t + (r - r^f)S_t \Delta + \sigma S_t \Delta W_t \tag{13.35}$$

The next step would be to use a random number generator to obtain N sequences of standard normal random variables $\{\in_i^j, i = 1, \ldots, k, j = 1, \ldots, N\}$ and then calculate the N simulated trajectories using the discretized SDE:

$$S_{t_i}^j = S_{t_{i-1}}^j + (r - r^f)S_{t_{i-1}}^j \Delta + \sigma S_{t_{i-1}}^j \sqrt{\Delta} \in_i^j \tag{13.36}$$

where the superscript j denotes the jth simulated trajectory and where $\Delta = t_i - t_{i-1}$.

Once the paths $\{S_{t_i}^j\}$ are obtained, the arbitrage-free value of the digital call option premium C (t) that pays R at expiration can be found by using the equality

$$C(t) = \mathrm{Re}^{-r(T-t)} E_t^{\tilde{P}}\left[I_{\{S_T > K\}}\right] \tag{13.37}$$

where the symbol $I_{\{S_T > K\}}$ is the indicator function that determines whether at time T, the S_T exceeds K or not:

$$I_{\{S_T > K\}} = \begin{cases} 1 & \text{IF } S_T > K \\ 0 & \text{Otherwise} \end{cases} \tag{13.38}$$

This means that $I_{\{S_T > K\}}$ equals 1 if the option expires in-the-money; otherwise it is 0. According to the expected payoff in Eq. (13.37), the arbitrage-free $C(t)$ depends on the value of $E_t^{\tilde{P}}\left[I_{\{S_T > K\}}\right]$. The latter can be written as

$$E_t^{\tilde{P}}\left[I_{\{S_T > K\}}\right] = \mathrm{Prob}(S_T > K) \tag{13.39}$$

Thus

$$C(t) = \mathrm{Re}^{-r(T-t)} \mathrm{Prob}(S_T > K) \tag{13.40}$$

---

[8]Discretization of stochastic differential equations is a nontrivial exercise and there are optimal ways of doing these. Here, we ignore such numerical complications. Interested readers can consult Kloeden and Platen (2011).

This equation is easy to interpret. The value of the digital option is equal to the risk-neutral probability that S_T will exceed K times the present value of the constant payoff R.[9]

Under these conditions, the role played by the Monte Carlo method is simple. We generate N paths for the exchange rate starting from the current observation S_t, and then calculate the *proportion* of paths that would end up above the level K. Once this tally is made, denoting this number by m, the arbitrage-free value of the option will be

$$C(t) = e^{-r(T-t)}R \ (\text{Prob}(S_T > K)) \tag{13.41}$$

$$\cong e^{-r(T-t)}R\left(\frac{m}{N}\right) \tag{13.42}$$

Thus, in this case the Monte Carlo method is used to calculate a special expected value, which is the risk-neutral probability of the event $\{S_T > K\}$. The following section discusses two examples.

13.2.3 PATH DEPENDENCY

In the examples discussed thus far, we used the Monte Carlo method to generate *trajectories* for an underlying risk S_t, yet considered only the time T values of these trajectories in calculating the desired quantity $C(S_t, t)$. The other elements of the trajectory were not directly used in pricing.

This changes if the asset under consideration makes interim payouts or is subject to some other restrictions as in the case of barrier options. When $C(S_t, t)$ denotes the price of a barrier call option with barrier H, the option may knock in or out depending on the event $S_u < H$ *during* the period $u \in [t, T]$. Consider the case of a down-and-out call. In pricing this instrument, once a Monte Carlo trajectory is obtained, the *whole* trajectory needs to be used to determine if the condition $S_u < H$ is satisfied by S_u^j during the entire trajectory. This is one example of the class of assets that are path dependent and hence require direct use of entire Monte Carlo trajectories.

We now provide two more examples of the application of the Monte Carlo procedure. In the first case, the procedure is applied to a vanilla digital option, and in the second example, we show what happens when the option is a down-and-out call.

EXAMPLE

Consider pricing a digital option written on S_t, the EUR/USD exchange rate with the same structure as in the first example. The digital euro call has strike $K = 1.091$ and pays \$100 if it expires in-the-money. The parameters are the same as before:

$$r = 2\%, \quad r^f = 3\%, \quad t_0 = 0, \quad t = 5 \text{ days}, \quad S_{t_0} = 1.09, \quad \sigma = 0.10 \tag{13.43}$$

---

[9]The interest rate differential governs arbitrage-free dynamics, but the discounting needs to be done using the domestic rate only.

The paths for S_t are given by

Path	Day 1	Day 2	Day 3	Day 4	Day 5
1	1.0937	1.0965	1.0981	1.1085	1.1060
2	1.0893	1.0875	1.0780	1.0850	1.092
3	1.0927	1.08946	1.0917	1.086	1.0710

The digital call expires in-the-money if $1.091 < S_T^j$. There are two incidences of this event in the previous case, and the estimated risk-neutral probability that the option expires in-the-money is 2/3. The option value is calculated as

$$C(t) = \text{Exp}\left(-0.02\frac{5}{365}\right)\frac{2}{3}[100] \tag{13.44}$$

$$C(t) = \$66.6 \tag{13.45}$$

Now, consider what happens if we add a knock-out barrier $H = 1.08$. The digital call knocks out if S_t falls below this barrier before expiration.

EXAMPLE

All parameters are the same as in the first example, and the paths are given by

Path	Day 1	Day 2	Day 3	Day 4	Day 5
1	1.0937	1.0965	1.0981	1.1085	1.1060
2	1.0893	1.0875	1.0780	1.0850	1.092
3	1.0927	1.08946	1.0917	1.086	1.0710

The digital knock-out call requires that $1.091 < S_T^j$ and that the trajectory never falls below 1.08. Thus, there is only one incidence of this in this case and the value of the option is calculated as

$$C(t) = \text{Exp}\left(-0.02\frac{5}{365}\right)\frac{1}{3}[100] \tag{13.46}$$

$$C(t) = \$33.3 \tag{13.47}$$

Hence, the digital option is cheaper. Also, note that in the case of vanilla call, only the terminal values were used to calculate the option value, whereas in the case of the knock-out call, the entire trajectory was needed to check the condition $H < S_t$.

13.2.4 DISCRETIZATION BIAS AND CLOSED FORMS

The examples on the Monte Carlo used discrete approximations of SDEs. Assuming that the arbitrage-free dynamics of an asset price S_t can be described by a geometric SDE,

$$dS_t = rS_t dt + \sigma S_t dW_t \quad t \in [0, \infty) \tag{13.48}$$

we selected an appropriate time interval Δ, and ignoring continuous compounding, discretized the SDE

$$S_{t+\Delta} = (1 + r\Delta)S_t + \sigma S_t (\Delta W_t) \tag{13.49}$$

Equation (13.49) is only an *approximation* of the true continuous time dynamics given by Eq. (13.48).

For some special SDEs, we *can* sample the exact S_t. In such special cases, the stochastic differential equation for S_t can be "solved" for a closed form. The geometric process shown in Eq. (13.48) is one such case. We can directly obtain the value of S_T using the closed-form formula:

$$S_T = S_{t_0} e^{r(T-t_0) - (1/2)\sigma^2(T-t_0) + \sigma\left(W_T - W_{t_0}\right)} \tag{13.50}$$

The term $\left(W_T - W_{t_0}\right)$ will be normally distributed with mean zero and variance $T - t_0$. Hence, by drawing replicas of this random variable, we can obtain exact replicas for S_T at any T, $t_0 < T$. It turns out that even in the case of a mean-reverting model, such closed-form formulas are available and lend themselves to Monte Carlo pricing. However, in general, we may have to use discretized SDEs that may contain a discretization *bias*.[10]

13.2.5 REAL-LIFE COMPLICATIONS

Obviously, Monte Carlo becomes a complex approach once we go beyond simple examples. Difficulties arise, yet significant improvements can be made in regard to (i) how to select random numbers with computers, (ii) how to trick the system, such that the greatest accuracy can be obtained in the shortest time, and (iii) how to reduce the variance of the calculated prices with a given number of random selections. For these questions, other sources should be considered; we will not discuss them given our focus on financial engineering.[11]

13.3 APPLICATION 2: CALIBRATION

Calibrating a model means selecting the model parameters such that the observed arbitrage-free benchmark prices are duplicated by the use of this model. In this section we give two examples for this procedure. Since we already discussed several examples of how the fundamental theorem can be applied to SDEs, in this section we concentrate instead on tree models. As the last section has shown, calibration can be done using Monte Carlo and the SDEs as well.

---

[10]Kloeden and Platten (2011) discuss how such biases can be minimized. Aït-Sahalia (1996) discusses this bias within a setting of interest rate derivatives and shows how continuous time SDEs can be utilized.
[11]For interested readers, an excellent introductory source on these issues is Ross et al. (2012).

13.3.1 CALIBRATING A TREE

The BDT model is a good example for procedures that extract information from market prices. The model calibrates future trajectories of the spot rate r_t. The BDT model illustrates the way arbitrage-free dynamics can be extracted from liquid and arbitrage-free asset prices.[12]

The basic idea of the BDT model is that of any other calibration methodology. Let it be implicit binomial trees, estimation of state prices implicit in asset prices, or estimation of risk-neutral probabilities. The model assumes that we are given a number of benchmark arbitrage-free zero-coupon bond prices and a number of relevant volatility quotes in these markets. These volatility quotes can come from liquid caps and floors or from swaptions that are discussed in Chapters 16 and 17, respectively. The procedure evolves in three steps. First, arbitrage-free benchmark securities' prices and the relevant volatilities are obtained. Second, from these data the arbitrage-free dynamics of the relevant variable are extracted. Finally, *other* interest-sensitive securities are priced using these arbitrage-free dynamics.

This section illustrates the procedure using a three-period *binomial tree*. To simplify the notation and concentrate on understanding the main ideas, this section assumes that the time intervals Δ in the tree equal 1 year, and that the day-count parameter δ in a LIBOR setting equals 1 as well. The reader can easily generalize this simple example.

13.3.2 EXTRACTING A LIBOR TREE

Suppose we have arbitrage-free prices of three default-free benchmark zero-coupon bonds $\{B(t_0, t_1), B(t_0, t_2), B(t_0, t_3)\}$. Also suppose we observe reliable volatility quotes σ_i $i = 0, 1, 2$ for the LIBOR rates $L_{t_0}, L_{t_1}, L_{t_2}$.

First note that σ_0 is by definition equal to 0, because time t_0 variables have already been observed at time t_0. Next, assume that we have the following data:

$$\sigma_1 = 15\% \tag{13.51}$$

$$\sigma_2 = 20\% \tag{13.52}$$

$$B(t_0, t_1) = 0.95 \tag{13.53}$$

$$B(t_0, t_2) = 0.87 \tag{13.54}$$

$$B(t_0, t_3) = 0.79 \tag{13.55}$$

From these data, we extract information concerning the future *arbitrage-free* behavior of the LIBOR rates L_{t_i}. We first need some pricing functions that tie the arbitrage-free bond prices to the dynamics of the LIBOR rates. These pricing functions are readily available from the fundamental theorem.

13.3.2.1 Pricing functions

Consider the fundamental theorem written for times t_0 and t_3. Suppose there are k states of the world at time t_3 and consider the matrix equation discussed in Chapter 11:

$$S_{kx1} = D_{kxk} Q_{kx1} \tag{13.56}$$

---

[12]The current convention in fixed income has evolved well beyond the BDT approach in different directions. On the one hand, there is the forward LIBOR model, and on the other hand, there are the trinomial interest rate models.

Here, S is a $(k \times 1)$ vector of arbitrage-free asset prices at time t_0, D is the payoff matrix at time t_3, and Q is the $(k \times 1)$ vector of positive state prices at time t_3.

Suppose the first asset is a 1-year LIBOR-based deposit and the second asset is the bond $B(t_0, t_3)$, which matures and pays 1 dollar at t_3. Then, the first two rows of the matrix equation in Eq. (13.56) will be as follows:

$$\begin{pmatrix} 1 \\ B(t_0, t_3) \end{pmatrix} = \begin{pmatrix} \dfrac{[(1+L_{t_0})(1+L_{t_1})(1+L_{t_2})]^1}{1} & \cdots & \dfrac{[(1+L_{t_0})(1+L_{t_1})(1+L_{t_2})]^k}{1} \\ \cdots & \cdots \end{pmatrix}$$
$$\times \begin{pmatrix} Q^1 \\ \cdots \\ \cdots \\ \cdots \\ Q^k \end{pmatrix}$$

(13.57)

where $[(1+L_{t_0})(1+L_{t_1})(1+L_{t_2})]^i$ represents the return to the savings account investment in the ith state of time t_3. We can write the second row as

$$B(t_0, t_3) = \sum_{i=1}^{k} Q^i$$

(13.58)

Normalizing by the savings account, this becomes

$$B(t_0, t_3) = \sum_{i=1}^{k} \frac{[(1+L_{t_0})(1+L_{t_1})(1+L_{t_2})]^i}{[(1+L_{t_0})(1+L_{t_1})(1+L_{t_2})]^i} Q^i$$

(13.59)

Relabeling the risk-neutral probabilities

$$\tilde{p}_i = [(1+L_{t_0})(1+L_{t_1})(1+L_{t_2})]^i Q^i$$

(13.60)

gives

$$B(t_0, t_3) = \sum_{i=1}^{k} \frac{1}{[(1+L_{t_0})(1+L_{t_1})(1+L_{t_2})]^i} \tilde{p}_i$$

(13.61)

Thus, we obtain the pricing equation for the t_3 maturity bond as:

$$B(t_0, t_3) = E_{t_0}^{\tilde{P}} \left[\frac{1}{(1+L_{t_0})(1+L_{t_1})(1+L_{t_2})} \right]$$

(13.62)

Proceeding in a similar way, we can obtain the pricing equations for the two remaining bonds:

$$B(t_0, t_1) = E_{t_0}^{\tilde{P}} \left[\frac{1}{(1+L_{t_0})} \right]$$

(13.63)

$$B(t_0, t_2) = E_{t_0}^{\tilde{P}} \left[\frac{1}{(1+L_{t_0})(1+L_{t_1})} \right]$$

(13.64)

The first equation is trivially true, since L_{t_0} is known at time t_0.

13.3.3 OBTAINING THE BDT TREE

In this particular example, we have three benchmark prices and two volatilities. This gives five equations:

$$B(t_0, t_1) = E_{t_0}^{\tilde{P}}\left[\frac{1}{(1 + L_{t_0})}\right] \tag{13.65}$$

$$B(t_0, t_2) = E_{t_0}^{\tilde{P}}\left[\frac{1}{(1 + L_{t_0})(1 + L_{t_1})}\right] \tag{13.66}$$

$$B(t_0, t_3) = E_{t_0}^{\tilde{P}}\left[\frac{1}{(1 + L_{t_0})(1 + L_{t_1})(1 + L_{t_2})}\right] \tag{13.67}$$

$$\text{Vol}(L_{t_1}) = \sigma_1 \tag{13.68}$$

$$\text{Vol}(L_{t_2}) = \sigma_2 \tag{13.69}$$

Once we specify a model for the dynamics of L_{t_i}, we can solve these equations to obtain the arbitrage-free paths for L_{t_i}.

13.3.3.1 Specifying the dynamics

We now obtain this arbitrage-free dynamics. Following the tradition in tree models, we simplify the notation and use the index $i = 0, 1, 2, 3$ to denote "time," and the letters u and d to represent the up and down states at each node. First note that we have five equations and, hence, we can at most, get five pieces of independent information from these equations. In other words, the specified dynamic must have at most five *unknowns* in it. Consider the following three-period binomial tree:

$$
\begin{array}{ccccc}
& & & \nearrow & L_2^{uu} \\
& & \nearrow L_1^u & & \\
& & & \searrow & L_2^{ud} \\
L_0 & & & & \\
& \searrow & & \nearrow & L_2^{du} \\
& & L_1^d & & \\
& & & \searrow & L_2^{dd}
\end{array}
\tag{13.70}
$$

The dynamic has seven unknowns, namely $\{L_0, L_1^u, L_1^d, L_2^{ud}, L_2^{du}, L_2^{dd}, L_2^{uu}\}$. That is two *more* than the number of equations we have. At least two unknowns must be eliminated by imposing additional restrictions on the model. These will come from the specification of variances, as we will now see.

13.3.3.2 The variance of L_i

The spot LIBOR rate L_i, $i = 0, 1, 2$ has a binomial specification. This means that at any node, the spot rate can take one of only two possible values. Thus, the *percentage* variance of L_i, conditional on state j at "time" i, is given by[13]

$$Var(L_i|j) = E\tilde{P}[\ln(L_i) - \ln(\bar{L}_i)^2|j] \tag{13.71}$$

---

[13]We calculate the percentage volatility because caps/floors markets quote volatility this way, by convention.

where

$$\ln \bar{L}_i = E\tilde{P}[\ln(L_i)|j] \tag{13.72}$$

is the conditional expected value of L_i.

We now make two additional assumptions. The first assumption is for notational purposes only. We let $j = u$, and hence assume that we are in an "up" state. The outcome for $j = d$ will be similar.

Second, we let

$$p_i^u = \frac{1}{2} \quad \forall i \tag{13.73}$$

$$p_i^d = \frac{1}{2} \quad \forall i \tag{13.74}$$

That is to say, we assume that the up-and-down *risk-neutral* probabilities are constant over the life of the tree and that they are equal. We will see that this assumption, which at first may look fairly strong, is actually not a restriction. Using these assumptions and the binomial nature of the LIBOR rate, we can immediately calculate the following:[14]

$$E\tilde{P}[\ln(L_i) \mid j = u] = p^u \ln(L_i^{uu}) + (1 - p^u)\ln(L_i^{ud}) \tag{13.75}$$

$$= \frac{1}{2}\left[\ln(L_i^{uu}) + \ln(L_i^{ud})\right] \tag{13.76}$$

Replacing this in Eq. (13.71) gives

$$Var(L_i \mid j = u) = E\tilde{P}\left[\left(\ln(L_i) - \frac{1}{2}\left[\ln(L_i^{uu}) + \ln(L_i^{ud})\right]\right)^2 \mid j = u\right] \tag{13.77}$$

Simplifying and regrouping, we obtain

$$Var(L_i \mid j = u) = \left(\frac{1}{2}\left[-\ln(L_i^{uu}) + \ln(L_i^{ud})\right]\right)^2 \tag{13.78}$$

This means that the volatility at time i, in state u, is given by

$$\sigma_i^u = \frac{1}{2}\ln\left[\frac{L_i^{uu}}{L_i^{ud}}\right] \tag{13.79}$$

The result for the down state will be similar:

$$\sigma_i^d = \frac{1}{2}\ln\left[\frac{L_i^{du}}{L_i^{dd}}\right] \tag{13.80}$$

These volatility estimates are functions of the possible values that the LIBOR rate can take during the subsequent period. Hence, given the market quotes on LIBOR volatilities, these formulas can be solved backward to obtain L_i^{uu}, L_i^{ud}. We will do this next.

---

[14]As usual, the first u in the superscript denotes the direction of the node for which the calculation is made, and the second superscript denotes where the LIBOR rate will go from there.

13.3.4 CALIBRATING THE TREE

The elements of the tree can now be calibrated to the observed prices. Using the assumptions concerning (i) the binomial nature for the process L_i, (ii) that $p^u = p^d = 1/2$, and (iii) that the tree is recombining, we get the following five equations:

$$B(t_0, t_1) = \frac{1}{(1 + L_0)} \tag{13.81}$$

$$B(t_0, t_2) = \frac{1}{2}\frac{1}{(1 + L_0)(1 + L_1^u)} + \frac{1}{2}\frac{1}{(1 + L_0)(1 + L_1^d)} \tag{13.82}$$

$$B(t_0, t_3) = \frac{1}{4}\left[\frac{1}{(1 + L_0)(1 + L_1^u)(1 + L_2^{uu})}\right] + \frac{1}{4}\left[\frac{1}{(1 + L_0)(1 + L_1^u)(1 + L_2^{ud})}\right]$$
$$+ \frac{1}{4}\left[\frac{1}{(1 + L_0)(1 + L_1^d)(1 + L_2^{du})}\right] + \frac{1}{4}\left[\frac{1}{(1 + L_0)(1 + L_1^d)(1 + L_2^{dd})}\right] \tag{13.83}$$

$$\frac{1}{2}\ln\left[\frac{L_1^u}{L_1^d}\right] = 0.15 \tag{13.84}$$

$$\frac{1}{2}\ln\left[\frac{L_2^{uu}}{L_2^{ud}}\right] = 0.20 \tag{13.85}$$

Of these equations, the first and second are straightforward. We just applied the risk-neutral measures to price the benchmark bonds. When weighted by these probabilities, the values of future payoffs discounted by the LIBOR rates become Martingales and hence, equal the current price of the appropriate bond, see Figure 13.1.

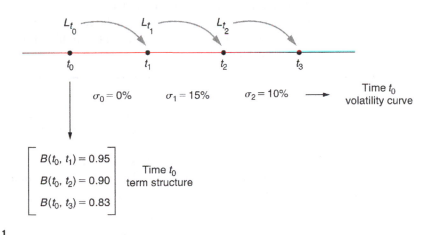

FIGURE 13.1

LIBOR rates and present value of future payoffs.

The third equation represents the pricing function for the bond that matures at time $t = 3$. It is interesting to see what it does. According to the tree used here, there are four possible paths the LIBOR rate can take during $t = 0, 1, 2$. These are

$$\{L_0, L_1^u, L_2^{uu}\} \tag{13.86}$$

$$\{L_0, L_1^u, L_2^{ud}\} \tag{13.87}$$

$$\{L_0, L_1^d, L_2^{du}\} \tag{13.88}$$

$$\{L_0, L_1^d, L_2^{dd}\} \tag{13.89}$$

Due to the way probabilities, p^u and p^d, are picked in this model, each path is equally likely to occur. This gives the third equation.

The last two equations are simply the volatilities *at each node*. We see that as the volatilities depend only on the time index i,

$$\frac{1}{2}\ln\left[\frac{L_2^{uu}}{L_2^{ud}}\right] = 0.20 \tag{13.90}$$

$$\frac{1}{2}\ln\left[\frac{L_2^{du}}{L_2^{dd}}\right] = 0.20 \tag{13.91}$$

which means that

$$L_2^{uu} L_2^{dd} = L_2^{ud} L_2^{du} \tag{13.92}$$

This adds another equation to the five listed earlier and makes the number of unknowns equal to the number of equations. Under the further assumption that the tree is recombining, we have

$$L_2^{uu} L_2^{dd} = (L_2^{ud})^2 \tag{13.93}$$

Now Eqs. (13.81)–(13.85) and (13.92)–(13.93) can be solved for the seven unknowns $\{L_0, L_1^u, L_1^d, L_2^{uu}, L_2^{du}, L_2^{ud}, L_2^{dd}\}$.

The simplest way to solve these equations is to start from $i = 0$ and work forward, since the system is recursive. It is trivial to obtain L_0 from the first equation. The second and fourth equations give L_1^u, L_1^d, and the remaining three equations give the last three unknowns. There is one caveat. The system of equations (13.81)–(13.85) is not linear. Hence, a nonlinear hill-climbing solution procedure must be used to determine the unknowns.

EXAMPLE

The situation is shown in the figure below. There are three periods. Hence, we have three discounts given by the corresponding zero-coupon bond prices and three volatilities. The first volatility is zero, since we do know the value of L_0.

The system in Eqs. (13.81)–(13.85) can be solved recursively. First, we solve for L_0, then for L_1^u and L_1^d, and last for the time $t = 2$ LIBOR rates. The nonlinear equations solved using Mathematica yield the following results:

$$\tag{13.94}$$

We now discuss how BDT trees that give arbitrage-free paths for LIBOR rates or other spot rates can be used.

13.3.5 USES OF THE TREE

Arbitrage-free trees have many uses. (i) We can price baskets of options written on the LIBOR rates L_i. These are called caps and floors and are very liquid. (ii) We can use the tree to price swaps and related derivatives. (iii) Finally, we can use the tree to price *forward* caps, floors, and swaps. We discuss one example below.

13.3.5.1 Application: pricing a cap

A caplet is an option written on a particular LIBOR rate L_{t_i}. A cap rate, L_K, is selected as a strike price, and the buyer of the caplet is compensated if the LIBOR rate moves above L_K, see Figures 13.2 and 13.3. The expiration date is t_i, and the settlement date is t_{i+1}. A caplet then "caps" the interest cost of the buyer. A sequence of consecutive caplets written on $L_{t_i}, L_{t_{i+1}}, \ldots, L_{t_{i+\tau}}$ forms a τ period cap. Similarly, a sequence of consecutive floorlets forms a floor. An interest rate floor is a derivative contract in which the buyer receives payments at the end of each period in which the interest rate is below the agreed strike price.

Suppose we have the following caplet to price:

- The t_i are such that $t_i - t_{i-1} = 12$ months.
- At time t_2, the LIBOR rate L_{t_2} will be observed.
- A notional amount N is selected at time t_0. Let it be given by

$$N = \$1 \text{ million} \tag{13.95}$$

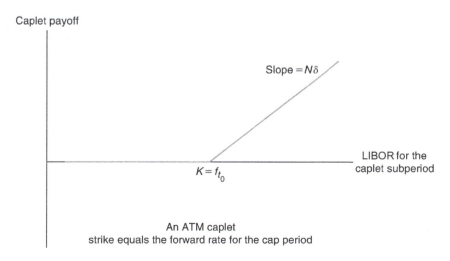

Caplet payoff

Slope $= N\delta$

LIBOR for the
caplet subperiod

$K = f_{t_0}$

An ATM caplet
strike equals the forward rate for the cap period

FIGURE 13.2

Caplet payoff.

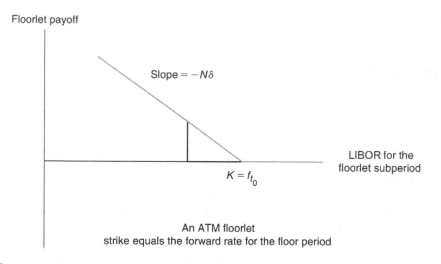

FIGURE 13.3

Floorlet payoff.

- If the LIBOR rate L_{t_2} is in excess of the *cap rate* $L_K = 6.5\%$, the client will receive payoff:

$$C(t_3) = \frac{N(L_{t_2} - L_K)}{100}$$

(13.96)

at time t_3. Otherwise the client is paid nothing.
- For this "insurance," the client pays a premium equal to $C(t_0)$.

The question is how to determine an arbitrage-free value of the caplet premium $C(t_0)$. The fundamental theorem says that the expected value of the expiration date payoff, discounted by the risk-free rate, will equal $C(t_0)$ if we evaluate the expectation using the risk-neutral probability. That is to say, remembering that we have $\delta = 1$,

$$C(t_0) = E_{t_0}^{\tilde{P}} \left[\frac{C(t_3)}{(1 + L_{t_0})(1 + L_{t_1})(1 + L_{t_2})} \right]$$

(13.97)

with expiration payoff

$$C(t_3) = N \max \left[\frac{(L_{t_2} - L_K)}{100}, 0 \right]$$

(13.98)

The pricing of the caplet is done with the BDT tree determined previously. In the example, the tree had four possible trajectories, each occuring with probability 1/4. Using these we can calculate the caplet price.

According to the BDT tree, the caplet ends in-the-money in three of the four trajectories. Calculating the possible payoffs in each case and then dividing by the discount factors, we get the numerical equivalent of the expectation in Eq. (13.98).

$$C_0 = 0.25 \left[\frac{53000}{(1.0526)(1.0639)(1.118)} + \frac{144400}{(1.0526)(1.0473)(1.0793)} + \frac{14400}{(1.0526)(1.0639)(1.0793)} \right]$$

$$= \$16,587$$

We should emphasize that under these circumstances the discount factors are random variables. They *cannot* be taken out of the expectation operator. Also, the center node, which is recombining and, hence, leads to the same value for L_2^{ud} and L_2^{du}, still requires different discount factors since the average interest rate is different across the two middle trajectories.

13.3.5.2 *Some assumptions of the model*

It may be worthwhile to summarize some of the assumptions that were used in the previous discussion.

- The BDT approach is an example of a one-factor model, since the short rate, here represented by the LIBOR rate L_i, is assumed to be the only variable determining bond prices. This means that bond prices are perfectly correlated.
- The distribution of interest rates is lognormal in the limit.
- We made several simplifying assumptions concerning the framework. There were neither taxes, nor any trading costs.

  Needless to say, the procedure also rests on the premise that the original data are arbitrage-free.

13.3.5.3 *Remarks*

The BDT approach may be considered simplistic. Yet, until the advent of the Forward LIBOR Model that we will introduce in the next chapter, market professionals preferred to stay with the BDT approach given the more sophisticated alternatives. A simple model may not fit reality exactly, but may have three important advantages.

1. If the model depends on few parameters, then few parameters have to be determined and the chance of error is less.
2. If the model is simple, a trader or risk manager will accumulate some personal experience in how to adjust for weaknesses of the model.
3. Simple models whose weaknesses are well known and well tried may be better than more sophisticated models with no track record.

We will see that another model with known weaknesses, namely the Black–Scholes model, is preferred by traders for similar reasons.

13.3.6 REAL-WORLD COMPLICATIONS

The BDT model as used in the previous example is, of course, based on symbolic parameters, such as two states, readily available pure discount bond prices, and so on. And as mentioned earlier, it rests on several restrictive assumptions.

In a real-world application, the following additions to the example discussed above need to be made. (i) *Day-count conventions* need to be checked and corrected for, (ii) *settlement* may be done at time $t = 2$, then further discounting may be needed from $t = 3$ to $t = 2$, and (iii) in market applications, caps consisting of several caplets instead of a single caplet are priced.

13.4 APPLICATION 3: QUANTOS

The first two examples of the application of the fundamental theorem shown thus far were essentially numerical. The pricing of quanto contracts constitutes another application of the fundamental theorem. This requires a conceptual discussion. It is a good example of how the techniques introduced in Chapter 12 can be used in *modeling*. The section is also intended to complete the discussion of the financial engineering aspects of quantoed assets that we started in Chapter 10.

A *quantoed foreign asset* makes future payoffs in the domestic currency at a known exchange rate. An exchange rate, x_t, is chosen at initiation, to settle the contract at time T. For example, using quantos, a dollar-based investor could benefit from the potential upside of a foreign stock market, while eliminating the implicit currency exposure to exchange rate movements.

13.4.1 PRICING QUANTOS

The following application of the fundamental theorem starts with pricing a *quanto forward*. Let S_t^* be a foreign stock denominated in the foreign currency. Let x_t be the exchange rate defined as the number of *domestic currency*, per 1 unit of *foreign currency*. The fundamental theorem can be used with the domestic risk-neutral measure $\tilde{P}$ to obtain the time t_0 value of the forward contract:

$$V(t_0) = e^{-r(T-t_0)} E_{t_0}^{\tilde{P}} x_{t_0} \left[S_T^* - F_{t_0} \right] \tag{13.99}$$

F_{t_0} is the time T forward value of the foreign stock. It is measured in foreign currency. Setting the $V(t_0)$ equal to zero gives the forward value F_{t_0}:

$$F_{t_0} = E_{t_0}^{\tilde{P}}[S_T^*] \tag{13.100}$$

Thus, in order to calculate F_{t_0} we need to evaluate the expectation of the foreign currency denominated S_T^* under the *domestic* risk-neutral measure $\tilde{P}$:

$$E_{t_0}^{\tilde{P}}[S_T^*]. \tag{13.101}$$

The fact that the state prices, Q^i, in the fundamental theorem are denominated in the domestic currency, whereas S_t^* is denominated in the foreign currency, makes this a nontrivial exercise.

But, if used judiciously, the fundamental theorem can still be exploited for obtaining the expectation in Eq. (13.101). To maintain continuity, we use the simple framework developed in Chapter 12. In particular, we assume that there are only two periods, t_0 and T, with n states of the world at time T. The notation remains the same.

Consider the matrix equation of the fundamental theorem for three assets. The first is the domestic savings account B_t which starts at 1 dollar and earns the domestic annual rate r. The second is a foreign savings account, B_t^* which starts with 1 unit of the foreign currency and earns the foreign interest rate r^*. These interest rates are assumed to be constant. The foreign currency has dollar value x_{t_0} at time t_0. Finally, we have the foreign stock, $S_{t_0}^*$.

Putting these into the matrix equation implied by the fundamental theorem we get

$$\begin{pmatrix} 1 \\ x_{t_0} \\ x_{t_0} S_{t_0}^* \end{pmatrix} = \begin{pmatrix} 1 + r(T - t_0) & \cdots & 1 + r(T - t_0) \\ x_T^1[1 + r^*(T - t_0)] & \cdots & x_T^n[1 + r^*(T - t_0)] \\ x_T^1 S_T^{*1} & \cdots & x_T^n S_T^{*n} \end{pmatrix} \begin{pmatrix} Q^1 \\ \cdots \\ \cdots \\ \cdots \\ Q^k \end{pmatrix} \qquad (13.102)$$

Here, x_T^i and S_T^{*i} have i superscripts because their time T value depends on the state that is realized at that time. This system involves domestic state prices, and therefore the value of the foreign stock $S_{t_0}^*$ is converted into domestic currency by multiplying with x_{t_0}. Q^i, $i = 1, \ldots, n$ are the state prices assumed to be known and positive.

We start with two results that are obtained by the following methods shown in Chapter 12. Define the domestic risk-neutral measure $\tilde{P}$ as

$$\tilde{p}_i = (1 + r(T - t_0))Q^i \qquad (13.103)$$

Then, from the third row of Eq. (13.102) we obtain the equality,

$$x_{t_0} S_{t_0}^* = \frac{1}{1 + r(T - t_0)} \sum_{i=1}^n x_T^i S_T^{*i} \tilde{p}_i \qquad (13.104)$$

This means that

$$x_{t_0} S_{t_0}^* = \frac{1}{1 + r(T - t_0)} E_{t_0}^{\tilde{P}}\left[x_T S_T^*\right] \qquad (13.105)$$

Using the second row of the system in Eq. (13.102), we obtain a similar equality for the exchange rate. After switching from Q^i to the risk-neutral probabilities $\tilde{P}$,

$$x_{t_0} = \frac{1 + r^*(T - t_0)}{1 + r(T - t_0)} E_{t_0}^{\tilde{P}}[x_T] \qquad (13.106)$$

We use Eqs. (13.105) and (13.106) in calculating the desired quantity, $E_{t_0}^{\tilde{P}}[S_T^*]$. We know from elementary statistics that

$$\text{Cov}(S_T^*, x_T) = E_{t_0}^{\tilde{P}}[S_T^* x_T] - E_{t_0}^{\tilde{P}}[S_T^*]E_{t_0}^{\tilde{P}}[x_T] \qquad (13.107)$$

Rearranging, we can write:

$$E_{t_0}^{\tilde{P}}\left[S_T^*\right] = \frac{E_{t_0}^{\tilde{P}}[S_T^* x_T] - \mathrm{Cov}(S_T^*, x_T)}{E_{t_0}^{\tilde{P}}[x_T]} \tag{13.108}$$

We substitute in the numerator from Eq. (13.105) and in the denominator from Eq. (13.106) to obtain

$$E_{t_0}^{\tilde{P}}\left[S_T^*\right] = \frac{[1 + r(T - t_0)]x_{t_0} S_{t_0}^* - \mathrm{Cov}(S_T^*, x_T)}{x_{t_0}[(1 + r(T - t_0))/(1 + r*(T - t_0))]} \tag{13.109}$$

We prefer to write this in a different form using the correlation coefficient denoted by ρ, and the percentage annual volatilities of x_t and S_t^* denoted by σ_x, σ_s, respectively. Let

$$\mathrm{Cov}(S_T^*, x_T) = \rho \sigma_x \sigma_s (x_{t_0} S_{t_0}^*)(T - t_0) \tag{13.110}$$

The expression in Eq. (13.109) becomes

$$E_{t_0}^{\tilde{P}}\left[S_T^*\right] = \frac{1 + r*(T - t_0)}{1 + r(T - t_0)}[1 + (r - \rho \sigma_x \sigma_s)(T - t_0)]S_{t_0}^* \tag{13.111}$$

We can approximate this as[15]

$$E_{t_0}^{\tilde{P}}[S_T^*] \cong [1 + (r* - \rho \sigma_x \sigma_s)(T - t_0)]S_{t_0}^* \tag{13.112}$$

This gives the foreign currency denominated price of the quanto forward in the domestic currency:

$$F_{t_0} \cong [1 + (r* - \rho \sigma_x \sigma_s)(T - t_0)]S_{t_0}^* \tag{13.113}$$

The present value of this in domestic currency will be the spot value of the quanto:

$$V_{t_0} = x_{t_0} \frac{1}{1 + r(T - t_0)}[1 + (r* - \rho \sigma_x \sigma_s)(T - t_0)]S_{t_0}^* \tag{13.114}$$

We can also write this relationship by reinterpreting the interest rates as continuously compounded rates:

$$V_{t_0} = e^{-r(T - t_0)} e^{(r* - \rho \sigma_x \sigma_s)(T - t_0)} x_{t_0} S_{t_0}^* \tag{13.115}$$

According to this expression, the value of the quanto feature depends on the sign of the correlation between exchange rate movements and the value of the foreign stock. If this correlation is positive, then the quanto feature is negatively priced. If the correlation is negative, the quanto feature has positive value.[16] If the correlation is zero, the quanto feature has zero value.

---

[15] We are using the approximation

$$\frac{1}{1 + z} = 1 - z$$

for small z. In the approximation, we ignore all terms of order $(T - t_0)^2$ and higher.
[16] Suppose the correlation is positive. Then, when foreign stock's value goes up, in general, the foreign currency will also go up. The quanto eliminates this opportunity from the point of view of a stockholder, and hence, has a negative value and the quantoed asset is cheaper.

13.4.2 THE PDE APPROACH

Our next example shows how the fundamental theorem can be used to obtain partial differential equations (PDE) for quanto instruments. The treatment will be in continuous time and is essentially heuristic. Consider the same two-currency environment. The domestic and foreign savings deposits are denoted by B_t and B_t^*, respectively. The corresponding *continuously compounded* rates are assumed to be constant, for simplicity, at r and r^*. This means that the savings account values increase incrementally according to the following (ordinary) differential equations:

$$dB_t = rB_t dt \quad t \in [0, \infty) \tag{13.116}$$

$$dB_t^* = r^* B_t^* dt \quad t \in [0, \infty) \tag{13.117}$$

Let x_t be the exchange rate expressed as the domestic currency price of 1 unit of foreign currency. x_t satisfies the SDE:

$$dx_t = \mu_x x_t dt + \sigma_x x_t dW_{1t} \quad t \in [0, \infty) \tag{13.118}$$

under the appropriate Martingale measure.

First we obtain the exchange rate dynamics under the B_t normalization. Note that B_t^* is a traded asset and its price in domestic currency is $x_t B_t^*$. According to the results obtained in Chapter 12, with B_t normalization, and the corresponding risk-neutral measure $\tilde{P}$, the ratio

$$\frac{x_t B_t^*}{B_t} \tag{13.119}$$

should behave as a Martingale. This means that the drift of the implied dynamics should be zero. Taking total derivatives,[17]

$$E_t^{\tilde{P}} \left[d \frac{x_t B_t^*}{B_t} \right] = E_t^{\tilde{P}} \left[\frac{B_t^*}{B_t} dx_t + \frac{x_t}{B_t} dB_t^* - \frac{x_t B_t^*}{B_t^2} dB_t \right] = 0 \tag{13.120}$$

Replacing from Eqs. (13.116), (13.117), and (13.118), we obtain[18]

$$\frac{B_t^*}{B_t} \mu_x x_t dt + \frac{x_t}{B_t} r^* B_t^* dt - \frac{x_t B_t^*}{B_t^2} rB_t dt = \frac{x_t B_t^*}{B_t} \left[\mu_x + r^* - r \right] dt \tag{13.121}$$

In order for this drift to be zero, we must have, under $\tilde{P}$, at all times:

$$\mu_x + r^* - r = 0 \tag{13.122}$$

Replacing this drift in Eq. (13.118) gives the exchange rate dynamics under the $\tilde{P}$:

$$dx_t = (r - r^*)x_t dt + \sigma_x x_t dW_{1t} \quad t \in [0, \infty) \tag{13.123}$$

Next, consider the $\tilde{P}$ dynamics of the foreign stock S_t^*.

$$dS_t^* = \mu_s S_t^* dt + \sigma_x S_t^* dW_{2t} \quad t \in [0, \infty) \tag{13.124}$$

---

[17]We are using differentials inside an expectation operator. We emphasize that this is heuristic since stochastic differentials are only symbolic ways of expressing some limits.

[18]For readers familiar with stochastic calculus, the second-order terms from Ito's Lemma are zero since x_t enters the formula linearly. Also, B_t is deterministic.

Under the B_t normalization, the domestic currency value of the foreign stock should behave as a Martingale. Applying Ito's Lemma:

$$E_t^{\tilde{P}}\left[d\frac{x_t S_t^*}{B_t}\right] = E_t^{\tilde{P}}\left[\frac{S_t^*}{B_t}dx_t + \frac{x_t}{B_t}dS_t^* - \frac{x_t S_t^*}{B_t^2}dB_t + \frac{dx_t dS_t^*}{B_t}\right] = 0 \tag{13.125}$$

Replacing the differentials and simplifying, we obtain

$$\frac{S_t^*}{B_t}(r - r^*)x_t + \frac{x_t}{B_t}\mu_s S_t^* - \frac{x_t S_t^*}{B_t^2}rB_t + \frac{\rho\sigma_x\sigma_s x_t S_t^*}{B_t}$$

$$= \frac{x_t S_t^*}{B_t}\left[(r - r^*) + \mu_s - r + \rho\sigma_x\sigma_s\right] = 0 \tag{13.126}$$

In order for this drift to be zero we must have, under $\tilde{P}$, at all times:

$$\mu_S = r^* - \rho\sigma_x\sigma_s \tag{13.127}$$

This gives the arbitrage-free stock price dynamics:

$$dS_t^* = (r^* - \rho\sigma_x\sigma_s)S_t^* dt + \sigma_x S_t^* dW_{2t} \quad t \in [0, \infty) \tag{13.128}$$

These dynamics imply that:

$$E_t^{\tilde{P}}[S_T^*] = e^{(r^* - \rho\sigma_x\sigma_s)(T-t)}S_t^* \tag{13.129}$$

as derived in the previous section. Here the interest rates r and r^* should be interpreted as continuously compounded rates. In the previous section, they were actuarial rates for the period $T - t_0$.

13.4.2.1 A PDE for quantos

Finally, using these results we can obtain a PDE for an arbitrary quanto asset written on a risk associated with a foreign economy. Let this foreign currency denominated asset be denoted by S_t^*. Let V_t denote the time t value of the quanto,

$$V(t) = x_t V(S_t^*, \ t) \tag{13.130}$$

$V(.)$, being a pricing function of the asset, needs to be determined. x_t is the initial exchange rate written in the quanto contract and, hence, $V(t)$ is expressed in domestic currency terms. Under B_t normalization, $V(t)$ should behave as a Martingale. Applying Ito's Lemma we obtain:[19]

$$E_t^{\tilde{P}}\left[d\frac{V(t)}{B_t}\right] = E_t^{\tilde{P}}\left[\frac{V_t}{B_t}dt + \frac{V_s}{B_t}dS_t^* - \frac{V}{B_t^2}dB_t + \frac{1}{2}\frac{V_{ss}\sigma_s^2(S_t^*)^2}{B_t}dt\right] = 0 \tag{13.131}$$

Replacing the stochastic differentials and simplifying yields the implied PDE,

$$V_t + (r^* - \rho\sigma_x\sigma_s)S_t^* V_s + \frac{1}{2}V_{ss}\sigma_s^2(S_t^*)^2 - rV = 0 \tag{13.132}$$

---

[19]In this expression, V_t is the partial derivative of $V(.)$ with respect to t and should not be confused with $V(t)$. Also, Chapter 9 contains a brief appendix that discusses Ito's Lemma. For related heuristics, see Hirsa and Neftci (2013). For a formal treatment, see Øksendal (2010).

with the terminal condition:

$$V(T) = x_t V(S_T^*, T) \tag{13.133}$$

We apply this PDE to two special cases.

13.4.3 QUANTO FORWARD

Suppose we know that a quanto forward has the value,

$$V(t) = q(t)S_t^* \tag{13.134}$$

but that the time-dependent function $q(t)$ is unknown. The PDE derived in the previous section can be used to solve for $q(t)$. Differentiating Eq. (13.134), we get the partial derivatives:

$$V_t = \frac{\partial q(t)}{\partial t} S_t^* = \dot{q} S_t^* \tag{13.135}$$

$$V_s = q(t) \tag{13.136}$$

$$V_{ss} = 0 \tag{13.137}$$

We replace these in the PDE for the quanto,

$$\dot{q} S_t^* + (r^* - \rho \sigma_x \sigma_s) q(t) S_t^* - r q(t) S_t^* = 0 \tag{13.138}$$

with the terminal condition,

$$V(T) = x_t S_T^* \tag{13.139}$$

Eliminating the common S_t^* terms, this ordinary differential equation can be solved for $q(t)$:

$$q(t) = x_t e^{(r^* - \rho \sigma_x \sigma_s)(T-t)} \tag{13.140}$$

This is the same result as obtained earlier.

13.4.4 QUANTO OPTION

Suppose the payoff $V(T)$ of a quanto asset relates to the payoff of a European call on a foreign stock S_t^*:

$$V_T = x_t \max[S_T^* - K^*, 0] \tag{13.141}$$

Here K^* is a foreign currency denominated strike price and T is the expiration date. Then the PDE derived in Eq. (13.132) can be solved using the equivalence with the Black–Scholes formula to obtain the pricing equation for a European quanto call:

$$C(t) = x_t \left[S_t^* e^{(r^* - r - \rho \sigma_x \sigma_s)(T-t)} N(b_1) - K^* e^{-r(T-t)} N(b_2) \right] \tag{13.142}$$

where

$$b_1 = \frac{(\ln S_t^* / K^*) + (r^* - \rho \sigma_x \sigma_s + 0.5 \sigma_s^2)(T - t)}{\sigma_s \sqrt{T - t}} \tag{13.143}$$

$$b_2 = b_1 - \sigma_s \sqrt{T - t} \tag{13.144}$$

The value of the call will be measured in domestic currency.

13.4.4.1 Black–Scholes and dividends

We now explain how to trick the PDE in Eq. (13.132) in order to arrive at the Black–Scholes type formula for the simple quantoed equity option shown above. To do this, we need the equivalent of the Black–Scholes formula in the case of a constant rate of dividends paid by the underlying stock during the life of the option.

Standard derivations in the Black–Scholes world will give the European call premium on a stock, S_t, that pays dividends at a constant rate Q as,

$$C(t) = e^{-Q(T-t)} S_t N(\tilde{d}_1) - K e^{-(r)(T-t)} N(\tilde{d}_2) \tag{13.145}$$

with

$$\tilde{d}_1 = \frac{(\ln S_t/K) + (r - Q + 0.5\sigma_s^2)(T-t)}{\sigma_s \sqrt{T-t}} \tag{13.146}$$

$$\tilde{d}_s = \tilde{d}_1 - \sigma_s \sqrt{T-t} \tag{13.147}$$

where S_t is the dividend paying stock.

Now, note that we can write the PDE in Eq. (13.138) as

$$V_t + (r - Q)S_t^* V_s + \frac{1}{2} V_{ss} \sigma_s^2 (S_t^*)^2 - rV = 0 \tag{13.148}$$

where Q is treated as a dividend yield, and is given by

$$Q = r - r^* + \rho \sigma_x \sigma_s \tag{13.149}$$

We can then use this Q in the standard Black–Scholes formula with a known dividend yield to get the quantoed call premium.

13.4.5 HOW TO HEDGE QUANTOS

Quanto contracts require dynamic hedging. The dealer would form a portfolio made of the underlying foreign asset, the foreign currency (or, better, an FX-forward), and the domestic lending and borrowing. The weights of this portfolio would be adjusted dynamically, so that the portfolio replicates the changes in value of the quanto contract. The trading gains (losses) realized from these hedge adjustments form the basis for the quanto premium or discount.

13.4.6 REAL-LIFE CONSIDERATIONS

The discussion of quantoed assets in this section has been in a simple, abstract, and unrealistic world. We used the following assumptions, among others: (i) the underlying processes were assumed to be lognormal, so that the implied SDEs were *geometric*. (ii) The correlation coefficient and the volatility parameters were assumed to be *constant* during the life of the option. (iii) Similarly, *interest rates* were assumed to be constant, although the corresponding exchange rate was stochastic.

These assumptions are not satisfied in most real-world applications. Especially important for quantos, the correlation coefficients between exchange rates and various risk factors are known to be quite unstable. The models discussed in this section therefore need to be regarded as a conceptual application of the fundamental theorem. They do not provide an algorithm for pricing real-world quantos.

13.5 CONCLUSIONS

This chapter dealt with three applications of the fundamental theorem of asset pricing. In general, a financial engineer needs to use such approaches when static replication of the assets is not possible. Mark-to-market requirements or construction of new products often requires calculating arbitrage-free prices internally without having recourse to synthetics that can be put together using liquid prices observed in the markets. The methods outlined in this chapter show some standard ways of doing this.

SUGGESTED READING

There are several sources the reader may consult to learn more on the methods introduced in this chapter via some simple examples. One of our preferred sources is **Clewlow and Strickland** (1998), which provides some generic codes for computer applications as well. A recent book that deals with the topic of Monte Carlo in finance is **Jackel** (2002). The series of articles referenced in **Avellaneda et al.** (2001) provides an in-depth discussion of calibration issues. Finally, the original **Black et al.** (1990) model is always an illuminating reading on the BDT model. For quanto assets, and related discussion, consider **Hull** (2014). **Wilmott** (2006) is very useful for learning further application of the techniques presented here.

EXERCISES

1. (Black–Derman–Toy model.) You observe the following default-free discount bond prices $B(t, T_i)$, where time is measured in years:

$$B(0, 1) = 95, B(0, 2) = 93, B(0, 3) = 91, B(0, 4) = 89 \qquad (13.150)$$

These prices are assumed to be arbitrage-free. In addition, you are given the following cap-floor volatilities:

$$\sigma(0, 1) = 0.20, \sigma(0, 2) = 0.25, \sigma(0, 3) = 0.20, \sigma(0, 4) = 0.18 \qquad (13.151)$$

where $\sigma(t, T_i)$ is the (constant) volatility of the LIBOR rate L_{T_i} that will be observed at T_i with tenor of 1 year.

a. Using the Black–Derman–Toy model, calibrate a binomial tree to these data.

b. Suppose you are given a bond call option with the following characteristics. The underlying, $B(2, 4)$, is a two-period bond, expiration $T = 2$, strike $K_B = 93$. You know that the BDT tree is a good approximation to arbitrage-free LIBOR dynamics. What is the *forward* price of $B(2, 4)$?

c. Calculate the arbitrage-free value of this call option using the BDT approach.

2. (Exchange rates and LIBOR rates.) You know that the euro/dollar exchange rate e_t follows the real-world dynamics:

$$de_t = \mu dt + 0.15 e_t dW_t \qquad (13.152)$$

The current value of the exchange rate is $e_o = 1.1015$. You also know that the price of a 1-year USD discount bond is given by

$$B(t, t+1)^{US} = 98.93 \qquad (13.153)$$

while the corresponding euro-denominated bond is priced as

$$B(t, t+1)^{EU} = 98.73 \qquad (13.154)$$

Both of these prices are arbitrage-free and there is no credit risk.
a. What are the 1-year LIBOR rates in these two currencies at time t?
b. What are the continuously compounded interest rates r_t^{US}, r_t^{EUR}?
c. Obtain the arbitrage-free dynamics of the e_t. In particular, state clearly whether we need to use continuously compounded rates or LIBOR rates to do this.
d. Is there a continuous time dynamic that can be written using the LIBOR rates?

3. (European option.) Consider again the data given in the previous question.
a. Use $\Delta = 1$ year to discretize the system.
b. Generate five sets of standard normal random numbers with five random numbers in each set. How do you know that these five trajectories are arbitrage-free?
c. Calculate the value of the following option using these trajectories. The strike is 0.95, the expiration is 3 years, and the European style applies.

4. (European FX option.) Suppose you know that the current value of the peso–dollar exchange rate is 3.75 pesos per dollar. The yearly volatility of the Mexican peso is 20%.
   The Mexican interest rate is 8%, whereas the US rate is 3%. You will price a dollar option written on the Mexican peso. The option is of European style and has a maturity of 270 days. All processes under consideration are known to be geometric.
a. Price this option using a standard Monte Carlo model. You will select the number of series, the size of the approximating time intervals, and other parameters of the Monte Carlo exercise.
b. Now assume that Mexico's foreign currency reserves follow a geometric SDE with a volatility of 10% and a drift coefficient of 5% a year. The current value of reserves is USD7 billion. If reserves fall below USD6 billion, there will be a one-shot devaluation of 100%. Is this information important for pricing the option? Explain.
c. Use importance sampling to reprice the option. Your pricing is supposed to incorporate the risk of devaluation.

EXCEL EXERCISES

5. (European Option.)
   Write a VBA program to simulate $M = 100$ stock prices using a Monte Carlo technique to calculate the prices of European Call and Put options based on the following data:
   • $S(0) = 100; K = 105; T = 1; r = 8\%; \sigma = 50\%$

Gradually increase the value of M and report the observed resulting price of the options.

6. (Digital Currency Options.) Write a VBA program to simulate $M = 100$ stock prices using a Monte Carlo technique to calculate the prices of digital call and put options FX options as discussed in the text. Use the following parameters:
    - $S(0) = \$1.54$; $K = \$1.58$; $T = 1$; $r = 8\%$; $r_f = 6\%$; $\sigma = 30\%$; payoff $R = \$10$
      Gradually increase the value of M and report the observed price of the options.

7. (Barrier Option.) Write a VBA program to simulate $M = 100$ stock prices using a Monte Carlo technique to calculate the price of Barrier down-and-out and down-and-in call options based on the following data:
    - $S(0) = 100$; $K = 110$; $T = 1$; $r = 8\%$; $\sigma = 50\%$; $H = 90$
      Gradually increase the value of M and report the observed price of the options.

MATLAB EXERCISES

8. (European Options.) Write a MATLAB program to document the efficiency of a Monte Carlo approach to the estimation of European Call and Put option prices based on the following data:
    - $S(0) = 100$; $K = 105$; $T = 1$; $r = 8\%$; $\sigma = 30\%$
      Plot a graph of estimated prices as a function of the number of stock price simulations.

9. (Barrier Option.) Write a MATLAB program to document the efficiency of a Monte Carlo approach to the estimation of the price of Down-and-Out and Down-and-In Call options based on the following data:
    - $S(0) = 100$; $K = 110$; $T = 1$; $r = 8\%$; $\sigma = 30\%$; $H = 90$
    - Plot a graph of estimated prices as a function of the number of stock price simulations.

10. (Digital Currency Option.) Write a MATLAB program to observe the efficiency of Monte Carlo technique to estimate the price of Digital Call and Put price with the following data:
    - $S(0) = \$1.54$; $K = \$1.58$; $T = 1$; $r = 8\%$; $r_f = 6\%$; $\sigma = 30\%$; $R = \$10$
      Plot the graph of estimated price v/s the no. of stock price simulation.

11. (BDT Model Calibration.) Write a MATLAB program to calibrate the BDT model based on the following data on bond prices and implied volatilities
    - $B(t_0, t_1) = 0.95$; $B(t_0, t_2) = 0.93$; $B(t_0, t_3) = 0.91$; $B(t_0, t_4) = 0.89$
    - $\sigma(0,1) = 20\%$; $\sigma(0,2) = 25\%$; $\sigma(0,3) = 20\%$; $\sigma(0,4) = 18\%$
      Draw the LIBOR tree based on the output results.

FIXED INCOME ENGINEERING

14.1 INTRODUCTION

This chapter extends the discussion of swap-type instruments and outlines a simple framework for fixed-income security pricing. Term structure modeling is treated within this framework. The chapter also introduces the recent models that are becoming a benchmark in this sector.

Until the late 1990s, short-rate modeling was the most common approach in pricing and risk-managing fixed-income securities. The publication in 1992 of the Heath—Jarrow—Merton (HJM) approach enabled arbitrage-free modeling of multifactor-driven term structure models, but markets continued to use short-rate modeling. A few years ago the situation changed. The Forward LIBOR or Brace—Gatarek—Musiela (BGM) Model published in 1997 became the market standard for pricing and risk management.

During the GFC, the LIBOR rate rose significantly and diverged from the overnight indexed swap (OIS) rate raising questions about whether LIBOR is an appropriate riskless discount. In Chapter 24, we will discuss recent models and market developments such as the use of OIS curve as a risk-free curve for discounting. The use of the OIS curve implies that more than one zero curve must be modeled for the purpose of pricing derivatives. The LIBOR curve is used to determine payoffs and the OIS zero curve is used for discounting. This chapter will discuss the Forward LIBOR Model and we will deal with more complex and recent issues such as approaches that simultaneously or separately model the LIBOR and the OIS curve in Chapter 24.

This chapter will approach the issues from a practical point of view using *swap markets* and *swap derivatives* as a background. We are interested in providing a framework for analyzing the mechanics of swaps and swap derivatives, for decomposing them into simpler instruments, and for constructing synthetics. Recent models of fixed-income modeling can then be built on this foundation very naturally.

It is worth reviewing the basic principles of swap engineering laid out in Chapter 4. First of all, swaps are almost always designed such that their value at initiation is zero. This is a characteristic of modern swap-type "spread instruments," and there is no surprise here. Second, what makes the value of the swap equal to zero is a spread or an interest rate that is chosen with the *purpose* that the initial value of the swap vanishes. Third, swaps encompass more than one settlement date. This means that whatever the value of the swap rate or swap spread, these will in the end be some sort of "average of shorter term floating rates or spreads." This not only imposes simple arbitrage conditions on relevant market rates but also provides an opportunity to trade the volatility associated with such averages through the use of options on swaps. Since swaps are very liquid, they form an excellent underlying for *swaptions*. Swaptions, in turn, are related to interest rate volatilities for the underlying subperiods, which will relate to cap/floor volatilities. This structure is conducive to

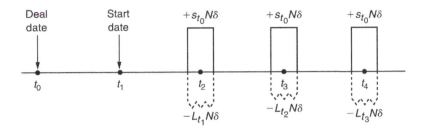

FIGURE 14.1

Three-period forward swap.

designing and understanding more complex swap products such as constant maturity swaps (CMS). The CMS is used as an example for showing the advantages of the Forward LIBOR Model.

Finally, the chapter will further use the developed framework to illustrate the advantages of measure change technology. Switching between various T-forward measures, we show how convexity effects can be calculated.

Most of the discussion will center on a three-period swap first, and then generalize the results. We begin with this simple example, because with a small number of cash flows the analysis becomes more manageable and easier to understand. Next, we lay out a somewhat more technical framework for engineering fixed-income instruments. Eventually, this is developed into the Forward LIBOR Model. Within our framework, measure changes using Girsanov-type transformations emerge as fundamental tools of financial engineering. The chapter discusses how measures can be changed *sequentially* during a numerical pricing exercise as was done in the simulation of the Forward LIBOR Model. These tools are then applied to CMS, which are difficult to price with traditional models.

14.2 A FRAMEWORK FOR SWAPS

We work with *forward* fixed payer interest rate swaps and their "spot" equivalent. These are vanilla products in the sense that contracts are predesigned and homogeneous. They are liquid, the bid−ask spreads are tight, and every market player is familiar with their properties and related conventions.

To simplify the discussion we work with a three-period forward swap, shown in Figure 14.1. It is worth repeating the relevant parameters again, given the somewhat more technical approach the chapter will adopt.

1. The notional amount is N, and the tenor of the underlying LIBOR rate is δ, which represents a proportion of a calendar year. As usual, if a year is denoted by 1, then δ will be 1/4 in the case of 3-month LIBOR.
2. The swap *maturity* is three periods. The swap ends at time $T = t_4$. The swap contract is signed at time t_0 but starts at time t_1, hence the term *forward* swap is used.[1]

---

[1] In the case of the spot swap that we will use, the swap will start at time t_0 and settle three times.

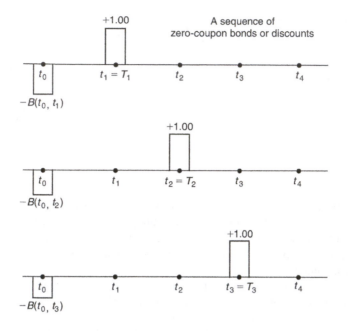

FIGURE 14.2

Payoff diagrams for three default-free pure discount bonds.

3. The dates $\{t_1, t_2, t_3\}$ are *reset dates* where the relevant LIBOR rates L_{t_1}, L_{t_2}, and L_{t_3} will be determined.[2] These dates are δ time units apart.

4. The dates $\{t_2, t_3, t_4\}$ are *settlement dates* where the LIBOR rates L_{t_1}, L_{t_2}, and L_{t_3} are used to exchange the floating cash flows, $\delta N L_{t_i}$ against the fixed $\delta N s_{t_0}$ at each t_{i+1}. In this setup, the time that passes until the start of the swap, $t_1 - t_0$, need not equal δ. However, it may be notationally convenient to assume that it does.

　　Our purpose is to provide a systematic framework in which the risk management and pricing of such swaps and the instruments that build on them can be done efficiently. That is, we discuss a technical framework that can be used for running a swap and swap derivatives book.

　　Swaps are one major component of a general framework for fixed-income engineering. We need *two* additional tools. These we introduce using a simple example again. Consider Figure 14.2, where we show payoff diagrams for three default-free pure *discount bonds*. The current price, $B(t_0, T_i)$, of these bonds is paid at t_0 to receive 1 dollar in the same currency at maturity dates $T_i = t_i$. Given that these bonds are default-free, the time t_i payoffs are *certain* and the price $B(t_0, T_i)$ can be considered as the value today of 1 dollar to be received at time t_i. This means they are, in fact, the relevant discount factors, or in market language, simply *discounts* for t_i. Note that as

$$T_1 < T_2 < T_3 < T_4 \tag{14.1}$$

[2]That is, determined by some objective and predefined authority such as ICE Benchmark Administration Limited (IBA).

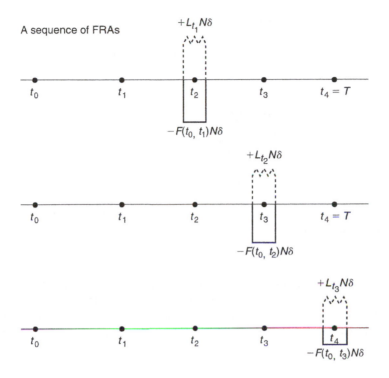

FIGURE 14.3

Cash flow diagrams of three FRAs paid in arrears.

bond prices must satisfy, regardless of the slope of the yield curve:[3]

$$B(t_0, T_1) > B(t_0, T_2) > B(t_0, T_3) > B(t_0, T_4) \tag{14.2}$$

These prices can be used as discount factors to calculate *present values* of various cash flows occurring at future settlement dates t_i. They are, therefore, quite useful in successive swap settlements and form the second component in our framework.

The third component of the fixed-income framework is shown in Figure 14.3. Here, we have the cash flow diagrams of three forward rate agreements (FRAs) paid in arrears. The FRAs are, respectively, $t_1 \times t_2$, $t_2 \times t_3$, and $t_3 \times t_4$. For each FRA, a floating (random) payment is made against a known (fixed) payment for a net cash flow of

$$\left[L_{t_i} - F(t_0, t_i) \right] N\delta \tag{14.3}$$

at time t_{i+1}. Here, the $F(t_0, t_i)$ is the forward rate of a fictitious forward loan contract signed at time t_0. The forward loan comes into effect at t_i and will be paid back at time $t_{i+1} = t_i + \delta$. We note that the fixed payments $N\delta \, F(t_0, t_i)$ are not the same across the FRAs. Although all FRA rates are

---

[3]As seen earlier, if we short *one* longer-term bond to fund a long position in *one* short-term bond, we would not have enough money.

known at time t_0, they will, in general, not equal each other or equal the payment of the fixed swap leg, $\delta N s_{t_0}$.

We can now use this framework to develop some important results and then apply them in financial engineering.

14.2.1 EQUIVALENCE OF CASH FLOWS

The first financial engineering rule that we discuss in this chapter is associated with the perceived *equivalence* of cash flows. In Figure 14.3, there is a strip of floating cash flows:

$$\left\{ L_{t_1} N\delta, L_{t_2} N\delta, L_{t_3} N\delta \right\} \tag{14.4}$$

and, given observed liquid prices, the market is willing to exchange these random cash flows against the known (fixed) cash flows:

$$\{ F(t_0, t_1) N\delta, F(t_0, t_2) N\delta, F(t_0, t_3) N\delta \} \tag{14.5}$$

According to this, if these FRAs are liquid at time t_0, the known cash flow sequence in Eq. (14.5) is perceived by the markets as the *correct* exchange against the unknown, floating payments in Eq. (14.4). If we then consider the swap cash flows shown in Figure 14.1, we note that exactly the same floating cash flow sequence as in Eq. (14.4) is exchanged for the known and fixed swap leg

$$\left\{ s_{t_0} N\delta, s_{t_0} N\delta, s_{t_0} N\delta \right\} \tag{14.6}$$

The settlement dates are the same as well. In both exchanges, neither party makes any upfront payments at time t_0. We can therefore combine the two exchanges at time t_0 and obtain the following result.

The market is willing to exchange the fixed and *known* cash flows

$$\left\{ s_{t_0} N\delta, s_{t_0} N\delta, s_{t_0} N\delta \right\} \tag{14.7}$$

against the variable *known* cash flows:

$$\{ F(t_0, t_1) N\delta, F(t_0, t_2) N\delta, F(t_0, t_3) N\delta \} \tag{14.8}$$

at no additional time t_0 compensation.

This has an important implication. It means that the time t_0 values of the two cash flow sequences are the same. Otherwise, one party would demand an initial cash payment. Given that the cash flows are known as of time t_0, their equivalence provides an *equation* that can be used in pricing, as we will see next. This argument will be discussed further using the Forward LIBOR Model.

14.2.2 PRICING THE SWAP

We have determined two *known* cash flow sequences the market is willing to exchange at no additional cost. Using this information, we now calculate the time t_0 values of the two cash flows. To do this, we use the second component of our framework, namely, the discount bond prices given in Figure 14.2.

Suppose the pure discount bonds with arbitrage-free prices $B(t_0, t_i)$, $i = 1, 2, 3, 4$ are liquid and actively traded. We can then use $\{B(t_0, t_2), B(t_0, t_3), B(t_0, t_4)\}$ to value cash flows settled at times t_2, t_3, and t_4, respectively.[4] In fact, the time t_0 value of the sequence of cash flows,

$$\{F(t_0, t_1)N\delta, F(t_0, t_2)N\delta, F(t_0, t_3)N\delta\} \tag{14.9}$$

is given by multiplying each cash flow by the *discount* factor that corresponds to that particular settlement date and then adding. We use the default-free bond prices as our discount factors and obtain the value of the fixed FRA cash flows

$$
\begin{aligned}
&B(t_0, t_2)F(t_0, t_1)N\delta + B(t_0, t_3)F(t_0, t_2)N\delta + B(t_0, t_4)F(t_0, t_3)N\delta \\
&= [B(t_0, t_2)F(t_0, t_1) + B(t_0, t_3)F(t_0, t_2) + B(t_0, t_4)F(t_0, t_3)]N\delta
\end{aligned} \tag{14.10}
$$

The time t_0 value of the fixed swap cash flows can be calculated similarly

$$B(t_0, t_2)s_{t_0}N\delta + B(t_0, t_3)s_{t_0}N\delta + B(t_0, t_4)s_{t_0}\delta N = [B(t_0, t_2) + B(t_0, t_3) + B(t_0, t_4)]\delta Ns_{t_0} \tag{14.11}$$

Now, according to the argument in the previous section, the values of the two cash flows must be the same.

$$
\begin{aligned}
&[B(t_0, t_2)F(t_0, t_1) + B(t_0, t_3)F(t_0, t_2) + B(t_0, t_4)F(t_0, t_3)]\delta N \\
&= [B(t_0, t_2) + B(t_0, t_3) + B(t_0, t_4)]\delta Ns_{t_0}
\end{aligned} \tag{14.12}
$$

This equality has at least two important implications. First, it implies that the value of the swap at time t_0 is zero. Second, note that equality can be used as an *equation* to determine the value of *one* unknown. As a matter of fact, pricing the swap means determining a value for s_{t_0} such that the equation is satisfied. Taking s_{t_0} as the unknown, we can rearrange Eq. (14.12), simplify, and obtain

$$s_{t_0} = \frac{B(t_0, t_2)F(t_0, t_1) + B(t_0, t_3)F(t_0, t_2) + B(t_0, t_4)F(t_0, t_3)}{B(t_0, t_2) + B(t_0, t_3) + B(t_0, t_4)} \tag{14.13}$$

This pricing formula can easily be generalized by moving from the three-period setting to a vanilla (forward) swap that makes n payments starting at time t_2. We obtain

$$s_{t_0} = \frac{\sum_{i=1}^{n} B(t_0, t_{i+1})F(t_0, t_i)}{\sum_{i=1}^{n} B(t_0, t_{i+1})} \tag{14.14}$$

This is a compact formula that ties together the three important components of the fixed-income framework we are using in this chapter.

---

[4]The fact that we are using default-free discount bonds to value a private party cash flow indicates that we are abstracting from all counter-party or credit risk.

14.2.2.1 Interpretation of the swap rate

The formula that gives the arbitrage-free value of the (forward) swap has a nice interpretation. For simplicity, revert to the three-period case. Rewrite Eq. (14.13) as

$$s_{t_0} = \frac{B(t_0, t_2)}{[B(t_0, t_2) + B(t_0, t_3) + B(t_0, t_4)]} F(t_0, t_1)$$

$$+ \frac{B(t_0, t_3)}{[B(t_0, t_2) + B(t_0, t_3) + B(t_0, t_4)]} F(t_0, t_2) \qquad (14.15)$$

$$+ \frac{B(t_0, t_4)}{[B(t_0, t_2) + B(t_0, t_3) + B(t_0, t_4)]} F(t_0, t_3) \qquad (14.16)$$

According to this expression, we see that the "correct" (forward) swap rate is a *weighted average* of the FRA paid-in-arrears rates during the life of the swap:

$$s_{t_0} = \omega_1 F(t_0, t_1) + \omega_2 F(t_0, t_2) + \omega_3 F(t_0, t_3) \qquad (14.17)$$

The weights are given by

$$\omega_i = \frac{B(t_0, t_{i+1})}{[B(t_0, t_2) + B(t_0, t_3) + B(t_0, t_4)]} \qquad (14.18)$$

and add up to one:

$$\omega_1 + \omega_2 + \omega_2 = 1 \qquad (14.19)$$

This can again be generalized for a (forward) swap that makes n payments:

$$s_{t_0} = \sum_{i=1}^{n} \omega_i F(t_0, t_i) \qquad (14.20)$$

with

$$\sum_{i=1}^{n} \omega_i = 1 \qquad (14.21)$$

Thus, the (forward) swap rate is an *average paid-in-arrears FRA rate*. We emphasize that this is true as long as the FRAs under consideration are paid in arrears. There are, on the other hand, so-called LIBOR-in-arrears FRAs where a convexity adjustment needs to be made for the argument to hold.[5]

It is important to realize that the weights $\{\omega_i\}$ are obtained from pure discount bond prices, which, as shown in Chapters 3 and 13, are themselves functions of forward rates:

$$B(t_0, t_i) = \frac{1}{\prod_{j=0}^{i-1}(1 + \delta F(t_0, t_j))} \qquad (14.22)$$

---

[5]We repeat the difference in terminology. One instrument is *paid-in-arrears*, the other is *LIBOR-in-arrears*. Here, the LIBOR of the settlement time t_i is used to determine the time t_i cash flows. With paid-in-arrears FRAs, the LIBOR of the previous settlement date, t_{i-1}, is used.

According to these formulas, three important components of our pricing framework—the swap market, the FRA market, and the bond market—are interlinked through nonlinear functions of forward rates. The important role played by the forward rates in these formulas suggests that obtaining *arbitrage-free dynamics* of the latter is required for the pricing of all swap and swap-related derivatives. The Forward LIBOR Model does exactly this. Because this model is set up in a way as to fit market conventions, it is also practical.

However, before we discuss these more advanced concepts, it is best to look at an example. In practice, swap and FRA markets are liquid and market makers readily quote the relevant rates. The real-world equivalents of the pure discount bonds $\{B(t_0, t_i)\}$, on the other hand, are not that liquid, even when they exist.[6] In the following example, we sidestep this point and assume that such quotes are available at all desired maturities. Even then, some important technical issues emerge, as the example illustrates.

EXAMPLE

Suppose we observe the following paid-in-arrears FRA quotes:

Term	Bid–Ask
0×6	4.05–4.07
6×12	4.15–4.17
12×18	4.32–4.34
18×24	4.50–4.54

Also, suppose the following treasury strip prices are observed:

Maturity	Bid–Ask
12 months	96.00–96.02
18 months	93.96–93.99
24 months	91.88–91.92

We can ask two questions. First, are these data arbitrage-free so that they can be used in obtaining an arbitrage-free swap rate? Second, if they are, what is the implied forward swap rate for the period that starts in 6 months and ends in 24 months?

The answer to the first question can be checked by using the following arbitrage equality, written for discount bonds with par value $100, as market convention suggests:

$$B(t_0, t_i) = \frac{100}{\prod_{j=0}^{i-1}(1 + \delta F(t_0, t_j))} \tag{14.23}$$

---

[6]In the United States, the instruments that come closest to these discount bonds are *treasury strips*. These are cash flows stripped from existing US treasuries, and there are a fair number of them. However, they are not very liquid and, in general, the market quoted prices cannot be used as substitutes for $B(t_0, t_i)$, for various technical reasons.

where the value of δ will be 1/2 in this example. Substituting the relevant forward rates from the preceding table, we indeed find that the given discount bond prices satisfy this equality. For example, for $B(0,2)^{\text{ask}}$ we have

$$B(0,2)^{\text{ask}} = \frac{100}{(1 + 0.5(0.0405))(1 + 0.5(0.0415))} = 96.02 \tag{14.24}$$

The relevant equalities hold for other discount bond prices as well. This means that the data are arbitrage free and can be used in finding an arbitrage-free swap rate for the above-mentioned forward start swap.

Replacing straightforwardly in

$$s_{t_0}^{\text{ask}} = \omega_1^{\text{ask}} F(t_0, t_1)^{\text{ask}} + \omega_2^{\text{ask}} F(t_0, t_2)^{\text{ask}} + \omega_3^{\text{ask}} F(t_0, t_3)^{\text{ask}} \tag{14.25}$$

$$\omega_i^{\text{ask}} = \frac{B(t_0, t_i+1)^{\text{ask}}}{[B(t_0, t_2)^{\text{ask}} + B(t_0, t_3)^{\text{ask}} + B(t_0, t_4)^{\text{ask}}]} \tag{14.26}$$

we find

$$\omega_1^{\text{ask}} = \frac{96.02}{[96.02 + 93.99 + 91.92]} = 0.341 \tag{14.27}$$

$$\omega_2^{\text{ask}} = \frac{93.99}{[96.02 + 93.99 + 91.92]} = 0.333 \tag{14.28}$$

$$\omega_3^{\text{ask}} = \frac{91.92}{[96.02 + 93.99 + 91.92]} = 0.326 \tag{14.29}$$

The asking swap rate is

$$s_{t_0}^{\text{ask}} = (0.341)4.17 + (0.333)4.34 + (0.326)4.54 = 4.34 \tag{14.30}$$

Similarly, we can calculate the bid rate:

$$s_{t_0}^{\text{bid}} = (0.341)4.15 + (0.333)4.32 + (0.326)4.50 = 4.32 \tag{14.31}$$

It is worth noting that the weights have approximately the same size.

We now consider further financial engineering applications of the fixed-income framework outlined in this section.

14.2.3 SOME APPLICATIONS

The first step is to consider the synthetic creation of swaps within our new framework. Our purpose is to obtain an alternative synthetic for swaps by manipulating the formulas derived in the previous section. In Chapter 4, we discussed one way of replicating swaps. We showed that a potential synthetic is the simultaneous shorting of a particular coupon bond and buying of a proper floating rate bond. This embodies the classical approach to synthetic swap creation, and it will be the starting point of the following discussion.

14.2.3.1 Another formula

We have already derived a formula for the (forward) swap rate, s_{t_0}, that gives an arbitrage-free swap value:

$$s_{t_0} = \frac{[B(t_0,t_2)F(t_0,t_1) + B(t_0,t_3)F(t_0,t_2) + B(t_0,t_4)F(t_0,t_3)]}{[B(t_0,t_2) + B(t_0,t_3) + B(t_0,t_4)]} \qquad (14.32)$$

Or, in the general form,

$$s_{t_0} = \frac{\sum_{i=1}^{n} B(t_0,t_{i+1})F(t_0,t_i)}{\sum_{i=1}^{n} B(t_0,t_{i+1})} \qquad (14.33)$$

Now, we would like to obtain an alternative way of looking at the same swap rate by modifying the formula. We start the discussion with the arbitrage relation between the discount bond prices, $B(t_0, t_i)$, and the forward rates, $F(t_0, t_i)$, obtained earlier in Chapter 3:

$$1 + \delta F(t_0,t_i) = \frac{B(t_0,t_i)}{B(t_0,t_{i+1})} \qquad (14.34)$$

Rearranging

$$F(t_0,t_i) = \frac{1}{\delta}\left[\frac{B(t_0,t_i)}{B(t_0,t_{i+1})} - 1\right] \qquad (14.35)$$

We now substitute this expression in Eq. (14.32) to obtain

$$s_{t_0} = \frac{1}{\delta[B(t_0,t_2) + B(t_0,t_3) + B(t_0,t_4)]}\left\{B(t_0,t_2)\left[\frac{B(t_0,t_1)}{B(t_0,t_2)} - 1\right]\right. \qquad (14.36)$$

$$\left. + B(t_0,t_3)\left[\frac{B(t_0,t_2)}{B(t_0,t_3)} - 1\right] + B(t_0,t_4)\left[\frac{B(t_0,t_3)}{B(t_0,t_4)} - 1\right]\right\} \qquad (14.37)$$

Simplifying the common $B(t_0, t_i)$ terms on the right-hand side, we get

$$s_{t_0} = \frac{1}{\delta\sum_{i=1}^{3} B(t_0,t_{i+1})}([B(t_0,t_1) - B(t_0,t_2)]$$

$$+ [B(t_0,t_2) - B(t_0,t_3)] + [B(t_0,t_3) - B(t_0,t_4)] \qquad (14.38)$$

$$= \frac{1}{\delta\sum_{i=1}^{3} B(t_0,t_{i+1})}[B(t_0,t_1) - B(t_0,t_4)] \qquad (14.39)$$

We can try to recognize what this formula means by first rearranging,

$$\delta s_{t_0}[B(t_0,t_2) + B(t_0,t_3) + B(t_0,t_4)] = [B(t_0,t_1) - B(t_0,t_4)] \qquad (14.40)$$

and then regrouping:

$$B(t_0,t_1) - \left[s_{t_0}\delta B(t_0,t_2) + s_{t_0}\delta B(t_0,t_3) + B(t_0,t_4)\left(1 + \delta s_{t_0}\right)\right] = 0 \qquad (14.41)$$

The equation equates two cash flows. $B(t_0, t_1)$ is the value of 1 dollar to be received at time t_1. Thus, the position needs to be *long* a t_1-maturity discount bond. Second, there appear to be coupon *payments* of constant size, δs_{t_0} at times t_2, t_3, t_4 and then a payment of 1 dollar at time t_4.[7] Thus, this seems to be a short (forward) position in a t_4-maturity coupon bond with coupon rate s_{t_0}.

To summarize, this particular forward fixed receiver interest rate swap is equivalent to

$$\text{Fixed-payer forward swap} = \{\text{Buy } t_1 \text{ discount bond, forward sell } t_4\text{-maturity coupon bond}\} \tag{14.42}$$

This synthetic will replicate the value of the forward swap. Note that the floating cash flows do not have to be replicated. This is because, in a forward swap, the floating cash flows are related to deposits (loans) that will be made in the future, at interest rates to be determined then.

14.2.3.2 Marking to market

We can use the same framework for discussing mark-to-market practices. Start at time t_0. As discussed earlier, the market is willing to pay the known cash flows

$$\left\{s_{t_0}N\delta, s_{t_0}N\delta, s_{t_0}N\delta\right\} \tag{14.43}$$

against the random cash flows

$$\left\{L_{t_1}N\delta, L_{t_2}N\delta, L_{t_3}N\delta\right\} \tag{14.44}$$

Now, let a short but noninfinitesimal time, Δ, pass. There will be a *new* swap rate $s_{t_0} + \Delta$, which, in all probability, will be different than s_{t_0}. This means that the market is now willing to pay the *new* known cash flows

$$\left\{s_{t_0+\Delta}N\delta, s_{t_0+\Delta}N\delta, s_{t_0+\Delta}N\delta\right\} \tag{14.45}$$

against the *same* random cash flows:

$$\left\{L_{t_1}N\delta, L_{t_2}N\delta, L_{t_3}N\delta\right\} \tag{14.46}$$

This implies that the value of the original swap, written at time t_0, is nonzero and is given by the difference:

$$\left[s_{t_0+\Delta}N\delta - s_{t_0}N\delta\right]\left[B(t_0 + \Delta, t_2) + B(t_0 + \Delta, t_3) + B(t_0 + \Delta, t_4)\right] \tag{14.47}$$

This can be regarded as the profit and loss for the fixed *payer*. At time $t_0 + \Delta$, the floating payment to be received has a value given by Eq. (14.47), and the actual floating payments would cancel out.[8] We can apply the same reasoning using the FRA rates and calculate the mark-to-market value of the original swap from the difference:

$$\left(N\delta\left[\sum_{i=1}^{n}\omega_{it_0}F(t_0, t_i)\right] - N\delta\left[\sum_{i=1}^{n}\omega_{i(t_0+\Delta)}F(t_0 + \Delta, t_i)\right]\right)\sum_{i=1}^{n}B(t_0 + \Delta, t_{i+1}) \tag{14.48}$$

---

[7]These payments are discounted to the present and this introduces the corresponding bond prices to the expression in the brackets.

[8]What one "plugs in" for unknown LIBOR rates in Eq. (14.46) does change. But we are valuing the swap from the fixed leg and we consider the fixed payments as compensation for random, unknown floating rates. These random variables remain the same.

This way of writing the expression shows the profit and loss from the point of view of a fixed *receiver*. It should be noted that here the weights w_i have time subscripts, since they will change as time passes. Thus, managing a swap book will depend nonlinearly on the forward rate dynamics.[9]

14.3 TERM STRUCTURE MODELING

The framework outlined in this chapter demonstrates the links between swap, bond, and FRA markets. We will now discuss the term structure implications of the derived formulas. The set of formulas we studied implies that, given the necessary information from *two* of these markets, we can, in principle, obtain arbitrage-free prices for the remaining market.[10] We discuss this briefly, after noting the following small, but significant, modification. Term structure models concern forward rates *as well as* spot rates. As a matter of fact, traditional yield curve construction is done by first obtaining the spot yields and then moving to forward rates.

Following this tradition, and noting that *spot swaps* are more liquid than *forward* swaps, in this section we let $s_{t_0}^n$ denote the *spot* swap rate with maturity n years. Without loss of generality, we can assume that swap maturities are across years $n = 1, \ldots, 30$, so that the longest dated swap is for 30 years.[11] The discussion will be conducted in terms of spot swap rates.

14.3.1 DETERMINING THE FORWARD RATES FROM SWAPS

Given a sufficient number of arbitrage-free values of observed spot swap rates $\left\{ s_{t_0}^n \right\}$ and using the equalities

$$s_{t_0}^n = \frac{\sum_{i=0}^{n-1} B(t_0, t_{i+1}) F(t_0, t_i)}{\sum_{i=0}^{n-1} B(t_0, t_{i+1})} \tag{14.49}$$

and

$$\frac{B(t_0, t_i)}{B(t_0, t_{i+1})} = (1 + \delta F(t_0, t_i)) \tag{14.50}$$

We can obtain all forward rates, for the case $\delta = 1$. By substituting the $B(t_0, t_i)$ out from the first set of equations, we obtain n equations in n forward rates.[12] In the case of $\delta = 1/4$ or $\delta = 1/2$, there are more unknown $F(t_0, t_i)$ than equations, if traded swap maturities are in years. Under these conditions, the t_i would run over quarters whereas the superscript in $s_{t_0}^n$, $n = 1, 2, \ldots$ will be in years. This is due to the fact that swap rates are quoted for annual intervals, whereas the settlement dates would be quarterly or semiannual. Some type of interpolation of swap rates or modeling will be required, which is common even in traditional yield curve calculations.

---

[9]Again, these forward rates need to be associated with paid-in-arrears FRAs or forward loans.

[10]This assumes that all maturities of the underlying instruments trade actively, which is, in general, not the case.

[11]Swaps start to trade from 2 years and on. A 1-year swap against 1-year LIBOR would, in fact, be equivalent to a trivial FRA.

[12]Remember that $F(t_0, t_0)$ equals the current LIBOR for that tenor and is a trivial forward rate.

14.3.2 DETERMINING THE $B(t_0, t_i)$ FROM FORWARD RATES

Now, if the forward rates $\{F(t_0, t_i)\}$ and the current LIBOR curve are provided by markets or are obtained from $\{s_{t_0}^n\}$ as in our case, we can use the formula

$$B(t_0, t_{i+1}) = \frac{1}{\prod_{j=0}^{i}(1 + \delta F(t_0, t_j))} \tag{14.51}$$

to calculate the arbitrage-free values of the relevant pure discount bond prices. In each case, we can derive the values of $B(t_0, t_i)$ from the observed $\{F(t_0, t_i)\}$ and $\{s_{t_0}^n\}$. This procedure would price the FRAs and bonds *off* the swap markets. It is called the *curve algorithm*.

14.3.3 DETERMINING THE SWAP RATE

We can proceed in the opposite direction as well. Given arbitrage-free values of forward rates, we can, in principle, use the same formulas to determine the swap rates. All we need to do is (i) calculate the discount bond prices from the forward rates and (ii) substitute these bond prices and the appropriate forward rates in our formula:

$$s_{t_0}^n = \frac{\sum_{i=0}^{n-1} B(t_0, t_{i+1})F(t_0, t_i)}{\sum_{i=0}^{n-1} B(t_0, t_{i+1})} \tag{14.52}$$

Repeating this for all available $s_{t_0}^n$, $n = 1, \ldots, 30$, we obtain the arbitrage-free *swap curve* and discounts. In this case, we would be going from the spot and forward LIBOR curve to the (spot) swap curve.

14.3.4 REAL-WORLD COMPLICATIONS

There are, of course, several real-world complications to going back and forth between the forward rates, discount bond prices, and swap rates. Let us mention three of these. First, as mentioned in the previous section, in reality swaps are traded for *yearly* intervals and the FRAs or Eurodollar contracts are traded for 3-month or 6-month tenors. This means that if we desired to go from swap quotes to quotes on forward rates using these formulas, there will be the need to interpolate the swap rates for portions of a year.

Second, observed quotes on forward rates do *not* necessarily come from paid-in-arrears FRAs. Market-traded FRAs settle at the time the LIBOR rate is observed, not at the end of the relevant period. The FRA rates generated by these markets *will* be consistent with the formulas introduced earlier. On the other hand, some traders use interest rate futures, and, specifically, Eurocurrency futures, in hedging their swap books. Futures markets are more *transparent* than the FRA markets and have a great deal of liquidity. But the forward rates determined in futures markets require *convexity adjustments* before they can be used in the swap formulas discussed in this chapter.

Third, the liquidity of FRA and swap rates depends on the maturity under consideration. As mentioned earlier, FRAs are more liquid for the shorter end of the curve, whereas swaps are more liquid at the longer end. This means that it may not be possible to go from FRA rates to swap rates for maturities over 5 years. Similarly, for very short maturities there will be no observed quotes for swaps.

14.3.4.1 Remark

Another important point needs to be mentioned here. In this chapter, for simplicity, our treatment has followed the pre-GFC convention in academic work of using the term "LIBOR" as if it relates to a default-free loan. In practice, LIBOR rates L_{t_i} are not risk free. Admittedly, this was implicit in the fixing process of the LIBOR rate. The banks that contributed to the rate had, in general, ratings of AA or AA−, and the interest rate that they paid reflected this level of credit risk. However, the probability of such highly rated banks defaulting was considered small and therefore market participants and academics considered LIBOR a reasonable proxy for the risk-free rate. Before the GFC, LIBOR was a key indicator of what banks were willing to lend to each other, and 3-month LIBOR in particular emerged as a key benchmark in the interest rate swaps market. As discussed in Chapter 3 during the GFC, the LIBOR rate rose dramatically and diverged from the OIS as fears of counter-party risk became acute. This led to the realization that LIBOR could not be used as a measure of the risk-free rate. Following the GFC, market participants started using the OIS as the discount rate for the pricing of (collateralized) interest rate swaps. In June 2010, LCH.Clearnet announced that it would use the OIS instead of LIBOR to discount its $218 trillion interest rate swap portfolio. The intuition is that in fully collateralized swaps, financial counterparties are exposed to overnight risk to other financial counterparties only and not the risk of the full term of a particular loan or instrument. Academic models have been developed to incorporate credit and counter-party risk more explicitly into derivatives pricing models. Such models and recent market practice are discussed in detail in Chapter 24.

Thus, if a financial engineer follows the procedures described here, the resulting curve will be the *swap curve* and not the treasury or sovereign curve. This swap curve will be "above" the sovereign or treasury curve, and the difference will be the curve for the swap spreads.

14.4 TERM STRUCTURE DYNAMICS

In the remainder of this chapter, we will see that the Forward LIBOR Model is the correct way to approach term structure dynamics. The model is based on the idea of converting the dynamics of each forward rate into a Martingale using some properly chosen forward measure. According to the linkages between sectors shown in this chapter, once such dynamics are obtained, we can use them to generate dynamics for other fixed-income instruments.

Most of the derivation associated with the Forward LIBOR Model is an application of the fundamental theorem of asset pricing discussed in Chapter 12. Thus, we continue to use the same finite-state world discussed in Chapter 12. The approach is mostly straightforward. There is only *one* aspect of Forward LIBOR or swap models that makes them potentially difficult to follow. Depending on the instruments, arbitrage-free dynamics of *different* forward rates may have to be expressed under the *same* forward measure. The methodology then becomes more complicated. It requires a judicious sequence of Girsanov-style measure changes to be applied to forward rate dynamics in some recursive fashion. Otherwise, arbitrage-free dynamics of individual forward rates would not be correctly represented.

The Girsanov theorem is a powerful tool, but it is not easy to conceive such successive measure changes. Doing this within a discrete framework, in a discrete setting, provides a great deal of *motivation* and facilitates understanding of arbitrage-free dynamics. This is the purpose behind the second part of this chapter.

14.4.1 THE FRAMEWORK

We adopt a simple discrete framework and then extend it to general formulas. Consider a market where instruments can be priced and risk managed in discrete times that are δ apart:

$$t_0 < t_1 < \cdots < t_n = T \tag{14.53}$$

with

$$t_i - t_{i-1} = \delta \tag{14.54}$$

Initially, we concentrate on the first three times, t_0, t_1, and t_2 that are δ apart. In this framework, we consider four simple fixed-income securities:

1. A default-free zero-coupon bond $B(t_0, t_2)$ that matures at time t_2.
2. A default-free zero-coupon bond that matures one period later, at time t_3. Its current price is expressed as $B(t_0, t_3)$.
3. A savings account that pays (in-arrears) the discrete-time simple rate L_{t_i} observed at time t_i. Therefore, the savings account payoff at t_2 will be

$$R_{t_2} = \left(1 + \delta L_{t_0}\right)\left(1 + \delta L_{t_1}\right) \tag{14.55}$$

   Note that the L_{t_1} observed from the initial time t_0 will be a random variable.
4. An FRA contracted at time t_0 and settled at time t_2, where the buyer receives/pays the differential between the fixed rate $F(t_0, t_1)$ and the floating rate L_{t_1} at time t_2. We let the notional amount of this instrument equal 1 and abbreviate the forward rate to F_{t_0}. The final payoff can be written as

$$\left(L_{t_1} - F_{t_0}\right)\delta \tag{14.56}$$

These assets can be organized in the following payoff matrix D for time t_2 as in Chapter 12, assuming that at every t_i, from every node there are only two possible movements for the underlying random process. Denoting these movements by u, d, we can write[13]

$$D = \begin{bmatrix} R_{t_2}^{uu} & R_{t_2}^{ud} & R_{t_2}^{du} & R_{t_2}^{dd} \\ 1 & 1 & 1 & 1 \\ B_{t_2}^{uu} & B_{t_2}^{ud} & B_{t_2}^{du} & B_{t_2}^{dd} \\ \delta\left(F_{t_0} - L_{t_1}^{u}\right) & \delta\left(F_{t_0} - L_{t_1}^{u}\right) & \delta\left(F_{t_0} - L_{t_1}^{d}\right) & \delta\left(F_{t_0} - L_{t_1}^{d}\right) \end{bmatrix} \tag{14.57}$$

where the $B_{t_2}^{ij}$ is the (random) value of the t_3 maturity discount bond at time t_2. This value will be state dependent at t_2 because the bond matures one period later, at time t_3. Looked at from time t_0, this value will be random. Clearly, with this D matrix we have simplified the notation significantly. We are using only four states of the world, expressing the forward rate $F(t_0, t_2)$ simply as F_{t_0}, and the $B(t_2, t_3)^{ij}$ simply as $B_{t_2}^{ij}$.

---

[13]This table can be regarded as the second step in a nonrecombining binomial tree.

If the FRA, the savings account, and the two bonds do not admit any arbitrage opportunities, the fundamental theorem of asset pricing permits the following linear representation:

$$
\begin{bmatrix} 1 \\ B(t_0, t_2) \\ B(t_0, t_3) \\ 0 \end{bmatrix} = \begin{bmatrix} R_{t_2}^{uu} & R_{t_2}^{ud} & R_{t_2}^{du} & R_{t_2}^{dd} \\ 1 & 1 & 1 & 1 \\ B_{t_2}^{uu} & B_{t_2}^{ud} & B_{t_2}^{du} & B_{t_2}^{dd} \\ \delta\left(F_{t_0} - L_{t_1}^u\right) & \delta\left(F_{t_0} - L_{t_1}^u\right) & \delta\left(F_{t_0} - L_{t_1}^d\right) & \delta\left(F_{t_0} - L_{t_1}^d\right) \end{bmatrix} \begin{bmatrix} Q^{uu} \\ Q^{ud} \\ Q^{du} \\ Q^{dd} \end{bmatrix}
\tag{14.58}
$$

where $\{Q^{ij}, i, j = u, d\}$ are the four state prices for period t_2. Under the no-arbitrage condition, the latter exist and are *positive*

$$
Q^{ij} > 0
\tag{14.59}
$$

for all states i, j.[14]

This matrix equation incorporates the ideas that (i) the fair market value of an FRA is zero at initiation, (ii) 1 dollar is to be invested in the savings account originally, and (iii) the bonds are default free. They mature at times t_2 and t_3. $R_{t_2}^{i,j}$, finally, represents the gross returns to the savings account as of time t_2. Because the interest rate that applies to time t_i is paid in arrears, at time $t_i + \delta$, we can express these gross returns as functions of the underlying LIBOR rates in the following way:

$$
R_{t_2}^{uu} = R_{t_2}^{ud} = \left(1 + \delta L_{t_0}\right)\left(1 + \delta L_{t_1}^u\right)
\tag{14.60}
$$

and

$$
R_{t_2}^{dd} = R_{t_2}^{du} = \left(1 + \delta L_{t_0}\right)\left(1 + \delta L_{t_1}^d\right)
\tag{14.61}
$$

We now present the LIBOR market model and the associated measure change methodology within this simple framework. The framework can be used to conveniently display most of the important tools and concepts that we need for fixed-income engineering. The first important concept that we need is the *forward measure* introduced in Chapter 12.

14.4.2 NORMALIZATION AND FORWARD MEASURE

To obtain the t_2 and t_3 forward measures, it is best to begin with a *risk-neutral probability*, and show why it is not a good working measure in the fixed-income environment described earlier. We can then show how to *convert* the risk-neutral probability to a desired forward measure explicitly.

14.4.2.1 Risk-neutral measure is inconvenient

As usual, define the risk-neutral measure $\{\tilde{p}_{ij}\}$ using the *first row* of the matrix equation:

$$
1 = R_{t_2}^{uu} Q^{uu} + R_{t_2}^{ud} Q^{ud} + R_{t_2}^{du} Q^{du} + R_{t_2}^{dd} Q^{dd}
\tag{14.62}
$$

---

[14]As usual, we are eliminating the time subscript on the state prices, since it is clear by now that we are dealing with time t_2 payoffs.

Relabel

$$\tilde{p}_{uu} = R_{t_2}^{uu} Q^{uu} \tag{14.63}$$

$$\tilde{p}_{ud} = R_{t_2}^{ud} Q^{ud} \tag{14.64}$$

$$\tilde{p}_{du} = R_{t_2}^{du} Q^{du} \tag{14.65}$$

$$\tilde{p}_{dd} = R_{t_2}^{dd} Q^{dd} \tag{14.66}$$

The $\{\tilde{p}_{ij}\}$ then have the characteristics of a probability distribution, and they can be exploited with the associated Martingale equality.

We know from Chapter 12 that, under the condition that every asset's price is arbitrage free, $\{Q^{ij}, i, j = u, d\}$ exist and are all positive, and $\tilde{p}_{ij}$ will be the risk-neutral probabilities. Then, by using the last row of the system in Eq. (14.58) we can write the following equality:

$$0 = \left[\delta\left(F_{t_0} - L_{t_1}^u\right)\frac{1}{R_{t_2}^{uu}}\right]\tilde{p}_{uu} + \left[\delta\left(F_{t_0} - L_{t_1}^u\right)\frac{1}{R_{t_2}^{ud}}\right]\tilde{p}_{ud} + \left[\delta\left(F_{t_0} - L_{t_1}^d\right)\frac{1}{R_{t_2}^{du}}\right]\tilde{p}_{du}$$
$$+ \left[\delta\left(F_{t_0} - L_{t_1}^d\right)\frac{1}{R_{t_2}^{dd}}\right]\tilde{p}_{dd} \tag{14.67}$$

Here, $\left(F_{t_0} - L_{t_1}^i\right)$, $i = u, d$ are "normalized" so that Q^{ij} can be replaced by the respective $\tilde{p}_{ij}$. Note that in this equation, F_{t_0} is determined at time t_0, and can be factored out. Grouping and rearranging, we get

$$F_{t_0} = \frac{\left([L_{t_1}^u(1/R_{t_2}^{uu})]\tilde{p}_{uu} + [L_{t_1}^u(1/R_{t_2}^{ud})]\tilde{p}_{ud} + [L_{t_1}^d(1/R_{t_2}^{du})]\tilde{p}_{du} + [L_{t_1}^d(1/R_{t_2}^{dd})]\tilde{p}_{dd}\right)}{\left((1/R_{t_2}^{uu})\tilde{p}_{uu} + (1/R_{t_2}^{ud})\tilde{p}_{ud} + (1/R_{t_2}^{du})\tilde{p}_{du} + (1/R_{t_2}^{dd})\tilde{p}_{dd}\right)} \tag{14.68}$$

This can be written using the expectation operator

$$F_{t_0} = \frac{1}{E_{t_0}^{\tilde{P}}[1/R_{t_2}]} E_{t_0}^{\tilde{P}}\left[\frac{L_{t_1}}{R_{t_2}}\right] \tag{14.69}$$

According to this last equality, if R_{t_2} is a random variable and is not independent of L_{t_1},[15] it cannot be moved outside the expectation operator. In other words, for general i,

$$F(t, t_i) \neq E_t^{\tilde{P}}[L_{t_i}] \quad t < t_i \tag{14.70}$$

That is to say, under the risk-neutral measure, $\tilde{P}$, the forward rate for time t_i is a *biased* "forecast" of the future LIBOR rate L_{t_i}. In fact, it is not very difficult to see that the $E_t^{\tilde{P}}[L_{t_i}]$ is the *futures* rate that will be determined by, say, a Eurodollar contract at time i. The "bias" in the forward rate, therefore, is associated with the convexity adjustment.

Another way of putting it is that F_t is *not* a Martingale with respect to the risk-neutral probability $\tilde{P}$, and that a discretized stochastic difference equation that represents the dynamics of F_t will, in general, have a trend:

$$F_{t+\Delta} - F_t = a(F_t, t)F_t\Delta + \sigma(F_t, t)F_t[W_{t+\Delta} - W_t] \tag{14.71}$$

where $a(F_t, t)$ is the non-zero expected rate of change of the forward rate under the probability $\tilde{P}$.

---

[15]Although we know, in general, that this will not be the case since these are returns to short-term investments in Eurocurrency markets.

The fact that F_t is *not* a Martingale with respect to probability $\tilde{P}$ makes the risk-neutral measure an inconvenient working tool for pricing and risk management in the fixed-income sector. Before we can use Eq. (14.71), we need to *calibrate* the drift factor $a(.)$. This requires first obtaining a functional form for the drift under the probability $\tilde{P}$. The original HJM article *does* develop a functional form for such drifts using *continuously compounded* instantaneous forward rates. But, this creates an environment quite different from LIBOR-driven markets and the associated actuarial rates L_{t_i} used here.[16]

On the other hand, we will see that in the interest rate sector, arbitrage-free drifts become much easier to calculate if we use the Forward LIBOR Model and switch to appropriate forward measures.

14.4.2.2 The forward measure

Consider defining a new set of probabilities for the states under consideration by using the default-free zero-coupon bond that matures at time t_2. First, we present the simple case. Use the second row of the system in Eq. (14.58):

$$B(t_0, t_2) = Q^{uu} + Q^{ud} + Q^{du} + Q^{dd} \tag{14.72}$$

and then divide every element by $B(t_0, t_2)$. Renaming, we get the forward t_2 measure $\tilde{P}^{t_2}$

$$1 = \tilde{p}^{t_2}_{uu} + \tilde{p}^{t_2}_{ud} + \tilde{p}^{t_2}_{du} + \tilde{p}^{t_2}_{dd} \tag{14.73}$$

where the probability of each state is obtained by scaling the corresponding Q^{ij} using the time t_0 price of the corresponding bond:

$$\tilde{p}^{t_2}_{ij} = \frac{Q^{ij}}{B(t_0, t_2)} \tag{14.74}$$

It is important to index the forward measure with the superscript, t_2, in these fixed-income models, as *other* forward measures would be needed for other forward rates. The superscript is a nice way of keeping track of the measure being used. For some instruments, these measures have to be switched sequentially.

As discussed in Chapter 12, using the t_2-forward measure we can price any asset C_t, with time t_2 payoffs $C^{ij}_{t_2}$

$$C_{t_0} = \left[B(t_0, t_2)C^{uu}_{t_2}\right]\tilde{p}^{t_2}_{uu} + \left[B(t_0, t_2)C^{ud}_{t_2}\right]\tilde{p}^{t_2}_{ud} + \left[B(t_0, t_2)C^{du}_{t_2}\right]\tilde{p}^{t_2}_{du} + \left[B(t_0, t_2)C^{dd}_{t_2}\right]\tilde{p}^{t_2}_{dd} \tag{14.75}$$

This implies that, for an asset that settles at time T and has no other payouts, the *general* pricing equation is given by

$$C_t = B(t, T)E^{\tilde{P}^T}_t[C_T] \tag{14.76}$$

where $\tilde{P}^T$ is the associated T-forward measure and where C_T is the time T payoff. According to this equality, it is the ratio

$$Z_t = \frac{C_t}{B(t, T)} \tag{14.77}$$

---

[16]Further, new technical problems appear that make the continuous compounding numerically unstable.

which is a Martingale with respect to the measure $\tilde{P}^T$. In fact, $B(t, T)$ being the *discount* factor for time T, and, hence, being ≤ 1, Z_t is nothing more than the *T-forward value* of the C_t. This means that the forward measure $\tilde{P}^T$ operates in terms of Martingales that are measured in time T dollars. The advantage of the forward $\tilde{P}^T$ measure becomes clear if we apply the same transformation to price the FRA as was done earlier for the case of the risk-neutral measure.

14.4.2.3 Arbitrage-free SDEs for forward rates

To get arbitrage-free dynamics for forward rates, we now go back to the simple model in Eq. (14.58). Divide the fourth row of the system by $B(t_0, t_2)$ and rearrange,

$$\frac{F_{t_0}}{B(t_0, t_2)} \left[Q^{uu} + Q^{ud} + Q^{du} + Q^{dd} \right] = \frac{L^u_{t_1}}{B(t_0, t_2)} Q^{uu} + \frac{L^u_{t_1}}{B(t_0, t_2)} Q^{ud}$$
$$+ \frac{L^d_{t_1}}{B(t_0, t_2)} Q^{du} + \frac{L^d_{t_1}}{B(t_0, t_2)} Q^{dd} \tag{14.78}$$

Now, as done in Chapter 12, substitute the t_2-forward measure into this equation using the equality:

$$\tilde{p}^{t_2}_{ij} = \frac{1}{B(t_0, t_2)} Q^{ij}$$

The equation becomes

$$F_{t_0} = [L^u_{t_1}] \tilde{p}^{t_2}_{uu} + [L^u_{t_1}] \tilde{p}^{t_2}_{ud} + [L^d_{t_1}] \tilde{p}^{t_2}_{du} + [L^d_{t_1}] \tilde{p}^{t_2}_{dd} \tag{14.79}$$

Extending this to the general case of m discrete states

$$F_{t_0} = \sum_{i=1}^{m} L^i_{t_1} \tilde{p}^{t_2}_i \tag{14.80}$$

This is clearly the expectation

$$F_{t_0} = E^{\tilde{p}^{t_2}}_{t_0} [L_{t_1}] \tag{14.81}$$

This means that, under the measure $\tilde{P}^{t_2}$, the forward rate for the period $[t_1, t_2]$ *will* be an *unbiased estimate* of the corresponding LIBOR rate.

Consequently, switching to the general notation of (t, T), the process

$$F_t = F(t, T, T + \delta) \tag{14.82}$$

will be a Martingale under the $(T + \delta)$-forward measure $\tilde{P}^{T+\delta}$. *Assuming that the errors due to discretization are small*, its dynamics can be described by a (discretized) SDE over small intervals of length Δ [17]

$$F_{t+\Delta} - F_t = \sigma_t F_t \Delta W_t \tag{14.83}$$

---

[17] See Suggested Reading at the end of the chapter for a source on discretization errors and their relevance here.

where W_t is a Wiener process under the measure $\tilde{P}^{T+\delta}$. ΔW_t is the Wiener process increment:

$$\Delta W_t = W_{t+\Delta} - W_t \tag{14.84}$$

This (approximate) equation has *no* drift component since, by arbitrage arguments, and writing for the general t, T, we have

$$1 + \delta F(t, T) = \frac{B(t, T)}{B(t, T + \delta)} \tag{14.85}$$

It is clear from the normalization arguments of Chapter 12 that, under the measure $\tilde{P}^{T+\delta}$ and normalization by $B(t, T + \delta)$, the ratio on the right-hand side of this equation is a Martingale with respect to $\tilde{P}^{T+\delta}$. This makes the corresponding forward rate a Martingale, so that the implied SDE will have no drift.

However, note that the forward rate for the period $[T - \delta, T]$ given by

$$1 + \delta F(t, T - \delta) = \frac{B(t, T - \delta)}{B(t, T)} \tag{14.86}$$

is *not* a Martingale under the *same* forward measure $\tilde{P}^{T+\delta}$. Instead, this forward rate is a Martingale under its *own* measure $\tilde{P}^{T}$ which requires normalization by $B(t, T)$. Thus, we get a critical result for the Forward LIBOR Model:

> Each forward rate $F(t, T_i)$ admits a Martingale representation under *its own* forward measure $\tilde{P}^{T_i+\delta}$.

This means that each forward rate dynamics can be approximated individually by a difference equation with no drift given the proper normalization. The only parameter that would be needed to characterize such dynamics is the corresponding forward rate volatility.

14.4.3 ARBITRAGE-FREE DYNAMICS

The previous section discussed the dynamics of forward rates under their *own* forward measure. We now show what happens when we use *one* forward measure for *two* forward rates that apply to two consecutive periods. Then, one of the forward rates has to be evaluated under a measure different from its own, and the Martingale dynamics will be broken. Yet, we will be able to obtain the new drift.

To keep the issue as simple as possible, we continue with the basic model in Eq. (14.58), except we add one more time period so that we can work with *two* nontrivial forward rates and their respective forward measures. This is the simplest setup within which we can show how *measure change* technology can be implemented. Using the forward measures introduced earlier and shown in Figure 14.4, we can now define the following forward rate dynamics for the *two* forward LIBOR processes $\{F(t_0, t_1), F(t_0, t_2)\}$ under consideration. The first will be a Martingale under the normalization with $B(t_0, t_2)$, whereas the second will be a Martingale under the normalization with $B(t_0, t_3)$. This means that $\tilde{P}^{t_2}$ and $\tilde{P}^{t_3}$ are the forward LIBOR processes' "own" measures.

Altogether, it is important to realize that during the following discussion we are working with a very simple example involving only four time periods, $t_0, t_1, t_2,$ and t_3. We start with the arbitrage-free

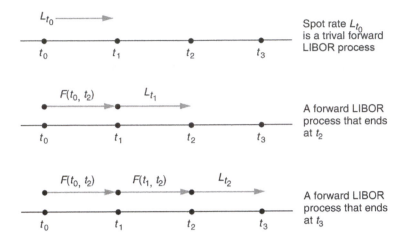

FIGURE 14.4

Spot and forward LIBOR process.

"dynamics" of the forward rate $F(t_0, t_2)$. In our simplified setup, we will observe only two future values of this forward rate at times t_1 and t_2. These are given by

$$F(t_1, t_2) - F(t_0, t_2) = \sigma_2 F(t_0, t_2) \Delta W_{t_1}^{t_3}$$

$$F(t_2, t_2) - F(t_1, t_2) = \sigma_2 F(t_1, t_2) \Delta W_{t_2}^{t_3} \qquad (14.87)$$

The superscript in $\Delta W_{t_i}^{t_3}$, $i = 1, 2$, indicates that the Wiener process increments have zero mean under the probability $\tilde{P}^{t_3}$. These equations show how the "current" value of the forward rate $F(t_0, t_2)$ first changes to become $F(t_1, t_2)$ and then ends up as $F(t_2, t_2)$. The latter is also L_{t_2}.

For the "nearer" forward rate $F(t_0, t_1)$, we need only one equation[18] defined under the normalization with the bond $B(t_0, t_2)$ (i.e., the $\tilde{P}^{t_2}$ measure) and the associated zero drift:

$$F(t_1, t_1) - F(t_0, t_1) = \sigma_1 F(t_0, t_1) \Delta W_{t_1}^{t_2} \qquad (14.88)$$

Similarly, the superscript in $\Delta W_{t_1}^{t_2}$ indicates that this Wiener process increment has zero mean under the probability $\tilde{P}^{t_2}$. Here, the $F(t_1, t_1)$ is also the LIBOR rate L_{t_1}. We reemphasize that each dynamic is defined under a different probability measure. Under these *different* forward measures, each forward LIBOR process behaves like a Martingale.[19] Consequently, there are no drift terms in either equation.

Fortunately, as long as we can work with these equations *separately*, no arbitrage-free *drift terms* need to be estimated or calibrated. The only parameters we need to determine are the volatilities of the two forward rates: σ_2 for the forward rate $F(t_0, t_2)$ and σ_1 for the forward rate $F(t_0, t_1)$.[20]

[18]This forward rate process will terminate at t_2.
[19]Again, we are assuming that the discretization bias is negligible.
[20]Note that according to the characterization here, the volatility parameters are not allowed to vary over time. This assumption can be relaxed somewhat, but we prefer this simple setting, since most market applications are based on constant volatility characterization as well.

In fact, each Wiener increment has a zero expectation under its own measure. For example, the Wiener increments of the two forward rates will satisfy, for time $t_0 < t_1$

$$E_{t_0}^{\tilde{P}^{t_2}} \left[\Delta W_{t_1}^{t_2} \right] = 0 \qquad (14.89)$$

and

$$E_{t_0}^{\tilde{P}^{t_3}} \left[\Delta W_{t_1}^{t_3} \right] = 0 \qquad (14.90)$$

Yet, when we evaluate the expectations under $\tilde{P}^{t_2}$, we get

$$E_{t_0}^{\tilde{P}^{t_2}} \left[\Delta W_{t_1}^{t_2} \right] = 0 \qquad (14.91)$$

and

$$E_{t_0}^{\tilde{P}^{t_2}} \left[\Delta W_{t_1}^{t_3} \right] = \lambda_{t_0}^{t_2} \Delta \neq 0 \qquad (14.92)$$

Here, $\lambda_{t_0}^{t_2}$ is a *mean correction* that needs to be made because we are evaluating the Wiener increment under a measure different from its own forward measure $\tilde{P}^{t_3}$. This, in turn, means that the dynamics for $F(t_0, t_2)$ lose their Martingale characteristic.

We will now comment on the second moments, variances, and covariances. Each Wiener increment is assumed to have the same variance under the two measures. The Girsanov theorem ensures that this is true in continuous time. In discrete time, this holds as an approximation. Finally, we are operating in an environment where there is only *one* factor.[21] So, the Wiener process increments defined under the two forward measures *will* be exactly correlated if they belong to the same time period. In other words, although their means are different, we can assume that, approximately, their *covariance* would be Δ:

$$E^{\tilde{P}^{t_3}} \left[\Delta W_t^{t_3} \Delta W_t^{t_2} \right] = E^{\tilde{P}^{t_2}} \left[\Delta W_t^{t_3} \Delta W_t^{t_2} \right] = \Delta \qquad (14.93)$$

Similar equalities will hold for the variances as well.[22]

14.4.3.1 Review

The results thus far indicate that for the pricing and risk managing of equity-linked assets, the risk-neutral measure $\tilde{P}$ may be quite convenient since it is easily adaptable to lognormal models where the arbitrage-free drifts are simple and known functions of the risk-free interest rate. As far as equity products are concerned, the assumption that short rates are constant is a reasonable approximation, especially for short maturities. Yet, for contracts written on future values of interest rates (rather than on asset prices), the use of the $\tilde{P}$ leads to complex arbitrage-free dynamics that cannot be captured easily by Martingales and, hence, the corresponding arbitrage-free drift terms may be difficult to calibrate.

Appropriate forward measures, on the other hand, result in Martingale equalities and lead to dynamics convenient for the calculation of arbitrage-free drifts, even when they are not zero. Forward (and swap) measures are the proper working probabilities for fixed-income environments.

---

[21] As a reminder, a one-factor model assumes that all random processes under consideration have the same unpredictable component up to a factor of proportionality. In other words, the correlation coefficients between these processes would be one.

[22] These relations will hold as Δ goes to zero.

14.4.4 A MONTE CARLO IMPLEMENTATION

Suppose we want to generate Monte Carlo "paths" from the two discretized SDEs for two forward rates, $F(t_i, t_1)$ and $F(t_i, t_2)$,

$$F(t_i, t_1) - F(t_{i-1}, t_1) = \sigma_1 F(t_{i-1}, t_1) \Delta W_{t_i}^{t_2} \qquad (14.94)$$

$$F(t_i, t_2) - F(t_{i-1}, t_2) = \sigma_2 F(t_{i-1}, t_2) \Delta W_{t_i}^{t_3} \qquad (14.95)$$

where $i = 1, 2$ for the second equation and $i = 1$ for the first.

It is easy to generate *individual* paths for the two forward rates separately by using these Martingale equations defined under their own forward measures. Consider the following approach.

Suppose volatilities σ_1 and σ_2 can be observed in the market. We select two random variables $\{\Delta W_1^3, \Delta W_2^3\}$ from the distribution

$$\Delta W_i^3 \sim N(0, \Delta) \qquad (14.96)$$

with a pseudo-random number generator, and then calculate, sequentially, the randomly generated forward rates in the following order, starting with the *observed* $F(t_0, t_2)$

$$F(t_1, t_2)^1 = F(t_0, t_2) + \sigma_2 F(t_0, t_2) \Delta W_1^3 \qquad (14.97)$$

$$F(t_2, t_2)^1 = F(t_1, t_2)^1 + \sigma_2 F(t_1, t_2)^1 \Delta W_2^3 \qquad (14.98)$$

where the superscript on the left-hand side indicates that these values are for the first Monte Carlo trajectory. Proceeding sequentially, all the terms on the right-hand side will be known. This gives the first simulated "path" $\{F(t_0, t_2), F(t_1, t_2)^1, F(t_2, t_2)^1\}$. We can repeat this algorithm to obtain M such paths for potential use in pricing.

What does this imply for the other forward LIBOR process $F(t_0, t_1)$? Can we use the *same* randomly generated random variable ΔW_1^3 in the Martingale equation for $F(t, t_1)$, and obtain the first "path" $\{F(t_0, t_1), F(t_1, t_1)^1\}$ from

$$F(t_1, t_1) = F(t_0, t_1) + \sigma_1 F(t_0, t_1) \Delta W_1^3 \qquad (14.99)$$

The answer is no. As mentioned earlier, the Wiener increments $\{\Delta W_{t_1}^{t_2}\}$ have zero mean only under the probability $\tilde{P}^{t_2}$. But, the first set of random variables was selected using the measure $\tilde{P}^{t_3}$. Under $\tilde{P}^{t_2}$, these random variables do not have zero mean, but are distributed as

$$N(\lambda_{t_0}^{t_2} \Delta, \Delta) \qquad (14.100)$$

Thus, if we use the same ΔW_1^3 in Eq. (14.99), then we need to correct for the term $\lambda_{t_0}^{t_2} \Delta$. To do this, we need to calculate the numerical value of $\lambda_{t_0}^{t_2}$.

Once this is done, the dynamics for $F(t_0, t_1)$ can be written as

$$F(t_1, t_1) = F(t_0, t_1) - \sigma_1 F(t_0, t_1)(\lambda_{t_0}^{t_2} \Delta) + \sigma_1 F(t_0, t_1) \Delta W_{t_1}^{t_3} \qquad (14.101)$$

To see why this is so, take the expectation under $\tilde{P}^{t_2}$ on the right-hand side and use the information in Eq. (14.100):

$$E_{t_1}^{\tilde{P}^{t_2}} \left[F(t_0, t_1) - \sigma_1 F(t_0, t_1)(\lambda_{t_0}^{t_2} \Delta) + \sigma_1 F(t_0, t_1) \Delta W_{t_1}^{t_3} \right] = F(t_0, t_1) \qquad (14.102)$$

Thus, we get the correct result under the $\tilde{P}^{t_2}$, after the mean correction. It is obvious that we need to determine these correction factors before the randomly generated Brownian motion increments can be used in all equations.

Yet, note the following simple case. If the instrument under consideration has additive cash flows where each cash flow depends on a single forward rate, then individual zero-drift equations *can* be used separately to generate paths. This applies for several liquid instruments. For example, FRAs and especially swaps have payment legs that depend on one LIBOR rate only. Individual zero-drift equations can be used for valuing each leg separately, and then these values can be added using *observed* zero-coupon bond prices, $B(t, T_i)$. However, this cannot be done in the case of CMSs, for example, because *each settlement leg will depend nonlinearly on more than one forward rate.*

We now discuss further how mean corrections can be conducted so that all forward rates are projected using a single forward measure. This will permit pricing instruments where individual cash flows depend on more than one forward rate.

14.5 MEASURE CHANGE TECHNOLOGY

We introduce a relatively general framework and then apply the results to the simple example shown previously. Basically, we need three previously developed relationships. We let t_i obey

$$t_0 < t_1 < \cdots < t_n = T \tag{14.103}$$

with

$$t_i - t_{i-1} = \delta \tag{14.104}$$

denote settlement dates of some basic interest rate swap structure and limit our attention to forward rates for successive forward loans contracted to begin at t_i, and paid at t_{i+1}. An example is shown in Figure 14.4.

- *Result 1*

  The forward rate at time t, for a LIBOR-based forward loan that starts at time t_i and ends at time $t_i + \delta$, denoted by $F(t, t_i)$, admits the following arbitrage relationship:

$$1 + F(t, t_i)\delta = \frac{B(t, t_i)}{B(t, t_{i+1})} \quad t < t_i \tag{14.105}$$

  where, as usual, $B(t, t_i)$ and $B(t, t_{i+1})$ are the time t prices of default-free zero-coupon bonds that mature at times t_i and t_{i+1}, respectively.

  The left side of this equality is a gross forward return. The right side, on the other hand, is a traded asset price, $B(t, t_i)$, normalized by another asset price, $B(t, t_{i+1})$. Hence, the ratio will be a Martingale under a proper measure—here, the forward measure denoted by $\tilde{P}^{t_{i+1}}$.

- *Result 2*

  In a discrete state setting with k states of the world, assuming that all asset prices are arbitrage free, and that the time t_i state prices Q^j, $j = 1, \ldots, k$, with $0 < Q^j$ exist, the time t_i values of the forward measure $\tilde{P}^{t_i}$ are given by[23]

$$\tilde{p}_1^{t_i} = \frac{1}{B(t, t_i)} Q^1, \tilde{p}_2^{t_i} = \frac{1}{B(t, t_i)} Q^2, \cdots, \tilde{p}_k^{t_i} = \frac{1}{B(t, t_i)} Q^k \qquad (14.106)$$

  These probabilities satisfy:

$$\tilde{p}_1^{t_i} + \tilde{p}_2^{t_i} + \cdots + \tilde{p}_k^{t_i} = 1 \qquad (14.107)$$

  and

$$0 < \tilde{p}_j^{t_i} \quad \forall j$$

  Note that the proportionality factors used to convert Q^j into $\tilde{p}_j^{t_i}$ are equal across j.
- *Result 3*

  In the same setting, the time t_i-probabilities associated with the t_{i+1} forward measure $\tilde{P}^{t_{i+1}}$ are given by

$$\tilde{p}_1^{t_{i+1}} = \frac{B(t_i, t_{i+1})^1}{B(t, t_{i+1})} Q^1, \tilde{p}_2^{t_{i+1}} = \frac{B(t_i, t_{i+1})^2}{B(t, t_{i+1})} Q^2, \ldots, \tilde{p}_k^{t_{i+1}} = \frac{B(t_i, t_{i+1})^k}{B(t, t_{i+1})} Q^k \qquad (14.108)$$

  where the $B(t_i, t_{i+1})^j$ are the state-dependent values of the t_{i+1}-maturity bond at time t_i. Here, the bond that matures at time t_{i+1} is used to normalize the cash flows for time t_i. Since the maturity date is t_{i+1}, the $B(t_i, t_{i+1})^j$ are not constant at t_i. The factors used to convert $\{Q^j\}$ into $\tilde{P}^{t_{i+1}}$ cease to be constant as well.

We use these results in discussing the mechanics of measure changes. Suppose we need to price an instrument whose value depends on *two* forward LIBOR processes, $F(t, t_i)$ and $F(t, t_{i+1})$, simultaneously. We know that each process is a Martingale and obeys an SDE with zero drift under its *own* forward measure.

Consider a one-factor setting, where a single Wiener process causes fluctuations in the two forward rates. Suppose that in this setting, starting from time t, with $t < t_i$, $i = 1, \ldots, n$, a small time interval denoted by h passes with $t + h < t_i$. By imposing a Gaussian volatility structure, we can write down the *individual* discretized arbitrage-free dynamics for two successive forward rates $F(t, t_i)$ and $F(t, t_{i+1})$ as

$$F(t + h, t_i) - F(t, t_i) = \sigma^i F(t, t_i) \Delta W_{t+h}^1 \qquad (14.109)$$

and

$$F(t + h, t_{i+1}) - F(t, t_{i+1}) = \sigma^{i+1} F(t, t_{i+1}) \Delta W_{t+h}^2 \qquad (14.110)$$

Changes in these forward rates have zero mean under their own forward measure and, hence, are written with zero drift. This means that the unique *real-world* Wiener process W_{t+h} is now denoted by ΔW_{t+h}^1 and ΔW_{t+h}^2 in the two equations. These are normally distributed, with mean zero and variance h *only* under their own forward measures, $\tilde{P}^{t_{i+1}}$ and $\tilde{P}^{t_{i+2}}$.

---

[23]The reader will note the slight change in notation, which is dictated by the environment relevant in this section.

The superscript in W_{t+h}^1 and W_{t+h}^2 expresses the t_{i+1} and t_{i+2} forward probability measures, respectively.[24] Finally, note how we simplify the characterization of volatilities and assume that they are constant over time.

The individual Martingale dynamics are very convenient from a financial engineering point of view. The respective drift components are zero and, hence, they need not be modeled during pricing. The only major task of the market practitioner is to get the respective volatilities σ^i and σ^{i+1}.

However, some securities' prices may depend on more than one forward rate in a nonlinear fashion and their value may have to be calculated as an expectation under *one single* measure. For example, suppose a security's price, S_t, depends on $F(t, t_i)$ and $F(t, t_{i+1})$ through a pricing relation such as:

$$S_t = E_t^{\tilde{P}}[g(F(t, t_i), F(t, t_{i+1}))] \tag{14.111}$$

where $g(.)$ is a known nonlinear function. Then, the expectation has to be calculated under *one* measure only. This probability can be *either* the time t_{i+1} or the time t_{i+2} forward measure. We then have to choose a forward rate equation with Martingale dynamics and carry out a mean correction to get the correct arbitrage-free dynamics for the other. The forward measure of one of the Martingale relationships is set as the *working probability distribution*, and the other equation(s) is obtained in terms of this unique probability by going through successive measure changes. We discuss this in detail below.

14.5.1 THE MECHANICS OF MEASURE CHANGES

We have the following expectations concerning ΔW_{t+h}^1 and ΔW_{t+h}^2, defined in Eqs. (14.109) and (14.110)

$$E_t^{\tilde{P}^{t_{i+1}}}[\Delta W_{t+h}^1] = 0 \tag{14.112}$$

$$E_t^{\tilde{P}^{t_{i+2}}}[\Delta W_{t+h}^2] = 0 \tag{14.113}$$

Under their own forward measure, each Wiener increment has zero expectation. If we select $\tilde{P}^{t_{i+2}}$ as our working measure, one of these equalities has to change. We would have[25]

$$E_t^{\tilde{P}^{t_{i+2}}}[\Delta W_{t+h}^1] = \lambda_t h \tag{14.114}$$

$$E_t^{\tilde{P}^{t_{i+2}}}[\Delta W_{t+h}^2] = 0 \tag{14.115}$$

The value of λ_t gives the correction factor that we need to use in order to obtain the correct arbitrage-free dynamics, if the working measure is $\tilde{P}^{t_{i+2}}$. Calculating this factor implies that we can change measures in the dynamics of $F(t, t_i)$.

We start with the original expectation:

$$E_t^{\tilde{P}^{t_{i+1}}}[\Delta W_{t+h}^1] = \sum_{j=1}^k \Delta W_{t+h}^{1j} \tilde{p}_j^{t_{i+1}} = 0 \tag{14.116}$$

---

[24]Remember that the time t_i forward rate will have a time t_{i+1} forward measure as its own measure.

[25]In the general case where there are m forward rates, all equations except one will change.

where $\tilde{p}_j^{t_i+1}$ are the probabilities associated with the individual states $j = 1, \ldots, k$. Now, using the identity,

$$\frac{B(t, t_{i+2})B(t_i, t_{i+2})^j}{B(t, t_{i+2})B(t_i, t_{i+2})^j} \equiv 1 \tag{14.117}$$

we rewrite the expectation as

$$E_t^{\tilde{p}^{t_i+1}}\left[\Delta W_{t+h}^1\right] = \sum_{j=1}^{k} \Delta W_{t+h}^{1j} \frac{B(t, t_{i+2})B(t_i, t_{i+2})^j}{B(t, t_{i+2})B(t_i, t_{i+2})^j} \equiv \tilde{p}_j^{t_i+1} \tag{14.118}$$

We regroup and use the definition of the t_{i+1} and t_{i+2} forward measures as implied by Result 3

$$\tilde{p}_j^{t_i+1} = \frac{B(t_i, t_{i+1})^j}{B(t, t_{i+1})} Q^j \tag{14.119}$$

and

$$\tilde{p}_j^{t_i+2} = \frac{B(t_i, t_{i+2})^j}{B(t, t_{i+2})} Q^j \tag{14.120}$$

Rescaling the Q^j using appropriate factors, Eq. (14.118) becomes

$$\sum_{j=1}^{k}(\Delta W_{t+h}^{1j})\left[\frac{B(t, t_{i+2})B(t_i, t_{i+1})^j}{B(t, t_{i+1})B(t_i, t_{i+2})^j}\right]\tilde{p}_j^{t_i+2} = 0 \tag{14.121}$$

Note that the probabilities switch as the factors that were applied to the Q^j changed. The superscript in W_{t+h}^1 does not change.

The next step in the derivation is to try to "recognize" the elements in this expectation. Using Result 1, we recognize the equality

$$1 + \delta F(t_i, t_{i+1})^j = \frac{B(t_i, t_{i+1})^j}{B(t_i, t_{i+2})^j} \tag{14.122}$$

Replacing, eliminating the j-independent terms, and rearranging gives

$$\sum_{j=1}^{k}(\Delta W_{t+h}^{1j})(1 + \delta F(t_i, t_{i+1})^j)\tilde{p}_j^{t_i+2} = 0 \tag{14.123}$$

Now, multiplying through, this leads to

$$\sum_{j=1}^{k}(\Delta W_{t+h}^{1j})\tilde{p}_j^{t_i+2} = -\left(\sum_{j=1}^{k}(\Delta W_{t+h}^{1j})F(t_i, t_{i+1})^j\tilde{p}_j^{t_i+2}\right)\delta \tag{14.124}$$

We can write this using the conditional expectation operator,

$$E_t^{\tilde{p}^{t_i+2}}[\Delta W_{t+h}^1] = -E_t^{\tilde{p}^{t_i+2}}[\Delta W_{t+h}^1 F(t_i, t_{i+1})]\delta. \tag{14.125}$$

In the last expression, the left-hand side is the desired expectation of the ΔW_{t+h}^1 under the new probability $\tilde{P}^{t_i+2}$. This expectation will *not* equal zero if the right-hand-side random variables are correlated. This correlation is nonzero as long as forward rates are correlated. To evaluate the mean of ΔW_{t+h}^1 under the new probability $\tilde{P}^{t_i+2}$, we then have to calculate the covariance.

Let the covariance be given by $-\lambda_t h$. We have,

$$\delta E_t^{\tilde{P}^{i+2}}[\Delta W_{t+h}^1 F(t_i, t_{i+1})] = -\lambda_t h \tag{14.126}$$

Using λ_t, we can switch probabilities in the $F(t, t_i)$ dynamics. We start with the original Martingale dynamics:

$$F(t + h, t_i) = F(t, t_i) + \sigma^i F(t, t_i)\Delta W_{t+h}^1 \tag{14.127}$$

Switch by adding and subtracting $\sigma^i F(t, t_i) \lambda_t h$ to the right-hand side and regroup:

$$F(t + h, t_i) = F(t, t_i) - \sigma^i F(t, t_i)\lambda_t h + \sigma^i F(t, t_i)[\lambda_t h + \Delta W_{t+h}^1] \tag{14.128}$$

Let

$$\Delta W_{t+h}^2 = [\lambda_t h + \Delta W_{t+h}^1] \tag{14.129}$$

We have just shown that the expectation of the right-hand side of this expression equals zero under $\tilde{P}^{i+2}$. So, under the $\tilde{P}^{i+2}$ we can write the new dynamics of the $F(t, t_i)$ as

$$F(t + h, t_i) = F(t, t_i) - \sigma^i F(t, t_i)\lambda_t h + \sigma^i F(t, t_i)\Delta W_{t+h}^2 \tag{14.130}$$

As can be seen from this expression, the new dynamics have a non-zero drift and $F(t, t_i)$ is *not* a Martingale under the new measure. Yet, this process is arbitrage free and easy to exploit in Monte Carlo type approaches. Since both dynamics are expressed under the same measure, the set of equations that describe the dynamics of the two forward rates can be used in pricing instruments that depend on these forward rates. The same pseudo-random numbers can be used in the two SDEs. Finally, the reader should remember that the discussion in this section depends on the discrete approximation of the SDEs.

14.5.2 GENERALIZATION

A generalization of the previous heuristic discussion leads to the Forward LIBOR Model. Suppose the setting involves n forward rates, $F(t_0, t_i)$, $i = 0, \ldots, n - 1$, that apply to loans which begin at time t_i, and end at $t_{i+1} = t_i + \delta$. The $F(t_0, t_0)$ is the trivial forward rate and is the spot LIBOR with tenor δ. The terminal date is t_n.

Similar to the discussion in the previous section, assume that there is a single factor.[26] Using the t_{i+1} forward measure we obtain arbitrage-free Martingale dynamics for each forward rate $F(t, t_i)$:

$$dF(t, t_i) = \sigma^i F(t, t_i)W_t^{i+1} \quad t \in [0, \infty) \tag{14.131}$$

The superscript in W_t^{i+1} implies that[27]

$$E_t^{\tilde{P}^{i+1}}[dW_t^{i+1}] = 0 \tag{14.132}$$

These arbitrage-free dynamics are very useful since they do not involve any interest rate modeling and are dependent only on the correct specification of the respective volatilities. However, when

---

[26]The multifactor model and an extensive discussion of the Forward LIBOR Model can be found in many texts. Musiela and Rutkowski (1998), Brigo and Mercurio (2006), and Rebonato (2002) are some examples.

[27]The use of a stochastic differential here is heuristic.

more than one forward rate determines a security's payoff in a nonlinear fashion, the process may have to be written under a unique working measure.

Suppose we chose $\tilde{P}^{t_n}$ as the working measure.[28] The heuristic approach discussed in the previous section can be generalized to obtain the following arbitrage-free *system* of SDEs that involve recursive drift corrections in a one-factor case:

$$dF(t, t_i) = - \left[\sigma^i F(t, t_i) \sum_{j=i}^{n-1} \frac{\delta \sigma^j F(t, t_j)}{1 + F(t, t_j)\delta} \right] dt + \sigma^i F\left[(t, t_i)dW_t^{t_n}\right] \quad t \in [0, \infty) \tag{14.133}$$

where superscript in $dW_t^{t_n}$ indicates that the working measure is $\tilde{P}^{t_n}$. The equations in this system are expressed under this forward measure for $i = 1, \ldots, n$. Yet, only the last equation has Martingale dynamics

$$dF(t, t_{n-1}) = \sigma^{n-1} F(t, t_{n-1}) dW_t^{t_n} \quad t \in [0, \infty) \tag{14.134}$$

All other SDEs involve successive correction factors given by the first term on the right side. It is important to realize that all terms in these factors can be observed at time t. The dynamics do not need a modeling of actual drifts.

14.6 AN APPLICATION

The forward measure and measure change technology are relevant for the pricing of many instruments. But there is one instrument class that has recently become quite popular with market participants and that can be priced with this technology. These are CMSs. They have properties that would illustrate some subtleties of the methods used thus far. In order to price them, forward rates need to be projected *jointly*.

First, we present a reading that illustrates some of the recent interest in this instrument class.

EXAMPLE

European institutional investors mindful of rising interest rates have been turning to constant maturity swap products to hedge their long-dated liabilities over the past few months. Whereas mainstream interest rate swaps tend to reference LIBOR, CMS pay-out structures are linked to the 10-year swap rate instead—a useful hedge for investors with long-dated liabilities such as insurance companies.

As the 10-year euro swap cratered from a peak of 3.77% in April 2011 to a low of 1.44% in early May this year, these firms piled into CMS structures to shield themselves against falling interest rates. The spectre of the US Federal Reserve scaling back its asset purchases was enough to put paid to this trend. By the end of June, the 10-year swap rate had rocketed to above 2% and CMS products are now gaining traction as a hedge against further rises.

"Buying of CMS caps has picked up in the first half of the year because of the change of sentiment in the market, which has put rising rates to the forefront of investors' minds," said Ivan Jossang, head of EMEA rates structuring at Morgan Stanley.

Caps have traditionally been popular with investors due to the relative simplicity of the structure. Take the example of an investor buying a cap with a 3% strike. If rates rise to 4% by the time the cap expires, the investor will be paid 100 bp on the notional it originally agreed. "CMS is the main light exotic product that

---

[28]Sometimes this is called the *terminal measure*.

survived the crisis, and is the most important building block for pay-offs in rates," said Adrian Bracher, European head of rates structuring at Credit Suisse.

As with many exotic products, CMS flourished in the run-up to the financial crisis and the market has since shrunk. There is no longer any CMS market to speak of in Swiss francs and sterling, with dollars and euros remaining the most liquid currencies. [...]

French insurance companies are without doubt the most prolific CMS users, and not just because of the proximity of the euro swaps curve to sovereign yields, against which many of their liabilities are benchmarked. They also offer savings products to policyholders comparable to bank deposits, which need hedging if rates go up or down. [...]

Bank hedging of CMS trades has seriously impacted on the swaptions market, making the 10-year euro swaption trade at a premium to options on the longer end of the curve. This has led some of the more dynamic Dutch and Danish pension funds—which use swaptions to hedge their liabilities—to move into the comparatively cheaper 20-year and 30-year contracts, according to Jossang.

CMS has also come into play on the asset side of the balance sheet. As 10-year swap rates are higher that 3-month Euribor, CMS can act as a useful yield enhancement tool, often overlaid on government bonds or even private placements from corporates if their treasurers can be persuaded to come on board.

(Thomson Reuters IFR Issue 1996, *CMS makes comeback to hedge rate rises*

by Christopher Whittall, August 14, 2013)

CMSs are instruments that build on the plain vanilla swaps in an interesting way. In a vanilla swap, a fixed swap rate is exchanged against a floating LIBOR that is an interest rate relevant for that particular settlement period *only*. In a CMS, this will be generalized. The fixed leg is exchanged against a floating leg, but the floating leg is *not* a "one-period" rate. It is itself a multi-period swap rate that will be determined in the future.

There are many versions of such exchanges, but as an example we consider the following. Suppose one party decides to pay 4% during the next 3 years against receiving a *2-year swap rate* that will be determined at the beginning of each one of those years. The future swap rates are unknown at time t_0 and can be considered as floating payments, except they are not floating payments that depend on the perceived volatility for that particular year *only*. They are themselves averages of 1-year rates. Clearly, such swaps have significant nonlinearities and we cannot do the same engineering as in the case of a plain vanilla swap.

An example of CMS is shown in Figure 14.5. The reader can see that what is being exchanged at each settlement date against a fixed payment is a floating rate that is a function of *more than one* forward rate. Under these conditions it is impossible to project individual forward rates using individual zero-drift stochastic differential equations defined under different forward measures. Each leg of the CMS depends on more than one forward rate and these need to be projected *jointly*, under a single measure.

14.6.1 ANOTHER EXAMPLE OF MEASURE CHANGE

This section provides another example to measure change technology from the FRA markets. Paid-in-arrears FRAs make time t_{i+1} payoffs:

$$N\delta[F(t_0, t_i) - L_{t_i}] \tag{14.135}$$

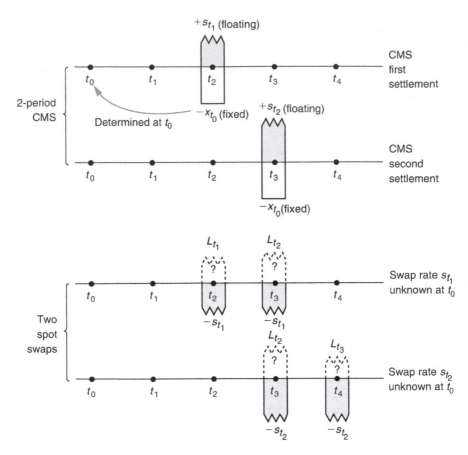

FIGURE 14.5

Example of a CMS.

The market-traded FRAs, on the other hand, settle at time t_i according to:

$$\frac{N\delta[F(t_0, t_i) - L_{t_i}]}{(1 + \delta L_{t_i})} \tag{14.136}$$

Finally, we have LIBOR-in-arrears FRAs that settle according to

$$N\delta[F(t_0, t_i) - L_{t_i}] \tag{14.137}$$

at time t_i. As we saw in Chapter 10, the LIBOR-in-arrears FRA payoffs settle in a "nonnatural" way, since L_{t_i}-related payments would normally be received or paid at time t_{i+1}.

We now show that the paid-in-arrears FRA and market-traded FRAs lead to the same forward rate. First, remember that under the $\tilde{P}^{t_{i+1}}$ forward measure for paid-in-arrears FRAs, we have:

$$F(t_0, t_i) = E_{t_0}^{\tilde{P}^{t_{i+1}}}[L_{t_i}] \tag{14.138}$$

That is to say, the FRA rate $F(t_0, t_i)$ is the average of possible values the LIBOR rate might take:

$$F(t_0, t_i) = \sum_{j=1}^{k} L_{t_i}^j \tilde{p}_j^{t_{i+1}} \tag{14.139}$$

where j represents possible states of the world, which are assumed to be discrete and countable.

Now, consider the settlement amount of market-traded FRAs:

$$\frac{N\delta[F(t_0, t_i) - L_{t_i}]}{(1 + \delta L_{t_i})} \tag{14.140}$$

Would the forward rate implied by this contract be the same as the paid-in-arrears FRAs?

The answer is yes. Using the measure change technology, we discuss how this can be shown. The idea is to begin with the expectation of this settlement amount under the $\tilde{P}^{t_i}$ measure and show that it leads to the same forward rate. Thus, begin with

$$E_{t_0}^{\tilde{p}^{t_i}} \left[\frac{N\delta[F(t_0, t_i) - L_{t_i}]}{(1 + \delta L_{t_i})} \right] \tag{14.141}$$

Setting this equal to zero, and rearranging, leads to the pricing equation

$$F(t_0, t_i) = \frac{E_{t_0}^{\tilde{p}^{t_i}} [N\delta L_{t_i}/(1 + \delta L_{t_i})]}{E_{t_0}^{\tilde{p}^{t_i}} [N\delta/(1 + \delta L_{t_i})]} \tag{14.142}$$

Now we switch measures on the right-hand side of Eq. (14.142). We have two expectations and we will switch measures in both of them. But first, let $N = 1$ and, similarly, $\delta = 1$.

Consider the numerator:

$$E_{t_0}^{\tilde{p}^{t_i}} \left[\frac{L_{t_i}}{(1 + L_{t_i})} \right] = \sum_{j=1}^{k} \frac{L_{t_i}^j}{\left(1 + L_{t_i}^j\right)} \tilde{p}_j^{t_i} \tag{14.143}$$

We know that for time t_i

$$\tilde{p}_j^{t_i} = \frac{1}{B(t_0, t_i)} Q^j$$

$$\tilde{p}_j^{t_{i+1}} = \frac{B(t_i, t_{i+1})^j}{B(t_0, t_{i+1})} Q^j \tag{14.144}$$

Thus:

$$\tilde{p}_j^{t_i} = \frac{1}{B(t_0, t_i)} \frac{B(t_0, t_{i+1})}{B(t_i, t_{i+1})^j} \tilde{p}_j^{t_{i+1}} \tag{14.145}$$

Replacing the right-hand side in Eq. (14.143), we get

$$\sum_{j=1}^{k} \frac{L_{t_i}^j}{\left(1 + L_{t_i}^j\right)} \frac{1}{B(t_0, t_i)} \frac{B(t_0, t_{i+1})}{B(t_i, t_{i+1})^j} \tilde{p}_j^{t_{i+1}} \tag{14.146}$$

In this expression, we recognize

$$\frac{B(t_0, t_{i+1})}{B(t_0, t_i)} = \frac{1}{1 + F(t_0, t_i)} \tag{14.147}$$

and

$$\frac{1}{B(t_i, t_{i+1})^j} = 1 + L_{t_i}^j \tag{14.148}$$

Using these we get the equivalence:

$$\sum_{j=1}^{k} \frac{L_{t_i}^j}{\left(1 + L_{t_i}^j\right)} \frac{1}{B(t_0, t_i)} \frac{B(t_0, t_{i+1})}{B(t_i, t_{i+1})^j} \tilde{p}_j^{t_{i+1}} = \sum_{j=1}^{k} \frac{L_{t_i}^j}{\left(1 + L_{t_i}^j\right)} \frac{1}{(1 + F(t_0, t_i))} \left(1 + L_{t_i}^j\right) \tilde{p}_j^{t_{i+1}} \tag{14.149}$$

Simplifying the common terms on the right-hand side reduces to

$$\sum_{j=1}^{k} \frac{L_{t_i}^j}{1 + F(t_0, t_i)} \tilde{p}_j^{t_{i+1}} \tag{14.150}$$

This we immediately recognize as the expectation:

$$E_{t_0}^{\tilde{p}^{i+1}} \left[\frac{L_{t_i}^j}{1 + F(t_0, t_i)} \right] \tag{14.151}$$

Now, consider the denominator in Eq. (14.142)

$$E_{t_0}^{\tilde{p}^i} \left[\frac{1}{(1 + L_{t_i})} \right] = \sum_{j=1}^{k} \frac{1}{\left(1 + L_{t_i}^j\right)} \tilde{p}_j^{t_i} \tag{14.152}$$

Using Eq. (14.144), we switch to the $\tilde{P}^{t_{i+1}}$ measure:

$$\sum_{j=1}^{k} \frac{1}{\left(1 + L_{t_i}^j\right)} \tilde{p}_j^{t_i} = \sum_{j=1}^{k} \frac{1}{\left(1 + L_{t_i}^j\right)} \frac{1}{B(t_0, t_i)} \frac{B(t_0, t_{i+1})}{B(t_i, t_{i+1})^j} \tilde{p}_j^{t_{i+1}} \tag{14.153}$$

Use the equivalences in Eq. (14.144), substitute:

$$\sum_{j=1}^{k} \frac{1}{\left(1 + L_{t_i}^j\right)} \frac{1}{B(t_0, t_i)} \frac{B(t_0, t_{i+1})}{B(t_i, t_{i+1})^j} \tilde{p}_j^{t_{i+1}} = \sum_{j=1}^{k} \frac{1}{\left(1 + L_{t_i}^j\right)} \frac{1}{(1 + F(t_0, t_i))} \left(1 + L_{t_i}^j\right) \tilde{p}_j^{t_{i+1}} \tag{14.154}$$

Note that, again, the random $(1 + L_{t_i}^j)$ terms conveniently cancel, and on the right-hand side we obtain:

$$= \sum_{j=1}^{k} \frac{1}{1 + F(t_0, t_i)} \tilde{p}_j^{t_{i+1}}$$

$$= \frac{1}{(1 + F(t_0, t_i))}$$

Putting the numerator and denominator together for general N and δ gives

$$F(t_0, t_i) = \frac{E_{t_0}^{\tilde{p}^i} \left[N \delta L_{t_i} / \left(1 + \delta L_{t_i}\right) \right]}{E_{t_0}^{\tilde{p}^i} \left[N \delta / \left(1 + \delta L_{t_i}\right) \right]} = \frac{E_{t_0}^{\tilde{p}^{i+1}} \left[N \delta \left(N_{t_i}^j / (1 + F(t_0, t_i) \delta) \right) \right]}{(N \delta / (1 + F(t_0, t_i) \delta))} \tag{14.155}$$

We simplify the common terms to get

$$F(t_0, t_i) = E_{t_0}^{\tilde{p}^{i+1}} \left[L_{t_i}^j \right] \tag{14.156}$$

Hence, we obtained the desired result. The FRA rate of paid-in-arrears FRAs is identical to the FRA rate of market-traded FRAs and is an unbiased predictor of the LIBOR rate L_{t_i}, under the right forward measure.

We conclude this section with another simple example.

EXAMPLE

We can apply the forward measure technology to mark-to-market practices as well. The paid-in-arrears FRA will settle at time t_{i+1} according to

$$\left[L_{t_i} - F(t_0, t_i)\right]N\delta \tag{14.157}$$

What is the value of this contract at time t_1, with $t_0 < t_1 < t_i$?

It is market convention to replace the random variable L_{t_i} with the corresponding forward rate of time t_1. We get

$$[F(t_1, t_i) - F(t_0, t_i)]N\delta \tag{14.158}$$

which, in general, will be nonzero. How do we know that this is the correct way to mark the contract to market? We simply take the time t_1 expectation of:

$$\left[L_{t_i} - F(t_0, t_i)\right]N\delta \tag{14.159}$$

with respect to the natural forward measure of t_{i+1}

$$E_{t_1}^{\tilde{P}^{i+1}}\left[L_{t_i} - F(t_0, t_i)\right]N\delta = [F(t_1, t_i) - F(t_0, t_i)]N\delta \tag{14.160}$$

where we use the fact that under the $\tilde{P}^{i+1}$, the $F(t_1, t_i)$ is an unbiased estimate of L_{t_i}.

14.6.2 PRICING CMS

Pricing CMS is known to involve convexity adjustments. Staying within the context of the simple framework used in this chapter, the industry first obtains the $t_1 \times t_2$ and $t_2 \times t_3$ swaption volatilities. Then, knowing that the swap is a Martingale under the "annuity" measure treated in Chapter 17, various transformations under specific assumptions are performed and then the convexity correction to the forward swap rate is estimated. In other words, the industry calculates the ϵ_t in the equation

$$cms_t = s_t^f + \epsilon_t \tag{14.161}$$

where cms_t is the CMS rate, s_t^f is the relevant forward swap rate, and ϵ_t is the convexity correction.

It is straightforward to price CMS using the forward LIBOR dynamics discussed earlier and then use successive measure changes for the required mean corrections. Because CMS offer a good example for such an application, we show a simple case.

Consider a two-period forward CMS where a fixed CMS rate x_{t_0} is paid at times t_2 and t_3 against the floating two-period cash swap rate at these times. The present value of the cash flows under the $\tilde{P}^{t_3}$ forward probability is given by

$$0 = E_{t_0}^{\tilde{P}^{t_3}}\left[\left(x_{t_0} - s_{t_1}\right) \frac{1}{\left(1 + L_{t_0}\delta\right)\left(1 + L_{t_1}\delta\right)} + \left(x_{t_0} - s_{t_2}\right) \right.$$

$$\left. \frac{1}{\left(1 + L_{t_0}\delta\right)\left(1 + L_{t_1}\delta\right)\left(1 + L_{t_2}\delta\right)} \right] N \tag{14.162}$$

where the settlement interval is assumed to be one and N is the notional swap amount. The s_{t_1} and s_{t_2} are the two 2-period swap rates unknown at time t_0. They are given by the usual spot swap formula shown in Eq. (14.49).

Setting $\delta = 1$, and rearranging this equation, we obtain

$$x_{t_0} = \frac{E_{t_0}^{\tilde{P}^{t_3}}\left[s_{t_1}\left(1/\left(1 + L_{t_0}\right)\left(1 + L_{t_1}\right)\right) + s_{t_2}\left(1/\left(1 + L_{t_0}\right)\left(1 + L_{t_1}\right)\left(1 + L_{t_2}\right)\right) \right]}{E_{t_0}^{\tilde{P}^{t_3}}\left[\left(1/\left(1 + L_{t_0}\right)\left(1 + L_{t_1}\right)\right) + \left(1/\left(1 + L_{t_0}\right)\left(1 + L_{t_1}\right)\left(1 + L_{t_2}\right)\right) \right]} \tag{14.163}$$

Hence, to find the value of the CMS rate x_{t_0}, all we need to do is write down the dynamics of the forward LIBOR processes, $F(t_0, t_1)$ and $F(t_0, t_2)$, under the *same* forward measure $\tilde{P}^{t_3}$ as done earlier, and then select Monte Carlo paths.

It is clear that proceeding in this way and obtaining Monte Carlo paths from the arbitrage-free forward LIBOR dynamics requires calibrating the respective volatilities σ^i. But once this is done, and once the correction factors are included in the proper equations, the Monte Carlo paths can be selected in a straightforward manner. The CMS rate can then be calculated from

$$x_{t_0} = \frac{\sum_{j=1}^{M}\left[s_{t_1}^j\left(1/\left(1 + L_{t_0}^j\right)\left(1 + L_{t_1}^j\right)\right) + s_{t_2}^j\left(1/\left(1 + L_{t_0}^j\right)\left(1 + L_{t_1}^j\right)\left(1 + L_{t_2}^j\right)\right) \right]}{\sum_{j=1}^{M}\left[\left(1/\left(1 + L_{t_0}^j\right)\left(1 + L_{t_1}^j\right)\right) + \left(1/\left(1 + L_{t_0}^j\right)\left(1 + L_{t_1}^j\right)\left(1 + L_{t_2}^j\right)\right) \right]} \tag{14.164}$$

where the swap rates $s_{t_i}^j$ themselves depend on the same forward rate trajectories and, hence, can be calculated from the selected paths.

The same exercise can be repeated by starting from perturbed values of volatilities and initial forward rates to obtain the relevant Greeks for risk-management purposes as well.

14.7 IN-ARREARS SWAPS AND CONVEXITY

Although an overwhelming proportion of swap transactions involve the vanilla swap, in some cases parties transact the so-called *LIBOR-in-arrears swap*. In this section, we study this instrument because it is a good example of how Forward LIBOR volatilities enter pricing *directly* through convexity adjustments.

But first we need to clarify the terminology. In a vanilla swap, the LIBOR rates L_{t_i} are assumed to "set" at time t_i whereas the floating *payments* are made in arrears at times t_{i+1}.[29] In the case of an

[29]It is important to remember that, in reality, there is another complication. The LIBOR is set, by convention, 2 business days before time t_i and the payment is made at t_{i+1}. Here we are ignoring this convention because it really does not affect the understanding of the instruments and pricing, while making the formulas easier to understand. The reader can incorporate such real-life modifications in the formulas given below.

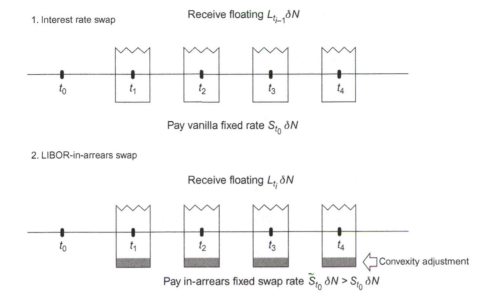

FIGURE 14.6

Interest rate swap versus LIBOR-in-arrears swap.

in-arrears swap, the payment days are kept the same, but the time t_{i+1} settlement will use the LIBOR rate $L_{t_{i+1}}$ that has *just* been observed at time t_{i+1}[30] to determine the floating payment. Thus, in a sense the *setting* of the LIBOR rate is in arrears, hence the name of the in-arrears swap.[31] The difference between LIBOR resets is illustrated in and the difference between vanilla interest rate swaps and LIBOR-in-arrears swaps is shown in Figure 14.6. LIBOR-in-arrears swaps *set* the LIBOR rates in arrears.

The simple modification of paying the $L_{t_{i+1}}$ observed at time t_{i+1} rather than the previously observed L_{t_i} makes a significant difference in pricing. We will work with a simple case of a two-period (forward) swap first, and then give the generalized formulas.

14.7.1 VALUATION

The valuation of the *fixed leg* of the in-arrears swap is the same as that of the vanilla swap, except of course the swap coupons are different. Let the s_{t_0} be the *vanilla* swap rate fixed at time t_0, N and δ be the notional amount and accrual parameters, respectively. Then the fixed leg payments are easy to value:

$$
\begin{aligned}
\text{Fixed-Leg}_{t_0} &= B(t_0, t_2)s_{t_0}\delta N + B(t_0, t_3)s_{t_0}\delta N \\
&= [B(t_0, t_2) + B(t_0, t_3)]s_{t_0}\delta N
\end{aligned}
\tag{14.165}
$$

---

[30]See the previous footnote concerning the two business days convention.
[31]Although in case of vanilla swaps, the payments are also in arrears.

It is clear that in case of the in-arrears swap, we will have a similar expression:

$$\text{Fixed-Leg}_{t_0} = [B(t_0, t_2) + B(t_0, t_3)]is_{t_0}\delta N \tag{14.166}$$

where the is_{t_0} is the swap rate of the in-arrears swap.

The difference in valuations emerge in the *floating leg*. Note that in the case of the in-arrears swap, the expected value under the P^{t_2} forward measure of the floating rate payments would be

$$\text{Floating-Leg}_{t_0} = E_{t_0}^{P^{t_2}}[B(t_0, t_2)L_{t_2}]\delta N + E_{t_0}^{P^{t_2}}[B(t_0, t_3)]\delta N \tag{14.167}$$

Multiply and divide by $(1 + L_{t_2}\delta)$ and $(1 + L_{t_3}\delta)$, respectively, we obtain

$$\text{Floating-Leg} = E_{t_0}^{P^{t_2}}\left[B(t_0, t_2)L_{t_2}\frac{(1 + L_{t_2}\delta)}{(1 + L_{t_2}\delta)}\right]\delta N + E_{t_0}^{P^{t_2}}\left[B(t_0, t_3)L_{t_3}\frac{(1 + L_{t_3}\delta)}{(1 + L_{t_3}\delta)}\right]\delta N \tag{14.168}$$

But in the case of finite-state random quantities, we have the usual correspondence between the t_3 and t_2 forward measures:

$$p_i^{t_2}\frac{B(t_2, t_3)^i}{B(t_0, t_3)}B(t_0, t_2) = p_i^{t_3} \tag{14.169}$$

Also, by definition

$$B(t_2, t_3)^i = \frac{1}{\left(1 + L_{t_2}^i\delta\right)} \tag{14.170}$$

This means that after regrouping, changing measures from P^{t_2} to P^{t_3}:

$$E_{t_0}^{P^{t_2}}\left[B(t_0, t_2)L_{t_2}\frac{(1 + L_{t_2}\delta)}{(1 + L_{t_2}\delta)}\right]\delta N = E_{t_0}^{P^{t_3}}\left[B(t_0, t_3)L_{t_2}\left(1 + L_{t_2}\delta\right)\right]\delta N \tag{14.171}$$

Changing the measure from P^{t_3} to P^{t_4}, a similar set of equations gives

$$E_{t_0}^{P^{t_3}}\left[B(t_0, t_3)L_{t_3}\frac{(1 + L_{t_3}\delta)}{(1 + L_{t_3}\delta)}\right] = E_{t_0}^{P^{t_4}}\left[B(t_0, t_4)L_{t_3}\left(1 + L_{t_3}\delta\right)\right] \tag{14.172}$$

Thus the valuation of the floating leg becomes

$$\text{Floating-Leg}_{t_0} = \left[E_{t_0}^{P^{t_3}}\left[B(t_0, t_3)L_{t_2}\left(1 + L_{t_2}\delta\right)\right] + E_{t_0}^{P^{t_4}}\left[B(t_0, t_4)L_{t_3}\left(1 + L_{t_3}\delta\right)\right]\right]\delta N \tag{14.173}$$

The right-hand side can be expanded to

$$\text{Floating-Leg}_{t_0} = B(t_0, t_3)\left[E_{t_0}^{P^{t_3}}\left[L_{t_2}\right] + E_{t_0}^{P^{t_3}}\left[\left(L_{t_2}\delta\right)^2\right]\right]\delta N + B(t_0, t_4)\left[E_{t_0}^{P^{t_4}}\left[L_{t_3}\right]\right.$$

$$\left. + E_{t_0}^{P^{t_4}}\left[\left(L_{t_3}\delta\right)^2\right]\right]\delta N \tag{14.174}$$

But the forward rates are Martingales with respect to their own measures. So,

$$F_{t_0}^{t_2} = E_{t_0}^{P^{t_3}}\left[L_{t_2}\right] \tag{14.175}$$

$$F_{t_0}^{t_3} = E_{t_0}^{P^{t_4}}\left[L_{t_3}\right] \tag{14.176}$$

and

$$E_{t_0}^{P^{t_3}}\left[L_{t_2}^2\right] = \left(F_{t_0}^{t_2}\right)^2 e^{\int_{t_0}^{t_2} \sigma(u)_{t_2}^2\, du} \tag{14.177}$$

$$E_{t_0}^{P^{t_4}}\left[L_{t_3}^2\right] = \left(F_{t_0}^{t_3}\right)^2 e^{\int_{t_0}^{t_3} \sigma(u)_{t_2}^2\, du} \tag{14.178}$$

So we get the final result as:

$$
\begin{aligned}
\text{Floating-Leg}_{t_0} = {} & B(t_0, t_3)\left[F_{t_0}^{P^{t_2}} + \delta\left(F_{t_0}^{t_2}\right)^2 e^{\int_{t_0}^{t_2} \sigma(u)_{t_2}^2\, du}\right] \\
& + B(t_0, t_4)\left[F_{t_0}^{t_3} + \delta\left(F_{t_0}^{t_3}\right)^2 e^{\int_{t_0}^{t_3} \sigma(u)_{t_3}^2\, du}\right]\delta N
\end{aligned}
\tag{14.179}
$$

This can be expressed as the floating leg of a vanilla swap plus an adjustment called the *convexity adjustment*:

$$
\begin{aligned}
\text{Floating-Leg}_{t_0} = {} & \left[B(t_0, t_3)F_{t_0}^{t_2} + B(t_0, t_4)F_{t_0}^{t_3}\right]\delta N + \left[B(t_0, t_3)\delta\left(F_{t_0}^{t_2}\right)e^{\int_{t_0}^{t_2} \sigma(u)_{t_2}^2\, du}\right. \\
& \left. + B(t_0, t_4)\delta\left(F_{t_0}^{t_3}\right)^2 e^{\int_{t_0}^{t_3} \sigma(u)_{t_3}^2\, du}\right]\delta N
\end{aligned}
\tag{14.180}
$$

For small settlement intervals, Δ and constant volatilities the approximation becomes

$$\left[B(t_0, t_3)\delta\left(F_{t_0}^{t_2}\right)^2 e^{\Delta \sigma_{t_2}^2} + B(t_0, t_4)\delta\left(F_{t_0}^{t_3}\right)^2 e^{\Delta \sigma_{t_3}^2}\right]\delta N \tag{14.181}$$

Note that the second bracketed term in the convexity adjustment is positive. This makes the value of the floating rate payments in the in-arrears swap be *greater* than the value of the floating payments in the vanilla swap. The consequence of this is that the in-arrears fixed swap rate denoted by $\tilde{s}_t$ is bigger than the vanilla fixed rate

$$s_{t_0} < \tilde{s}_{t_0} \tag{14.182}$$

A number of comments can be made. First note that the volatilities can be obtained from the corresponding caplet volatilities. Second, note that the value of the in-arrears swap does not depend on the correlation between various forward rates. Third, the volatilities are likely to be different than the swaption volatilities.

14.7.2 SPECIAL CASE

A special case of this is if we look at a single period in-arrears swap. Then we get a relation between forward rates and futures rates.

A Eurodollar contract leads to an exchange of f_{t_0} for L_{t_i} at time t_i. The forward contract leads to an exchange of F_{t_0} for $L_{t_{i-1}}$ at t_i. So this is the one-period replica of the comparison we just made.

This means

$$f_{t_0}^{t-i} - F_{t_0}^{t-i} = \delta\left(F_{t_0}\right)^2 e^{\int_{t_0}^{t_i} \sigma(u)_{t_i}^2\, du} \tag{14.183}$$

Directly from Eq. (14.177) of the previous section. The right-hand side is known as the convexity adjustment that needs to be applied when going from futures to forward rates. Note that the futures *price* of the contract will then be smaller than the forward price.[32]

14.8 CROSS-CURRENCY SWAPS

A cross-currency swap can serve different purposes. It can be used for hedging, that is to reduce the exposure to exchange rate changes. Alternatively, one can use it to exploit arbitrage opportunities between different rates.

The following reading illustrates some of the recent issues related to cross-currency swaps.

EXAMPLE

It has been a rollercoaster couple of years for the Australian cross-currency basis swap [...], which has ripsawed to unprecedented levels primarily as a result of large irregular capital dislocations during the global financial crisis. On October 10, 2008, a shortage of US dollars, coupled with major rehedging of Japanese power reverse dual currency (PRDC) notes, caused the five-year AUD/USD cross-currency basis swap to drop to −50 basis points—quite a move for a swap that has traditionally traded in a fairly predictable range of around six to 10 bp. Its 10-year equivalent suffered an equally volatile fate, hitting −47 bp on the same date.

By September 2009, this situation had completely reversed. Kangaroo bond issuance had all but ground to a halt, while Australian banks—buoyed by AA credit ratings and government guarantees—were able to comfortably tap offshore markets to shore up their balance sheets. [...]

These extreme levels of volatility have attracted the attention of a number of hedge funds. "The hedge fund community has been active in the basis swap market for some time. By nature, it is a function of opportunity," says Anthony Robson, head of rates for Australian and New Zealand dollars at Barclays Capital in Sydney.

Those opportunities, he says, have occurred regularly in the past two years. "In 2008 and early 2009, the yield curve was steeply negative, which reflected the sheer illiquidity at that time, severe global funding issues and bank hedging need to receive in the long-end around 30 years. For hedge funds looking for the basis swap to widen, that presented attractive opportunities to pay. For example, they could pay a five-year basis swap, starting in five years' time and pick up 60 bp, benefiting from the negative-shaped curve and the very rapid change in levels," he says, adding: "At the moment, and especially late last year, the basis swap is quite steeply positive, so there are certainly receivers of the forward basis spread."

Dealers say that since hitting its peak in late 2009, interest from hedge fund managers in the basis swap has noticeably picked up. Betting on the basis to trend back towards its historical average, they have put on a range of forward-starting basis swaps to extract value from the shape of the curve. "A fund manager could receive a five-year basis swap on a hold-to-maturity basis, yielding say, +31.5 bp," says Ian Martin, head of global rates for Australia and New Zealand at Deutsche Bank in Sydney, who says hedge funds have been particularly involved in two- to five-year maturities. "By hedging the first year's cash flows at +5.5 bp, the manager can create a forward-starting four-year swap at 39 bp. Assuming the swap tightens in line with its historical average, they would benefit from carry and capital appreciation as the trade rolls down the curve."

Since the fourth quarter of 2009, following a surge in kangaroo issuance, the three- to seven-year basis swap tenors have come back in substantially from their late 2009 peaks, with the five-year AUD/USD cross-currency swap falling to +32 bp on February 17. Ten-year and longer tenors, however, have not fallen as

---

[32]Because we subtract a bigger term.

sharply. That caused the inversion of the long-end of the basis to remain pronounced and has led Deutsche Bank to recommend a butterfly spread to take advantage of the inversion of the curve. "While the basis swap spread slope has returned to fairly normal long-run levels, the five-year/10-year/15-year butterfly is currently near the wides reached in February last year. We think this butterfly (receiving 10-year basis against paying the five-year and 15-year) is one of the better trade opportunities in basis swap spreads at present, and we recommend entering the trade," the bank wrote in a research note dated February 17.

('Betting on basis' by Wietske Blees, Asia Risk, 19 April 2010. URL: http://www.risk.net/asia-risk/ feature/1600830/betting-basis).

As an example of the use of a cross-currency swap for hedging, consider a European company that intends to buy US dollar bonds and would like to hedge the US dollar exchange rate risk. The European company could buy a cross-currency swap from a US-based bank. As a result of the swap, the European company would pay in euros and receive a (fixed) US dollar cash flow. These US dollar cash flows from the swap could be used to buy the US dollar bond. The valuation of a cross-currency swap is similar to that of an interest-rate swap but one difference relates to the exchange of notional. A cross-currency swap has two principal amounts, one for each currency. The initial principals can be exchanged or not. The nonexchange of the initial amounts is a minor issue. But eliminating the exchange of the final amounts changes the pricing structure significantly. The exchange rate used to determine the principals is the prevailing spot rate.

With an interest rate swap, there is no exchange of principal at either the start or end of the transaction, as both principal amounts are the same and therefore net out. For a cross-currency swap, it is essential that the parties agree to exchange principal amounts at maturity. The exchange of principal at the start is optional.

Like all swaps, a cross-currency swap can be replicated using on-balance sheet instruments, in this case with money market deposits or FRNs denominated in different currencies. This explains the necessity for principal exchanges at maturity as all loans and deposits also require repayment at maturity.[33]

The initial exchange can be replicated by the bank by entering into a spot exchange transaction at the same rate quoted in the cross-currency swap. Actually, all foreign exchange forwards can be described as cross-currency swaps as they are agreements to exchange two streams of cash flows in different currencies. Many banks manage long-term foreign exchange forwards as part of the cross-currency swap business, given the similarities.

Like FX forwards, the cross-currency swap exposes the user to foreign exchange risk. The swap leg the party agrees to pay is a liability in one currency, and the swap leg they have agreed to receive is an asset in the other currency. One of the users of cross-currency swaps are debt issuers. In the Eurobond markets, issuers sell bonds in the currency with the lowest cost and swap their exposure to the desired currency using a cross-currency swap.

14.8.1 PRICING

At the inception of the swap, the present value of one leg must be equal to the present value of the other leg at the then-prevailing spot rate. Using this simple logic, it would seem natural that cash

---

[33]While the corporate can elect not to exchange principal at the start, the bank needs to.

flows of LIBOR (flat) payments in one currency could be exchanged for cash flows of LIBOR (flat) payments in another currency.

In reality this is not true, and there is a constant spread. By convention cross-currency swap spreads are quoted on the currency side against USD LIBOR: USD LIBOR versus Foreign Currency LIBOR plus a spread. The spread can be positive or negative. Before 2007 the EUR/USD cross-currency swap spread was close to zero, but during the GFC it reached 120 bp. The reasons for the spread are twofold. First is the daily demand—supply imbalances that are always possible. There may be more demand for paying a LIBOR in a particular currency and this will lead to a positive (basis) spread to be paid. Consider US companies, for example, that issue a loan in euros. The companies will exchange principal and coupons payments in EUR for principal and coupons payments in USD. If the cross-currency basis is negative, the companies will have to pay LIBOR but will receive Euribor minus the basis. Such corporate supply and hedging activity can be a potential driver cross-currency swap basis. If the corporate issuance in EUR is high and most of it is swapped into USD, there could be a positive or spread to exchange EUR payments in USD payments. Such corporate activity, however, cannot explain the huge swings in the spread during the GFC.

The second effect that leads to positive spreads is credit risk. Counterparties may not have the same credit risk and the currency swap spread may then reflect this. For example, if one party is paying the Philippine peso equivalent of the LIBOR rate against USD LIBOR, then this party is likely to have a higher credit risk. So the party will pay a higher spread. During the GFC, an excess demand of USD by non-US banks drove the EUR/USD cross-currency swap down dramatically in 2008. In 2011, US Money-Market funds withdrew funding to European banks which also led to a decline in the spread. These examples illustrate that the spread reflects the relative creditworthiness of banks across different countries.[34]

14.8.2 CONVENTIONS

The usual convention for quoting the currency swap spread, also called the *basis*, is to quote it relative to the USD LIBOR.

14.9 DIFFERENTIAL (QUANTO) SWAPS

Financial Accounting Standard (FAS) 133 and the International Accounting Standard (IAS) 39 set the accounting rules for derivatives for the United States and for European companies, respectively. According to these rules, a derivative position will have to be marked-to-market and included in income statements unless it qualifies for *hedge accounting*.[35]

A quanto (differential) swap is a special type of cross-currency swap. It is an agreement where one party makes payments in, say, USD LIBOR and receives payments, say, in Euro LIBOR. However, the *important* point is that both parties make payments in the *same* currency. In other

---

[34]The following BIS document provides details on recent cross-currency swap basis movements: http://www.bis.org/publ/qtrpdf/r_qt0803h.pdf.

[35]Qualifying for hedge accounting is a lengthy and costly process. At the end the qualification is still random. However, some vanilla instruments such as vanilla swaps can qualify relatively easily.

words, the quanto swap value is an exposure on pure play of international interest rate differentials and has no foreign exchange risk.

Quanto swap popularity depends on the relative shapes of the forward curves in the two underlying money markets. Quanto swaps become "cheap" if one of the two forward curves is lower at the short end and higher at the long end.

14.9.1 BASIS SWAPS

This discussion is limited to US dollar markets. The particular interest rates discussed below will change if other currencies are considered, since basis swaps are directly related to the business environment in an economy. A *basis swap* is a floating–floating interest rate swap in which cash flows based on one floating reference rate are exchanged against the cash flows based on a different floating reference rate.

In a basis swap, one party will often pay USD LIBOR and will receive *another money-market rate*. Most bank liabilities are in fact LIBOR based, but assets are not. A corporation that deals mainly in the *domestic* US economy may be exposed to commercial paper (CP) rates; a bank may be exposed to T-bill rates, and another to the *prime rate*. Basis swaps could then be used to protect the party with respect to changes in these different money market rates.

The most common types of basis swaps are Fed Funds against LIBOR,[36] T-bill rates against LIBOR, CP rates against LIBOR, and the prime rate against LIBOR.[37] Which basis swap a client picks depends on his or her business. For example, a party that has concerns about credit squeezes can use Fed Funds-LIBOR basis swaps or the T-bill-LIBOR basis swaps. During a credit squeeze, a flight to safety will make the basis swap spread increase. On the other hand, the Prime-LIBOR basis swap can hedge the exposures of those players involved in credit card, auto, or consumer loans.

14.10 CONCLUSIONS

This chapter was devoted to the connections between the swap, FRA, and bond markets. Our discussion led us to the issue of constructing a satisfactory yield curve, which is the fundamental task of a financial engineer. Two main tools were introduced in the chapter. The first was the *T*-forward measures and the second was the related measure change technology. This permitted setting up convenient arbitrage-free dynamics for a sequence of forward rates. These dynamics were then used as a tool for calculating the desired quantities using the formulas that connect swap rates, forward rates, and their derivatives.

The next topic was the Forward LIBOR Model. Here, the essential idea was to obtain sequential correction factors to express the dynamics of various forward rates under a single forward measure.

---

[36]The Fed Funds market consists of overnight lending of free reverses kept at the Federal Reserve between high-quality banks. The quality of the banks and the sort tenor implies that Fed Funds rate will be lower than, say, LIBOR.

[37]The prime rate is not an interbank rate. It applies to the best retail clients.

SUGGESTED READING

The standard readings for this chapter will make interesting reading for a financial engineer. **Brace et al.** (1997) and **Jamshidian** (1997) are the fundamental readings. **Miltersen et al.** (1997) is another important reference. **Glasserman and Zhao** (2000) is a good source on the discretization of BGM models. Finally, the text by **Brigo and Mercurio** (2006) provides a comprehensive treatment of all this material. **Rebonato** (2002) is a good introduction.

EXERCISES

1.   You are given the following quotes for liquid FRAs paid in arrears. Assume that all time intervals are measured in months of 30 days.

Term	Bid/Ask
3×6	4.5–4.6
6×9	4.7–4.8
9×12	5.0–5.1
12×15	5.5–5.7
15×18	6.1–6.3

You also know that the current 3-month LIBOR rate is 4%.
   a. Calculate the discount bond prices $B(t_0, t_i)$, where $t_i = 6, 9, 12, 15$, and 18 months.
   b. Calculate the yield curve for maturities $0-18$ months.
   c. Calculate the swap curve for the same maturities.
   d. Are the two curves different?
   e. Calculate the par yield curve.
   f. Calculate the zero-coupon yield curve.

2.   You are given the following quotes for liquid swap rates. Assume that all time intervals are measured in years.

Term	Bid/Ask
2	6.2–6.5
3	6.4–6.7
4	7.0–7.3
5	7.5–7.8
6	8.1–8.4

You know that the current 12-month LIBOR rate is 5%.
a. Calculate the FRA rates for the next 5 years, starting with a 1×2 FRA.
b. Calculate the discount bond prices $B(t_0, t_i)$, where $t_i = 1, \ldots, 6$ years.
c. Calculate the yield curve for maturities of 0–18 months.
d. Calculate the par yield curve.

3. Going back to the data given in Exercise 2, calculate the following:
a. The bid–ask on a forward swap that starts in 2 years with maturity in 3 years. The swap is against 12-month LIBOR.
b. The forward price of a coupon bond that will be delivered at time 2. The bond pays coupon 7% and matures in 2 years.

EXCEL EXERCISES

4. Write a VBA program to determine the bond price and swap rates from the set of forward rates given below and plot the yield curves along with the swap curve. Comment on their characteristics.

Term (Month)	Forward Rate
0	4.00%
3	4.50%
6	4.70%
9	5.00%
12	5.50%
15	6.10%
18	6.40%
21	6.46%
24	7.00%

5. Write a VBA program to determine the bond price and forward rates from the set of swap rates below with the term of 1 year as the input. Assume that the current LIBOR rate is 5.00%. Plot the yield curves along with swap curve to observe their characteristic.

Term (Years)	Swap Rate
2	6.20%
3	6.40%
4	7.00%
5	7.50%
6	8.10%
7	8.70%

6. Write a VBA program to determine the FRA rates and swap rates from the set of bond prices below as the input and plot the yield curves along with swap curve. Comment on their characteristics.

Time (Month)	Bond Price
0	100.00
3	99.01
6	97.91
9	96.77
12	95.58
15	94.28
18	92.86
21	91.40
24	89.95
27	88.40

7. Write a VBA program to determine the price of an interest rate swap which makes a floating payment every 6 months at the rate equal to USD LIBOR rate against a fixed rate of 5.10% for the notional amount of $10,000. Use the data given in the "Input" worksheet for carrying out the calculation.

Time (Years)	Forward Rate
0.00	5.70%
0.50	5.18%
1.00	5.83%
1.50	5.10%
2.00	5.75%
2.50	5.74%
3.00	5.28%
3.50	5.65%
4.00	5.53%

8. (Cross-Currency Swap) Write a VBA program to determine the cross-currency swap rate based on the information below. The fixed payment is made in USD in the exchange for the floating payment (in USD) rate equal to the GBP LIBOR at every 3 months. Also include the exchange of notional amount £10,000 against payment in dollars at their corresponding exchange spot rate during initiation and maturity of the swap in the swap rate calculation.
    Notional amount (GBP) = £10,000, notional amount (USD) = $20,000

Time (Month)	Forward Rate (USD)	Exchange Rate (USD/GBP)	LIBOR Rate (GBP)
0	4.00%	2.0000	5.70%
3	4.50%	2.0225	5.18%
6	4.70%	2.0150	5.83%
9	5.00%	1.9875	5.10%
12	5.50%	1.9750	5.75%
15	6.10%	1.9685	5.74%
18	6.30%	1.9480	5.28%
21	6.36%	1.9325	5.65%
24	6.40%	1.9300	5.53%

MATLAB EXERCISE

9.   a. Write a MATLAB program to determine the bond price and swap rate from the given set of FRA rates and calculate the yield, par yield, and zero-coupon yield. The input data are to be read from the "Forward" worksheet of the "Data.xlsx" file on the chapter webpage. Now plot all the curves simultaneously and comment on their characteristics.

   b. Write a MATLAB program to determine the bond price and FRA rate from the given set of forward swap rates and calculate the yield, par yield, and zero-coupon yield. The input data are to be read from the "Swap" worksheet of "Data.xlsx" file. Now plot all the curves simultaneously and comment on their characteristics.

   c. Write a MATLAB program to determine the FRA price and swap rate from the given set of bond prices and calculate the yield, par yield, and zero-coupon yield. The input data are to be read from the "Bond" worksheet of "Data.xlsx" file. Now plot all the curves simultaneously and comment on their characteristics.

TOOLS FOR VOLATILITY ENGINEERING, VOLATILITY SWAPS, AND VOLATILITY TRADING

15

CHAPTER OUTLINE

15.1 INTRODUCTION

Liquid instruments that involve *pure* volatility trades are potentially very useful for market participants who have natural exposure to various volatilities in their balance sheet or trading book. The classical option strategies discussed in Chapter 11 have serious drawbacks in this respect. When a trader takes a position or hedges a risk, he or she expects that the random movements of the underlying would have a *known* effect on the position. The underlying may be random, but the *payoff function* of a well-defined contract or a position has to be known. Payoff functions of most *classical* volatility strategies are not invariant to underlying risks, and most volatility instruments turn out to be *imperfect* tools for isolating this risk. Even when traders' anticipations come true, the trader may realize that the underlying volatility payoff functions have changed due to movements in *other* variables. Hence, classical volatility strategies cannot provide satisfactory hedges for volatility exposures. The reason for this and possible solutions are the topics of this chapter.

Traditional volatility trades used to involve buying and selling portfolios of call and put options, straddles or strangles, and then possibly *delta*-hedging these positions. But such volatility positions were not *pure* and this led to a search for volatility tools whose payoff function would depend on the volatility risk *only*. This chapter examines two of the pure volatility instruments that were developed—variance and volatility swaps. Volatility swaps are forward contracts on future realized (stock) volatility. Variance contracts are similar contracts on variance, the square of future volatility. Variance and volatility swaps are interesting for at least two reasons: First, volatility is an important risk for market practitioners, and ways of hedging and pricing such risks have to be understood. Second, the discussion of volatility swaps constitutes a good example of the basic principles that need to be followed when devising new instruments.

The chapter uses *variance* swaps instead of volatility swaps to conduct the discussion. Although markets in general use the term *volatility*, it is more appropriate to think in terms of *variance*, the square of volatility. Variance is the second centered moment of a random variable, and it falls more naturally from the formulas used in this chapter. For example, volatility (i.e., standard deviation) instruments require convexity adjustments, whereas variance instruments in general do not. Thus, when we talk about *vega*, for example, we refer to *variance vega*. This is the sensitivity of the option's price with respect to σ^2, not σ. In fact, in the heuristic discussion in this chapter, at times the term volatility and variance are used interchangeably.

15.2 VOLATILITY POSITIONS

Volatility positions can be taken with the purpose of hedging a volatility exposure or speculating on the future behavior of volatility. These positions require instruments that isolate volatility risk as well as possible. To motivate the upcoming discussion, we introduce two examples that illustrate traditional volatility positions.

15.2.1 TRADING VOLATILITY TERM STRUCTURE

We have seen several examples for strategies associated with shifts in the interest rate term structure. They were called *curve steepening* or *curve flattening* trades. It is clear that similar positions can be taken with respect to *volatility* term structures as well. Volatilities traded in markets come with different maturities. As with the interest rate term structure, we can buy one "maturity" and sell another "maturity," as the following example shows.[1]

> **EXAMPLE**
>
> [A] dealer said he was considering selling short-dated 25-delta euro puts/dollar calls and buying a longer-dated straddle. A three-month straddle financed by the sale of two 25-delta one-month puts would have cost 3.9% in premium yesterday.
>
> These volatility plays are attractive because the short-dated volatility is sold for more than the cost of the longer-maturity options.

In this particular example, the anticipations of traders concern not the level of an asset price or return, but, instead, the volatility associated with the price. Volatility over one interval is bought using the funds generated by selling the volatility over a different interval.

Apparently, the traders think that *short-dated* euro volatility will fall *relative to* the long-dated euro volatility. The question is to what extent the positions taken will meet the traders' needs, *even when their anticipations are borne out*. We will see that the payoff function of this position is not invariant to changes in the underlying euro/dollar exchange rate.

---

[1]The term "arbitrage" is used here in the sense financial markets use it and not in the sense of academic analysis. The following arbitrage positions may have zero cost and have a relatively high probability of succeeding, but the gains are by no means risk-free.

15.2.2 TRADING VOLATILITY ACROSS INSTRUMENTS

Our second example is from the interest rate sector and involves another "arbitrage" position on volatility. The trader buys the volatility of one risk and sells a related volatility on a different risk. This time, the volatilities in question do not belong to different time periods, but instead are generated by different *instruments*.

EXAMPLE

Dealers are looking at the spreads between euro cap-floor straddle and swaption straddle volatility to take advantage of a 5% volatility difference in the 7-year area. Proprietary traders are selling a two-year cap-floor straddle starting in six years with vols close to 15%. The trade offers a good pick-up over the five-year swaption straddle with volatility 10%. This compares with a historical spread closer to 2%.

Cap-floors and swaptions are instruments on interest rates. There are both similarities and differences between them. We will study them in more detail in the next chapter. Selling a cap-floor straddle will basically be *short* interest rate volatility. In the example, the traders were able to take this position at 15% volatility. On the other hand, buying a swaption amounts to a *long* position on volatility. This was done at 10%, which gives a *volatility spread* of about 5%. The example states that the latter number has historically been around 2%. Hence, by selling the spread the traders would benefit from a future narrowing of differences between the volatilities of the two instruments.

This position's payoff is not invariant to interest rate trajectories. Even when volatilities behave as anticipated, the path followed by the level of interest rates may result in unexpected volatility.[2] The following discussion intends to clarify why such positions on volatility have serious weaknesses and require meticulous risk management. We will consider pure volatility positions later.

15.3 INVARIANCE OF VOLATILITY PAYOFFS

In previous chapters, convexity was used to isolate *volatility* as a risk. In Chapters 9 and 10, we showed how to convert the volatility of an underlying into "cash," and with that took the first steps toward volatility engineering.

Using the methods discussed in Chapters 9 and 10, a trader can hedge and risk-manage exposures with respect to volatility movements. Yet, these are positions influenced by variables *other* than volatility. Consider a speculative position taken by an *investor*.

Let S_t be a risk factor and suppose an investor *buys* S_t volatility at time t_0 for a future period denoted by $[t_1, T]$, T being the expiration of the contract. As in every long position, this means that

---

[2]We must point out that there are differences between cap-floor volatilities and swaption volatilities. In fact, this 4% spread may very well be due to these factors. Also, such positions become even more complicated with the existence of a volatility smile.

the investor is anticipating an increase in *realized* volatility during this period. If realized volatility during $[t_1, T]$ exceeds the volatility "purchased" at time t_0, the investor will benefit. Thus far this is not very different from other long positions. For example, a trader buys a stock and benefits if the stock price goes up. He or she can buy a fixed receiver swap and benefit if the swap value goes up (the swap rate goes down). Similarly, in our present case, we receive a "fixed" volatility and benefit if the actual volatility exceeds this level.

By buying call or put options, straddles, or strangles, and then *delta*-hedging these positions, the trader will, in general, end up with a long position that benefits if the realized volatility increases, as was shown in Chapters 9 and 10. Yet, there is one major difference between such volatility positions and positions taken on other instruments such as stocks, swaps, forward rate agreements (FRAs), and so on. Consider Figure 15.1a, that shows a long position on a stock funded by a money market loan. As the stock price increases, the position benefits by the amount $S_{t_1} - S_{t_0}$. This potential payoff is known and depends *only* on the level of S_{t_1}. In fact, it depends on S_t *linearly*. In Figure 15.1b, we have a *short*-dated discount bond position. As the yield decreases, the position gains. Again, we *know* how much the position will be making or losing, depending on the movements in the yield, y_t, if convexity gains are negligible.

A volatility position taken via, say, straddles, is fundamentally different from this as the payoff diagram will move depending on the path followed by variables *other* than volatility. For example, a change in (i) interest rates, (ii) the underlying asset price, or (iii) the level of *implied* volatility may lead to different payoffs at the same *realized* volatility level.

Variance (volatility) swaps, on the other hand, are pure volatility positions. Potential gains or losses in positions taken with these instruments depend *only* on what happens to realized volatility until expiration. This chapter shows how volatility engineering can be used to set up such contracts and to study their pricing and hedging. We begin with imperfect volatility positions.

15.3.1 IMPERFECT VOLATILITY POSITIONS

In financial markets, a *volatility position* is often interpreted to be a static position taken by buying and selling straddles, or a dynamically maintained position that uses straddles or options. As

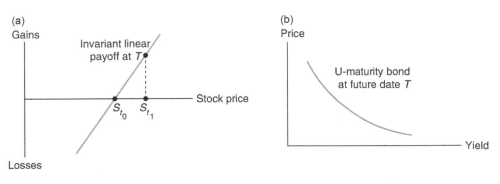

FIGURE 15.1

Long position on a stock (funded by a money market loan) and long discount bond position.

mentioned previously, these volatility positions are not the right way to price, hedge, or risk-manage volatility exposure. In this section, we go into the reasons for this. We consider a simple position that consists of a dynamically hedged single-call option.

15.3.1.1 A dynamic volatility position

Consider a volatility exposure taken through a dynamically maintained position using a plain vanilla call. To simplify the exposition, we impose the assumptions of the Black–Scholes world where there are no dividends; the interest rate, r, and implied volatility, σ, are constant; there are no transaction costs; and the underlying asset follows a geometric process. Then the arbitrage-free value of a European call $C(S_t, t)$ will be given by the Black–Scholes formula[3]:

$$C(S_t, t) = S_t \int_{-\infty}^{d_1} \frac{1}{\sqrt{2\pi}} e^{-\frac{1}{2}x^2} dx - e^{-r(T-t)} K \int_{-\infty}^{d_2} \frac{1}{\sqrt{2\pi}} e^{-\frac{1}{2}x^2} dx \quad (15.1)$$

where S_t is the spot price, and K is the strike. The d_i, $i = 1, 2$, are given by

$$d_i = \frac{\log(S_t/K) \pm (1/2)\sigma^2(T-t) + r(T-t)}{\sigma\sqrt{T-t}} \quad (15.2)$$

For simplicity, and without loss of generality, we let

$$r = 0 \quad (15.3)$$

This simplifies some expressions and makes the discussion easier to follow.[4]

Now consider the following simple experiment. A trader uses the Black–Scholes setting to take a dynamically hedged *long* position on implied volatility. Implied volatility goes up. Suppose the trader tracks the gains and losses of the position using the corresponding variance-*vega*. What would be this trader's possible gains in the following specific case? Consider the following simple setup.

1. The parameters of the position are as follows:

$$\text{Time to expiration} = 0.1 \quad (15.4)$$

$$K = S_{t_0} = 100 \quad (15.5)$$

$$\sigma = 20\% \quad (15.6)$$

Initially we let $t_0 = 0$.

2. The trader expects an increase in the *implied* volatility from 20% to 30%, and considers taking a *long* volatility position.

3. To *buy* into a volatility position, the trader borrows an amount equal to 100 $C(S_t, t)$, and buys 100 calls at time t_0 with funding cost $r = 0$.[5]

4. Next, the position is *delta*-hedged by short-selling C_s units of the underlying per call to obtain the familiar exposure shown in Figure 15.2.

---

[3]See Chapter 10 for a derivation.
[4]This is a useful assumption for discussing volatility trading.
[5]Remember that an identical position could be taken by buying puts. We take calls simply as an example.

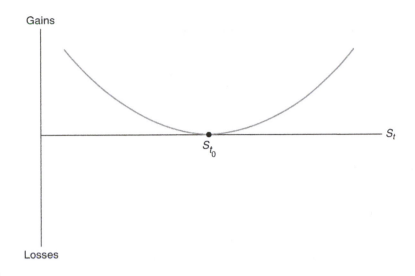

FIGURE 15.2

Delta-hedged long call position.

In this example, there are about 1.2 months to the expiration of this option, the option is at-the-money, and the initial implied volatility is 20%.

It turns out that in this environment, even when the trader's anticipations are borne out, the payoffs from the volatility position may vary significantly, depending on the path followed by S_t. The implied volatility may move from 20% to 30% as anticipated, but the position may not pay the expected amount. The following example displays the related calculations.

EXAMPLE

We can calculate the relevant Greeks and payoff curves using Mathematica or MATLAB. First, we obtain the initial price of the call as

$$C(100, t_0) = 2.52 \qquad (15.7)$$

Multiplying by 100, the total position is worth $252. Then, we get the implied delta of this position by first calculating the S_t-derivative of $C(S_t, t)$ evaluated at $S_{t_0} = 100$, and then multiplying by 100:

$$100 \left(\frac{\partial C(S_t, t)}{\partial S_t} \right) = 51.2 \qquad (15.8)$$

Hence, the position has $+51$-delta. To hedge this exposure, the trader needs to short 51 units of the underlying and make the net delta exposure approximately equal to zero.

Next, we obtain the associated gamma and the (variance) vega of the position at t_0. Using the given data, we get

$$\text{Gamma} = 100 \left[\frac{\partial^2 C(S_t, t)}{\partial S_t^2} \right] = 6.3 \tag{15.9}$$

$$\text{Variance vega} = 100 \left[\frac{\partial C(S_t, t)}{\partial \sigma^2} \right] = 3152 \tag{15.10}$$

The change in the option value, given a change in the (implied) variance, is given approximately by

$$100[\partial C(S_t, t)] \cong (3152)\partial \sigma^2 \tag{15.11}$$

This means that, everything else being constant, if the implied volatility rises *suddenly* from 20% to 30%, the instantaneous change in the option price will depend on the product of these numbers and is expected to be

$$100[\partial C(S_t, t)] \cong 3152(0.09 - 0.04) \tag{15.12}$$

$$= 157.6 \tag{15.13}$$

In other words, the position is expected to gain about $158, if everything else remained constant.

The point is that the trader was long implied volatility, expecting that it would increase, and it did. So if the volatility *does* go up from 20% to 30%, is this trader guaranteed to gain the $157.6? Not necessarily. Let us see why not.

Even in this simplified Black−Scholes world, the (variance) *vega* is a function of S_t, t, r, as well as σ^2. Everything else is not constant and the S_t may follow any conceivable trajectory. But, and this is the important point, when S_t changes, this in turn will make the *vega* change as well. The following table shows the possible values for variance-*vega* depending on the value assumed by S_t, within this setting.[6]

S_t	Vega
80	0.0558
90	7.4666
100	31.5234
110	10.6215
120	0.5415

---

[6]The numbers in the table need to be multiplied by 100.

Thus, if the expectations of the trader are fulfilled, the implied volatility increases to 30%, but, at the same time, if the underlying price moves *away from the strike*, say to $S_{t_1} = 80$, the same calculation will become approximately:

$$\text{Vega}(\partial\sigma^2) \cong 5.6(0.09 - 0.04) \tag{15.14}$$

$$= 0.28 \tag{15.15}$$

Hence, instead of an anticipated gain of $157.6, the trader could realize almost no gain at all. In fact, if there are costs to maintaining the volatility position, the trader may end up losing money. The reason is simple: as S_t changes, the option's sensitivity to implied volatility, namely the *vega*, changes as well. It is a function of S_t. As a result, the outcome is very different from what the trader was originally expecting.

For a more detailed view on how the position's sensitivity moves when S_t changes, consider Figure 15.3 where we plot the partial derivative:

$$100 \frac{\partial \text{ variance vega}}{\partial S_t} \tag{15.16}$$

Under the present conditions, we see that as long as S_t remains in the vicinity of the strike K, the trader has some exposure to volatility changes. But as soon as S_t leaves the neighborhood of K, this exposure drops sharply. The trader may think he or she has a (variance) volatility position, but, in fact, the position costs money, and may not have any significant variance exposure as the underlying changes right after the trade is put in place. Thus, such classical volatility positions are imperfect ways of putting on volatility trades or hedging volatility exposures.

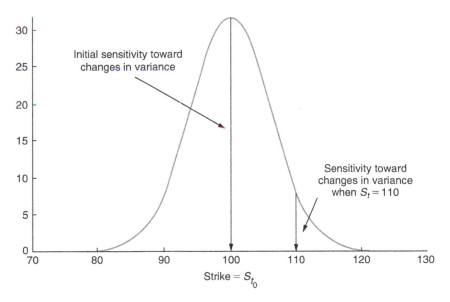

FIGURE 15.3

Vega as a function of the price of the underlying.

15.3.2 **VOLATILITY HEDGING**

The outcome of such volatility positions may also be unsatisfactory if these positions are maintained as a hedge against a *constant* volatility exposure in another instrument. According to what was discussed, movements in S_t can make the hedge *disappear* almost completely and the trader may hold a naked volatility position in the end. An institution that has volatility exposure may use a hedge only to realize that the hedge may be *slipping* over time due to movements unrelated to volatility fluctuations.

Such slippage may occur for more reasons than just a change in S_t. In reality, there are also (i) smile effects, (ii) interest rate effects, and (iii) shifts in correlation parameters in some instruments. Changes in these can also cause the classical volatility payoffs to move away from initially perceived levels.

15.3.3 **A STATIC VOLATILITY POSITION**

If a *dynamic delta*-neutral option position loses its exposure to movements in σ^2 and, hence, ceases to be useful as a hedge against volatility risk, do static positions fare better?

A classic position that has volatility exposure is buying (selling) ATM straddles. Using the same numbers as above, Figure 15.4 shows the joint payoff of an ATM call and an ATM put struck at $K = 100$. This position is made of two plain vanilla options and may suffer from a similar defect. The following example discusses this in more detail.

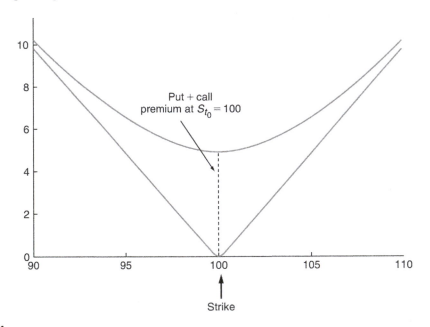

FIGURE 15.4

Joint payoff of an ATM call and ATM put.

EXAMPLE

As in the previous example, we choose the following numerical values:

$$S_{t_0} = 100, \quad r = 0, \quad T - t_0 = 0.1 \tag{15.17}$$

The initial volatility is 20%, which means that

$$\sigma^2 = 0.04 \tag{15.18}$$

We again look at the sensitivity of the position with respect to movements in some variables of interest. We calculate the variance vega of the portfolio:

$$V(S_t, t) = 100\{\text{ATM Put} + \text{ATM Call}\} \tag{15.19}$$

by taking the partial:

$$\text{Straddle vega} = 100 \frac{\partial V(S_t, t)}{\partial \sigma^2} \tag{15.20}$$

Then, we substitute the appropriate values of S_t, t, σ^2 in the formula. Doing this for some values of interest for S_t, we obtain the following sensitivity factors:

S_t	Vega
80	11
90	1493
100	6304
110	2124
120	108

According to these numbers, if S_t stays at 100 and the volatility moves from 20% to 30%, the static position's value increases approximately by

$$\partial \text{ Straddle} \cong 6304(0.09 - 0.04) \tag{15.21}$$

$$= 315.2 \tag{15.22}$$

As expected, this return is about twice as big as in the previous example. The straddle *has* more sensitivity to volatility changes. But, the option's responsiveness to volatility movements is again not constant, and depends on factors that are external to what happens to volatility. The table shows that if S_t moves to 80, then even when the trader's expectation is justified and volatility moves from 20% to 30%, the position's mark-to-market gains will go down to about 0.56.

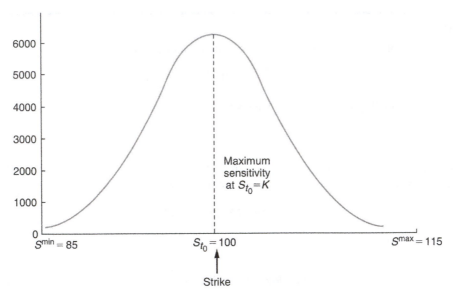

FIGURE 15.5

Straddle's sensitivity with respect to implied volatility for different values of S_t.

Figure 15.5 shows the behavior of the straddle's sensitivity with respect to implied volatility for different values of S_t. We see that the volatility position is again not invariant to changes in external variables. However, there is one major difference from the case of a dynamically maintained portfolio. Static non-*delta*-hedged positions using straddles will benefit from *actual* (realized) movements in S_t. For example, if S_t stays at 80 until expiration date T, the put leg of the straddle would pay 20 and the static volatility position would gain. This is regardless of how the *vega* of the position changed due to movements in S_t over the interval $[t_0, T]$.

15.4 PURE VOLATILITY POSITIONS

The key to finding the right way to hedge volatility risk or to take positions in it is to isolate the "volatility" completely, using existing liquid instruments. In other words, we have to construct a *synthetic* such that the value of the synthetic changes *only* when "volatility" changes. This position should not be sensitive to variations in variables other than the underlying volatility. The exposure should be invariant. Then, we can use the synthetic to take volatility exposures or to hedge volatility risk. Such volatility instruments can be quite useful.

First, we know from Chapters 12 and 13 that by using options with different strikes we can essentially create *any* payoff that we like—if options with a broad range of strikes exist and if markets are complete. Thus, we should, in principle, be able to create pure volatility instruments by using judiciously selected option portfolios.

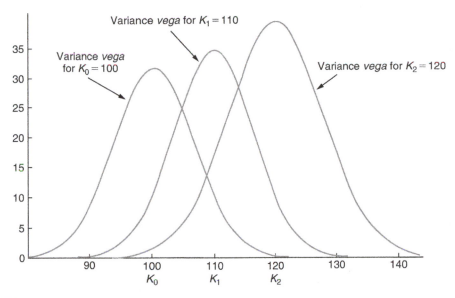

FIGURE 15.6

Vega of three plain vanilla European call options.

Second, if an option position's *vega* drops suddenly once S_t moves away from the strike, then, by combining options of *different* strikes appropriately, we may be able to obtain a *portfolio* of options whose *vega* is more or less insensitive to movements in S_t. Heuristically speaking, we can put together small portions of smooth *curves* to get a desired horizontal *line*.

When we follow these steps, we can create pure volatility instruments. Consider the plot of the *vega* of three plain vanilla European call options, two of which are out-of-the-money. The options are identical in all respects, except for their strike. Figure 15.6 shows an example. Three σ^2 sensitivity factors for the strikes $K_0 = 100$, $K_1 = 110$, $K_2 = 120$ are plotted. Note that *each* variance *vega* is very sensitive to movements in S_t, as discussed earlier. Now, what happens when we consider the portfolio made of the *sum* of all three calls? The sensitivity of the portfolio,

$$V(S_t, t) = \{C(S_t, t, K_0) + C(S_t, t, K_1) + C(S_t, t, K_2)\} \tag{15.23}$$

again varies as S_t changes, but less. So, the direction taken is correct except that the previous portfolio did not optimally combine the three options. In fact, according to Figure 15.6, we should have combined the options by using different *weights* that depend on their respective strike price. The more out-of-the-money the option is, the higher should be its weight, and the more it should be present in the portfolio.

Hence, consider the new portfolio where the weights are inversely proportional to the square of the strike K,

$$V(S_t, t) = \frac{1}{K_0^2} C(S_t, t, K_0) + \frac{1}{K_1^2} C(S_t, t, K_1) + \frac{1}{K_2^2} C(S_t, t, K_2) \tag{15.24}$$

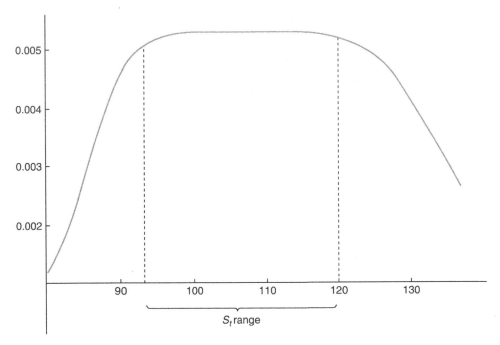

FIGURE 15.7

Variance vega of a portfolio of options with weights inversely proportional to the square of the strike price.

The variance *vega* of this portfolio that uses the parameter values given earlier, is plotted in Figure 15.7. Here, we consider a suitable $0 < \epsilon$, and the range

$$K_0 - \epsilon < S_t < K_2 + \epsilon \tag{15.25}$$

Figure 15.7 shows that the *vega* of the portfolio is approximately constant over this range when S_t changes. This suggests that more options with different strikes can be added to the portfolio, weighting them by the corresponding strike prices. In the example below we show these calculations.

EXAMPLE

Consider the portfolio

$$V(S_t, t) = \left[\frac{1}{80^2} C(S_t, t, 80) + \frac{1}{90^2} C(S_t, t, 90) + \frac{1}{100^2} C(S_t, t, 100) \right. \tag{15.26}$$

$$\left. + \frac{1}{110^2} C(S_t, t, 110) + \frac{1}{120^2} C(S_t, t, 120) \right] \tag{15.27}$$

This portfolio has an approximately constant vega for the range

$$80 - \epsilon < S_t < 120 + \epsilon \tag{15.28}$$

By including additional options with different strikes in a similar fashion, we can lengthen this section further.

We have, in fact, found a way to create synthetics for volatility positions using a portfolio of liquid options with varying strikes, where the portfolio options are weighted by their respective strikes.

15.4.1 PRACTICAL ISSUES

In our attempt to obtain a pure volatility instrument, we have essentially followed the same strategy that we have been using all along. We constructed a *synthetic*. But this time, instead of matching the cash flows of an instrument, the synthetic had the purpose of matching a particular *sensitivity* factor. It was put together so as to have a constant (variance) *vega*.

Once a constant *vega* portfolio is found, the payoff of this portfolio can be expressed as an approximately linear function of σ^2

$$V(\sigma^2) = a_0 + a_1\sigma^2 + \text{small} \tag{15.29}$$

with

$$a_1 = \frac{\partial V(\sigma^2, t)}{\partial \sigma^2} \tag{15.30}$$

as long as S_t stays within the range

$$S^{\min} = K_0 < S_t < K_n = S^{\max} \tag{15.31}$$

Under these conditions, the volatility position will look like any other long (or short) position, with a positive slope a_1.

The portfolio with a constant (variance) *vega* can be constructed using vanilla European calls and puts. The rules concerning synthetics discussed earlier apply here also. It is important that elements of the synthetic be liquid; therefore, liquid calls and puts have to be selected. The previous discussion referred only to calls. Practical applications of the procedure involve puts as well. This brings us to two somewhat complicated issues. The first has to do with the smile effect. The second concerns liquidity.

15.4.1.1 The smile effect

Suppose we form a portfolio at time t_0 that has a constant *vega* as long as S_t stays in a reasonable range

$$S^{\min} < S_t < S^{\max} \tag{15.32}$$

Under these conditions, the portfolio consists of options with different "moneyness" properties, and the volatility parameter in the option pricing formulas may depend on K if there is a volatility smile. In general, as K decreases, the implied $\sigma(K)$ would increase for constant S_t. Under these conditions, the trader needs to accurately determine the smile and the way to model it before the portfolio is formed.

15.4.1.2 Liquidity problems

From the preceding it follows that we need to select out-of-the-money options for the synthetic since they are more liquid. But as time passes, the moneyness of these options changes and this affects their liquidity. Those options that become in-the-money are now less liquid. Other options that were not originally included in the synthetic become more liquid. Even though the replicating portfolio was static, the illiquidity of the constituent options may become a drawback in case the position needs to be unwound.

15.5 VARIANCE SWAPS

One instrument that has invariant exposure to fluctuations in (realized) variance is the *variance swap*. In this section, we introduce this concept and in the next, we provide a simple framework for studying it.

A variance swap is, in many ways, just like any other swap. The parties exchange *floating* risk against a risk fixed at the contract origination. In this case, what is being swapped is not an interest rate or a return on an equity instrument, but the variances that correspond to various risk factors.

15.5.1 USES AND USERS OF VARIANCE SWAPS

Before we study the hedging and pricing of variance swaps it is instructive to think about what the potential uses of variance swaps are.

15.5.1.1 Uses of variance swaps

We can distinguish several uses and users of variance swaps. First, some users may be interested in variance swaps as a way of directionally trading variance levels. This could be for hedging or speculative purposes. Market participants who would like to speculate on the future levels of stock or index variance or volatility can go long or short realized variance with a variance swap. The advantage of such a position over trading and hedging options is that the variance swap is not contaminated by other risk exposures. If an investor expects a sharp decline in political and financial uncertainty after a forthcoming election, for example, a short position in a variance swap may be a good way to express this view and profit from it.

Second, some investors may use the variance swap to trade the spread between realized and implied variance levels. As we will see below, the fair strike price F_{t_0} in a variance swap is related to the level of current-implied volatilities for options with the same expiration as the swap. Therefore, by unwinding the swap before expiration, a market participant can trade the spread between realized and implied volatility.

15.5.1.2 Market participants with an implicit volatility exposure to hedge

We can think of at least three groups of market participants with an implicit volatility exposure to hedge. Some event-driven hedge funds engage in merger arbitrage and take positions that assume that the spread between an acquirer and takeover target will narrow. The risk is, however, that if overall market volatility increases the merger may be less likely and the spread may widen leading to loses for merger arbitrage hedge funds.

Some institutional investors that follow active benchmarking strategies regularly rebalance their portfolios. In volatile periods such rebalancing may need to be more frequent, thus leading to higher portfolio turnover and transaction costs. Independent from transaction costs, higher volatility may also lead to increased tracking error for portfolio managers who are judged against a benchmark. As asset prices move quickly, they may not be able to track the benchmark as effectively as when volatility is low.

Standard equity funds could be argued to be short volatility since there is a negative empirical correlation between equity market levels and volatility as we will see below. Thus, as global equity correlations rise, diversification across securities and countries becomes less effective. The GFC represents a recent example of many such turbulent periods during which most assets fall, but variances tends to rise, thus leading to a potential hedge.

The three market participants listed above could all benefit potentially from long positions in variance swaps to hedge their exposures and reduce portfolio risk.

In the following section, we move to a more technical discussion of variance swaps. However, we emphasize again that the discussion will proceed using the variance rather than the volatility as the underlying.

15.5.2 A FRAMEWORK FOR VARIANCE SWAPS

Let S_t be the underlying price. The time-T_2 payoff $V(T_1, T_2)$ of a variance swap with a notional amount, N, is given by the following:

$$V(T_1, T_2) = \left[\sigma_{T_1,T_2}^2 - F_{t_0}^2\right](T_2 - T_1)N \tag{15.33}$$

where σ_{T_1,T_2} is the *realized* volatility rate of S_{tb} during the interval $t \in [T_1, T_2]$, with $t < T_1 < T_2$. It is similar to a "floating" rate, and will be observed only when time T_2 arrives. The F_{t_0} is the "fixed" S_t volatility *rate* that is quoted at time t_0 by markets. This has to be multiplied by $(T_2 - T_1)$ to get the appropriate volatility for the contract period. N is the notional amount that needs to be determined at contract initiation. At time t_0, the $V(T_1, T_2)$ is unknown. The swap is set so that the time-t_0 "expected value" of the payoff, denoted by $V(t_0, T_1, T_2)$ is zero. At initiation, no cash changes hands:

$$V(t_0, T_1, T_2) = 0 \tag{15.34}$$

Thus, variance swaps are similar to a vanilla (interest rate) swap in that a "floating" $\sigma_{T_1,T_2}^2(T_2 - T_1)N$ is received against a "fixed" $(T_2 - T_1)F_{t_0}^2N$.

The cash flows implied by a variance swap are shown in Figure 15.8. The contract is initiated at time t_0, and the start date is T_1. It matures at T_2. The "floating" volatility (variance) is the total volatility (variance) of S_t during the entire period $[T_1, T_2]$. F_{t_0} has the subscript t_0, and, hence, has to be determined at time t_0.

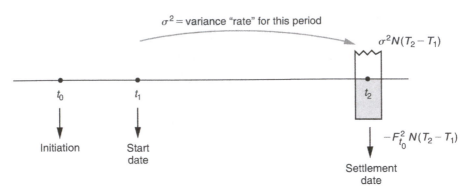

FIGURE 15.8

Variance swap cash flows.

15.5.2.1 Real-world example of a variance swap

Below we provide a real-world example of a variance swap.

EXAMPLE: VARIANCE SWAP ON S&P500—SPX INDICATIVE TERMS AND CONDITIONS

Instrument:	Swap
Variance buyer:	TBD (e.g., JPMorganChase)
Variance seller:	TBD (e.g., Investor)
Denominated currency:	USD
Vega amount:	100,000
Variance amount:	3125 (determined as vega amount/(strike $\times$ 2))
Underlying:	S&P500 (Bloomberg ticker: SPX index)
Strike price K_{vol}:	16
Equity amount:	$T + 3$ after the observation end date, the Equity Amount will be calculated and paid in accordance with the following formula:

Final Equity payment = variance amount $\times$ (final realized volatility^2 − strike price^2)

If the Equity Amount is positive the Variance Seller will pay the Variance Buyer the Equity Amount. If the Equity Amount is negative the Variance Buyer will pay the Variance Seller an amount equal to the absolute value of the Equity Amount, where

$$\text{Final realized volatility} = \sqrt{\frac{252 \times \sum_{t=1}^{t=N} (\ln(P_t/P_{t-1}))^2}{\text{Expected\_}N}} \times 100$$

and

> *Expected_N* = [number of days], being the number of days which, as of Trade Date, are expected to be Scheduled Trading Days in the Observation Period
> P_0 = The Official Closing of the underlying at the Observation Start Date
> P_t = Either the Official Closing of the underlying in any observation date *t* or, at Observation End Date, the Official Settlement Price of the Exchange-Traded Contract.
>
> **(Bossu et al. (2005))**

The terms and conditions of the variance swap contract show that if realized variance is higher than the strike price K_{vol}, the variance buyer will receive a payment from the variance seller. Similar to other derivative contracts, variance swaps can be used for hedging or speculation. Let's consider an example in which the contract is used for hedging purposes. The variance buyer buys the variance swap as insurance, that is, for the purpose of hedging against unexpected increases in variance. Why would the variance buyer want to hedge against variance increases? The variance buyer (i.e., JPMorganChase, here) might have, for example, issued structured equity products with a capital guarantee to retail investors and now wants to hedge the structured product by means of a variance swap. The capital guarantee protects the retail investor against losses. The structured product could consist of an equity market exposure and a capped downside. JPMorganChase could use a long position in a put option on the index, or if it is cheaper, a long position in a variance swap to generate the cash flow to pay for the capital guarantee. When stock markets plunge variances and correlations tend to increase. The variance contract term sheets show a strike price K_{vol} of 16, that is a volatility of 16. A stock market plunge might occur in a scenario when the realized volatility is (substantial) above 16. The variance buyer (JPMorganChase) could buy, that is go long, the variance swap. Who would be the counterparty? The "Investor" could be a hedge fund for example. We could economically interpret the transaction as JPMorganChase buying insurance from the hedge fund against unexpected increases in variance.

In line with market practice the variance notional is defined in volatility terms:

$$\text{Variance notional} = \frac{\text{vega notional}}{2 \times K_{vol}}$$

The vega amount is \$100,000 (which implies a variance amount of \$3125(=\$100,000/ (2×16)). The implication of the above adjustment is that when the realized volatility is 1 "vega" or 1 volatility point above the strike at maturity, the payoff is approximately equal to the Vega Notional.

In the variance swap term sheet, it is interesting to observe that returns are calculated on a logarithmic basis and that the mean return is assumed to be zero for simplicity which means that the payoff is perfectly additive: 6-month + 6-month variance = 1-month variance.

If the index realizes a 20% volatility over the next year, for example, the Variance Buyer—the long position—will *receive \$450,000 = 3125 × (Realized volatility^2 − K_{vol}^2) = 3125 × (20^2 − 16^2)*. This could be interpreted as a *realized* insurance *payout* from the hedge fund to JPMorganChase. Why would the hedge fund enter into such a transaction as a counterparty? As we will see below, historically, the implied variance (the fixed leg) has been above the realized variance (floating leg) *on average.* For example, if the index only realizes 12%, the long will *pay $350,000 = 3125 ×* *(Realized volatility^2 − K_{vol}^2) = 3125 × (12^2 − 16^2)* to the seller. Here JPMorganChase would pay the investor or hedge fund $350,000.[7] This could be interpreted as an *insurance premium.* In the past over certain time periods it was possible for a hedge fund to earn a substantial risk or insurance premium by selling variance swaps or being short variance swaps. This is by no means a riskless trade, however, since in crises times, the realized volatility tends to exceed and rise above the implied volatility.

Figure 15.9 shows the floating and fixed leg for a S&P500 variance swap contract from 1996 to 2014. The contract is for 30-day variance. The floating leg is based on the realized variance and

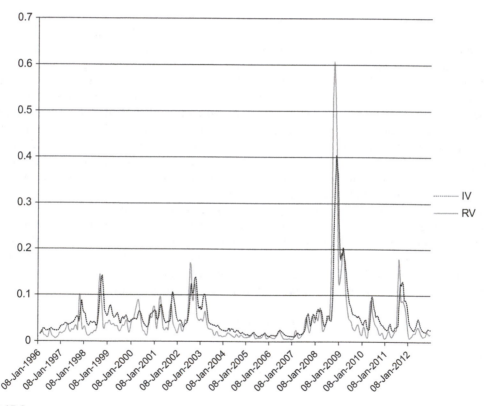

FIGURE 15.9

Floating and fixed leg for a S&P500 variance swap contract.

---

[7]Note that in the above example, the average exposure for a realized volatility being 4% away from the strike is $400,000 or four times the vega amount, as required.

the fixed leg is based on the implied variance.[8] As we can see the implied variance was above the realized variance for most of the pre-GFC and post-GFC period. This highlights the insurance premium that a variance seller could have pocketed for selling variance swaps.

However, the variance seller would have made large losses during the GFC. As the figure illustrates, on September 29, 2008, the realized variance reached a level of 0.70 while implied variance was 0.13. Many trading desks and hedge funds that were short variance in 2008 suffered substantial losses and some were forced to close. Note that variance swaps were traded as OTC contracts and no historical data from exchange traded variance contracts are currently available. Therefore, the plot is based on synthetic variance swap contracts based on S&P500 index options and Optionmetrics volatility surface data.

Figure 15.9 also illustrates an important feature of volatility, which is its tendency to revert to the mean. If we were to compare the volatility plot to the realized return of the S&P500 we would also see that equity volatility is negatively correlated with the equity index and tends to rise in times of elevated uncertainly or risk.

15.5.2.2 Real-world conventions

As the term sheet illustrates it is important for counterparties to agree the procedure for calculating the realized volatility, in particular, the sources and the frequency of the price of the underlying as well as the annualization factor used in moving from daily to annual volatilities. It also matters whether the standard deviation is calculated by subtracting the sample mean from each return, or by assuming a zero return. The zero mean method has theoretical advantages since it corresponds most closely to the contract that can be replicated by means of option portfolios.

15.5.2.3 Floating leg

Variance (volatility) positions need to be taken with respect to a well-defined time interval. After all, the volatility *rate* is like an interest rate: It is defined for a specific time interval. Thus, we subdivide the period $[T_1, T_2]$ into equal subintervals, say, days:

$$T_1 = t_1 < t_2 \cdots < t_n = T_2 \tag{15.35}$$

with

$$t_i - t_{i-1} = \delta \tag{15.36}$$

and then define the realized variance for period δ as

$$\sigma_{t_i}^2 \delta = \left[\frac{S_{t_i} - S_{t_{i-1}}}{S_{t_{i-1}}} - \mu\delta \right]^2 \tag{15.37}$$

where $i = 1, \ldots, n$.[9] Here, μ is the expected rate of change of S_t during a year. This parameter can be set equal to zero or any other estimated mean. Regardless of the value chosen, μ needs to be carefully defined in the contract. If μ is zero, then the right-hand side is simply the squared returns during intervals of length δ.

---

[8]To create a smooth plot we transformed the original monthly time series by taking the 12-month moving average of the two time series.

[9]Of course, there are many other ways to define these "short-period" volatilities. Some of the recent research uses the estimated variance of daily price changes during a trading day, for example.

Adding the marginal variances for successive intervals, $\sigma^2_{T_1,T_2}$ is equal to

$$\left(\sigma^2_{T_1,T_2}\right)(T_2 - T_1) = \sum_{i=1}^{n} \left[\frac{S_{t_i} - S_{t_{i-1}}}{S_{t_i}} - \mu\delta\right]^2 \tag{15.38}$$

Thus, $\sigma^2_{T_1,T_2}$ represents the realized percentage variance of the S_t during the interval $[T_1, T_2]$. If the intervals become smaller and smaller, $\delta \to 0$, the last expression can be written as

$$\left(\sigma^2_{T_1,T_2}\right)(T_2 - T_1) = \int_{T_1}^{T_2} \left[\frac{1}{S_t} dS_t - \mu dt\right]^2 \tag{15.39}$$

$$= \int_{T_1}^{T_2} \sigma_t^2 dt \tag{15.40}$$

This formula defines the realized variance. It is a random variable at time t_0, and can be viewed as the floating leg of the swap. Obviously, such floating variances can be defined for any interval in the future and can then be exchanged against a "fixed" leg.

15.5.2.4 *Determining the fixed variance*

Determining the fixed volatility, F_{t_0}, will give the fair value of the variance swap at time t_0. How do we obtain the numerical value of F_{t_0}? We start by noting that the variance swap is *designed* so that its fair value at time t_0 is equal to zero. Accordingly, the $F_{t_0}^2$ is that number (variance), which makes the fair value of the swap equal zero. This is a basic principle used throughout the text and it applies here as well.

We use the fundamental theorem of asset pricing and try to find a proper arbitrage-free measure $\tilde{P}$ such that

$$E_{t_0}^{\tilde{P}}\left[\sigma^2_{T_1,T_2} - F_{t_0}^2\right](T_2 - T_1)N = 0 \tag{15.41}$$

What could this measure $\tilde{P}$ be? Suppose markets are complete.

We assume that the continuously compounded risk-free spot rate r is constant. The random process $\sigma^2_{T_1,T_2}$ is, then, a nonlinear function of S_u, $T_1 \le u \le T_2$, only:

$$\sigma^2_{T_1,T_2}(T_2 - T_1) = \int_{T_1}^{T_2} \left[\frac{1}{S_t} dS_t - \mu dt\right]^2 \tag{15.42}$$

Under some conditions, we can use the normalization by the money market account and let $\tilde{P}$ be the risk-neutral measure.[10] Then, from Eq. (15.41), taking the expectation inside the brackets and arranging, we get

$$F_{t_0}^2 = E_{t_0}^{\tilde{P}}\left[\sigma^2_{T_1,T_2}\right] \tag{15.43}$$

This leads to the pricing formula

$$F_{t_0}^2 = \frac{1}{T_2 - T_1} E_{t_0}^{\tilde{P}}\left[\int_{T_1}^{T_2} \left[\frac{1}{S_t} dS_t - \mu dt\right]^2\right] \tag{15.44}$$

---

[10] We remind the reader that this contract will be settled at time T_2.

Therefore, to determine $F_{t_0}^2$ we need to evaluate the expectation under the measure $\tilde{P}$ of the integral of σ_t^2. The discrete time equivalent of this is given by

$$F_{t_0}^2 = \frac{1}{T_2 - T_1} E_{t_0}^{\tilde{P}} \left[\sum_{i=1}^{n} \left[\frac{S_{t_i} - S_{t_{i-1}}}{S_{t_{i-1}}} - \mu\delta \right]^2 \right] \tag{15.45}$$

Given a proper arbitrage-free measure, it is not difficult to evaluate this expression. One can use Monte Carlo or tree methods to do this once the arbitrage-free dynamics are specified.

15.5.3 A REPLICATING PORTFOLIO

The representation using the risk-neutral measure can be used for pricing. But how would we hedge a variance swap? To create the right hedge, we need to find a replicating portfolio. We discuss this issue using an alternative setup. This alternative has the side advantage of the financial engineering interpretation of some mathematical tools being clearly displayed. The following model starts with Black–Scholes assumptions.

The trick in hedging the variance swap lies in isolating σ_{T_1,T_2}^2 in terms of observable (traded) quantities. This can be done by obtaining a proper synthetic. Assume a diffusion process for S_t:

$$dS_t = \mu(S_t, t)S_t dt + \sigma(S_t, t)S_t dW_t \quad t \in [0, \infty) \tag{15.46}$$

where W_t is a Wiener process defined under the probability $\tilde{P}$. The diffusion parameter $\sigma(S_t, t)$ is called local volatility. Now consider the nonlinear transformation:

$$Z_t = f(S_t) = \log(S_t) \tag{15.47}$$

We apply Ito's Lemma to set up the dynamics (i.e., the SDE) for this new process Z_t:

$$dZ_t = \frac{\partial f(S_t)}{\partial S_t} dS_t + \frac{1}{2} \frac{\partial^2 f(S_t)}{\partial S_t^2} \sigma(S_t, t)^2 S_t^2 dt \quad t \in [0, \infty) \tag{15.48}$$

which gives

$$d\log(S_t) = \frac{1}{S_t} \mu(S_t, t)S_t dt - \frac{1}{2S_t^2} \sigma(S_t, t)^2 S_t^2 dt + \sigma(S_t, t)dW_t \quad t \in [0, \infty) \tag{15.49}$$

where the S_t^2 term cancels out on the right-hand side. Collecting terms, we obtain

$$d\log(S_t) = \left[\mu(S_t, t) - \frac{1}{2}\sigma(S_t, t)^2 \right] dt + \sigma(S_t, t)dW_t \tag{15.50}$$

Notice an interesting result. The dynamics for dS_t/S_t and $d\log(S_t)$ are almost the same *except* for the factor involving $\sigma(S_t, t)^2 dt$. This means that we can subtract the two equations from each other and obtain

$$\frac{dS_t}{S_t} - d\log(S_t) = \frac{1}{2}\sigma(S_t, t)^2 dt \quad t \in [0, \infty) \tag{15.51}$$

This operation has *isolated* the instantaneous percentage local volatility on the right-hand side. But, what we need for the variance swap is the *integral* of this term. Integrating both sides we get

$$\int_{T_1}^{T_2} \left[\frac{1}{S_t} dS_t - d\log(S_t) \right] = \frac{1}{2} \int_{T_1}^{T_2} \sigma(S_t, t)^2 dt \tag{15.52}$$

We now take the integral on the left-hand side,

$$\int_{T_1}^{T_2} d \log(S_t) = \log(S_{T_2}) - \log(S_{T_1}) \tag{15.53}$$

We use this and rearrange to obtain the result:

$$2\left[\int_{T_1}^{T_2} \frac{1}{S_t} dS_t\right] - 2 \log\left(\frac{S_{T_2}}{S_{T_1}}\right) = \int_{T_1}^{T_2} \sigma(S_t, t)^2 dt \tag{15.54}$$

We have succeeded in isolating the percentage total variance for the period $[T_1, T_2]$ on the right-hand side. Given that S_t is an asset that trades, the expression on the left-hand side replicates this variance.

15.5.4 THE HEDGE

The interpretation of the left-hand side in Eq. (15.54) is quite interesting. It will ultimately provide a hedge for the variance swap. In fact, the integral in the expression is a good example of what Ito integrals often mean in modern finance. Consider

$$\int_{T_1}^{T_2} \frac{1}{S_t} dS_t \tag{15.55}$$

How do we interpret this expression?

Suppose we would like to maintain a long position that is made of $1/S_t$ units of S_t held during each infinitesimally short interval of size dt, and for all t. In other words, we purchase $1/S_t$ units of the underlying at time t and hold them during an infinitesimal interval dt. Given that at time t, S_t is observed, this position can easily be taken. For example, if $S_t = 100$, we can buy 0.01 units of S_t at a total cost of 1 dollar. Then, as time passes, S_t will change by dS_t and the position will gain or lose dS_t dollars for every unit purchased. We readjust the portfolio, since the S_{t+dt} will presumably be different, and the portfolio needs to be $1/S_{t+dt}$ units long.

The resulting gains or losses of such portfolios during an infinitesimally small interval dt are given by the expression[11]

$$\frac{1}{S_t}(S_{t+dt} - S_t) = \frac{1}{S_t} dS_t \tag{15.56}$$

Proceeding in a similar fashion for all subsequent intervals dt, over the entire period $[T_1, T_2]$, the gains and losses of such a dynamically maintained portfolio add up to

$$\int_{T_1}^{T_2} \frac{1}{S_t} dS_t \tag{15.57}$$

The integral, therefore, represents the *trading gains or losses* of a dynamically maintained portfolio.[12]

---

[11]The use of dt here is heuristic.

[12] In fact, this interpretation can be generalized quite a bit. Often the stochastic integrals in finance have a structure such as

$$\int_{T_1}^{T_2} f(S_t) dS_t$$

These can be interpreted as trading gains or losses of *dynamically* maintaining $f(S_t)$ units of the asset that have price S_t.

The second integral on the left-hand side of Eq. (15.52)

$$\int_{T_2}^{T_2} d \log(S_t) = \log(S_{T_2}) - \log(S_{T_1})$$

(15.58)

is taken with respect to time t, and is a standard integral. It can be interpreted as a *static* position. In this case, the integral is the payoff of a contract written at time T_1, which pays, at time T_2, the difference between the unknown log (S_{T_2}) and the known $\log(S_{T_1})$. This is known as a *log contract*. The long and short positions in this contract are logarithmic functions of S_t.

In a sense, the left-hand side of Eq. (15.54) provides a hedge of the variance contract. If the trader is short the variance swap, he or she would also maintain a dynamically adjusted long position on S_t and be short a static log contract. This assumes complete markets.

15.6 REAL-WORLD EXAMPLE OF VARIANCE CONTRACT

Variance (volatility) swaps are clearly useful for taking positions with variance (volatility) exposure and hedging. But, each time a new market is born, there are usually further developments beyond the immediate uses. We briefly mention some further applications of the notions developed in this chapter.

First of all, the F_t^2, which is the fixed leg of the variance swap, can be used as a benchmark in creating new products. It is important to realize, however, that this price was obtained using the risk-neutral measure and that it is *not* necessarily an unbiased forecast of future variance for the period $[T_1, T_2]$. Just like the FRA market prices, the F_t will include a risk premium. Still, it is the proper price on which to write volatility options.

The pricing of the variance swap does not necessarily give a volatility that will equal the implied volatility for the same period. Implied volatility comes with a smile and this may introduce another wedge between F_t and the ATM volatility.

Finally, F_t^2 should be a good indicator for risk-managing volatility exposures and also options books.

Volatility trading, volatility hedging, and arbitraging are still a relatively new area. Below, we will see some recent difficulties that volatility products encountered during the GFC and new positions associated with them.

15.7 VOLATILITY AND VARIANCE SWAPS BEFORE AND AFTER THE GFC—THE ROLE OF CONVEXITY ADJUSTMENTS?

As we have seen in Section 15.4, *in theory*, a market maker could sell a variance swap and hedge it by means of a static hedge with a portfolio of put and call options across a range of strike prices, weighted inversely proportional to the squared strike price.

15.7.1 THE DIFFICULTY OF HEDGING VARIANCE SWAPS IN PRACTICE

In practice, options across the whole range of required strike prices and maturities are either not available or not very liquid. Index options tend to be more liquid than individual stock options. As a result, dealers and market makers before the GFC concentrated on a limited number of strike prices near the spot level. Since the 1990s the over-the-counter market for variance swaps on indices and individual equities developed. However, what many dealers painfully learned during the GFC when markets dropped and volatility spiked is that liquidity is not constant and it tends to disappear in crisis times. In September 2008, as markets plunged, market makers, who tried to rebalance their hedges in accordance with their models, found that liquidity dried up in listed option markets and it was difficult and expensive to hedge exposures. The drop in liquidity was more severe for single name options than for index options. Dealers who were short single-stock variance swaps suffered the most accordingly.

EXAMPLE

When volatility rises by the amount it did in late 2008 and you square it to calculate the payout of a variance swap, it's a nasty product to be short of—and a lot of dealers were short single-stock variance. The hedges they had put on just didn't perform in any way approaching how they would have wanted. Although those positions are theoretically hedgeable with listed options, it assumes that wherever the stock price is, you can buy strikes down to zero, but that is never the case.

(a quote of Dean Curnutt, president of Macro Risk Advisors, a New York-based
derivatives asset management firm, and previously head of institutional equity
derivatives sales at Bank of America, Structured Products,
April 1, 2010, www.risk.net/1595196)

One of the features of variance swaps is that they exhibit a constant vega irrespective of the level of the spot price. Since this vega increases linearly with the level of volatility, the dealer needs to buy more options to hedge the swap as volatility increases. This is possible in normal liquid markets, but in crises times when markets fall and risk aversion increases it is difficult to find liquidity options.

The above example not only illustrates practical difficulties in hedging variance swaps but also the different risk profiles of volatility and variance swaps. The payoff of a variance swap is, as we saw earlier, equal to the difference between the realized variance (that is the square of the volatility) and a preagreed strike level, multiplied by the vega notional. Thus, the payoff of a variance swap is convex in volatility. What does this imply for an investor that is long a variance swap? It means that the investor will benefit from increased gains and reduced losses compared with an alternative exposure to volatility by means of a volatility swap. Similarly, for an investor that is short a variance swap it implies that losses are magnified compared to a volatility swap. Why would someone want to short a variance swap, then? The answer is that, the fair strike of a variance swap is higher than that of a volatility swap. This makes the selling of variance swaps more attractive compared to volatility swaps since the seller can sell at a higher price. Some investors that are long volatility prefer variance swaps to volatility swaps due to the additional convexity and the resulting higher profit potential when volatility suddenly increases. This investor clientele often chooses single-stock variance trades over index variance trades since volatility moves in single stocks tend to be more pronounced than in indices.

The following reading illustrates one of the lessons learned from the GFC in equity derivatives.

EXAMPLE

"The convexity in variance swaps was an appealing proposition for clients that tended to be long volatility. For the sell side, it involved a big bet on liquidity in vanilla options being continuously available to provide some degree of hedge. In retrospect, variance is not a product that works for the sell side in very volatile markets, [...].

(**Structured Products, April 1, 2010, www.risk.net/1595196**)

As the example illustrates, theoretical models assume that continuous hedging is possible in practice. This assumption does not hold in reality and especially in volatile markets hedging becomes difficult and expensive. Some market participants have therefore argued that the variance swap market grew too fast and did not pay attention to the underlying complexities associated with liquidity and hedging costs.

15.7.2 CONVEXITY AND THE DIFFERENCE BETWEEN VARIANCE AND VOLATILITY SWAPS

15.7.2.1 Source of the convexity adjustment

As we have seen in Section 15.5.4, variance emerges naturally from hedged options trading. However, there is no simple replication strategy for synthesizing a volatility swap despite the preference for volatility quotes over variance quotes by many market participants. To move from variance swaps to volatility swaps prices a convexity adjustment is required. In other words, from a contingent claims or derivatives perspective, variance is the primary underlying and all other volatility payoffs, such as volatility swaps, should be viewed as derivative securities on the variance as underlying. Intuitively, volatility is the square root of variance and therefore a nonlinear function of variance. As a result volatility is theoretically and practically more difficult to hedge and value.

To explain the difference between volatility and variances swap pricing, let's look at a naïve strategy that approximates a volatility swap by statically holding a suitably chosen variance contract. In order to make the variance and volatility payoffs agree in value and volatility sensitivity (the first derivative with respect to σ), we can choose the following approximation:

$$\sigma - K_{vol} \approx \frac{1}{2K_{vol}}(\sigma^2 - K_{vol}^2) \tag{15.59}$$

The above states that $(1/2K_{vol})$ variance contracts with a strike K_{vol} can be used to approximate a volatility swap with a notional of ($\$1$/vol point) for realized volatilities near K_{vol}. This naïve approximation would also mean that the fair price of future volatility (the stroke for which the volatility swap has zero value) is simply the square root of the fair variance, that is, $K_{vol} = \sqrt{K_{var}}$. Figure 15.10 plots the left-hand side and the right-hand side of Eq. (15.59) for different values of realized volatility and a strike price $K_{vol} = 16\%$. The payoff of the volatility swap is represented by the dashed line and the payoff of the variance swap is represented by the solid parabola. The figure shows that the actual volatility swap and the corresponding approximation based on a variance swap different considerably when future realized volatility is far away from the strike price. We see that the naïve approximation

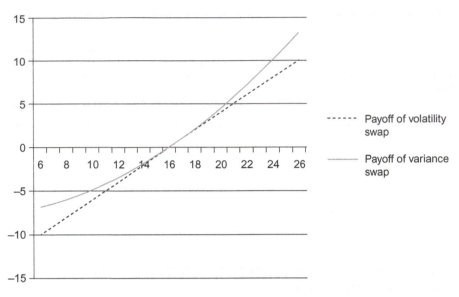

FIGURE 15.10

Payoff of variance and volatility swaps.

represented by Eq. (15.59) does not work well since it leads to variance swap payoffs that are always greater than the volatility swap payoff. The difference between the two payoffs is called the convexity bias:

$$\text{Convexity bias} \approx \frac{1}{2K_{\text{vol}}}(\sigma^2 - K_{\text{vol}}^2) \qquad (15.60)$$

The implication of this is that due to the square term, which is always positive, the fair delivery price for volatility K_{vol} would lead to higher payoffs from a long variance swap than a long volatility swap position. One approach to correct for this is to lower the fair strike K_{vol} for a volatility contract so that it is lower than the square root of the fair stroke for a variance contract K_{var}: $K_{\text{vol}} < \sqrt{K_{\text{var}}}$. This could be graphically represented by a shift in the dashed line in Figure 15.10 to the left, so that it does not always lie below the parabola.

So far we have discussed the static replication of variance swap by means of a volatility swap. In theory, the mismatch and convexity bias mentioned above could be addressed by dynamically trading variance swap contracts during the life of the volatility swap. The intuition for this is similar to the replication of a nonlinear stock option payoff by means of delta-hedging using the linear underlying stock price. In practice, unfortunately, there is no liquid market for variance swaps that would allow such dynamic trading of the underlying. If a liquid variance swap market was to develop in the future, then our financial engineering principles provide us with guidance on how such a dynamic replication could be implemented. When we discussed the replication of call and put options we saw that the optimal hedge ratio depends on the assumed future volatility of the underlying (stock). Here the dynamic replication of a volatility swap also requires a model for the volatility of volatility. In other works, the replication strategy would require holding continuous *variance-delta* equivalent of variance contracts to hedge the volatility derivative. Such a practical implementation would require an arbitrage-free model for the stochastic evaluation of the volatility surface. The next chapter will describe the volatility surface in more detail.

15.7.2.2 *The role of convexity in the volatility trading market during the GFC*

In response to the large losses on variance swaps, some banks, in 2009, decided to offer simpler volatility products to clients. In response to hedge fund client demand in Asia in the second quarter of 2009, BNP Paribas brought back volatility swaps which first traded in the late 1990s. Since the volatility swap has a payoff that depends on the volatility and not the variance, it does not have the same convexity as the variance swap. The downside is that it gives less of a profit to investors that long the volatility swap when volatility rises significantly.

Another downside of volatility swaps is that they do not have the same theoretical replication as variance swaps.

One of the few banks that continued to offer variance swap was Societe Generale. The bank was able to do this since it reduced risk in early 2008, earlier and by more than many of its competitors. Thus it was better placed to continue to offer variance swaps when the crisis worsened. As we discussed earlier long only clients find variance swaps attractive due to the convexity and additional profit and loss potential that goes with it. From the dealer or market maker perspective variance swaps are theoretically easier to hedge than volatility swaps since they can be replicated with vanilla options, although dealers use different hedging models and practices.

Innovation in the variance trading has continued and in March 2009, SG CIB started offering an American variance swap. This product embeds an option that allows investors to close their swap position before maturity at its realized value only. The benefit to SG CIB is that the product gives it the right to exist in the trade, which means that it can sell volatility to the client at a lower price.

In 2009, market participants were very risk averse and often too scared to trade volatility as an asset class. For those willing to take volatility risk, the American variance swap offered an alternative in a very uncertain environment.

The above reading illustrates that the market for volatility works like any other market in the sense that it responds to supply and demand. Since 2009, market conditions have improved and volatility has dropped to precrisis levels. While the Vix reached a level of 80 on November 20, 2008, it then fell to around 40 by the end of March 2009 and traded around 11 by June 2014. If volatility was to remain at these pre-GFC levels, then variance swaps may be viewed again more favorably than volatility swaps again, since the breakeven points on the upside, beyond which variance are more profitable than volatility swaps, will fall. The return on the variance swaps for a market maker will depend on the level of volatility and the liquidity of stock options required to hedge variance positions.

15.7.3 INTRODUCTION TO VOLATILITY AS AN ASSET CLASS

Acceptance of implied volatility and implied correlation as an asset class is growing. The main players are (i) institutional investors, (ii) hedge funds, and (iii) banks. This increased liquidity facilitates the engineering of structured products with embedded volatility.

EXAMPLE

It appears that institutional investors are migrating to four types of strategies for going long exposure to implied volatility. Among the largest institutional investors, variance swap-based strategies are the most popular.

Variance swaps offer the easiest and most liquid way to get exposure to volatility. Institutional investors and hedge funds are the target audience for the new services offered in this field. These services enable customers to trade variance swaps through Bloomberg terminals. Volumes in the inter-dealer market and with clients prompted the move. Variance swaps are used to go long or short volatility on an index or equity with a selected maturity.

Pure volatility instruments, such as volatility swaps and variance swaps, make sense for institutional investors, because volatility is both a diversifier on the downside and a global hedge on an equity portfolio.

Institutional investors such as pension funds and insurance companies clearly need to diversify. While they are moving to other asset classes, such as hedge funds, they also do not want to reduce their exposure to equity markets, particularly if there is a good chance of equity markets performing well. With this in mind, they are increasingly turning to long-term volatility strangles.

The main external driver of the current ongoing rise in volatility is M & A activity.

Requests for forward volatility strategies to hedge structured products are also on the rise, particularly among private banks. These strategies fit their needs, as dealers sold a lot of forward volatility certificates and warrants to them last year.

(Thomson Reuters IFR, 2004)

15.7.4 POST-GFC REGULATION, STANDARDIZATION AND EXCHANGE TRADED VOLATILITY PRODUCTS

In addition to supply and demand the market for volatility is also influenced by regulation. OTC derivatives regulation in Europe and the United States is pushing markets toward standardization and exchange trading.

15.7.4.1 Variance futures

The properties of the CBOE Volatility Index, or VIX, are now relatively well understood.[13] The Vix has a negative correlation with the S&P500 and this correlation tends to increase in crisis times. Thus the VIX can be used as a hedge or a "*tail risk*" hedge against extreme negative events. Unfortunately, the VIX is just a mathematical formula and not an investable index. It is not tradable and cannot be purchased. Fortunately there are several ways to benefit from a rise in the VIX by using exchange traded instruments: VIX futures, variance futures, VIX options, S&P options and volatility-related exchange traded funds (ETFs) and exchange traded notes (ETNs). There is also a myriad of over-the-counter products linked to the VIX.

On December 10, 2012, the CBOE listed the S&P 500 Variance futures.

The contract is *not* a variance swap since it depends on the realized variance of the S&P500, but it does allow investors to speculate on the realized S&P500 volatility. Figure 15.11 shows the open interest and volume of the S&P500 variance futures from 2012 until 2013.

Below is a term sheet for the S&P500 Variance Futures

---

[13]See http://www.cboe.com/micro/VIX/vixintro.aspx for more details about the VIX.

Description:	S&P 500 Variance futures are exchange-traded futures contracts based on the realized variance of the S&P 500 Composite Stock Price Index (S&P 500). The final settlement value for the contract will be determined based on a standardized formula for calculating the realized variance of the S&P 500 measured from the time of initial listing until expiration of the contract. The standard formula inputs for discount factor and daily interest rate are determined by the Exchange.
Contract size:	The contract multiplier for the S&P 500 Variance futures contract is $1 per variance unit. One contract equals one variance unit.
Trading hours:	8:30 a.m.–3:15 p.m. (Chicago time).
	All Orders, quotes, cancellations and Order modifications for S&P 500 Variance futures during trading hours must be received by the Exchange by no later than 3:14:59 p.m. Chicago time and will be automatically rejected if received by the Exchange during trading hours after 3:14:59 p.m. Chicago time.
Contract months:	The Exchange may list contract months on S&P 500 Variance futures that correspond to the listed contract months for S&P 500 Index options listed on Chicago Board Options Exchange, Incorporated (CBOE).

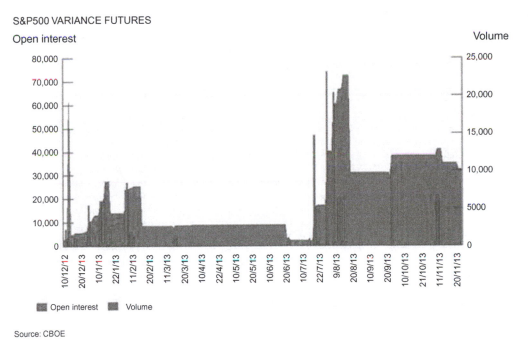

S&P500 VARIANCE FUTURES

Source: CBOE

FIGURE 15.11

Open interest and volume of the S&P500 variance futures.

In addition to the S&P500 VIX and Variance futures, the CBOE offers futures on several other US equity indices such as

- VSW CBOE Short-Term Volatility Index Futures (VSW)
- VX CBOE S&P 500 Volatility Index (VIX) Futures

- VU CBOE Russell 2000 Volatility Index (RVX) Futures
- VN CBOE Nasdaq-100 Volatility Index (VXN) Futures

emerging markets indices,

- VXEM CBOE Emerging Markets ETF
- Volatility Index (VXEEM) Security Futures
- VXEW CBOE Brazil ETF Volatility Index (VXEWZ) Security Futures

and commodities

- CBOE Gold ETF Volatility Index (GVZ) Security Futures
- CBOE Crude Oil ETF Volatility Index (OVX) Security Futures.

This shows the widespread acceptance of volatility as an asset class. It also highlights the fact that market participants are become more discerning regarding different types of volatility exposure depending on country and asset class.

15.7.4.2 Variance swaps

Variance swaps have been traded over-the-counter for almost 20 years, but recently, there have been plans to list variance swaps on exchanges. The following example discusses plans by Eurex to move over-the-counter variance swaps on to exchanges.

EXAMPLE

"Plans to move over-the-counter variance swaps on to exchanges are in full swing in Europe as the Eurex derivatives exchange prepares to list futures-style contracts tracking EuroStoxx 50 variance in mid-2014.

The exchange is not the first to attempt a listed alternative that produces the same pay-off as the widely traded OTC product. The CBOE launched S&P 500 variance futures a year ago (see "Replicating variance"), but the contract has struggled to gain traction. Since the launch last October, open interest has built up to about 32,000 contracts and volume remains lumpy with relatively few days of active trading.

Despite the inauspicious start for the US version, some traders remain optimistic there could be support for a listed version in Europe, where many asset managers looking to trade volatility do not have OTC approval, while the VStoxx futures have just a fraction of the liquidity seen in VIX contracts.

"The demand is there," said Rory Hill, EMEA co-head of equity derivatives at Citigroup. "After the crisis, closing down and novating OTC risk became an issue. Now, anything that's OTC, institutions and funds want to try to get listed. Client flows are now pushing volumes on to exchanges and even banks themselves don't want OTC exposure any more."

Sluggish demand for the CBOE product has been attributed to a variety of factors—not least the ample liquidity available in the US OTC variance swaps, where 10-year contracts can trade tighter than six-month maturities in Europe.

"The S&P variance OTC market is very healthy and very liquid, so there hasn't been a natural need. Most of the clients trading the product are already set up to trade OTC with the top 10 dealers, so their needs are already fulfilled," said Mark Chen, head of equity derivatives index trading at Citigroup in New York.

[...]

A number of dealers and hedge funds have already expressed interest in supporting the new product, subject to the right fee structure and incentives. But the real driver is likely to come from regulatory pressure as swaps are pushed into central clearing on a product-by-product basis.

"What could change things is when the regulatory requirement to clear the OTC products through CCPs goes live. Then we might see some clients move to the listed platform," said Bornhauser.

> *Variance swap activity stalled to a near standstill in the wake of the crisis—Citi's Hill estimates that volume plummeted to around 20% of its 2006–07 peak, but has since returned to as much as 50%, though that is concentrated in index trades.*
>
> *"There's still some single-stock vol swap exposure but single-stock variance swaps died a death. Most of the variance swap flow is concentrated"*
>
> **(Thomson Reuters IFR 2011, November 23, 2013)**

Regulation restricts the ability of certain investors to buy OTC products. The costs of hedging and funding complex products has also increased in banks, thus explaining the increased attraction of exchange traded assets.

15.7.5 THE HEDGE

In Section 15.5, we discussed the potential uses and users of variance swaps. But who were the counterparties in variance swaps in the run up to the GFC? It turns out that variance swaps were not just attractive to investors but also for dealers in investment banks with large structured products businesses.

EXAMPLE

> *Variance swaps became a very critical part of the risk recycling process for dealers that were running structured products books and had a short correlation exposure to offset. That was a big driver of the growth of the market in the early part of the last decade.*
>
> **(Structured Products, April 1, 2010, www.risk.net/1595196)**

The above example refers to dealers in large investment banks who had sold structured products to their clients. Equity structured products, for example, often involved a downside protection to the client. The product would thus pay when markets fell. When markets fall, correlations tend to rise sharply. This would leave the seller of such products short correlation. Thus one way of hedging such structured product exposure was to go long correlation by means of variance swaps. This is akin to buying protection on unexpected increases in correlations. Who would sell such protection? Before the financial crisis many hedge funds went short correlation. Thus, a hedge fund would go short a variance swap on an index and go long a variance swap on the constituent stock. This created a short position for the hedge fund and a long position for the dealer. We discuss correlation trading and correlation swaps in more detail in the next chapter.

15.8 WHICH VOLATILITY?

This chapter dealt with *four* notions of volatility. These must be summarized and distinguished clearly before we move on to discussion of the volatility smile in the next chapter.

When market professionals use the term "volatility," chances are they refer to Black–Schole's *implied* volatility. Otherwise, they will use terms such as *realized* or *historical* volatility. *Local volatility* and *variance swap volatility* are also part of the jargon. Finally, *cap-floor volatility* and *swaption volatility* are standard terms in financial markets.

As explained in Chapter 9, *implied volatility* is simply the value of σ that one would plug into the Black–Scholes formula to obtain the fair market value of a plain vanilla option as observed in the markets. For this reason, it is more correct to call it *Black–Scholes implied vol* or *Black volatility* in the case of interest rate derivatives. It is quite conceivable for a professional to use a different formula to price options, and the volatility implied by this formula would naturally be different. The term implied volatility is, thus, a formula-dependent variable.

We can attach the following definitions to the term "volatility."

- First, there is the class of *realized volatilities*. This is closest to what is contained in statistics courses. In this case, there is an observed or to-be-observed data set, a "sample," $\{x_1, \ldots x_n\}$, which can be regarded as a realization of a possibly vector-stochastic process, x_t, defined under some real-world probability P. The process x_t has a second moment

$$\sigma_t = \sqrt{E_t^P[(x_t - E_t^P[x_t])^2]} \tag{15.61}$$

We can devise an estimator to estimate this σ_t. For example, we can let

$$\hat{\sigma}_t = \sqrt{\frac{\sum_{i=0}^{m}(x_{t-i} - \bar{x}_t^m)^2}{m}} \tag{15.62}$$

where $\bar{x}_t^m$ is the *m*-period sample mean:

$$\bar{x}_t^m = \frac{\sum_{i=0}^{m} x_{t-i}}{m} \tag{15.63}$$

Such volatilities measure the actual real-world fluctuations in asset prices or risk factors. One example of the use of this volatility concept was shown in this chapter. The σ_t^2 defined earlier represented the floating leg of the variance swap discussed here.

- The next class is *implied volatility*.[14] There is an observed market price. The market practitioner has a pricing formula (e.g., Black–Scholes) or procedure (e.g., implied trees) for this price. Then, implied volatility is that "volatility" number, or series of numbers, which must be plugged into the formula in order to recover the fair market price. Thus, let $F(S_t, t, r, \sigma_t, T)$ be the Black–Scholes price for a European option written on the underlying S_t, with interest rates r and expiration T. At time t, σ_t represents the implied volatility if we solve the following equation (nonlinearly) for σ_t:

$$F(S_t, t, r, \sigma_t, T) = \text{observed price} \tag{15.64}$$

---

[14]This definition could be a little misleading since these days most traders quote volatility directly and then calculate the market price of options implied by this volatility quote.

This implied volatility may differ from the realized volatility significantly, since it incorporates any *adjustments* that the trader feels he or she should make to expected realized volatility. Implied volatility may be systematically different than realized volatility if volatility is *stochastic* and if a *risk premium* needs to be added to volatility quotes. Violations of Black–Scholes assumptions may also cause such a divergence.

- *Local volatility* is used to represent the function $\sigma(.)$ in a stochastic differential equation:

$$dS(t) = \mu(S,t)dt + \sigma(S,t)S_t dW_t \quad t \in [0, \infty) \tag{15.65}$$

However, local volatility has a more specific meaning. Suppose options on S_t trade in all strikes, K, and expirations T, and that the associated arbitrage-free prices, $\{C(S_t, t, K, T)\}$, are observed for all K, T. Then the function $\sigma(S_t, t)$ is the local volatility, if the corresponding SDE successfully replicates all these observed prices either through a Monte Carlo or PDE pricing method.

In other words, *local volatility* is a concept associated with calibration exercises. It can be regarded as a generalization of Black–Scholes implied volatility. The implied volatility replicates a *single* observed price through the Black–Scholes formula. The local volatility, on the other hand, replicates an entire *surface* of options indexed by K and T, through a pricing method. As a result, we get a *volatility surface* indexed by K and T, instead of a single number as in the case of Black–Scholes implied volatility.

- Finally, in this chapter we encountered the *variance swap volatility*. This referred to the expectation of the average future squared deviations. But, because the expectation used the risk-neutral measure, it is different from real-world volatility.

Discussions of the volatility smile relate to these volatility notions. The implied volatility is obviously of interest to most traders but it cannot exist independently of realized volatility. It is natural to expect a close relationship between the two concepts.[15] Also, as volatility trading develops, more and more instruments are written that use the realized volatility as some kind of underlying risk factor for creating new products. The variance swap was only one example.

15.9 CONCLUSIONS

This chapter provided a brief introduction to a sector that may, in the future, play an even more significant role in financial market strategies. Our purpose was to show how we can isolate the volatility of a risk factor from other related risks, and then construct instruments that can be used to trade it. An important point should be emphasized here. The introductory discussion contained in this chapter deals with the case where the volatility parameter is a function of time and the underlying price only. These methods have to be modified for more complex volatility specifications.

---

[15]For an interesting comparison of the ability of different volatility measures to forecast S&P100 volatility see Blair et al. (2001).

SUGGESTED READING

Rebonato (2000) and (2002) are good places to start getting acquainted with the various notions of volatility. **Bossu et al.** (2005) provide a good introduction to variance swaps. Some of the material in this chapter comes directly from **Demeterfi et al.** (1999), where the reader will find proper references to the literature as well. The important paper by **Dupire** (1992) and the literature it generated can be consulted for local volatility.

EXERCISES

1. Read the quote carefully and describe how you would take this position using volatility swaps. Be precise about the parameters of these swaps.
    a. How would you price this position? What does pricing mean in this context anyway? Which price are we trying to determine and write in the contract?
    b. In particular, do you need the correlations between the two markets?
    c. Do you need to know the smile before you sell the position?
    d. Discuss the risks involved in this volatility position.
    Volatility Swaps
       A bank is recommending a trade in which investors can take advantage of the wide differential between Nasdaq 100 and S&P500 longer-dated implied volatilities.

       Two-year implied volatility on the Nasdaq 100 was last week near all-time highs, at around 45.7%, but the tumult in tech stocks over the last several years is largely played out, said [a] global head of equity derivatives strategy in New York. The tech stock boom appears to be over, as does the most eye-popping part of the downturn, he added. While there will be selling pressure on tech companies over the next several quarters, a dramatic sell-off similar to what the market has seen over the last six months is unlikely.

       The bank recommends entering a volatility swap on the differential between the Nasdaq and the S&P, where the investor receives a payout if the realized volatility in two years is less than about 21%, the approximate differential last week between the at-the-money forward two-year implied volatilities on the indices. The investor profits here if, in two years, the realized two-year volatility for the Nasdaq has fallen relative to the equivalent volatility on the S&P.

       It might make sense just to sell Nasdaq vol, said [the trader], but it's better to put on a relative value trade with the Nasdaq and S&P to help reduce the volatility beta in the Nasdaq position. In other words, if there is a total market meltdown, tech stocks and the market as a whole will see higher implied vols. But volatility on the S&P500, which represents stocks in a broader array of sectors, is likely to increase substantially, while volatility on the Nasdaq is already close to all-time highs. A relative value trade where the investor takes a view on the differential between the realized volatility in two years time on the two indices allows the investor to profit from a fall in Nasdaq volatility relative to the S&P.

The two-year sector is a good place to look at this differential, said [the trader]. Two years is enough time for the current market turmoil, particularly in the technology sector, to play itself out, and the differential between two-year implied vols, at about 22% last week, is near all-time high levels. Since 1990, the realized volatility differential has tended to be closer to 10.7% over long periods of time.

[The trader] noted that there are other means of putting on this trade, such as selling two-year at-the-money forward straddles on Nasdaq volatility and buying two-year at-the-money forward straddles on S&P vol. (Derivatives Week (now part of GlobalCapital), October 30, 2000)

2. The following reading deals with another example of how spread positions on volatility can be taken. Yet, of interest here are further aspects of volatility positions. In fact, the episode is an example of the use of knock-in and knock-out options in volatility positions.

    a. Suppose the investor sells short-dated (1-month) volatility and buys 6-month volatility. In what sense is this a naked volatility position? What are the risks? Explain using volatility swaps as an underlying instrument.

    b. Explain how a 1-month break-out clause can hedge this situation.

    c. How would the straddles gain value when the additional premium is triggered?

    d. What are the risks, if any, of the position with break-out clauses?

    e. Is this a pure volatility position?

    Sterling volatility is peaking ahead of the introduction of the euro next year. A bank suggests the following strategy to take advantage of the highly inverted volatility curve. Sterling will not join the euro in January and the market expects reduced sterling positions. This view has pushed up one-month sterling/Deutsche mark vols to levels of 12.6% early last week. In contrast, six-month vols are languishing at under 9.2%. This suggests selling short-dated vol and buying six-month vols. Customers can buy a six-month straddle with a one-month break-out clause added to replicate a short volatility position in the one-month maturity. This way they don't have a naked volatility position. (Based on an article in Derivatives Week (now part of GlobalCapital).)

MATLAB EXERCISES

3. (Variance swap price). Write a MATLAB program to illustrate the variation of the price of an (equity) variance swap with respect to the change in the mean return of the underlying stock and also show this on a graph. Use the following parameters for the specification of the stochastic process followed by the underlying.
    - $S(0) = 100$; $T = 1$; $r = 8\%$; $\sigma = 30\%$; $\mu = 10\%$

4. (Volatility swap price). Write a MATLAB program to show the invariance of the volatility payoff of a delta hedged long call for the various frequency of delta adjustment until the time of expiry of the call. Use the following data:
    - $S(0) = 100$; $K = 105$; $T = 1$; $r = 8\%$; $\sigma = 30\%$ and realized volatility $= 50\%$

    Plot the graph for the volatility payoff for 20 simulations for the adjustment frequencies 10, 20, 50, 100, and 500.

CORRELATION AS AN ASSET CLASS AND THE SMILE

CHAPTER OUTLINE

Principles of Financial Engineering.

16.1 INTRODUCTION TO CORRELATION AS AN ASSET CLASS

In the previous chapter, we saw that volatility as an asset class is firmly established. Developments in financial engineering did not stop at volatility and have continued their natural progression to correlation.

Equity correlation measures how much stock prices tend to move together. Correlation links the volatility of an index to the volatilities of component stocks. As an approximation, we can write:

$$\text{Index volatility} \approx \sqrt{\text{correlation}} \times \text{average single stock volatility}$$

This formula shows that correlation is an important driver of index volatility. Historically, more than half of the changes in index volatility are driven by changes in correlation. This shows that correlation exposure is a fundamental component of volatility. Correlation risk is also an important driver of returns on options and other assets such as hedge funds. How can we use the principles of financial engineering applied in previous chapters to correlation? It turns out that it is quite straight-forward to back out the level of forward-looking correlation being priced by the market by calculating the relative levels of implied index and implied single-stock volatilities. As with implied index volatility, this *implied correlation* tends to trade at a premium to that realized correlation. Figure 16.1 shows the realized and implied correlation for the S&P500.

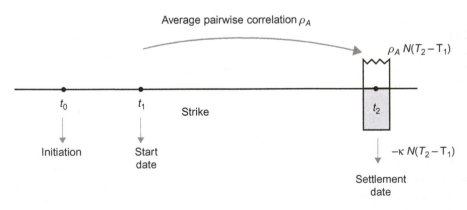

FIGURE 16.1

Cash flows in a correlation swap.

16.1.1 REASONS FOR TRADING CORRELATION

The reasons for trading correlation are similar to the reasons for trading volatility:

- Since implied correlation has historically been above realized correlation, a short correlation (swap) position can be a good source of return. The hedge fund industry and several hedge fund strategies are exposed to correlation risk as Buraschi et al. (2014) show.
- Correlation is a major component of volatility and similar to volatility it is negatively correlated with equity index returns. Therefore, correlation trading can provide diversification benefits.
- Since correlation is the residual from index and single-stock volatility, correlation trading can be used to express views and take positions regarding the relative moves of index versus single-stock volatility.

16.1.2 CORRELATION TRADING VEHICLES

There are two main vehicles for trading correlation, namely variance dispersion trades and correlation swaps.[1] Correlation swaps provide direct exposure to correlation. The payoff of a correlation swap is the difference between the strike of the swap and the subsequent realized average correlation of a pre-agreed basket of stocks or index. Although correlation swaps provide the purest correlation risk exposure, they are less liquid than dispersion trades.

In recent years, the most common vehicle for trading correlation has been the dispersion trade. The trade involves taking opposing positions in the volatility of an index (or basket) and the volatility of its constituents. Since a dispersion trade is typically implemented by means of variance swaps, it is called a *variance dispersion trade*. If variance swaps are used to benefit from unexpectedly higher correlations, the trade consists of a short position in an index variance swap and long positions in single stock variance swaps. Sometimes, for simplicity, options are used instead of variance swaps to generate index and stock-level exposure to volatility.

Acceptance of correlation as an asset class is growing as the following example illustrates.

EXAMPLE

Correlation trading diversifies:

> With correlation at historic highs across many indices, traders continue to see investors rush to monetise the widely-held view that correlation will fall over the next 12 months as macro-economic drivers become less prevalent in shaping investor sentiment, triggering the return of alpha strategies and increased stock dispersion. After effectively shutting down amid heightened uncertainty in the wake of Lehman Brothers' bankruptcy, correlation trading has been gaining traction in recent months.
>
> "The correlation market has been re-opened for more than a year now and although the dispersion trade is a standard trade, investors are now tracking every kind of correlation, both within and between different asset classes," said Stephane Mattatia, global head of financial engineering and advisory at SGCIB global equity flow.

---

[1]Several simple option derivative products such as basket options and quantos are implicitly exposed to correlation risk but they are not pure correlation instruments.

> In recent months, macro hedge funds have made substantial profits by taking leveraged bets on correlation between a variety of asset classes including S&P 500 versus euro-dollar, which was highly correlated following the crisis and broke in September 2009. The strong correlation between gold and euro-dollar, which was also widely traded, broke earlier this year.
>
> In equity dispersion markets, traders are seeing greater diversity of client types as numerous pension funds become active players, and a greater diversity of trade types is being implemented.
>
> While macro hedge funds continue to favour the dispersion trade—buying volatility swaps on separate stocks while selling volatility swaps on the index—options strategies such as dispersion straddles have increased in popularity as liquidity in more traditional products declines.
>
> "Funds are now building dispersion books and the key is diversification," said Mattatia. "They are trading all instrument types including straddles, variance swaps and volatility swaps, and looking across all indices."
>
> **(Thomson Reuters IFR 1850, 'Correlation trading diversifies', September 11, 2010)**

The above reading illustrates that correlation trading now extends across asset classes. Pension funds are mentioned as participants in correlation trading. They typically go long correlation swaps, which implies that they benefit when correlation unexpectedly increases. Macro hedge funds on the other hand are reported to be short correlation, by selling index and buying single stock variance swaps, thus benefiting when markets remain calm and correlations remain relatively low. To understand how correlation exposure can be created it is instructive to start with relatively simple correlation derivatives linked to one asset class, such as equities. Next we examine the replication and construction of dispersion traders and correlation swaps.

The purest way to trade correlation is directly via a correlation swap. A correlation swap provides exposure to the *average pairwise correlation* of a predetermined basket of stocks. Average pairwise correlation ρ_A is a simple equally weighted average of the correlations between all pairs of distinct stocks in an index:

$$\rho_A = \frac{2}{N(N-1)} \sum_{i<j} \rho_{ij}$$

where ρ_{ij} is the pairwise correlation of the ith and jth stocks in a basket or index whose variance is given by $\sigma_I^2 = \sum_i \omega_i^2 \sigma_i^2 + 2\sum_{i<j} \omega_i \omega_j \sigma_i \sigma_j \rho_{ij}$, where ω_i and σ_i are respectively the weight and the volatility of the ith stock in the index.

The swap strike is the level of (average pairwise) correlation that is bought or sold. It is typically scaled by a factor of 100 (that is a correlation of 0.5 is quoted as a strike of 50). Similar to the variance swap (discussed in the previous chapter), the payout of a correlation swap is the notional amount multiplied by the difference between the swap strike and the subsequent realized average pairwise correlation on the basket of underlyings:

$$\text{Payout} = \text{notional} \times (\text{realized average pairwise correlation} - \text{strike})$$

Figure 16.1 shows the cash flows in a typical correlation swap cash flow. Let's examine a correlation swap example.

EXAMPLE: *CORRELATION SWAPS*

Suppose that on January 6, 2014, a hedge fund sells a 6-month correlation swap on an equally weighted basket of stocks consisting of the constituents of the Eurostoxx 50 index. The counter-party is assumed to be a pension fund. The strike price is 40 and the notional amount is €10,000. This means that the hedge fund is short the correlation swap and the pension fund is long the swap. One way to interpret the transaction is to say that, for a fee, the hedge fund sells insurance against unexpected increases of the correlation to the pension fund.

After 6 months, on July 6, 2014, we calculate the realized 6-month average pairwise correlation of the stocks in our basket as 0.3. Then the buyer of the correlation swap, the pension fund, pays the following amount to the seller of the pension fund:
Notional $\times (55 - 42) = €10,000(40 - 30) = €100,000$.

In the above example, the hedge fund received a premium for bearing the risk of unexpected increases in correlations. A good proxy for the correlation strike, also known as implied correlation, is the square of the ratio between the implied index volatility and the average of the equity component implied volatilities (as implied volatility is a good estimate of the forward-realized volatility, we could then also infer that implied correlation is a good estimate of the forward-realized correlation).

Historically the implied correlation, the strike, has been above the realized correlation. This is illustrated in Figure 16.2 where the implied correlation is represented by the dashed line and the solid line represents the realized correlation.

Although correlation swaps and variance swaps appear similar, the stochastic process followed by correlation and variance is different and therefore correlation and variance swaps allow the expression of different views in the form of directional trades or hedges as the following reading illustrates.

EXAMPLE

Implied correlation hits multi-year lows

Equity correlation has fallen in recent weeks alongside rising equity markets and plummeting volatility. The result has been a renewed focus on dispersion trades and other strategies that seek to profit from increased differentiation between stocks, sectors and global indices.

"For investors, dispersion is a great story in the current environment and an interesting market parameter as it isn't a directional view. It's still very tricky to express a directional view on one market," said Arnaud Jobert, executive director, equity derivatives structuring at JP Morgan.

Despite a long-term rising trend, driven in part by a growing use of index and exchange traded products, implied correlation has dipped across global indices as well as at the sector and single stock level over the past year.

And although it remains historically high, the measure is now trending back towards pre-crisis levels. Implied correlation across global indices hit multi-year lows last week as equities ramped up their January rally with the S&P 500 hitting 1500 for the first time since 2007.

Over the course of just last week, two-year implied levels on a variety of index pairs including EuroStoxx 50/Nikkei 225 and S&P 500/Nikkei 225 fell by more than five points.

Meanwhile, the CBOE's S&P 500 implied correlation index (January 2014 maturity), which estimates correlation between the index constituents based on options prices of the S&P 500 index itself and its 50 largest stocks, has fallen from 70 to 60 since the end of December, having traded in the 80s at the start of 2012.

Macro shift

Equity correlation pushed to record highs during the crisis as volatility spiked and investors fled equities for the safer haven of fixed income markets, leaving macro factors as the only real drivers of equity performance. But even while volatility declined to historic lows, correlation remained relatively high as abundant liquidity injected by worldwide central bank policy ensured that macro drivers outweighed other factors even as stock prices began to stage a rebound.

However, that looks set to change as investors increasingly seek to express views on macro divergence as regions move towards different parts of their economic cycles.

(Thomson Reuters IFR 1968, January 26, 2013)

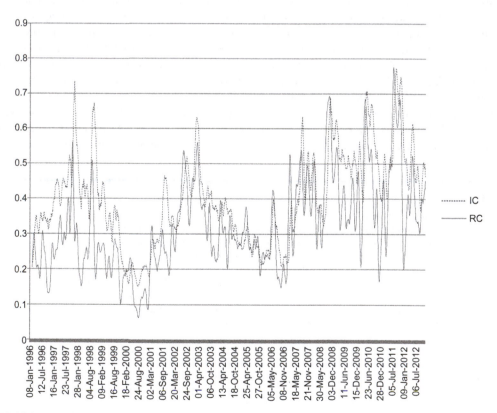

FIGURE 16.2

Implied and realized correlation over time.

16.2 **VOLATILITY AS FUNDING**

For market professionals and hedge funds, the issue of how to *fund* an investment is as important as the investment itself. After all, a hedge fund would look for the "best way" to borrow funds to carry a position. The best way may sometimes carry a negative interest. In other words, the hedge fund would make money from the investment *and* from the funding itself.

The normal floating LIBOR funding one is accustomed to think about is "risk-free,"[2] but at the same time may not always carry the lowest funding cost. In principle, volatility could be used as a funding source. A practitioner could for example go short volatility and long another asset such as a fixed-income asset. To fund the position, why not select an appropriate *volatility*, sell options of value N, and then *delta* hedge these option positions? In fact, this would fund the bond position with volatility. We analyze it below.

First we know from Chapter 9 that *delta*-hedged short option positions are convex exposures that will pay the *gamma*. These payouts are unknown initially. As market volatility is observed, the hedge is dynamically adjusted, and depending on the market volatility the hedge fund will face a cash outflow equal to *gamma*. To the hedge fund this is similar to paying floating money market interest rates.

Note one difference between loan cash flows and volatility cash flows. In volatility funding, there is no payback of the principal N at the end of the contract. In this sense, N is borrowed and then paid back gradually over time as *gamma* gains. One example is provided below from the year 2005.

> **EXAMPLE**
>
> Merrill notes "one of the most overcrowded trades in the market has been to take advantage of the long term trading range," by selling volatility and "earning carry via mortgage-backed securities."

Market professionals use options as funding vehicles for their positions. The main problem with this is that in many cases option markets may not have the depth needed in order to sell large chunks of options. If such selling depresses prices (i.e., volatility), then this idea may be hard to implement no matter how attractive it looks at the outset.

16.3 **SMILE**

Options were introduced as volatility instruments in Chapter 9. This is very much in line with the way traders think about options. We showed that when we deal with options as volatility instruments mathematically we arrived at the same formula, in this case the same partial differential equation (PDE) as the Black–Scholes PDE. *Mathematically* the approach was identical to the

---

[2]See the section on the zero in finance in Chapter 4.

standard textbook treatment that considers options as directional instruments.[3] Yet, although the interpretation in Chapter 9 is more in line with the way traders and option markets think, in that discussion there was still a major missing component.

It turns out that everything else being the same, an out-of-the-money put or call has a higher implied volatility than an ATM call or put. This effect, alluded to several times up to this point, is called the *volatility smile*. However, in order to discuss the smile in this chapter we adopt still another interpretation of options as instruments.

The discussion in Chapter 9 showed that the option price (after some adjustments for interest receipts and payments) is actually related to the *expected gamma* gains due to volatility in the underlying. The interpretation we use in this chapter will show that these expected gains will depend on the option's *strike*. One cost to pay for this interesting result is the need for a different mathematical approach. The advantage is that the smile will be the *natural* outcome. A side advantage is that we will discuss a dynamic hedging strategy other than the well-known *delta*-hedging. In fact, we start the chapter with a discussion of options from a more "recent" point of view which uses the so-called *dirac delta* functions. It is perhaps the best way of bringing the smile explicitly in option pricing.

16.4 DIRAC *DELTA* FUNCTIONS

Consider the integral of the Gaussian density with mean K given below

$$\int_{-\infty}^{\infty} \frac{1}{\sqrt{2\pi\beta^2}} e^{-(1/2)(x-K)^2/\beta^2} dx = 1 \tag{16.1}$$

where β^2 is the "variance" parameter. Let $f(x)$ denote the density:

$$f(x) = \frac{1}{\sqrt{2\pi\beta^2}} e^{-(1/2)(x-K)^2/\beta^2} \tag{16.2}$$

We will use $f(x)$ as a mathematical *tool* instead of representing a probability density associated with a financial variable. To see how this is done, suppose we consider the values of β that sequentially go from 1 toward 0. The densities will be as shown in Figure 16.3. Clearly, if β is very small, the "density" will essentially be a spike at K, but still will have an area under it that adds up to 1.

Now consider the "expectations" calculated with such an $f(x)$. Let $C(x_t)$ be a random value that depends on the random variable x_t, indexed by the time t. Then we can write

$$E[C(x_t)] = \int_{-\infty}^{\infty} C(x_t) f(x_t) dx_t \tag{16.3}$$

---

[3]On one hand, in this textbook approach, calls are regarded as a bet in increasing prices, and put a bet on decreasing prices. This, however, would be true under the risk-adjusted probability and leaves the wrong impression that calls and puts are different in some sense. On the other hand, the volatility interpretation shows that the calls and puts are in fact the same from the point of view of volatility.

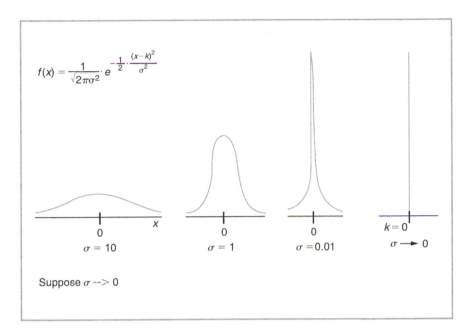

FIGURE 16.3

Density as β goes to zero.

Now we push β toward zero. The density $f(x_t)$ will become a spike at K. This means that all values of $C(x_t)$ will be multiplied by a probability of almost zero, except the ones around $x_t = K$. After all, at the limit $f(.)$ is nonzero only around $x_t = K$. Thus at the limit we obtain

$$\lim_{\beta \to 0} \int_{-\infty}^{\infty} C(x_t)f(x_t)\mathrm{d}x_t = C(K) \qquad (16.4)$$

The integral of the product of a function $C(x_t)$ and of $f(x_t)$ as β goes to zero *picks up* the value of the function at $x_t = K$.

Hence we define the *Dirac delta function* as

$$\delta_K(x) = \lim_{\beta \to 0} f(x, K, \beta) \qquad (16.5)$$

Remember that β determines how close $f(x)$ is to a spike at K. The integral can then be rewritten as

$$\int_{-\infty}^{\infty} C(K)\delta_K(x)\mathrm{d}x_t = C(K) \qquad (16.6)$$

This integral shows the most useful property of dirac *delta* function for our purposes. Essentially, the dirac *delta* picks up the value of $C(x_t)$ at the point $x_t = K$. We now apply this property to option payoffs at expiration.

16.5 APPLICATION TO OPTION PAYOFFS

The major advantage of the dirac *delta* functions, interpreted as the limits of distributions, is in differentiating functions that have points that cannot be differentiated in the usual sense. There are many such points in option trading. The payoff at the strike K is one example. Knock-in, knock-out barriers is another example. Dirac *delta* will be useful for discussing derivatives at those points.

Before we proceed, for simplicity we will assume in this section that interest rates are equal to zero:

$$r_t = 0 \tag{16.7}$$

We also assume that the underlying S_T follows the risk-neutral SDE, which in this case will be given by

$$dS_T = \sigma(S_t)S_t dW_t \tag{16.8}$$

Note that with interest rates being zero, the drift is eliminated and that the volatility is *not* of the Black–Scholes form. It depends on the random variable S_t. Let

$$
\begin{aligned}
f(S_T) &= \max[S_T - K, 0] \\
&= (S_t - K)^+
\end{aligned}
\tag{16.9}
$$

be the vanilla call option payoff shown in Figure 16.4. The function is not differentiable at $S_T = K$, yet its first-order derivative is like a step function. More interestingly, the *second*-order derivative can be interpreted as a dirac *delta* function. These derivatives are shown in Figures 16.4 and 16.5.

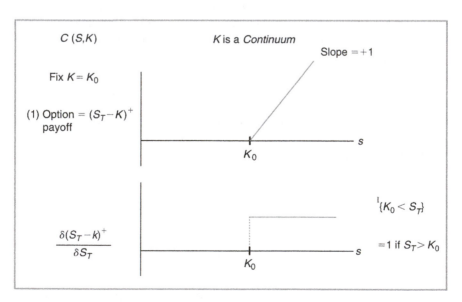

FIGURE 16.4

Call option payoff and first derivative.

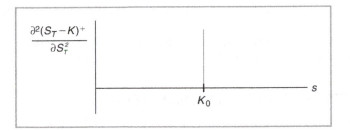

FIGURE 16.5

Second derivative and dirac delta function.

Now write the equivalent of Ito's Lemma in a setting where functions have kinks as in the option payoff case. This is called *Tanaka's formula* and essentially extends Ito's Lemma to functions that cannot be differentiated at all points. We can write

$$d(S_t - K)^+ = \frac{\partial(S_t - K)^+}{\partial S_t} dS_t + \frac{1}{2} \frac{\partial^2(S_t - K)^+}{\partial S_t^2} \sigma(S_t)^2 dt \tag{16.10}$$

where we define

$$\frac{\partial(S_t - K)^+}{\partial S_t} = 1_{s_t > K} \tag{16.11}$$

$$\frac{\partial^2(S_t - K)^+}{\partial S_t^2} = \delta_K(S_t) \tag{16.12}$$

Taking integrals from t_0 to T we get:

$$(S_T - K)^+ = (S_{t_0} - K)^+ + \int_{t_0}^T 1_{s_t > K} \, dS_t + \frac{1}{2} \int_{t_0}^T \frac{\partial^2(S_t - K)^+}{\partial S_t^2} \sigma(s_t)^2 dt \tag{16.13}$$

where the first term on the right-hand side is the time value of the option at time t_0 and is known with certainty. We also know that with zero interest rates, the option price $C(S_{t_0})$ will be given by

$$C(s_{t_0}) = E_{t_0}^{\tilde{P}}(S_T - K)^+ \tag{16.14}$$

Now, using the risk-adjusted probability $\tilde{P}$, (i) apply the expectation operator to both sides of Eq. (16.13), (ii) change the order of integration and expectation, and (iii) use the property of dirac *delta* functions in eliminating the terms valued at points other than $S_t = K$. We obtain the characterization of the option price as:

$$E_t^P[(S_T - K)^+] = (S_{t_0} - K)^+ + \int_{t_0}^T \sigma(K)^2 \phi_t(K) dt \tag{16.15}$$

$$= C(S_{t_0})$$

where $\phi_t(.)$ is the continuous density function that corresponds to the risk-adjusted probability of S_t.[4] This means that the time value of the option depends (i) on the intrinsic value of the option,

---

[4]We assume that a density exists.

(ii) on the *time spent* around K during the life of the option, and (iii) on the *volatility at that strike*, $\sigma(K)$.

The main point for us is that this expression shows that the option price depends *not* on the overall volatility, but on the volatility of S_t *around* K. This is exactly what the notion of volatility smile is.

16.5.1 AN INTERPRETATION OF DYNAMIC HEDGING

There are many dynamic strategies that replicate an option's final payoff. The best known is *delta* hedging. In *delta* hedging, the financial engineer will buy or sell the *delta* $= D_t$ units of the underlying, borrow any necessary funds, and adjust D_t as the underlying S_t moves over time. As $t \rightarrow T$, the expiration date, this will duplicate the option's payoff. This is the case because, as the time value goes to zero the option price merges with $(S_T - K)^+$.

However, there is an alternative dynamic hedging procedure that is similar to the approach adopted in the previous section. The dynamic hedging technique, called stop-loss strategy, is as follows.

In order to *replicate* the payoff of the long call, hold *one* unit of S_t if $K < S_t$. Otherwise hold *no* S_t. This strategy requires that as S_t crosses level K, we keep adjusting the position as soon as possible. Either buy one unit of S_t or sell the S_t immediately as S_t crosses the K *from left to right* or *from right to left*, respectively. The P/L of this position is given by the term

$$\frac{1}{2}\int_{t_0}^{T} \frac{\partial^2 (S_{t_0} - K)^+}{\partial S_t^2} \sigma(S_t)^2 dt \tag{16.16}$$

Clearly the switches at $S_t = K$ cannot be done instantaneously at zero cost. The trader is moving with time Δ while the underlying Wiener process is moving at a *faster* rate $\sqrt{\Delta}$. These adjustments are shown in Figures 16.6 and 16.7. The resulting hedging cost is the options value.

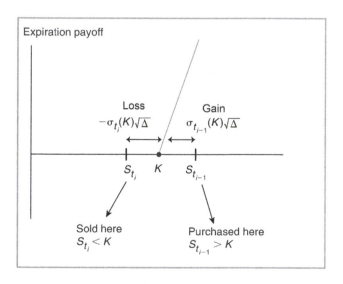

FIGURE 16.6

Hedging strategy adjustment—call option.

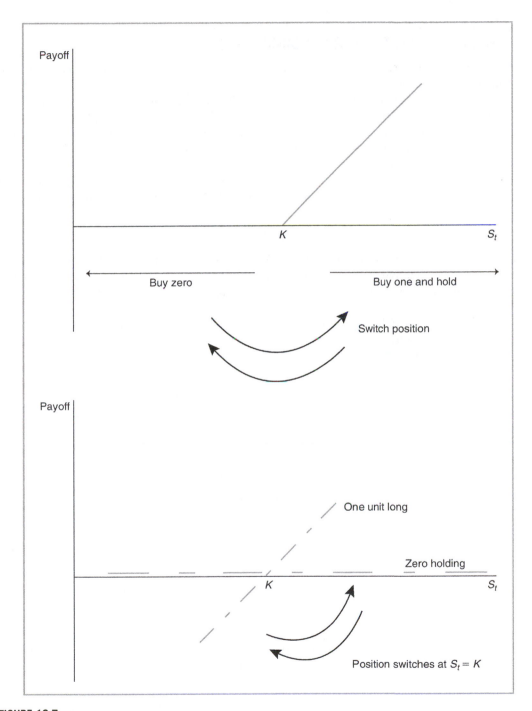

FIGURE 16.7

Hedging strategy adjustment and positions.

16.6 BREEDEN–LITZENBERGER SIMPLIFIED

The so-called Breeden–Litzenberger Theorem is an important result that shows how one can back out risk-adjusted probabilities from liquid arbitrage-free option prices. In this section, we discuss a trader's approach to Breeden–Litzenberger. This approach will show the theoretical relevance of some popular option strategies used in practice. Below, we use a simplified framework which could be generalized in a straightforward way. However, we will not generalize these results, but instead in the following section use the dirac *delta* approach to prove the Breeden–Litzenberger theorem.

Consider a simple setting where we observe prices of four liquid European call options, denoted by $\{C_t^1, \ldots, C_t^4\}$. The options all expire at time T with $t < T$. The options have strike prices denoted by $\{K^1 < \ldots < K^4\}$ with

$$K^i - K^{i-1} = \Delta K \tag{16.17}$$

Hence, the strike prices are equally spaced. Apart from the assumption that these options are written on the same underlying S_t which does not pay dividends, we make no distributional assumption about S_t. In fact the volatility of S_t can be stochastic and the distribution is not necessarily log normal.

Finally we use the LIBOR rate L_t to discount cash flows to be received at time T. The discount factor will then be given by

$$\frac{1}{(1 + L_t \delta)} \tag{16.18}$$

Next we define a simple probability space. We assume that the strike prices define the four *states of the world* where S_T can end up. Hence the state space is discrete and is assumed to be made of only four states $\{\omega^1, \ldots, \omega^4\}$.[5]

$$\omega^i = K^i \tag{16.19}$$

We then have four risk-adjusted probabilities associated with these states defined as follows:

$$p^1 = P(S_T = K^1) \tag{16.20}$$

$$p^2 = P(S_T = K^2) \tag{16.21}$$

$$p^3 = P(S_T = K^3) \tag{16.22}$$

$$p^4 = P(S_T = K^4) \tag{16.23}$$

The arbitrage-free pricing of Chapter 11 can be applied to these vanilla options:

$$C_t^i = \frac{1}{(1 + L_t \delta)} E_t^{\tilde{P}} \left[(S_T - K, 0)^+ \right] \tag{16.24}$$

The straightforward application of this formula using the probabilities p^i gives the following pricing equations, where possible payoffs are weighed by the corresponding probabilities (Figure 16.8).

---

[5]The following discussion can continue unchanged by assuming n discrete states.

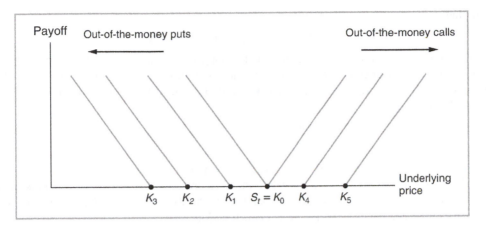

FIGURE 16.8

Payoff diagrams for puts and calls with different moneyness.

$$C_t^1 = \frac{1}{(1 + L_t\delta)} \left[p^2 \Delta K + p^3 (2\Delta K) + p^4 (3\Delta K) \right] \tag{16.25}$$

$$C_t^2 = \frac{1}{(1 + L_t\delta)} \left[p^3 (\Delta K) + p^4 (2\Delta K) \right] \tag{16.26}$$

$$C_t^3 = \frac{1}{(1 + L_t\delta)} \left[p^4 (\Delta K) \right] \tag{16.27}$$

Next we calculate the first differences of these option prices.

$$C_t^1 - C_t^2 = \frac{1}{(1 + L_t\delta)} \left[p^2 \Delta K + p^3 (\Delta K) + p^4 (\Delta K) \right] \tag{16.28}$$

$$C_t^2 - C_t^3 = \frac{1}{(1 + L_t\delta)} \left[p^3 (\Delta K) + p^4 (\Delta K) \right] \tag{16.29}$$

Finally, we calculate the *second* difference and obtain the following interesting result:

$$(C_t^1 - C_t^2) - (C_t^2 - C_t^3) = \frac{1}{(1 + L_t\delta)} p^2 \Delta K \tag{16.30}$$

Divide by ΔK twice to obtain

$$\frac{\Delta^2 C}{\Delta K^2} = \frac{1}{(1 + L_t\delta)\Delta K} p^2 \tag{16.31}$$

where

$$\Delta^2 C = (C_t^1 - C_t^2) - (C_t^2 - C_t^3) \tag{16.32}$$

This is the well-known Breeden–Litzenberger result in this very simple environment. It has interesting implications for the options trader.

Note that

$$(C_t^1 - C_t^2) - (C_t^2 - C_t^3) = (C_t^1 - C_t^3) - 2C_t^2 \tag{16.33}$$

In other words, this is an option position that is long two wings and short the center twice. In fact this is a butterfly centered at K_2. It turns out that the arbitrage-free market value of this butterfly multiplied by the $(1 + L_t\delta)\Delta K$ gives the risk-adjusted probability that the underlying S_t will end up at state K_2. Letting $\Delta K \to 0$ we get

$$\frac{\partial^2 C}{\partial K^2} = \frac{1}{(1 + L_t\delta)}\phi(S_T = K) \tag{16.34}$$

where $\phi(S_T = K)$ is the (conditional) risk-adjusted *density* of the underlying at time T.[6]

This discussion illustrates one reason why butterflies are traded as vanilla instruments in option markets. They yield the probability associated with their center. Below we prove the Breeden–Litzenberger result using the dirac *delta* function.

16.6.1 THE PROOF

The idea behind the Breeden–Litzenberger result has been discussed before. It rests on the fact that by using liquid and arbitrage-free options prices we can back out the risk-adjusted probabilities associated with various states of the world in the future. The probabilities will relate to the future values of the underlying price, S_T.

The theorem asserts that (i) if a continuum of European vanilla option prices exists for all $0 \le K$ and (ii) if the function giving $C(S_t, K)$ is twice differentiable with respect to K, then we have

$$\frac{\partial^2 C}{\partial K^2} = \frac{1}{(1 + L_t\delta)}\phi(S_T = K) \tag{16.35}$$

where $\phi(S_T = K | S_{t_0})$ is the conditional risk-adjusted density of S_T. In other words, if we had a continuum of vanilla option prices, we could obtain the risk-adjusted density with a straightforward differentiation. We now prove this using the dirac *delta* function $\delta_K(S_T)$.

Apply the twice differential operator to the definition of both sides of the arbitrage-free price $C(S_t, K)$. By definition, this means

$$\frac{\partial^2}{\partial K^2}C(S_t, K) = \frac{1}{(1 + L_t\delta)}\frac{\partial^2}{\partial K^2}\int_0^\infty (S_T - K)^+\phi(S_T)dS_T \tag{16.36}$$

Assuming that we can interchange the operators and realizing that $\phi(S_T)$ does not depend on K we obtain

$$\frac{\partial^2}{\partial K^2}C(S_t, K) = \frac{1}{(1 + L_t\delta)}\int_0^\infty \frac{\partial^2}{\partial K^2}(S_T - K)^+\phi(S_T)dS_T \tag{16.37}$$

---

[6] Remember that if the density at x_0 is $f(x_0)$, then $f(x_0)dx$ is the probability of ending around x_0. In other words we have

$$p^2 \sim \phi(S_T = K^2)\Delta K$$

But

$$\frac{\partial^2}{\partial K^2}(S_T - K)^+ = \delta_K(S_T) \tag{16.38}$$

is a dirac *delta*, which means that

$$\frac{\partial^2}{\partial K^2}C(S_t, K) = \frac{1}{(1 + L_t\delta)}\int_0^\infty \delta_K(S_T)\phi(S_T)\mathrm{d}S_T \tag{16.39}$$

The previous discussion and Eq. (16.4) tells us that in this integral $\phi(S_T)$ is being multiplied by zero everywhere except for $S_T = K$. Thus,

$$\frac{\partial^2}{\partial K^2}C(S_t, K) = \frac{1}{(1 + L_t\delta)}\phi(S_T = K) \tag{16.40}$$

To recover the risk-adjusted density just take the second partial of the European vanilla option prices with respect to K. This is the Breeden–Litzenberger result.

16.7 A CHARACTERIZATION OF OPTION PRICES AS *GAMMA* GAINS

The question then is, how does a trader "characterize" an option using these hedging gains? First of all, in liquid option markets the order flow determines the price and the trader does not have to go through a pricing exercise. But still, can we use these trading gains to represent the frame of mind of an options trader?

The discussion in the previous section provides a hint about this issue. The trader buys or sells an option with strike price K. The cash needed for this transaction is either borrowed or lent. Then the trader *delta* hedges the option. Finally, this hedge is adjusted as the underlying price fluctuates *around the initial S_{t_0}.*

According to this, the trader could add the (discounted) future gains (payouts) from these hedge adjustments and this would be the true *time-value* of the option, besides interest or other expenses. The critical point is that these future gains need to be calculated at the initial *gamma*, evaluated at the initial s_{t_0}, and adjusted for passing time.

We can explain this statement. First, for simplicity assume interest rates are equal to zero. We then let the price of the vanilla call be denoted by $C(S_t, t)$. Then by definition we have

$$C(S_{t_0}, T) = \text{Max}[S_{t_0} - K, 0] \tag{16.41}$$

This will be the future value of the option if the underlying ended up at S_{t_0} at time T. Now, this value is equal to the initial price plus how much the time value has changed between t_0 and T,

$$C(S_{t_0}, T) = C(S_{t_0}, t_0) + \int_{t_0}^T \frac{\partial C}{\partial t}\Big|_{S_t=S_{t_0}}\mathrm{d}t \tag{16.42}$$

Now, we know from the Black–Scholes PDE that

$$\frac{\partial C}{\partial t} = \frac{1}{2}\frac{\partial^2 C}{\partial S_t^2}\sigma_t^2(S_t, t) \tag{16.43}$$

Substituting and reorganizing Eq. (16.42) above becomes

$$C(S_{t_0}, t_0) = \text{Max}[S_{t_0} - K, 0] + \int_0^T \frac{1}{1} \frac{\partial^2 C(S_{t_0}, t)}{\partial S_t^2} \sigma_t^2(S_{t_0}, t) dt \qquad (16.44)$$

Note that on the right-hand side, the integral is evaluated at the constant S_{t_0} so we don't need to take expectations.

According to this, the trader is valuing the option at time t_0 by adding the intrinsic value and the *gamma* gains evaluated with a *gamma* at S_{t_0} and a volatility centered at S_{t_0}. Still, the time t changes and this will change the *gamma* over time. Thus, it is important to realize that the trader is *not* valuing the option by looking at the expected value of the future *gamma* gains evaluated at *random* future S_t. The *gamma* is evaluated at the initial S_{t_0}.

16.7.1 RELATIONSHIP TO TANAKA'S FORMULA

The discussion above is also consistent with the option interpretation obtained using Tanaka's formula. Consider the value of the option as shown above, again

$$C(S_{t_0}, t_0) = \text{Max}[S_{t_0} - K, 0] + \int_{t_0}^T \frac{1}{2} \frac{\partial^2 C(S_{t_0}, t)}{\partial S_t^2} \sigma_t^2(S_{t_0}, t) dt \qquad (16.45)$$

Now we know from Breeden–Litzenberger that

$$\frac{\partial^2 C(S_{t_0}, t)}{\partial S_t^2} \sigma_t^2(S_{t_0}, t) = \Phi(S_{t_0}, t | S_{t_0}) \qquad (16.46)$$

where $\Phi(S_{t_0}, t | S_{t_0})$ is the risk-adjusted density of S_t at time t. Substituting this in the option value

$$C(S_{t_0}, t_0) = \text{Max}[S_{t_0} - K, 0] + \int_{t_0}^T \frac{1}{2} \sigma_t^2(S_{t_0}, t) \Phi(S_{t_0}, t | S_{t_0}) dt \qquad (16.47)$$

This is the same equation we obtained by using dirac *delta* functions along with the Tanaka formula. The second term on the right-hand side was called *local time*. In this case, the local time is evaluated for the ATM option with strike $K = S_{t_0}$.

16.8 INTRODUCTION TO THE SMILE

Markets trade *many* options with the same underlying, but different strike prices and different expirations. Does the difference in strike price between options that are identical in every other aspect have any important implications?

At first, the answer to this question seems to be no. After all, vanilla options are written on an underlying, with say, price S_t, and this price will have only *one* volatility at any time t, regardless of the strike price K_i. Hence, it appears that, regardless of the differences in the strike price, the implied volatility of options written on the same underlying, with the same expiration, should be the same.

Yet, this first impression is wrong. In reality, options that are identical in every respect, except for their strike, in general, have *different* implied volatilities. Overall, the more out-of-the-money a call (put) option is, the higher is the corresponding implied volatility. This well-established empirical fact is known as the volatility *smile*, or volatility *skew*, and has major implications for hedging, pricing, and marking-to-market of many important instruments. In the remainder of this chapter, we discuss the volatility smile and skews using caps and floors as vehicles for conveying the main ideas. This will indirectly give us an opportunity to discuss the engineering of this special class of convex instruments.

From this point and on, in this chapter we will use the term *smile* only. This will be the case even when the smile is, in fact, a one-sided *skew*. However, whenever relevant, we will point out the differences.

16.9 PRELIMINARIES

The volatility smile has important implications for trading, hedging, and pricing financial instruments. To illustrate how far things have come in this area, we look at a position taken with the objective of benefiting from abnormal conditions regarding the volatility smile.

We can trade stocks, bonds, or, as we have seen before, the slope of the yield curve. We may, for example, expect that the long-term yields will decline *relative to* the short-term yields. This is called a flattening of the yield curve and it invites curve-flattening strategies that (short) sell short maturities, and buy long maturities. This can be done using cash instruments (i.e., bonds) or swaps.

In any case, such trades have become routine in financial markets. A more recent relative value trade relates to the volatility smile. Consider the following episode.

EXAMPLE

Over the last month, European equity options traders have seen interest by contrarian investors, namely hedge funds, in buying at-the-money volatility and selling out-of-the-money volatility to take advantage of a skew in volatility levels in certain markets.

> . . . the skew trade involves an investor buying at-the-money vol and selling out-of-the-money vol. Due to supply/demand pressures, the level of out-of-the-money vol sometimes rises higher than normal. In other words, the spread between out. and at-the-money volatility increases, causing a so-called skew. Investors put on the trade in anticipation of the skew dissipating.
>
> [A trader] explained that along with the bull run of US and European equity markets has come a sense of unease among some investors regarding a downturn. Many have thus sought protection via over-the-counter put contracts. Because out-of-the-money puts are usually cheaper than at-the-money puts, many investors have opted for the former. The heavy volume has caused out-of-the-money vol levels to rise. Many investors want crash protection but today puts are too expensive. So, instead of buying today at 100 they buy puts at 80.

(Based on an article in Derivatives Week (now part of GlobalCapital))

According to this example, equity investors that had heavily invested during the "stock market bubble" of the 1990s were looking for *crash protection*. They were long equities and would have suffered significant losses if markets crashed. Instead of selling the stocks that they owned, they bought put options. With a put an investor has the right to *sell* the underlying stocks at a predetermined price, say, K. If market price declines below K, the investor would have some protection.

According to the reading, the large number of investors who were willing to buy puts increased, first, the at-the-money (ATM) volatility.[7] The ATM options became expensive. To lower the cost of insurance sought, investors instead bought options that were, according to the reading, 20% out-of-the-money. These options were cheaper. But as more and more investors bought them, the out-of-the-money volatility started to increase relative to ATM volatility of the same option series. This led to an abnormally steep skew.[8]

The reading suggests that this "abnormal" skew may have attracted some hedge funds who expected the abnormality to disappear in the longer run. According to one theory, these funds sold out-of-the-money volatility and bought ATM volatility. This position will make money if the skew *flattens* and out-of-the-money volatility decreases *relative to* the ATM volatility.[9] As this example shows, the skew, or smile, should be considered as an integral part of financial markets activity. However, as we see in this section, its existence causes several complications and difficulties in financial modeling and in risk management.

16.10 A FIRST LOOK AT THE SMILE

The volatility smile can be a confusing notion, and we need to discuss some preliminary ideas before getting into the mechanics of pricing and market applications. It is well known that the Black–Scholes assumptions are not very realistic. And yet, the Black–Scholes formula is routinely used by options traders, although these traders know better than anybody else that the assumptions behind the model are problematic. One of the major Black–Scholes assumptions, for example, is that volatility is *constant* during the life of an option. How can a trader still use the Black–Scholes formula if the realized volatility is known to fluctuate significantly during the life of the option?

If this Black–Scholes assumption is violated, wouldn't the price given by the Black–Scholes formula be "wrong," and, hence, the volatility implied by the formula be erroneous? This question needs to be carefully considered. In the end, we will see that there are really no inconsistencies in traders' behavior. We can explain this as follows.

1. First, note that the Black–Scholes formula is *simple* and depends on a small number of parameters. In fact, the only major parameter that it depends on is the volatility, σ. A simple

---

[7]This is, of course, somewhat circular. If there is a fear of a crash, one would normally expect such an increase in the volatility anyway.

[8]These out-of-the-money puts would still be cheaper in monetary terms when compared with ATM puts. Only, the volatility that they imply would be higher. Another way of saying this is that if we plugged the ATM volatility into the Black–Scholes formula for these out-of-the-money options, they would end up being even cheaper.

[9]However, a similar effect would be observed if investors were unwinding their previous insurance bought when markets were at, say, 120 and buying new insurance at $K = 80$. This is equivalent to rolling over their protection. Hence, it is difficult to tell what the real driving force behind this observation is, namely, whether it is due to speculative relative value plays or simply rolling over the positions.

formula has some advantages. It is easy to understand and remember. But, more importantly, it is also easy to realize *where* or *when* it may go wrong. A simple formula permits developing ways to correct for any inaccuracies *informally* by making subjective adjustments during trading. The Black–Scholes formula has one parameter, and it may be easier to remember how to "adjust" this parameter to cover for the imperfections of the formula.[10]

2. An important aspect of the Black–Scholes formula is that it has become a *convention*. In other words, it has become a *standard* among professionals and also in computer platforms. The formula provides a way to connect a volatility quote to a dollar value attached to this quote. This way traders use the *same* formula to put a dollar value on a volatility number quoted by the market. This helps in developing common platforms for hedging, risk managing, and trading volatility.

3. Thus, once we accept that the use of the Black–Scholes formula amounts to a convention, and that traders differ in their selection of the value of the parameter σ, then the critical process is no longer the option price, but the volatility. This is one reason why in many markets, such as caps, floors, and swaptions markets, the volatility is quoted *directly*.

    One way to account for the imperfections of the Black–Scholes assumptions would be for traders to adjust the volatility parameter.

4. However, the convention creates new risks. Once the underlying is the volatility process, another issue emerges. For example, traders could add a *risk premium* to quoted volatilities. Just like the risk premium contained in asset prices, the quotes on volatility may incorporate a risk premium.

The volatility smile and its generalization, the *volatility surface*, could then contain a great deal of information concerning the implied volatilities and any arbitrage relations between them. Trading, pricing, hedging, and arbitraging of the smile thus become important.

16.11 WHAT IS THE VOLATILITY SMILE?

Consider the Black–Scholes world with vanilla European call and put options written on the equity price (index), S_t, that expire on the same date T. Let K_i denote the ith strike of the option series and σ_i the *constant* Black–Scholes instantaneous (implied) volatility coefficient for the strike K_i. Finally, let r be the constant risk-free rate.

The Black–Scholes setting makes many assumptions beyond that of constant volatility. In particular, the underlying equity does not make any dividend payments, and there are no transaction costs, tax issues, or regulatory costs. Finally, S_t is assumed to follow the *geometric* stochastic differential equation (SDE)

$$dS_t = \mu S_t \, dt + \sigma S_t \, dW_t \quad t \in [0, \infty] \tag{16.48}$$

---

[10]In the theory of prediction, there is the notion of *parsimony*. During a prediction exercise it is costly to have too many parameters because errors are more likely to occur. The notion applies to the numerical calculation of complex options prices. If a model has fewer parameters to be calibrated, the likelihood of making mistakes decreases.

where W_t is a Wiener process defined under the probability P. Here, the parameter μ may also depend on S_t. The crucial assumption is that the *diffusion* component is given by σS_t. This is the assumption that we are concerned with in this chapter. The Black–Scholes setting assumes that the absolute volatility during an infinitesimally small interval dt is given (heuristically) by[11]

$$\sqrt{E_t^P[(dS_t - \mu S_t \, dt)^2]} = \sigma S_t \sqrt{dt} \tag{16.49}$$

Thus, for a small interval, Δ, we can write the *percentage* volatility approximately as

$$\frac{\sqrt{E_t^P[(\Delta S_t - \mu S_t \Delta)^2]}}{S_t} \cong \sigma\sqrt{\Delta} \tag{16.50}$$

According to this, as S_t changes, the percentage volatility during intervals of length Δ remains approximately constant.

In this environment, a typical put option's price is given by the Black–Scholes formula:

$$P(S_t, K, \sigma, r, T) = -S_t N(-d_1) + Ke^{-r(T-t)}N(-d_2) \tag{16.51}$$

with

$$d_1 = \frac{\log(S_t/K) + ((1/2)\sigma^2 + r)(T - t)}{\sigma\sqrt{T - t}} \tag{16.52}$$

$$d_2 = d_1 - \sigma\sqrt{(T - t)} \tag{16.53}$$

Suppose the markets quote implied volatility, σ. To get the monetary value of an option with strike K_i, the trader will put the current values of S_t, t, and r and the quoted value of the implied volatility σ_i at which the trade went through, in this formula. According to this interpretation, the Black–Scholes formula is used to assign a dollar value to a quoted volatility. Conversely, given the correct option price $P(.)$, the implied volatility, σ_i, for the K_i put could be extracted.

We can now define the volatility smile within this context. Consider a series of T-expiration, *liquid*, and arbitrage-free out-of-the-money put option prices indexed by the strike prices K_i, denoted respectively by P_{K_i}:

$$P_{K_1}, \ldots, P_{K_n} \tag{16.54}$$

for

$$K_n < \cdots < K_1 < K_0 = S_t \tag{16.55}$$

According to this, the K_0 put is ATM and, as K_i decreases, the puts go deeper out-of-the-money, see Figure 16.9.

Then, given the (bid or ask) option prices, we can use the Black–Scholes formula *backward* and extract σ_i that the trader used to conclude the deal on P_{K_i}. If the assumptions of the Black–Scholes world are correct, all the implied volatilities would turn out to be the same

$$\sigma_{K_0} = \sigma_{K_1} = \cdots = \sigma_{K_n} = \sigma \tag{16.56}$$

---

[11]Here, dS_t is an infinitesimally small change in the price. It is only a symbolic way of writing small changes, and the expectation of such infinitesimal increments can only be heuristic.

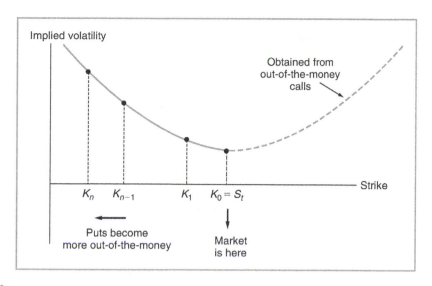

FIGURE 16.9

Volatility smile: implied volatility against strike price.

since the put options would be identical except for their strike price. Thus, in a market that conforms to the Black–Scholes world, the traders would use the *same* σ in the Black–Scholes put formula to obtain each $P_{K_i}, i = 0, \ldots, n$. Going backward, we would then recover the same constant σ from the prices.[12]

Yet, if we conducted this exercise in reality with observed option prices, we would find that the implied volatilities would satisfy

$$\sigma_{K_0} < \sigma_{K_1} < \cdots < \sigma_{K_n} \tag{16.57}$$

In other words, the more out-of-the-money the put option is, the higher the corresponding implied volatility. As a result, we would obtain a "smiley" curve.

We can also use the implied vols from progressively out-of-the-money call options and obtain, depending on the underlying instrument, the second half of the smile, as shown in Figure 16.10.

16.11.1 SOME STYLIZED FACTS

Volatility smiles observed in reality seem to have the following characteristics,

1. Options written on *equity* indices yield, in general, a nonsymmetric *one-sided* "smile" as shown in Figure 16.10a. For this reason, they are often called *skews*.

---

[12]This exercise requires that the put values were indeed obtained at the same time t and were identical in all other aspects except for K_i.

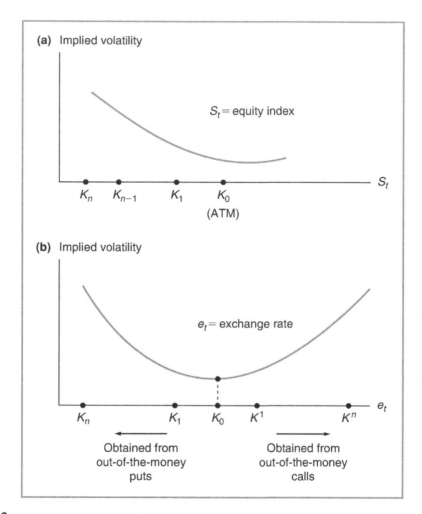

FIGURE 16.10

Skew for equity indices and typical symmetric FX smile.

2. The *FX markets* are quite different in this respect. They yield a more or less *symmetric* smile, as in Figure 16.10b. However, the smile will rarely be exactly symmetric and it is routine practice in foreign exchange markets to trade this asymmetry using risk reversals.
3. Options on various *interest rates* yield a more *monotonous* one-sided smile than the equity indices. The fact that "smile" patterns vary from market to market would suggest, on the surface, that there are different explanations involved.

It is also natural to think that the *dynamics* of the smile vary depending on the sector. This point is relevant for risk management, running swaption, cap/floor books, and volatility trading. But, before we discuss it, we consider an example.

EXAMPLE

Table 16.1 displays all the options written on the S&P100 index with a very short expiration. These data were obtained from live quotes early in the morning, so few trades had passed. However, the option bid–ask quotes were live, in the sense that reasonably sized trades could be conducted on them.

When the data were gathered, the underlying was trading at 589.14. We use 12 out-of-the-money puts and 9 out-of-the-money calls to obtain the Black–Scholes implied volatilities.

The interest rate is taken to be 1.98%, and the time to expiration was 8/365. Using these values and the bid prices for the options given in the table, the equations

$$C(S_t, K_i, r, T, \sigma_i) = C_i \qquad (16.58)$$

$$P(S_t, K_j, r, T, \sigma_j) = P_j \qquad (16.59)$$

were solved for the implied vols of calls $\{\sigma_i\}$ and the implied vols of puts $\{\sigma_i\}$, C_i and P_j being observed option prices.

Table 16.1 OEX Options with January 18, 2002, Expiration

Calls	Bid	Ask	Vol	Puts	Bid	Ask	Vol
Jan 550	39.5	41.5	0	Jan 550	0.45	0.75	0
Jan 555	34.8	36.3	0	Jan 555	0.65	0.95	0
Jan 560	30	31.5	0	Jan 560	0.9	1.2	0
Jan 565	25.2	26.7	0	Jan 565	1.25	1.55	0
Jan 570	20.6	22.1	0	Jan 570	1.8	2.1	0
Jan 575	16.3	17.8	0	Jan 575	2.3	3	0
Jan 580	13	13.5	0	Jan 580	3.4	4.1	2
Jan 585	9.1	9.8	0	Jan 585	5	5.7	5
Jan 590	6.1	6.8	50	Jan 590	7.6	7.9	5
Jan 595	4.1	4.5	12	Jan 595	10.1	10.8	25
Jan 600	2.5	2.8	3	Jan 600	13.1	14.5	0
Jan 605	1.2	1.5	0	Jan 605	17.2	18.7	0
Jan 610	0.55	0.85	1	Jan 610	21.7	23.2	0
Jan 615	0.25	0.55	0	Jan 615	26.6	28.1	0
Jan 620	0.2	0.35	1	Jan 620	31.4	32.9	0
Jan 625	0.05	0.2	0	Jan 625	36.3	37.8	0
Jan 630	0	0.15	0	Jan 630	41	43	0
Jan 635	0	0.1	0	Jan 635	46	48	0
Jan 640	0	0.1	0	Jan 640	51	53	0
Jan 645	0	0.1	0	Jan 645	56.5	57.5	0
Jan 650	0	0.1	0	Jan 650	60.5	63.5	0
Jan 660	0	0.05	0	Jan 660	70.5	73.5	0
Jan 680	0	0.05	0	Jan 680	90.5	93.5	0

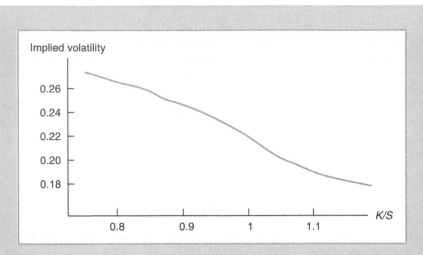

FIGURE 16.11

Implied volatility of short-dated OEX options against moneyness.

The resulting vols were plotted against K_i/S_t in Figure 16.11. We see a pronounced smile. For example, the January 400 put, which traded at about 32% out-of-the-money, had a volatility of about 26%, while the ATM option traded at an implied volatility of 18.5%.

OEX options are of American style and this issue was ignored in the example above. This would introduce an upward bias in the calculated volatilities. This bias is secondary for our purpose, but in real trading should be corrected. One correction has been suggested by Barone-Adesi and Whaley (1987).

16.11.2 HOW CAN WE DEFINE MONEYNESS?

The way a smile is plotted varies from one market to another. The implied volatility, denoted by σ_i, always appears on the y-axis. Unless stated otherwise, we extract this volatility from the Black–Scholes formula in equity or FX markets, and from the Black formula in the case of interest rates. The implied volatilities are then treated as if they were *random* and *time varying*.

What to put on the horizontal axis is a more delicate question and eventually depends on how we define "moneyness" of an option. Sometimes the smile is plotted against *moneyness* measured by the ratio of the strike price to the current market price, K_i/S_t. If the smile is a function of how much the option is out-of-the-money only, then this normalization will stabilize the smile in the sense that as S_t changes, the smile for that particular option series may be more or less invariant. But there are almost always factors other than moneyness that affect the smile, and some practitioners define moneyness differently.

For example, sometimes the smile is plotted against $K_i e^{-r(T-t)}/S_t$. For short-dated options, this makes little difference, since $r(T-t)$ will be a small number. For longer-dated options, the difference is more relevant. By including this discount factor, market practitioners hope to eliminate the effect of the changes in the remaining life of the option.

Sometimes the horizontal axis represents the option's *delta*. FX traders take the size of *delta* as a measure of moneyness. This practice can be challenged on the grounds that an option's *delta* depends on more variables than just moneyness. It also depends, for example, on the instantaneous implied volatility. Yet, as we will see later, there are some *deltas* at which volatility trading is particularly liquid in FX markets.

The reader should note that the smile in Figure 16.9 is a plot of the implied volatility against the *strike* only. Figures 16.12 and 16.13 are plots of the implied volatility against the *delta*. These curves

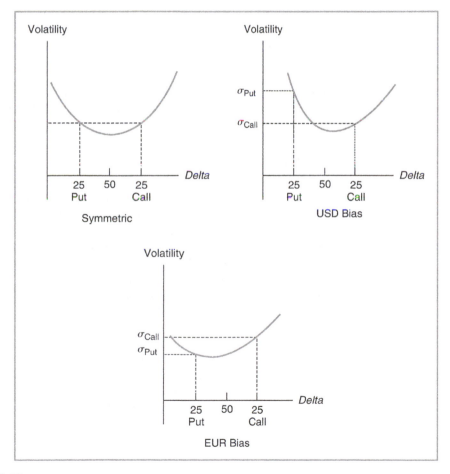

FIGURE 16.12

Three different smile shapes for FX markets.

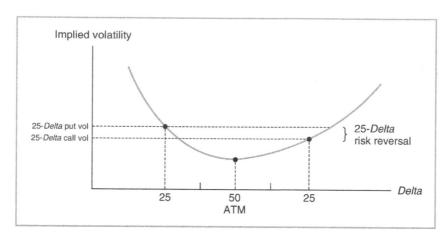

FIGURE 16.13

Risk reversals and curvature of smile.

relate to a particular time t and expiration date T. As the latter change, the smile will, in general, shift. It is quite important to know how changes in time t and expiration date T affect the smile.

16.11.3 REPLICATING THE SMILE

The volatility smile is a plot of the implied volatility of options that are alike in all respects except for their moneyness. The basic shape of the volatility smile has two major characteristics. The first relates to the extent of symmetry in the smile. The second is about how "pronounced" the curvature is. There are good approximations for measuring both characteristics.

First, consider the issue of symmetry. Figure 16.12 shows three smiles for FX markets. One is symmetric, the other two are asymmetric with different *biases*. If the smile is symmetric, the volatilities across similarly out-of-the-money puts and calls will be the same. This means that if a trader buys a call and sells a put with the same moneyness, the structure will have zero value. Such positions were called risk reversals in Chapter 11. A symmetric smile implies that a zero cost risk reversal can be achieved by buying and selling similarly out-of-the-money options. In the case of asymmetric smiles, puts and calls with similar moneyness are sold at *different* implied volatilities, and this is labeled a *bias*. Thus, a risk reversal is one way of measuring the bias in a volatility smile.

The way risk reversals measure the symmetry of the volatility smile is shown in Figure 16.13. We use the *delta* of the option as a measure of its moneyness on the x-axis. ATM options would have a *delta* of around 50 and would be in the "middle" of the x-axis. The volatility of the *25-delta* risk reversal gives the difference between the volatilities of a 25-*delta* put and a 25-*delta* call as indicated in the graph. We can write this as

$$\sigma(25\text{-}delta\ RR\ \text{spread}) = \sigma(25\text{-}delta\ \text{put}) - \sigma(25\text{-}delta\ \text{call}) \tag{16.60}$$

where $\sigma(25\text{-}delta\ RR\ \text{spread})$, $\sigma(25\text{-}delta\ \text{put})$, and $\sigma(25\text{-}delta\ \text{call})$ indicate, respectively, the implied volatilities of a risk reversal, a 25-*delta* put, and a 25-*delta* call.

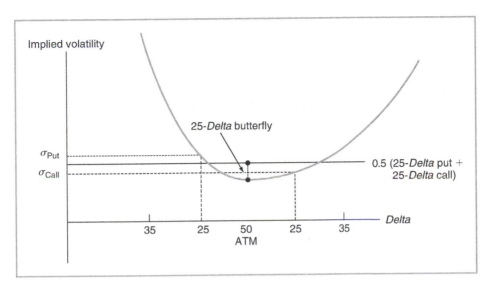

FIGURE 16.14

Butterfly measures curvature of smile.

The curvature of the smile can be measured using a *butterfly*. Consider the sale of an ATM put and an ATM call along with the purchase of one *25-delta* out-of-the-money put and a *25-delta* out-of-the-money call. This butterfly has a payoff diagram that should be familiar from Chapter 11. The position consists of buying two symmetric out-of-the-money volatilities and selling two ATM volatilities. If there were no smile effects, these volatilities would all be the same and the net volatility position would be zero. On the other hand, the more pronounced the smile becomes, the higher the out-of-the-money volatilities would be relative to the ATM volatilities, and the net volatility position would become more and more positive. Figure 16.14 shows how a butterfly measures the magnitude of the curvature in a smile. The following equality holds:

$$\sigma(25\text{-}delta \text{ butterfly spread}) = \sigma(25\text{-}delta \text{ put}) + \sigma(25\text{-}delta \text{ call}) - 2\sigma(\text{ATM}) \qquad (16.61)$$

where the σ(25-*delta* butterfly spread) and σ(ATM) are the butterfly and the ATM implied volatilities, respectively.

16.11.3.1 Contractual equations

Chapter 3 dealt with contractual equations for simple assets. The equalities discussed in the preceding paragraphs now permit considering quite different types of contractual equations. In fact, we can rearrange equalities shown in Eqs. (16.60) and (16.61) to generate some contractual equations for out-of-the-money implied volatilities:

$$\boxed{\begin{array}{c}\text{25-}delta \text{ put} \\ \text{volatility}\end{array}} = \boxed{\text{ATM volatility}} + \frac{1}{2}\boxed{\begin{array}{c}\text{25-}delta \text{ RR} \\ \text{volatility spread}\end{array}} + \frac{1}{2}\boxed{\begin{array}{c}\text{25-}delta \text{ butterfly} \\ \text{volatility spread}\end{array}} \qquad (16.62)$$

$$\boxed{\begin{array}{c}\text{25-}\textit{delta}\text{ put}\\ \text{volatility}\end{array}} = \boxed{\text{ATM volatility}} - \frac{1}{2}\boxed{\begin{array}{c}\text{25-}\textit{delta}\text{ RR}\\ \text{volatility spread}\end{array}} + \frac{1}{2}\boxed{\begin{array}{c}\text{25-}\textit{delta}\text{ butterfly}\\ \text{volatility spread}\end{array}} \qquad (16.63)$$

These equalities can be used to determine out-of-the-money volatilities in the case of vanilla options. For example, if ATM, *RR*, and butterfly volatilities are liquid, we can use these equations to "calculate" 25-*delta* call and put volatilities. However, it has to be noted that for exotic options, adjusting the volatility parameter this way will not work. This issue will be discussed at the end of the chapter.

16.12 SMILE DYNAMICS

There are at least two types of smile "dynamics." In the first, we would fix the time parameter t and consider options with longer and longer expirations, T. In the second case, we would keep T constant, but let time t pass and study how changes in various factors affect the volatility smile. In particular, we can observe *if* changes in S_t affect the smile when the moneyness K_i/S_t is kept constant.

We first keep t fixed and increase T. We consider *two* series of options that trade at the same time t. Both series have comparable strikes, but one series has a relatively longer maturity. How would the smiles implied by the two series of options with expirations, say, T_1, T_2, compare with each other?

The second question of interest is how the smile of the *same* option series moves over time as S_t changes. In particular, would the smile be a function of the ratio K_i/S_t only, or would it also depend on the level of S_t over and above the moneyness?

The answers to these questions change depending on which underlying asset is considered. This is because there is more than one explanation for the existence of the smile, and for different sectors, different explanations seem to prevail. Thus, before we analyze the smile dynamics and its properties any further, we need to discuss the major explanations advanced for the existence of the volatility smile.

16.13 HOW TO EXPLAIN THE SMILE

The volatility smile is an empirical phenomenon that violates the assumptions of the Black–Scholes world. At the same time, the volatility smile is related to the implied volatilities obtained *from* the Black–Scholes formula. This may give rise to confusion. The smile suggests that the Black–Scholes formula is not valid, while at the same time, the trader obtains the smile using the very same Black–Scholes formula. Is there an internal inconsistency?

The answer is no. To clarify the point, we use an analogy that is unrelated to the present discussion, but illustrates what market conventions are. Consider the 3-month LIBOR rate L_t. What is the present value of, say, $100 that will be received in 3 months' time? We saw in Chapter 3 that all we need to do is calculate the ratio:

$$\frac{100}{(1 + L_t \frac{1}{4})} \qquad (16.64)$$

An economist who is used to a different decompounding may disagree and use the following present value formula:

$$\frac{100}{(1 + L_t)^{\frac{1}{4}}} \qquad (16.65)$$

Who is right? The answer depends on the market convention. If L_t is quoted under the condition that formula (16.64) be used, then formula (16.65) would be wrong if used with the *same* L_t. However, we can always calculate a new L_t^* using the equivalence:

$$\frac{100}{(1 + L_t \frac{1}{4})} = \frac{100}{(1 + L_t^*)^{\frac{1}{4}}} \qquad (16.66)$$

Then, the formula

$$\frac{100}{(1 + L_t^*)^{\frac{1}{4}}} \qquad (16.67)$$

used with L_t^* would *also* yield the correct present value. The point is, the market is quoting an interest rate L_t with the condition that it is used with formula (16.64). If for some odd reason a client wants to use formula (16.65), then the market would quote L_t^* instead of L_t. The result would be the same since, whether we use formula (16.64) with L_t, as the market does, or formula (16.65) with L_t^*, we would obtain the same present value. In other words, the question of which formula is correct depends on *how* the market quotes the variable of interest.

This goes for options also. The Black–Scholes formula may be the wrong formula if we substitute one particular volatility, but may give the right answer if we use another volatility. And the latter may be different than the real-world volatility at that instant. But traders can still use a particular volatility to obtain the right option price from this "wrong" formula, just as in the earlier present value example. This particular volatility, when associated with the Black–Scholes formula, may give the correct value for the option even though the assumptions leading to the formula are not satisfied.

Thus, suppose the arbitrage-free option price obtained under the "correct" assumptions is given by

$$C(S_t, t, T, K, \sigma_t^*, \theta_t) \qquad (16.68)$$

where K is the strike price, T is the expiration date, and S_t is the underlying asset price. The (vector) variable θ_t represents all the other parameters that enter the "correct" formula and that may not be taken into account by the Black–Scholes world. For example, the volatility may be stochastic, and some parameters that influence the volatility dynamics may indirectly enter the formula and be part of θ_t.[13] The critical point here is the meaning that is attached to σ_t^*. We assume for now that it is the correct instantaneous volatility as of time t.

---

[13]In the case of a mean-reverting stochastic volatility model, we will have

$$d\sigma_t = \lambda(\sigma_0 - \sigma_t)dt + \kappa\sigma_t dW_t$$

where σ_0, κ, and λ are, respectively, the long-run average volatility, the volatility of the volatility, and the speed of mean reversion. The θ_t in formula (16.68) will include λ, σ_0, and the, possibly time varying, γ_t.

The (correct) pricing function in Eq. (16.68) may be more complex and may not even have a closed-form solution in contrast to the Black–Scholes formula, $F(S_t, t, \sigma)$. Suppose traders ignore Eq. (16.68) but prefer to use the formula $F(S_t, t, \sigma)$, even though the latter is "wrong." Does this mean traders will calculate the wrong price?

Not necessarily. The "wrong" formula $F(S_t, t, \sigma)$ can very well yield the same option price as C $(S_t, t, K, \sigma_t^*, \theta_t)$ if the trader uses in $F(S_t, t, \sigma)$, another volatility, σ, such that the two formulas give the same correct price:

$$C(S_t, t, K, \sigma_t^*, \theta_t) = F(S_t, t, \sigma) \qquad (16.69)$$

Thus, we may be able to get the correct option price from the "unrealistic" Black–Scholes formula if we proceed as follows:

1. We quote the K_i-strike option volatilities σ_i directly at every instant t, *under the condition that the Black–Scholes formula be used* to obtain the option value. Then, liquid and arbitrage-free markets will supply "correct" observations of the ATM volatility σ_0.[14]
2. For out-of-the-money options, we use the Black–Scholes formula with a new volatility denoted by $\sigma((S/K_i), S)$, and

$$\sigma\left(\frac{S}{K_i}, S\right) = \sigma_0 + f\left(\frac{S}{K_i}, S\right) \qquad (16.70)$$

where $f(.)$ is, in general, positive and implies a smile effect. The adjustment made to the ATM volatility, σ_0, is such that when $\sigma((S/K_i), S)$ is used in the Black–Scholes formula, it gives the correct value for K_i strike option:

$$F\left(S_t, t, K_i, \sigma_0 + f\left(\frac{S}{K_i}, S\right)\right) = C(S_t, t, K_i, \sigma_t^*, \theta_t) \qquad (16.71)$$

The adjustment factor $f((S/K_i), S)$ is determined by the trader's experience, knowledge, and the trading environment at that instant. The relationships between risk reversal, butterfly, and ATM volatilities discussed in the previous section can also be used here.[15]

The trader, thus, *adjusts* the volatility of the non-ATM options so that the wrong formula gives the correct answer, even though what is used in the Black–Scholes formula may not be the "correct" instantaneous realized volatility of the S_t process.

The $f((S/K_i), S)$ is, therefore, an adjustment required by the imperfections of the Black–Scholes formula in adequately representing the real-world environment. The upshot is that when we plot $\sigma((S/K_i), S)$ against K_i/S or K_i we get a smile, or a skew curve, depending on the time and the sector we are working with.

For what types of situations should the volatilities be adjusted? At least three inconsistencies of the Black–Scholes assumptions with the real world can be corrected for by adjusting the volatilities across the strike K_i. The first is the lognormal process assumption. The second is the fact that if asset prices fall dramatically during a relatively short period of time, this could increase the

---

[14]Especially FX markets quote such implied volatilities and active trading is done on them.
[15]For convenience, the t subscripts are ignored in these formulas.

"fear factor" and volatility would increase. The third involves the organizational and regulatory assumptions concerning financial markets. We discuss these in more detail next.

16.13.1 CASE 1: NONGEOMETRIC PRICE PROCESSES

Suppose the underlying obeys the true risk-neutral dynamics described by the SDE:

$$dS_t = rS_t \, dt + \sigma S_t^\alpha \, dW_t \quad t \in [0, \infty) \tag{16.72}$$

With $\alpha = 1$, S_t would be lognormal. Everything else being conformable to the Black–Scholes world, there would be no smile in the implied volatilities.

The case of $\alpha < 1$ would require an adjustment to the volatility coefficient used in the Black–Scholes formula as the strike changes. This is true, since, unlike in the case of $\alpha = 1$, now the percentage volatility is dependent on the level of S_t. We divide by S_t to obtain

$$\frac{dS_t}{S_t} = r \, dt + \sigma S_t^{\alpha-1} dW_t \quad t \in [0, \infty) \tag{16.73}$$

The percentage volatility is given by the term $\sigma S_t^{\alpha-1}$. This percentage volatility will be a decreasing function of S_t if $\alpha < 1$. As S_t declines, the *percentage* volatility increases. Thus, the trader needs to use higher implied volatility parameters in the Black–Scholes formula for put options with lower and lower strike prices. This means that the more out-of-the-money the put option is, the higher the volatility used in the Black–Scholes formula must be.

This illustrates the idea that although the trader knows that the Black–Scholes world is far from reality, the volatility is adjusted so that the original Black–Scholes framework is preserved and that a "wrong" formula can still give the correct option value.

16.13.2 CASE 2: POSSIBILITY OF CRASH

Suppose a put option series has an expiration of 2 months. All options are identical except for their strike. They run from ATM to deep out-of-the-money. Suppose also that the current level of S_t is 100. The liquid put options have strikes 90, 80, 70, and 60.

Here is what the 90-strike option implies. If the option expires in-the-money, then the market would have fallen by at least 10% in 2 months. This is a big fall, perhaps, but not a disaster. In contrast, if the 60-strike put expires in-the-money, this would imply a 40% drop in 2 months. This is clearly an unusual event, and unusual events lead to sudden spikes in volatility. Thus, the 60-strike option is relatively more associated with events that are labeled as crises and, everything else the same, this option would, in all likelihood, be in-the-money when the volatility is very high. But when this option becomes in-the-money, its *gamma*, which originally is close to zero, will also be higher. Thus, the trader who sells this option would have higher cash payouts due to *delta* hedge adjustments. So, to compensate for these potentially higher cash payments, the trader would use higher and higher vol parameters in put options that are more and more out-of-the-money, and, hence, are more and more likely to be associated with a crisis situation.

This explanation is consistent with the smiley shapes observed in reality. Note that in FX markets, sudden drops *and* sudden increases would mean higher volatility because in each case *one* of the observed currencies could be falling dramatically. So the smile will be more or less symmetric. But in the case of equity markets, a sudden increase in equity prices may be an important event, but not a crisis at all. For traders (excluding the shorts) this is a "happy" outcome, and the volatility may not increase much. In contrast, when asset prices suddenly crash, this increases the fear factor and the volatilities may spike. Thus, in equity markets the smile is expected to be mostly one-sided if this explanation is correct. It turns out that empirical data support this contention. Out-of-the-money equity puts have a smile, but out-of-the-money equity calls exhibit almost no smile.

EXAMPLE

Consider Table 16.2 which displays the prices of options with June 2002 expiry, on January 10, 2002, and ignore issues related to Americanness or any possible payouts. These data are collected at the same time as those discussed in the earlier example. In this case, the options are longer dated and expire in about 6 months. First, we obtain the volatility smile for these data.

The data are collected at the same instant, and since the current value of the underlying index is the same in each case, the division by S_{t_0} is not a major issue, but we still prefer to graph the volatility smile against the K/S.

We extract ask prices for the 8 out-of-the-money puts and consider the 600-put as being in-the-money. This way we can calculate nine implied vols. The price data that we use are shown in Table 16.2. We consider first the out-of-the-money put asking prices listed in the sixth column of this table. This will give nine prices.

Table 16.2 OEX Options with June 21, 2002, Expiry (Collected 9:46 CBOT on January 10, 2002)

Calls	Bid	Ask	Puts	Bid	Ask
Jun 440	153.4	156.4	Jun 440	4.2	4.8
Jun 460	134.8	137.8	Jun 460	5.6	6.3
Jun 480	116.7	119.7	Jun 480	7.4	8.1
Jun 500	99.2	102.2	Jun 500	9.9	10.6
Jun 520	82.6	85.6	Jun 520	12.9	14.4
Jun 540	67.2	69.7	Jun 540	17.2	18.7
Jun 560	52.7	55.2	Jun 560	22.7	24.2
Jun 580	39.8	41.8	Jun 580	29.3	31.3
Jun 600	28.6	30.6	Jun 600	38.3	40.3
Jun 620	19.9	21.4	Jun 620	49.5	51.5
Jun 640	12.8	14.3	Jun 640	62.2	64.7
Jun 660	8	8.7	Jun 660	76.9	79.9
Jun 680	4.7	5.4	Jun 680	93.7	96.7
Jun 700	2.55	3.2	Jun 700	111.6	114.6

Ignoring other complications that may exist in reality, we use the Black—Scholes formula straightforwardly with

$$S_{t_0} = 589.15, r = 1.90\%, t = \frac{152}{365} = 0.416 \tag{16.74}$$

We solve the equations

$$P(589.15, K_i, 1.90, \sigma_{K_i}, 0.416) = P_{K_i} \quad i = 1, \ldots, 9 \tag{16.75}$$

and obtain the nine implied volatilities σ_{K_i}. Using Mathematica, we obtain the following result, which shows the value of K_i/S and the corresponding implied vols for out-of-the-money puts:

$\frac{K}{S}$	Vol
0.74	0.26
0.78	0.26
0.81	0.26
0.84	0.25
0.88	0.25
0.91	0.24
0.95	0.23
0.98	0.22
1.01	0.21

This is shown in Figure 16.15. Clearly, as the moneyness of the puts decreases, the volatility increases. Option market makers will conclude that, if in 6 months, US equity markets

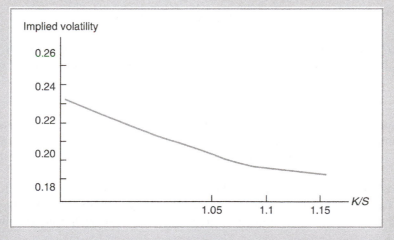

FIGURE 16.15

Implied volatility of OEX options against moneyness.

were to drop by 25%, then the fear factor would increase volatility from 21% to 26%. By selecting the seven out-of-the-money call prices, we get the implied vols for out-of-the-money calls.

$\frac{K}{S}$	Vol
0.98	0.23
1.01	0.22
1.05	0.21
1.08	0.20
1.12	0.19
1.15	0.19
1 18	0.18

Here, the situation is different. We see that as moneyness of the calls decreases, the volatility also decreases.

Option market makers may now think that if, in 6 months, US equity markets were to increase by 20%, then the fear factor would decrease and so would volatility.

The fear of a crash that leads to a smile phenomenon can, under some conditions, be represented analytically using the so-called jump processes. We discuss this modeling approach next.

16.13.2.1 Modeling crashes

Consider again the standard geometric Brownian motion case:

$$dS_t = rS_t \, dt + \sigma S_t \, dW_t \quad t \in [0, \infty) \tag{16.76}$$

W_t is a Wiener process under the risk-neutral probability $\tilde{P}$. Now, keep the volatility parameterization the same, but instead, add a jump component as discussed in Lipton (2002). For example, let

$$dS_t = rS_t \, dt - \sigma S_t \, dW_t + S_t[(e^j - 1)dJ_t - \lambda m \, dt] \quad t \in [0, \infty) \tag{16.77}$$

Some definitions are needed regarding the term $(e^j - 1)dJ_t - \lambda m dt$. The j is the size of a *random* logarithmic jump. The size of the jump is not related to the occurrence of the jump, which is represented by the term dJ_t. If the jump is of size zero, then $(e^j - 1) = 0$ and the jump term does not matter.

The term dJ_t is a Poisson-type process. In general, at time t, it equals zero. But, with "small" probability, it can equal one. The probability of this happening depends on the length of the interval we are looking at, and on the size of the *intensity coefficient* λ. The jump can heuristically be modeled as follows:

$$dJ_t = \begin{cases} 0 & \text{with probability}(1 - \lambda \, dt) \\ 1 & \text{with probability } \lambda \, dt \end{cases} \tag{16.78}$$

where $0 < dt$ is an infinitesimally short interval. Finally, m is the expected value of $(e^j - 1)$:

$$E_t^{\tilde{P}}[(e^j - 1)] = m \qquad (16.79)$$

Thus, we see that, for an infinitesimal interval we can heuristically write

$$E_t^{\tilde{P}}[(e^j - 1)dJ_t] = E_t^{\tilde{P}}[(e^j - 1)]E_t^{\tilde{P}}[dJ_t] \qquad (16.80)$$

$$= m[0.(1 - \lambda \, dt) + 1.\lambda \, dt] \qquad (16.81)$$

$$= m\lambda \, dt \qquad (16.82)$$

According to this, the expected value of the term $(e^j - 1)dJ_t - \lambda m dt$ is zero.

This jump-diffusion model captures some crash phenomena. Stock market crashes, major defaults, 9/11-type events, and currency devaluations can be modeled as rare but discrete events that lead to jumps in prices.

The way these types of jumps create a smile can be heuristically explained as follows. In a world where the Black–Scholes assumptions hold, with a geometric S_t process, a constant volatility parameter $\tilde{\sigma}$, and *no* jumps, the volatility trade yields the arbitrage relation:

$$\frac{1}{2}C_{ss}\tilde{\sigma}^2 S^2 + C_t + rC_s S - rC = 0 \qquad (16.83)$$

With a jump term added to the geometric process as in Eq. (16.77), the corresponding arbitrage relation becomes

$$\frac{1}{2}C_{ss}^* \sigma^2 S^2 + C_t^* + (r - \lambda m)C_s^* S - rC^* + \lambda E_t^{\tilde{P}}\left[C^*(Se^j, t) - C^*(S, t)\right] = 0 \qquad (16.84)$$

where $\tilde{P}$ is the risk-neutral probability. Suppose we decide to use, as a convention, the Black–Scholes formula, but believe that the true PDE is the one in Eq. (16.83). Then, we would select $\tilde{\sigma}$ such that the Black–Scholes formula yields the same option value as the other PDE would yield.

For example, out-of-the-money options will have much smaller *gammas*, C_{ss}. If the expected jump is negative, then $\tilde{\sigma}$ will be bigger, and the more out-of-the-money the options are. As the expiration date T increases, C_{ss} will increase and the smile will be less pronounced.

16.13.3 OTHER EXPLANATIONS

Many other effects can cause a volatility smile. One is *stochastic volatility*. Consider a local volatility specification using

$$dS_t = \mu S_t \, dt + \sigma S_t^\alpha \, dW_t \quad t \in [0, \infty) \qquad (16.85)$$

with, say, $\alpha < 1$. In this specification, percentage volatility *will* be stochastic since it depends on the random variable S_t. But often this specification does not express what is meant by models of stochastic volatility. What is captured by stochastic volatility is a situation where an additional Wiener process dB_t, possibly correlated with dW_t, affects the dynamics of percentage volatility. For example, we can write

$$dS_t = \mu S_t \, dt + \sigma_t S_t \, dW_t \quad t \in [0, \infty) \qquad (16.86)$$

$$d\sigma_t = a(\sigma_t, S_t)dt + \kappa \sigma_t \, dB_t \quad t \in [0, \infty) \qquad (16.87)$$

where κ is the parameter representing the (constant) percentage volatility of the volatility of S_t. In this model, the volatility itself is driven by some random increments that originate in the volatility market *only*. These shocks are only partially correlated with the innovation terms dW_t affecting the price data.

It can be shown that stochastic volatility generates a volatility smile. In fact, with stochastic volatility, we can perform an analysis similar to the PDE with a jump process (Lipton, 2002). The result will essentially be similar. However, it is important to emphasize that, everything else being the same, this model may be incomplete in the sense that there may not be enough instruments to hedge the risks associated with dW_t and dB_t completely, and form a risk-free, self-financing portfolio. The jump-diffusion model discussed in the previous section may entail the same problem. To the extent that the jump part and the diffusion part are affected by different processes, the model may not be complete.

16.13.3.1 Structural and regulatory explanations

Tax effects (Merton, 1976) and the capital requirements associated with carrying out-of-the-money options in options books may also lead to a smile in implied volatility. We briefly touch on the second point.

The argument involves the concept of *gamma*. A negative *gamma* position is considered to be more risky, the more out-of-the-money the option is. Essentially, negative *gamma* means that the market maker has sold options and *delta* hedged them, and that he or she is paying the *gamma* through the rebalancing of this hedge. If the option is deep out-of-the-money, *gamma* would be close to zero. Yet, if the option suddenly becomes in-the-money, the *gamma* could spike, especially if the option is about to expire. This may cause significant losses. Out-of-the-money options, therefore, involve substantial risk and require more capital. Due to such costs, the market maker may want to sell the out-of-the-money option at a higher price than warranted by the ATM volatility.

16.14 THE RELEVANCE OF THE SMILE

The volatility smile is important in financial engineering for at least *three* reasons.

First, if we associate a volatility smile with all the risk factors, and if this smile shifts randomly over time, then we may be able to *trade* it, take spread positions, and arbitrage it. The smile dynamics, thus, imply new opportunities for a market professional.

The second reason for the relevance of the volatility smile is that it may contain important information about the dynamics of the underlying realized volatility processes. With a volatility smile, pricing and hedging may become much more complicated, especially if the instrument has characteristics of an exotic option. Is volatility constant or a stochastic process? If the latter is the case, then what type of stochastic process is it? Are there jumps or is a process with Wiener-type increments a sufficiently good approximation? These questions are important for risk management and pricing.

Third, the creation of new products and synthetics must pay attention to the causes of the smile. If the smile is the result of conventions and practices adopted by market professionals rather than resulting from the underlying volatility processes, we must take these conventions into account. We now discuss these issues in more detail and provide some examples to the uses of the volatility smile.

16.15 **TRADING THE SMILE**

The volatility smile is actively traded to a different extent in different sectors. The smile is an integral part of daily trading in the FX sector. Here, market practitioners routinely quote risk reversals, which relate to the symmetry in exchange rate volatility and butterflies related to the curvature of the smile. Traders trade and arbitrage these effects. The volatility smile is also traded in the equity sector. Traders arbitrage volatility across stock market indices, and in doing this, sometimes trade the smile indirectly. At other times, this trade is direct. The smile relating to a risk may be too steep and is expected to flatten. The trader then sells the deep out-of-the-money options and buys those that are closer to being ATM. In the interest rate sector, volatility smile is mainly traded due to its risk management and hedging implications for cap/floor positions and swaption books.

Smiles can be of interest to investors who may want to take positions on the slope and the curvature of the volatility smile, thinking that the market has under- or overemphasized one of the underlying parameters. In the following example, traders are putting together *skew swaps* that will trade *realized* skews against *implied* skews.

EXAMPLE

As the skew in volatility between out-of-the-money puts and calls on Standard & Poor's 500 index has grown, street traders are looking to capture discrepancies between the realized and implied skews of the options. One trader in New York noted interest in a skew swap on the S&P500 from hedge funds trading volatility. The swaps—which traders believe would be a first—would offer the realized skew of puts and calls in return for the implied skew.

Currently, the S&P skew is above 30—if strikes on puts and calls are moved by 10%, the volatility would [increase by] 3%, explained one structurer. This compares with a level of 15 at the beginning of October, which is in line with the historical levels of 15–20.

One structurer who had tried to put together a skew swap noted that there is no mathematical formula that can capture implied skews for any period of time. He also admitted to being stumped by hedging the product. "To hedge this, we would have to close every night with a vol swap on the deal and that can't always be done," he said. A rival noted that one popular trade to capture flattening skews is selling out-of-the-money puts and calls and buying ATM puts and calls.

(Derivatives Week (now part of GlobalCapital), November 1998)

One interesting point in this reading is that, at least in this particular case, the observed smile (skew) is characterized by multiplying a *linear* relationship with a slope of 3. According to the traders, if moneyness decreases by 10%, volatility increases by 3%. Traders expect this relationship to be around 15 during normal times. Hence, the smile is expected to flatten.

16.16 **PRICING WITH A SMILE**

Pricing and hedging are fairly closely related activities, at least in abstract settings. Once an asset is replicated with liquid securities, the price of the asset is the cost of the replicating portfolio plus some profit margin. At several points in the previous chapters, we saw that assets can be replicated using a series of options with different strike prices. This was the method applied for finding a hedge for a volatility swap in Chapter 15, for example. The replicating portfolio was made of a weighted sum of relevant options with the same characteristics except for their strikes. In Chapter 12, we saw that option portfolios could replicate statically almost any future payoff function. Again, these options were similar except for their strike prices.

The potential use of options with different strikes makes the volatility smile a crucial parameter in forming hedging portfolios and in pricing complex instruments. In fact, if implied volatilities depend on the moneyness of the option, then the volatility parameters used in formulas for replicating portfolios would automatically change as the markets move and the moneyness of the replicating options changes. The critical point is that this will be true even if the underlying *realized* volatility remains the same. This section presents two examples of this phenomenon.

The first involves the class of interest rate derivatives known as caps and floors. These are among the most widely traded instruments. Their hedging and pricing are influenced in a crucial way if there is a volatility smile. The second involves the pricing and hedging of exotic options. Here also, the methodology and market practices crucially depend on the existence of the volatility smile.

16.17 EXOTIC OPTIONS AND THE SMILE

The second major category of instruments where the existence of the volatility smile can change pricing and hedging practices significantly is exotic options. In this section, we consider a simple knock-out call that is representative of the main ideas we want to convey. Due to the contractual equation between vanilla options, knock-out calls, and knock-in calls, our discussion immediately extends to knock-in calls as well. At the end of this section, we briefly discuss digital options and how the existence of the volatility smile affects them.

16.17.1 A HEDGE FOR A BARRIER

Knock-out calls were discussed in Chapters 9 and 11. As a reminder, a simple knock-out call is similar to a European vanilla call with strike K and expiration T, written on the underlying S_t, except that the option will cease to exist if, during the life of the option, S_t falls below barrier H:

$$S_t < H \quad t \in [t_0, T] \tag{16.88}$$

The price of a knock-out barrier approaches the price of a vanilla call as the option becomes more in-the-money. However, as the underlying approaches the barrier, the value of the knock-out will approach zero.

There are several ways of hedging knock-out options used by practitioners. Here, we explain a hedge that (i) has nice financial engineering implications and (ii) shows clearly the important role played by the smile.

Suppose we bought the corresponding vanilla call *and* sold a carefully chosen out-of-the money vanilla put with strike K^*, $K^* < K$, with a very precise objective. We want the put and the call to be such that, as S_t approaches the barrier H, the portfolio's value becomes zero. This portfolio, which is, in fact, a type of risk reversal, approximately replicates the knock-out option. If S_t moves away from H, the put becomes more out-of-the-money, and its value will decline. Then the portfolio looks more and more like a vanilla call. This is what the knock-out option accomplishes anyway. On the other hand, as S_t approaches H, the put becomes more valuable. If it is carefully chosen, at H the value of the put position can equal the value of the vanilla call and the portfolio would have zero value. This is, again, what the knock-out option accomplishes near H. As S_t falls below H, the portfolio has to be liquidated. Thus, the portfolio of

$$\{\text{Short } x \text{ units of } K^* - \text{Put, Long one } K\text{-Call}\} \tag{16.89}$$

replicates the knock-out call if x and K^* are appropriately selected. One way to do this is to use the "symmetry" principle.

Suppose there is no smile effect and that all options with different strikes that belong to a series have the *same* volatility. Then we can choose x and K^* as follows. We want the value of x units of the K^* put to equal the value of the vanilla call when $H \leftarrow S_t$. This can be achieved by choosing K^* such that

$$K^* \cdot K = H^2 \tag{16.90}$$

The prices of these options are assumed to satisfy

$$\frac{K^* \text{ put}}{K \text{ call}} = \sqrt{\frac{K^*}{K}} \tag{16.91}$$

This means that if x is chosen so that

$$x = \frac{K}{K^*} \tag{16.92}$$

then the value of x units of these K^* puts would equal the value of the K call as S_t approaches H. As a result, the portfolio would replicate the knock-out call, except that once the barrier is hit, the portfolio needs to be liquidated.

16.17.2 EFFECTS OF THE SMILE

Consider first the effect of a stable volatility smile on this procedure. If the smile does not shift over time, then it is easy to incorporate the effect into the previous replicating portfolio. Suppose the smile is downward sloping over $[K^*, K]$. Then, one could plug different volatility parameters in Black–Scholes formulas for each vanilla option. The same K^* put selected earlier would be relatively more valuable than in the case of a flat smile. This means that the knockout option could be sold at a cheaper price. If the smile was upward sloping over the range, then the reverse would be true and the knockout would be more expensive. Hence, the smile has a direct effect that needs to be taken into account in the pricing and hedging of the knockout.

There is a second effect of the smile as well. Suppose the smile is *not* stable during the life of the option, and that it shifts as time passes. Then the logic of replication would fall apart since a smile that shifts over time would make the relative values of the call and the put differ from the originally intended ratio as S_t approaches H. Given that in most markets the smile *is* unstable over time, the hedging technique by this replication is questionable. The pricing of the knock-out would also be unreliable.

16.17.2.1 An example of technical difficulties

We can look at the complications introduced by the volatility smile using knock-out pricing formulas in case all standard assumptions are satisfied. The pricing formula for knock-outs was given in Chapter 9. In particular, the price of a down-and-out call written on a stock S_t, satisfying all standard assumptions, and paying no dividends, was given by

$$C^b(t) = C(t) - J(t) \tag{16.93}$$

where $C(t)$ is the value of a vanilla call, given by the standard Black–Scholes formula, and where $J(t)$ is the "discount" that needs to be applied because the option may die if S_t falls below H during the life of the contract. The formula for $J(t)$ was

$$J(t) = S_t \left(\frac{H}{S_t}\right)^{(2(r-(1/2)\sigma^2)/\sigma^2)+2} N(c_1) - Ke^{-r(T-t)}\left(\frac{H}{S_t}\right)^{2(r-(1/2)\sigma^2)/\sigma^2} N(c_2) \tag{16.94}$$

where

$$c_{1,2} = \frac{\log(H^2/S_tK) + (r \pm \frac{1}{2}\sigma^2)(T-t)}{\sigma\sqrt{T-t}} \tag{16.95}$$

Note that just like the Black–Scholes formula, this pricing function contains a *single* volatility parameter σ. In the plain vanilla case, this parameter could be manipulated to make the formula yield the correct answer, even when the underlying assumptions do not hold. Thus, in a smile environment with nonconstant volatilities, the trader could use *one* value for volatility and make the formula yield the correct answer. In fact, this was used in pricing caps and floors even though the volatilities of the individual forward rates that are relevant to these instruments were different.

Unfortunately, this approach is not guaranteed to work for the down-and-out call pricing formula given here if a volatility smile exists and if this smile is continuously (and stochastically) shifting over time. In fact, there may not be a single *well-behaved* volatility value to replace in Eq. (16.94) to obtain the correct price of the down-and-out call. Even when the realized and ATM implied volatilities remain the same, if the slope of the smile changes, the price of the down-and-out call may change as in the previous argument. In fact, if the smile becomes less negatively sloped, the short put component of the replicating portfolio will become relatively less expensive and the value of the down-and-out call may increase. Hence, in the case of exotic options, the relationship between volatility adjustments and the correct option price may become much more complex as smile effects become significant.

16.17.2.2 Pricing exotics

Actual trading takes place in the presence of a volatility smile, and the prices of exotic options need to be set so as to take into account the future costs and benefits of adjusting the hedge to the exotic option. With the presence of smile, as time passes, the *vega* hedge of an exotic option book needs to be adjusted for the reasons explained earlier. As this happens, depending on the net position of the option book, the trader may realize some net cash gains or losses. The present value of these "expected" cash gains and losses needs to be incorporated into the market price. At the end, the market price of the exotic may be higher or lower than the theoretical price indicated by Eq. (16.94).

16.17.3 THE CASE OF DIGITAL OPTIONS

Chapter 11 showed that a theoretical replication of a European digital call with strike K and expiration T would be to buy $1/h$ units of the vanilla call with strike K and to sell $1/h$ units of the vanilla call with strike $K + h$. In this case h would be the minimum tick in a futures market.

If there is a volatility smile, then the prices of these vanilla calls would need to be adjusted since they have different strikes and, therefore, different volatilities. Of course, if h is small, this volatility difference would be small as well, but then $1/h$ would be large and the position would involve buying and selling a large number of vanilla calls. With such numbers, small variations in volatilities can make a difference to the end result.[16]

16.17.4 ANOTHER APPLICATION: RISK REVERSALS

One of the most liquid ways of trading the smile is using risk reversals from FX markets. Consider options written on an exchange rate e_t. Fix the expiration at T, and arrange the puts and calls by their strike price K_i. Then calculate these options' *deltas* and consider a grid of reasonable *deltas*. We use the options' *deltas* to represent the moneyness.

A typical smile for these exchange rate options will then look like the one shown in Figure 16.13. It is a "symmetric" curve and is plotted in a two-dimensional graph having the percentage volatility on the vertical axis and the option's *delta* on the x-axis. In particular, consider the 25, 50, and 75-*delta* options.[17]

We look at the following example.

EXAMPLE

Activity in the dollar/yen foreign exchange markets over the past fortnight has emphasized the severe complexities associated with pricing exotic options. More importantly, it has provided sophisticated option houses with an opportunity to test their pricing theories against each other and against their less-advanced competitors...

However, it was not the decline in the spot rate itself that provided the interest for options dealers but the resulting risk-reversal position. Risk-reversal is an expression of the directional preference in the market. If spot is expected to fall, as in the case of dollar/yen, then there will be greater demand for puts relative to calls, and so the volatility trader will pay a higher price for the puts than the calls. The upshot of this, said one commentator, is that volatility is not constant, as assumed by the standard Black–Scholes option-pricing model, but instead changes according to the option delta.

One-month dollar/yen risk reversal shot up to nearly 3 last week and has continued to hover at levels not seen since the summer of 1995. This extraordinary situation enabled sophisticated traders to find value in the pricing of their so-called naive counterparts.

"A lot of banks must have learned a lot about risk-reversal over the past few days," said one trader. According to market insiders, the less advanced houses failed to adequately account for the effects of risk-reversal in pricing and hedging exotic structures such as knock-out and path-dependent options.

---

[16]In addition, note that this hedge requires selling and buying volatilities and, hence, is subject to variations in the volatility bid–ask spreads.

[17]Practitioners multiply the Black–Scholes *delta* by 100 in their daily usage of this concept.

> *They also asserted that simple off-the-shelf option pricing software was unable to cope with pricing exotic derivatives in a risk-reversal scenario. These packages did not allow the user to enter different volatilities for different deltas and so failed to capture the nuances of exotics pricing.*
>
> *However, other commentators argued that this too was an over-simplification and that other "third-order" effects came into play when hedging certain types of options in very high risk-reversal scenarios. They added that the third-order effects meant the barrier described might not be overpriced by the Black–Scholes user, merely mispriced.*
>
> **(Thomson Reuters IFR Issue 1188, June 1997)**

As this reading illustrates, risk reversals are creations of the volatility smile in FX markets and are heavily traded. But, as indicated in the reading, market practitioners involved in risk-reversal trading are clearly dealing with the underlying smile dynamics. The dynamics can become very complex. Pricing and hedging such positions on exotic options may become much more difficult.

16.18 CONCLUSIONS

The volatility smile is a fascinating topic in finance. Yet, it is also a complex phenomenon and more research needs to be done on its causes and on the ways to model it. This chapter offered a simple introduction. However, it has illustrated some of the essential points associated with this topic.

SUGGESTED READING

Allen and Granger (2005) provide an excellent overview of techniques for trading equity correlation. The text by **Brigo and Mercurio** (2006) deals with the volatility smile in the interest rate sector. For equity smiles, the reader should consult the papers by **Derman et al.** (1994) and (1996), at a minimum. A comprehensive treatment of the volatility smile is provided in **Lipton** (2002). **Taleb** (1996) is the source that we used to discuss most market practices. There is also a flurry of papers dealing with empirical and theoretical issues involving the volatility smile. One recent source is **Johnson and Lee** (2003). The cited sources contain further references on the previous research.

EXERCISES

1. Consider the following table displaying the bid–ask prices for all options on the OEX index passed on January 10, 2002, at 9:46. These options have February 22, 2002, expiry and at the time of data collection, the underlying was at 589.14.

Calls	Bid	Ask	Vol	Puts	Bid	Ask	Vol
Feb 400	188.9	191.9	0	Feb 400	0.05	0.2	0
Feb 420	169	172	0	Feb 420	0.1	0.4	0
Feb 440	149.2	152.2	0	Feb 440	0.25	0.55	0
Feb 460	129.4	132.4	0	Feb 460	0.45	0.75	0
Feb 480	109.6	112.6	0	Feb 480	0.8	1.1	0
Feb 500	90.2	92.7	0	Feb 500	1.4	1.7	0
Feb 520	71	73.5	0	Feb 520	2.5	2.8	0
Feb 530	61.6	64.1	0	Feb 530	2.8	3.5	0
Feb 540	52.4	54.9	0	Feb 540	3.7	4.4	0
Feb 550	43.8	45.8	0	Feb 550	4.9	5.6	1
Feb 560	35.4	37.4	0	Feb 560	6.6	7.3	0
Feb 570	27.9	29.4	0	Feb 570	8.9	9.6	0
Feb 580	20.8	22	0	Feb 580	11.8	12.8	0
Feb 590	14.8	15.8	0	Feb 590	15.8	16.8	1
Feb 600	10	10.7	1	Feb 600	20.8	22	0
Feb 610	6.1	6.8	0	Feb 610	27.1	28.6	0
Feb 615	4.6	5.3	0	Feb 615	31	32	0
Feb 620	3.4	4.1	0	Feb 620	34.3	36.3	0
Feb 630	1.9	2.2	0	Feb 630	42.8	44.8	0
Feb 640	0.9	1.2	0	Feb 640	52	54	0
Feb 650	0.4	0.7	0	Feb 650	61.4	63.9	0
Feb 660	0.15	0.45	0	Feb 660	71.2	73.7	0
Feb 680	0	0.25	0	Feb 680	90.8	93.8	0
Feb 700	0	0.2	100	Feb 700	110.8	113.8	0

a. Using the out-of-the-money ask prices for the puts, calculate the implied volatility for the relevant strikes. Plot the volatility smile against K/S.

b. Using the out-of-the-money bid prices for the puts, calculate the implied volatility for the relevant strikes. Plot the volatility smile against K/S. Are the bid−ask spreads for these vols constant?

c. Using the out-of-the-money ask prices for the calls, calculate the implied vol for the relevant strikes. Plot the volatility smile against K/S. When you put this figure together with the out-of-the-money put volatilities, do you obtain a smile or a skew?

EXCEL EXERCISE

2. (Stop-loss hedge). Write a VBA program to show the stop-loss hedged portfolio adjustments and cash flows for 100 long calls from the dealers' perspective with the following data:

$$S(0) = 100; K = 100; T = 1; \quad r = 8\%; \sigma = 30\%; \& \text{ realized volatility} = 50\%$$

MATLAB EXERCISES

3. Write a MATLAB program to observe the volatility payoff of stop loss hedged long call position and measure the performance of this strategy of replicating long call position when the frequency of adjustment is increased during the time interval until the call expires. Use the following data for the calculation

$$S(0) = 100; K = 100; T = 1; \quad r = 8\%; \sigma = 30\%; \& \text{ realized volatility} = 50\%$$

Plot the graph for the volatility payoff for 30 instances for these sets of adjustment frequency [10, 20, 50, 100, 500]. Also show the performance of the method on the graph as a plot of mean return of the hedging strategy versus frequency of adjustment.

4. (Stochastic volatility) Write a MATLAB program to observe the implied volatility smile for the option based on the stock price having stochastic volatility. Use the following data for the calculation.

$$S(0) = 100; K = 100; T = 0.5; \quad r = 8\%; \sigma = 30\%; \sigma_\sigma = 10\%, r_\sigma = 10\%$$

Plot the graph of the estimated implied volatility against the moneyness of the option.

5. (Nongeometric Brownian motion) Write a MATLAB program to observe the implied volatility smile for the option based on the stock price following the dynamics of nongeometric Brownian motion. Use the following data for the calculation.

$$S(0) = 100; K = 100; T = 0.5; \quad r = 8\%; \sigma = 30\%; \alpha = 0.8$$

Plot the graph of the estimated implied volatility against the moneyness of the option.

6. (Stock price crashes and jumps)

Write a MATLAB program to observe the implied volatility smile for the option based on the stock price having jumps (particularly crashes). Use the following data for the calculation.

$$S(0) = 100; K = 100; T = 0.5; \quad r = 8\%; \sigma = 30\%; \lambda = 0.3; \alpha_J = -2\%; \sigma_J = 5\%$$

Plot the graph of the estimated implied volatility against the moneyness of the option.

7. (VIX calculation) Using the S&P 500 option data on the chapter website (collected on July 4, 2013 for options expiring on July 25, 2013), calculate the VIX index value following the procedure discussed in the book. Take the risk-free interest rate value to be 5% for the purpose of calculation. Also highlight the strike price K_0 and show the contribution by strike for all the options.

CAPS/FLOORS AND SWAPTIONS WITH AN APPLICATION TO MORTGAGES

CHAPTER OUTLINE

17.1 INTRODUCTION

In the previous chapters, we discussed convexity and options as well as their applications in the form of volatility trading. The US mortgage market provides us with another application of a long convexity position as a result of the prepayment option of US homeowners. We carefully explain the value of the prepayment option and its effect on mortgage agencies and discuss potential approaches to hedge this risk. As we will see the structure of the US mortgage market and the presence of a mortgage prepayment option, in particular, leads to optionality and convexity in payoffs.

Swap markets rank among the world's largest in notional amount and are among the most liquid. The same is true for swaption markets. Why is this so?

There are many uses for swaps. Borrowers "arbitrage" credit spreads and borrow in currencies that yield the lowest all-in-cost. This, in general, implies borrowing in a currency other than the one the borrower needs. The proceeds, therefore, need to be swapped into the needed currency. Theoretically, this operation can be done by using a single currency swap where floating rates in different currencies are exchanged. Currency swaps discussed in Chapter 6 are not interest rate swaps, but the operation would involve vanilla interest swaps as well. Most issuers would like to *pay* a fixed long-term coupon. Thus, the process of swapping the proceeds into a different currency requires, first, swapping the fixed coupon payments into floating in the same currency, and then, through currency swaps, exchanging the floating rate cash flows. Once the floating rate payments in the desired currency are established, these can be further swapped into fixed payments in the same currency. Thus, a new issue requires the use of *two* plain vanilla interest rate swaps in different currencies coupled with a vanilla currency swap. Hence, new bond issues are one source of liquidity for vanilla interest rate swaps.

Balance sheet management of interest exposure is another reason for the high liquidity of swaps. The asset and liability interest rate exposure of financial institutions can be adjusted using interest rate swaps and swaptions. If a loan is obtained in floating rates and on-lent in fixed rate, then an interest rate swap can be entered into and the exposures can be efficiently managed. FRAs, swaps, caps, floors and swaptions are traded on many liquid currencies and their existence in these countries is not chiefly driven by mortgage market driven.

In some countries and cases, the uses of swap markets can be matched by the needs of *mortgage-based activity*. It appears that a major part of the swaption and a significant portion of plain vanilla swap trading in the US are due to the requirements of the mortgage-based financial strategies. In this chapter we focus on the US mortgage market for the purpose of illustrating the link between option markets and the market for the underlying. Mortgages have prepayment clauses and this introduces convexities in banks' fixed-income portfolios. This convexity can be hedged using swaptions, which creates liquidity in the swaption market. On the other hand, swaptions are positions that need to be dynamically hedged. This hedging can be done with forward swaps as the underlying. This leads to further swap trading. Mortgage markets are huge and this activity can sometimes dominate the swap and swaption markets.

In this chapter, we use the mortgage sector as an example to study the financial engineering of swaptions. We use this to introduce caps/floors and swaptions. The chapter also presents a simple discussion of the *swap measure*. This constitutes another example of measure change technology introduced in Chapter 12 and further discussed in Chapters 13 and 14. This chapter provides an example that puts together most of the tools used throughout the text. We start by first reviewing the essentials of the mortgage sector.

17.2 THE MORTGAGE MARKET

Lenders such as mortgage bankers and commercial banks *originate* mortgage loans by lending the original funds to a home buyer. This constitutes the primary mortgage market. Primary market lenders are mortgage banks, savings and loan institutions, commercial banks, and credit unions.

These lenders then group similar mortgage loans together and sell them to Fannie Mae or Freddie Mac-type *agencies*. This is part of the secondary market which also includes pension funds, insurance companies, and securities dealers.

Agencies buy mortgages in at least two ways. First, they pay "cash" for mortgages and hold them on their books.[1] Second, they issue mortgage-backed securities (MBSs) *in exchange* for pools of mortgages that they receive from lenders. The lenders can in turn either hold these MBSs on their books, or sell them to investors. To the extent that mortgages are converted into MBSs, the purchased mortgage loans are *securitized*.[2] Agencies guarantee the payment of the principal and interest. Hence, the "credit risk" is borne by the issuing institution.[3] We now discuss the engineering of a mortgage deal and the resulting positions. As usual, we use a highly simplified setting.

17.2.1 THE LIFE OF A TYPICAL MORTGAGE

The process from home buying activity in the US to hedging swaptions is a long one, and it is best to start the discussion by explaining the mechanics of primary and secondary mortgage markets. The implied cash flow diagrams eventually lead to swaption positions. The present section is intended to clarify the sequence of cash flows generated by this activity. We will see that, at the end, the prepayment right amounts to holding a short position on a *swaption*. This is from the point of view of an institution that holds the mortgage and finances it with a straight fixed rate loan. Essentially, the institution sold an American-style option on buying a fixed payer swap at a predetermined rate. The option is exercised when the future mortgage rates fall below a certain limit.[4]

The interrelationships between home buying, mortgage lending, and agency activity are shown in Figures 17.1−17.3. We go step by step and then put all the positions together in Figure 17.3 to obtain the consolidated position of the mortgage warehouse. We prefer to deal with a relatively simple case which can then be generalized. Some of these generalizations are straightforward, others involve considerable problems.

We consider the following setup. A *balloon mortgage* is issued at time t_0.[5] The mortgage holder pays only interest and returns the principal at maturity. The mortgage principal is N, the rate is c_{t_0}, and the mortgage matures at t_4. The t_i, $i = 1, \ldots, 4$, is measured in years.[6] We emphasize that, with such a balloon mortgage, the periodic payments are made of interest only and no amortization takes place. This simplifies the engineering and permits the use of plain vanilla interest rate swaps.

The home buyer makes four interest payments and then pays the N at time t_4 if he or she does *not* prepay during the life of the mortgage. On the other hand, if the mortgage rate c_t falls below a

---

[1] The lending institutions then use the cash in making further mortgage loans.

[2] We discuss the securitization process in further detail in a subsequent chapter.

[3] These agencies have direct access to treasury borrowing and there is a perception in the markets that this is an implicit government guarantee.

[4] However, the party who is long this option is not a financial institution and may not exercise this American-style option at the right time, in an optimal fashion. This complicates the pricing issue further.

[5] In the United States, balloon (payment) mortgages are named after the final "balloon," or disproportionately large, payment that they require towards the end of the mortgage. This type of mortgage specifies that borrowers make regular payments for a specific interval and then pay off the remaining balance that remains within a short period of time or at maturity.

[6] Of course, mortgage payments are monthly in general, but this assumption simplifies the notation, which is cumbersome. It can be easily generalized.

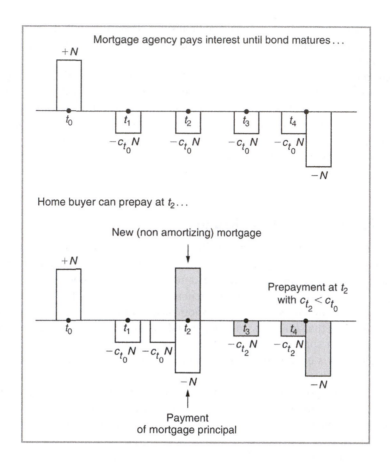

FIGURE 17.1

Mortgage agency and home buyer cash flows.

threshold level at a future date, the home buyer refinances. We assume that the home buyer can prepay only at time t_2. When the loan is paid at time t_2, it is replaced by a new mortgage of two periods[7] that carries a lower rate. This situation is shown in Figure 17.1.

The bank borrows at the floating LIBOR rate, L_{t_i}, and lends at the fixed rate, c_{t_0}. Then the mortgage is sold to an agency such as Fannie Mae or to some other mortgage warehouse. The bank will only be an intermediary and earns a fee for servicing mortgage payments.

The interesting position is that of the agency that ends up with the mortgage. The agency buys the mortgage and puts it on its books. This forms the asset side of its balance sheet. The issue is how these secondary market purchases are funded. In reality, there is more than one way. Some of the mortgages bought in the secondary market are kept on the books, and funded by issuing *agency securities* to investors. Other mortgages are packaged as MBS, and resold to end investors.

[7]In reality, the prepayment option is valid for all future time periods and, hence, the prepayment time is random.

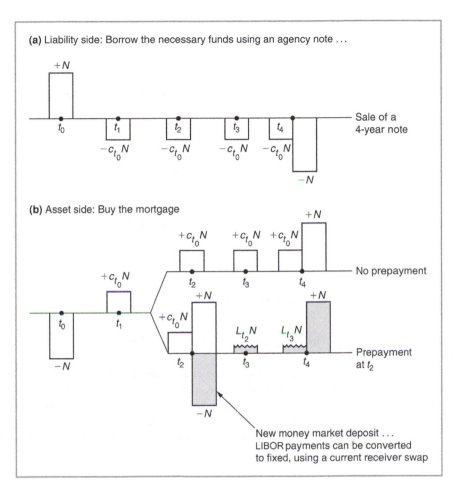

(a) Liability side: Borrow the necessary funds using an agency note . . .

$+N$

t_0 t_1 t_2 t_3 t_4

$-c_{t_0}N$ $-c_{t_0}N$ $-c_{t_0}N$ $-c_{t_0}N$

$-N$

Sale of a
4-year note

(b) Asset side: Buy the mortgage

$+N$

$+c_{t_0}N$ $+c_{t_0}N$ $+c_{t_0}N$

t_2 t_3 t_4

No prepayment

$+c_{t_0}N$

t_0 t_1

$-N$

$+c_{t_0}N$

$+N$

t_2

$+c_{t_0}N$

$+N$

$L_{t_2}N$ $L_{t_3}N$

t_3 t_4

$-N$

Prepayment
at t_2

New money market deposit . . .
LIBOR payments can be converted
to fixed, using a current receiver swap

FIGURE 17.2

Mortgage agency asset and liability side and prepayment risk.

Suppose in our case the agency uses a 4-year *fixed coupon note* to secure funding at the rate c_{t_0}. The liabilities of the agency are shown in Figure 17.2. The asset side is shown in the lower part of this figure. The critical point is the *prepayment issue.* According to Figure 17.2b, the agency faces a prepayment risk only at time t_2.[8]

If the home buyer does not refinance, the agency receives four interest payments of size $c_{t_0}N$ each, then, at time t_4 receives the principal. We assume that the home buyer never defaults.[9] If, on the other hand, the home buyer pays the "last" interest payment $c_{t_0}N$ and prepays the principal earlier at time t_2, the agency places these funds in a floating rate money market account until time t_4. The rate would be L_{t_i}. Alternatively, the agency can get into a fixed receiver swap and receive the

---

[8]According to our simplifying assumption, the home buyer can prepay only at time t_2.
[9]In any case default risk is borne by agencies of the issuing bank.

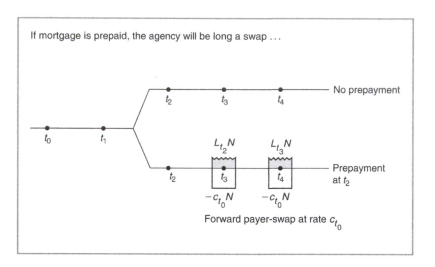

FIGURE 17.3

Prepayment results in long fixed payer forward swap position.

swap rate beginning at time t_3. This swap rate is denoted by s_{t_2}. This explains the cash flows faced by the agency in case the home buyer prepays.

The asset and liability positions of the agency are consolidated in Figure 17.3. We see that all cash flows, except the ones for the prepayment case, cancel. Figure 17.3 shows that if the home buyer prepays, the agency will be *long* a fixed payer forward swap at the predetermined swap rate c_{t_0}. If no prepayment occurs, the agency will have a flat position.

Now assume for simplicity, and without much loss of generality, that the swap rates and mortgage rates are equal for all t:

$$c_t = s_t \tag{17.1}$$

Then, the final position of the agency amounts to the following contract.

The agency has *sold* the right to enter into a fixed receiver swap at the predetermined rate c_{t_0} to a counterparty. The swap notional is N, the option is European,[10] and the exercise date is t_2. The underlying cash swap has a life of 2 years. The agency is short this option, and would be a *fixed payer* in case the option is exercised.

This contract is a *swaption.* The agency has sold an option on a swap contract and is short convexity. The home buyer is the counterparty, and he or she is long convexity. The option is exercised if $s_{t_2} \leq c_{t_0}$. The option premium is Sw_{t_0}.[11] It is paid at time t_0.[12] According to this, we are ignoring any transaction costs or fees.

---

[10]This is the case since we assumed that the home buyer is allowed to prepay only at t_2. In reality, home buyers can prepay any time during the mortgage duration. This would make a Bermudan-style option.

[11]We are ignoring the spreads between mortgage rates, swap rates, and Fannie Mae cost of debt capital.

[12]In the US market, this could be paid during the initiation of the mortgage as Bermudan "points" for example.

17.2.2 HEDGING THE POSITION

The important point of the exercise in the previous section is that, after all the dust settles, the agency faces prepayment risk and this results in a short swaption position. The *short* European swaption position can be hedged in several ways. One option is direct dynamic hedging. The hedger would (i) first calculate the *delta* of the swaption, and then, (ii) buy *delta* units of a 2-year forward-receiver swap with start date t_2.[13]

After following these steps, the agency will have a short convexity position, which is also a short position on volatility. If volatility increases, the agency suffers losses. In order to eliminate this risk as well, the agency has to buy convexity from the market. Another way to hedge the short swaption position is to issue callable bonds, and buy the convexity directly from the investors.

> **EXAMPLE**
>
> *Fannie Mae Callable Benchmark Notes will have a notional of at least US$500m and a minimum reopening size of US$100m. There will be at least one issue per month and four main structures will be used: a five-year, non-call two; a five-year, non-call three; a 10-year, non-call three; and a 10-year, non-call five. The volume and regular issuance pattern of the program is expected to depress volatility.*
>
> *Agencies have always had a natural demand for volatility and this has traditionally been provided by the banking community. But with the invention of the callable benchmark issuance program, agencies have instead started to source volatility from institutional investors.*
>
> **(Thomson Reuters IFR, Issue 1291)**

Finally, as an alternative to eliminate the short swaption position, the agency can buy a similar swaption from the market.

17.2.3 ASSUMPTIONS BEHIND THE MODEL

The discussion thus far has made some simplifying assumptions, most of which can be relaxed in a straightforward way. Some assumptions are, however, of a fundamental nature. We list these below.

1. It was assumed that the home buyer could finance *only* at time t_2. The resulting swaption was, therefore, of *European* style. Normally, home buyers can finance at all times at t_1, t_2, and t_3. This complicates the situation significantly. The same cash flow analysis can be done, but the underlying swaption will be of *Bermudan* style. It will be an option that can be exercised at dates t_1, t_2, or t_3, and the choice of the exercise date is left to the home buyer. This is *not* a trivial extension. Bermudan options have no closed-form pricing formulas, and pricing them requires more advanced techniques. The presence of an interest rate volatility smile complicates things further.

2. The reader can, at this point, easily "add" derivatives to this setup to create mortgages seen during the credit crisis of 2007−2008. We assumed, unrealistically, that there was no *credit*

---

[13]Remember that, by convention, buying a swap means paying fixed, so this is equivalent to selling a swap.

risk. The home buyer, or the agency, had zero default probability in the model discussed. This assumption was made for simplifying the engineering. As the credit crisis of 2007−2008 shows, mortgages can have significant credit risk, especially when they are not of "vanilla" type.

3. The mortgage was assumed to be of *balloon* style. Normally, mortgages amortize over a fixed period of, say, 15 or 30 years. This changes the structure of cash flows shown in Figures 17.1−17.3. The assumption can be relaxed by moving from the standard plain vanilla swaps to *amortizing swaps.* This may lead to some interesting financial engineering issues which are ignored in the present text.

Finally, there are the issues of transaction costs, fees, and a more general specification of the maturities and settlement periods. These are relatively minor extensions and do not change the main points of the argument.

17.2.4 TWO RISKS

According to the preceding discussion, there are *two* risks associated with prepayments. One is that the bank or the issuing agency will be caught long a payer swap in the case of prepayment. If this happens, the agency would be paying a fixed rate, c_{t_0}, for the remainder of the original mortgage maturity, and this c_{t_0} will be *more* than the prevailing fixed receiver swap rate.[14] The agency will face a negative carry of $s_\tau - c_{t_0} < 0$ for the remaining life of the original mortgage, where τ is the random prepayment date.

The second risk is more complicated. The agency does not know *when* the prepayment will happen. This means that, when mortgages are bought, the agency has to estimate the *expected maturity* of the mortgage. Then, based upon this expected maturity, notes with fixed maturity are sold. The liability side has, then, a more or less predictable duration,[15] whereas the asset side has an average maturity that is difficult to calculate. The example below illustrates how such problems can become very serious.

EXAMPLE

As rates rally, mortgage players become more vulnerable to call risk, which leads to a shortening of their asset duration, as occurred in September. "The mortgage supply into the street has been very heavy, so there has been an overall buying of volatility to hedge convexity risk from both Freddie and Fannie," says the chief strategist of a fixed-income derivatives house. The refinancing boom has already caught out some mortgage players. A New Jersey-based arbitrage fund specializing in mortgage securities, announced losses of around $400 million at the end of October. Analysts say the fund had not adequately hedged itself against the rate rally.

While some speculated that criticism of Fannie Mae's duration gap—the difference between the maturity of its assets and its liabilities—in the media would lead speculators to position themselves to benefit from the agency's need to hedge, this was not a serious contributor to the increase in volatility.

---

[14]This is the case since prepayment means the interest rates have fallen and that the fixed receiver swap rate has moved below c_{t_0}.

[15]Or, average maturity.

> *However, with the big increase in its mortgage portfolio, Fannie Mae, and indeed Freddie Mac, which has also been building up its mortgage portfolio, now faces the increased threat of extension risk—the risk of mortgage assets lengthening in duration as yields increase and prepayments decrease.*
> *Fannie Mae's notional derivatives exposure stood at $594.5 billion at the end of June, while Freddie reported a total derivatives exposure of $1.1 trillion.*
>
> **(Based on an article in Derivatives Week (now part of GlobalCapital).)**

Of the two risks associated with trading and *warehousing* mortgages, the first is an issue of hedging *convexity* positions, whereas the second involves the Americanness of the implicit prepayment options.

17.3 SWAPTIONS

Mortgages form the largest asset class that households own directly or indirectly. The total mortgage stock is of similar size as the total US Treasury debt. Most of these mortgages have prepayment clauses, and this leads to massive short swaption positions. As a result, Bermudan swaptions become a major asset class. In fact, mortgages are not the only instrument with prepayment clauses. Prepayment options exist in many other instruments. Callable and putable bonds also lead to contingent, open-forward swap positions. These create similar swaptions exposures and can be fully hedged only by taking the opposite position in the relevant swaption market. Given the important role played by swaptions in financial markets, we need to study their modeling. This section deals with technical issues associated with European swaptions.[16]

A swaption can be visualized as a generalization of caplets and floorlets. The caplet selects a floating LIBOR rate, L_{t_i}, for *one* settlement period, such that, if L_{t_i} is greater than the cap rate κ written in the contract, the caplet buyer receives compensation proportional to the difference. We can generalize this. Select a floating swap rate for n periods. If the time t_1 value of this swap rate denoted by s_{t_1} is greater than a fixed strike level, $f_{t_0} = \kappa$, the buyer of the instrument receives payments proportional to the difference. Thus, in this case both the "underlying interest rate" and the strike rate cover more than one period and they are forward swap rates. A fixed payer swaption can then be regarded as a generalization of a caplet. Similarly, a fixed receiver swaption can be regarded as a generalization of a floorlet.

Let us look at this in more detail. Consider a European-style vanilla call option with expiration date t_1. Instead of being written on equity or foreign exchange, let this option be written on a plain vanilla *interest rate swap*. The option holder has the right to "buy" the underlying at a selected strike price, but this underlying is now a swap. In other words, the option holder has the right to enter into a *payer swap* at time t_1, at a predetermined swap rate κ. The strike price itself can be specified either in terms of a swap rate or in terms of the value of the underlying swap.

---

[16]Pricing and risk management of Bermudan swaptions are much more complex and will not be dealt with here. Rebonato (2002) is a good source for related issues.

Let t_0 be the trade date. We simplify the instrument and consider a two-period *forward* interest rate swap that starts at time t_1, with $t_0 < t_1$, and that exchanges two fixed payments against the 12-month LIBOR rates, L_{t_1} and L_{t_2}. Let the *forward* fixed payer swap rate for this period be denoted by f_{t_0}. The *spot* swap rate, s_{t_1}, will be observed at time t_1, while the forward swap rate is known at t_0.

The buyer of the swaption has thus purchased the right to buy, at time t_1, a fixed payer (spot) swap, with nominal value N and payer swap rate f_{t_0}. If the strike price, κ, equals the ongoing forward swap rate for the time t_1 swap, this is called an ATM swaption. A forward fixed payer swap is contracted at time t_0. The forward swap rate f_{t_0} is set at time t_0, and the swap will start at time t_1. The corresponding *spot* swap can be contracted at time t_1. Thus, the swap rate s_{t_1} is unknown as of t_0.

The ATM swaption involves the following transactions. No cash changes hands at time t_1. The option premium is paid at t_0. If at time t_1 the spot swap rate, s_{t_1}, turns out to be higher than the strike rate f_{t_0}, the swaption holder will enter a fixed payer swap at the previously set rate f_{t_0}. If the time t_1 spot swap rate s_{t_1} is below f_{t_0}, the option holder has an incentive to buy a new swap from the market since he or she will pay less. The swaption expires unexercised. Thus, ignoring the bid−ask spreads, suppose, at time t_1,

$$f_{t_0} < s_{t_1} \tag{17.2}$$

the 2-year interest rate swap that pays s_{t_1} against 12-month LIBOR will have a zero time t_1 value. This means that at time t_1, the swaption holder can decide to pay f_{t_0}, and receive s_{t_1}. The time t_1 value of the resulting cash flows can be expressed as[17]

$$\left[\frac{s_{t_1} - f_{t_0}}{(1 + L_{t_1})} + \frac{s_{t_1} - f_{t_0}}{(1 + L_{t_1})(1 + L_{t_2})} \right] N \tag{17.3}$$

Note that at time t_1, L_{t_1} will be known, but L_{t_2} would still be unknown. But, we know from earlier chapters that we can "replace" this random variable by the forward rate for the period $[t_2, t_3]$ that is known at time t_1. The forward rate $F(t_1, t_2)$ is an unbiased estimator of L_{t_2} under the forward measure $\tilde{P}^{t_3}$:

$$F(t_1, t_2) = E_{t_1}^{\tilde{P}^{t_3}} \left[L_{t_2} \right] \tag{17.4}$$

Then, the time t_1 value of a swap entered into at a predetermined rate f_{t_0} will be

$$\left[\frac{s_{t_1} - f_{t_0}}{(1 + L_{t_1})} + \frac{s_{t_1} - f_{t_0}}{(1 + L_{t_1})(1 + F(t_1, t_2))} \right] N \tag{17.5}$$

This is zero if the swaption expires at-the-money, at time t_1, with

$$s_{t_1} = f_{t_0} \tag{17.6}$$

Thus, we can look at the swaption as if it is an option written on the value of the cash flows in Eq. (17.5) as well. The ATM swaption on a fixed payer swap can then either be regarded as an option on the (forward) swap rate f_{t_0}, or as an option on the *value* of a forward swap with strike price $\kappa = 0$.

---

[17] $\delta = 1$, since we assume 12-month settlement intervals.

17.3.1 A CONTRACTUAL EQUATION

As in the case of caps and floors, we can obtain a contractual equation that ties swaptions to forward swaps.

$$
\boxed{\begin{array}{c}\text{Forward fixed payer}\\\text{swap at rate } f_{t_0}\end{array}} = \boxed{\begin{array}{c}\text{Payer swaption}\\\text{with strike } \kappa\end{array}} - \boxed{\begin{array}{c}\text{Receiver swaption}\\\text{with strike } \kappa\end{array}}
\tag{17.7}
$$

We interpret this contractual equation the following way. Consider a forward payer swap with rate f_{t_0} that starts at time t_1 and ends at time t_{n+1}. The forward swap settles n times in between and makes (receives) the following sequence of cash payments in arrears:

$$
\left\{ (f_{t_0} - L_{t_1})N\delta, \ldots, (f_{t_0} - L_{t_n})N\delta \right\}
\tag{17.8}
$$

This is regardless of whether $f_{t_0} < s_{t_1}$ or not.

A payer swaption, on the other hand, makes these payments *only* if $f_{t_0} < s_{t_1}$. To receive the cash flows in Eq. (17.8) when $f_{t_0} > s_{t_1}$, one has to buy a receiver swaption.

17.4 PRICING SWAPTIONS

The pricing and risk management of swaptions can be approached from many angles using various *working measures*. For example, we have already seen in Chapters 4 and 14 that the time t_0 forward swap rate $f(t_0, t_1, T)$ for a forward swap that begins at time t_1 and ends at time T will be a weighted average of the forward rates $F(t_0, t_i, t_{i+1})$. If we adopt this representation, a possible working measure would be the time T forward measure. The Martingale dynamics of all forward rates under this measure could be obtained, and swaptions and various other swap derivatives could be priced with it.

However, we adopt an alternative approach. For some pricing and risk-management problems, using the *swap measure* as the working probability may be more convenient. This gives us an opportunity to discuss this interesting class of working measures in a very simple context. Therefore, we will obtain the pricing functions for European swaptions using the swap measure.

17.4.1 SWAP MEASURE

The measures considered thus far were obtained using normalization by the price of a *single* asset. Under some conditions, we may want to use normalization by the value of a stream of payments instead of a single asset. One well-known case is the *swap measure*. We will discuss the swap measure in a simple, two-period model with finite states. The model has the same time and cash flow structure discussed earlier in this chapter. We start with the definition of an *annuity*.

Consider a contract that pays δ dollars at every δ units of time. The payment dates are denoted by t_i,

$$
t_i - t_{i-1} = \delta
\tag{17.9}
$$

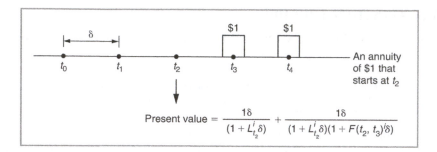

FIGURE 17.4

Present value of annuity.

For example, δ can be 1/2 and the payments would be made every 6 months. The payments continue for N years. This would be an annuity and we could use its time t value to normalize time t asset prices. This procedure leads us to the swap measure.

Suppose at time t_2 there is a finite number of states of the world indexed by $i = 1, 2, \ldots, n$. In this context suppose an annuity starts at time t_2 and makes two payments of 1 dollar at times t_3 and t_4, respectively. Hence, in this case, $\delta = 1$. This situation is shown in Figure 17.4. Obviously, for simplicity, we eliminated any associated credit risks. The annuity is assumed to be default-free. In reality, such contracts do have significant credit risk as was seen during the GFC.

The present value of the payments at time t_2 is a random variable when considered from time t_0 and is given by the expression:[18]

$$PV_{t_2}^i = \frac{1\delta}{\left(1 + L_{t_2}^i \delta\right)} + \frac{1\delta}{\left(1 + L_{t_2}^i \delta\right)(1 + F(t_2, t_3)^i \delta)} \tag{17.10}$$

Here, $L_{t_2}^i$ is the time t_2 LIBOR rate in state i and $F(t_2, t_3)^i$ is the forward rate at time t_2, in state i, on a loan that starts at time t_3. The loan is paid back at time t_4. We are using these rates in order to calculate the time t_2 present value of the payments that will be received in times t_3 and t_4. These values are state dependent and $PV_{t_2}^i$ is therefore indexed by i.

Using the default-free discount bonds $B(t, t_3)$ and $B(t, t_4)$ that mature at times t_3 and t_4, we can write the $PV_{t_2}^i$ as

$$PV_{t_2}^i = (\delta)B(t_2, t_3)^i + (\delta)B(t_2, t_4)^i \tag{17.11}$$

In this representation, the right-hand side default-free bond prices are state dependent, since they are measured at time t_2.

Suppose two-period interest rate swaps trade actively. A spot swap that *will* be initiated at time t_2 is shown in Figure 17.5. The spot swap rate $s_{t_2}^i$ for this instrument is unknown at time t_0, and this is implied by the i superscript. As of time t_0, markets are assumed to trade a forward swap, with a rate denoted by $f_{t_0}^i$ that corresponds to $s_{t_2}^i$.

---

[18]Although $\delta = 1$ in our case, we prefer to discuss the formulas using a symbolic δ.

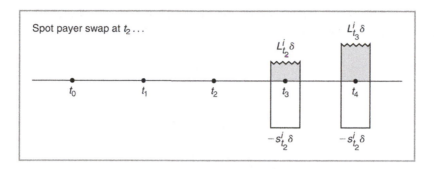

FIGURE 17.5

Spot swap initiated at date t_2.

Now consider the time t_0 market value of a two-period forward swap contract that starts at time t_2. In the context of the fundamental theorem of Chapter 12, we can write a matrix equation containing the annuity, the forward swap contract, and the European fixed payer swaption Sw_t that delivers the same two-period swap, if $s_{t_2} < f_{t_0}$. Putting these assets together in the simplified matrix equation of Chapter 12, we obtain:

$$
\begin{bmatrix} B(t_0,t_3)\delta + B(t_0,t_4)\delta \\ 0 \\ Sw_{t_0} \end{bmatrix} = \begin{bmatrix} \cdots & \dfrac{1\delta}{(1+L^i_{t_2}\delta)} + \dfrac{1\delta}{(1+L^i_{t_2}\delta)(1+F(t_2,t_3)^i\delta)} & \cdots \\ \cdots & \dfrac{1}{(f_{t_0}-S^i_{t_2})\delta} & \cdots \\ \cdots & \dfrac{1}{Sw^i_{t_2}} & \cdots \end{bmatrix} \begin{bmatrix} \cdots \\ Q^i \\ \cdots \end{bmatrix} \quad (17.12)
$$

Here $\{Q^i, i=1, \ldots, n\}$ are the n state prices for the period t_2. Under the no-arbitrage condition, these exist, and they are *positive*

$$ Q^i > 0 \quad (17.13) $$

for all states i.[19] The last row of the matrix equation shows the swaption's current value as a function of Q^i and the state-dependent expiration values of the swaption. In the states of the world where $f_{t_0} > s^i_{t_2}$, the swaption will expire out-of-the-money and the corresponding $Sw^i_{t_2}$ will be zero. Without loss of generality, let these be the first $m < n$, values:

$$ Sw^i_{t_2} = 0 \quad i=1,\ldots,m \quad (17.14) $$

For the remaining $n-m$ states, the expiration value of the swaption is

$$ Sw^i_{t_2} = \frac{(f_{t_0}-s^i_{t_2})\delta}{(1+L^i_{t_2}\delta)} + \frac{(f_{t_0}-s^i_{t_2})\delta}{(1+L^i_{t_2}\delta)(1+F(t_2,t_3)^i\delta)} \quad s^i_{t_2} < f_{t_0} \quad (17.15) $$

[19]The matrix equation is written using only the first three rows. Other assets are ignored. However, the existence of Q^i requires the existence of n other assets that can serve as a basis.

The first row of the matrix equation shows the value of the annuity. We obtain this row from the arbitrage-free prices of the corresponding default-free discount bonds with par value $1:

$$B(t_2, t_3)^i = \frac{1}{\left(1 + L_{t_2}^i \delta\right)} \tag{17.16}$$

and

$$B(t_2, t_4)^i = \frac{1}{\left(1 + L_{t_2}^i \delta\right)(1 + F(t_2, t_3)^i \delta)} \tag{17.17}$$

The quantity $B(t_0, t_3)\delta + B(t_0, t_4)\,\delta$ is, therefore, the present value of the annuity payment. The same quantity was previously called PV01. In fact, the present discussion illustrates how PV01 comes to center stage as we deal with streams of fixed-income cash flows.

The first row of the matrix equation gives:

$$B(t_0, t_3)\delta + B(t_0, t_4)\delta = \sum_{i=1}^{n} \left[\frac{1\delta}{\left(1 + L_{t_2}^i \delta\right)} + \frac{1\delta}{\left(1 + L_{t_2}^i \delta\right)(1 + F(t_2, t_3)^i \delta)} \right] Q^i \tag{17.18}$$

Now we switch measures and obtain a new type of working probability known as the *swap measure*. Dividing the first row by the left-hand side we have:

$$1 = \sum_{i=1}^{n} \frac{1}{B(t_0, t_3)\delta + B(t_0, t_4)\delta} \left[\frac{1\delta}{\left(1 + L_{t_2}^i \delta\right)} + \frac{1\delta}{\left(1 + L_{t_2}^i \delta\right)(1 + F(t_2, t_3)^i \delta)} \right] Q^i \tag{17.19}$$

We define the measure $\tilde{p}_i^s$ using this expression:

$$\tilde{p}_i^s = \frac{1}{B(t_0, t_3)\delta + B(t_0, t_4)\delta} \left[\frac{1\delta}{\left(1 + L_{t_2}^i \delta\right)} + \frac{1\delta}{\left(1 + L_{t_2}^i \delta\right)(1 + F(t_2, t_3)^i \delta)} \right] Q^i \tag{17.20}$$

Note that as long as $Q^i > 0$ is satisfied, we will have:

$$\tilde{p}_i^s > 0 \quad i = 1, \ldots, n \tag{17.21}$$

and

$$1 = \sum_{i=1}^{n} \tilde{p}_i^s \tag{17.22}$$

Thus, $\tilde{p}_i^s$ have the properties of a well-defined discrete probability distribution. We call this the *swap measure*. We will show two interesting applications of this measure. First, we will see that under this measure, the corresponding forward swap rate behaves as a Martingale and has very simple dynamics. Second, we will see that by using this measure, we can price European swaptions easily.

17.4.2 THE FORWARD SWAP RATE AS A MARTINGALE

In order to see why forward swap rates behave as a Martingale under the appropriate swap measure, consider the second row of the matrix equation in Eq. (17.12). Since the forward swap has a value of zero at the time of initiation, we can write:

$$0 = \sum_{i=1}^{n} (f_{t_0} - s_{t_2}^i)\delta \left[\frac{1}{\left(1 + L_{t_2}^i \delta\right)} + \frac{1}{\left(1 + L_{t_2}^i \delta\right)(1 + F(t_2, t_3)^i \delta)} \right] Q^i \tag{17.23}$$

We can express this as:

$$0 = \sum_{i=1}^{n} \left(f_{t_0} - s_{t_2}^i\right) \delta \left[\frac{1}{\left(1 + L_{t_2}^i \delta\right)} + \frac{1}{\left(1 + L_{t_2}^i \delta\right)\left(1 + F(t_2, t_3)^i \delta\right)} \right] \left[\frac{B(t_0, t_3)\delta + B(t_0, t_4)\delta}{B(t_0, t_3)\delta + B(t_0, t_4)\delta} \right] Q^i \tag{17.24}$$

Now, we use the definition of the swap measure given in Eq. (17.20) and relabel, to get:

$$0 = \sum_{i=1}^{n} \left(f_{t_0} - s_{t_2}^i\right) [B(t_0, t_3)\delta + B(t_0, t_4)\delta] \tilde{p}_i^s \tag{17.25}$$

After taking the constant term outside the summation and eliminating, this becomes:

$$0 = \sum_{i=1}^{n} \left(f_{t_0} - s_{t_2}^i\right) \tilde{p}_i^s \tag{17.26}$$

or again:

$$f_{t_0} = \sum_{i=1}^{n} s_{t_2}^i \tilde{p}_i^s \tag{17.27}$$

which means:

$$f_{t_0} = E_{t_0}^{\tilde{P}^s}[s_{t_2}] \tag{17.28}$$

Hence, under the *swap measure*, the forward swap rate behaves as a Martingale. The dynamics of the forward swap rate under this measure can then be written using the SDE:

$$df_t = \sigma f_t dW_t \quad t \in [t_0, T] \tag{17.29}$$

where we assumed a particular diffusion structure and where T is the general swap maturity. Given the forward swap rate volatility σ, it would be straightforward to generate Monte Carlo trajectories from these dynamics.

17.4.3 SWAPTION VALUE

In order to obtain a pricing function for the European swaption, we consider the third row of the matrix equation given in Eq. (17.12). We have:

$$Sw_{t_0} = \sum_{i=1}^{n} Sw_{t_2}^i Q^i \tag{17.30}$$

We use the definition of the swap measure $\tilde{P}^s$

$$\tilde{p}_i^s = \frac{1}{[B(t_0, t_3)\delta + B(t_0, t_4)\delta]} \left[\frac{1\delta}{\left(1 + L_{t_2}^i \delta\right)} + \frac{1\delta}{\left(1 + L_{t_2}^i \delta\right)\left(1 + F(t_2, t_3)^i \delta\right)} \right] Q^i \tag{17.31}$$

and rewrite the equality in Eq. (17.30):

$$Sw_{t_0} = \sum_{i=m+1}^{n} Sw_{t_2}^i \left[\frac{[B(t_0, t_3)\delta + B(t_0, t_4)\delta]}{\left(1\delta / \left(1 + L_{t_2}^i \delta\right)\right) + \left(1\delta / \left(1 + L_{t_2}^i \delta\right)\left(1 + F(t_2, t_3)^i \delta\right)\right)} \right] \tilde{p}_i^s \tag{17.32}$$

Note that we succeeded in introducing the swap measure on the right-hand side. Now, we know from Eq. (17.15) that the non-zero expiration values of the swaption are given by

$$Sw_{t_2}^i = \frac{(f_{t_0} - s_{t_2}^i)\delta}{(1 + L_{t_2}^i \delta)} + \frac{(f_{t_0} - s_{t_2}^i)\delta}{(1 + L_{t_2}^i \delta)(1 + F(t_2, t_3)^i \delta)} \qquad s_{t_2}^i < f_{t_0} \tag{17.33}$$

Substituting these values in Eq. (17.32) gives,

$$Sw_{t_0} = \sum_{i=m+1}^{n} \left[\frac{(f_{t_0} - s_{t_2}^i)\delta}{(1 + L_{t_2}^i \delta)} + \frac{(f_{t_0} - s_{t_2}^i)\delta}{(1 + L_{t_2}^i \delta)(1 + F(t_2, t_3)^i \delta)} \right]$$
$$\left[\frac{[B(t_0, t_3) + B(t_0, t_4)]\delta}{(\delta/(1 + L_{t_2}^i \delta)) + (\delta/(1 + L_{t_2}^i \delta)(1 + F(t_2, t_3)^i \delta))} \tilde{p}_i^s \right] \tag{17.34}$$

We see the important role played by the swap measure here. On the right-hand side of this expression, the state-dependent values of the annuity will cancel out for each i and we obtain the expression:

$$Sw_{t_0} = [B(t_0, t_3) + B(t_0, t_4)] \sum_{i=m+1}^{n} \left[(f_{t_0} - s_{t_2}^i)\delta \tilde{p}_i^s \right] \tag{17.35}$$

This is equivalent to

$$Sw_{t_0} = [B(t_0, t_3) + B(t_0, t_4)] E_{t_0}^{\tilde{P}^s} \left[\max \left[(f_{t_0} - s_{t_2})\delta, 0 \right] \right] \tag{17.36}$$

Using this pricing equation and remembering that the forward swap rate behaves as a Martingale under the proper swap measure, we can easily obtain a closed-form pricing formula for European swaptions.

There are many ways of formulating market practice. One way to proceed is as follows. The market assumes that:

1. The forward swap rate f_t follows a geometric (log-normal) process

$$df_t = \mu(f_t, t)dt + \sigma f_t dW_t \tag{17.37}$$

which can be converted into a Martingale by using the swap measure $\tilde{P}^s$

$$df_t = \sigma f_t d\tilde{W}_t \tag{17.38}$$

2. Black's formula can be applied to obtain the value

$$f_{t_0} N(d_1) - \kappa N(d_2) \tag{17.39}$$

with

$$d_1 = \frac{\log(f_t/\kappa) + (1/2)\sigma^2(t_2 - t_0)}{\sigma\sqrt{t_2 - t_0}} \tag{17.40}$$

$$d_2 = d_1 - \sigma\sqrt{t_2 - t_0} \tag{17.41}$$

where t_2 is the expiration of the swaption contract discussed in this section.

3. Finally, the value of the European swaption Sw_t is obtained by discounting this value using the present value of the annuity. For time t_0, this will give:

$$Sw_{t_0} = [B(t_0, t_3)\delta + B(t_0, t_4)\delta] [f_{t_0} N(d_1) - \kappa N(d_2)] \tag{17.42}$$

In other words, swaptions are priced by convention using Black's formula and then discounting by the value of a proper annuity.

17.4.4 REAL-WORLD COMPLICATIONS

There may be several complications in the real world. First of all, the market quotes the swaption volatilities σ directly, and the preceding formula is used to put a dollar value on a European swaption. However, most real-world applications of swaptions are of Bermudan nature and this introduces Americanness to the underlying option. There are no closed formulas for Bermudan swaptions.

Second, just as in the case of caps and floors, the existence of the volatility smile introduces significant complications to the pricing of swaptions, especially when they are of the *Bermudan* type.[20] If the swap curve is not flat, then relative to a single strike level, different forward swap rates have different moneyness characteristics. Pricing Bermudan swaptions would then become more complicated.

17.5 MORTGAGE-BASED SECURITIES

MBS is a very important market, especially in the United States. MBS is also called "mortgage pass-through certificates." They essentially take the cash flows involving interest and amortization of a principal paid by home buyers, and then pass them through to an investor who buys the MBS. In MBS, a selected pool of mortgages serves as the underlying asset. An MBS investor receives a "pro-rata share of the cash flows" generated by the underlying mortgages.

> **EXAMPLE**
>
> A typical Fannie Mae MBS will have the following characteristics.
>
> Each pool of mortgages has a pass-through rate, which forms the coupon passed on to the investor. This is done on the 25th day of a month following the month of the initial issue.
>
> The pass-through rate is lower than the interest rate on the underlying mortgages in the pool. (We assumed this away in the present chapter.)
>
> This interest differential covers the fee paid to Fannie Mae, and the fee paid to the servicing institution for collecting payments from homeowners and performing other servicing functions.
>
> The lender that delivers the mortgages for securitization can retain servicing of the loans.
>
> During pooling, Fannie Mae makes sure that the mortgage rates on the underlying loans fall within 250 basis points of each other.
>
> MBSs are sold to investors through securities dealers.
>
> Certificates are issued in book-entry form and are paid by wire transfer.

---

[20]A Bermudan swaption has more than one well-defined expiration date. It is not of American style, but it can only be exercised at particular, set dates.

Fannie Mae's paying agent is the Federal Reserve Bank of New York. The centralized payment simplifies accounting procedures since investors receive just one monthly payment for various MBSs that they hold.

The certificates will initially represent a minimum of $1000 of the unpaid principal of the pooled mortgages. However, as time passes, this principal is paid gradually due to amortization and the remaining principal may go down.

Such activity by Fannie Mae-type institutions has several important consequences. First of all, the MBS and the Fannie Mae agency securities are issued in large sizes and normally form a liquid asset class. Further, the MBS and some of the agency securities contain implicit call and put options, and this creates a large class of convex assets. Important hedging, arbitraging, and pricing issues emerge. Second, institutions such as Fannie Mae become huge players in derivatives markets which affect the liquidity and functioning of swap, swaption, and other interest rate option markets. Third, the issue of *prepayment* requires special attention from the part of market participants. Fourth, as institutions such as Fannie Mae bear the credit risk in the mortgages, they separate credit risk from market risk of mortgages. The credit crisis of 2007—2008 has shown that there is a delicate pricing issue here. It is difficult to estimate this credit risk. We will deal with the issue of credit risk in subsequent chapters. Finally, the structure of the balance sheets and the federal government connection that such institutions have may create market gyrations from time to time.

17.6 CAPS AND FLOORS

Caps and floors are important instruments for fixed-income professionals. Our treatment of caps and floors will be consistent with the approach adopted in this text from the beginning. First, we would like to discuss, not the more liquid spot caps and floors that start immediately, but instead the so-called forward caps and floors. These instruments have a future start date and are perhaps less liquid, but they are more appropriate for learning purposes. Second, we will follow our standard practice and generate these instruments from instruments that have already been discussed.

Again, we keep the framework as simple as possible and leave the generalization to the reader. Consider the two-period *forward fixed payer swap* shown in Figure 17.6a. In this swap, there will be two settlements

$$(f_{t_0} - L_{t_1})N\delta \quad \text{and} \quad (f_{t_0} - L_{t_2})N\delta \tag{17.43}$$

where L_{t_i}, f_{t_0}, and N are the relevant LIBOR rate, forward swap rate, and the forward swap notional, respectively. δ is the reset interval. The forward swap starts at time t_1.[21]

We now consider a particular decomposition of this forward swap. First, note that depending on whether $L_{t_1} > f_{t_0}$ or not, the swap party either *receives* a payment at time t_2, or *makes* a payment. Thus, the first cash flow to be settled at time t_2 can be decomposed into two *contingent* cash flows

---

[21] f_{t_0} is a forward rate that applies to the spot swap rate s_{t_1}. The latter will be observed at time t_1. f_{t_0}, on the other hand, is known as of t_0.

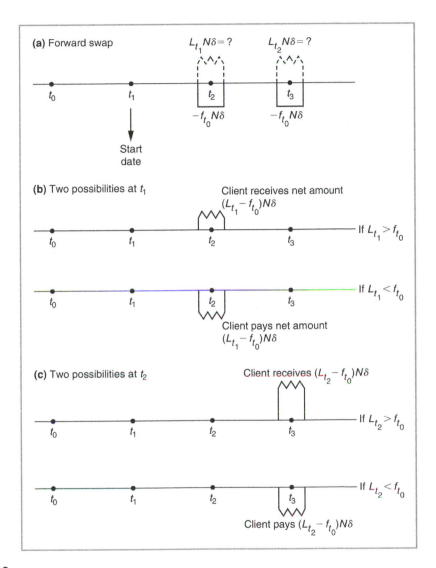

FIGURE 17.6

Forward fixed payer swap.

depending on whether $L_{t_1} < f_{t_0}$ or $L_{t_1} > f_{t_0}$. The same can be done for the second cash flow to be settled at time t_3. Again, the cash flow can be split into two contingent payments.

These contingent cash flows can be interpreted as payoffs of some kind of interest rate option. Consider the cash flows of time t_2 in Figure 17.6b. Here, the client receives $(L_{t_1} - f_{t_0})N\delta$ if $L_{t_1} > f_{t_0}$, otherwise he or she receives nothing. Thus, this cash flow will replicate the payoff of an option with expiration date t_1, and settlement date t_2 that is written on the LIBOR rate L_{t_1}. The client receives the appropriate difference at time t_2 if the LIBOR observed at time t_1 exceeds a *cap level* κ, which in this case is $\kappa = f_{t_0}$. Thus, the top part of Figure 17.6b is like an insurance against movements of LIBOR rates above level κ. This instrument is labeled a *caplet* and its time t price is denoted by Cl^{t_1}. The client is *long* a caplet.

The lower portion of Figure 17.6b is similar but shows insurance against a drop of the LIBOR rate below the *floor level* κ. In fact, the cash flow here is a payment of $(f_{t_0} - L_{t_1})N\delta$ if $L_{t_1} < f_{t_0}$. Otherwise, the client pays nothing. This cash flow replicates the payoff of an option with expiration t_1 and settlement t_2 that is written on the LIBOR rate. The client pays the appropriate difference at time t_2, if the LIBOR observed at time t_1 falls below *floor* κ, which in this case, is $\kappa = f_{t_0}$. We call this instrument a *floorlet* and denote its price by Fl^{t_1}. Clearly, the client is *short* the floorlet in this case.

The treatment of the time t_3 settled caplet-floorlets is similar. The cap and floor levels are again the same:

$$\kappa = f_{t_0} \tag{17.44}$$

but exercise times, settlement times, and the underlying LIBOR rate L_{t_2} are different. By putting the two caplets together, we get a two-period forward interest rate cap, which starts at time t_1 and ends at t_3

$$\text{Cap} = \{Cl^{t_1}, Cl^{t_2}\} \tag{17.45}$$

Similarly, the two floorlets can be grouped as a two-period forward interest rate floor:

$$\text{Floor} = \{Fl^{t_1}, Fl^{t_2}\} \tag{17.46}$$

These forward caps and floors can be extended to n periods by putting together n caplets and floorlets defined similarly.

Figure 17.6 shows that we obtained a contractual equation. By adding Figures 17.6b and c vertically, we get back the original forward swap. Thus, we can write the contractual equation:

Forward fixed payer swap at rate $f_{t_0} = \kappa$	=	Forward cap with cap rate κ	−	Forward floor with floor rate κ	(17.47)

This contractual equation shows that caps, floors, and swaps are closely related instruments.

17.6.1 PRICING CAPS AND FLOORS

This discussion will be conducted using the same two-period cap and floor discussed earlier. One obvious conclusion from the engineering shown here is that the caplets that make up the cap can be priced separately and their values added, after discounting them properly. This is how a caplet is typically priced; we take the Cl^{t_1} written on L_{t_1}.

First, let the $F(t, t_1)$, $t < t_1$, be the forward rate that corresponds to L_{t_1}, observed at t. More precisely, $F(t, t_1)$ is the interest rate decided on at time t on a loan that will be made at time t_1 and paid back at time t_2. Market practice *assumes* that the forward LIBOR rate $F(t, t_1)$ can be represented as a Martingale with *constant* instantaneous percentage volatility σ:

$$dF(t, t_1) = \sigma F(t, t_1)dW_t \quad t \in [0, \infty) \tag{17.48}$$

with initial point $F(t_0, t_1)$.

Second, the market assumes that the arbitrage-free value of a t_2-maturity default-free pure discount bond price, denoted by $B(t_0, t_2)$, can be calculated.[22]

Third, it is shown that the time t_0 value of the caplet Cl^{t_1} is given by the formula

$$Cl_{t_0}^{t_1} = B(t_0, t_2) E_{t_0}^{\tilde{P}^{t_2}} \left[\max \left[L_{t_1} - \kappa, 0 \right] N\delta \right] \tag{17.49}$$

where, in this case, κ also equals the forward swap rate f_{t_0}. Hence, this is an ATM cap. N is the national amount. Here, the expectation is taken with respect to the forward measure $\tilde{P}^{t_2}$ and the normalization is done with the t_2-maturity pure discount bond, $B(t_0, t_2)$.

Finally, remembering that under the measure $\tilde{P}^{t_2}$, the forward LIBOR rate $F(t_0, t_1)$ is an unbiased estimate of L_{t_1}

$$F(t_0, t_1) = E_{t_0}^{\tilde{P}^{t_2}} \left[L_{t_1} \right] \tag{17.50}$$

a closed-form formula can be obtained. This is done by applying Black's formula to the $F(t, t_1)$ process, which has a zero drift term. Then the expectation

$$E_{t_0}^{\tilde{P}^{t_2}} \left[\max \left[L_{t_1} - \kappa, 0 \right] N\delta \right] \tag{17.51}$$

can be calculated to obtain Black's formula:[23]

$$Cl_{t_0}^{t_1} = B(t_0, t_2)[F(t_0, t_1)N(h_1) - \kappa N(h_2)]\delta N \tag{17.52}$$

where

$$h_1 = \frac{\log(F(t_0, t_1)/\kappa) + (1/2)\sigma^2(t_1 - t_0)}{\sigma\sqrt{t_1 - t_0}} \tag{17.53}$$

$$h_2 = \frac{\log(F(t_0, t_1)/\kappa) + (1/2)\sigma^2(t_1 - t_0)}{\sigma\sqrt{t_1 - t_0}} - \sigma\sqrt{(t_1 - t_0)} \tag{17.54}$$

Note that both $F(t_0, t_1)$ and f_{t_0} are rates that relate to amounts calculated in t_2 dollars. The discount bond price $B(t_0, t_2)$ is then a market-determined "expected" discount rate for this settlement.

The caplet that expires at time t_2, Cl^{t_2}, will be priced similarly, except that this time the expectation has to be taken with respect to the time t_3-forward measure $\tilde{P}^{t_3}$, and the underlying forward rate will now be $F(t_0, t_2)$ and not $F(t_0, t_1)$.[24] Solving the pricing formula, we get

$$Cl_{t_0}^{t_2} = B(t_0, t_3)[F(t_0, t_2)N(g_1) - \kappa N(q_2)]\delta N \tag{17.55}$$

where

$$q_1 = \frac{\log(F(t_0, t_2)/\kappa) + (1/2)\sigma^2(t_2 - t_0)}{\sigma\sqrt{t_2 - t_0}} \tag{17.56}$$

$$q_2 = \frac{\log(F(t_0, t_2)/\kappa) + (1/2)\sigma^2(t_2 - t_0)}{\sigma\sqrt{t_2 - t_0}} - \sigma\sqrt{(t_2 - t_0)} \tag{17.57}$$

---

[22]Note that this discount bond matures at settlement time t_2 and not at caplet expiration t_1. This is necessary to discount the random payments that will be received at time t_2.

[23]Here $N(.)$ represents a probability, whereas N at the end is the notional amount.

[24]L_{t_2} will *not* be a Martingale with respect to $\tilde{P}^{t_2}$, and the drift of the corresponding SDE will not be zero.

It is important to realize that the same constant percentage volatility was used in the two caplet formulas even though the two caplets expired at *different* times. In fact, the first caplet has a life of $[t_0, t_1]$, whereas the second caplet, $Cl_{t_0}^{t_2}$, has a longer life of $[t_0, t_2]$. Also, note that despite the strike price being the same, the two caplets have different money.

Finally, to get the value of the cap itself, the market professional simply adds the two caplet prices:

$$\text{Cap}_{t_0} = B(t_0, t_2)[F(t_0, t_1)N(h_1) - \kappa N(h_2)] + B(t_0, t_3)[F(t_0, t_2)N(g_1) - \kappa N(g_2)] \qquad (17.58)$$

where h_i and g_i are given as noted earlier. The parameters N, δ are set equal to one in this formula. Otherwise, the right-hand side needs to be multiplied by N, δ as well.

17.6.2 A SUMMARY

It is worth summarizing some critical steps of the cap/floor pricing convention. First, to price each caplet, the market uses normalization by means of a discount bond that matures at the settlement date of that particular caplet and obtains the forward measure that corresponds to that caplet. Second, the market uses an SDE with a zero drift since the corresponding forward LIBOR rate is a Martingale under this measure. Third, the percentage volatility for *all* the caplets is assumed to be constant and identical across caplets. Finally, the use of proper forward measures makes it possible to calculate the expectations for the random discount factors and the caplet payoffs, separately. Black's formula gives the caplet prices which are added using the appropriate liquid discount bond prices.

17.6.3 CAPLET PRICING AND THE SMILE

The previous summary should make clear why the volatility smile is highly relevant for pricing and hedging caps/floors. The latter are made of several caplets and floorlets and, hence, are baskets of options written on *different* forward rates denoted by $F(t_0, t_i)$, $i = 1, \ldots, n$.

As long as the yield curve is not flat, the forward rates $F(t_0, t_i)$ will be different from each other. Yet, each cap or floor has only *one* strike price κ. This means that, relative to this strike price, we will have

$$\frac{F(t_0, t_1)}{\kappa} \neq \frac{F(t_0, t_2)}{\kappa} \neq \cdots \frac{F(t_0, t_n)}{\kappa} \qquad (17.59)$$

In other words, as long as the yield curve slopes upward or downward, the ratios

$$\frac{F(t_0, t_i)}{\kappa} \qquad (17.60)$$

will be different, implying different moneyness for each caplet/floorlet.

Now, suppose there is a volatility smile. If, under these conditions, each caplet and floorlet were sold *separately*, the option trader would substitute a *different* volatility parameter for each, conformable with the smile in the corresponding Black's formula. But caplets or floorlets are bundled into caps and floors. More important, a *single* volatility parameter is used in Black's formula.

This has at least two implications. First, we see that the volatility parameter associated with a cap is some weighted average of the caplet volatilities associated with options that have different moneyness characteristics. Second, even when the realized and ATM volatilities remain the same, a movement of the yield curve would change the moneyness of the underlying caplets (floorlets) and, through the smile effect, would change the volatility parameter that needs to be used in the corresponding Black's formulas. In other words, marking cap/floor books to the market may become a delicate task if there is a volatility smile.

17.7 CONCLUSIONS

Swaptions play a fundamental role in economic activity and in world financial markets. This chapter has shown a simple example that was, however, illustrative of how swap measures can be defined and used in pricing swaptions.

SUGGESTED READING

Rebonato (2002) as well as **Brigo and Mercurio** (2006) are two extensive sources that the reader can consult after reading this chapter. Further references can also be found in these sources. There is a growing academic and practical literature on this topic. Two technical articles that contain references are **Andersen and Andreasen** (2000) and **Pedersen** (1999).

EXERCISES

1. Consider the following statement:

    One prop trader noted that cap/floor volatility should be slightly higher than swaptions. Corporates buy caps and investors sell swaptions through callable bonds, said one London-based prop trader. The market is structurally short caps and long swaptions.

    a. Swaptions are options to get into swaps in a predetermined data, at a predetermined rate. Explain why, according to this reading, cap/floor volatility should be higher than swaption volatility.
    b. What are some plausible reasons for the market to be structurally short of caps and long on swaptions?
    c. What would this statement mean in terms of hedging and risk management?

2. The reading below deals with some typical swaption strategies and the factors that originate them. First, read it carefully.

    Lehman Brothers and Credit Suisse First Boston are recommending clients to buy long-dated swaption vol ahead of the upcoming US Federal Open Markets Committee meeting. In the trade, the banks are recommending clients to buy long-dated swaption straddles, whose value increases if long-dated swaption vol rises.

> *Mortgage servicers and investors are keen to hedge refinancing risk with swaptions as interest rates fall, meaning that long-dated swaption vol should rise over the next several months, said an official at CSFB in New York.*
>
> *The traditional supply of swaption vol has been steadily decreasing over the last year, said (a trader) in New York. Agencies supply vol to dealers mainly via entering swaps with embedded options on the back of issuing callable notes. Over the past year, demand for callable notes has been decreasing due to high interest rate volatility, and lower demand from banks, one of the larger investors in callables. This decreased supply in vol for dealers would point to intermediate- and longer-dated swaption vol rising.*
>
> *Lehman is recommending a relative value trade in which investors sell short-dated swaption volatility, such as options, to enter 10-year swaps in one year, and buy long-dated swaption vol, such as the option to enter a 10-year swap in five years. The trade is constructed with at-the-money swaption straddles, weighted in such a way as to make this a play on the slope of the vol curve, which at current levels is more inverted compared to historical levels, noted (the trader). Short-dated swaption vol should fall as the Fed's plans become clearer in the next several months, and the market is assuaged concerning the direction of interest rates. If the Fed cuts interest rates 50 basis points when it meets this week, the market should breathe a sigh of relief, and short-dated vol should fall more than longer-dated vol, he added.*
>
> *CSFB recommends buying long-dated vol. The option to enter a five-year swap in five years had a premium of about 570 basis points at press time. At its peak this month, this level was 670 bps, said the official at CSFB. During a similar turning point in the US economy in 1994–1995, vol for a similar swap was nearly 700 bps. This week's meeting might provide an instant lift to the position. If the Fed cuts rates by more or less than the expected 50 bps, long-dated swaption vol could rise in reaction, he added. Even if the Fed cuts by 50 basis points, swaption vol could rise as mortgage players hedge. An offsetting factor here is the fact that following a Fed meeting, short-dated vol typically falls, sometimes dragging longer-dated vol down with it. (Derivatives Week (now part of GlobalCapital) November 2001).*

First, some comments on the mechanics mentioned in the reading. The term *Agencies* is used for semiofficial institutions such as Fannie Mae that provide mortgage funds to the banking sector. These agencies sell bonds that contain imbedded call options. For example, the agency has the right to call the bond at par at a particular time. The owners of these bonds are thus option writers. The agency is an option buyer. To hedge these positions they also need to be option sellers.

a. Using cash flow diagrams, show the positions that agencies would take due to issuing callable bonds.

b. Why would the short-term vol decrease according to the market? Why would long-term vol increase?

c. What do you think an at-the-money swaption straddle is? Show the implied cash flow diagrams.

d. If the expectations concerning the volatilities are realized, how would the gains be cashed in?

e. What are the differences between the risks associated with the positions recommended by CSFB versus the ones recommended by Lehman?

3.   Answer the questions related to the following case study.

CASE STUDY: DANISH MORTGAGE BONDS

Danish mortgage bonds (DMBs) are liquid, volatile, and complex instruments. They are traded in fundamentally sound legal and institutional environments and attract some of the best hedge funds. Here, we look at them only as an example that teaches us something about swaps, swaptions, options, and the risks associated with modeling household behavior.

Background Material

To answer the questions that follow, you need to have a good understanding of these notions and instruments:

1. Plain vanilla interest-rate swaps
2. Swaptions
3. Prepayment risk of an MBS
4. The relationship between volatility and options
5. LIBOR and simple LIBOR instruments
6. Government bond markets versus high-yield bonds.

Questions

Part I

a. Define the following concepts: MBS, prepayment risk, implicit option, and negative convexity.
b. Show how you can hedge your DMB positions in the swap and swaption markets and then earn a generous arbitrage. Why would this add liquidity to the Danish markets?
c. Show how you can do this using Bermudan options.
d. Explain carefully if this is true arbitrage. Are there any risks?

Part II

Now consider the second reading.

e. What is a corridor structure?
f. What does the reading mean by "dislocations"?
g. What are "balance-protected swaps"?
h. Explain the purpose of the dislocation trades.

Reading 1

Arbitrageurs Swoop on Danish Mortgage Market

Yield-starved banks and hedge funds have been scooping up long-dated Danish mortgage bonds, hedging parts of the exposure in the swap and swaption markets, and earning a generous arbitrage (1). But the strategy is not without risk. The structures involved are heavily dependent on complex statistical modelling techniques.

The Danish mortgage bond market has been undergoing something of a revolution. Securities traditionally bought by Danish pension funds or investment managers are now being snapped up by sophisticated foreign banks and hedge funds. The international players are trying to arbitrage between mortgage, credit- and other interest-rate markets. This activity has also generated significant added volume in the Danish krone and Deutsche mark swap and swaption markets. (1)

The exact size and nature of the international activity is unknown, but it is believed to be large and growing. "Foreign involvement has gone through the roof," said one structurer at a London-based bank. Banks believed to be at the forefront of activity include Barclays Capital, Bankers Trust, and Merrill Lynch International.

Outside interest, said market professionals, had been stoked by a decreasing number of profit opportunities in both Europe and the US European investors have seen Southern European bond yields plummet ahead of EMU and have had to look for new sources of return. US hedge funds have turned to Denmark as US mortgage market arbitrage opportunities have started to run dry.

The European activity has focused on Denmark's mortgage bonds because they are seen as the only real arbitrage alternative (2) to US securities. Seven dedicated private companies, Nykredit, Real Kredit, BRF Kredit, Danske Kredit, Unikredit, Totalkredit, and DLR, continue to issue into a market now worth DKr900bn. Mortgage-backed markets are developing in other sectors of Europe—though slowly. Deutsche Bank last week launched the first German mortgage securitization, Haus I, and ABN AMRO have twice securitized Dutch mortgages (see Asset-Backed Securities).

Arbing MBS

Foreign activity in the Danish mortgage bond market (2) concentrates on the valuation of embedded prepayment clauses— options that enable the borrower to buy back the bonds at par in order to take out the pre-payment risks associated with the underlying collateral pool. Derivative professionals analyze pre-payment options in terms of traditional non-mortgage backed swaptions and realize any relative value by trading swaptions against mortgage bonds. (2)

In one version of the trade, the speculator buys the mortgage bonds and transacts a Bermudan receivers swaption, either in Danish kroner or in (closely correlated) Deutsche marks. This cancels out the prepayment risk and leaves relatively pure interest rate and credit exposure (3).

Some players go even further and transact both a swap and a swaption against the mortgage bonds, thus creating a true arbitrage (3). They pay fixed on the swap to eliminate the interest rate risk and buy receivers swaptions to offset the inbuilt optionality. They are left with simple LIBOR plus returns and a locked-in profit, subject to credit risk.

Very similar activity forms the backbone of the US mortgage bond market. Sophisticated US houses such as Goldman Sachs, Morgan Stanley, and Salomon Brothers have for years generated favorable arbitrage profits by examining mortgage bonds in terms of other instruments (3).

But such potential profits, whether in the US or in Denmark, are subject to considerable risks. First, there are problems involved in pricing long-dated Bermudan swaptions. The product, according to swaption professionals, is sensitive to the tilt and shape of the yield curve—features not captured by simple one-factor swaption pricing models.

Second, there are also difficulties involved with analyzing the exact size of the prepayment risks. "Prepayment concerns have consumed mortgage desks in the US for the last few years. And many operations have blown up because they made the wrong assumptions," said one structurer at a prominent Wall Street investment bank.

"There's always a rumor that one house or another has got their risk modelling wrong," said another New York-based mortgage derivative professional. The problem for banks and hedge funds is that the size of prepayment is governed by the size of individual Danish property owners' refinancing activity. And this is a function of both current interest rate conditions and the statistical properties of individual house-owners.

As a result, the science of valuing mortgage bonds revolves partly on building statistical models that analyze factors other than interest rate market conditions. If rates fall in Denmark, it is likely that some mortgage holders will choose to redeem their high-rate mortgages and take out a new lower-rate loan. This will lead to the borrower prepaying some of its mortgage bonds. But not every homeowner reacts in such a way. Some may never refinance; some have a greater inclination to refinance at certain times of the year; some refinance more in certain parts of the country, and so on.

Looking to the future, mortgage bond arbitrage activity may soon be spreading to the rest of Europe. The Danish government is considering cooling the economy by restricting the flow of 30-year issues. (5) This will depress liquidity in the market and make arbitrage more difficult. Meanwhile, investment banks are talking to authorities in Sweden, Germany, the Netherlands, and the UK about changes to make their own mortgage bond markets more efficient. (Thomson Reuters IFR May 1998)

Reading 2

This reading refers to Part II of the case study.

Dynamic Funds Eye Dislocation Opportunities

With short-term yields continuing to fall across Europe, money market funds are increasingly targeting higher yields through structured products. Generally, they are lifting returns by selling volatility through corridor structures. (1) More specifically, they are taking advantage of hedge fund-induced market dislocations in Danish mortgage markets and sterling swap spread markets. "In the last few weeks funds have substituted a little credit risk for a bit more market risk," said one market professional at a major European bank. He added that funds saw the increasing confidence in financial markets as an opportunity to generate higher yields from market dislocations.

In particular, dealers said funds were looking to buy Danish mortgage bonds together with balance-protected swaps (2). The balance-protected swap guarantees the bond buyer the coupons but still leaves the holder with duration risk— the risk that the instrument will have a shorter duration if prepayment increases. With unswapped Danish mortgage bonds, the holder is exposed to the risk that prepayment rates increase and that coupons levels are reduced.

In addition to specific dislocation-related trades (3), market professionals said there was a general trend for funds to move away from standard commercial paper and asset-swap investments towards higher-returning, more structured products. Whereas traditional funds buy floating-rate instruments and are only exposed to the credit risk of the instrument, dynamic money market funds buy instruments that are exposed to interest rate and volatility risks.

One structurer last week said the dynamic fund sector had been growing for some time, and added that some European funds had more funds under management in dynamic instruments than traditional instruments. "There's lots of excitement surrounding money market funds and their attempts to churn yield," he said.

Typical dynamic money market fund trades are corridors, (4) with popular recent trades being based on two LIBOR rates remaining within the band. One bank structurer said he had recently traded a note that offered higher coupons, provided both US dollar LIBOR and French franc LIBOR remained within set limits. Capital repayment was guaranteed, but coupon payments were contingent. By buying structures such as these, funds are in effect going short LIBOR volatility and using the earned premium to enhance their received coupon levels. (5) The size of the enhanced coupon typically depends on the width of the corridor, with extra yields of up to 200 bp generated by a tight corridor and additional yields of around 50 bp delivered by a wider corridor. Dealers said funds preferred to be more cautious and favored the wider corridor, lower yield products.

EXCEL EXERCISES

4.   (Interest Rate Cap Pricing). Write a VBA program to determine the price of an interest rate cap which makes payments if the floating USD LIBOR rate is above the fixed level of 5.00%. Use the data given in the "Cap Pricing Input" worksheet on the chapter webpage for the calculation.

5.   (Interest Rate Floor Pricing). Write a VBA program to determine the price of interest rate floor which makes the payment if the floating USD LIBOR rate is below the fixed level of 4.60%. Use the data given in the "Floor Pricing Input" worksheet on the chapter webpage for the calculation.

CREDIT MARKETS: CDS ENGINEERING

CHAPTER OUTLINE

18.1 INTRODUCTION

In the previous chapters, we examined the application of financial engineering to forwards, futures, swaps, and options. In this discussion, for simplicity we abstracted away from issues related to credit risk. In the present and the following chapters, we will discuss financial engineering applications related to credit risk, credit derivatives, and credit structured products.

Credit derivatives have had a revolutionary effect on financial engineering. This is true in (at least) two respects. First, liquid credit derivatives permit stripping, pricing, and trading the *last* major component in financial instruments, namely the credit risk. With credit derivatives, synthetics for almost *any* instrument can be built.[1] Second, and as important, is the special role played by credit quants and financial engineers. Major broker dealers started organizing the credit market after the development of major instruments such as options, swaps, constant maturity swaps, and swaptions was complete. New knowledge and skills were already in place. Credit markets were developed and instruments were structured using this expertise. At the end, most of the new innovations were put in place in credit markets and in structured credit. Hence, it is important to study the credit instruments, not only because they form a huge market but also because without them many of the new financial engineering techniques would not be understood properly.

Liquid credit derivative markets extend the creation of synthetics to assets with default risk. They also permit pricing and trading default correlation. With credit default swaps (CDS) and the index products, pricing *default risk* and *default correlation* is left to the markets. In contrast, traditional approach to credit risk uses *ad hoc* estimates of *credit curves* and tries to model the spread dynamics.

This chapter deals with three sets of instruments. The first is the fundamental building block, CDS. We will see that CDSs are a natural extension of liquid fixed-income instruments. Next, we move to index products and their tranches. Tranche trading leads to calibration of default correlation and has implications both for finance and for business cycle analysis. Finally, as the third component of credit markets we look at various structured credit products. Structured credit is one area where new innovation takes place at a brisk rate. The chapter will also briefly review credit derivatives other than CDSs.

---

[1]Without credit derivatives, creating *exact* synthetics for non-AAA-rated instruments would be possible, but it would be imperfect. A synthetic that does not use credit derivatives would require some effort in modeling credit spreads and would be *ad hoc* to some extent. The principle that is applied throughout this text is that pricing, hedging, and risk management should be based on liquid and tradable securities' prices as much as possible. With credit derivatives, the *ad hoc* modeling aspects are minimized, and the model parameters can be calibrated to liquid markets.

18.2 TERMINOLOGY AND DEFINITIONS

First, we need to define some terminology. The credit sector is relatively new in modern finance, although an *ad hoc* treatment of it has existed as long as banking itself.[2] Some of the terms used in this sector come from swap markets, but others are new and specific to the credit sector. The following list is selective.

1. *Reference name.* The issuer of a debt instrument on which one is buying or selling default insurance.
2. *Reference asset.* The instrument on which credit risk is traded. Note that the credit sector adopts a somewhat more liberal definition of the basis risk. A trader may be dealing in loans but may hedge the credit risk using a bond issued by the same credit.
3. *Credit event.* Credit risk is directly or indirectly associated with some specific events (e.g., defaults or downgrades). These are important, discrete events, compared to market risk where events are relatively small and continuous.[3] The underlying credit event needs to be defined carefully in credit derivative contracts. The industry differentiates between *hard* credit events such as bankruptcy versus *soft* credit events such as restructuring.[4] We discuss this issue later in this chapter.
4. *Protection buyer, protection seller.* Protection buyer is the entity that *buys* a credit instrument such as a CDS. This entity will make periodic payments in return for compensation in the event of default. A *protection seller* is the entity that sells the CDS.
5. *Recovery value.* If default occurs, the payoff of the credit instrument will depend on the recovery value of the underlying asset at the moment of default. This value is rarely zero; some positive amount will be recoverable. Hence, the buyer needs to buy protection over and above the recoverable amount. Major rating agencies such as Moody's or Standard and Poor's have recovery rate tables for various credits which are prepared using past default data. As a result of the 2009 CDS standardization, the recovery rate is now fixed by convention at 40%. This removed some earlier ambiguity related to the calculation of default probabilities and CDS mark-to-market from the quoted spread.
6. *Credit indices.* This is the most liquid part of the credit sector. A credit index is put together by first selecting a pool of reference names and then taking the arithmetic average of the CDS rates for the names included in the portfolio. There are economy-wide credit indices with investment grade and speculative grade ratings, as well as indices for particular sectors. *iTraxx* for Europe and Asia and *CDX* for the United States and Emerging Markets are the most liquid credit indices.
7. *Tranches.* Given a portfolio of reference names, it is not known at the outset which name will default, or for that matter whether there will be defaults at all. Under these conditions the structure may decide to sell, for example, the risks associated with the first 0−3% of the

---

[2]Credit derivatives and structured products based on CDS are relatively recent. The credit default swaps themselves are several decades old and were pioneered by JP Morgan in 1994.

[3]Wiener versus Poisson-type events provide two theoretical examples.

[4]The idea is that the default probability of a company that restructures the debt is quite different from a company that has defaulted or signaled that it will default.

defaults. In a pool of 100 names, the risk of the first three defaults would then be transferred to another investor. The investor would receive periodic payments for bearing this risk. Similarly, the structurer may sell the risk associated with 3−6% of the defaults, etc. We discuss credit indices and tranches in detail in Chapter 21.

8. *E-trading.* Credit indices are typically traded OTC and, as a result of post-GFC regulation, centrally cleared. Most trading is electronic and certain US trades are required to be executed through an electronic platform called Swaps Execution Facility since October 2013. Exchange-traded futures on credit indices have been launched.[5]

9. *Standardization.* In April 2009, ISDA implemented the standardization of CDS contracts. This event is also referred to as the CDS "Big Bang." The objective of this initiative was to facilitate CDS settlement by reducing uncertainty and making credit event management more operationally efficient. The standardization had three main parts: auction hardwiring, standardizing trading conventions, and central clearing. We will discuss these in detail in Section 18.4.

The credit sector has many other sector-specific terms that we will introduce during our discussion.

18.2.1 TYPES OF CREDIT DERIVATIVES

Crude forms of credit derivatives have existed since the beginning of banking. These were not liquid, did not trade, and, in general, did not possess the desirable properties of modern financial instruments, like swaps, that facilitate their use in financial engineering. Banking services such as a *letter of credit, banker's acceptances*, and *guarantees* are precursors of modern credit instruments and can be found in the balance sheet of every bank around the world.

Broadly speaking, there are *three* major categories of credit derivatives.

1. *Credit event*-related products make payments depending on the occurrence of a mutually agreeable event. The CDS is the major building block here.
2. Credit *index* products that are used in trading portfolios of credit. Obviously, such indices would come with their own derivatives such as options, futures, and forwards.[6] An example would be an option written on the iTraxx Europe index.
3. The structured credit products and the index tranches.

The difference between the value of risky and riskless debt is typically expressed in terms of a *credit spread*. The spread is the difference between the (promised) yield on the risky bond and the yield on riskless bonds. Credit risk can be broadly grouped into two different categories: on one hand, *credit deterioration*. Widening of the underlying credit spread can indicate how credit deteriorates. On the other hand, *default risk*. This is a separate risk from credit deterioration, although it is certainly correlated with it. Default products trade default risk by separating it from credit deterioration risk.

---

[5]Some market participants cannot trade swaps due to internal restrictions or a lack of an ISDA agreement. In such cases, credit index futures provide an alternative to obtaining exposure to or hedge macro credit risk.

[6]In 2007, Eurex launched futures contracts on three iTraxx indices but trading in these contracts never took off. ICE incorporated the lessons from this episode into its product design and in June 2013 launched futures on four Markit CDX and iTraxx indices. Unlike the Eurex and CME products, the ICE contracts do not offer default protection for the investor, but a forward view of credit spreads.

As mentioned above, banks have issued letters of credit, guarantees, and insurance. The major distinguishing characteristic of these traditional instruments is that they transfer default risk *only*. They do not, in particular, transfer market risk or the risk of credit deterioration. Essentially, a payment is made when default occurs. With these products, no compensation changes hands when the underlying credit deteriorates. New credit *default* products share some of the properties of these old instruments. Some of the new features of credit contracts are as follows:

1. The payout is dependent on an *event* rather than an underlying price, similar to insurance products and unlike other derivatives. The dependence of a payoff on an event leads to new techniques and instruments.
2. The existence of an event leads to the issue of *recovery value*. How to determine (model) the value of an asset in case of default is now easy. Throughout this chapter, we will use the assumption that the recovery rate is constant and known at a level R.
3. The process of *settling* credit contracts is more complex than in other markets. In the case of *physical delivery* of the underlying, this does not present a major problem. The protection seller will be the legal owner of the defaulted instrument and may take necessary legal steps for recovery. But if the contract is cash-settled, then neither party has legal recourse to the borrower unless the party owns the underlying credit directly. For this reason, the industry prefers physical delivery, and a large majority of default swaps settle this way.

We will address the additional characteristics of default products when we study CDSs in more detail. In the following section, we will look at the most liquid credit derivatives in more detail and study the financial engineering of CDSs.

18.3 CREDIT DEFAULT SWAPS

The major building block of the credit sector is the CDS, introduced in the first chapter as an example in the swap family. It is, however, a major category. A typical default swap from the point of view of a protection *seller* is shown in Figure 18.1. The CDS seller of a particular credit denoted by i receives a preset coupon called the CDS rate. The CDS expires at time T. The credit protection seller in a CDS is referred to as selling the CDS while the protection buyer buys the CDS. The protection buyer pays a fixed rate (called the CDS rate) periodically until the earlier of default and the fixed maturity date. The CDS spread is denoted by $cds_{t_0}^j$ and is set at time t_0. A payment of $cds_{t_0}^j \, \delta N$ is made at every t_i. The j represents the reference name. If no default occurs until T, the contract expires without any other payments. On the other hand, if the name j defaults during $[t_0, T]$, the CDS seller has to compensate the counterparty by the insured amount, N dollars. Against this payment of cash, the protection buyer has to deliver eligible debt instruments with *par value N* dollars. These instruments will be from a *deliverable basket* and are clearly specified in the contract at time t_0. Obviously, one of these instruments will, in general, be cheapest-to-deliver in the case of default, and all players may want to deliver that particular underlying.

Note that the discussion so far is simplified since we assume that there is no upfront payment at the initiation of the CDS. This is in contrast to recent market conventions following CDS standardization which have moved from a running spread basis to an upfront basis. We will discuss these

Credit default swap with default possibility at t_3 only

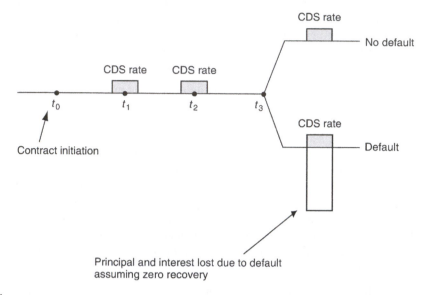

FIGURE 18.1

Credit default swap.

market conventions later in detail. Later in this chapter, we will consider additional properties of the default swap market that a financial engineer should be aware of. At this point, we discuss the engineering aspects of this product. This is especially important because we will show that a default swap will fall naturally as the residual from the decomposition of a typical risky bond. In fact, we will take a risky bond and decompose it into its components. The key component will be the default swap. This natural function played by default swaps partly explains their appeal and their position as the leading credit instrument.

We discuss the creation of a default swap by using a specific example. The example deals with a special case, but illustrates almost all the major aspects of engineering credit risk. Many current practices involving index CDSs, basket default swaps, synthetic collateralized debt obligations (CDOs) and credit linked notes (CLNs), and other popular credit instruments can be traced to the discussion provided next.[7]

Independently, this section can be seen as another example of engineering cash flows. We show how the static replication methods change when default risk is introduced into the picture. Essentially, the same techniques are used. But the creation of a satisfactory synthetic becomes possible only if we add CDSs to other standard instruments.

---

[7]CDOs and CLNs are described in Chapter 21. Synthetic CDOs were popular before the GFC and are experiencing renewed interest.

18.3.1 CREATING A CDS

The steps we intend to take can be summarized as follows. We take a bond issued by the reference name j that has *default risk* and then show how the cash flows of this bond can be decomposed into simpler, more liquid constituents. Essentially we decompose the bond risk into two—one depending on market volatility only, the other depending on the reference home's likelihood of default. CDSs result naturally from this decomposition.

Our discussion leads to a new type of *contractual equation* that will incorporate credit risk. We then use this contractual equation to show how a CDS can be created, hedged, and priced in theory. The contractual equation also illustrates some of the inherent difficulties of the hedging and pricing process in practice. At the end of the section, we discuss some practical hedging and pricing issues.

18.3.2 DECOMPOSING A RISKY BOND

We keep the example simple in order to illustrate the fundamental issues more clearly. Consider a "risky" bond, purchased at time t_0, subject to default risk. The bond does not contain any implicit call and put options and pays a coupon of c_{t_0} annually over 3 years. The bond is originally sold at par.[8]

We make two further *simplifying assumptions* which can be relaxed with little additional effort. These assumptions do not change the essence of the engineering, but significantly facilitate the understanding of the credit instrument. First, we assume that, in the case of default, the recovery value equals the *known* constant R. Second, and without much loss of generality, we assume that the default occurs *only* at settlement dates t_i. Finally, to keep the graphs tractable we assume that settlement dates are annual, and that the maturity of the bond is $T = 3$ years.

Figure 18.2 shows the cash flows implied by this bond. The bond is initially purchased for 100, three coupon payments are made, and the principal of 100 is returned *if* there is no default. On the

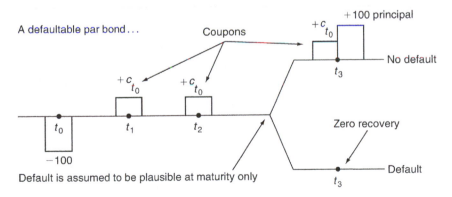

FIGURE 18.2

A defaultable par bond.

---

[8]The latter is an assumption made for convenience and is rarely satisfied in reality. Bonds sold at a discount or premium need significant adjustments in their engineering as discussed below. However, these are mostly technical in nature.

other hand, if there is default, the bond pays nothing. The dependence on default is shown with the fork at times t_i. At each settlement date, there are two possibilities and the claim is *contingent* on these.

How do we reverse engineer these cash flows and convert them into liquid financial instruments? We answer this question in steps.

First, we need to introduce a useful trick that will facilitate the application of static decomposition methods to defaultable instruments. We do this in Figure 18.3. Remember that our goal is to *isolate* the underlying default risk using a *single* instrument. This task will be greatly simplified if we add *and* subtract the amount $(1 + c_{t_0})N$ to the cash flows in the case of default at times t_i. Note that this does not change the original cash flows. Yet, it is useful for isolating the inherent CDS, as we will see.

Now we can discuss the decomposition of the defaultable bond. The bond in Figure 18.2 contains three different types of cash flows:

1. Three coupon payments on dates t_1, t_2, and t_3. We strip these fixed cash flows and place them in Figure 18.3b. Although the coupons are risky, we can still extract three default-free coupon payments from the bond cash flows due to the trick used. To get the default-free coupon payments, we simply pick the positive $(c_{t_0})100$ at the default state for times t_i of Figure 18.3a. Note that this leaves the negative $(c_{t_0})100$ in place.
2. Initial and final payment of 100 as shown in Figure 18.3c. Again, adding and subtracting 100 is used to obtain a default-free cash flow of 100 at time t_3. These two cash flows are then carried to Figure 18.3c. As a result, the negative payment of 100 in the default state of times t_i remains in Figure 18.3a.
3. All remaining cash flows are shown in Figure 18.3d. These consist of the negative cash flow $(1 + Co_{t_0})100$ that occurs in the time t_3 default state. This is detached and placed in Figure 18.3d.

The next step is to convert the three cash flow diagrams in Figure 18.3b−d into recognizable and, preferably, liquid contracts traded in the markets. Remember that to do this, we need to add and subtract arbitrary cash flows to those in Figure 18.3b−d while ensuring that the following three conditions are met:

- For each cash flow added, we have to *subtract* the same amount (or its present value) at the same t_i from one of the Figures 18.3b, 18.3c, or 18.3d.
- These new cash flows should be introduced to make the resulting instruments as liquid as possible.
- When added back together, the modified Figure 18.3b−d should give back the original bond cash flows in Figure 18.3a. This way, we should be able to recover the cash flows of the defaultable bond.

This process is displayed in Figure 18.4. The easiest cash flows to convert into a recognizable instrument are those in Figure 18.3b. If we add floating LIBOR-based payments, L_{t_i} at times t_1, t_2, and t_3, these cash flows will *look like* a fixed receiver interest rate swap (IRS). This is good because swaps are very liquid instruments. However, one additional modification is required. The fixed receiver swap rate, s_{t_0}, is less than the coupon of a par bond issued at time t_0, since the bond can default while the swap is subject only to a counterparty risk. Thus, we have

$$s_{t_0} \leq c_{t_0} \tag{18.1}$$

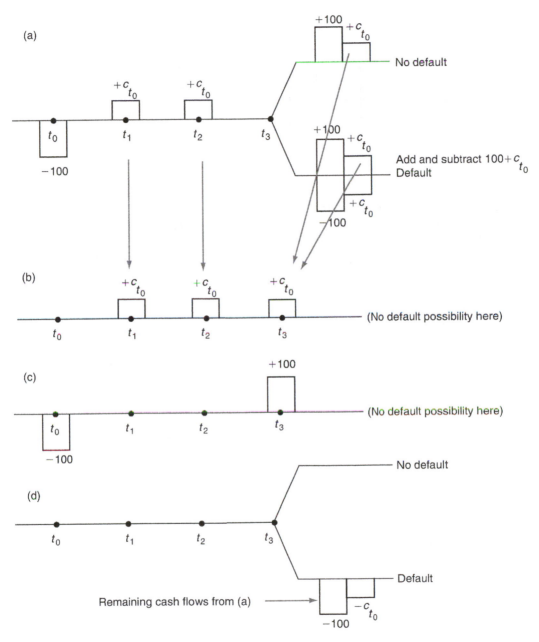

FIGURE 18.3

Decomposition of a defaultable bond.

(a)

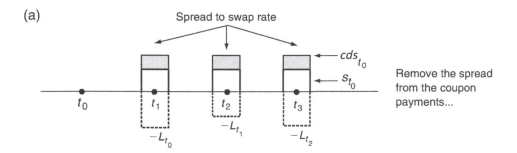

Spread to swap rate

cds_{t_0}

s_{t_0}

Remove the spread
from the coupon
payments...

(b) Place the spreads on the cash flows
in Figure 18.3d

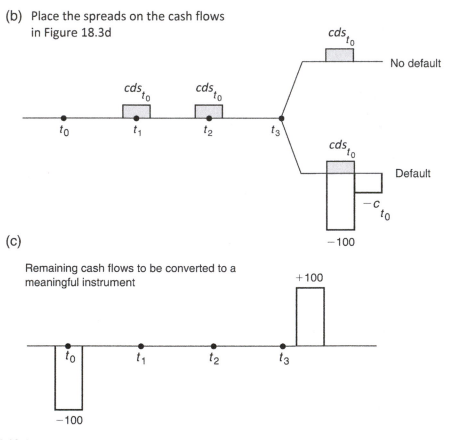

cds_{t_0}

No default

cds_{t_0}

cds_{t_0}

cds_{t_0}

Default

$-c_{t_0}$

-100

(c)

Remaining cash flows to be converted to a
meaningful instrument

$+100$

-100

FIGURE 18.4

Synthetic credit default swap.

The difference, denoted by cds_{t_0},

$$cds_{t_0} = c_{t_0} - s_{t_0} \qquad (18.2)$$

is the *credit spread over the swap rate*. This is how much a credit has to pay over and above the swap rate due to the default possibility. Note that we are defining the credit spread as a spread over the corresponding swap rate and *not* over that of the treasuries. This definition falls naturally from cash flow decompositions.[9]

Thus, in order for the cash flows in Figure 18.4a to be equivalent to a receiver swap, we need to subtract c_{t_0} from each coupon as is done in Figure 18.4a. This will make the fixed receipts equal the swap rate:

$$c_{t_0} - cds_{t_0} = s_{t_0} \qquad (18.3)$$

The resulting cash flows become a true *IRS*.

The next question is where to *place* the counterparts of the cash flows c_{t_0} and L_{t_i} that we just introduced in Figure 18.4a. After all, unless the same cash flows are placed *somewhere* else with opposite signs, they will not cancel out, and the resulting synthetic will not reduce to a risky bond.

A natural place to put the LIBOR-based cash flows is shown in Figure 18.5. Nicely, the addition of the LIBOR-related cash flows converts the figure into a *default-free money market deposit* with tenor δ. This deposit will be rolled over at the going floating LIBOR rate. Note that this is also a liquid instrument.[10]

The final adjustment is how to compensate the reduction of c_{t_0}'s by the credit spread cds_{t_0}. Since the first two instruments are complete, there is only one place to put the compensating cds_{t_0}'s. We add the cds_{t_0} to the cash flows shown in Figure 18.3d, and the result is shown in Figure 18.4b. This is the critical step, since we now have obtained a *new* instrument that has fallen

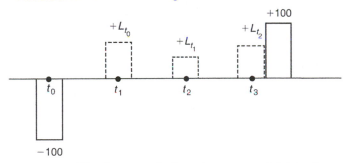

Add LIBOR-based cash flow to figure 18.3c...

...They become a floating rate money market deposit or a floating rate note (FRN)

FIGURE 18.5

LIBOR cash flows become deposit.

[9]We should also mention that AAA credits have sub-Libor funding cost.
[10]Alternatively, we can call it a floating rate note (FRN).

naturally from the decomposition of the risky bond. Essentially, this instrument has potentially three receipts of cds_{t_0} dollars at times t_1, t_2, and t_3. But if *default* occurs, the instrument will make a compensating payment of $(1 + c_{t_0})100$ dollars.[11]

To make sure that the decomposition is correct, we add Figures 18.4a,b and 18.5 vertically and see if the original cash flows are recovered. The vertical sum of cash flows in Figures 18.4a,b and 18.5 indeed replicates exactly the cash flows of the defaultable bond.

The instrument we have in Figure 18.4b is equivalent to *selling* protection against the default risk of the bond. The contract involves collecting *fees* equal to cds_{t_0} at each t_i until the default occurs. Then the protection buyer is compensated for the loss. On the other hand, if there is no default, the fees are collected until the expiration of the contract and no payment is made. We call this instrument a *CDS*.

18.3.3 A SYNTHETIC

The preceding discussion shows that a defaultable bond can be decomposed into a portfolio made up of (i) a fixed receiver IRS, (ii) a default-free money market deposit, and (iii) a CDS. The use of these instruments implies the following contractual equation:

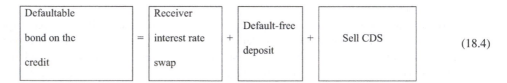

$$\begin{array}{|c|}\hline \text{Defaultable} \\ \text{bond on the} \\ \text{credit} \\ \hline \end{array} = \begin{array}{|c|}\hline \text{Receiver} \\ \text{interest rate} \\ \text{swap} \\ \hline \end{array} + \begin{array}{|c|}\hline \text{Default-free} \\ \text{deposit} \\ \hline \end{array} + \begin{array}{|c|}\hline \text{Sell CDS} \\ \hline \end{array} \qquad (18.4)$$

By manipulating the elements of this equation using the standard rules of algebra, we can obtain synthetics for every instrument in the equation. In the next section, we show two applications.[12]

18.3.4 USING THE CONTRACTUAL EQUATION

As a first application, we show how to obtain a *hedge* for a long or short CDS position by manipulating the contractual equation. Second, we discuss the implied pricing and the resulting real-world difficulties.

18.3.4.1 Creating a synthetic CDS

First, we consider the way a CDS would be hedged. Suppose a market maker sells a CDS on a certain name. In other words, the market maker provides protection or *sells* the CDS and the credit. How would the market maker hedge this position while it is still on his or her books?

---

[11]According to this, in the case of default, the total *net* payment becomes $(1 + c_{t_0})100 - cds_{t_0}100$.

[12]The market convention is to refer to the counterparty paying the fixed rate as called the *payer* (while receiving the floating rate), and the counterparty receiving the fixed rate as the "receiver" (while paying the floating rate).

To obtain a hedge for the CDS, all we need to do is to manipulate the contractual equation obtained above. Rearranging, we obtain the following equation for selling a CDS

$$
\boxed{\begin{array}{c}\text{Defaultable bond}\\[4pt]\text{issued by the credit}\end{array}} - \boxed{\begin{array}{c}\text{Receiver}\\[4pt]\text{interest rate}\\[4pt]\text{swap}\end{array}} - \boxed{\begin{array}{c}\text{Default-free}\\[4pt]\text{deposit}\end{array}} = \boxed{\text{Sell CDS}} \qquad (18.5)
$$

Remembering that a negative sign implies the opposite position in the relevant instrument, we can write the formal synthetic for buying a CDS as

$$
\boxed{\begin{array}{c}\text{Buy}\\[4pt]\text{CDS on the}\\[4pt]\text{credit}\end{array}} = \boxed{\begin{array}{c}\text{Short a}\\[4pt]\text{defaultable}\\[4pt]\text{bond}\\[4pt]\text{on credit}\end{array}} + \boxed{\begin{array}{c}\text{Receiver}\\[4pt]\text{interest rate}\\[4pt]\text{swap}\end{array}} + \boxed{\begin{array}{c}\text{Default-free}\\[4pt]\text{deposit}\end{array}} \qquad (18.6)
$$

The market maker who sold such a CDS provided protection needs to take the *opposite* position of the left-hand side of Eq. (18.1). This hedge corresponds to the right-hand side of Eq. (18.6). That is to say, the credit derivatives dealer who sold a CDS will hedge the CDS position by creating a synthetic opposite CDS position. This is achieved by first *shorting* the risky bond, depositing the received 100 in a default-free deposit account, and contracting a payer swap. This and the sale of the CDS will then "cancel" out. The market maker will make money on the bid−ask spread.

18.3.4.2 Negative basis trades

The second application of the contractual equation is referred to as *negative basis trades*. Negative basis trades are an important position frequently taken by the traders in credit markets. A discussion of the trade provides a good example of how the contractual equation defining the CDS contracts changes in the real world.

The contractual equation that leads to the creation of a CDS can be used to construct a synthetic CDS that can be used against the actual one. As we saw in Chapter 7, the basis typically refers to the difference between the spot (cash) price of an asset and its future's price (derivative). In the credit derivatives market, traders refer to the basis as the difference between the CDS premium of a given bond (cds_{t_0}) and the bond spread $(c_{t_0} - s_{t_0})$ for the same debt issuer and with similar, if not exactly equal, maturities. Since CDS are derivatives, market participants in credit markets refer to the CDS leg of a negative basis trade as *synthetic*, and the bond leg as *cash*.[13]

---

[13]We noted earlier that the fixed rate in an interest swap rate differs from the rate on Treasury securities with the same maturity. The interpretation of the negative basis spread trade is slightly different if the bond spread is measured relative to a swap rate and not against Treasuries with a similar maturity. For now, let's assume that s_{t_0} represents the fixed rate in an interest rate swap.

Essentially, with a negative basis trade one would take out a floating rate loan to buy a bond that pays the par-yield c_{t_0} while at the same time, buying insurance on the same bond and entering a fixed payer IRS. Clearly such a position has no default risk and makes sense if the coupon minus the bond risk premium $(c_{t_0} - s_{t_0})$ is greater than the CDS premium payments (cds_{t_0})

$$s_{t_0} + \text{cds}_{t_0} < c_{t_0} \qquad (18.7)$$

This is in fact the case of *negative basis*. Here, as in the derivation of the contractual equation above, we assume that the rate on the floating rate loan is the same as the LIBOR rate in the IRS. The above interpretation of the negative basis trade follows straight from the contractual equations (18.5) and (18.6).

In practice, there is sometimes an alternative interpretation of the negative basis trade, since the bond spread can also be measured as the yield on the defaultable bond minus the yield of a Treasury security with a similar maturity. In other words, s_{t_0} is taken to be the yield on a Treasury security and not the fixed rate from an IRS. The negative basis trade in that case can be interpreted as shorting the Treasury security and using the cash from this short position to buy a defaultable bond while at the same time taking out insurance purchasing a CDS on the same bond. Since the short *finances* the long, there is no need for a loan.

Normally, the insurance on a default risky bond should be "slightly" higher than the credit risk spread one would obtain from the bond. This *basis* is in general positive. Otherwise, if the bond spread is larger than the CDS premium, then one would buy the bond *and* buy insurance on it. This would be a perfect arbitrage. This is exactly what a *negative basis* is. The reading below suggests when and how this may occur.

EXAMPLE

A surge of corporate bond and structured finance issuance this past month has pushed risk premiums on credit default swaps down and those on investment-grade corporate bonds up, nearly erasing the difference between the two asset classes. If this trend continues, it has the potential to create new trading opportunities for investors who can take positions in both bonds and derivatives. Analysts are expecting to see more so-called "negative basis trades" as a result.

January has been a particularly active month in the US primary corporate bond market, with over $60 billion issued in investment-grade debt alone. This supply helps to widen risk premiums on corporate bonds. At the same time, there has been no shortage of synthetic collateralized debt obligations (CDOs), which has helped narrow CDS spreads. A corporate credit default swap contract features a seller of protection and a buyer of protection. The seller is effectively long that company's debt while the buyer is short. When a synthetic CDO is created, dealers sell a certain amount of credit protection, which helps compress CDS risk premiums.

Typically, credit default swap risk premiums trade at wider levels than comparable bond risk premiums. This is partly because it is easier to take a short position via a CDS rather than a bond. Another reason that CDS risk premiums trade wider is the cheapest-to-deliver option. In the event of a bankruptcy, the seller of protection in a CDS contract will make a cash payment to the buyer in exchange for the bonds of the bankrupt company. However, the seller of protection will receive the cheapest-to-deliver bond from the protection buyer.

Once the basis between a CDS risk premium and cash bond reverses and becomes negative, it can become advantageous to buy the bond and buy protection through a credit default swap. In such a trade, known as a negative basis trade, the investor has hedged out their credit risk, but is earning more on their bond position than they are paying out on their CDS position.

(Thomson Reuters IFR, January 2006)

The above reading also illustrates the terminology used in CDS trades. In the context of IRSs and volatility swaps, we defined the counterparty that was long the swap as the party that paid the first interest rate or volatility. In the contexts of CDS, one can describe the buyer (seller) of a CDS as being short (long) the underlying bond or credit.[14] To the extent one refers to the buyer of the CDS as being short, this can be counterintuitive. To avoid any confusion we refer to the buyer (seller) of a CDS as the protection buyer (seller). We will deal with CDOs and their construction and pricing in detail in Chapter 21.

Negative basis trades become a possibility due to leveraged buyout (LBO) activity as well. During an LBO the buyer of the company issues large amounts of debt, which increases the debt to equity ratio on the balance sheet. Often, rating agencies downgrade such LBO targets several notches which makes LBO candidates risky for the original bond holder. Bond holders, hearing that a company is becoming an LBO target, may sell before the likely LBO; bond prices may drop suddenly and the yields may spike. On the other hand, the CDS rates may move less since LBO is not necessarily going to increase the *default* probability. As a result, the basis may momentarily turn negative.

18.3.5 MEASURING CREDIT RISK OF CASH BONDS

Many credit and fixed-income strategies involve arbitraging between cash and synthetic instruments. CDS is the synthetic way of taking an exposure to the default of a single name credit. It is a clean way of trading default. But, cash bonds contain default risk as well as interest rate and curve risk. How would we strip from a defaultable bond yield the component that is being paid due to default risk? In other words, how do we obtain the equivalent of the CDS rate from a defaultable bond *in reality?*

This question needs to be answered if we are to take arbitrage positions between cash and derivatives; for example, when we have to make a decision whether cash bonds are *too expensive* or not relative to the CDS of the same name.

At the outset the question seems unnecessary, since we just developed a contractual equation for the CDS. We showed that if we combined the cash bond and a vanilla IRS accordingly, then we would obtain a synthetic FRN that paid LIBOR plus a spread. The spread would equal the CDS rate. In other words, take the par yield of a par bond, subtract from this the comparable IRS rate, and get the equivalent of the CDS rate included in the bonds yield.

This is indeed true, except for one major problem. The formula is a good approximation only if all simplifying assumptions are satisfied and if the cash bonds are selling *at par*. It is only then that we can straightforwardly put together an *asset swap* and strip the credit spread.[15] If the bond is not selling for par, we would need further adjustments.

There are two major methods used to obtain a measure of the default risk contained in a bond that does not trade at par. The first is to calculate the *asset swap spread* and the second is to calculate the so-called *Z-spread*. We discuss these two practical concepts next and give examples.

---

[14]We use the term *long credit* as describing the position of the CDS protection seller since the seller is exposed to the credit risk. In the context of variance swaps, we also refer to the variance swap seller as being long volatility.

[15]There are also some day-count adjustments that we assumed away.

18.3.5.1 Asset swap

An asset swap spread is one way of calculating the credit spread associated with a default risky bond. Essentially it converts the risky yield into a *LIBOR plus credit spread*, using an IRS.

In order to create a position equivalent to selling protection, we must buy the defaultable bond and get in a payer IRS. Note that the IRS will be a par swap here, while the bond might be selling for a discount or premium.[16] So adjustments are needed in reality since the bond sells either at a discount or premium.[17]

In the asset swap, the bond cash flows are discounted using the corresponding zero-coupon swap rates. A spread (called the asset swap spread) is added to (subtracted from) the bond cash flows so that the resulting bond price equals the market price.

The formula for calculating the asset swap spread is as follows: first use the *zero-coupon* swap curve to calculate the discount factors. By definition, *zero-coupon swap curve* discounts are obtained using the corresponding forward rates f_{t_j}, and the equation,

$$\left(1+s_{t_0}^i\delta\right)^i = \prod_{j=0}^{i-1}\left(1+f_{t_j}\delta\right)$$ (18.8)

Then calculate the discount factors

$$B(t_0, t_i) = \frac{1}{\left(1+s_{t_0}^i\delta\right)^i}$$ (18.9)

Once this is done, form the *annuity factor*, A.

$$A = \sum_{i=1}^n B(t_0, t_i)\delta$$ (18.10)

Next, calculate the value of the bond cash flows using these factors:

$$\tilde{P}_{t_0} = \sum_{i=1}^n \frac{c_{t_i}}{\left(1+s_{t_0}^i\delta\right)^i}$$ (18.11)

The asset swap spread is $\tilde{a}_{t_0}$ that solves the equation:

$$\tilde{P}_{t_0} - P_{t_0} = \tilde{a}_{t_0}\sum_{i=1}^n B(t_0, t_i)\delta$$ (18.12)

Thus, $\tilde{a}_{t_0}$ is how much one needs to be paid extra in order to compensate for the difference between the market price and the theoretical price implied by the default-free swap curve. This is the case, since if there were no default risk the value of the bond would be $\tilde{P}_{t_0}$ which is greater than the market price P_{t_0}. $\tilde{a}_{t_0}$ is a measure that converts this price differential into an additional spread.

---

[16]Remember how swaps were engineered. We added an FRN position to a par bond paying the swap rate s_{t_0}. This leads to the par swaps traded in the markets.

[17]In an asset swap, the IRS and the bond position are written on separate tickets. If there is default, the swap position needs to be closed separately and this may involve an extra exposure. The swap position may be making or losing money when default occurs.

We can show how asset swaps are structured using this last equation. A par bond paying the swap rate s_{t_0} satisfies the equation:

$$100 = \sum_{i=1}^{n} B(t_0, t_i) s_{t_0} \delta + B(t_0, t_n) 100 \qquad (18.13)$$

Thus after adding 100 to both sides, the previous equation can be written as:

$$\tilde{P}_{t_0} - \left(P_{t_0} - 100 \right) = \left(s_{t_0} + \tilde{a}_{t_0} \right) \sum_{i=1}^{n} B(t_0, t_i) \delta + B(t_0, t_n) 100 \qquad (18.14)$$

where the second term on the left-hand side is the upfront payment (receipt) upf_{t_0}

$$\text{upf}_{t_0} = \left(P_{t_0} - 100 \right) \qquad (18.15)$$

The latter is negative if the bond is selling at a discount and positive if the bond is at a premium.

We interpret this equation as follows. Suppose the bond is trading at a discount, then an upfront payment of upf_{t_0} plus an IRS contracted at a rate $s_{t_0} + \tilde{a}_{t_0}$ is equivalent to the present value of the coupon payments of the bond. In other words, a par swap rate has to be *augmented* by $\tilde{a}_{t_0}$ in order to compensate for the bond's default risk. Note that during this exercise we worked with *risk-free discount factors*. This will change with the next notion.

18.3.5.2 The Z-spread

The so-called Z-spread is another way of calculating the credit spread. It gives a result similar to the asset swap spread but is not necessarily the same.

In order to calculate the Z-spread, the cash flows generated by a default risk bond will be discounted by the *zero-coupon swap rate* s_{t_0} augmented by a spread z_{t_0}, so that the sum equals the market price of the bond. That is to say we have

$$P_{t_0} = \sum_{i=1}^{n} \frac{\text{cf}_i}{\left(1 + \left(s_{t_0}^i + z_{t_0} \right) \delta \right)^i} \qquad (18.16)$$

We solve this equation for the unknown z_{t_0}. Here cf_{t_i} are the cash flows received at time t_i and are made of coupon payments c_{t_i} and possibly of the principal.

According to this, the zero-coupon swap curve is adjusted in a parallel fashion so that the present value of the cash flows equals the bond price. The Z-spread is the amount of parallel movement in the zero-coupon swap curve needed to do this.

Note that during the calculation of the Z-spread we worked with a measure of *risky discount factors*.[18] The major difference between the Z-spread and the asset swap spread arises from the discount rates used. Asset swaps use zero-coupon swap rates whereas Z-spread uses zero-coupon swap rates *plus* the Z-spread.

In this sense, the Z-spread method uses a stream of risky discounts from the *whole risky curve* to adjust the future cash flows of the bond. The asset swap spread uses a *single* maturity swap rate to measure the credit risk. Although the Z-spread is better suited to risky discount

---

[18]Although the risky discount factors are calculated differently from the risky DV01 we will see in this chapter.

factors, markets prefer to use the asset swaps as a measure of credit risk. The reason is that the markets do *not* quote the credit spread as a spread to *zero-coupon* swap rates. The credit spread is quoted as a *spread to a par swap rate* because the par swaps are much more liquid than the zero-coupon swaps.

18.4 REAL-WORLD COMPLICATIONS

Credit markets and credit derivatives trading are inherently more complicated and heterogeneous than most other markets, and one faces an unusual number of real-life complications that theoretical models may not account for. In this section, we look at some of the real-life aspects of CDS contracts.

Contractual equation (18.6) provides a natural hedge for the CDS and shows one way of pricing it. Similar contractual equations may provide usable hedges and pricing methods for some bread-and-butter instruments with negligible credit risk, but for CDSs these equations are essentially *theoretical*. The simplified approach discussed above may sometimes misprice the CDS and the hedge obtained earlier may not hold. There may be several reasons why the *benchmark spreads* on this credit may deviate significantly from the CDS rates. We briefly discuss some of these reasons.

1. In the preceding example, the CDS had a maturity of 3 years. What if the particular credit had no outstanding 3-year bonds at the time the CDS was issued? Then the pricing would be more complicated and the benchmark spread could very well deviate from the CDS rate.
2. Even if similar maturity bonds exist, these may not be very liquid, especially during times of high market volatility. Then, it would be natural to see discrepancies between the CDS rates and the benchmark spreads.
3. The tax treatment of corporate bonds and CDSs is different, and this introduces a wedge between the corresponding spread and the CDS rate.
4. As mentioned earlier, CDSs result in physical delivery in the case of default. But this delivery is from a basket of deliverable bonds. This means that the CDS contains a *delivery option*, which was not built into the contractual equation presented earlier.

In reality, another important issue arises. The construction of the synthetic shown above used a money market account that was assumed to be risk free. In general, such money market accounts are almost never risk free and the deposit-accepting institution will have a default risk. This introduces another wedge between the theoretical construction and actual pricing. This additional credit risk that creeps into the construction can, in principle, be eliminated by buying a new CDS for the deposit-accepting institution.

The following reading illustrates some of the real-world complications, such as liquidity issues, in the context of credit curve strategies. Standard fixed-income strategy can also be applied in the credit sector.

Just as in the case of standard fixed-income products, the credit curve allows for curve flattener and curve steepener trades. The idea is self explanatory and follows the fixed-income swap positions applied to the iTraxx curve.

EXAMPLE

Consider Hutchison Whampoa credit-curve flattener via bond-basis play, exploiting value in long end while protecting against any general market deterioration. Buy Hutch 2027 bonds (cheapest on curve), buy 5-year CDS protection. Spread now 82 bps (asset-swap bond spread 131 bps versus CDS at 49 bps). Target compression to 65 bps in coming months. Exit if spread widens to 90 bps.

   Trade primarily a play on long-end bonds: buying protection at shorter end offers hedge in case of bad news on Hutch or general bond downturn. CDS cheaper alternative to selling short-end bond given lack of repo liquidity; using asset swap avoids fixed-rate risk at short end. Spread should slowly grind tighter as more investors switch out of shorter Hutch bonds (2010, 2011, 2014 maturities) where value has already been squeezed out into undervalued longer end.

This trade can be interpreted as primarily a play on long-end bonds. The reason why the protection is bought at the shorter maturity end is that it offers a hedge in case of bad news on Hutch or general bond downturn in the near term. The reason why a CDS was chosen over shorting a bond is that the CDS is cheaper than the alternative of selling a short-end bond given the *relative* lack of repo liquidity in the corporate bond market. The use of an asset swap avoids fixed-rate risk at the short end. Spread should slowly grind tighter as more investors switch out of shorter Hutch bonds (2010, 2011, 2014 maturities) where value has already been squeezed out into undervalued longer end.

18.4.1 CDS STANDARDIZATION AND 2009 "BIG BANG"

On April 8, 2009, a *"Big Bang"* occurred in the market for CDS contracts and the way in which they are traded. The changes affected both contracts and conventions. The standardization had three main parts: auction hardwiring, standardizing trading conventions, and central clearing. We discuss the main elements of these below.[19]

18.4.1.1 Auction hardwiring

ISDA released a supplement and a protocol for credit and succession events in 2009. This led to the creation of regional Determination Committees of dealers and investors to arbitrate when a credit event arises. The committee's decisions are binding. If a committee decides on a specific credit event an automatic and mandatory CDS auction settlement process takes place.

18.4.1.2 Standardization coupons and trading conventions

The standardization of North American Corporate CDS saw the introduction of fixed coupons of 100—500 bp. For more than a decade after they were first introduced most single-name CDS contracts traded with coupons in *all-running* format, that is single-name CDS traded on a *running*

---

[19]For further details, see http://www.markit.com/assets/en/docs/markit-magazine/issue-4/60-cds-big-bang.pdf as well as the end of chapter readings.

spread or *par spread* basis which meant that the protection buyer pays for protection by making premium (or spread) payments to the protection seller. Consider a 5-year CDS with a $10 million notional, for example. If we assume that the running spread was 200 bp, the protection buyer paid $200,000 a year (typically in quarterly installments).

Under the running spread convention, no money was exchanged upfront which implied that CDS positions were implicitly leveraged. Following the April 2009 "Big Bang" in the CDS market, the market moved to an *points upfront* basis. Nowadays contracts have fixed coupons of either 100 bp or 500 bp. The upfront payments are made at initiation and are equal to the present value of the difference between the current market credit spread and the fixed coupon. As an example, if a contract with a fixed coupon of 100 bp is trading at 200 bp, the protection buyer will make an upfront payment equal to the present value of the difference between 200 bp and 100 bp. The payment made in the opposite direction of the market spread is below the coupon. Note that the most liquid indices such as CDX and iTraxx have traded on such a points upfront basis for a long time. These market conventions do not require a major modification of CDS pricing approaches since a running spread contract can be viewed as a special case of an upfront and fixed-coupon contract where the upfront is zero and the fixed coupon is equal to the par spread.

How is the fixed coupon set for CDS indices? When the new series of the index is launched, the fixed coupon, which is also known as a deal spread, since it is expressed as a spread, is determined.

The move to the points upfront basis for single-name CDS has three advantages. First, it will make netting between single name and index positions easier. Such positions are common as we will see below in the context of index arbitrage trades. Second, the more convenient netting also facilitates the recent move of CDS contracts to central clearing. Third, it also reduces market participant's exposures to sudden market movements or *jump* risk which we discuss in more detail in the context of CDS unwinding later.

18.4.1.3 Central clearing

The changes above facilitated the introduction of central clearing for CDS. As discussed below in the context of unwinding of CDS, central clearing helps reduce counterparty risk in the CDS market. Central clearing for CDS in the United States and Europe began in 2009.

18.4.2 RESTRUCTURING

Another real-life complication deals with the definition of default itself. Credit events are normally *failure to pay* and *bankruptcy*. Any nonpayments of interest or principal would count as the former and any type of formal bankruptcy would count as the latter. In our theoretical engineering, we used this second definition of defaults. Yet, in reality, most single-name CDSs also consider restructuring as an additional default event. Further complicating the picture is the type restructuring, summarized below.

There are four types of restructuring clauses in the CDS contracts.[20] The first is simply no-restructuring, *No R*. In this case, any structuring would not constitute a credit event. It is important

---

[20]The differences between the clauses related to the maturity of the deliverable obligations and transferability of deliverable obligations. See 2014 ISDA Credit Definitions, for details.

to note that credits have come to trade based on market-defined conventions. Normally, high-yield CDS typically trade No R. This is especially true of the CDX indexes in the United States. Markit iTraxx indices trade with Modified-Modified Restructuring except for the Sub-Financials which trade with Restructuring.

The second type is modified restructuring, *Mod R*. This creates new conditions for a credit event.[21] North American investment grade names trade with a "Modified" restructuring convention. *Mod R* clauses arose historically for these North American investment-grade credits because of the needs of hedgers of bank loan portfolios.

The third is the modified modified restructuring, *Mod Mod R*. In this case, the maturity limits on the deliverable bonds or loans are somewhat different. There is an exception that the bonds (loans) with a maturity of more than 30 months but no more than 60 months can be delivered. European CDS contracts typically traded with a Mod-Mod R convention. The Mod-Mod R convention in Europe stems from the fact that in Europe bankruptcy laws make it difficult for borrows to file in many jurisdictions.

The fourth type is *Old R* or *Old Restructuring*. Old R clauses imply that there is no limit on the maturity of the deliverable obligation and no tranching in the auction postcredit event. Western European sovereign CDS typically contain the clause Old R.

Obviously CDS contracts with restructuring will have higher CDS spreads than the contracts of the same name, without restructuring. Note the key difference: in a bankruptcy or failure-to-pay-type credit event the price differences of deliverable bonds will be relatively small. In a restructuring, the deliverable bonds could have very different values depending on the maturity. The protection buyer has the option to deliver the cheapest bond and hence this option could be very expensive.

18.4.3 FIXED RECOVERY CDS

CDSs with fixed recovery rates are called *Fixed Recovery CDS* and are also known as *Digital CDS* or *Binary CDS*. In contrast to standard CDSs which require a valuation following a credit event (such as default), digital CDS simply specify payment of a fixed dollar payoff. At inception of the CDS, the payoff amount is agreed based on the severity of the default event. Remember that in a standard CDS the payoff is equal to the notional amount of the CDS minus the postdefault value of the insured assets.

Digital CDSs have several advantages. They have lower costs, more precise focus on the credit risk, and greater transparency. They will not be subject to difficulties associated with the auctioning process after a default event.

Essentially, the digital CDSs eliminate the random recovery rates associated with the conventional CDS. In fact, recent events suggest the possibility that the recovery rates associated with CDSs and synthetic CDOs have higher volatility than do recovery rates implied by corporate bond defaults. This led credit derivatives arrangers to offer a range of structural features to lessen this variability. One approach is a fixed recovery rate that is applied to all credit events, as in a digital CDS.

---

[21]Here there is a condition that on a restructuring event, the maximum maturity of the delivered bonds is no more than 30 months after the event, although there are exceptions.

However, CDS can also have disadvantages. There is a potential moral hazard associated with digital CDS since, in a digital CDS, dealers may have an incentive to more often call *soft* credit events, events that fall within the ISDA definition of default, but outside that of the rating agencies.

18.4.4 A NOTE ON THE ARBITRAGE EQUALITY

The simple contractual equation derived earlier suggests that we should have

$$\text{cds}_{t_0} - s_{t_0} = \text{cds}_{t_0} \tag{18.17}$$

under ideal conditions.

Yet, in reality, even when bonds are selling close to par, we in general observe

$$c_{t_0} - s_{t_0} < \text{cds}_{t_0} \tag{18.18}$$

Under these conditions instead of buying credit protection on the issuer, the client would simply short the bond and get in a receiver swap. This will provide the same protection against default, and, at the same time, cost less. So why would clients buy CDS instead? In fact, such inequalities can be caused by many different factors, briefly listed below.

1. CDS protection is "easy" to buy. On the other hand, it is "costly" to short bonds. One has to first go to the repo market to find such bonds, and repo has the mark-to-market property. With CDS protection, there is no such inconvenience.
2. Shorting a bond is risky because of the possibility of a short squeeze. If too many players are short the bond, the position may have to be covered at a much higher price.
3. Some bonds may be very hard to find when a sudden need for protection arises.
4. Also, as discussed earlier, a delivery option premium is included in the CDS rate.

These factors may cause the theoretical hedge to be different from the CDS sold to clients. Finally, it should be noted that when the probability of default becomes significant CDS dealers may suddenly move their prices out and stop trading. In more precise terms, the market moves from trading default toward trading the *recovery*. This is done by quoting the implied *upfronts* (UPF) instead of spreads.

18.5 CDS ANALYTICS

We will discuss the main quantitative tools used in CDS pricing and hedging using a 3-year, single-name CDS. The idea is to illustrate the way (risk-adjusted) *probability of default* is obtained and used. Also, we would like to determine the so-called *risky DV01* and *risky annuity* factors. These factors are used in obtaining hedge ratios during the CDS pricing.

Let cds_t be rate of a single-name CDS at time t. Let R denote the fixed recovery rate, and N be the notional amount. As usual $B(t_0, t_i)$ with $t_0 < t_i$ represents the default-risk free pure discount bond prices at time t_0. The bonds mature at times t_i and have par value of \$1. First we develop the notion of the default probability p_t at time t.

18.6 DEFAULT PROBABILITY ARITHMETIC

Modeling the occurrence and the timing of default events can be quite complex. The market, however, gravitates to some simple and tradable notions as we have seen before with options and implied volatility. In this section, we discuss the arithmetic behind the trading of default probability.

Imagine default as an event that happens at a random time τ, starting from some time t_0. How should we model the probabilities associated with such events?

We consider the *exponential distribution*, which can be described as a probability distribution describing the *waiting time* between events. In this case the event is the default and the waiting time is time until default. Thus we are dealing with modeling the random times until some events occur.

Let τ be the occurrence time of a default. The density function of an *exponentially distributed* random variable, $f(\tau)$, is given by

$$d(\tau) = \lambda e^{-\lambda \tau} \tag{18.19}$$

This is one way to model default occurrence and timing.

We see that as the parameter λ gets bigger, the probability that the event will occur earlier goes up. Hence this parameter governs when the default event is likely to occur. It is called the *intensity* associated with the random event.

The market does not like to use the exponential distribution. Taking the second-order Taylor series approximation of the $f(\tau)$ around the point $\tau_{t_0} = 0$ we get

$$f(\tau) \cong \lambda - \lambda^2 \tau + \frac{\lambda^2 \tau^2}{2} + o(\tau)^3 \tag{18.20}$$

Thus the probability that default will occur in a small interval Δ immediately following τ_{t_0} is approximately given by

$$f(\tau_{t_0}) \Delta \cong \lambda \Delta \tag{18.21}$$

Integrating the density $f(\tau)$ from 0 to some T we get the corresponding probability distribution function (PDF)

$$\begin{aligned} P(\tau < T) &= \int_0^T \lambda e^{-\lambda \tau} d\tau \\ &= 1 - e^{-\lambda T} \end{aligned} \tag{18.22}$$

The first-order Taylor series approximation of the PDF is then given by

$$1 - e^{\lambda \tau} \cong \lambda \tau \tag{18.23}$$

According to this, the probability that default occurs during a period Δ is approximately proportional to λ. The probability that the event will occur within 1 year is obtained by replacing T in the PDF by 1. We obtain

$$P(\tau \leq 1) \cong \lambda \tag{18.24}$$

Hence the parameter λ can be looked at as the approximate constant rate of default probability. The market likes to trade *annual* default probabilities, assuming that the corresponding probability is constant over various trading maturities. Obviously as time passes and quotes change, the corresponding default probability will also change. Hence it is best to use the subscript t_0 to denote a probability that is written in an instrument at time t_0, and use $p_{t_0} \cong \lambda_{t_0}$ as the default probability written in a contract at inception time t_0.

Looking at this from a different way: suppose $0 < \Delta$ is a small time interval. The default event is represented by a random variable d_t that assumes the values of 0 or 1, depending on whether during $[t, t + \Delta]$ the credit defaults or not.

$$d_t = \begin{cases} 0 & \text{No default} \\ 1 & \text{Default} \end{cases} \tag{18.25}$$

Now we make the assumption that the probability of default follows the equation

$$\text{Prob}(d_t = 1) \cong p_t \Delta \tag{18.26}$$

which says that the probability that the credit defaults during a small interval Δ depends on the length of the interval and on a parameter p_t called the "intensity." According to this, the probability that default occurs by time t will be given by

$$\text{Prob}(\tau < t) = 1 - e^{-\int_0^t p_s \, ds} \tag{18.27}$$

Assuming p_s is constant gives

$$\text{Prob}(\tau < \Delta) = 1 - e^{-p\Delta} \tag{18.28}$$

We have the Taylor series approximation around $t = 0$

$$e^{-p\Delta} = 1 - p\Delta + \frac{1}{2} p^2 \Delta^2 \cdots \tag{18.29}$$

or

$$1 - e^{-p\Delta} \cong p\Delta \tag{18.30}$$

According to this, the probability that the credit defaults during 1 year will equal p.

18.6.1 THE DV01'S

Working with a CDS of maturity $T = 3$ years, let p_{t_0} be the risk-adjusted default probability associated with a CDS contract of maturity 3 years. We will ignore the t_0 subscript and write this parameter simply as p. The CDS rate is cds_{t_0} and is observed in the markets. The δ_i is the time as a proportion of the year between two consecutive settlement dates.

Using this we can write the *initial* value of the CDS at inception time t_0 as follows. First, if during the 3 years the name does not default, the present value of the cash flows denoted by PV_{ND} will be

$$PV_{ND} = \left[B(t_0, t_1)(1 - p)\delta_1 \text{cds}_{t_0} + B(t_0, t_2)(1 - p)^2 \delta_2 \text{cds}_{t_0} + B(t_0, t_2)(1 - p)^3 \delta_3 \text{cds}_{t_0} \right] N \tag{18.31}$$

On the other hand, the name can default during years 1, 2, or 3. The expected present value of the accrued premium denoted by PV_{AP} if a case default event occurs will be

$$\begin{aligned} PV_{AP} = [&B(t_0, t_0 + \Delta_1)\Delta_1 p \times \text{cds}_{t_0} + B(t_0, t_1 + \Delta_2)\Delta_2(1 - p)p \times \text{cds}_{t_0} \\ &+ B(t_0, t_2 + \Delta_3)\Delta_3(1 - p)^2 p \times \text{cds}_{t_0}]N \end{aligned} \tag{18.32}$$

Three comments might help here. First, Δ_i are the parameters that determine the prorated spreads that will be received. If the name defaults right after the settlement date then that particular Δ_i will be close to 0. If the default is right before the next settlement date, Δ_i will be close to δ_i.[22] Assuming that the expected default time is the middle of the settlement period gives

$$
\mathrm{PV_{AP}} = \frac{1}{2}[B(t_0, t_0 + \Delta_1)\delta_1 p \times \mathrm{cds}_{t_0} + B(t_0, t_1 + \Delta_2)\delta_2(1-p)p \times \mathrm{cds}_{t_0}
$$

$$
+ B(t_0, t_2 + \Delta_3)\delta_3(1-p)^2 p \times \mathrm{cds}_{t_0}]N
$$

(18.33)

The $\frac{1}{2}$ comes from the expected default time during an interval $[t_i, t_{i+1}]$ as given by a uniform distribution on the interval $[0,1]$, the height of the uniform density being dt.

The expected value of the compensation for the cash payouts during default will be given by

$$
\mathrm{PV_D} = B(t_0, t_0\Delta_1)p(1-R)N + B(t_0, t_1 + \Delta_2)(1-p)p(1-R)N
$$

$$
+ B(t_0, t_2 + \Delta_3)(1-p)^2(1-R)N
$$

(18.34)

The expected payments and receipts should be equal under the *risk-adjusted* probability and the CDS rate cds_{t_0} must satisfy the equation

$$
\mathrm{PV_{ND}} = \mathrm{PV_D} - \mathrm{PV_{AP}}
$$

(18.35)

Generalizing from the 3-year maturity to n settlement dates, we can write

$$
\left[\sum_{i=1}^{n} B(t_0, t_i)(1-p)^i \delta_i \mathrm{cds}_{t_0}\right]N + \left[\frac{1}{2}\sum_{i=1}^{n} B\left(t_0, t_{i-1} + \frac{1}{2}\delta_i\right)(1-p)^{i-1} p\delta_i \mathrm{cds}_{t_0}\right]N
$$

$$
= \sum_{i=1}^{n} B\left(t_0, t_{i-1} + \frac{1}{2}\delta_i\right)(1-p)^{i-1} p(1-R)N
$$

(18.36)

Using Eq. (18.36), we now define two important concepts. The first is the *risky annuity* factor. Suppose the investor receives \$1 at all t_i. The payments will stop with a default. For that period the investor receives only a prorated premium. The risky annuity factor, denoted by $\tilde{A}$, is the value of this defaultable annuity. It is obtained by letting $\mathrm{cds}_{t_0}N = 1$ on the right-hand side of the expression in Eq. (18.36)

$$
\tilde{A} = \left[\sum_{i=1}^{n} B(t_0, t_i)(1-p)^i \delta_i\right] + \left[\frac{1}{2}\sum_{i=1}^{n} B\left(t_0, t_{i-1} + \frac{1}{2}\delta_i\right)(1-p)^{i-1} P\delta_i\right]
$$

(18.37)

---

[22]For example, $\frac{1}{4}$ if the CDS settles quarterly.

The second concept is the *risky* DV01. This is given by letting the cds_{t_0} change by one basis point.[23] The CDS DV01 is also known as a *credit delta* or *spread delta*. Often the traders use the $\tilde{A}$ as the risky DV01. Yet, there is a difference between the two concepts. The risky *DV01* is how much the value of the CDS changes if one increases cds_{t_0} by 0.0001. In general, this is not going to equal $\tilde{A}$, although it will be close to it depending on the shape of the curve and the change in spreads. The reason is following the relationship between the probability of default and the CDS spread cds_{t_0}[24]

$$(1 - R)p_{t_0} = \text{cds}_{t_0}. \tag{18.38}$$

If DV01 is the value of a stream of payments to a $d\text{cds}_{t_0} = 1$, then note that we have

$$dp = \frac{1}{(1 - R)} \tag{18.39}$$

In other words, the risky annuity factor $\tilde{A}$ will change due to *two* factors. Both cds_{t_0} and p_{t_0} would change. This means that the risky DV01 is a nonlinear function of cds_{t_0} and that there will be a convexity effect. Most market participants ignore this effect and consider the annuity factor $\tilde{A}$ as a good approximation of DV01. Yet, if the CDS spreads are moving in a volatile environment then the two sensitivity measures would differ.

18.6.2 UNWINDING A CDS

There are three essential ways to unwind a CDS transaction. The first two are similar to the transactions we routinely see in futures markets. The third is a bit different.

The most common way of unwinding a CDS position is to offset the position with another CDS or by getting in an offsetting position in the underlying assets.

A second way to unwind a CDS position is by terminating the contract and pay (or receive from) the counterparty the present value of the remaining CDS cash flows. The trick here is to remember that in a credit event, the cash flows would terminate early, hence, the calculation of the PV should take this into account. This is a good example for the use of risky annuities and risky DV01s.

Finally, assigning the contract to another dealer is the third way of terminating the contract. Below we discuss an example for the use of DV01 by terminating the contract before maturity. We will calculate the upfront cash payment or receipt.

---

[23]See Chapter 3 for an introduction to DV01. Credit delta measures the dollar present value changes for each basis point shift in the credit curve. The DV01 is quoted as if one basis point change is given by $d\text{cds}_{t_0} = 1$. In reality the basis point change would be 0.0001, but the market quotes it after multiplying by 10,000. As usual the decimal point is disliked by the trader during the trading process and eliminated from the quotes altogether.

[24]This relationship is also known as the *credit triangle*. The equation assumes that defaults occur at the settlement dates only. Otherwise it is only an approximation. See Hull et al. (2005). We will obtain this relationship in a slightly different context in Chapter 22.

EXAMPLE

Theoretically, one can unwind a CDS position by getting into an offsetting position or by receiving or paying the PV of the contract. However, in practice, if the CDS in question is substantially in-the-money, then it may be difficult to find a counterparty who will be willing to pay a substantial upfront for a position that can be attained with no upfront cash. This means that the original owner of the contract may have to accept a PV significantly lower to entice the counterparty to take over the position. In other words, there will be a haircut issue. For example:

An investor bought 3-year ABC protection on October 29, 2007 at 122 bps on $100 notional. Three days later on November 1, the spread is at 140 bps. This means that the original CDS will be in-the-money by an amount.

If there is no credit event this means an annuity of

$$\frac{(140 - 122)}{10,000} 100 \, m = \$180,000 \tag{18.40}$$

to be paid annually at times t_1, t_2, t_3. On the other hand, there may be a credit event and the coupon payments may stop. Assuming that this credit event can occur only at the end of the year there are three possible cash flow paths.

$$\{180, 180, 180\}, \{180, 180\}, \{180\} \tag{18.41}$$

The payments during a default event will approximately cancel each other out assuming that the same cheapest-to-deliver bond is involved during the delivery.

Suppose the recovery rate is 40%. In order to determine the present value (PV) of these cash flows we use risk-adjusted probabilities of default p obtained from the CDS spread at time t_0

$$p = \frac{142}{10,000(1 - 0.4)} = 0.023 \tag{18.42}$$

Clearly, we also need the corresponding risky zero bond prices, $\tilde{B}(t_0, t_i)$. Suppose they are given by

$$\tilde{B}(t_0, t_1) = 0.92 \tag{18.43}$$

$$\tilde{B}(t_0, t_2) = 0.86 \tag{18.44}$$

$$\tilde{B}(t_0, t_3) = 0.79 \tag{18.45}$$

Hence we can write the present value of the coupon payments if no default occurs,

$$\begin{aligned} DV01 = [((\tilde{B}(t_0, t_1) + \tilde{B}(t_0, t_2) + \tilde{B}(t_0, t_3))(1 - p)^3 \\ + (\tilde{B}(t_0, t_1) + \tilde{B}(t_0, t_2))(1 - p)^2 + (\tilde{B}(t_0, t_1))(1 - p))] \times 0.18 \times (100) \end{aligned} \tag{18.46}$$

where p^i, $i = 1, 2, 3$ are the probabilities associated with each annuity path. Plugging in the numbers we obtain

$$DV01 = 87.4775 \tag{18.47}$$

Note that we needed to use the default risky discounts and the corresponding DV01 because the default will change the timing of the cash flows.

18.6.3 UPFRONT PAYMENTS AND CDS UNWINDING

As we discussed earlier, in 2009 the single-name CDS also moved to upfront payments.

18.7 PRICING SINGLE-NAME CDS

The basic approach to pricing CDS is relatively simple, but there are real-world market conventions that have to be taken into account in practice. We first present a simplified CDS pricing example before reviewing market conventions and how they affect pricing.

18.7.1 A SIMPLIFIED CDS VALUATION EXAMPLE

We can use some of the tools reviewed above to value a CDS. Essentially the approach is to choose the CDS spread *cds* in such a way as to set the PV of the expected payments by the protection seller (the protection leg) to the PV of the payments made by the protection buyer (the premium leg). This was captured by Eq. (18.35). The two main inputs in this calculation are the survival probabilities and the discount rates. The survival probabilities should be risk-neutral probabilities and they can be obtained from bond prices or asset swaps discussed in Section 18.3. Of course, once market prices or CDS spreads are given, the survival probabilities can also be backed out from CDS markets, similar to how implied volatilities are backed out from option prices.

Let's consider the following example of a 5-year CDS with a notional amount $N = \$1$. We assume that default can occur in the middle of each year. The riskless rate is assumed to be 4% and the yield curve is flat. If one uses the default intensity approach outlined in Section 18.6 with a hazard rate of $\lambda = 5\%$, then one can calculate the probability of survival for each year from Eq. (18.22). The default intensity approach is also referred to as a reduced form approach to modeling default. We distinguish it from the structural approach to default modeling discussed in Chapter 19.

For the end of the first year, the equation leads to a probability of default (PD) of $P(\tau < T) = 1 - e^{-\lambda T} = 1 - e^{-5\% \times 1} = 0.0488$. Therefore the probability of survival is $1 - 0.0488 = 0.9512$. We can similarly calculate the probability of survival (PS) until the end of years 1, 2, 3, 4, and 5. This is shown in column 2 of Table 18.1. The second column shows the expected premium payment times the CDS spread *cds*. It is calculated as $PS \times cds \times$ Notional amount.

The discount factor in this example is based on continuous discounting. The discount factor for year 1 is $e^{-0.04 \times 1} = 0.9608$. The present value of the expected premium payment at the end of the first year in Table 18.1 is $0.9512 \times 0.9608 \times cds \times 1 = 0.9139 \times cds$. Note that this present value does not take into account the accrual payments (Eq. (18.32)).

We also need to add the accrued premiums in the event of a default. Table 18.2 shows their calculation in the first five columns. The probability of default during year $T + 1$ is calculated as the probability of survival until the end of year T minus the probability of survival until the end of year $T + 1$. For example of date $T = 1.5$, we calculate the probability of survival until year $T = 1.5$

Table 18.1 Simplified CDS Valuation Example (PV of Premium Leg)

Time (Years)	Probability of Survival (PS)	Expected Payment	Discount Factor	PV of Expected Payment
1	0.9512	0.9512 × cds	0.9608	0.9139 × cds
2	0.9048	0.9048 × cds	0.9231	0.8353 × cds
3	0.8607	0.8607 × cds	0.8869	0.7634 × cds
4	0.8187	0.8187 × cds	0.8521	0.6977 × cds
5	0.7788	0.7788 × cds	0.8187	0.6376 × cds
Total				3.8479 × cds

Table 18.2 Simplified CDS Valuation Example (PV of Protection Leg and Accrued Payments)

Time (Years)	Probability of Default (PD)	Expected Accrual Payment	Discount Factor	PV of Expected Accrual Payment	R	Expected Payoff	PV of Expected Payoff
0.5	0.0488	0.0244 × cds	0.9802	0.0239 × cds	0.4	0.0293	0.0287
1.5	0.0464	0.0232 × cds	0.9418	0.0218 × cds	0.4	0.0278	0.0262
2.5	0.0441	0.0221 × cds	0.9048	0.0200 × cds	0.4	0.0265	0.0240
3.5	0.0420	0.0210 × cds	0.8694	0.0182 × cds	0.4	0.0252	0.0219
4.5	0.0399	0.0200 × cds	0.8353	0.0167 × cds	0.4	0.0240	0.0200
				0.1006 × cds			0.1208

to be $0.9512 - 0.9048 = 0.0464$. Since we expect to receive the CDS premium *cds* at date $T = 1.5$, the expected accrual payment is $0.464 \times \mathbf{0.5} \times \mathbf{cds}$. If we multiply the expected accrual payments times the discount factor we obtain a present value of all expected accrual payments of $0.1006 \times \mathbf{cds}$.

If we add the expected accrual payments and the expected premium payments, we obtain a value of $3.8479 \times \mathbf{cds} + 0.1006 \times \mathbf{cds} = 3.9585 \times \mathbf{cds}$.

Finally, lets consider the protection leg payments. The last three columns of Table 18.2 show the calculations. We assume a standard recovery rate of $R = 40\%$. The expected payoff for date $T = 0.5$, e.g., $(1 - R) \times N \times \mathbf{0.0488} = \mathbf{0.6} \times \mathbf{1} \times \mathbf{0.0488} \times \mathbf{0.9802} = \mathbf{0.0287}$. One can use programs such as Matlab or VBA to calculate the spread *cds* that sets the present value of the premium leg and the protection leg equal to each other so that $3.9585 \times \mathbf{cds} = 0.1208$. If we do that we find a value of $cds = 0.0306$ or 306 basis points. In other words, given the parameters in this example a CDS spread of 306 basis points would set the present value of the premium leg payments equal to the protection leg payments.

18.7.2 REAL-WORLD COMPLICATIONS

The above example is very much a simplified and unrealistic treatment of CDS valuation. It ignores many real-world complications such as coupon, day-count conventions, interest rate curves, and upfront payments. CDS premiums are paid typically on a quarterly basis and since the "CDS big bang", CDS include upfront payments. The choice of the benchmark yield curve and riskless rate is important in practice. The swap yield curve or the LIBOR curve could be used as the basis for the discount factors, but as we will discuss in Chapter 24 the LIBOR curve has some issues due to counterparty risk. White (2013) provides detailed CDS pricing formulae that take into account the ISDA model and real-world conventions.[25] Since the CDS market has moved from the spread convention for single-name contracts to a fixed coupon and upfront payments, following standardization it is important for market participants to be able to match the upfront payment amounts and be able to translate upfront quotations to spread quotations and vice versa in a standardized manner. Therefore, ISDA created the CDS Standard model and made the underlying source code for CDS calculations freely available at http://www.cdsmodel.com/cdsmodel/.

18.7.3 LESSONS FROM THE GFC FOR CDS PRICING

Historically financial innovation has played an important role in economic growth, funding economic activity and risk sharing, but occasionally it also turns sour. Derivatives are one example of financial innovation and derivatives debacles occur with an disturbingly high frequency.[26] Such debacles have led to unexpected losses on the part of unsophisticated or misinformed counterparties such as corporates, municipalities, or banks, but sometimes they lead to systemic risk. The insurance company AIG was nearly brought down by losses on CDS positions during the GFC. AIG had a AAA credit rating and was therefore considered a safe counterparty to sell CDS protection. However, the size of its portfolio grew disproportionately large and reached around $500 billion dollars in notional value in 2008. AIG also provided insurance against the default of Lehman Brothers which famously went under in September of 2008. One of the mistakes that AIG made was that, in contrast to the reserves it held for its traditional insurance business, it never correctly reserved for the CDS that it was selling. The irony was that CDS were first developed within a unit of AIG Financial Products which was started in 1987.

18.8 COMPARING CDS TO TRS AND EDS

To appreciate the nuances of CDS it is helpful to compare them to different but related products.

18.8.1 TOTAL RETURN SWAPS VERSUS CREDIT DEFAULT SWAPS

We encountered total return swap (TRS) applications in Chapter 4 and discussed equity swaps as an example of TRS principle. A TRS trades *default, credit deterioration*, and *market risk* simultaneously. Therefore, unlike CDS, TRS are not a pure credit derivative.

It is instructive to compare them with CDSs. In the case of a CDS, a protection buyer owns a bond issued by a credit and would like to buy insurance against *default*. This is done by making

---

[25]The document is available at: http://developers.opengamma.com/quantitative-research/Pricing-and-Risk-Management-of-Credit-Default-Swaps-OpenGamma.pdf.

[26]See Jacque (2010) for a comprehensive and very readable history of such debacles.

constant periodic payments during the maturity of the contract to the protection seller. It is similar to, say, fire insurance. A constant amount is paid, and if during the life of the contract the bond issuer defaults, the protection seller compensates the protection buyer for the loss and the contract ends. The compensation is done by paying the protection buyer the face value of, say, 100, and then, in return, accepting the delivery of a deliverable bond issued by the credit. In brief, CDSs are instruments for trading defaults only.[27]

A TRS has a different structure. Consider a bond or any arbitrary risky security issued by a credit. This security makes two types of payments. First, it pays a coupon interest. Second, there will be associated capital gains (appreciation in asset price) and capital losses (depreciation in asset price), which include default in the extreme case. In a TRS, the *protection seller* pays any depreciation in the asset price during periodic intervals to the protection buyer. Default is included in these payments, but it is not the only component. In general, assets gain or lose value for many reasons, and this does not mean the issuer has defaulted or will default. Nevertheless, the protection seller will compensate the protection buyer for these losses as well.

However, in a TRS, the protection seller's payments will not stop there. The protection seller will also make an additional payment linked to LIBOR plus a spread.

The *protection buyer*, on the other hand, will make periodic payments associated with the appreciation and the coupon of the underlying asset. Normally, asset prices appreciate and pay coupons more often than decline, but this is compensated by the LIBOR plus any spread received.

The TRS structure is equivalent to the following operation. A market participant buys an asset, S_t, and funds this purchase with a LIBOR-based loan. The loan carries an interest rate, L_{t_i}, and has to be rolled over at each t_i. The market participant is rated A and has to pay the credit spread d_{t_0} known at time t_0. S_t has periodic (coupon) payouts equal to c. The market participant's net receipts at time t_{i+1} would, then, be the following:

$$\left(\Delta S_{t_{i+1}} + c\right) - \left(L_{t_i} + d_{t_0}\right)S_{t_0}\Delta \tag{18.48}$$

where $\Delta S_{t_{i+1}}$ is the appreciation or depreciation of the asset price during the period, $\Delta = [t_i,\ t_{i+1}]$. The c is paid during Δ. The payments are in-arrears.

A TRS swap is equivalent to this purchase of a risky asset with LIBOR funding. Except, in this particular case, instead of going ahead with this transaction, the market participant can simply sign a TRS with a proper counterparty. This will make him or her a protection seller. Banks may prefer these types of TRS contracts to lending to market practitioners.

18.8.2 EDS VERSUS CDS

Equity default swaps (EDS) are strictly speaking equity derivatives, but they have similarities with CDS. EDS have been marketed by dealers with mixed success thus far.[28] The EDS emulate CDS. In a CDS, the reference asset is a bond, and the protection is provided against the default of other credit event. In an EDS, the reference asset is a company's stock, and protection is provided against a dramatic decline in the company's stock price.

---

[27]Maturity is typically 5 years in the case of most corporate credits.
[28]JP Morgan pioneered the first EDS in 2003.

An EDS is "exercised" when a company's stock price S_t falls below a prespecified barrier H. Normally this barrier will be 30—40% below the current stock price level. If $S_t < H$ happens, then an "equity event" similar to a credit event takes place.

If a credit event occurs, the EDS terminates, and the protection seller makes a specified payment to the protection buyer. The payment is calculated as

$$\text{notional amount}(1 - \text{recovery rate}) \tag{18.49}$$

Note, however, that hitting an equity barrier H, no matter how distant it is, is more likely than a credit event. After all, a company's stock price can fall dramatically without the company going bankrupt. An EDS shares similarities with CDS and with options. An EDS is similar to a deep out-of-the-money, long-dated American digital put. However, there is a difference in that the option premium is paid in installments and these stop when the option is exercised. A better analogy is between the EDS and the CDS. The reason is that typically when a CDS is triggered the stock price of the underlying typically dramatically collapses and might fall to zero at the same time as the reference asset, that is the bond, drops in value. One of the advantages of EDS over CDS is that the trigger events are less ambiguous. It is generally clear whether a stock price has fallen below a certain threshold, while in a CDS there is some ambiguity about what constitutes a credit event or default, as we will see in our discussion of sovereign CDS below, for example.

Again, similar to CDSs, dealers can, and have tried to, put together CDOs of the EDSs. Such CDOs can get rated. Normally, however, it is difficult for dealers to get a big enough tranche of such a CDO rated higher than A.[29]

> It is worth adding that the EDS structure is very similar to a deep out-of-the-money put option written on the stock. In both cases there is a barrier, namely the strike price in the case of the option, such that the option buyer receives a cash payment. The major difference is perhaps the expiration date of the EDS which can be much longer.

18.9 SOVEREIGN CDS

The cash flow engineering approach that we used to create synthetic CDS in Section 18.3 applies to any reference asset. In a corporate CDS, the reference asset is a corporate bond. In a sovereign CDS, the reference asset is sovereign debt. There are some special considerations associated with sovereign CDS which we want to highlight and therefore we discuss sovereign CDS in this section. In economic terms, sovereign CDS are also becoming more important. In terms of gross notional amounts the *sovereign* CDS market in 2012 was 11% of the whole CDS market, but the sovereign CDS market has been growing while the *single-name* CDS market has been declining since the GFC. We discuss reasons for the decline in the single-name CDS market in Chapters 21 and 22.

---

[29]Rating agencies normally stress test instruments by putting them through scenarios. There are three basic scenarios: *normal*, *stress*, and *crash*. It is very difficult for a CDO of EDS to stay highly rated during a stock market crash scenario.

A sovereign CDS can in principle protect investors against losses on sovereign debt if a country restructures or defaults on its debt. Sovereign CDS have four main uses: hedging, speculation, basis trading, and credit risk management.

First, sovereign CDS are also used as *proxy hedges* for other types of credit risk, such as financial and nonfinancial corporate bonds. An investor that holds Italian government bonds can buy an Italian sovereign CDS to hedge the credit risk in the Italian government bond.

Second, sovereign CDS can be used to speculate. A market participant could buy or sell sovereign CDS on a *naked* basis, that is without an offsetting position in the underlying reference asset. A hedge fund, for example, with a view that Japanese sovereign credit ratings will improve could buy a sovereign CDS on Japanese government debt. CDS differ from traditional insurance in that in principle a purchase of a CDS does not need to own the reference asset. Expressing a view about the evolution of a country's credit rating could be achieved by using other financial markets such as cash bond markets or interest rate futures, but these alternatives reflect other types of risk in addition to sovereign credit risk.

Naked CDS positions are banned in some markets. As a result of the rising influence of the sovereign CDS market, concerns have been voiced about whether speculative uses of sovereign CDS could be destabilizing. Since 2012, as part of the EU regulation "Short Selling and Certain Aspects of Credit Default Swaps" naked CDS on the debt of European Economic Area countries are banned. Liquidity in the sovereign CDS market had started to decrease ahead of the ban, but proving that the ban *caused* liquidity to decrease is made difficult by various other events and policy announcements at the same time. Some market participants have closed even *covered* CDS positions following the ban because of ambiguity in the hedge rules in the regulation. There are alternative instruments that can be used to hedge sovereign risk such as some corporate CDS contracts and bond futures but these proxy hedges are likely to be more expensive and less precise.

Third, as we saw in Section 18.3, CDS, together with IRSs and FRNs can be used to replicate the cash flows of the underlying asset. Therefore, sovereign CDS can be used for basis trading, that is exploiting mispricing between the sovereign CDS and the underlying government bond. If the sovereign CDS spread is narrower than the credit spread of the underlying debt, that is if the basis is negative, arbitrageurs may be able to profitably buy the debt and buy CDS protection.

Fourth, sovereign CDS are important credit risk management tools and sovereign CDS premia are widely used as market indicators of credit risk. Figure 18.6 shows an example of how CDS spreads can be used as risk and default probability indicators. The figure shows a screenshot from Bloomberg which shows the implied 1-year default probability in percent for a range of reserve currency and nonreserve currency sovereign credits on July 2, 2014. The default probabilities are derived from the CDS spreads with are reported in the columns to the right. The 5-year CDS spread for Greece is 455 bps or 4.55% and the implied annual default probability for Greece on that day is 9.5%. Note however that this probability is lower than the highest probability that had been reached historically, which according to the screenshot was around 27%.

Dealer banks act as market makers and dominate the sovereign CDS market on the buy and the sell side. The banks' exposure to sovereign risk arises from their direct holdings of sovereign debt as well as counterparty credit risk associated with their derivatives trades with countries. Historically, sovereigns did not post collateral on a mark-to-market basis related to their OTC derivatives positions in interest rate and cross-currency derivatives for example. As a result dealer

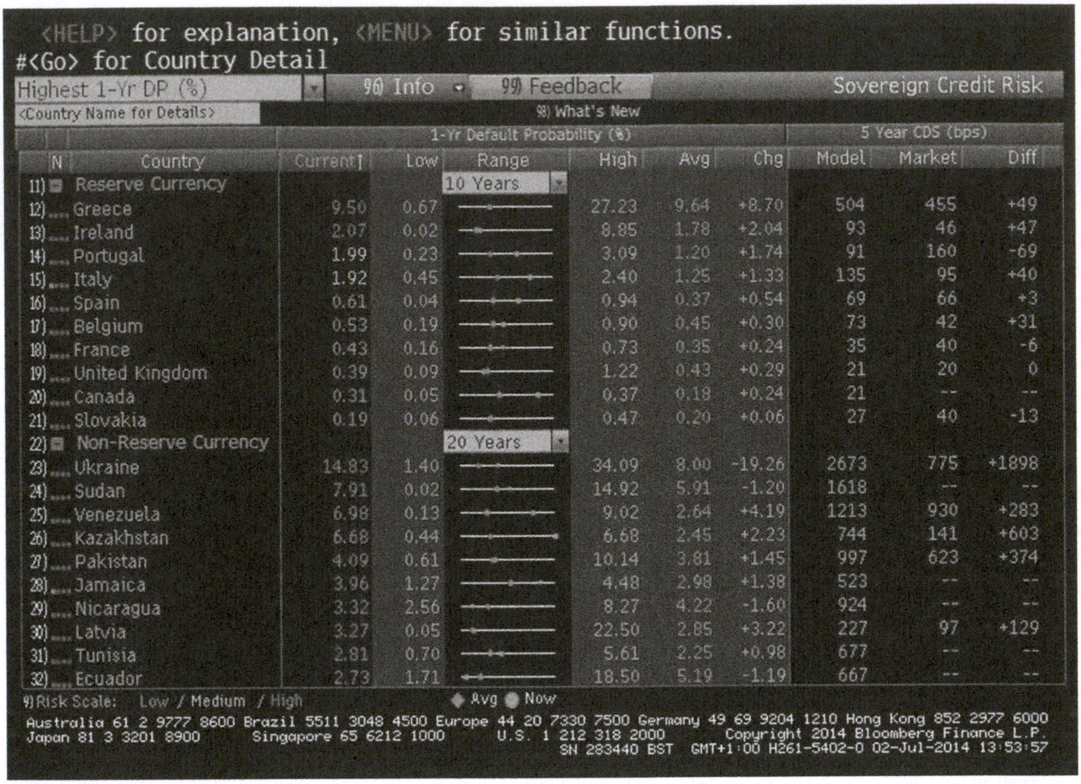

<HELP> for explanation, <MENU> for similar functions.
#<Go> for Country Detail

Highest 1-Yr DP (%) | 96) Info | 99) Feedback | Sovereign Credit Risk

<Country Name for Details> | 98) What's New

N	Country	1-Yr Default Probability (%) Current	Low	Range	High	Avg	Chg	5 Year CDS (bps) Model	Market	Diff
11)	Reserve Currency			10 Years						
12)	Greece	9.50	0.67		27.23	9.64	+8.70	504	455	+49
13)	Ireland	2.07	0.02		8.85	1.78	+2.04	93	46	+47
14)	Portugal	1.99	0.23		3.09	1.20	+1.74	91	160	-69
15)	Italy	1.92	0.45		2.40	1.25	+1.33	135	95	+40
16)	Spain	0.61	0.04		0.94	0.37	+0.54	69	66	+3
17)	Belgium	0.53	0.19		0.90	0.45	+0.30	73	42	+31
18)	France	0.43	0.16		0.73	0.35	+0.24	35	40	-6
19)	United Kingdom	0.39	0.09		1.22	0.43	+0.29	21	20	0
20)	Canada	0.31	0.05		0.37	0.18	+0.24	21	--	--
21)	Slovakia	0.19	0.06		0.47	0.20	+0.06	27	40	-13
22)	Non-Reserve Currency			20 Years						
23)	Ukraine	14.83	1.40		34.09	8.00	-19.26	2673	775	+1898
24)	Sudan	7.91	0.02		14.92	5.91	-1.20	1618	--	--
25)	Venezuela	6.98	0.13		9.02	2.64	+4.19	1213	930	+283
26)	Kazakhstan	6.68	0.44		6.68	2.45	+2.23	744	141	+603
27)	Pakistan	4.09	0.61		10.14	3.81	+1.45	997	623	+374
28)	Jamaica	3.96	1.27		4.48	2.98	+1.38	523	--	--
29)	Nicaragua	3.32	2.56		8.27	4.22	-1.60	924	--	--
30)	Latvia	3.27	0.05		22.50	2.85	+3.22	227	97	+129
31)	Tunisia	2.81	0.70		5.61	2.25	+0.98	677	--	--
32)	Ecuador	2.73	1.71		18.50	5.19	-1.19	667	--	--

9) Risk Scale: Low / Medium / High ◆ Avg ● Now

Australia 61 2 9777 8600 Brazil 5511 3048 4500 Europe 44 20 7330 7500 Germany 49 69 9204 1210 Hong Kong 852 2977 6000
Japan 81 3 3201 8900 Singapore 65 6212 1000 U.S. 1 212 318 2000 Copyright 2014 Bloomberg Finance L.P.
SN 283440 BST GMT+1:00 H261-5402-0 02-Jul-2014 13:53:57

FIGURE 18.6

Sovereign CDS and implied default probabilities.

banks have counterparty risk exposure on these OTC contracts in which sovereigns are counterparties and sovereign CDS can be used to hedge the counterparty risk.

Argentina is an interesting name to be associated with the CDS market because of the large size of the default and the ongoing legal proceedings regarding Argentina's debt. The following deals with Argentina, where the CDS rate was around 40% for 1 year around the default period.

EXAMPLE

One-year Argentina credit default swap mid-levels hit 4,000 bp late last week, though the highest trade in the sovereign is thought to have been a one-year deal at 2,350 bp early in the week.

Derivatives market-makers were cautiously quoting default swap prices on an extremely wide bid/offer spread (the two-year Argentina mid rose to around 3,900 bp), but mostly concentrated on balancing cash market hedges, which did not prove easy.

Dealers who have sold protection also consulted their lawyers to plot tactics in the event that Argentina defaults, or restructures its debt. It is likely that more than US$1bn of credit default protection on Argentina has traded in the last few years, which could result in the biggest default swap payout yet, if there is a *clear-cut default or debt restructuring. There is plenty of scope for disagreement on whether or not the payout terms of swaps have been met, however, depending on how any debt restructuring is handled by the Argentine authorities.*

Pricing default swaps when a payout trigger could be hours away is an art, not a science. Late last week traders were working from the closing price on Thursday of Argentina's FRBs of 63.5, which was the equivalent of 3,060 bp over LIBOR, then adding a 30—40% basis for the theoretical risk of writing a default swap, as opposed to the asset swap value of a bond trade. For much of this year, traders have been using a default against asset swap basis of around 10% of the total spread for deals in Latin American sovereigns.

(Thomson Reuters IFR, July 2001)

Following the events described in the reading above, Argentina defaulted on a total of USD 93 billion of external debt on December 26, 2001. The following article summarizes the lengthy legal proceedings and impact on CDS markets in the subsequent years up to 2014 and provides valuable lessons.

EXAMPLE

ISDA Asked If Event Clause Triggered on Argentina Debt Swaps

The International Swaps & Derivatives Association said it was asked to rule whether a clause in credit-default swaps on Argentina has been triggered after the government said it won't make bond interest payments.[. . .]

Argentina is negotiating with creditors, who refused to accept restructured debt after its default in 2001, following a U.S. Supreme Court ruling last week requiring full payment. The ruling blocks interest payments on restructured bonds until holdout creditors are paid. President Cristina Fernandez de Kirchner's government said it's unable to pay all claims.[. . .]

The government owes $900 million in interest on June 30 for bonds issued as part of restructurings in 2005 and 2010. The New York judge's order requires Argentina to pay creditors, including billionaire Paul Singer's Elliott Management Corp., $1.5 billion on defaulted debt before it's allowed to make the interest payments.

Fernandez says the nation could owe as much as $15 billion if forced to pay all holders of defaulted bonds on the same terms, which would deplete more than half of its foreign reserves.[. . .]

The government defaulted on a record $95 billion debt in 2001, replacing the defaulted bonds with new ones at a discount in two restructurings. Holdouts have fought for full payment on the defaulted bonds. Argentina had vowed never to pay the holdouts, calling them "vultures" and refusing to pay U.S. court judgments in their favor.

Argentina's debt is the world's most expensive to insure, according to data compiled by Bloomberg. It cost $3.4 million upfront and $500,000 annually to protect $10 million of Argentina's debt for five years, signaling a 64 percent chance of default within that time, according to CMA.

There were 2,602 credit-default swaps contracts covering a net $906 million of Argentina's debt outstanding as of June 13, according to the Depository Trust & Clearing Corp.

(Bloomberg, June 23, 2014)

The legal wrangling regarding the payment of creditors by Argentina in 2013—2014 led to gyrations in the CDS spread and implied default probability. Figure 18.7 shows the annual probability of default for Argentina implied by 5-year CDS spreads and under the assumption of a 40% recovery rate. In the

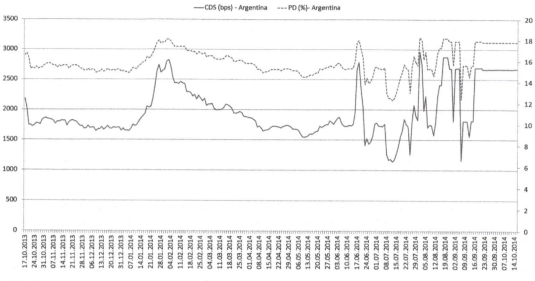

FIGURE 18.7

Argentina probability of default and CDS spread.

figure, the dashed line and y-axis on the right show the default probability while the solid line and y-axis on the left show the CDS spread. In August 2014, ISDA determined that Argentina's failure to pay its sovereign bonds was a credit event, triggering a payout to holders of credit default swap.

The sovereign debt stress in the euro area following the GFC has again raised concerns about the reliability and usefulness of sovereign CDS. Between June 2005 and April 2013, there were 103 CDS credit events but only two sovereign CDS credit events in which settlements were publicly documented. The March 2012 Greece debt exchange was an example that highlighted the potential complexity of sovereign CDS credit event triggering and settlement. About €200 million in Greek government bonds (GGBs) were exchanged against new GGBs making this the largest sovereign restructuring event in history. European governments had concerns about the effect of a Greek debt default on European banks. Therefore, governments attempted to delay the triggering of the associated credit event. These actions however undermined confidence in the CDS market and raise questions about the effectiveness of sovereign CDS. The effectiveness of CDS protection depends on two main things. The first is whether the event responsible for the losses triggers the CDS payout. The second is whether the payout offsets the losses if the CDS is triggered. The April 2013 IMF Global Financial Stability report notes that it is a fortunate coincidence that the new GGBs were trading at about 22% of par going into the CDS settlement, which was the same price that the old GGBs were trading at before the exchange. This implied that the settlement resulted in sovereign CDS payouts roughly in line with losses incurred in the debt exchange. Since this result for a fortunate coincidence for debt and CDS holders, the associated uncertainty led to a rethinking of alternative settlement mechanisms such as delivering a package of new instruments in proportion to the instruments that they replace.

18.10 CONCLUSIONS

This chapter is only a very brief introduction to this important class of credit derivatives. We saw that CDS play a key role in completing the methodology of financial engineering. We developed a contractual equation for CDS and applied it to negative basis trades. We discussed real-world complications related to recovery clauses and drew lessons about the importance of credit event triggers and settlement following recent sovereign defaults and sovereign CDS triggers.

SUGGESTED READING

Several recent books deal with this new sector. For a good theoretical background and some empirical work, **Duffie and Singleton** (2003) is very useful. **Bielecki and Rutkowski** (2001) is more mathematically involved, but excellent. The **Barclays Capital Standard Corporate CDS Handbook** 2010 provides a good summary of the main changes in the CDS market after the GFC. The monthly Risk publication, **Credit**, is also good reading on market activity. We only covered a simple example of CDS pricing in this chapter. For a detailed description of CDS pricing under real-world market conventions see **White** (2013). **Hull** (2014) and **O'kane** (2008) are also good references. The classic reading here is **Merton** (1974). **Giesecke** (2002) is a good survey on pricing. The reader should also consult the very good source, **Schönbucher** (2003). The April 2013 **IMF Global Financial Stability Report** reviews recent issues in sovereign CDS markets.

EXERCISES

1. This exercise deals with value-at-risk calculations for credit portfolios. Using the data on a corporate financial statement, answer the following questions:
   a. How would you calculate the default probabilities?
   b. How can one obtain the migration matrix for a credit?
   c. How can one obtain the joint migration probabilities for the relevant credits in a bank's portfolio?

2. You are given two risky bonds with the following specifications:
   Bond A
   a. Par: 100
   b. Currency: USD
   c. Coupon: 10
   d. Maturity: 4 years
   e. Callable after 3 years
   f. Credit: AA−
   Bond B
   a. Par: 100
   b. Currency: EUR
   c. Coupon: LIBOR + 78 bp
   d. Maturity: 5 years
   e. Credit: AAA

You will be asked to transform Bond A into Bond B by acquiring some proper derivative contracts. Use cash flow diagrams and be precise.

- Show how you would use a currency swap to switch into the right currency.
- Show how you would use an IRS to switch to the needed interest rate.
- Is there a need for using a swaption contract? Can the same be accomplished using forward caps and floors?
- Finally, show *two* ways of using credit derivatives to switch to the desired credit quality.

3. Consider the CDS pricing example in Section 18.7. Assume that hazard rate is 3% instead of 5% but all other input parameters remain the same. Calculate the value of the CDS by finding the CDS spread *cds* that sets the expected present value of the protection leg payments equal to the expected present value of the premium leg payments as in the text.

4. In the CDS pricing example in the text we assumed a hazard rate to derive the CDS spread and to price the contract. Now assume that the hazard rate is unknown, but assume that you can observe a CDS spread of 200 bp in the market for this credit. The recovery rate is still 40%, the maturity is 5 years, the riskless rate is 4% and the yield curve is flat. Assumptions about the timing of defaults and accrued premia are unchanged. Calculate the implied hazard rate.

5. **a.** Consider the following quote from Reuters:

    The poor correlation between CDS and cash in Swedish utility Attentat (VTT.XE) is an anomaly and investors can benefit by setting up negative basis trades, says ING. 5-yr CDS for instance has tightened by approx. 5 bp since mid-May while the Attentat 2010 is actually approx. 1 bp wider over the same period.

    Buy the 2010 bond and CDS protection at approx midas + 27 bp. (MO)
    - **i.** Display this position on a graph with cash flows exactly marked.
    - **ii.** Explain the logic of this position.
    - **iii.** Explain the numbers involved. In particular, suppose you have 100 to invest in such a position, what would be the costs and expected returns?
    - **iv.** What other parameters may have caused such a discrepancy?

6. Consider the following news from Reuters:
    1407 GMT [Dow Jones] LONDON—According to a large investment bank investors can boost yields using the following strategies:
    - **a.** *In the strategy, sell 5-yr CDS on basket of Greece (9 bp), Italy (8.5 bp), Japan (4 bp), Poland (12 bp), and Hungary (16 bp), for 34 bp spread. Buy 5-yr protection on iTraxx Europe at 38 bp to hedge.*
        Trade gives up 4 bp but will benefit if public debt outperforms credit.
    - **b.** *To achieve neutral or positive carry, adjust notional amounts—for example in the first trade, up OECD basket's notional by 20% for spread neutral position.*
    - **c.** *Emerging market basket was 65% correlated with iTraxx in 2005, hence use the latter as hedge.*
        - **i.** Explain the rationale in item (a). In particular, explain why the iTraxx Xover is used as a hedge.

    ii. Explain how you would obtain positive carry in (b).

    iii. What is the use of the information given in statement (c)?

7. Consider the following news from Reuters:

HVB Suggests Covered Bond Switches

0843 GMT [Dow Jones] LONDON—Sell DG Hyp 4.25% 2008s at 6.5 bp under swaps and buy Landesbank Baden-Wuerttemberg(LBBW) 3.5% 2009s at swaps-4.2 bp, HVB says. The LBBW deal is grandfathered and will continue to enjoy state guarantees; HVB expects spreads to tighten further in the near future.

    a. What is a German Landesbank? What are their ratings?

    b. What is the logic behind this credit strategy?

    c. Can you take the same position using CDSs? Describe how.

8. Explain the logic behind the two following strategies using cash flow diagrams.

    a. *WestLB mortgage Pfandbriefe trade too tight. Sell the WestLB 3% 2009s at 5.4 bp under swaps and buy the zero risk weighted Land Berlin 2.75% 2010s at 2.7 bp under. (TMA)*

    b. The following quote deals with implied forward rates in the credit sector. Using proper diagrams explain what the trade is.

Implied forward CDS levels look high because shorter-dated CDS are currently too cheap to 5-year, says BNP Paribas. Using the iTraxx Main curve as reference gives a theoretical 3-year forward curve that shows 6-month and 1-year CDS both at 60 bp. "In 3-years time, we find that 6-month and 1-year CDS are very unlikely to be trading above 60 bp."

Take advantage through the 3-5-10-year barbell, buying iTraxx 3-year at 20.75 bp for EUR20M, selling iTraxx 5-year at 38 bp for EUR25M, and buying iTraxx 10-year at 61.25 bp for EUR7M.

The trade has a yearly carry of EUR32,000 for a short nominal exposure of EUR2M.

ENGINEERING OF EQUITY INSTRUMENTS AND STRUCTURAL MODELS OF DEFAULT

CHAPTER OUTLINE

19.1 INTRODUCTION

Fixed-income instruments involve *payoffs* that are, in general, known and "fixed." They also have set *maturity* dates. Putting aside the credit quality of the instrument, fixed-income assets have relatively simple cash flows that depend on a known, small set of variables and, hence, risk factors. There are also well-established and quite accurate ways to calculate the relevant term structure. Finally, there are several liquid and efficient fixed-income derivatives markets such as swaps, forward rate agreements (FRAs), and futures, which simplify the replication and pricing problems existing in this sector.

There is no such luxury in equity analysis. The underlying asset, which is often a stock or a stock index, does not have a set maturity date. It depends on a nontransparent, idiosyncratic set of risks, and the resulting cash flows are complex. The timing and the size of future cash flows may not be known. There are also complex issues of growth, investment, and management decisions that further complicate the replication and pricing of equity instruments. Finally, relatively few related derivatives markets are liquid and usable for a replication exercise.

Yet, the general principles of option pricing and replication can be applied to value equity. In this chapter, we extend the methods introduced earlier to equity and equity-linked products. We also discuss the engineering applications of some products that are representative of this sector.

First, we show how the equity of a company can be viewed and priced as an option on the assets of the firm with a strike price equal to the present value of the debt. This approach is based on the so-called Merton (1974) *structural model of default*. It is an elegant and useful application of financial engineering principles and importantly it allows us to link the pricing of credit and equity instruments. Similar to the Black–Scholes model that we have used in earlier chapters to either price an option based on a given set of input parameters or to back out an implied volatility based on observed market price of the option, the Merton (1974) model can be applied in three main different ways. First, given information about a firm's assets, its capital structure and debt value we can determine the model implied price of the equity. Second, in the *equity-to-credit* approach we can plug observed market prices of equity into the model to back out the implied credit spread and probability of default for the debt. The third application uses observed market prices for the equity and observed credit spreads (from bond or CDS markets) to back out an implied volatility for the equity.

These wide-ranging uses have important practical uses. The following example highlights the practical relevance of structural models of default in today's markets.

EXAMPLE

[...] leading investors say Merton models are now so frequently used that they are actually driving the credit market. Take Barclays Global Investors (BGI) in San Francisco, with $780 billion under management. BGI has been using Merton models − an in-house model as well as those available on the market − for a number of years, but its head of US fixed income, Peter Wilson, says the model is now becoming so prevalent with other investors that it is starting to dictate which way credit markets will move.

(Source: Risk Magazine, November 1, 2003, 'Barra joins crowded market for Merton models', www.risk.net/1497515/)

As markets for credit derivatives developed and market prices for credit spreads became more widely available the Merton model became more widely used in the financial sector and many banks integrated their equity and fixed-income divisions to a greater extent and located them on the same trading floor. Previously equity and fixed-income divisions were often found on different trading floors. The underlying intellectual foundation for this reorganization lies in structural models of default. Furthermore, these models can be used for so-called capital structure arbitrage strategies. Such strategies use the Merton model to identify mispricing between credit and equity markets and take positions in equity, bond, or CDS markets to exploit it.

Given the complexity of equity markets and their risks, as we will see, there are however also limitations associated with the basic Merton model and its applications.

In the previous chapter, we had discussed credit risk, the probability of default of a risky bond, and credit default swaps. We had seen that credit spreads and therefore an implied probability of default can be backed out from both corporate bond spreads as well as CDS spreads. The Merton model provides us with a *third* source of information for the probability of default, since the model can be used to calculate it if the market value of equity and other parameters about the capital structure are observed.

The simple Merton model has been extended significantly over the last 40 years and industry versions such as the KMV or the CreditGrades model are widely used. We discuss how hedge funds pursue so-called capital structure arbitrage strategies that are based on structural models of default.

The second purpose of this chapter is to discuss the engineering of some popular equity instruments such as convertible bonds and *hybrid equity products.*

The plan of this section is as follows. First, we discuss how financial engineering and option pricing principles can be applied to value a company's equity. We discuss applications and limitations of the model. Second, we consider *convertible bonds* and their relative, warrant-linked bonds. Convertible bonds are another example of bonds with embedded options. In Chapter 17, we saw the first example of a bond with an embedded option in the form of callable bonds. In the next chapter, we will discuss structured products including equity structured products and so-called reverse convertibles.

We begin by applying option replication and financial engineering principles to a company's equity.

19.2 WHAT IS EQUITY?

Equity holders have a claim to the residual value of a company after creditors have been paid. The traditional approach to equity valuation is to view the price of equity as the present discounted value of future cash flows. This approach is covered in most standard corporate finance textbooks and, therefore, we will briefly summarize it without discussing it further.[1] However, this approach, which discounts cash flows under the real-world probability measure and uses a risky discount rate, does not require any financial engineering. It turns out that there is also a second alternative perspective which views equity as an option on the assets of the firm and which can be used to calculate its value under the risk-neutral measure. The second interpretation is the one that represents an elegant application of financial engineering principles such as option replication and pricing that we discussed in previous chapters and therefore we focus on it in this chapter. From a practitioner perspective, there is another important reason why the option replication approach to equity valuation is a useful framework. The reason is that it creates a link between the value of equity and bonds and thus between equity and bond markets. In fact, with the existence of liquid credit derivative markets, it can even be used to connect equity, bond, and CDS markets. The reason for these linkages is that equity can be viewed as an option on the assets of the firm with a strike price equal to the present value of the debt. This leads to a closed-form expression that links equity and bond prices as well as equity prices and credit spreads. In this *equity-to-credit* approach, equity market values can be used to back out default probabilities of the debt.

Bonds are contracts that promise the delivery of known cash flows, at known dates. Sometimes these cash flows are floating, but the dates are almost always known, and with floating rate instruments, pricing and risk management is less of an issue. Finally, the owner of a bond is a *lender* to the institution that issues the bond. This means a certain set of covenants would exist.

Stocks, on the other hand, entitle the holder to some *ownership* of the company that issues the instrument.[2] Thus, the position of the equity holder is similar to that of a partner of the company, benefiting directly from increasing profits and getting hurt by losses. One of the differences compared to a partnership is however that shareholders of public companies have limited liability.

---

[1]Damadoran (2012) provides one of the most comprehensive overviews of different equity valuation approaches including the Dividend Discount Model and multiples valuation.

[2]Not all stocks are like this. There is Euro-equity, where the asset belongs to the bearer of the security and is not registered. In this case, the owner is anonymous, and, hence, it is difficult to speak of an owner. Yet, the owner still has access to the cash flows earned by the company, although he or she has no voting rights and, hence, cannot influence how the company should be run. This justifies the claim that the Euro-stock owner is not a "real" owner of the company.

19.3 EQUITY AS THE DISCOUNTED VALUE OF FUTURE CASH FLOWS

The traditional approach before option pricing models were developed was to view equity as the present discounted value of cash flows to equity holder. The stock or share prices corresponded to the equity value per share. These approaches postulate that expected future cash flows properly discounted should equal the current price S_t. Here it is important to note that the discount rate that is used to discount cash flows is risky and is used under the real-world probability measure. One variant of this approach is the Dividend Discount Model, for example, which takes dividends to be the cash flows that equity holds receive.

The corporation has future dividends *per share* denoted by DPS_t. We can buy one unit of S_t to get the title for future dividends. We then use the *real-world probability* P and the real-world discount rate k_t that apply to the dollar dividends paid out by this company to write an equation that is similar to the representation for the coupon bond price:[3]

$$S_t = E_t^P \left[\sum_{i=1}^{\infty} \frac{DPS_{t+i}}{\prod_{j=1}^{i}(1 + k_{t+j})} \right] \qquad (19.1)$$

It is worth re-emphasizing that, in this expression, we are using the real-world probability. Thus, the relevant discount rate will differ from the risk-free rate:

$$k_t \neq r \qquad (19.2)$$

How is k_t obtained in practice? Various models such as the Capital Asset Pricing Model (CAPM), the Arbitrage Pricing Theory (APT), and others have been proposed. These models form the theoretical foundation for valuation and practitioners use different single or multifactor models to calculate the expected return and discount rate in practice.

19.4 EQUITY AS AN OPTION ON THE ASSETS OF THE FIRM

19.4.1 MERTON'S STRUCTURAL MODEL OF DEFAULT

In Chapter 9, we introduced the Black—Scholes option pricing model to price an option on a stock. Can the option pricing approach also be used to price equity? It turns out that the answer is affirmative. Merton (1974) proposed a seminal *structural model of default* in which the equity of a firm is viewed as an option on the assets of the firm with a strike price equal to the (face) value of the debt. As the name indicates, the typical application of the model does not lie in the pricing of the equity, but rather in backing out information about the debt (including the default probability) from observed market prices for equity. As a result structural models of default have a large range of important practical applications including the modeling, pricing, and forecasting of default risk.[4] The following example illustrates the practical relevance of the *structural* approach.

---

[3] In the following formula, t is in years.
[4] Structural models differ from reduced-form models in that in the latter model the default process is modeled directly without a model of the capital structure of the firm.

EXAMPLE

The Tequila crisis, the Asian financial crisis, Brazil, Russia, Enron... it seems most large credit events are followed by critical post mortems on the effectiveness of credit rating agencies. Investors seeking better default indicators than produced by rating analysts have turned to a variety of sources, not least of which is the family of so-called Merton model tools, perhaps the most well known of which is KMV's.

(Risk Magazine, February 1, 2002, 'Battle of the supermodels', www.risk.net/1506552)

The GFC provided further recent examples of corporate defaults that *preceded* downgrades by rating agencies. Merton-type models, of course, are not infallible and like all models they have their strengths and weaknesses. One application of structural models of default is in the form of so-called *capital structure arbitrage*. As we will see capital structure arbitrage consists of exploiting potential misvaluation between the debt and the equity issued by a given company.

The Merton (1974) model is an excellent example of the financial engineering and option valuation concepts that we discussed in previous chapters. In the previous chapter, we discussed credit risk, the probability of default of a risky bond and credit default swaps. We saw that credit spreads and therefore an implied probability of default can be backed out from both corporate bond spreads as well as CDS spreads. The Merton model provides us with a *third* source of probability of default information. All three sources of default probabilities use market prices and therefore provide us with *risk-neutral* default probabilities. These probabilities can be used to calculate expected cash flows in a risk-neutral world with credit risk. We saw in Chapter 12 that according to the fundamental theorem of asset pricing, a derivative's price is the discounted expected value of the future payoff under the risk-neutral measure. Since structural models of default tie together the values of bond and equity markets they allow financial engineers and market participants to exploit potential arbitrage and relative value opportunities between equity, bond, and CDS markets. We will discuss examples of such arbitrage strategies below.

The Merton model has spawned a large literature that has extended the basic model in many different directions.[5] We will first describe the basic model and then discuss applications before pointing to recent extensions. Since it is the simplest case, we review the Merton model in the context of the valuation of a risky zero-coupon bond, but Merton (1974) also considered the cases of coupon-bearing and callable debt.

19.4.2 PAYOFFS TO BOND AND EQUITY HOLDERS

A standard company balance sheet is a summary of the financial position of a company at a given point in time and consists of three main parts: assets, liabilities, and equity. Equity holders therefore have a residual claim on the firm after creditors have been paid. Figure 19.1a illustrates a stylized balance sheet. The stock market allows us to observe the market value of equity. Therefore the balance sheet in the figure can be interpreted as reflecting *market* as opposed to *book* values of assets, debt, and equity. For (medium and large) companies that publicly issue bonds we can observe market values for the debt in the form of bond prices.[6]

---

[5]We refer to the Merton (1974) model and related structural models as *the Merton model* in this chapter.
[6]If traded bonds are not available in practice, then the book value of debt can be used as an imperfect approximation of the market value.

(a) Firm's balance sheet (market values)

Assets	Liabilities
	Debt
	Equity

Use traded bond
and equity values
if available

(b) Payments to debt and equity holders under different scenarios

	Scenario 1: $A_T > F$	Scenario 2: $A_T < F$
Debt holders	F	A_T
Equity holders	$A_T - F$	0

(c) Payoff profile of debt and equity positions

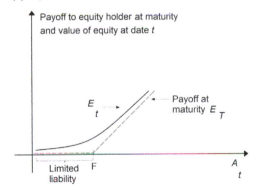

Payoff to equity holder at maturity
and value of equity at date t

E_t

Payoff at
maturity E_T

Limited F
liability

A_t

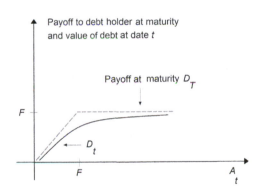

Payoff to debt holder at maturity
and value of debt at date t

F

Payoff at maturity D_T

D_t

F

A_t

(d) Decomposition of payoff to debt holders into short put option position

		Scenario 1: $A_T > F$	Scenario 2: $A_T < F$
	Debt holders	F	A_T
Portfolio of	Riskless debt	F	F
	-put option	0	$-(A_T - F)$

FIGURE 19.1

Equity as option on firm's assets.

19.4.2.1 Payments to debt and equity holders under different scenarios

Consider a firm that has only issued two types of securities: equity and zero-coupon bonds, that is the debt. How can we apply option valuation techniques to the equity of such a firm that issues only one type of debt in the form of a zero-coupon bond with face value F? The value of the firm at date t is equal to the (market) value of the firm's assets and therefore we denote it by A_t. Let's first consider the relationship between the value of the firm, denoted by A_t, the value of the equity E_t, and the value of the debt D_t. Debt holders are promised a payment of F at maturity date T. Now, consider two possible scenarios:

1. $A_T > F$: If A_T, the assets of the firm at maturity T exceed the promised value of the payment to bond holders F, then equity holders will be paid the residual amount $A_T - F$ and the equity is worth $E_T = A_T - F$. In this scenario, there is no default.
2. $A_T < F$: If A_T, the assets of the firm at maturity T are below the amount promised to debt holders, the firm defaults. In this case, there is no residual amount remaining for equity holders and the equity is worth zero E_T. The equity holders have limited liability and as a result their claim cannot be worth less than zero. These two scenarios are illustrated in Figure 19.1b.

We can plot the payoffs to equity, E_T, and the payoff to bond holders, D_T, against the value of the firm's asset at maturity T, A_T. This is represented by the dashed lines in Figure 19.1c. As we can observe the payoff profile of equity resembles that of a call option on the assets of the firm, A_T. The figure also indicates the limited liability feature. It implies that equity holders can walk away from the firm's debt in exchange for a payoff of zero. The strike price is the face value of the zero-coupon debt. The payoff profile for the bond or debt holders, in contrast is reminiscent of a short put position. Since before maturity at date $t < T$ the options also have time value, the curved black lines plot the value of the equity and the debt at dates before maturity. The value of the equity and debt has some time value before maturity. The black lines represent the value of equity E_t and debt D_t before maturity.

Figure 19.1c illustrates the basic intuition for why in the Merton model, the equity is priced as a European call option on the firm's assets while the bond holders' claim is equivalent to risk-free debt in addition to a short put position. The payoff of the creditors at date T can be written as:

$$D(A_T, T) = \min(A_T, F) = F - (F - A_T)^+. \tag{19.3}$$

This is illustrated in Figure 19.1d. The decomposition of the payoff to the bond holders shows that the bond holders are short a put option written on the assets of the borrowing firm with a strike price equal to the face value of the debt F. The put option implicit in the creditor payoff is sometimes referred to as a *default* or *credit put*.

We can also apply the put–call parity relationship, introduced in Chapter 11, to derive the value of the firm and the value of risky debt. Remember that the put–call parity relationship states that a portfolio of a call minus a put is equal to the value of a portfolio containing the underlying minus a riskless bond.

$$C_t - P_t = S_t - e^{-r(T-t)}K \tag{19.4}$$

In the context of the Merton model the underlying assets (S_t) are the firm's assets, that is $S_t = A_t$. The face value of the bond is F, that is $K = F$, so that the put–call parity relationship becomes

$$C_t - P_t = S_t - e^{-r(T-t)}K = A_t - \text{Riskless Bond}(t) \tag{19.5}$$

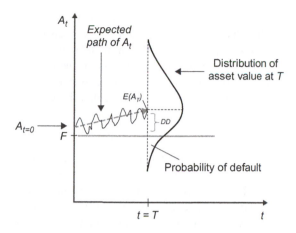

FIGURE 19.2

Distribution of firm value in Merton model.

Equation (19.5) can be rewritten in terms of the assets A_t of the firm as follows:

$$A_t = \text{Riskless Bond}(t) + C_t - P_t \tag{19.6}$$

We also know that the value of the assets equals the (risky) debt plus the equity:

$$A_t = E_t + D_t \tag{19.7}$$

Since the underlying assets in Eq. (19.6) are the assets of the firm, we can set Eqs. (19.6) and (19.7) equal to each other to get

$$\text{Riskless Bond}(t) + C_t - P_t = E_t + D_t \tag{19.8}$$

Now we note that equity in the Merton model is a call option on the assets of the firm ($C_t = E_t$) so that we can write Eq. (19.8) as follows:

$$\text{Riskless Bond}(t) + E_t - P_t = E_t + D_t \tag{19.9}$$

and therefore we subtract C_t from both sides to get

$$\text{Riskless Bond}(t) - P_t = D_t \tag{19.10}$$

This leads to the following contractual equation:

Value of risky debt	=	Riskless bond	−	Value of put option on firm's assets	(19.11)

19.4.2.2 Bond and equity value in the Merton model

One of the convenient results of the Merton model and subsequent papers using this methodology is that closed-form solutions for the value of debt and equity can be obtained. To do so, we need to make an assumption about the evolution of the firm's assets over time and its relationship to the firm's liabilities. Figure 19.2 illustrates one such possible path followed by the firm's assets A_t over time.

The dashed blue line represents the path of the firm's expected asset value over time. The curve at time T represents the probability distribution of A_t at time T. The distribution illustrates the fact that there is uncertainty about the evolution of A_t. The Merton model makes specific distributional assumptions about the evolution of assets over time.[7] The face value of the debt is represented by F. Default occurs if the assets A_T at time T fall below the face amount of the debt F. The difference between $E(A_T)$ and F is called *distance to default (DD)*. We assume that the value of the assets of the firm A_t follows a geometric Brownian motion.

$$dA_t = A_t(\mu dt + \sigma dW_t) \tag{19.12}$$

where σ is the asset volatility, μ is the expected continuously compounded return on A_t, and W_t represents a geometric Brownian motion. The basic model makes several restrictive assumptions which we discuss later and which have been relaxed by subsequent research. Here we highlight one of the assumptions that is implicit in our discussion above, namely that default can only occur at date T and there are no distress costs associated with default, which implies that bond holders receive A_T in case of default. This assumption allows us to model the option feature as a European option with maturity date T.

As we saw in Figure 19.1c, the firm's equity value at time T is given by

$$E_T = \max(A_T - F, 0) \tag{19.13}$$

We can apply the Black–Scholes European call option pricing formula[8] to derive the value of the firm's equity at date t. If we make the assumption that one can trade the firm's assets A_t, then, from Chapter 12, we know that $e^{-rt}A_t$ is a martingale under the risk-neutral measure $\tilde{P}$ which leads to the following:

$$E_t = E^{\tilde{P}}[e^{-r(T-t)}(A_T - F)^+] = BSCall(A_t, F, r, \sigma, T - t) \tag{19.14}$$

The Black–Scholes formula for a European call is

$$E_t = A_t N(d_1) - Fe^{-r(T-t)}N(d_2) \tag{19.15}$$

where

$$d_1 = \frac{ln(A_t/F) + (r + (1/2)\sigma^2)(T - t)}{\sigma\sqrt{T - t}}$$

and

$$d_2 = d_1 - \sigma\sqrt{T - t}$$

The value of the debt at date t is

$$D_t = A_t N(-d_1) + e^{-r(T-t)}FN(d_2) \tag{19.16}$$

An important insight from the Merton model is that the spread between the risky debt and the risk-free debt is the value of the put option. Credit spreads are therefore a function of the input

---

[7]The original Merton (1974) paper uses V_t, instead of A_t, to denote the firm value and assumes that V_t followed a diffusion process.
[8]The Black–Scholes formula is also known as the Black–Scholes–Merton formula since it was developed by Fisher Black, Robert Merton and Myron Scholes in 1973.

parameters of the model such as the maturity of the debt, the leverage, the strike price of the put, and the business risk of the firm.

Let the yield on a riskless bond be represented by r. Furthermore, if we denote the yield on a risky zero-coupon bond by R, then the price of a risky bond at date t as a function of the yield and the face value is

$$D_t = e^{-R(T-t)}F \qquad (19.17)$$

After taking logs and solving for R, the yield R can be written as

$$R = -\left(\frac{1}{T-t}\right)ln\left(\frac{D_t}{F}\right) \qquad (19.18)$$

We can plug the expression for D_t from Eq. (19.16) into Eq. (19.18) to obtain the yield R of a risky bond:

$$R = -\frac{1}{T-t}ln\left[\frac{A_t N(-d_1) + e^{-r(T-t)}FN(d_2)}{F}\right] \qquad (19.19)$$

$$= -\frac{1}{T-t}ln\left[e^{-r(T-t)}\left(\frac{A_t}{e^{-r(T-t)}F}N(-d_1) + N(d_2)\right)\right] \qquad (19.20)$$

$$= -\frac{1}{T-t}\left\{-r(T-t) + ln\left[\left(\frac{1}{L}N(-d_1) + N(d_2)\right)\right]\right\} \qquad (19.21)$$

$$= r - \frac{1}{T-t}\left(ln\left[\left(\frac{1}{L}N(-d_1) + N(d_2)\right)\right]\right) \qquad (19.22)$$

where $L = e^{-r(T-t)}F/A_t$ is a measure of leverage. We can rewrite Eq. (19.22) in terms of the credit spread $R - r$:

$$R - r = -\frac{1}{T-t}\left(ln\left[\left(\frac{1}{L}N(-d_1) + N(d_2)\right)\right]\right) \qquad (19.23)$$

The expression in Eq. (19.23) is important and useful since it shows us how based on the input parameters one can calculate the implied credit spread from the Merton model. This equity implied credit spread can be compared to credit spreads from bond or CDS prices. The Merton model implies that as share prices increase, the credit spread decreases. Do we observe such a negative relationship in practice? Figure 19.3 provides an illustrative example. It plots the credit default spread for Goldman Sachs against the Goldman Sachs share price over the period 2007–2014 on a weekly basis. It is clear that the relationship is negative as predicted by the Merton model. This example does, of course, not represent proof of the practical usefulness of the Merton model in explaining credit spreads. Recent comprehensive academic studies find some support for the Merton model and for its input parameters being able to explain credit spreads, but these studies also suggest that factors outside the Merton model are important for the determination of credit spreads.

It is *a priori* not clear which financial market should price credit risk correctly, but a market participant can take the view that either the equity, the bond, or the CDS market prices credit risk correctly and then take long and short positions in two of these markets to try to exploit potential relative mispricing. We will describe the underlying rationale and examples of such trades in detail in one of the subsequent sections.

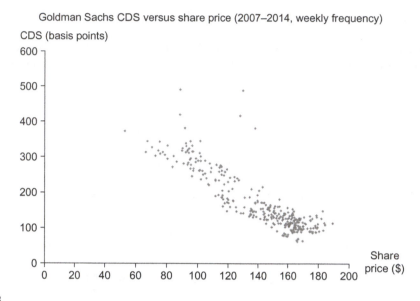

FIGURE 19.3

Goldman Sachs CDS versus share price (2007–2014).

19.4.3 PROBABILITY OF DEFAULT

One of the uses of Merton models in practice is as an indicator of default risk. The Merton model can be used to calculate the option implied risk-neutral probability of default, that is the risk-neutral probability that $A_T < F$. The risk-neutral probability of default is the probability that the put finishes in-the-money. The probability of default can be calculated as follows:

$$\text{Probability of Default} = P(A_T < F) = N(-d_2) \tag{19.24}$$

In the risk-neutral world, the drift of the firm value process is the risk-free rate r. Note that under the stochastic process in Eq. (19.12) for A_t the real-world probability of default is the probability that $A_T < F$ given that A_t has a drift μ and a volatility σ. The drifts μ and r may be different. Therefore, the risk-neutral probability of default in Eq. (19.24) may differ from the actual probability of default. On the one hand, if $\mu > r$, the firm value in the real world drifts upwards faster than in the risk-neutral world, which implies that the real-world probability of default would be lower than the risk-neutral probability. On the other hand, if $\mu < r$, the real-world probability of default would be higher than the risk-neutral probability of default. It is possible to derive closed-form solutions for the actual probability of default from the risk-neutral probability. Typically $\mu > r$ and therefore the real-world probability is below the risk-neutral probability. We can interpret the higher risk-neutral probability as consisting of the real-world probability plus a risk premium for uncertainty regarding the timing and magnitude of default. Note that in the basic Merton model, there is no uncertainty considering the timing which is not realistic.

The following example illustrates the application of the Merton model.

EXAMPLE

We assume that the value and volatility of assets is given and our objective is to calculate the current market value of equity and debt as well as the implied credit spread and default probability. Table 19.1 shows the input parameters.

Table 19.1 Input Parameters for Merton Model Example

Parameter	Value
Asset/firm value (A_t)	100
Face value of the debt (F)	75
Maturity (T)	1
(Riskless) interest rate (r)	2%
Asset volatility (σ)	20%

If we apply Eq. (19.15) to calculate the price of the equity E_0 at date $t = 0$ we obtain a value of $E_0 = 26.84$. Similarly, according to Eq. (19.16) the market value of debt is 73.16. The implied yield to maturity of the risky debt based on Eq. (19.18) is 2.49% and the resulting credit spread is 49. The value of the riskless debt is 73.51 and the difference between the value of the riskless debt and the risk debt is the value of put, which is 0.36. The default probability is given by Eq. (19.24) and is equal to 3.46% in this example.[9]

19.4.4 APPLICATION OF THE MERTON MODEL AND EQUITY-TO-CREDIT APPROACH

As Eq. (19.15) shows, structural models such as the Merton model are based on the unobserved variable A_t. For publicly traded companies, the share price and therefore the equity value is observable in the market. The usual approach to estimating the value of the firm's assets A_t and their volatility σ_A is to use Eq. (19.13) together with another equation that links the value and volatility of equity to the value and volatility of assets. If we apply Itô's formula to Eq. (19.15), we obtain:

$$E_t \sigma_E = A_t \sigma_A \frac{\partial E}{\partial A} = N(d_1) \sigma_V V_t \qquad (19.25)$$

Equations (19.15) and (19.25) represent a pair of simultaneous equations that can be solved for A_t and σ_A. Once we calculate A_t and σ_A it is also possible to calculate the probability of default $N(-d_2)$ based on Eq. (19.24). We can also obtain the market value of the debt which is equal to $A_t - E_t$. By comparing the current model implied market value to the present value of the promised debt payment, we can calculate the expected loss (EL). The end of chapter exercises provide a numerical example of an application of this approach.

One of the first firms to use Merton-type models in practice to estimate probabilities of default of a debt issuer from share price data and the firm's capital structure data was KMV (now Moody's

---

[9]The end of chapter exercises provide further examples.

KMV) in the 1980s. This approach is sometimes also referred to as *equity-to-credit* and its advantage over credit ratings is that it is based on real-time data from liquid markets. This makes it particularly useful for *predicting* defaults. Default probabilities derived from credit ratings issued by credit rating agencies on the other hand are updated relatively infrequently and there are many examples when large publicly listed companies were only downgraded by credit rating agencies *after* the companies had filed for bankruptcy and defaulted on their debt. Default probabilities can also be obtained from other sources. Four other potential sources of such information are (i) credit ratings, (ii) historical default rates, (iii) credit spreads implied by publicly traded bonds, and (iv) credit spreads implied by CDS. Merton-type models should therefore be viewed as complementing rather than replacing other models and approaches when investment decisions are taken. Although Merton-type models are popular in credit markets, market participants still continue to use other fundamental and quantitative tools instead of purely relying on the outputs of structural models. Sometimes structural models are viewed as useful for understanding the sources of risk, while other classes of models can be useful for predicting market values of bonds and equities. Merton-type models are used by many institutional asset managers to understand how, for example corporate bonds will perform, in different asset volatility scenarios. However, ultimately decisions are made by taking into account information from all models.

19.4.5 ASSUMPTIONS OF THE MERTON MODEL AND EXTENSIONS

The basic Merton model that we reviewed above has been extended significantly by both practitioners and academics over the last 40 years. Many of the more recent and advanced structural models implemented in practice are proprietary and there is no public information about their underlying assumptions. Published academic research has, however, also developed more complex structural models of default that fit the stylized empirical facts better than the basic model.

19.4.5.1 Extensions in published academic research

As part of the introduction of the basic Merton model, we mentioned a range of simplifying assumptions that the basic model is based on. For reasons of space, the list of assumptions provided above is incomplete. Sundaresan (2013) provides a comprehensive review of the basic Merton model, its assumptions and how subsequent structural models of default have relaxed these assumptions to fit stylized empirical facts better. One of the assumptions that is used in almost all papers in this literature is that trading in assets takes place continuously. Clearly this is an unrealistic assumption for all firms and all assets. While some very liquid exchange traded assets can be viewed to a first-order approximation as exhibiting continuous trading, the physical assets of many firms are very illiquid and no market may exist for some of them. In fact, the value of the firm's assets and its parameters are not even directly observable. The extensions that have been incorporated in richer, more complex structural models of default include (a) agency costs, (b) moral hazard, (c) bankruptcy codes, (d) renegotiations, (e) investments, (f) taxes, (g) jumps in the stochastic process followed by firm value, and (h) the dependence of default probabilities on the business cycle.

19.4.5.2 Evolution of structural models of default in industry

Although the details of models used by practitioners are not known, some published accounts indicate that several of the extensions such as (a)–(h) mentioned in the previous section have been incorporated into such models. Even the original KMV model developed in the 1980s contained some significant differences compared to the basic Merton model since it scaled the distance to default obtained from the basic Merton model to actual default probabilities using a proprietary default database.[10] Many software vendors, banks, and other private sector institutions developed their own Merton-type models. While the focus of early versions of the KMV model was on default probabilities, the development of credit derivatives markets in the 1990s led the industry to develop structural models of default to estimate bond and CDS spreads. One such example was *CreditGrades* launched by the *Riskmetrics* group (now part of MSCI) in 2002. The original CreditGrades model was based on the Black–Cox model, which relaxes some of the assumptions of the basic Merton model to allow default to occur prior to time T if the value of the firm's assets hit a predetermined default barrier. Models such as CreditGrades can generate theoretical (model-implied) CDS spreads and various *Greeks*. In practice, to produce realistic credit spreads and default probabilities, contemporary models have been extended to include such variables as convertible debt, preferred equity, and so on. One of the challenges of applying standard Merton-type models is that the capital structure of a company can suddenly change. It may be possible to use the model to state whether the equity is correctly priced relative to the credit for the current capital structure but it is almost impossible to reliably forecast how the capital structure will change in the future. For example, it is very difficult to predict when the capital structure will change due to a management decision. Consider the situation of a capital structure arbitrage fund that has a long credit position and a short equity position on the company. What would happen if the company unexpectedly carries out a leveraged buyout? The likely result is that the position will lose money as a result. Some arbitrageurs trade equity and credit to express a view about expected future changes in a company's capital structure.

The paragraph above discusses the usefulness of Merton-type models for capital structure arbitrage strategies. We will introduce these strategies in the following section.

19.5 CAPITAL STRUCTURE ARBITRAGE

19.5.1 EXAMPLES OF CAPITAL STRUCTURE ARBITRAGE TRADES

Capital structure arbitrage is a class of strategies used by market participants such as credit hedge funds and certain banks. The basic idea behind the strategy is to go long one security in a company's capital structure while at the same time going short another security in the same company's capital structure. Examples of capital structure arbitrage trades include

1. "Long the stock of company ABC and short the bonds of company ABC".
2. "Long equity of a company ABC and short the CDS on the debt issued by ABC".
3. "Go long subordinated debt and short senior bonds".

How could a structural model of default be used to provide a decision rule for the implementation of the trade in (1)? One could use a Merton-type model to calculate the implied credit spread and probability of default based on company ABC's stock price. Call this spread $s_{Merton, \ implied}$.

---

[10]See http://www.moodysanalytics.com/About-Us/History/KMV-History for further details.

If company ABC has publicly traded bonds one can use bond prices to calculate the bond implied credit spread $s_{Bond,\ implied}$.

If $s_{Merton,\ implied} < s_{Bond,\ implied}$, then $\varepsilon = s_{Bond,\ implied} - s_{Merton,\ implied} > 0$. Assume one takes the view that the equity market correctly prices the default risk while the bond market overestimates the risk. Furthermore, if one believes that the two markets should converge in their assessment of the default risk, then one could take a long bond and a short stock position. Implicit in this approach is the view that the pricing error ε declines (to 0) over time.

The signal for such trades can be the discrepancy between the Merton model implied bond spread derived from equity prices and the CDS spread.

The example in (2) above is based on a comparison of equity and CDS implied information and not equity and bond implied information as in the example in (1). As described in the previous chapter, CDS markets can be used to calculate a CDS implied credit spread $s_{CDS,\ implied}$ for company ABC. Consider the scenario where $s_{Merton,\ implied} > s_{CDS,\ implied}$. Now consider the case where a market participant believes that the CDS market is pricing the credit risk correctly while the equity market is not. One way to express this view is through a long position in the stocks of ABC while buying a CDS of company ABC. If, over time, the equity and the CDS market converge in their assessment of the credit risk, this position would make a profit.

The following reading illustrates that such strategies can be applied not just to individual companies but also to indices. In the example, the CDS market is seen to have interpreted the macroeconomic environment differently from the equity market.

EXAMPLE

JP Morgan Chase and Morgan Stanley's CDS index for Europe, Trac-x Europe, narrowed from 140 basis points in October 2002 to 105 bp during the war, while the Dow Jones Eurostoxx 50, which was at 2,275 in early October 2002, reached a low of 1,928 in March. The apparent discrepancy was largely due to the fact that a lot of European companies were reducing their debt levels at this time [...]. The credit market took the view that corporate debt reduction was a stronger indicator of company performance than the negative macroeconomic indicators that were pummelling stock levels.

(Risk Magazine, November 1, 2003, 'Barra joins crowded market for Merton models', www.risk.net/1497515)

The above reading illustrates that the narrowing of CDS spreads from 140 to 105 bp reflected a view in the CDS market that credit risk diminished as European companies reduced debt levels. Since equity holders are residual claimants to the assets of the company, such an interpretation should coincide with a rise in equity prices *if* the equity market agreed with the CDS market. But as the example illustrates the two markets disagreed in this case and the equity market actually fell from 2275 to 1928.

19.5.2 USING THE MERTON MODEL TO PROVIDE SIGNALS FOR CAPITAL STRUCTURE ARBITRAGE TRADES

The above examples illustrated the basic idea behind capital structure arbitrage trades. We now want to look in more detail at how the Merton model can be used to provide signals for such trades. Equation (19.23) above shows how we can back out implied credit spreads from observed share prices. For a given company, which we call XYZ, we could plot the Merton model implied credit spreads for a range of different share prices. Figure 19.4a shows such a relationship.[11] The axes in the figure correspond to those from the output of the basic Merton model. We see that a share price of around $30 implies a credit spread of 5% for XYZ. The relationship is downward sloping since a share price increase implies a lower probability of default and thus a lower credit spread. Conversely, if share prices decrease the probability of default increases.

Now imagine that for company XYZ we gather credit spreads from bond prices of CDS markets. For simplicity we assume that the data are from bond markets, but an analogous argument could be made if the data were from CDS markets. If we used CDS market data, any potential position to exploit the mispricing would involve buying or selling CDS, while below we explain how one can go long or short bonds to exploit the mispricing.

We superimpose these *observed* spreads on the relationship in Figure 19.4a. The result is shown in Figure 19.4b. The circles represent different hypothetical market observations. As the figure shows, the fit is quite good. Assume that on a given trading day a market participant plots the figure and observes share prices of around $25 and a credit spread of around 10% (1000 bp) as indicated by point P and the horizontal line passing through P. What would be the theoretical credit spread implied by the Merton model for a share price of $25? Figure 19.4b suggests that the Merton model would imply a credit spread closer to 5% (500 bp) if the share price is $25. This is illustrated by point P' and the dashed red line. Does this mean that the debt is overvalued or undervalued relative to the equity? How could one exploit any potential mispricing?

First we note that according to Figure 19.4b, point P corresponds to an observation that lies above the line indicating the theoretical relationship. This implies that the actual spread of 1000 bp (point P) is higher than the theoretical spread of 500 bp (point P'). The actual spread can be interpreted as being too high. If the actual spread is derived from bond data, this means that bonds are too cheap. If the spread is from CDS prices it means the CDS spread is too high relative to the model. From the perspective of the equity value, we observe that the actual share price is $25, whereas the Merton model would predict that the share price should be closer to $14 if the credit spread was 1000 bp, which is illustrated by the dashed black line. From this perspective, the equity is overvalued. This example shows that the Merton model can be used to identify *relative* mispricing. To exploit the relative mispricing it is not enough to only take one position in bonds, CDS, or equities, respectively. A long position in bonds and a short position in stocks would allow exploiting a potential correction in the relative mispricing between the bonds and equities of company XYZ. Therefore to exploit the observed mispricing we would go short XYZ equity and go long XYZ bonds in the hope that stock prices will fall and bond yields fall as predicted by the Merton model. The following example illustrates however that capital structure arbitrage trades can also lose money even if it is based on long/short positions:

---

[11]The figure is just for illustration and is not based on actual market data. In the end of chapter exercises, we will derive a similar relationship using real-world share prices and balance sheet data.

EXAMPLE

A number of capital structure arbitrageurs were caught out when stock levels plummeted in the run-up to the war in Iraq—the Dow Jones Eurostoxx 50, which was at 2,275 in early October 2002, reached a low of 1,928 in March. But credit default swap (CDS) spreads remained relatively tight, rather than widening as a Merton model-based credit system would predict. The Trac-x Europe CDS index narrowed from 140 basis points in October to 105 bp during the war.

(Risk Magazine, November 1, 2003, 'Barra joins crowded market for Merton models', www.risk.net/1497515/)

The above example is an illustration of the use of CDS instead of bond positions together with stock positions. Arbitrageurs were long equities and sustained losses as the index fell from 2275 to 1928. They are referred to as having been short credit, which typically means long the CDS or buying protection. Thus, as credit spread's tightened they would have suffered mark-to-market losses on the CDS position. The reason for these market moves was that the credit market believed that the ongoing corporate deleveraging was a more important determinant of company performance than the macroeconomic factors related to the war. The result was that arbitrageurs that were short credit and long equities suffered losses.

We can generalize the above conclusions and state that the capital structure arbitrage strategy and the decision which security to go long or short depends on which side of the line representing the theoretical relationship between the actual credit spread and share price observation falls. This is shown in Figure 19.4c.

How can we calculate the appropriate hedge ratio for the arbitrage strategy? We can derive the delta theoretically from the delta of equity with respect to the firm's asset value A_t. From Eq. (19.15) we know that

$$\frac{dE_t}{dA_t} = N(d_1) \tag{19.26}$$

and

$$\frac{dD_t}{dE_t} = \frac{d(A_t - E_t)}{dE_t} = \frac{dA_t}{dE_t} - 1 = \frac{1}{N(d_1)} - 1. \tag{19.27}$$

The hedge ratio dD_t/dE_t allows us to calculate how many shares to go short (or long) for each bond bought (or sold). Our discussion so far has been framed in terms of the relative pricing of equity and bonds. If instead of bond prices the *actual* credit spreads were derived from CDS prices, then to exploit the mispricing one would instead of going long bonds (in the hope that bond prices would rise and spreads would tighten) one would sell a CDS.

19.5.3 SOURCE OF PROFITS FROM CAPITAL

Figure 19.4 illustrated a decision rule based on the Merton model that can be used to exploit any potential mispricing between bonds (or CDS) and equity. Now we want to examine the main

(a) Illustrative example of relationship between credit spread and share price

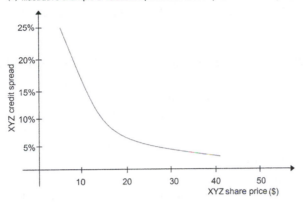

(b) Model implied relationship versus credit spreads observed in the market

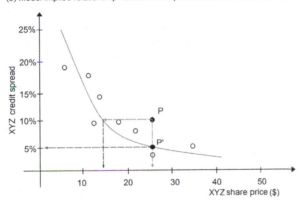

(c) Indictator for capital structure arbitrage strategy and decision to go long or short

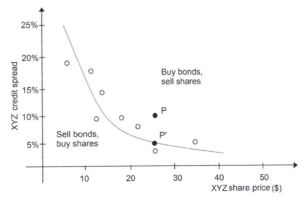

FIGURE 19.4

Credit spread versus share price.

sources of profit for the capital arbitrage strategy discussed above. It turns out that the profits can be viewed either from (a) the perspective of bond/CDS and equity prices correcting or (b) a more advanced perspective of volatility arbitrage.

19.5.3.1 Capital structure arbitrage as a mispricing of bonds (or CDS) and equity

Let's examine (a) first. To exploit the apparent arbitrage represented by point P we noted that a short stock and long bond position is optimal. Such a position will benefit if either the share price increases or the bond price rises (that is credit spreads tighten) or both occur simultaneously. There are however also other scenarios when the strategy will be profitable, for example, when the share price rises, but losses on the short stock position are outweighed by gains from the long bond position if credit spreads tighten to a sufficient extent.

We can distinguish three potential areas in the credit spread–share price space to discuss whether the strategy (i) will make money, (ii) will lose money, or (iii) will break even. Figure 19.5 illustrates the three areas. The starting point is a delta-hedged position at point P. The position consists of a long bond and short stock position. As the arrow which points towards the south-west indicates any move from point P towards the theoretical credit spread–share price relationship represents a correction of the apparent mispricing at point P and would make money for the strategy. A share price/spread move toward the north-east would on the other hand lead to losses. There is a third area however represented by the line that passes through point P. This line can be interpreted as a break-even area. Point P represents a delta-hedged position and any move in the credit spreads along the line corresponds to a movement as the Merton model would indicate *if* the current share price/spreads observation at point P *was* correct, that is if there was no mispricing.

19.5.3.2 Capital structure arbitrage from the perspective of mispriced volatility

As discussed, there is a second perspective (b) which interprets mispricing as mispricing of volatility rather than spreads and share prices. In our discussion of the application of the Merton model in Section 19.4.4, we showed how one can back out a credit spread from observed equity prices and volatility. This approach could be presented by the following relationship which suggests that using variables such as the equity value and volatility we can obtain the Merton model implied credit spread:

$$E_t, \sigma_E, F, r, \sigma_A, T - t \xrightarrow[\text{Merton–Model}]{} R - r_{\text{Merton}} \tag{19.28}$$

The alternative route is to start with actual credit spreads and then back out the implied equity volatility.

$$(R - r)_{\text{actual}}, E_t, F, r, \sigma_A, T - t \xrightarrow[\text{Merton–Model}]{} \sigma_E^{\text{actual}} \tag{19.29}$$

The procedure in Eq. (19.28) illustrates the use of theoretical/Merton model spreads. Figure 19.5b shows the lines that represent that theoretical spread–share price relationship for different volatilities. For example, at point P' if we plug in a share price of around \$25 we obtain a credit spread of 5% if we use an implied equity volatility of σ_E^{Merton}.

What would happen if we used the procedure in Eq. (19.29) and, together with the other parameters, plugged the share price of \$25 and the observed credit spread of 10% corresponding to point P into the Merton formula to obtain an implied equity volatility? In this case, we would obtain an equity volatility of σ_E^{actual}.

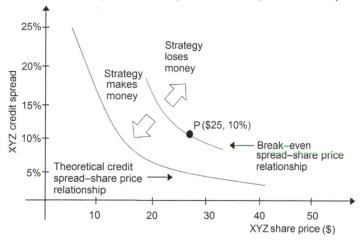

(a) Illustrative example of relationship between credit spread and share price

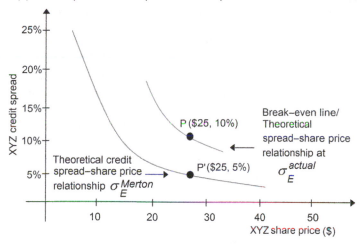

(b) Model implied relationship versus credit spreads observed in the market

FIGURE 19.5

Capital structure arbitrage profits.

Since the Merton model is based on the Black–Scholes formula, it is clear that to obtain a higher credit spread at the same stock price level of $25 one would need to use a higher implied volatility. Therefore, the volatility backed out using the combination of a $25 share price and 5% credit spread must be lower than the volatility obtained using a $25 share price and 10% credit spread. $\sigma_E^{\text{Merton}} < \sigma_E^{\text{actual}}$. In other words, the actual bond (or CDS) spreads being too high (compared to the Merton model implied ones for the same stock price) can be also interpreted as the bond (or CDS) spread implied equity volatility being high relative to what could be referred to as the *true* volatility. The profits from the strategy that is long XYZ bonds and short XYZ stocks can be reinterpreted as a result. The strategy will make money if the implied equity volatility corrects (that is decreases), that is by moves towards the theoretical credit spread—share relationship based on volatility σ_E^{Merton}.

What could be a source of the "true" equity market volatility be in the comparison above? How can we get an estimate of the input parameter σ_E. One potential source is equity option implied volatility. The following reading illustrates the use of the implied equity volatility perspective on capital structure arbitrage in practice. The reading refers to an asset manager that takes CDS prices and extracts from them an implied equity volatility which is then compared to the implied volatility in the equity options market. The equity volatilities also provide the sensitivities for the delta hedging.

EXAMPLE

On September 10, the CDS spread on Sun was around 70 bp above LIBOR. "This gives us a kind of implied equity volatility—when we use our capital structure model—of something like 45–50 volatility [basis] points," [. . .]

The Axa team then compared this value with the implied volatility value on Sun in the equity options market. On that day, the implied volatility of 3-month at-the-money options on Sun was around 65 bp. The 15–20 bp difference was the arbitrage opportunity. To monetise this, Axa bought CDS protection on Sun while buying Sun stock to delta hedge the position.

To work out the implied equity volatilities, [. . .] Axa relies on a proprietary equity-based firm-value model based on Merton theory [. . .]

(Risk Magazine, October 1, 2003, 'Capital structure builds a following', www.risk.net/1526277)

The Merton model was used in the example to back out σ_E^{actual} as in Eq. (19.29). The observed CDS spread in the market implied an equity volatility of 45–50 bp, which was lower than the equity option market implied volatility of 65 bp. It seems that the traders took the view that the equity option volatility was more correct. This view was expressed by buying CDS protection and hedging by buying stock. If the credit spread in the CDS market widens the protection buyer would make mark-to-market gains. If the spread tightens the long stock position would hopefully provide a hedge given the negative empirical relationship between the two.

19.6 ENGINEERING EQUITY PRODUCTS

19.6.1 PURPOSE

Companies raise capital by issuing debt or equity.[12] Suppose a corporation or a bank decides to raise funds by issuing *equity*. Are there more advantageous ways of doing this? Financial engineering offers several alternatives that may address the company's specific needs.

1. Some strategies may decrease the cost of equity financing.
2. Other strategies may result in modifying the composition of the balance sheet.
3. There are steps directed toward better timing for issuing securities depending on the direction of interest rates, stock markets, and currencies.

---

[12]There is also what is called the "mezzanine finance," which comes close to a combination of these two.

4. Finally, there are strategies directed toward broadening the investor base.

In discussing these strategies, we consider two basic instruments that the reader is already familiar with. First, we need a *straight coupon bond* issued by the corporation. The second instrument is an *option* written on the stock. The (call) option on the stock is of European style, has expiration date T and strike price K. The call is sold at a premium $C(t_0)$. Its payoff at time T is

$$C(T) = \max[S_T - K, \ 0] \tag{19.30}$$

These sets of instruments can be complemented by two additional products. In some equity-linked products, we may want to use a call option on the bond as well. The option will be European. In other special cases, we may want to add a credit default swap to the analysis. Many useful synthetics can be created from these building blocks. We start with the engineering of a *convertible bond* in a simplified setting.

19.6.2 CONVERTIBLES

In Chapter 17, we discussed *callable* bonds that convey an option to the issuer to redeem the bond issue early. Convertible bonds and convertible preferred stock are another example of securities with embedded options. In contrast to callable bonds they give an option to the holder of the security rather than the issuing firm. A *convertible* bond is a bond that incorporates an option to convert the principal into stocks. The principal can be converted to a predetermined number of stocks of the issuing company. Otherwise, the par value is received. It is clear that the convertible bond is a *hybrid product* that gives the bond holder exposure to the company stock in case the underlying equity appreciates significantly.

Consider the example of a convertible bond with a *conversion ratio* of 20 and a par value of $1000. This allows the convertible bond holder to convert one bond of par value $1000 into 20 shares of common stock. The *conversion price* is defined as the par value divided by the *conversion ratio* or $1000/20, that is $50 in this example. The bond holder evaluates whether it is worth foregoing bonds with a face value of $1000 or $50 per share. When is it optimal for the holder to convert the bond? This is the case when the present value of the bond's promised payments is <20 times the value of one share of stock. A convertible bond that is trading at $900 and has a conversion ratio of 20 might be worth converting if the stocks were selling above $45 since the value of 20 shares that the holder would receive for each bond is higher than $900. The bond's *conversion value* is defined as the value that it would have if the holder converted it into stock.

Can we say anything about the price of the convertible bond and how high it should be *relative* to the conversion value and the value of a straight bond? On the one hand, the convertible bond should trade at least as high as its conversion value. Otherwise one could buy the convertible bond, convert it and make an immediate profit. Such an arbitrage strategy could in reality be expected to push up the price to the conversion value or above.

On the other hand, the convertible bond is worth at least as much as a straight bond since the convertible consists of a straight bond plus a call option. The straight bond value provides a *bond floor*. Figures 19.6a illustrates the straight bond value. For bond issuers with a strong balance sheet, the straight debt value should be largely independent of the value of the stock since the default risk

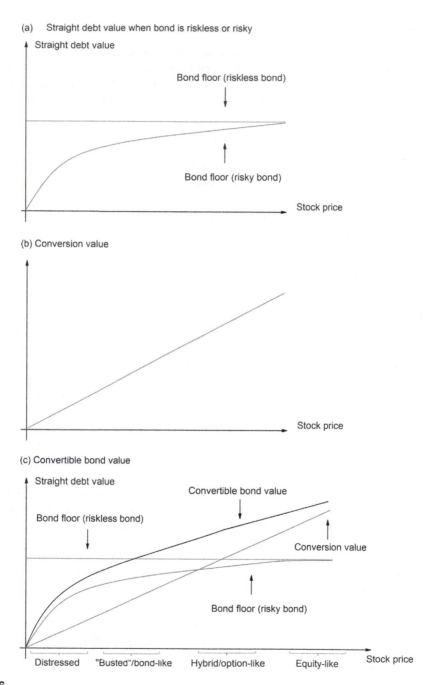

FIGURE 19.6

Convertible bond, conversion, and straight bond value.

is negligible. This is represented by the horizontal bond floor. If the bond is risky then as the stock price declines the bond value can be expected to decline and even become worth close to zero as it becomes distressed. Figure 19.6b illustrates the conversion value. Figure 19.6c shows that the convertible bond value is bounded from below by the straight (risky) bond value when the stock price makes conversion unlikely. In this case, the convertible is also referred to as being *busted*. It is bounded from below by the conversion value when the stock price is high enough to make conversion likely. When the stock price is significantly above the conversion price, the convertible behaves like equity. The difference between the convertible bond value and the conversion value is called the *conversion premium*. Some convertible bonds have a required conversion or redemption feature, in which case they are called *mandatory* convertibles.

We discuss the engineering of such a convertible bond under simplified assumptions. In the first case, we discuss a bond that has no default risk. This is illustrative, but unrealistic. All corporate bonds have some default risk. Sometimes this risk is significant. Hence, we redo the engineering, after adding a *default risk* in the second example.

19.6.3 SYNTHETIC CONVERTIBLE BONDS AND CASH FLOW ENGINEERING

From the above discussion, we can see that a convertible bond can be viewed as a portfolio of a straight bond plus a call option. This leads us to the following contractual equation.

$$
\boxed{\begin{array}{c}\text{Convertible}\\\text{bond}\end{array}} = \boxed{\begin{array}{c}\text{Straight}\\\text{bond}\end{array}} + \boxed{\begin{array}{c}\text{Call}\\\text{option}\end{array}}
\qquad (19.31)
$$

Figure 19.7 shows the composition of a convertible bond's cash flows into a defaultable bond and a call option.[13] There are four potential scenarios depending on whether the bond defaults or not and whether the option is exercised or not.

As the issuer sells an embedded option to convert its shares, the issuer expects to pay a lower fee or coupon. The buyer of the convertible is rewarded by the potential to participate in the upside of the company.

For a risky bond, we can also build on the discussion from the previous chapter on CDS which showed that a defaultable can be replicated by a portfolio consisting of a receiver interest rate swap, a default-free deposit and selling a CDS. Thus if we replace the straight bond in Eq. (19.31) by a portfolio we obtain the following contractual equation:

$$
\boxed{\begin{array}{c}\text{Convertible}\\\text{bond}\end{array}} = \boxed{\begin{array}{c}\text{Receiver}\\\text{interest}\\\text{rate swap}\end{array}} + \boxed{\begin{array}{c}\text{Default}\\\text{+ free}\\\text{deposit}\end{array}} + \boxed{\begin{array}{c}\text{Sell}\\\text{CDS}\end{array}} + \boxed{\begin{array}{c}\text{Buy call}\\\text{option}\end{array}}
\qquad (19.32)
$$

This contractual equation provides an interesting insight since it shows that the convertible bond contains default risk that can be hedged by buying a CDS.

---

[13]For simplicity, we assume that the bond is sold at par.

(a) A defaultable bond and a call option

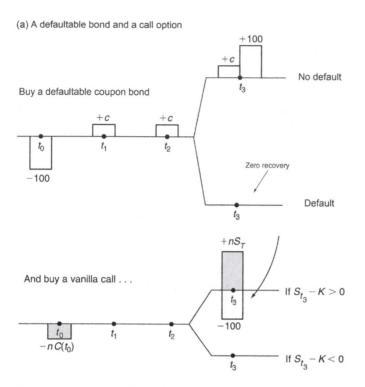

Buy a defaultable coupon bond

And buy a vanilla call . . .

(b) Cash flows of a convertible bond

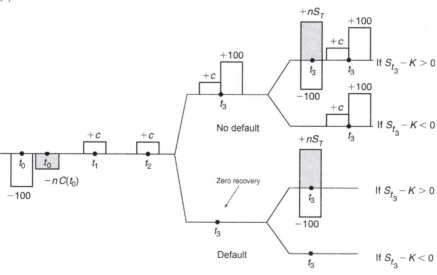

FIGURE 19.7

Convertible bond decomposition.

Table 19.2 Intel Corporation 2.95% Convertible Bonds Due in 2035	
Issuer	**INTEL CORP**
Par amount	$1000
Conversion ratio	34.9501
Conversion price	28.6122
Conversion value	1000
Coupon (fixed)	2.95%
(Conversion) Parity	110.547
Conversion premium (over conversion value)	15.703
Price of one Intel Share	31.63
Yield to maturity	1.51
Coupon Frequency	**Semiannual**
Price of the convertible bond	126.25
Call provision	None
Implied volatility	62.194
Delta	0.560
Vega	0.268
Theta	0.004
Psi	−2.092
Credit sensitivity	−3.539
Aggregate amount issued	1,600,000 (M)
Announcement date	03/30/2006
1st coupon date	06/14/2035
Convertible until	12/14/2035
Maturity date	12/15/2035
Book Runner	JPM

Source: *Bloomberg, July 15, 2014.*

19.6.4 REAL-WORLD EXAMPLE

It is instructive to examine a real-world example. Table 19.2 provides information on a convertible bond issued by Intel. The bond matures in 2035. The bond pays a 2.95% semiannual coupon. The reported conversion ratio is 34.9501 and the conversion price is 26.6122. If we multiply the conversion ratio times the conversion price we obtain a conversion value of 1000. The share price of Intel on that day was 31.63. The convertible bond is currently selling for 126.25 per $100 or 1262.5 per $1000 par value. The conversion parity is defined as the share price multiplied by the conversion ratio: $34.9501 \times 31.63 = 1105.47$. The conversion premium is defined as the price of the convertible minus the conversion parity, that is $1262.5 - 1105.47 = 157.03$ per $1000 or $15.703 per $100 as indicated in the table. The implied volatility is 62.194 and can be compared to the implied volatility of equity options on Intel shares on that day. The delta is the sensitivity of the convertible bond to changes in the share prices and it is 0.56 according to the table.

19.6.5 CONVERTIBLE BOND PRICING

19.6.5.1 Bond plus equity option approach

There are several issues that complicate the decomposition in Eq. (19.31) and any valuation based on it. These complications include dividend payments by the stock, a conversion price that may increase over time and any potential call options that allow the issuer to repurchase the bond early. Moreover, convertible bonds contain an option component with a stochastic strike price equal to the bond price.

19.6.5.2 Multifactor model

A further complicating factor relates to default risk. As we saw in Figure 19.6, the share price level is a key determinant of the convertible bond value. In principle, this means that we can use the Black—Scholes approach to value the convertible with the state variable being the share price and use a dynamic hedging strategy to value to embedded option. However, as Figure 19.6c shows, the credit risk of the bond also depends on the stock price. The most recent and sophisticated convertible bond pricing models incorporate the stochastic nature of interest rates and credit spreads. References to recent papers on convertible bond pricing are provided at the end of the chapter. Bloomberg's convertible bond pricing function, for example, incorporates not only time-varying interest rates, dividends, and volatility but also allows for jumps in the stock price which can be calibrated to CDS data.

19.6.6 CONVERTIBLE BOND ARBITRAGE

Historically some of the most active buyers of convertible bonds were hedge funds.[14] However, recent estimates suggest that the proportion of hedge funds as buyers of convertible bonds has fallen from 75% in 2008 to less than 40% in 2013.

Convertible arbitrage strategies are used by hedge funds in an attempt to generate risk-adjusted performance. Consider the following reading.

EXAMPLE

Convertible arb hedge funds in the U.S. are piling into the credit default swaps market. The step-up in demand is in response to the rise in investment-grade convertible bond issuance over the last month, coupled with illiquidity in the U.S. asset swaps market and the increasing credit sensitivity of convertible players' portfolios, said market officials in New York and Connecticut.

Arb hedge funds are using credit default swaps to strip out the credit risk from convertible bonds, leaving them with only the implicit equity derivative and interest rate risk. The latter is often hedged through futures or treasuries. Depending on the price of the investment-grade convertible bond, this strategy is often cheaper than buying equity derivatives options outright, said [a trader].

Asset swapping, which involves stripping out the equity derivative from the convertible, is the optimal hedge for these funds, said the [trader] as it allows them to finance the position cheaply, and removes interest rate risk and credit risk in one fell swoop. But with issuer-credit quality in the U.S. over the last 12 to 18 months declining, finding counterparties willing to take the other side of an asset swap has become more difficult...

(Based on an article in Derivatives Week (now part of GlobalCapital))

---

[14]See Choi et al. (2010), for example, for empirical evidence on this.

Convertible bonds are generally issued at a discount to their theoretical value. This implies that hedge fund managers can extract the undervalued component from the convertible using appropriate techniques. One of the most basic convertible bond arbitrage strategies is to buy *cheap* volatility. A convertible bond consists of a bond and an option as we saw earlier. A hedge fund manager can calculate the price of the embedded option or the equity volatility implied by the current price of the convertible. By comparing the implied volatility of the embedded option to historical volatility or the implied volatility of listed options on the same stock, it is possible to calculate whether the embedded option in the convertible bond is undervalued. The implied equity volatility is the volatility that needs to be entered into a convertible bond pricing model together with other input parameters in order to generate the observed convertible bond price. The implied volatility for the convertible bond issued by Intel, for example, was 62.194 according to the Bloomberg data in Table 19.2.

If the option embedded in the convertible is cheaper than a vanilla call option on the same stock, then a fund can go long the convertible bond in the hope that the undervalued embedded option will appreciate in value. A range of techniques can be employed to hedge against unwanted risks and isolate the desired equity, credit, event, or even currency exposure embedded in the securities.

The basic convertible bond arbitrage strategy is *long volatility*. This means that the strategy benefits from volatility in the underlying stock. The volatility exposure arises from two different sources. First, the embedded stock option implies that the buyer of the convertible bond is long gamma, that is the rate of change in the delta with respect to stock prices. Second, the strategy is long vega since the call option price rises if implied volatility rises. In the case of the Intel convertible discussed above, the vega was 0.268. If the implied volatility corrects and rises, this will lead to gains for the holder of the convertible bond. Convertible bonds are also exposed to more complex risk including credit risk. On the one hand, a deterioration in the creditworthiness of the issuer is likely to coincide with an increase in the volatility of the underlying stock which in turn affects the value of the embedded call option. The opposite is likely to happen if creditworthiness improves. On the other hand, a deterioration of the credit quality of the issuer directly affects the underlying risky bonds. As credit spreads tighten, the value of a convertible bond is likely to increase. As credit spreads widen its value decreases.

Equation 19.32 showed that a convertible bond can be decomposed into several constituents including a CDS position. Therefore, CDS can be used to hedge credit risk embedded in a long convertible bond position. The reason is that the mark-to-market value of an existing CDS position will change as the credit risk of the issuer changes. To establish the correct hedge ratio, we examine how sensitive the CDS is to changes in credit spreads relative to the sensitivity of the convertible bond. *Omnicron* measures the price impact of an increase in credit spreads. Typically this measure is rebased for 1 basis point increase in the level of credit spreads. Practitioners also call this sensitivity the *Credit DV01*.

The equity market risk can be hedged using short stock positions or put options. The credit market risk exposure is in principle best hedged using CDS. However, sometimes, equity market puts may be a cheaper hedging alternative than CDS. As the Merton model illustrated, equity market and credit markets are linked. One can calculate the sensitivity of put options and CDS to the underlying credit risk and use this sensitivity as a hedge ration and choose between a CDS or equity put option hedge depending on which one is cheaper.

19.6.7 COMPARING THE ROLE OF VOLATILITY IN CONVERTIBLE BOND ARBITRAGE AND CAPITAL STRUCTURE ARBITRAGE

As we saw in Section 19.5, in capital structure arbitrage we can interpret the straight bond as an equity derivative since there is a theoretical relationship between the bond and the equity. We can use a Merton-type model to calculate the equity volatility implied by the share price and the credit spread. If this implied equity volatility differs from what is viewed as the correct volatility, the market participant can buy or sell the bonds (or the CDS) and delta hedge the position. There is an important difference between the capital structure arbitrage strategy and the convertible bond arbitrage strategy since a convertible bond can be converted into equity while a straight bond cannot. This implies that there are theoretical arbitrage bounds for the convertible but no such bounds exist for straight debt based on the Merton-type models. This in addition to the restrictive assumptions of the Merton model imply that arbitrage strategies based on simple structural models of default may not predict actual movements in equity and credit markets well.

19.6.8 INCORPORATING MORE COMPLEX STRUCTURES

This section considers two variations of the basic convertible structure. First of all, the basic convertible can be modified in a way that will make the buyer operate in two *different currencies*. In fact, a dollar-denominated bond may be sold, but the underlying shares may be, say, French shares, denominated in Euros. This amounts, as we will see, to adding a call or put option on a *foreign currency*.

The second alternative is also important. The basic convertible can be made *callable*. This amounts to making the underlying debt issue a callable bond. It leads to adding a call option on the bond. Before we see how these are used, we consider some of the financial engineering issues in each case.

19.6.8.1 Exchange rate exposure

Suppose the convertible bond is structured in *two* currencies. A Thai company secures funding by selling a euro convertible in the Eurodollar market, and the debt component of the structure is denominated in dollars. So, the bonds have a par value of, say, $100. The conversion is into the shares of the firm, which trade, say, in Bangkok. The shares are baht denominated. We assume, unrealistically, that there is no default risk.

Because Thai shares trade in Thai exchanges and are quoted in Thai baht, the conversion price to be included in the convertible bond needs to specify *something* about the value of the exchange rate to be used during a potential conversion. Otherwise, the conversion rule will not be complete. That is to say, instead of specifying only the number of shares, n, and the conversion price, K, using the equality

$$100\$ = Kn \tag{19.33}$$

the conversion condition now needs to be

$$e_t 100\$ = Kn \tag{19.34}$$

where e_t is an exchange rate denoting the price of USD1 in terms of Thai baht at date t. This is needed since the original conversion price, K, will be in Thai baht, yet, the face value of the bond will be in USD. The bond structure can set a value for e_t and include it as a parameter in the contract. Often, this e_t will be the current exchange rate.

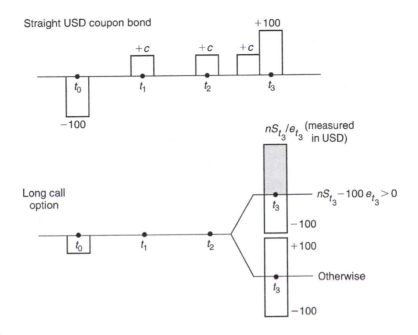

FIGURE 19.8

Convertible with currency conversion.

Now, suppose a Thai issuer has sold such a Euro convertible at e_t, the current exchange rate. Then, if Thai stocks rise *and* the exchange rate remains stable, the conversion will occur. Here is the important point. With this structure, at maturity, the Thai firm will meet its obligations by using its own *shares* instead of returning the original $100 to bond holders. Yet, if, in the meantime, e_t rises,[15] then, in spite of higher stock prices, the value of the original principal $100, when measured in Thai baht, may still be higher than the nS_T and the conversion may not occur. As a result, the Thai firm may face a significant dollar cash outflow.[16]

This shows that a convertible bond, issued in major currencies but written on domestic stocks, will carry an FX exposure. This point can be seen more clearly if we reconstruct this type of convertible and create its synthetic. This is done in Figure 19.8.

The top part of Figure 19.8 is similar to Figure 19.7. A straight coupon bond with coupon c matures at time t_3 and pays the principal $100. The difference is in the second part of the figure. Here, we have, as usual, the call option on the stock S_t. But S_t is denominated in baht and the call will be in-the-money—that is to say, the conversion will occur only if

$$nS_{t_3} > 100e_{t_3} \tag{19.35}$$

---

[15]That is to say, if the Thai baht is devalued, for example.

[16]This may be something that is occurring at a bad time, since if the currency is devalued, the international markets may not be receptive to rolling over the dollar debt of Thai corporates.

The idea in Figure 19.8 is the following. We would like to begin with a dollar bond and then convert the new call option into an option as in the case before. But, if the Thai baht collapses,[17] then the $100 received from the principal at maturity will be much more valuable than $S_{t_3} n / e_{t_0}$.

19.6.8.2 Making the convertibles callable

One can extend the basic convertible structure in a second way and add a call option on the underlying convertible bond. For example, if the bond maturity is T, then we can add an implicit option that gives the issuer the right to buy the bond back at time, U, $U < T$ at the price

$$\max [\$100, \ nS_U] \tag{19.36}$$

This way the company has the right to *force* the conversion and issue new securities at time U. Some corporations may find this a useful strategy.

With this type of convertible, forcing the conversion is the main purpose. Suppose the following two conditions are satisfied:

1. The share is trading at a higher price than the conversion price (i.e., the strike K).
2. The expected future dividends to be paid on the stock are lower than the current coupon of the convertible.

Then, if the convertible is callable, the issuer may force the conversion by calling the bond. This will convert a debt issue in the issuer's balance sheet into equity and affect some important ratios, in case these are relevant. Second, the immediate cash flow of the firm will improve.

19.6.8.3 Exchangeable bond

The basic convertible-warrant structures can be modified to meet further financial engineering needs. We can consider another example. Suppose the convertible bond, when it converts, converts into *another* company's security. This may be the case, for example, if company A has acquired an interest in company B. This way, the company can sell convertible bonds where the conversion is into company B's securities.

From a financial engineering point of view, the structure of this "exchangeable" is the same. Yet, the pricing and risk management are different because now there are *two* credits that affect the price of the bond: the credit of the company that issues the bond and the credit of the company this bond may convert.

Another difference involves the dilution of the shares of the target company. When a convertible is issued and converts at a later date, there may be dilution of the shares, yet, in an exchangeable the shares that are exchanged will come, in general, from the free float.

19.6.9 USING CONVERTIBLES

A convertible bond has some attractiveness from the point of view of end investors. For example, the investor who buys the convertible will have some exposure to the share price. If S_t increases significantly, the bond becomes a portfolio of shares. On the other hand, if the bond fails to

---

[17]This means that the Thai stock market is also down.

convert, the investor has at least some minimum cash flow to count on as income, and the principal is recovered (when there is no default).

But, our interest in this book is not with the investors, but rather, in the advantages of the product from an issuer's point of view. For what types of purposes can we use a convertible bond?

- The first consequence of issuing convertibles rather than a straight bond is that the convertible carries a lower coupon. Hence, it "seems" like the funds are secured at lower cost.
- More notably for a financial engineer, convertibles have interesting implications for balance sheet management. If an equity-linked capital is regarded as equity, it may have less effect on ratios such as debt to equity. But, in general, rating agencies would consider straight convertibles as debt rather than equity.
- Note that with a convertible, in case conversion occurs, the shares will be sold at a higher price than the original stock price at issue time.
- Finally, convertibles are bonds, and they can be sold in the Euro markets as Euro-convertibles. This way a new investor base can be reached.

We should also point out that convertibles, when combined with other instruments, may have significant and subtle tax advantages. The best way to show this is by looking at an example from the markets.

EXAMPLE

(ABC Capital) has entered into a total return swap on 154,000 shares of Cox Communications preferred stock exchangeable into shares of Sprint PCS, and a total return swap on 225,000 shares of Sprint PCS. In the Cox swap, the hedge fund pays three-month LIBOR plus 50 basis points and receives the return on the exchangeable preferred shares. In the Sprint swap, ABC pays the return on the stock and receives three-month LIBOR less 25 bps. Both total return swaps mature in about 13 months.

The total return swaps were entered into for tax reasons. ABC's positions are held by a Cayman Islands limited duration company. Because the Cayman Islands do not have a tax treaty with the U.S., income from these securities is withheld at the non-treaty rate of 30%. Entering the total return swaps ensures that ABC does not physically hold the securities, and, hence, is not subject to U.S. withholding.

The underlying position was put on as part of a convertible arb play. ABC bought the exchangeable preferred stock and is using the cash equity to delta hedge the implicit equity option. The market is undervaluing the exchangeable preferred shares, according to a trader, who noted that although these shares recently traded at USD76.50, the fund's models indicate they should be priced around USD87. The company's model is based in part on the volatility of the underlying stock, the credit quality of the issuer, and the features of the convertible. In this case, the market may be undervaluing the security because it is not pricing in all the features of the complicated preferreds and because of general malaise in the telecom sector.

(Derivatives Week (now part of GlobalCapital), November 2000)

This reading is also an example of how implicit options can be used to form arbitrage portfolios. However, there are many delicate points of doing this as were shown earlier.

19.6.10 WARRANTS

Warrants are *detachable* options linked to bonds. In this sense, they are similar to convertibles. But, from a financial engineering point of view, there are important differences.

1. The warrant is detachable and can be sold separately from the bond. Of course, a financial engineer can always detach the implicit option in a convertible bond as well, but still there are differences. The fact that the warrant is detachable means that the principal will always have to be paid at maturity.

   The number of warrants will not necessarily be chosen so as to give an exercise cash inflow that equals the cash outflow due to the payment of the principal. Thus, the investor can, in principle, end up with both the debt and the equity arising from the same issue.

2. The exchange rate used in a convertible is fixed. But, because there is no such requirement for a warrant and because the latter is detachable, this is, in general, not the case for a warrant. Hence, there is no implicit option on the exchange rate in the case of warrants. In this sense warrants are said to be relatively more attractive for strong currency borrowers, whereas convertibles are more attractive for weak currency borrowers.

3. Finally, because warrants are detachable, the warrant cannot be forced to convert. The bond can be called, but the conversion is not required.

We now move to another topic and look at securitizing cash flows. This can be regarded as an example of new product structuring.

19.7 CONCLUSIONS

In this chapter, we saw how we can apply the financial engineering and option pricing principles to value equity. Equity can be viewed as an option on the assets of the firm with a strike price equal to the debt value. Structural models of default thus establish a link between equity markets and bond or CDS markets. The development of credit derivative markets has accelerated the use of structural models of default in practice. Their applications range from the forecasting of default to capital structure arbitrage. Convertible bonds are hybrid products that can be converted from debt to equity. Convertible bond arbitrage strategies are based on exploiting cheap volatility while delta hedging the position. In the next chapter, we will review various structure products including reverse convertibles. Unlike convertible bonds, in reverse convertibles, the conversion option is conveyed to the issuer of the product and not the buyer. Both types of instruments have thus embedded options whose value depends on the implied volatility.

SUGGESTED READING

This chapter has provided a simple introduction to structural models of default and hybrid equity products. The reader may want to follow up on the discussion in two ways. **Das and Sundaram** (2010) provide an accessible summary and application of the **Merton** (1974) model. **Sundaresan** (2013) reviews recent developments in research on structural models of default. **Tepla** (2004) provides a nice example and case study of convertible and capital structure arbitrage strategies based on KBC Investment Management. The discussion of capital arbitrage profits in this chapter is based on the case study.

Ayache et al. (2003) provide a review of valuation of convertible bonds with credit risk. Several books deal with the current state of hybrid equity instruments. **Das** (2000) is one example. **Duffie and Singleton** (2003) is also a good reference. The reader may also want to learn more about the technical issues related to pricing, hedging, and risk-managing hybrid equity products.

Kat (2001) deals with some of the related issues. **Damadoran** (2012) is a good reference for discounted cash flow and other equity valuation approaches under the real-world probability measure which we do not discuss in this chapter.

EXERCISES

1. Explain why debt in the Merton (1974) model is viewed as an option.

2. Assume that company A has an asset volatility of 20%. The current value of its assets is $100 million and the face value of its 1-year maturity zero-coupon debt is $50 million. The risk-free rate of interest is 2%. Use the Merton (1974) model to calculate the value of the firm's equity. What is the value of the debt?

3. Consider company B which issued equity and zero-coupon bonds with a maturity of 1 year. Assume that the value of the firm is $100 and the value of the equity is $50 million. The risk-free rate is 2%. The equity volatility is 30%.
    a. What is the market value of debt and the implied credit spread?
    b. Assume that the company has 1 million shares outstanding. Plot the credit spread against the share price for a range of different share price values.
    c. Plot the hedge ratios for different values of the share price.

4. What variables and real-world complications are important in practice but ignored by the basic Merton model.

5. Consider two convertible bonds X and Y, which for simplicity are assumed to be riskless. The following table provides information about the two bonds. Calculate the conversion value. What is the yield to maturity of the convertible bonds based on the actual bond price?

	Convertible Bond X	Convertible Bond Y
Annual coupon	50	50
Time to maturity	5 years	5 years
Conversion ratio	30	40
Stock price	$25	$30
Conversion value	?	?
Yield on 5-year Treasuries	5%	5%
Value of straight debt	913.41	913.41
Actual convertible bond price	920	1207
Yield to maturity of convertible	?	?

Examine the information given in Table 19.2 in the text. Consider a trader that buys the convertible bond because it views the implied volatility as cheap and expects the implied volatility of the bond to correct. If the trader decides to hedge the long convertible bond position, how many shares does he have to short to be delta neutral?

ESSENTIALS OF STRUCTURED PRODUCT ENGINEERING

CHAPTER OUTLINE

20.1 INTRODUCTION

Structured products consist of packaging basic assets such as stocks, bonds, and currencies together with some derivatives. The final product obtained this way will, depending on the product, (i) have an *enhanced return* or *improved credit quality*, (ii) *lower costs* of asset–liability management for corporates, (iii) build in the *views* held by the clients, and (iv) often be *principal protected*.[1]

Households do not like to build their own cars, computers, or refrigerators themselves. They prefer to buy them from the *producers* who manufacture and assemble them. Every complex product has its own specialists, and it is more cost effective to buy products manufactured by these specialists. The same is true for *financial products*. Investors, corporates, and institutions need solutions for problems that they face in their lives. The packaging solutions for investors' and institutions' needs are called *structured* products. "Manufacturers," i.e., the "structurers," put these together and sell them to clients. Clients consist of investment funds, pension funds, insurance companies, and individuals. Clients may have *views* on the near- or medium-term behavior of equity prices, interest rates, or commodities. Structured products can be designed so that such clients can take positions according to their views in a convenient way.

Industrial goods such as cell phones and cars are constantly updated and improved. Again, the same is true for structured products. The views, the needs, or simply the risk appetite of clients change and the structurer needs constantly to provide new structured products that fit these new views. This chapter discusses the way financial engineering can be used to service retail clients' particular needs.

Financial engineering provides ways to construct any payoff structure desired by an investor. However, often these payoffs involve complex option positions, and clients may not have the knowledge, or simply the means, to handle such risks. Market practitioners can do this better. For example, many structured products offer principal protection or credit enhancements to investors. Normally, institutions that may not be allowed to invest in such positions due to regulatory reasons will be eligible to hold the structured product itself once principal protection is added to it. Providing custom-made products for clients due to differing views, risk appetite, or regulatory conditions is one way to interpret structured products and in general they are regarded this way. Hybrid securities that consist of several financial products such as a stock or bond plus a derivative are also referred to as structured notes.

However, in this book our main interest is to study financial phenomena from the manufacturer's point of view. This view provides a *second interpretation* of structured products. Investment banks deal with clients, corporates, and with each other. These activities require holding

---

[1]According to this, the investor (or the corporate) would not lose the principal in case the expectations turn out to be wrong. We deal with the so-called Constant Proportion Portfolio Insurance (CPPI) and its more recent version, Dynamic Proportion Portfolio Insurance (DPPI) in Chapter 23.

inventories, sourcing and outsourcing exposures, and maintaining books. However, due to market conditions, the instruments that banks are keeping on their books may sometimes become too costly or too risky, or sometimes better alternatives emerge. The natural thing to do is to sell these exposures to "others." Structured products may be one convenient way of doing this. Consider the following example. A bank would like to *buy volatility* at a reasonable price, but suppose there are not enough sellers of such volatility in the interbank market. Then a structured product can be designed so that the bank can buy volatility at a reasonable price from the retail investor.

In this interpretation, the structured product is regarded from the manufacturer's angle and looks like a tool in inventory or balance sheet management. A structured product is either an indirect way to sell some existing risks to a client or an indirect way to buy some desired risks from the retail client. Given that bank balance sheets and books contain a great deal of interest rate and credit risk-related exposures, it is natural that a significant portion of the recent activity in structured products relates to managing such exposures.

In this chapter, we consider two major classes of structured products. The first group is the new *equity*-, commodity- and FX-based structured products and the second is *LIBOR-based fixed-income* products. The latter are designed so as to benefit from expected future movements in the yield curve. We will argue that the general logic behind structured products is the same, regardless of whether they are LIBOR-based or equity-linked. Hence we try to provide a unified approach to structured products. In a later chapter, we will consider the *third* important class of structured products based on the occurrence of an *event*. This event may be a mortgage prepayment or, more importantly, a credit default. These will be discussed through *structured credit* products. Because credit is considered separately in a different chapter, during the discussion that follows it is best to assume that there is no credit risk.

20.2 PURPOSES OF STRUCTURED PRODUCTS

Structured products may have at least *four* specific objectives.

The first objective could be *yield enhancement*—to offer the client a higher return than what is normally available. This of course implies that the client will be taking additional risks or foregoing some gains in other circumstances. For example, the client gets an enhanced return if a stock price increases up to 12%. However, any additional gains would be forgone and the return would be *capped* at 12%. The value of this cap is used in offering an enhanced yield.

The second could be *credit enhancement*. In this case, the client will buy a predetermined set of debt securities at a lower default risk than warranted by their rating. For example, a client invests in a portfolio of 100 bonds with average rating BBB. At the same time, the client buys insurance on the first default in the portfolio. The cost of first debtor defaulting will be met by another party. This increases the credit quality of the portfolio to, say, BBB+.

The third objective could be to provide a desired payoff profile to the client according to the client's *views*. For example, the client may think that the yield curve will become steeper. The structurer will offer an instrument that gains value if this expectation is realized.

Finally, a fourth objective may be facilitating *asset/liability management needs* of the client. For example, a corporate treasurer who thinks that the cost of funds would increase in the future may decide to enter into a payer interest rate swap. The structurer will provide a modified swap structure that will protect against this eventuality at a smaller cost. In the following, we discuss these generalities using different sectors in financial markets.

20.2.1 EQUITY STRUCTURED PRODUCTS

First we take a quick glance at the history of equity structured products. This provides a perspective on the most common methodologies used in this sector. The first examples of structured products appeared in the late 1970s. One example was the *stop-loss* strategies. According to these, the risky asset holdings would automatically be liquidated if the prices fell through a target tolerance level. These were precursors of the CPPI techniques to be seen later in Chapter 23. They can also be regarded as precursors of barrier options.

Then, during the late 1980s, market practitioners started to move to principal protected products. Here the original approach was offering "zero coupon bond plus a call" structures. For example, with 5-year treasury rates at r_t, and with an initial investment of $N = 100$, the product would invest

$$\frac{N}{(1+r_t)^5} \tag{20.1}$$

into a discount bond with a 5-year maturity. The rest of the principal would be invested in a properly chosen call or put option. This simple product is shown in Figure 20.1.

At the simplest level, the guaranteed product consists of a zero-coupon bond and one or more options.[2] Suppose S_t denotes the value of an underlying security. This security can essentially be anything from stocks to credit index tranches or the value of some hedge fund investment. We can then write the following contractual equation:

$$
\boxed{\begin{array}{c}\text{Guaranteed product}\\ \text{with } S_t \text{ exposure}\end{array}} = \boxed{\begin{array}{c}\text{A zero-}\\ \text{coupon bond}\end{array}} + \boxed{\begin{array}{c}\text{Long option with } S_t\\ \text{exposure}\end{array}} \tag{20.2}
$$

Suppose an investor invests the amount $N = 100$ directly to a basket of options over a T-year maturity. Then, options being risk investments and investors having limited risk management capabilities, part of the principal may be lost if these options expire out-of-the-money. On the other hand, if the yield on a T-maturity zero-coupon bond is $r\%$ and if the same investor invests, at time t_0, a carefully chosen amount PV_{t_0} in this bond, the security will be worth 100 in 5 years:

$$PV_{t_0}\left(1+r_{t_0}\right)^T = N \tag{20.3}$$

The payoff of the zero-coupon bond is illustrated in Figure 20.1a as a function of the share, that is the underlying of the option. Thus the investor can allocate PV_{t_0} to buy a bond and will still have $N - PV_{t_0}$ to invest in (a basket of) options. Depending on the level of volatility, the level of r and

---

[2]Thus regulation of this sector can be visualized as demanding principal protection before options-related products are sold to retail investors.

(a) Payoff of a zero-coupon bond as a function of the share price

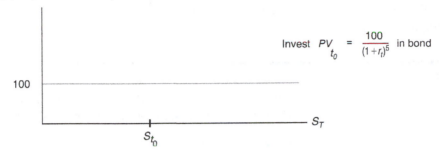

Invest $PV_{t_0} = \dfrac{100}{(1+r_t)^5}$ in bond

(b) Payoff of a long call option position

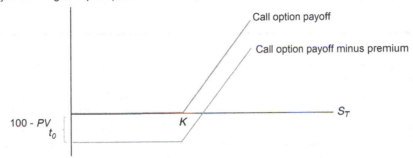

(c) Payoff of a long zero-coupon bond and long call option position

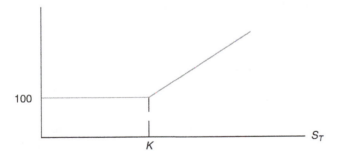

FIGURE 20.1

Zero-coupon bond and long call position.

the expiration dates under consideration, this residual will provide an exposure—the growth of S_t. In fact, let g_{t_i} be the percentage rate of change in S_t during the interval $[t_i, t_{i-1}]$,

$$g_{t_i} = \frac{S_{t_i} - S_{t_{i-1}}}{S_{t_{i-1}}} \tag{20.4}$$

Then the investor's exposure will be $\lambda g_{t_i} S_{t_0}$, where λ is the familiar *participation rate*. In the case of structured products, it is the bank that makes all these calculations, selects a structure with a high participation rate, and sells the principal protected security as a package to the investor.

According to this, in the simplest case the bank will buy a PV_{t_0} amount of the zero-coupon bond for every invested 100, and then options will be purchased with the rest of the principal, that is $100 - PV_{t_0}$ or $100 \times \left(1 - \frac{1}{(1+r_t)^5}\right)$.

EXAMPLE

An investor has the principal $N = 100$. The observed yield on a 5-year zero-coupon treasury is 4.50%. If the investor invests 80.2451 in this bond, the security will be worth 100 in 5 years:

$$80.2451(1 + .045)^5 = 100 \qquad (20.5)$$

Thus the investor can allocate 80.2451 to buy a bond and will still have 19.7549 to invest in options.[3]

In the above example, assume that the investor invests in S&P500 options. After buying the zero-coupon bond, the structurer can invest \$19.7549 in S&P 500 options and pay for administration, costs, and commissions. Assume that a 5-year S&P500 call option costs \$22.19. We also add \$2 for administration and margin costs. Thus the investor will benefit from a participation rate of 80%. The participation rate is calculated as $(19.7549 - 2)/22.19$. Consider two scenarios. In the optimistic scenario, we assume that the S&P500 goes up by 20% over the course of 5 years. In this case, the investor will achieve 16% ($=80\% \times 20\%$). The structured product would redeem at maturity at 116% of the principal (100% from the zero-coupon bond and a 16% gain from the option). In the pessimistic scenario, we assume the S&P500 would fall by 40%. In this case, the call option would expire worthless and the investor would receive 100% of his capital back.

The general idea is simple. The problems arise in implementation and in developing more refined ways of doing this. In practice, several problems can occur.

First, zero-coupon interest rates for a maturity of T may be too low. Then the zero-coupon bond may be too expensive and not enough "excess" may be left over to invest in options. For example, during the years 2011–2013, 5-year USD Treasury rates were around 1.25%. This leaves only:

$$100 - \frac{100}{(1+0.02)^5} = 6.02 \qquad (20.6)$$

to invest in the option basket. Once we factor out the fees paid for such products, which could be several percentage points, the amount that can be invested in the option goes down even more.[4] Depending on the level of volatility, such an investment may not be able to secure any meaningful participation rate.

Second, options on the underlying where exposure is desired may not exist. For example, considering hedge funds, there are few options traded on these risks. Yet, an investor may want exposure to hedge fund activity, or, say, to credit tranches without risking (part of) his or her principal.

---

[3]In real life there are also structurers' fees that need to come out of this amount. These fees are collected up front.

[4]A historical estimate of the average fees for structured products is provided by Thomson Reuters IFR, March 12, 2007. According to this source, the average fees range between 30 and 100 bps in Europe and between 60 and 150 bps in the United States.

Third, even if options on the underlying risk exist, the set of available maturities is limited and longer-term maturities are often illiquid. To address this, hedgers may have to resort to rolling a sequence of short maturity options which can result in a highly path-dependent strategy since the cost of the rolling strategy will depend on changing market conditions.

Fourth, irrespective of the level of interest rates, the options may be too expensive, depending on the level of volatility. This may, again, not secure a meaningful participation rate.

The following reading illustrates how the behavior of various market participants that buy and sell volatility affects implied volatility levels and thus the cost of options.

EXAMPLE

A resurgence in close-out activity has been triggered by a sharp rise in long-dated implied volatility during the first five months of 1998. Having declined significantly throughout November and December last year, three-year index volatility levels have risen steadily since January.

The trend closely mirrors changes in implied volatility experienced last year. Implied volatility levels first took off in 1997, driven by a surge in the market for guaranteed equity funds. Arrangers of the funds bought longer-dated options from banks to hedge the guarantee embedded in retail products. No natural sellers of long-dated volatility were available, and a short squeeze in volatility quickly developed, pushing up rates.

Market professionals asserted last week that the short squeeze was the result of a vicious circle. Banks closing-out their short option positions had pushed volatility even higher, which in turn has prompted other houses to close out their positions. "It's something of a chain reaction. I think most professionals are fairly concerned," said one head of equity derivatives trading in London.

Although in agreement over the severity of the volatility squeeze, market professionals were divided on its cause or solution. Some professionals alleged that bank risk controllers had exacerbated the volatility squeeze by setting too tight risk limits. "This is a typical example of accountants sticking to their guns, whatever happens. It's a unique situation which will not last and they should take account of that," said one.

History is Bunk?

Professionals also pointed to factors that they felt would eventually alleviate the demand/supply imbalance. Several market participants said they thought high volatility was a temporary phenomenon; they argued that volatility is mean-reverting and that implied volatility rates would soon descend towards (much lower) historical levels.

The discrepancy between the two views of volatility lies at the heart of conflicting views over the market's development. Implied volatility rates are calculated by feeding current option prices into an option model, and so are a function of the supply and demand in volatility. In contrast, historical volatility rates are calculated from previous equity market movements. Three-year FTSE 100 historical volatility levels are around 11%.

The yawning gap between historical and implied volatility rates has already created trading opportunities for unconventional suppliers of volatility. While dealers remain naturally short volatility, other trading firms have a more flexible approach to volatility rates. Hedge funds had already been seeking to sell long-dated volatility, with the view that implied levels would descend to historical rates.

> *Equity derivatives professionals asserted that lower participation rates were the direct result of the rise in implied volatility. Guaranteed products are a mixture of a long position in equities—usually achieved through one or more futures contracts—and an option used to provide a floor on possible losses. If the floor is more expensive, the upside offered to the buyer of the guaranteed product will be reduced to lower the overall cost of the product.*
>
> *According to several market professionals, the new lower participation rates are a function of the volatility squeeze and are thus here to stay. "I think participation rates will decline and so reduce demand for volatility. It's simply the market finding an equilibrium," said one equity derivatives marketer.*
>
> **(Thomson Reuters IFR, June 1998)**

The reading shows that high implied volatility levels increase the price of call options which are one of the building blocks in capital guaranteed products. The higher costs means that the participation rates offered by such products necessarily decline.

The capital guaranteed products described above were followed in the early 1990s with structures that essentially *complicated* the long option position. Some products started to "cap" the upside. The structure would consist of a discount bond, a long call with strike K_L and a *short* call with strike K_U, with $K_L < K_U$. This way, the premium obtained from selling the second call would be used to increase the participation rate, since more could be invested in the long option. This is shown in Figure 20.2. Figure 20.2a shows the zero-coupon bond payoff. Figure 20.2b and c shows the payoffs of a long call position with strike price K_L, a short call position with strike price K_U. Figure 20.2d shows the payoff of the portfolio consisting of the zero-coupon bond and the two call options. Other products started using Asian options. The gains of the index to be paid to the investor would be calculated as an *average* of the gains during the life of the contract.

Late 1990s started seeing *correlation* products. A *worst of* structure would pay at maturity, for example, 170% of the initial investment plus the return of a worst performing asset in a basket of say 10 stocks or commodities. Note that this performance could be negative, thus the investor could receive <170% return. However, such products were also principal protected and the investor would still recover the invested 100 in the worst case.

In the *best of* case, the investor would receive the return of the best performing stock or commodity given a basket of stocks or commodities. The *observation period* could be over the entire maturity or could be annual. In the latter case, the product would lock in the annual gains of the best-performing stock, which can be different every year. Mid-2000s brought several new versions of these equity-linked structured instruments including reverse convertibles which we discuss in more detail below, but first we consider the main tools underlying the products.

20.2.2 THE TOOLS

Equity structured products are manufactured using a relatively small set of *tools* that we will review in this section. We will concentrate on the main concepts and instruments: basically five main types of instruments and a major conceptual issue that will recur in dealing with equity structured products.

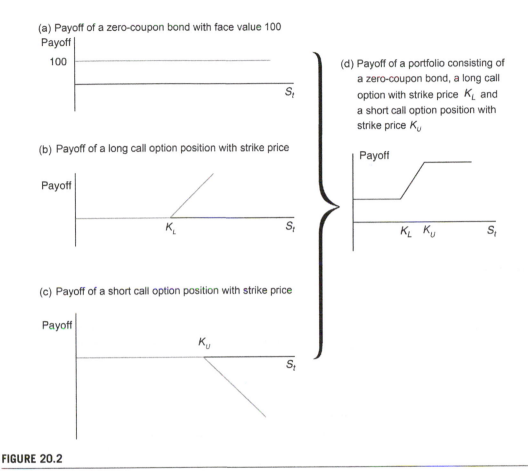

(a) Payoff of a zero-coupon bond with face value 100

(b) Payoff of a long call option position with strike price

(c) Payoff of a short call option position with strike price

(d) Payoff of a portfolio consisting of a zero-coupon bond, a long call option with strike price K_L and a short call option position with strike price K_U

FIGURE 20.2

Zero-coupon bond, long call and short call.

First there are *vanilla call* or *put* options. These were discussed in Chapters 9 and 11 and are not handled here. The second tool is *touch* or *digital* options, discussed in a later chapter, but we'll provide a brief summary below.

Touch or digital options are essentially used to provide payoffs (of cash or an asset) if some levels are crossed. Most equity structured products incorporate such *levels*. The third tool is reverse convertibles, which are conceptually related to convertible bonds with the important difference that they embed an option that is conveyed to the issuer and not the holder of the option.

The fourth tool is new; it is the so-called *rainbow* options. These are options written on the maximum or minimum of a *basket* of stocks. They are useful since almost all equity structured products involve payoffs that depend on more than one stock. The fifth tool is the *cliquet*. These options are important prototypes and are used in buying and selling *forward starting* options. Note that an equity structured product would naturally span over several years. Often the investor is offered returns of an index during a future year, but the initial index value during these future years

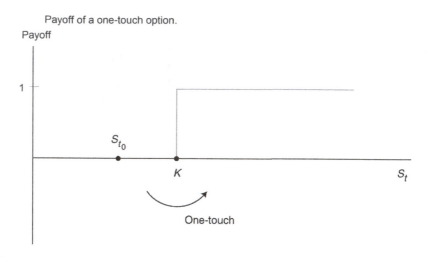

FIGURE 20.3

Payoff of a one-touch option.

would not be known. Hence, such options would have forward-setting strikes and would depend on *forward volatility*. Forward volatility plays a crucial role in pricing and hedging structured products, both in equity and in fixed-income sectors.[5]

20.2.2.1 Touch and digital options

Touch options are similar to the digital options introduced in Chapter 11. European digital options have payoffs that are step functions. Digital or binary options are contracts that pay out a fixed amount or nothing at expiration, depending on the settle price of the underlying asset. If, at the maturity date, a long digital option ends in-the-money, the option holder will receive a predetermined amount of cash, or, alternatively, a predetermined asset. As discussed in Chapter 11, under the standard Black–Scholes assumptions the digital option value will be given by the risk-adjusted probability that the option will end up in-the-money. In particular, suppose the digital is written on an underlying S_t and is of European style with expiration T and strike K. The payoff is R and risk-free rates are constant at r, as shown in Figure 20.3. Then the digital call price will be given by

$$C_t = e^{-r(T-t)}\tilde{P}(S_T > K)\ R \tag{20.7}$$

where $\tilde{P}$ denotes the proper risk-adjusted probability. Digital options are standard components of structured equity products and will be used below.

A *one-touch* option is a slightly modified version of the vanilla digital. A one-touch call is shown in Figure 20.3. The underlying with original price $S_{t_0} < K$ will give the payoff $R = \$1$ if (i) at expiration time T, the price of the underlying is above the strike price ($K < S_T$) and (ii) if the level K is breached only *once*.

[5]The equity structured products are often principal protected. A discussion of CPPI-type portfolio insurance which is relevant here will be considered later. We do not include the CPPI techniques in this chapter.

Otherwise the payoff will be zero. A previous chapter discussed a *double-no-touch* (DNT) option which is often used to structure *wedding cake* structures for FX markets.[6]

One advantage of digital options from the seller's point of view is that the maximum possible downside is known in advance. From the point of view of a bank or structure this implies that selling digital options is much more risk controlled and less negatively skewed as a strategy than a typical short volatility position. On the other hand, touch and digital options are less liquid than vanilla options since it is more difficult for market makers to hedge the digital options near the strike price around expiry. Some digital options are now exchange traded while they were previously OTC instruments. In 2008, the American Stock Exchange (Amex) and the CBOE launched exchange-traded European cash-or-nothing binary options. The more complicated tools are the *rainbow options* and the concept of *forward volatility*. We will discuss them in turn before we start discussing recent equity structured products.

20.2.2.2 Rainbow options

The term *rainbow options* is reserved for options whose payoffs depend on the trajectories of *more* than one asset price. Obviously, they are very relevant for equity products that have a basket of stocks as the underlying. The major class of such options is those that pay the *worst-of* or *best-of* the n underlying assets. Suppose $n = 2$; two examples are

$$\text{Min}[S_T^1 - K^1, S_T^2 - K^2] \tag{20.8}$$

where the option pays the smaller of the two price changes on two stocks and

$$\text{Max}[0, S_T^1 - K^1, S_T^2 - K^2] \tag{20.9}$$

where the payoff is the larger one and it is floored at zero. Needless to say the number of underlying assets n can be larger than 2, although calibration and numerical burdens make a very large n impractical.

20.2.2.3 Reverse convertibles

In the previous chapter, we saw that convertible bonds can be viewed as a portfolio of straight debt and an embedded call option. This typically implies that the issuer can issue the debt at a *lower* coupon than would be the case with straight debt alone. For a bond investor who is seeking to participate in the upside of the underlying equity or who is exploiting potentially undervalued volatility a convertible bond can be the right investment. However, for an investor who is seeking a high yield and is not concerned with participating in the upside a convertible bond is less attractive than the straight debt. How could a financial engineer create a product that pays a coupon that is *higher* than that of straight debt? The answer is that if the product embedded a short put position instead of a long call position, the buyer issuer of the instrument would receive a long put option from the buyer of the product and the issuer could in return compensate the buyer of the product with a higher coupon. In other words, the embedded put is financing the higher coupon. Such securities are called *reverse convertibles*. The put option is written on an underlying stock (or basket of

---

[6]A wedding cake is a portfolio of DNT options with different bases.

stocks) S_t and the conversion is not optional, but occurs automatically if S_t falls below a certain level K, which can be viewed as the strike price of the put option. The underlying stock or basket of stocks is referred to as *reference shares*.

Reverse convertibles have been mainly sold as structured products to retail investors. Reverse convertibles embed a put option that depends on the underlying stock volatility in a similar way that the embedded call option in a convertible bond depends on implied volatility. The fundamental difference between convertible bonds and reverse convertibles is, however, that convertible bonds offer participation in the *upside* of the underlying, while reverse convertibles offer participation in the *downside* of the underlying. Moreover, the conversion option is conveyed to the issuer, not the holder of the reverse convertible since the holder implicitly writes a put option on the underlying to the issuer. While the coupon payments in a reverse convertible may be considerably higher than the yields available in the bond market, reverse convertibles carry a higher risk than bonds because repayment of the principal amount is not guaranteed.

The payoff of a basic reverse convertible at maturity depends on two scenarios:

- Scenario 1: $S_T > K$. The underlying stock at maturity is above the strike price. In this case, the holder receives the coupon and 100% of the original investment.
- Scenario 2: $S_T < K$. The underlying stock at maturity is below the strike price. The holder receives a predetermined number of stocks.

For example, consider the buyer of a reverse convertible linked to the share price of ABC. If the stock price of ABC was initially £100, and in 1 year the stock price was £120 then scenario 1 obtains. The holder of a £1000 note would receive £100 for the 10% coupon, and the £1000 principal. Scenario 2 would occur if the stock price was £80 at the end of 1 year. Then the holder of the note would receive £100 for the coupon and £800 worth of stock. In other words, this would lead to a capital loss. Figure 20.4 illustrates the payoff profile of a typical reverse convertible. Figure 20.4a shows the payoff of a zero-coupon bond. Figure 20.4b shows the profit and loss from a short put option position with a strike price K. The write of the put option receives the put option premium if the underlying S_t remains above the strike price K. If we combine the zero-coupon bond with the short put position we obtain the profit and loss diagram for the reverse convertible which shows that the gain is higher than for a zero-coupon bond since the put option premium enhances the coupon.

Often the embedded option is not a simple option, but a knock-in option. This implies that the scenarios above are different in the sense that the condition is that the stock price S_t remains above the strike price at any point in time until maturity. The knock-in level is often set at 70−80% of the initial reference price.

When are reverse convertibles typically bought by investors? In a low interest rate and high market volatility environment reverse convertibles are popular since they provide a way for investors to receive an enhanced yield. However, the products' popularity does not mean that investors fully understand the price of the embedded put option that they are writing to the issuer of the product. If market volatility is high it is possible that the embedded put is very valuable and that the structure does not pass all of its value onto the buyer in the form of a coupon. In this case, the buyer takes on a large downside risk and may be surprised that in an equity market downturn the product leads to losses. Thus, investors should carefully compare the yield offered by the

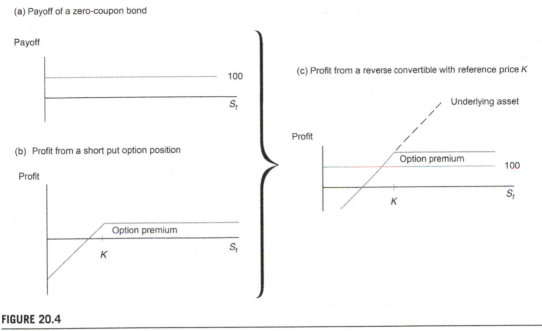

(a) Payoff of a zero-coupon bond

(b) Profit from a short put option position

(c) Profit from a reverse convertible with reference price K

FIGURE 20.4

Reverse convertible.

reverse convertible to money market rates, since if the two diverge significantly it may mean that the reverse convertible embeds significant risk.

In the 1990s, reverse convertibles were often issued with embedded short at-the-money put options. The downside of such products was that investors would suffer losses if the underlying was below the initial level. As investor demand waned in response to the market downturn following the bursting of the tech bubble, a new generation of products was developed that embedded a short at-the-money down-and-in put option. The barrier feature provided investors with additional protection since the put option would not be triggered unless the (down) barrier was reached. Such barrier reverse convertibles are popular structured products in Europe and in the United States. For the structurer, the barrier feature poses new challenges in practice, however, since the structure requires hedging long-dated equity barrier risks (with maturities between 3 and 5 years) and the Greeks of the products near the barrier tended to be unstable as discussed in the context of digital options above. Moreover, the large number of structured products and the relative illiquidity of the underlying equity market made hedging such products more difficult than is the case for FX products, for example.

Some structured products including reverse convertibles have embedded *call* features. Thus, such products have call and conversion features. Banks that issue structured products refer to these products with call features as *autocallable*, which is the abbreviation of *automatically callable*. This feature is often found in structured products with longer maturities. Such products are callable by the issuer if the reference asset is at or above its initial level on a specified observation date.

This is effectively an option for the issuer to redeem the product early. In this case, the investor receives the principal amount of their investment plus a predetermined premium. Similar to callable bonds, this call option conveyed to the issuer makes the product less attractive to the holder and therefore the yield on autocallables can be higher than for alternatives without this feature. Autocallable features are often found in capital guaranteed notes and barrier reverse convertibles.

Structured products do not just contain hedging risks from the perspective of the structurer or issuer, but also legal risks. There are many examples in the United States, Europe, and Asia/Pacific when buyers of structured products suffered losses and then went to court against the issuers of the products. Therefore, it is not just in the interest of the buyers but also in the interest of the structurers to understand the consequences of any embedded downside in the products. In some instances, banks settle law suits in order not to jeopardize future client relationships even if the products were properly marketed and sold. In 2013, IOSCO (the International Organization of Securities Commissions) published a report on the *Regulation of Retail Structured Products*. The objective of the report is to enhance investor protection. It outlines a range of regulatory options that securities regulators can use to regulate retail structured products. The following reading provides one example of regulatory issues and risks associated with structured products. The example is based on reverse convertibles discussed in the section.

EXAMPLE: *REVERSE CONVERTIBLES IN LIMBO*

Issuance of lira-denominated reverse convertibles ground to a halt in response to growing uncertainty over their tax status under Italian law. The Italian authorities are concerned that investors may buy the instruments in the belief that they are capital-protected fixed income instruments, when, in fact, they would be exposed to equity downside risk.

According to warrant market participants, about a month ago the Bank of Italy warned potential issuers of lira-denominated reverse convertibles that they might be classified as "abnormal securities." If classified as such, the coupon on the securities would be taxed at 27% instead of 12.5%—the rate for normal fixed income and derivative structures. Since then, lira reverse convertible issuance has dwindled as structurers await a decision on their status.

Market commentators said the Bank of Italy was concerned about the lack of principal protection in the structure. Reverse convertibles generate a yield considerably higher than that of vanilla bonds by embodying a short equity put position.

The investor receives a high coupon and normal bond redemption as long as a specific equity price is above a particular level at maturity. However, if the equity falls below the specified mark, then the investor is forced to receive the physical equity instead of the normal bond principal.

As a result, the buyer of the reverse convertible could end up with a long stock position at a low level, which would mean an erosion of initial principal. In contrast, the buyer of vanilla bond paper is assured of getting back the initial principal investment.

(Thomson Reuters IFR, May 1998)

The above reading illustrates the embedded downside risk in reverse convertible products. The issuer of the products will typically exercise the put option if the stock price is less than the strike price. As a result the bond holders or buyers of the reverse convertibles receive the stock under adverse conditions. The concern is that not all investors understand that even if the issuer does not default, the holders may suffer from substantial losses.

20.2.2.4 Cliquets

Cliquet options are frequently used in engineering equity and FX-structured products. They are also quite useful in understanding the deeper complexities of structured products. Cliquets are also known as ratchet options due to the resetting strikes in the structure.

A *cliquet* is a series of prepurchased options with forward setting strikes. The first option's strike price is known but the following options have unknown strike prices. The strike price of future options will be set according to where the underlying closes at the end of each future sub-period. The easiest case is at-the-money options. At the beginning of each observation period, the strike price will be the price observed for S_{t_i}. The number of reset periods is determined by the buyer in advance. The payout on each option is generally paid at the end of each reset period.

At reset dates, the option locks in the difference between the old and new strikes and pays it out as a profit. If the stock has moved in such a way during the preceding period that one of the component options expires out-of-the-money there will be no profit, and the investor will lose the premium corresponding to that period.

EXAMPLE

A five-year cliquet call on the S&P with annual resets is shown in Figure 20.5. Essentially the cliquet is a basket of five annual at-the-money spot calls.

The initial strike is set at, say, 1419, the observed value of the underlying at the purchase date. If at the end of the first year, the S&P closes at 1450, the first call matures in-the-money and the payout is paid to the buyer. Next, the call strike for the second year is reset at 1450 and so on.

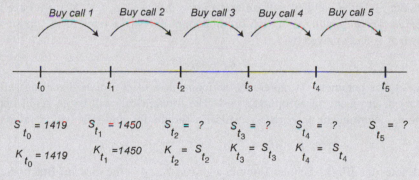

FIGURE 20.5

Cliquet.

To see the significance of a 5-year cliquet, consider two alternatives. In the first case, one buys a 1-year at-the-money call, then continues to buy new at-the-money calls at the beginning of future years four times. In the second case, one buys a 5-year cliquet. The difference between these is that the cost of the cliquet will be known in advance, while the premium of the future calls will be unknown at t_0. Thus a structurer will know at t_0 what the costs of the structured product will be only if he uses a cliquet.

Consider a 5-year maturity again. The chance that the market will close lower for 5 *consecutive* years is, in general, lower than the probability that the market will be down after 5 years. If the market is down after 5 years, chances are it will close higher in (at least) one of these 5 years. It is thus clear that a cliquet call will be more expensive than a vanilla at-the-money call with the same final maturity.

One of the selling points that sellers of cliquet options point to is that when volatility is expected to rise investors can lock in profits periodically rather than risk losing gains that have accumulated. An investor that takes the view that the underlying risky asset will go up or down over time, but does not have a view about the precise timing may see a cliquet structure as a better way of locking in profits. One recent example of a cliquet structure is market-linked certificates of deposit (CDs) in the US structured product market.

The important point is that cliquet needs to be priced using the implied *forward volatility* surface. Once this is done the cliquet premium will equal the present value of the premiums for the future options.

20.2.3 FORWARD VOLATILITY

Forward volatility is an important *concept* in structured product pricing and hedging. This is a complicated technical topic and can only be dealt with briefly here. Consider a vanilla European call written at time t_0. The call expires at T, $t_0 < T$ and has a strike price K. To calculate the value of this call we find an implied volatility and plug this into the Black–Scholes formula. This is called the Black–Scholes implied volatility.

Now consider a vanilla call that will *start at a later date* at t_1, $t_0 < t_1$. Yet, we have to price the option at time t_0. The expiration is at t_2. More important, the strike price of the option denoted by K_{t_1} is unknown at t_0 and is given by

$$K_{t_1} = \alpha S_{t_1} \tag{20.10}$$

where $0 < \alpha \leq 1$ is a parameter. It represents the moneyness of the forward starting call and hence is an important determinant of the option's cost. The forward call will be an ATM option at t_1 if $\alpha = 1$. Assuming deterministic short rates r, we can write the forward start option value at t_0 as

$$C\left(S_{t_0}, K_{t_1}, \sigma(t_0, \ t_1, \ t_2)\right) = e^{r(t_2 - t_0)} E_{t_0}^{\tilde{P}}\left[(S_{t_2} - \alpha S_{t_1})^+\right] \tag{20.11}$$

where $C(.)$ denotes the Black–Scholes formula and $\sigma(t_0, t_1, t_2)$ is the *forward* Black–Scholes volatility. The volatility is calculated at t_0 and applies to the period $[t_1, t_2]$. We can replace the (unknown) K_{t_1}, using Eq. (20.10) and see that the cliquet option price would depend only on the current S_{t_0} and on forward volatility.

Thus the pricing issue reduces to calculating the value of the forward volatility given liquid vanilla option markets on the underlying S_t. This task turns out to be quite complex once we go beyond very simple characterizations of the instantaneous volatility for the underlying process. We consider two special cases that represent the main ideas involved in this section. For a comprehensive treatment we recommend that the reader consult Gatheral (2006).

EXAMPLE: *DETERMINISTIC INSTANTANEOUS VOLATILITY*

Suppose the volatility parameter that drives the S_t process is time dependent, but is deterministic in the sense that the only factor that drives the instantaneous volatility σ_t is the time t. In other words we have the risk-neutral dynamics,

$$dS_t = rS_t + \sigma_t S_t dW_t \tag{20.12}$$

Then the implied Black–Scholes volatility for the period $[t_0, T_1]$ is defined as

$$\sigma_{BS}^{T_1} = \sqrt{\frac{1}{T_1 - t_0} \int_{t_0}^{T_1} \sigma_t^2 dt} \tag{20.13}$$

In other words, $\sigma_{BS}^{T_1}$ is the average volatility during period $[t_0, T_1]$. Note that under these conditions the variance of S_t during this period will be

$$\left(\sigma_{BS}^{T_1}\right)^2 (T_1 - t_0) \tag{20.14}$$

Now consider a longer time period defined as $[t_0, T_2]$ with $T_1 < T_2$ and the corresponding implied volatility

$$\sigma_{BS}^{T_2} = \sqrt{\frac{1}{T_2 - t_0} \int_{t_0}^{T_2} \sigma_t^2 dt} \tag{20.15}$$

We can then define the forward Black–Scholes variance as

$$\left(\sigma_{BS}^{T_2}\right)^2 (T_2 - t_0) - \left(\sigma_{BS}^{T_1}\right)^2 (T_1 - t_0) \tag{20.16}$$

Plug in the integrals and take the square root to get the forward implied Black–Scholes volatility from time T_1 to time T_2, $\sigma_{BS}^f (T_1, T_2)$

$$\sigma_{BS}^f(T_1, T_2) = \sqrt{\frac{1}{T_2 - T_1} \int_{T_1}^{T_2} \sigma_t^2 dt} \tag{20.17}$$

The important point of this example is the following: In case the volatility changes deterministically as a function of time t, the forward Black–Scholes volatility is simply the forward volatility. Hence it can be calculated in a straightforward way given a (deterministic) volatility surface. What intuition suggests is correct in this case. We now see a more realistic case with *stochastic* volatility where this straightforward relation between forward Black–Scholes volatility and forward volatility disappears.

EXAMPLE: *STOCHASTIC VOLATILITY*

Suppose S_t obeys

$$dS_t = rS_t + \sigma I_t S_t dW_t \tag{20.18}$$

where I_t is a zero-one process given by

$$I_t = \begin{cases} 0.30 & \text{with probability } 0.5 \\ 0.1 & \text{with probability } 0.5 \end{cases} \tag{20.19}$$

Thus, we have a stochastic volatility that fluctuates randomly (and independently of S_t) between high and low volatility periods. Then, the average variances for the periods $[t_0, T_1]$ and $[t_0, T_2]$ will be given respectively as

$$(\sigma^{T_1})^2 (T_1 - t_0) = \int_{t_0}^{T_1} E[(\sigma(I_t)dW_t)^2]$$
$$= (T_1 - t_0) \ (0.3)^2 \tag{20.20}$$

which implies that forward volatility will be 0.2.

Yet, the forward implied Black–Scholes volatility will not equal 0.2.

According to this, whenever instantaneous volatility is stochastic, calculating the Black–Scholes forward volatility will not be straightforward. Essentially, we would need to model this stochastic volatility and then, using Monte Carlo, price the vanilla options. From there we would back out the implied Black–Scholes forward volatility. The following section deals with our first example of equity structured products where forward volatility plays an important role.

20.2.4 **PROTOTYPES**

The examples of major equity structured products below are selected so that we can show the major methods used in this sector. Obviously, these examples cannot be comprehensive.

We first begin with a structure that imbeds a cliquet. The idea here is to benefit from fluctuations in forward equity prices. *Forward volatility* becomes the main issue. Next we move to structures that contain rainbow options. Here the issue is to benefit from the maxima or minima of stocks in a basket. The structures will have exposure to *correlation* between these stocks and the investor will be long or short correlation. Third, we consider Napoleon-type products where the main issue becomes hedging the forward volatility movements. With these structures, the volatility exposure will be convex and there will be a *volatility gamma*. If these dynamic hedging costs involving volatility purchases and sales are not taken into account at the time of initiation, the structure will be mispriced. Such dynamic hedging costs involving volatility exposures is another important dimension in equity structured products.

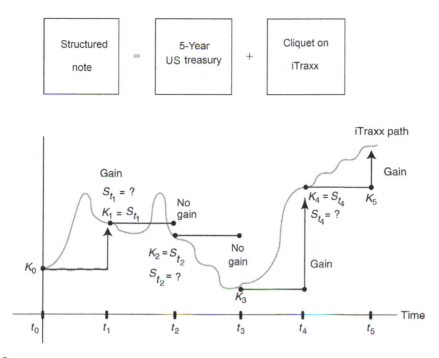

FIGURE 20.6

Structured product consisting of T-note and cliquet on iTraxx.

20.2.4.1 Case I: A structure with built-in cliquet

Cliquets are convenient instruments to structure products. Let S_t be an underlying like stock indices or commodities or FX. Let g_{t_i} be the annual rate of change in this underlying calculated at the end of year.

$$g_{t_i} = \frac{S_{t_i} - S_{t_{i-1}}}{S_{t_{i-1}}} \qquad (20.21)$$

where t_i, $i = 1, 2, \ldots, n$ are settlement dates. There is no loss of generality in assuming that t_i is denoted in years.

Suppose you want to promise a client the following: Buying a 5-year note, the client will receive the future annual returns $\lambda g_{t_i} N$ at the end of every year t_i. The $0 < \lambda$ is a parameter to be determined by the structurer. The annual returns are *floored* at zero. In other words, the annual payoffs will be

$$P_{t_i} = \text{Max}[\lambda g_{t_i} N, \ 0] \qquad (20.22)$$

λ is called the *participation rate*.

It turns out that this structure is less straightforward than appears at the outset. Note that the structurer is promising *unknown* annual returns, with a known coefficient λ at time t_0. In fact, this is a cliquet made of one vanilla option and four forward starting options. The forward starting options depend on forward volatility. The pricing should be done at the initial point t_0 after calculating the forward volatility for the intervals $t_i - t_{i-1}$. Figure 20.6 shows how one can use cliquets to structure this product. In this structured note, the structurer will take the principal N, deposit part of it in a 5-year Treasury note, and with the remainder buy a 5-year cliquet. The underlying risky assets in the example are assumed to be the iTraxx index, but it could also be an equity index such as the S&P500 or the MSCI World. As the figure shows the cliquet consists of five forward start options. The strike price of the first option is K_0 and since the price of the underlying is $S_{t1} > K_0$ at date t_1, the first option is exercised and pays out. The second option has a strike price of $K_1 = S_{t1}$ and at date t_2 it is out-of-the-money. The third option does not pay out either. The fourth and fifth options pay out. We would like to discuss this structure in more detail.

First let us incorporate the simple principal protection feature. Suppose 5-year risk-free interest rates are denoted by $r\%$. Then the value at time t_0 of a 5-year default-free Treasury bond will be given by

$$PV_{t_0} = \frac{100}{(1+r)^5} \tag{20.23}$$

Clearly this is less than 100. Then define the *cushion* Cu_{t_0}

$$Cu_{t_0} = 100 - PV_{t_0} \tag{20.24}$$

Note that

$$0 < Cu_{t_0} \tag{20.25}$$

and that these funds can be used to buy options. However, note that we cannot buy *any* option; instead we buy a *cliquet* since λ times the *unknown* annual returns are promised to the investor. The issue is how to price the options on these unknown forward returns at time t_0. To do this, forward volatility needs to be calibrated and substituted in the option pricing formula which, in general, will be Black–Scholes.

With this product, if the annual returns are positive the investor will receive λ times these returns. If the returns are negative, then the investor receives nothing. Note that even in a market where the long-run trend is downward, *some* years the investor may end up getting a positive return.

20.2.4.2 Case II: Structures with mountain options

Structures with payoffs depending on the maximum and minimum of a basket of stocks are generally denoted as *mountain options*. This type of exotic option combines features of basket options and range options and Société Générale was one of the first sellers of such options in the late 1990s. There are several examples. We consider a simple case for each important category.

20.2.4.2.1 Altiplano

Consider a basket of stocks with prices $\{S_t^1, \ldots, S_t^n\}$. A level K is set. For example, 70% of the initial price. The simplest version of an Altiplano structure entitles the investor to a "large" coupon if none of the S_t^i hits the level K during a given time period $[t_i, t_{i-1}]$. Otherwise, the investor will receive lower coupons as more and more stocks hit the barrier. Typically, once three to four stocks hit the barrier the coupon becomes zero. The following is an example.

EXAMPLE: *AN ALTIPLANO*

Currency: Eur; Capital guarantee: 100%; Issue price: 100.

Issue date: 01-01-2014; Maturity date: 01-01-2019

Underlying basket: {Pepsico, JP Morgan Chase, General Motors, Time Warner, Seven-Eleven}

Annual coupons:

Coupon = 15% if no stocks settle below 70% of its reference price on coupon payment dates.

Coupon = 7% if one stock settles below the 70% limit.

Coupon = 0.5% if more than one stock settles below the limit.

Figure 20.7 shows how we can engineer such a product. Essentially, the investor has purchased a zero-coupon bond and then sold five digital puts. The coupons are a function of the premia for the digitals. Figure 20.7a shows the zero-coupon bond and Figure 20.7b shows the payoff and profit of the short digital put position. The put option premium can be used to enhance the coupon. Clearly this product can offer higher coupons if the components of the reference portfolio have higher volatility.

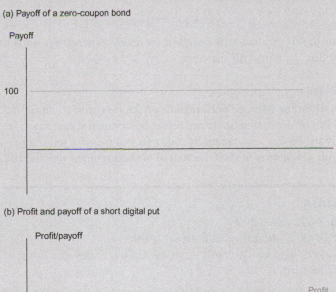

(a) Payoff of a zero-coupon bond

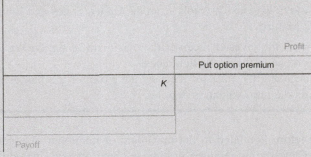

(b) Profit and payoff of a short digital put

FIGURE 20.7

Components of Altiplano.

This product has an important property that may not be visible at the outset. In fact, the Altiplano investor will be *long equity correlation*, whereas the issuer will be *short*. This property is similar to the pricing of CDO equity tranches and will be discussed in detail later. Here we consider two extreme cases.

Suppose we have a basket of k stocks S_t^i, $i = 1, 2, ..., k$. For simplicity suppose all volatilities are equal to σ. For all stocks under consideration, we define the annual probability of *not* crossing the level $K\,S_{t_0}$,

$$P\left(S_t^i, t \in [t_0, t_1] > K S_{t_0}^i\right) = (1 - p^i) \tag{20.26}$$

for all i and $t \in [t_0, T]$. Here the $(1 - p^i)$ measure the probability that the ith stock never falls below the level $K\,S_{t_0}$. For simplicity let all p^i be the same at p. Then *if* S_t^i, $i = 1, 2, ...$ are independent, we can calculate the probability of receiving the high coupon at the end of the first year as $(1 - p)^k$. Note that as k increases, this probability goes down.

Now go to the other extreme case and assume that the correlation between S_t^i becomes *one*. This means that all stocks are the same. The probability of receiving a high coupon becomes simply $1 - p$. This is the case since all of these stocks act identically; if *one* does not cross the limit, *none* will. Since $0 < p < 1$ with $k > 1$ we have

$$(1 - p)^k < (1 - p) \tag{20.27}$$

Thus, the investor in this product will *benefit* if correlation increases, since the investor's probability of receiving higher coupons will increase.

20.2.4.2.2 Himalaya

The *Himalaya* is a call on the average performance of the best stocks within the basket. In one version, throughout the life of the option, there are preset observation dates, say $t_1, t_2, ..., t_n$, at which the best performer within the basket is sequentially removed and the realized return of the removed stock is recorded. The payoff at maturity is then the sum of all best returns over the life of the product.

EXAMPLE: *A HIMALAYA*

Currency: Eur; Issue price: 100.
Issue date: 01-01-2013; Maturity date: 01-01-2019.
Underlying basket: 20 stocks possibly from the United States and Europe.
Redemption at Maturity:

If the basket rose, the investor receives the maximum of the basket of remaining securities observed on one of the evaluation dates.
If the basket declined, the investor receives the return of the basket of remaining securities observed on the last evaluation date.

In this case, the return is related to the maximum or minimum of a certain basket over some evaluation periods. Clearly, this requires writing rainbow options, including them in a structure, and then selling them to investors.

20.2.4.3 Case III: The Napoleon and Vega hedging costs

A *Napoleon* is a capital-guaranteed structured product which gives the investor the opportunity to earn a high fixed coupon each year, say, $c_{t_0} = 10\%$, *plus* the worst monthly performance in an underlying basket of k underlying stocks S_t^i. If k is large there will be a high probability that the worst performance is negative. In this case, the actual return could potentially be much less than the coupon c_{t_0}.

The importance of Napoleon for us is the implication of dynamic hedging that needs to accompany such products. The key issue is that Napoleon-type products cannot be hedged statically and require dynamic hedging. But the main point is that the dynamic hedging under question here is different than the one in plain vanilla options. In plain vanilla options, the practitioner buys and sells the underlying S_t^i to hedge the directional movements in the option price. This dynamic hedging results in *gamma* gains (losses). What is being hedged in Napoleon-type products is the volatility exposure. The practitioner has to buy and sell volatility dynamically.

These products have exposure to the so-called *volatility gamma*. The structurer needs to buy option volatility when volatility *increases* and sell it when volatility *decreases*. This is similar to the *gamma* gains of a vanilla option discussed in Chapter 9, except that now it is being applied to the *volatility* itself rather than the underlying price; hence the term *volatility gamma*. By buying volatility when vol is expensive and selling it when it is cheap, the structurer will suffer hedging costs. The expected value of these costs needs to be factored in the initial selling price, otherwise the product will be mispriced.

EXAMPLE: *NAPOLEON HEDGING COSTS*

Suppose there is a basket of 10 stocks $\{S_t^1, \ldots, S_t^{10}\}$ whose prices are monitored monthly. The investor is paid a return of 10% plus the worst monthly return among these stocks.

Suppose now volatility is very high with monthly moves of, say, 50%. Then a 1 percentage point change in volatility does not matter much to the seller since, chances are, one of the stocks will have a negative monthly return which will lower the coupon paid. Thus the seller has relatively little volatility exposure during high volatility periods.

If, on the other hand, volatility is very low, say, 9%, then the situation changes. A 1% move in the volatility will matter, leading to a high volatility exposure.

This implies that with low volatility the seller is long volatility, and with high volatility, the volatility exposure tends to zero. Hence the structurer needs to sell volatility when volatility decreases and buy it back when volatility increases in order to neutralize the vol of the position.

This is an important example that shows the need to carefully calculate future hedging costs. If volatility is *volatile*, Napoleon-type structured products will have volatility *gamma* costs to hedging costs that need to be incorporated in the initial price.

20.2.5 SIMILAR FX STRUCTURES

It turns out that cliquets, mountain options, Napoleons, or other structured equity instruments can all be applied to FX or commodity sectors by considering baskets of currencies of commodities instead of stocks. Because of this close similarity, we will not discuss FX and commodity structures in detail. Wystub (2006) is a very good source for this.

20.3 STRUCTURED FIXED-INCOME PRODUCTS

Structured fixed-income products follow principles that are similar to the ones based on equity or commodity prices. But the analysis of the principles of fixed income is significantly more complex for several reasons.

First, the main driving force behind the fixed-income structured products is the yield *curve*, which is a k-dimensional stochastic process. Equity or commodity indices are scalar-valued stochastic processes, and elementary structured products based on them are easy to price and hedge. Equity (commodity) products that are based on *baskets* would have a k-dimensional underlying, yet the arbitrage conditions associated with this vector would still be simpler. Second, the basis of fixed-income products is the LIBOR reference system, which leads to the LIBOR market model or swap models. In equity even when we deal with a vector process, there is no need to use similar models. Third, fixed-income markets are bigger than the equity and commodity markets combined. The very broad nature of fixed-income products' maturities and credits can make some maturities in fixed income much less *liquid*. Finally, the fixed-income structured products do have long maturities, whereas in equity or commodity-linked derivatives they are relatively short dated.

20.3.1 YIELD CURVE STRATEGIES

It is clear that most fixed-income structured products will deal with yield curve strategies. There aren't too many yield curve movements.

1. The yield curve may shift *parallel* to itself up or down—called the *level effect*.
2. The yield curve *slope* may change. This could be due to monetary policy changes or due to changes in inflationary expectations. The curve can steepen if the Central Bank lowers short-term rates or flatten if the Central Bank raises short-term rates. This is called the *slope effects*.
3. The "belly" of the curve may go up and down. This is in general interpreted as a *convexity* effect and is related to changes in interest rate volatility.

The next point is that many of these yield curve movements are at least partially *predictable*. After all, Central Banks often announce their future policies clearly to inform the markets. Structurers can use this information to put together constant maturity swap (CMS)-linked products that benefit from the expected yield curve movements. One can also add callability to enhance the yield further.[7]

---

[7]There is some possibility that investors are more interested in yield enhancement and are willing to tolerate some *duration uncertainty*. Accordingly, if a product is called before maturity, investors may not be too disappointed. In fact, for many structured products, many retail investors prefer that the product is called, and that they receive the first-year high coupon.

The reading below is one example of how the structurers look at yield curve strategies.

EXAMPLE

The popularity of CMS-linked structured notes derives from end users wanting to take advantage of the inverse sterling yield curve, which seems to have stabilized in the long end.

A typical structured note might be EUR5-50 million, with a 20-year maturity. It could pay a coupon of 8% for the first 5 years, and then an annual coupon based on the 10-year sterling swap rate, capped at 8% for the remainder of the note. It would be noncallable. The 10-year sterling swap rate was about 6.77% last week.

The long end of the sterling yield curve has likely stopped dropping because UK life insurers, who have been hedging guaranteed rate annuity products sold in the 1980s, have stopped scrambling for long-dated gilts. They have done so either because they no longer require further hedging, or because they have found more economical ways of doing so. If the long end fails to fall further, investors are more secure about receiving a long-term rate in a CMS, a trader said.

The sterling yield curve, which is flat for about three years, and then inverts, makes these products attractive for investors who believe the curve will disinvert at some point in the future, according to traders.

(Thomson Reuters IFR, January 31, 2000)

Hence it is clear that fixed-income structured products are heavy in terms of their involvement in LIBOR, swaption, and call/floor volatilities and their dynamics. Essentially, to handle them the structurer needs to have, at the least, a very good command of the forward LIBOR and swap models.

20.3.2 THE TOOLS

Some of the tools involved in designing and risk-managing structured products were discussed earlier. Digital and rainbow options and forward volatility were among these. Fixed-income structured products use additional tools. Two familiar tools are modified versions of *cap/floors* and *swaptions* and a third major tool is *CMS*. We review these briefly in this section.

A *digital caplet* is similar to a vanilla caplet. It makes a payment if the reference LIBOR rate exceeds a cap level. The difference is the payoff. While the vanilla caplet payoff may vary according to how much the LIBOR exceeds the level, the digital caplet would make a constant payment no matter what the excess is, given that the LIBOR rate is greater than the cap level.

A *Bermudan swaption* can be defined as an option on a swap rate s_t. The option can be exercised only at some specific dates t_1, t_2, ... When the option is exercised, the option buyer has the right to get in a payer (receiver) swap at a predetermined swap rate κ. The option seller has the obligation of taking the other side of the deal. Clearly with this product the option buyer receives swaps of different maturity as the exercise date changes.

CMS are fundamental elements of fixed-income structured products; hence, we review them separately.

20.3.3 CMS

A *constant maturity swap* is similar to a plain vanilla swap except for the definition of the floating rate. They were discussed in Chapter 14. There is a fixed payer or receiver, but the floating payments would *no longer* be LIBOR-referenced. LIBOR is a short-term rate with tenors of 1, 2, 3, 6, 9, or 12 months. It can only capture views concerning increasing or decreasing *short-term* rates. In a CMS, the floating rate will be *another vanilla swap rate*. This swap rate could have a maturity of 2 years, 3 years, or even 30 years. This way, instruments that benefit from increasing or decreasing long-term rates can also be put together. A 10-year CMS with a maturity of 2 years was shown in Figure 14.5. Note that there are variations on the CMS and it is possible that in a CMS, one party periodically pays a swap rate of a specific tenor (or the spread between swap rates of different specified tenors), known as the CMS rate, and in exchange receives a specified fixed *or* floating rate from the counterparty.

A special property of CMS should be repeated at this point. Note that at every reset date, the contract requires obtaining, say, a 10-year swap rate from some formal fixing process. This 10-year swap rate is normally valid for the next 10 years. Yet, in a CMS that settles semiannually, this rate will be used for the next 6 months *only*. At the next reset date, the new fixing will be used. Thus, the floating rate that we are using is not the "natural rate" for the payment period. In other words, denoting the 10-year floating swap rate by $s_{t_i}^{10}$ we have:

$$\frac{\left(1 + s_{t_i}^{10}\delta\right)}{\left(1 + L_{t_i}\delta\right)} \neq 1 \tag{20.28}$$

even though both rates are "floating." As long as the yield curve is upward sloping, the ratio will in fact be greater than one. But, in the case of a *vanilla* swap, each floating rate L_{t_i} is the natural rate for the payment period and we have:

$$\frac{\left(1 + L_{t_i}\delta\right)}{\left(1 + L_{t_i}\delta\right)} = 1 \tag{20.29}$$

This is true regardless of whether we have observed L_{t_i} or not. For this reason, the CMS require a *convexity adjustment*. This means, heuristically speaking, that the future unknown floating rates cannot simply be replaced by their forward equivalents. For example, if in 3 years we receive a 10-year floating swap rate s_t, during pricing we cannot replace this by the corresponding forward swap rate s_t^f. Instead we replace it with a forward swap rate adjusted for convexity.

20.3.4 YIELD ENHANCEMENT IN FIXED-INCOME PRODUCTS

Suppose an investor desires an enhanced return or a corporation wants a hedging solution at a lower cost. The general principle behind putting together such structured products is similar to equity products and is illustrated in the following contractual equation:

$$\boxed{\begin{array}{c}\text{Buy a}\\\text{standard}\\\text{asset}\end{array}} + \boxed{\begin{array}{c}\text{Sell one or}\\\text{more}\\\text{options}\end{array}} = \boxed{\begin{array}{c}\text{An asset}\\\text{with}\\\text{enhanced}\\\text{return}\end{array}} \tag{20.30}$$

As in equity structured products, in order to offer a return higher than the one offered by straight bonds, make the client sell one or more options. In fact, as long as the client properly understands the risks and is willing to bear them, the more expensive and more numerous the options are, the higher will be the return. If the client is a corporation and is looking for a cheaper hedge, selling an option would again lower the associated costs.

In structured fixed income products, there are at least two standard ways one can enhance yields.

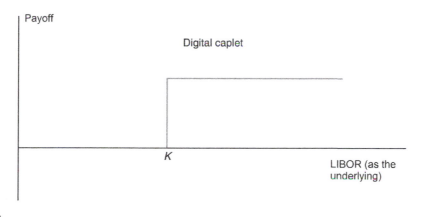

FIGURE 20.8

Digital caplet.

20.3.4.1 Method 1: Sell cap volatility

The first method to enhance yields is conceived so as to make the client sell cap/floor volatility. We saw in Chapter 17 that a caplet (floorlet) was an insurance written on a particular LIBOR rate that made a payment to the buyer of the option if the observed LIBOR rate went above (below) the cap level (floor level). A structurer could offer an investor a product that pays a higher coupon than would be available with the same credit risk and maturity if the product implicitly involves the investor writing a call option to the structurer. If interest rates are low and the yield curve is steep, investors may obtain above-market yields by betting against the market. Steep yield curves normally imply that interest rates are expected to rise. If an investor believes that interest rates will not rise as far as the market predicts and the investor is proven right, then the investor may be able to obtain an enhanced return. Of course, there is a price in the market for such views and the structurer could price the product based on the cost of a short caplet in the market plus a profit market. To specify a precise range of interest rates and payments associated with the option, the structurer can use digital caplets. Figure 20.8 shows the payoff of a long position in a digital caplet.

The structure will consider daily fixings of LIBOR and make coupon payments (enhanced by the implicit digital caplet-type premium) when a day's observation stays within a range, say [0,7%]. If the observed LIBOR exceeds that range for that day, no coupon is received.

This way, the client is short a digital caplet and is *selling* digital caplet volatility and he or she will receive an enhanced yield for bearing this risk. Such products are called Range Accrual Notes (RAN).[8] The client will earn interest for the proportion of the day's LIBOR observations that remain within the range. This feature would be suitable for a client who does not expect LIBOR rates to fluctuate significantly during the maturity period.

---

[8]As with other structured products different banks uses different terms to describe them. A RAN is also known as a corridor bond or note, a range floater, an accrual note, a LIBOR range note, a range accumulation note, a fairway bond or accretion bond index range note.

Let $F_t^{t_i}$ be the time-t 6-month *forward* rate associated with the LIBOR rate L_{t_i}. The associated settlement of the spot LIBOR is done, in-arrears, at time t_{i+1} and the day-count adjustment parameter is δ as usual.

Let the index $j = 1, 2, \ldots$ denote *days*. A typical caplet starts on day $t_i + (j - 1)$ and has an expiration 1 day later at $t_i + j$. Each caplet's payoff will depend on the selected reference rate that is followed daily. Often this would be the LIBOR rate at time $t_i + j$, L_{t_i+j}. Depending on this daily observation the caplets will expire in- or out-of-the-money. In other words, the seller collects *daily* fixings on the LIBOR rate and sees if the rate stayed within the *range* that day. If it does, there will be a digital payoff for that day (i.e., interest accrues); otherwise, no interest is earned for that particular day.

On the other hand, the actual *amount paid* will depend on *another* predetermined LIBOR rate. The rate applied to the payoff will be L_{t_i} + spread, settled at t_{i+1}.[9] According to this the return of the structured product return is a function of the payoffs of m digital options, where m is the number of calendar days in the payment period. Hence, the issue of *whether interest is earned or not* and the *payoff* depends on different LIBOR rates for each settlement period.

Symbolically, assuming that interest paid is constant at R, the jth day's payoff of the caplets can be written as

$$
\text{Pay}_{t_i+j} = \begin{cases} L_{t_i} \dfrac{1}{360} N & \text{if } L_{t_i+j} \geq L^{\max} \\ 0 & \text{if } L_{t_i+j} \leq L^{\max} \end{cases} \tag{20.31}
$$

where $L^{\max}$ is the upper limit of "range," N is the notional amount, and Pay_{t_i+j} is the daily payoff that depends on the jth observed LIBOR rate L_{t_i+j}. The investor will be *short* this caplet. How would this enhance the return?

Suppose there are m days during the interest payment period;[10] then the client is selling m digital caplets. For observation days these caplets are written on that day's LIBOR rate that we denoted by L_{t_i+j}. For weekends, the previous observation day's LIBOR is used. So these digital caplets can be regarded as an m-period digital *cap*, made of caplets with daily premiums c_{t_i+j}, if settled at the end of that day. The investor receives the daily premiums and pays off that day's payoff at every $t_i + j$, instead of collecting all the cap premiums at the contract conception t_0. The total value of these digital caplets at the payment time t_{i+1} will be given by

$$
C_{t_0} = \sum_{j=1}^{m} B(t_0, t_i + j) c_{t_i+j} \tag{20.32}
$$

where c_{t_i+j} are the caplet premiums for the option that starts at time $t_i + j$. Clearly these quantities are known at time t_0.[11] Note that this quantity is measured in time t_{i+j} dollars. Then the enhanced yield of the RAN settled at time t_{i+j} will be given by

$$
L_{t_i} + c_{t_i+j} \tag{20.33}
$$

---

[9]Or, alternatively, it could simply be a fixed rate, say R.

[10]For example, on a 30/360 day basis, and semiannual payment periods, we will have $m = 180$.

[11]Hence a more complicated notation could be useful. We can let the caplet premia be denoted by $c(t_0, t_i + (j - 1), t_i + j)$ which means that this is the premium calculated at time t_0, for a caplet that starts at $t_i + (j - 1)$ and expires at $t_i + j$.

At the time of inception t_0 of the note, the relevant LIBORs will be $\{L_{t_i}\}$ and these will be "equivalent" to s_{t_0}, the swap rate observed at the time of inception. So the enhanced yield can in fact be expressed by the constant R_{t_0}:

$$R_{t_0} = s_{t_0} + c_{t_i+j} \qquad (20.34)$$

If at time t_0 the structurer observes the (i) swap rate s_{t_0}, (ii) the forward volatilities of each digital cap c_j, and (iii) the discount factors $B(t_0, t_{i+1})$, then R_{t_0} can be calculated. Thus the investor will receive the $R_{t_0}N$ and will pay the payoffs of daily caplets that expire in-the-money. Naturally, all this assumes a correct calculation of the digital cap premium c_j. Here there are some small technical complications. However, before we get to these we look at an example.

EXAMPLE: *INTEREST RATE RANGE ACCRUAL/STRUCTURED DEPOSIT*

[Price quoted as of May 23, 2011]
    Customer View: US Dollar 3-month LIBOR will stay within the range of 0.25–0.60% (inclusive) in the coming 1 year
    Deposit Currency: Renminbi (CNY)
    Deposit Period: 1 year
    Deposit Amount: CNY1,000,000
    Interest Rate: US Dollar 3-month LIBOR (3M LIBOR)
    Reference Index
    Accrual in rate: 0.70% p.a.
    Accrual out rate: 0.00% p.a.
    Interest rate: Accrual in rate × (No. of days 3M LIBOR stays at or within the Accrual Range)/Total number of days + Accrual out rate × (No. of days 3M LIBOR stays outside the Accrual Range)/Total number of days
    Accrual range: 0.25–0.60% p.a.
    Interest period: Quarterly
    Interest payment: (Principal × Interest rate/4) for each interest period for each period
    Upon maturity: 100% of principal customer receives
                  (HSBC, https://www.hsbc.com.hk/1/2/hk/investments/sp/range-acc#example)

In the above example, there are some interesting variations since the deposit currency (CNY) differs from the currency underlying the interest rate reference index. This may be convenient for customers that deposit CNY but have reviews regarding US interest rates. Note that the range is not bounded from below by zero, but is instead bounded by $L^{min} = 0.25\%$ and $L^{max} = 0.60\%$. The accrual out rate is the rate received if the interest rate falls outside the range. This implies that the investor is not just short a digital cap, but also a digital floor.

Note that the underlying reference rate in the examples above that we use to determine the payoff of the digital caplets is 3-month or 6-month LIBOR rates, which are not the "natural rates" for the 1-day payoffs. Hence, the pricing of these digital caplets would require that a convexity adjustment is applied to the LIBOR rates, similar to CMS.

20.3.4.2 Method 2: Sell swaption volatility
Making a straight bond callable is the second way of enhancing yields. This will result in the investor being short *swaption volatility*.

The difference is important. In the first case one is writing a series of options on a *single* cash flow, namely the caplet payoff. But in the case of callable bonds, the investor will write options on all the cash flows *simultaneously* and will receive his principal 100 if the bond is called. Thus a swaption involves payoffs with baskets of cash flows. These cash flows will depend on different LIBOR rates. When the swaption is Bermudan, this is similar to selling several options (although dependent on each other) at the same time. Hence the Bermudan swaption will be more expensive and there will be more yield enhancement.

An investor that buys a callable LIBOR exotic has sold the issuer the right (but not the obligation) to redeem the notes at 100% of the face value at any given *call date*. A note that is callable just once (European) will have a lower yield than a comparable note with multiple calls (Bermudan). The question whether a callable note will be called or not depends on the initially assumed dynamics of the forward LIBOR rates versus the behavior of these forward rates and their volatilities in the future.

20.4 SOME PROTOTYPES

In this section, we discuss some typical fixed-income structured products and their engineering in detail using the tools previously introduced. We consider three typical structured products that are representative.

20.4.1 THE COMPONENTS

In order to engineer fixed-income structured products, the market practitioner will need a small number of components. These are

1. The relevance *discount curve* $B(t_0, t_i)$ in a certain currency. This will be used to discount future "expected" cash flows.[12]
2. A relevant *forward curve* in the same currency. This could be a forward LIBOR curve or a forward swap curve. Obviously, this can be obtained from the discount curve using relations such as[13]

$$(1 + F(t_0, t_i, t_k)\delta) = \frac{B(t_0, t_i)}{B(t_0, t_k)} \quad i \le k \tag{20.35}$$

This is equivalent to needing a market for vanilla swaps, i.e., a tradeable swap curve.
3. A market for CMS, since the structurer may want to receive or pay a *floating* rate that can be any point of the yield curve. A fixed CMS rate will be paid against this.[14]
4. A market for (Bermudan) swaptions if the structure is callable.
5. A market for caps/floors if the structure is of range accrual type.

---

[12]The expectation is with respect to some working probability measure.

[13]To review this go back to the arbitrage argument used to obtain the FRA rates. In this case, the FRA rate is defined more generally.

[14] Remember that the CMS rate is quoted as a spread to the vanilla swap rate or the relevant LIBOR rates:

$$s_{t_0}^{\text{cms}} = s_{t_0} + \text{spread}$$

We now show how some prototypes for fixed-income structured products can be manufactured using these components. The prototypes we discuss are *exotics* in the sense that the structurer cannot buy the note from some wholesale market and then sell it. The structurer has to *manufacture* the note. In other words, they are exotics because one side of the market does not exist and the structurer has to know how to price and hedge the product in-house.

20.4.2 CMS-LINKED STRUCTURES

There are (at least) two kinds of CMS-based products. Some link the coupon to a CMS *rate*. This would be similar to a floating rate note, but the floating rate would be a long-term rate this time. The second kind will be linked to a CMS *spread*. An investor buying CMS spread-linked structures will not be affected by parallel shifts in the yield curve. Rather, the buyer will be affected by the *slope* of the yield curve. Depending on whether the curve flattens or steepens, the buyer of the spread notes would benefit.

A note linked to a CMS rate enables investors to benefit from shifts in the long or short end of the yield curve over prolonged periods of time. The note pays a fixed coupon which goes up as the short end of the yield curve shifts upward over the lifetime of the product.[15] If made *callable*, the investor also receives enhanced returns. This is one of the most common types of structured fixed-income products. A straightforward example is shown below.

EXAMPLE: *A CMS-LINKED NOTE*

    Issuer: Bank ABC
    Tenor: 5 years
    Principal: Guaranteed at maturity at 100%
    Coupon:

    Year 1: 4.00%
    Year 2: CMS 10 + 2.5%
    Year 3: CMS 10 + 2.5%
    Year 4: CMS 10 + 2.5%
    Year 5: CMS 10 + 2.5%
  where CMS 10 refers to the 10-year constant maturity swap rate.

    Call: Callable on each coupon date.

Suppose an institutional investor expects 10-year rates to rise successively during the next 5 years. This would be equivalent to five consecutive shifts in the long end of the yield curve. Then the note provides a way to take exposure to this risk. We now discuss how CMS-linked products can be engineered using standard tools in fixed-income markets.

---

[15]If these shifts are predetermined and are written in the contract at contract initiation, then the instrument becomes a step-up note.

20.4.3 ENGINEERING A CMS-LINKED NOTE

In this section, we engineer a straightforward CMS-linked note. Suppose an investor has "a view" concerning the yield curve. For example, he or she expects the long end of the yield curve to shift up gradually during the next 5 years. For such an investor, the structurer would like to put together a portfolio of elementary assets that generates the promised risk-return characteristics of the relevant CMS-linked note with maturity $T = 5$ years. The example discussed above provides a good framework. For simplicity we assume annual interest payment dates. How do we engineer the CMS-linked note shown above?

1. First we select a CMS rate. For example, let the CMS rate be the floating rate of an m-period swap observed at every t_i (in-advance) and paid at time t_{i+1} (in arrears). There is no harm in thinking that $m = 10$ years.
2. Next we manufacture the high, known, first-year coupon c_{t_0}. There is no harm in thinking that $c_{t_0} = 5\%$ as in the above case. The question is, of course, how to manufacture such a high coupon. This value will be determined during pricing.
3. Next, offer a "floating" coupon for the following 4 years of the form

$$c_{t_i} = \text{cms}_{t_i}^m + \alpha_{t_0}^{t_i} \tag{20.36}$$

where $\text{cms}_{t_i}^m$ is the CMS rate of period t_i.
4. $\alpha_{t_i}^{t_i}$ will be known constants at time t_0 and will have to be determined during the pricing. Below we consider two different formulations for this term.
5. Make the note callable at call dates t_1, t_2, t_3, and t_4.

First, some general observations. The first-year coupon needs to be constant, since even with floating rate instruments the first coupon is always known. The investor is taking a position on "floating" rates, but the first floating rate will be observed at the purchase date. Second, the note is made callable. The structurer is making the investor sell an option, so that the returns can be enhanced by this option's premium. Third, the option is of Bermudan style, so it can be exercised four times at any of the four future dates. This option would naturally be more expensive than a vanilla swaption and hence the investor can be better compensated.

20.4.3.1 A contractual equation

Now we can put together a replicating portfolio, i.e., obtain a contractual equation for this note.

First, start with the floating CMS coupons $\text{cms}_{t_i}^m$. How can the structurer pay such coupons to the investor?[16] Here the answer is straightforward. The structurer gets into a 5-year *receiver* CMS at time t_0. We assume that in this CMS the floating CMS rate is exchanged against a floating LIBOR rate. In this swap the first-year coupon is fixed, but at every t_i, $i = 1, 2, 3, 4$ the going CMS rate $\text{cms}_{t_i}^m$ will be received by the structurer and that period's LIBOR L_{t_i} will be paid. The structurer will pass the $\text{cms}_{t_i}^m$ to the investor. So part of the coupon has now been constructed. This situation is shown in Figure 20.9. The source of the LIBOR payments will be discussed later.

What would α_{t_i} represent then? Calculate the premium of a 5-year Bermudan swaption—call it C_{t_0}. This swaption is on the *CMS rate* and can be exercised four times during the period $[t_0, T]$

---

[16] In other words, how would such payments be hedged?

Premium C_{t_0} from 5-year Bermudan option used to enhance coupon

$$C_{t_0} = B(t_0, t_1)\,\alpha_{t_0}^{t_1} + B(t_0, t_2)\,\alpha_{t_0}^{t_2} + B(t_0, t_3)\,\alpha_{t_0}^{t_3} + B(t_0, t_4)\,\alpha_{t_0}^{t_4} + B(t_0, t_5)\,\alpha_{t_0}^{t_5}$$

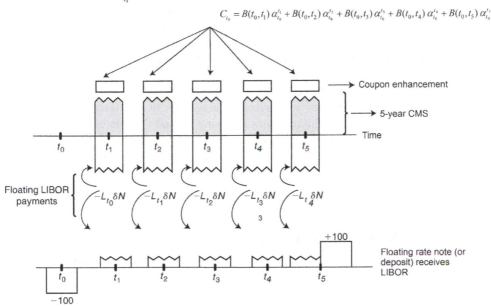

FIGURE 20.9

CMS-linked note.

annually at each t_i, $i = 1, 2, 3, 4, 5$. Allocate this premium to the four future years by choosing the $\alpha_{t_0}^{t_i}$ such that,

$$C_{t_0} = B(t_0, t_1)\,\alpha_{t_0}^{t_1} + B(t_0, t_2)\,\alpha_{t_0}^{t_2} + B(t_0, t_3)\,\alpha_{t_0}^{t_3} + B(t_0, t_4)\,\alpha_{t_0}^{t_4} + B(t_0, t_5)\,\alpha_{t_0}^{t_5} \tag{20.36}$$

This gives $\alpha_{t_0}^{t_i}$.

Finally, note that the structurer will be making LIBOR referenced payments to the CMS market maker. These payments will come from the original principal $N = 100$. The structurer will place the principal into a deposit account and receive floating LIBOR.

The replication is complete. The structurer buys a receiver a 5-year CMS on the 10-year CMS rate and sells a 5-year Bermudan swaption on the CMS rate that can be exercised annually. The original N received as principal is held in a deposit account.

The cash flows of this portfolio are shown in Figure 20.9. These cash flows are identical to the ones promised by the note. The structurer is essentially buying the portfolio, repackaging it, and then selling it to the client as a structured CMS-linked note. We can summarize such CMS-linked structures using contractual equations. For this particular CMS-linked note we have

$$
\boxed{\begin{array}{c}\text{Callable}\\\text{CMS-linked}\\\text{note}\end{array}}
=
\boxed{\begin{array}{c}\text{LIBOR}\\\text{deposit}\end{array}}
+
\boxed{\begin{array}{c}\text{Receive}\\\text{10-year CMS}\end{array}}
+
\boxed{\begin{array}{c}\text{Short}\\\text{Bermudan}\\\text{swaption}\end{array}}
\tag{20.37}
$$

Note that the replicating portfolio presents further opportunities to the structurer. The structurer may be in need to sell swaption volatility to other clients. Through this CMS-linked note the structurer is buying *swaption volatility* from retail clients. Hence, the note may be a good way of generating a needed supply of swaption volatility at an attractive price. The structurer will naturally sell the swaption at a higher (offer) price than the price of the swaption implicitly bought from the retail client.

A similar comment is valid for the CMS. The structurer may in fact be a CMS market maker and may be receiving fixed CMS rates and paying floating rates in a different deal. By marketing the CMS-linked note to the investor the structurer is paying a fixed CMS rate. This is like receiving the asked CMS rate and then paying the bid CMS rate.

In other words, the instruments that need to be purchased from the market may in fact already be in the books of the structurer. The structured product is then a good way of taking these risks, repackaging them, and then selling them, to retail clients.

20.4.4 ENGINEERING A CMS SPREAD NOTE

Suppose an investor expects that the yield curve will *steepen*. In that case, a *CMS-spread structure* is appropriate. First we consider the product itself.

This LIBOR exotic has three additional properties. First, the instrument is more complicated because the floating annual coupon will depend on *more* than one CMS rate, hence the "spread." Second, this spread will be offered to the retail investor after multiplying it with a *participation rate*. The participation rate has the potential of significantly enhancing the yields if the expectation turns out to be correct and if the product is not called. The spread in question can become negative. To prevent investors from paying negative coupons, such spread-related coupons are often *floored at zero*. Third, because the product is written on more than one CMS rate, the value of the structure will explicitly depend on the correlation between these rates.

The example below is typical of CMS spread notes.

EXAMPLE: *A CMS SPREAD NOTE*

Issuer: Bank ABC; Tenor: 5 years
Principal: Guaranteed at maturity at 100%
Coupons:

Year 1: 4.00%
Year 2–5: $17 \times (\text{CMS}^{10} - \text{CMS}^{20})$, max 22%, mm 0%

Call: Callable on each coupon date, t_i.

Why would an investor be interested in such a note? Suppose an institutional investor expects that the yield curve will steepen further. This can happen in two ways *at least*.

First, if at constant inflation and hence at constant long rates, the short end interest rates decline. A loosening of monetary policy by the Central Bank due to moderately weakening real economy may be one example. Second, short rates and Central Bank monetary policy may stay the same, but due either to inflationary pressures or strengthening economic activity the long rates may go up. These two cases represent two possible *views* and the spread note will provide one way to take an exposure toward such an event.

This product will offer higher rates if the yield curve keeps steepening gradually, as the coupon is dependent on the differential between the rates. Below we synthetically create a CMS spread note starting with more elementary instruments.

20.4.5 THE ENGINEERING

The CMS spread product has (at least) two novel financial engineering features: the product will illustrate the utilization of the *participation rate* and the way one has to *floor* the spreads. In addition, on the pricing side, we will see that correlation becomes an *explicit* additional risk.[17] Now we engineer the CMS spread note mentioned above. The first year coupon is already discussed; it comes from the first year LIBOR rate which is known, plus part of the option premium sold by the investor. The real novelty of the structure is in the coupons for years $i = 2, 3, 4, 5$. In fact, the coupons are of the form,

$$c_{t_i} = \max\left[\lambda\left(\text{cms}_{t_i}^m - \text{cms}_{t_i}^h\right) + \alpha_{t_i}, 0\right] \tag{20.38}$$

where $\text{cms}_{t_i}^m$ and $\text{cms}_{t_i}^h$ are two floating CMS rates observed at time t_i, with CMS maturities of m and h years, respectively. As shown in the example, there is no harm in thinking that $m = 10$ and $h = 2$ years, respectively. According to this, the coupon gets bigger or smaller depending on the difference between the 10-year and 2-year swap rates at times t_i in the future. At times t_i, we are taking snapshots of the swap curve and paying the client a coupon proportional to the *slope* of the curve. In this particular case, the client would get progressively higher coupons if the swap curve becomes steeper and steeper during the following 5 years. The note will benefit from progressive steepening *if* it is not called.

In order to engineer the coupons, first ignore the floor, and let the α represent a swaption premium allocated to the five settlement dates, as before. Now consider engineering the component,

$$\lambda\left(\text{cms}_{t_i}^m - \text{cms}_{t_i}^h\right) \tag{20.39}$$

These coupons can be replicated using the following position: pay 2-year CMS rate and receive 10-year CMS rate for 5 years during the t_1, t_2, t_3, t_4. This can be accomplished by getting in *two* CMS. The structurer buys λ units of the 10-year CMS and sells λ units of the 2-year CMS. For simplicity the cash flows for the first 2 years only are shown in Figure 20.10.

The figure incorporates the way CMS market quotes the CMS. It turns out that the market does this as a spread to the plain vanilla swap rate. Thus,

$$\text{cms}_{t_i}^{10} = s_{t_0,t_i}^5 + sp_{t_i}^{10,5} \tag{20.40}$$

and

$$\text{cms}_{t_i}^2 = s_{t_0,t_i}^5 + sp_{t_0}^{2,5} \tag{20.41}$$

---

[17]We say explicit, because implicitly one can claim that all CMS products have to deal with LIBOR rate correlations across the curve.

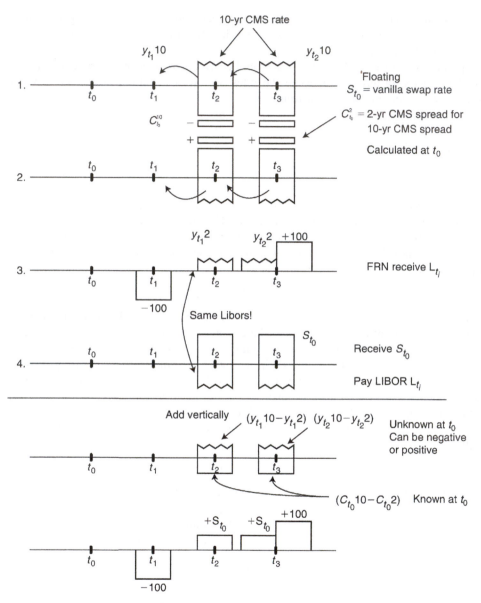

FIGURE 20.10

CMS spread product and two CMS.

where s_{t_0,t_i}^5 is the 5-year vanilla (forward) swap rate known at time t_0 for the case swap beginning at time t_i. $sp_{t_0}^{10,5}$ is the 10-year CMS *spread* for an instrument of 5-year maturity. The structurer will receive this. $sp_{t_0}^{2,5}$ on the other hand is the 2-year CMS *spread* for an instrument of 5-year maturity. The structurer will pay this. Note that at time t_0, both spreads are known for all t_i.

Adding together the components we have:

$$
\begin{aligned}
c_{t_i} &= \max\left[\lambda\left(cms_{t_i}^{10} - cms_{t_i}^{2}\right) + \alpha_{t_i}, 0\right] \\
&= \max\left[\lambda\left(\left(s_{t_0,t_i}^{5} + sp_{t_0}^{10,5}\right) - \left(s_{t_0,t_i}^{5} + sp_{t_0}^{2,5}\right)\right) + \alpha_{t_0}, 0\right] \\
&= \max\left[\lambda\left(sp_{t_0}^{10,5} - sp_{t_i}^{2,5}\right) + \alpha_{t_0}, 0\right]
\end{aligned}
\tag{20.42}
$$

This is the coupon. Note that at this point of the engineering the floating rates have dropped, and the only unknown on the right-hand side is the participation rate λ. Next we show how λ can be determined.

First remember that the principal N is received from the investor. This is placed in a deposit account that pays the LIBOR rates L_{t_i} in the future. At time t_0, one can get in a 5-year swap and convert these floating cash flows into a strip of *known* swap rate cash flows at the rate $s_{t_0}^{5}$. This means at every t_i the structurer will receive the known quantity $s_{t_0}^{5}$. This is shown in Figure 20.11 for a simplified 2-year maturity version of the 5-year maturity product. Consider the following:

$$
\lambda\left(sp_{t_0}^{10,5} - sp_{t_0}^{2,5}\right) = s_{t_0}^{5}
\tag{20.43}
$$

This is an equation where all quantities are known at t_0 except λ. Solve for λ and insert this number in the original coupon rate,

$$
c_{t_i} = \max\left[\lambda\left(cms_{t_i}^{m} - cms_{t_i}^{h}\right) + \alpha_{t_i}, 0\right]
\tag{20.44}
$$

In this expression, the unknowns are the CMS rates and this is the risk the client is assuming. To the structurer, however, the spread $cms_{t_i}^{m} - cms_{t_i}^{h}$ does not represent a risk since it can be hedged at a cost of $sp_{t_0}^{10,5} - sp_{t_0}^{2,5}$. The example below shows this simple calculation.

EXAMPLE

Suppose the 2-year and 10-year CMS trade at spreads of

$$
sp^{10,5} = 50 \text{ bps}
\tag{20.45}
$$

$$
sp^{2,5} = 20 \text{ bps}
\tag{20.46}
$$

The difference is 30 bps. Suppose also that the 5-year swap rate is 4.5%. Then λ will be given by

$$
\frac{450}{30} = 15
\tag{20.47}
$$

This explains the high participation rates. Even if the curve steepens by a small 30 bp, the investor can receive a coupon over 10%: α_{t_i} plus the 450 bp.

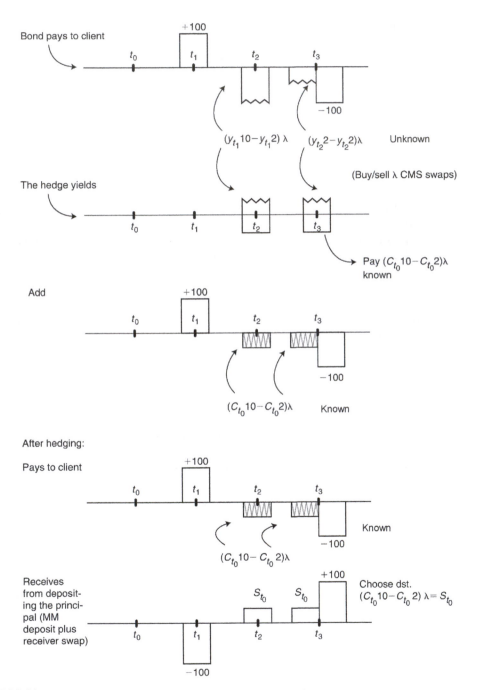

FIGURE 20.11

CMS spread product with participation rate.

The structurer has determined all the unknown parameters. Essentially the structurer will buy and sell two CMS with different reference rates, buy floors and sell swaptions on CMS rates to manufacture a synthetic of the 5-year CMS note. Then the structurer will repackage these as a structured note and sell it to clients.

20.4.5.1 A contractual equation

This characterization is shown in the contractual equation below.

$$(20.48)$$

As in the previous case, the synthetic structure can open several possibilities for the structurer. The structurer can buy swaption volatility, sell cap/floor volatility at advantageous rates from the retail client, and market them at better rates in the interbank market or to other clients. Again, as before, the structurer may in fact have some of the components of the synthetic on his books and the CMS spread note would be a good way of passing them along to other customers and removing them from the balance sheet.

Or, the structurer can take the exposure itself. The structurer can buy/sell all the components in the right-hand side of the contractual equation except the swaption. Note that, then, if the expectation turns out to be correct, the synthetic structure will have a positive value. But, the note is callable. The structurer will call the note and close the position on the hedge side with a good profit. It is partly for this reason that when the callable notes are likely to pay very high coupons they are in fact called. The investors basically receive the high initial coupon.

20.4.6 SOME OTHER STRUCTURES

A special case of structured fixed-income products is called *Target Redemption Note* (TARN). This security provides a sum of coupons until a *target level* is reached. The note then terminates early. TARNs may be quite popular with investors when interest rates are low, or more correctly when they are heading lower. The additional risk in TARNs is the uncertainty of termination date. Although this is like a callable note, there is a difference. The termination condition is *explicitly stated* in a TARN and can easily be priced using the LIBOR market model. On the other hand, callable products contain embedded Bermudan swaptions. It is much more difficult to determine when the option will be exercised (i.e., called). Some investors may prefer the more transparent way the TARNs are redeemed early. Like the others, the instrument is path dependent.

Another example is an *inverse floater*. As typical in such structured products, the first coupon is fixed. The subsequent coupons are set so as to depend on LIBOR inversely. When LIBOR rates decline, the coupon automatically increases. Often such coupons accumulate. If the accumulated coupons reach the target level, the note will be redeemed early. The client is paid the par value.

We can also give examples of structured products that are useful for asset−liability management. Consider a *trigger swap* that fixes borrowing costs for T years at a level lower than the current comparable market rates. In this sense, it is an asset−liability management tool. We discuss a simple variant. It is easy to complicate this simple prototype. Products such as trigger swaps belong in the category of fixed-income structured products although they are not marketed to investors. The clients are corporations and the product is useful in managing assets and liabilities. Still, the main idea is the same. The corporation has a *view* on the yield curve movements, or simply desires to lower hedging costs, which is the equivalent of yield enhancement.

20.5 CONCLUSIONS

We close this chapter with a comment on modeling structured products. Which model to use and how to calibrate the chosen models is clearly a crucial component of structured product trading. The structurer cannot buy these products in the wholesale market. The products need to be manufactured in-house. This requires extensive pricing and hedging efforts that will often depend on the model one is working with.

In equity structured products, versions of the stochastic volatility model are found to be quite effective and are widely used. For fixed-income versions of the forward LIBOR model that incorporate some volatility, smile needs to be used.

SUGGESTED READING

For LIBOR exotics consider **Piterbarg** (2005) and **Andersen and Piterbarg** (2010). **Wystup** (2006) is recommended for FX structured products. For the new equity structured products, **Gatheral** (2010) is required reading.

EXERCISES

1. Case Study: Reverse-Convertibles and Volatility Trading
    This case study shows another example of volatility trading and reverse convertibles.
    Read the case study below and answer the following questions.
    a. What is a reverse-convertible bond? How would you decompose this instrument? How would a corporate treasurer use reverse-convertible bonds?
    b. How are the market professionals using reverse-convertibles? Why is there a "flood"?
    c. What is a synthetic convertible?
    d. What is the other solution mentioned in the text? What are the possible risks behind this solution?

e. Finally, the regulators. Consider the case of French regulators and reverse-convertible bonds. Why would the regulators have an issue with these products? Do you think this is justified? Discuss briefly.

Reverse convertibles activity soared to new heights last week, driven by sustained high stock index volatility levels across Europe and stock market jitters. Warrant professionals highlighted the increasing number of structuring banks involved in the sector and the broadening range of issuance currencies.

These factors were evidence, they said, of the product's growing acceptance throughout the industry. Stock index volatilities across Europe have consistently been in the high twenties of percentage points since the beginning of the year.

Explaining the sudden surge in demand, equity derivatives professionals said reverse convertibles were an ideal means of taking advantage of current high volatility rates throughout Europe.

Others said interest in the product was spreading across Europe, whereas it had previously been confined to Switzerland and Germany. "There's been a flood. It's all over the place," enthused one German bank derivatives official.

Reverse convertibles are credit products with an embedded put option referenced to a particular quoted company stock. The face value of the bond is discounted to the equivalent value of the option premium. If the stock price breaches a certain minimum value threshold, the bond investor receives equity instead of a cash payout.

Credit Lyonnais Equity Derivatives was in the thick of the action, structuring three deals. Volkswagen was chosen as the underlying for an issue paying a 10% coupon; Telecom Italia was selected as the reference stock for an issue paying an 11% coupon; and ABN AMRO was the reference for a third.

If the stock price on the final date is greater than or equal to 95% of the initial price, then the investor receives 100% cash redemption. But if the price is below, then the investor receives one physical.

Thomson Reuters IFR, June 1998

2. Consider the swap and LIBOR curves available in Reuters or Bloomberg.
   a. Obtain the 3-month discount and forward curves
   b. Obtain the 2-year forward curve
   c. Find the components for the following note: maturity: 3 years
      Callable: each coupon payment date
      Payments: annual
      Coupons:
        Year 1: R_1
        Year 2: $\alpha_1 \times$ (2-year CMS) + previous coupon
        Year 3: $\alpha_2 \times$ (2-year CMS) + previous coupon
   Determine the unknowns R_1, α_1, α_2.
   d. Find the components for the following note:
      Maturity: 3 years
      Callable: each coupon payment date
      Payments: annual
      Coupons:
        Year 1: R_1
        Year 2: $\alpha \times$ [(3-year CMS) − (2-year CMS)] + β_1
        Year 3: $\alpha \times$ [(3-year CMS) − (2-year CMS)] + β_2
   Determine the unknowns R_1, α, β_i.
   e. In the latter case, when would $\beta_1 = \beta_2$?

3.   What follows is the description of a rather complex swap structured by a bank. The structure is sold for the purpose of liability management and involves an exotic option (digital cap) and a CMS component.

   At time 0 the bank and the client agree to exchange cash flows semiannually for 5 years according to the following rules:

   •   The bank pays semiannually, 6m-LIBOR on the notional amount. This is a vanilla swap.
   •   The client pays a coupon c_t according to the following formula:

$$c_t = c_{1t} - c_{2t} \qquad (20.49)$$

   where

$$c_{1t} = \begin{cases} \text{LIBOR} + 47 \text{ bp}\% & \text{if} \quad \text{LIBOR} < 4.85\% \\ 5.23\% & \text{if} \quad 4.85\% < \text{LIBOR} < 6.13\% \\ 2.98\% & \text{if} \quad \text{LIBOR} > 6.13\% \end{cases}$$
$$c_{2t} = \{\text{cms } (30Y) - (\text{cms } (10Y) + 198 \text{ bp})\}$$

   for the first 2 years. And

$$c_{2t} = 8\{\text{cms}(30Y) - (\text{cms}(10Y) + 198 \text{ bp})\}$$

   for the last 3 years.

   a.   Unbundle this LIBOR exotic into vanilla products the best you can.
   b.   Why would an investor demand this product? What would be his or her expectations?

4.   Show how you would engineer the following *Snowball Note*.

   Issuer: ABC bank
   Notional: $10 mio
   Tenor: 10 years; Principal: Guaranteed at maturity
   Coupon: Yr 1; Q1: 9.00%
   Q2: Previous Coupon + CMS10 4.65%
   Q3: Previous Coupon + CMS10 4.85%
   Q4: Previous Coupon + CMS10 5.25%
   Yr 2 Q1: Previous Coupon + CMS10 5.45%
   Yr 2 Q2−Q4: Previous Coupon + CMS10 5.65%
   Yr 3: Previous Coupon + CMS10 5.75%
   Yr 4−10: Previous Coupon
   Coupon subject to a minimum of 0%
   Call: Callable on each coupon

   a.   What is the view of the investor?
   b.   What are the risks?
   c.   Now forget about the call provision and calculate the coupons paid under the two following realizations of LIBOR rates:
     Realization 1 = 5.0, 6.0, 6.5, 7.0, 8.0, 9.0, 10.0
     Realization 2 = LIBOR stays at 3.5
   d.   How can you characterize these coupons using a swap? What type of swap is this?

    e. Suppose you have 8 annual FRAs quoted to you. How can you price these coupon payments?

    f. How do you generate the first-year coupon?

    g. Are the coupons floored at 0?

    h. Write a contractual equation representing this instrument.

5. Show how you would engineer the following *CMS spread note*.

    Issuer: ABC

    Notional: $10mio

    Tenor: 10 years

    Principal: Guaranteed at maturity

    Coupon:

    Yr 1: 11.50%

    Yr 2−10: $16 \times (CMS30 - CMS10)$, max of 30%, min 0%

    Call: Callable on each coupon date by the issuer

    a. What is the view of the investor?

    b. What are the risks?

CHAPTER OUTLINE

Principles of Financial Engineering.

739

21.1 INTRODUCTION

In this chapter, we introduce securitization, asset-backed securities (ABSs) and the modeling of collateralized debt obligations (CDOs) as well as other credit structured products. We explain the economic drivers of securitization, potential for misaligned interests, and the role of securitization in the run-up to the GFC. The securitization and CDO market has rebounded in recent years and we offer an outlook for these markets based on recent regulatory developments. The chapter also deals with credit indices, their implications in creating *tranched* securities.

Tranching a basket of credit-risky instruments makes the trading and pricing of credit correlation possible. Once we learn how to strip correlation from tranched products we can price it and then trade it. The issues are discussed within credit default context, but the techniques themselves are very general and buying and selling correlation is routinely applied in creating equity and FX-based structured products as well as we saw in earlier chapters. Yet, in this chapter we focus on *default* correlation and discuss the issue using the credit names included in the *iTraxx* or, alternatively, *CDX* indices.

The chapter also discusses more recent credit indices ABX and LCDX and credit structured products. Finally, we discuss one post-GFC financial innovation in the form of contingent convertibles (CoCos). CoCos are a hybrid product and we explain how valuation approaches from earlier chapters can be applied to the valuation of CoCos.

21.2 FINANCIAL ENGINEERING OF SECURITIZATION

Every business or financial institution is associated with a "credit" or, more precisely, a credit rating. If this entity issues a debt instrument to secure funding, then the resulting bonds, in general, have the same credit rating as the company. Yet, a company can also be interpreted as the receiver of future cash flows with different credit ratings. The value and the credit rating of the company will be a function of these future cash flows. Not all the receivables will have the same rating. For example, some cash flows may be owed by institutions with a dubious credit record, and these cash flows may not be received in the case of default or delinquency. Other cash flows may be liabilities of highly reputable companies, may carry a low probability of default, and may indeed be received with very high probability.

A debt issue will be backed by an average of these credits, since it is the average receivable cash flows that determine the probability that the bonds will be repaid at maturity. If the receivables of a company carry mostly a relatively high probability of default, then the company may experience difficulties in the future and, hence, may end up defaulting on the loan. Alternatively, the credit spread on the bond will increase and the investors will be subject to mark-to-market losses. All these possibilities reflect on the debt issued by this company and are factors in the determination of the proper cost of funding.

On the other hand, instead of issuing debt on the back of the average cash flows to be received in the future, the company can issue special types of bonds that are backed only by the *higher rated* portion of the receivables. Clearly, such receivables have a comparatively lower probability of default, and this makes the bonds carry a lower default probability. The funding cost will decrease significantly. The company has thus securitized a certain portion of the cash flows that are to be received in the future. In other words, securitization can be regarded as a way to issue debt and raise funds that have a higher rating than that of the company. It is also a way of repackaging various cash flows.

What are the critical aspects of such financial engineering? Essentially, various cash flows are to be analyzed and a proper selection is made so as to obtain an optimal basket. This is then sold to investors through special types of bonds.

Yet, besides the financial engineering aspects, securitization involves (i) legal issues, (ii) balance sheet considerations, and (iii) tax considerations. Securitization is a way of funding an operation. Instead of selling bonds or securing bank lines, the company issues ABSs. The option of securitization helps corporations and banks make decisions among the various funding alternatives.

21.2.1 CHOOSING CASH FLOWS

Consider Figure 21.1. Here, we show a bank that expects three different (random) cash flows in the future. The institutions that are supposed to pay these cash flows have different credit ratings. For example, the first series of cash flows, rated BBB, may represent credit card payments. The third could represent the random cash flows due to mortgages made in the past. Arguably, people pay better attention to the timely payment of mortgages than they do to the timely payments of unsecured credit card proceeds. Credit card defaults are much more common and plausible than mortgage delinquencies. Thus, the mortgage cash flows will be rated higher, say, with an A rating as shown in the figure.

Now, if the company is set to receive these three cash flows only, assuming similar liabilities, the company's average rating will perhaps be around BBB+. A corporate bond issued by the company will carry a BBB+ credit spread.

Consider two different ways of packaging the same cash flows. If the company "sells" cash flow 3 and backs a bond issue with this cash flow only, the probability of default will be much

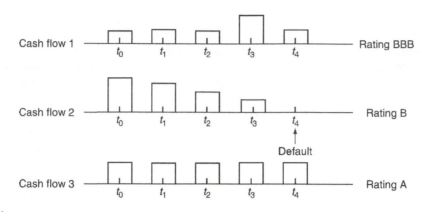

FIGURE 21.1

Cash flow characteristics of three risky cash flows.

lower and funding can be secured at a lower rate. A bond backed by cash flows 1 and 2 will have a lower credit rating than a bond based on cash flow 3, but depending on the diversification benefits of combining different cash flows and the attractiveness of offering a bigger pool of receivables in this second option, it may make sense for a company to combine cash flows 1 and 2 instead of selling separate bonds based on them.

Early examples of securitization can be found as far back as the eighteenth century but the first modern residential mortgage backed security (MBS) was created in 1970 by the U.S. Department of Housing and Urban Development. During the 1990s, banks moved into securitization of a wider range of assets and more complex structures. Loans from credit cards, mortgages, equity, and cars were packaged and sold to investors. A typical strategy was to (i) buy the loan from an *originator*, (ii) warehouse it in the bank while the cash flows stabilized and their credit quality was established, (iii) hedge the credit exposure during this warehousing period, and (iv) finally sell the loan to the investor in the form of ABS.

21.2.2 THE CRITICAL STEP: SECURING THE CASH FLOW

The idea of securitization is quite simple. Instead of borrowing against the average quality of the company's receivables, which is what happens if the company sells a straight corporate bond, the entity decides to borrow against a higher-quality *subset* of the receivables. In case of default, these receivables have a higher chance of being collected (recovered) and, hence, the cost of these funds will be lower.

But there is a critical step. How can the buyer of an ABS make sure that the receivables that are supposed to back the security are not used by the company for other purposes, and that, in the case of bankruptcy, these receivables will be there to cover losses?

The question is relevant, since after issuing the ABS security, it is still the original company that handles the business of processing new receivables (e.g., by issuing new mortgages), as well as the receipts of cash generated by such cash flows, and then uses them in the daily business of the firm. Clearly, there must be an additional mechanism that guarantees, at least partially, that these cash flows will be there in case of default.

A *bankruptcy remote SPV (special-purpose vehicle)* is one such mechanism used quite often to resolve such problems in practice. The idea is as follows. (i) The issuing company forms a SPV, which is a shell company, often independent of the parent company, and whose sole purpose is to act as a *vehicle* in structuring the ABS. (ii) Steps are taken to make the SPV *bankruptcy remote*. That is to say, the probability that the SPV itself defaults is zero (since it does not engage in any meaningful economic activity other than that of issuing the paper), and in case the original company goes bankrupt, the underlying cash flows remain in the hands of the SPV. (iii) The issuing company draws all the necessary papers so that, at least from a legal point of view, the cash flows are sold to the SPV. This is a *true sale at law*.

The idea is to transfer the right to these cash flows and guarantee them under the ABS as much as possible.[1] In fact, several SPVs with different purposes can be layered to make sure that the ABS has the desired characteristics.

1. An SPV may be needed for tax reasons.
2. Another SPV may be needed for balance sheet reasons.
3. Still another SPV may be needed to comply with other regulations.

Hence, one possible structure can be layered as shown in Figure 21.2. Note that here, the first SPV is a subsidiary of the company, so the company can "buy" the cash flows, and this is the reason for its existence. But if the SPVs keep the cash flows, these will still be on the balance sheet of the original company.

Finally, note the role played by the investment bank. The first three layers make up the ABS structure, and the investment bank still has to handle the original sale of the ABS. The structure clearly shows that the ABS has three important purposes, namely lowering the funding cost, managing the balance sheet, and handling tax and accounting restrictions.

21.2.3 SOME COMPARISONS

The first use of securitization concerns *funding costs*, as already discussed. Securitization is a form of funding. But we must add that it is also an unconstrained form of funding, and an *off-balance-sheet* form of liquidity for small and medium-sized companies. Finally, it is a diversified funding source. This way it can lower leverage. Securitization also implies less public disclosure.

Securitization is neither secured corporate financing, nor the sale of an asset. It is a hybrid, a combination of both that uses the well-accepted legal, regulatory, tax, and accounting concepts that already exist.

21.2.3.1 Loan sales

We should also compare whole loan sales versus securitization. Securitization is on a service-retained basis, whereas a loan sale is service released. The buyer of the loan would like to service

---

[1]However, the existence of such bankruptcy-remote SPVs is not sufficient to fully guarantee that the cash flows will be there to use. A bankruptcy is sometimes a chaotic event, involving difficult legal issues.

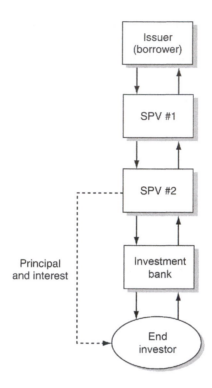

FIGURE 21.2

Securitization structure.

the loan himself or herself. Another point is the retention of credit and prepayment risk. In a loan sale, 100% of these risks are transferred. With securitization, some of these risks may be retained. A loan sale is also often done at a premium, whereas securitization issues are often around par.

In addition, there is a timing issue. In securitization, cash flows from assets are often invested in short-term investments and then transferred to the bond holders. Thus, the investor receives the payments later than the *servicer*. Finally, securitization sometimes uses credit enhancements, and this makes the paper somewhat more liquid.

21.2.3.2 Secured lending

Securitization is similar to secured financing, with one important difference. In an ABS, the issuing company is not liable for its ABSs. It is as if the company has not really "borrowed" the funds. A separate legal entity needs to be established to do the borrowing. Securitization is structured so that this entity becomes the legal owner of the asset. If the company defaults, the cash flows will not belong to the company, but to the SPV. This way the owners of the bonds have an *ownership* interest in the case of securitization, whereas in the case of secured lending, they only have a *security* interest.

21.3 ABSs VERSUS CDOs

This chapter introduces financial engineering applications that relate to ABSs and securitization. It turns out that securitization and hybrid asset creation are similar procedures with different objectives. From the *issuer's* point of view, one is a solution to balance sheet problems and it helps to reduce funding costs. From an *investor's* point of view, securitization gives access to payoffs the investor had no access to before and provides opportunities for better diversification. Hybrid assets, on the other hand, can be regarded as complex, ready-made portfolios.

A financial engineer needs to know how to construct an ABS. In fact, engineering is implicit in this asset class. The remaining tasks of pricing and risk managing are straightforward. A similar statement can be made about hybrid assets.

This book cannot deal with the technical issues concerning ABSs and CDOs. Still, credit indices should be put in context so that they can be compared to ABSs and CDOs. This is also a good opportunity to introduce the basic definitions of and the differences between the ABS- and CDO-type securities. A CDO is a security whose value is backed or *collateralized* by a pool of underlying fixed-income assets. The CDO pays a return which is based on payments from the performance of this pool of assets.

Imagine three different classes of defaultable securities. The first class is defaultable bonds. Except for U.S. Treasury bonds, all existing bonds are defaultable and fall into this category. The second class can be defined as "loans." There are several types of loans, but for our purposes here, we consider just the secured loans extended to businesses. Finally, there are loans extended to households, the most important of which are mortgages.

ABS securities can be defined by either using other assets such as loans, bonds, or mortgages, or more commonly using the stream of cash flows generated by various assets such as credit cards or student or equity loans. In this section, we consider the first kind. Figure 21.3 displays the way we can structure an ABS security. A basket of debt securities is divided into subclasses with different ratings, then the subclasses are placed "behind" different *classes* of the ABS security. This means that any cash flows or the corresponding shortfalls from the original debt instruments would be passed on to the investor who buys that particular class of ABS security. In contrast, if there are defaults, the investors receive an accordingly lower coupon or may even lose their principal. Note that according to Figure 21.3, classes of ABS securities have different ratings because they are backed by *different* debt instruments. Often in an ABS, the credit pool is made of loans such as credit cards, auto loans, home equity loans, and other similar consumer-related borrowing. When the underlying instrument is a mortgage, the ABS is called an MBS, which we examined in detail in Chapter 17. In each case, the rating of the ABS is determined by the rating of the loans that back it.[2]

Figure 21.4 shows a "cash" CDO, also called a *funded* CDO. Again a pool of credit instruments are selected. But they are classified in a very different way. The CDO classes called *tranches* are formed, not by classifying the underlying securities, but the risk in them. In fact, all CDO tranches will be backed by the *same* pool of securities. What distinguishes the tranches is the *subordination* of the default risk. The ABS categorizes the securities *themselves*. A CDO categorizes the *priority*

---

[2]There are other ways one can define the classes of ABS securities.

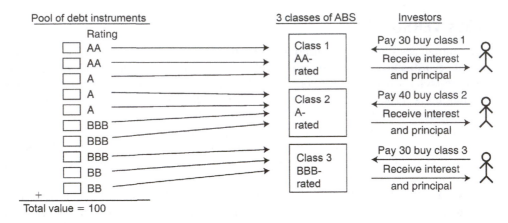

FIGURE 21.3

ABS structure.

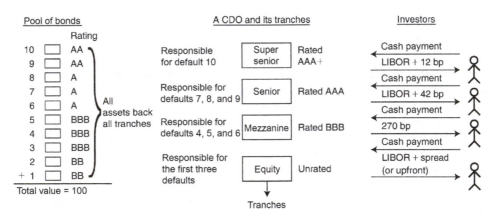

Note that these are cash bonds backing the CDO.

FIGURE 21.4

Cash CDO.

of payments during defaults. The first few defaults will be the first tranche, then if defaults continue the next tranche will suffer and so on.

21.3.1 TRANCHES AND SOME SECURITIZATION HISTORY

In the 1990s, the securitization process that started with ABS was then extended to CDOs, collateralized loan obligations (CLOs), and their variants. In packaging bonds, mortgages, loans, and ABS securities into CDO-type instruments, banks went through the following practice. The banks decided to keep the *first loss piece* called the *equity tranche*. For example, the bank took responsibility for the first, say, 3% of the defaults during the CDO maturity. This introduced some *subordination* to the securities sold. The responsibility for the next tranche, the mezzanine took the risk of the defaults between, say, 3% and 6% of the defaults in the underlying portfolio. This was less risky than the equity tranche and the paper could be rated, say, BBB. Institutional investors who are prevented by law to invest in speculative grade securities could then buy the mezzanine tranche and, indirectly, the underlying loans even if the underlying debt was speculative grade. Consider the following illustrative example. Take 100 bonds all rated B. This is a fairly speculative rating and many institutions by law are not allowed to hold such bonds. Yet, suppose we sell the risk of the first 10% of the defaults to some hedge fund, at which time the default risk on the remaining bonds may in fact become A. Institutional investors can then buy this risk. Hence, credit enhancement made it possible to sell paper that originally was very risky.

Then, the very high-quality tranches, called *senior* and *super senior*, were also kept on banks' books because their implied return was too small for many investors. As a result, the banks found themselves *long equity tranche* and *long senior and super senior tranches*. They were short the *mezzanine* tranche which paid a good return and was rated investment grade (IG). It turns out, as we will see in this chapter, that the equity tranche value is *positively* affected by the default correlation while the super senior tranche value is *negatively* affected. The sum of these positions formed the correlation books of the banks. Banks had to learn *correlation trading*, *hedging*, and *pricing* as a result.

As discussed below, the bespoke tranches evolved later into standard tranches on the credit indices. The *equity tranche* was the first loss piece. A protection seller on the equity tranche bears the risk of the first 0−3% of defaults in *standard tranches*. The *mezzanine tranche* bears the second highest risk. A protection seller on the tranche is responsible, by convention, for 3−6% of the defaults. The 6−9% tranche is called senior mezzanine. The *senior* and *super senior* iTraxx tranches bear the default risk of 9−12% and 12−22% of names, respectively. In this chapter, we study the pricing and the engineering of such tranches.

If the default risk comes from a pool of bonds, the CDO is called a collateralized bond obligation (CBO). If the underlying securities are loans, then it is called a CLO. The term CDO is more general and the securities it represents may be a mixture of all these. In fact, these underlying securities may include MBS or other ABS securities. Even tranches of other CDOs were sometimes included in a CDO before the GFC.

21.3.2 A COMPARISON OF ABSs AND CDOs

Consider an investor pays 100 in cash to buy the ABS and the CDO structures shown in Figures 21.3 and 21.4. By buying the ABS or the CDO the investor is, in a sense, buying the underlying securities as a pool. The underlying pool can be arranged so that the investor can choose among classes of ABS securities with different ratings. In the CDO, the *priority* of interest and principal payments determine the rating of the tranche. Thus different classes of ABS securities (at least as defined here) represent different underlying assets grouped according to their credit rating, whereas the tranches of CDOs are actually backed by the same underlying assets, while having different ratings due to how quickly they will be hit during successive defaults. Figure 21.4 shows that the cash flows in CDOs have a cascading effect between classes. This pattern is therefore often referred to as a *cash flow waterfall*.

21.3.2.1 Credit indices

Markets prefer to trade credit risk through standardized and transparent credit indices. We discussed the 2009 standardization of single-name credit default swap (CDS) contracts in Chapter 18. Credit indices are liquid benchmarks that are based on portfolios of reference names or single-name CDS contracts. The corresponding default risks are then repackaged by slicing the portfolio into various *standard tranches* bearing different risk and return characteristics.[3]

The indices allow investors to take a position on the broad corporate credit market with a single trade and provide liquid hedging instruments for institutions to hedge their complex credit exposures. Within the IG corporate credit names, there exist two major CDS index categories in the world: *CDX* and *iTraxx*. During the discussion in this chapter, we focus mostly on *iTraxx* indices.

The credit indices are constructed by Markit's subsidiary International Index Company (IIC) after a dealer liquidity poll. Market makers submit a list of names to the IIC based on the following criteria.

1. The entities have to be incorporated in Europe and have to have the highest CDS trading volume, as measured over the previous 6 months. Traded volumes for entities that fall under the same ticker, but trade separately in the CDS market, are summed to arrive at an overall volume for each ticker. The most liquid entity under the ticker qualifies for index membership.
2. The list of entities in the index is ranked according to trading volumes. IIC removes any entities rated below BBB (by S&P) and those on negative outlook.
3. The final portfolio is created using 125 names. It is assembled according to the following classification: 10 Autos, 30 Consumers, 20 Energy, 20 Industrials, 20 TMT, and 25 Financials. Each name is weighted equally in the overall and subindices.[4]

The markets prefer to trade mostly the subinvestment grade indices called *iTraxx* Crossover (XO) in Europe and CDX High Yield (HY) in the United States. These indices see a large majority of the trading, they contain fewer names, and the spreads are significantly higher.[5]

21.3.2.2 Synthetic CDOs

A credit index and the associated tranches are a synthetic version of a CDO. The underlying assets are related to the bonds of the reference portfolio names but they are *not* purchased! Hence, they are *unfunded*. Thus, with an index and most of the tranches, there is no initial cash payment or receipt involved.[6] While a cash CDO pays LIBOR plus some spread, the synthetic (unfunded) CDO pays just the spread because it involves no initial cash payments. In this sense, the relationship between an index (unfunded CDO) and a funded CDO is similar to the relationship between cash bonds versus the CDS written on that name.

Figure 21.5 shows how the index can be interpreted as a synthetic unfunded CDO. In this figure, the tranches are selected so that they conform to the standard *iTraxx* tranches. Synthetic CDOs

---

[3]A custom-made tranche is known as a *bespoke tranche*. Note that a bank or corporation trying to hedge a basket of loans would in general need or sell bespoke tranches.

[4]On July 24, 2014, Markit announced that the iTraxx Europe Index will expand from 125 to 130 names as of March 2015.

[5]For more information on the indices, the reader should visit www.markit.com which actually collects and processes the quotes for the indices.

[6]Only the most risky tranche is traded with *upfront*. Many CDS contracts require payment of an upfront free.

A credit index and the implied synthetic CDO. Note that there is no initial cash investment.

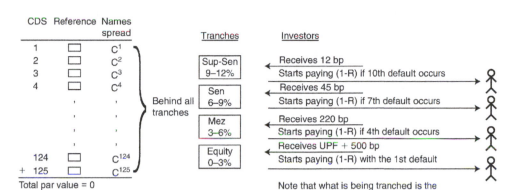

FIGURE 21.5

A credit index and the implied synthetic CDO.

were popular in the run-up to the GFC since they were cheaper and easier to create than traditional CDOs which as we saw were mostly created from mortgages. The following example illustrates recent developments in CDS and CDO markets.

EXAMPLE

The GFC led to a recession that started in December 2007 and ended in June 2009. Default rates following the recession peaked in 2010 according to S&P but subsequently declined to historical lows in 2012. As a result of the fall in defaults the demand for hedging default risk declined and the amount of single-name CDS outstanding fell. According to data from the BIS, at the end of 2012, the notional amount outstanding of single-name CDS was $14.3tn which represents a 57% decline from its $33.4tn peak in the first half of 2008. It is important to bear in mind that part of the decline can be attributed to investors compressing or clearing their swaps portfolios but this does not explain the majority of the decline.

   The shrinking of the single-name CDS markets may have been welcomed by some regulators and politicians but it also means that banks and other market participants have fewer hedging tools available. As is clear from Figure 21.5, the creation for synthetic CDOs contributed to the growth of the single-name CDS market before the GFC as CDOs sold credit protection in waves. Demand for CDS protection stemmed from banks that used swaps to offset their loan exposures. At the same time hedge funds were active market participants and "correlation traders" as they arbitraged discrepancies in the prices of CDS or different CDO tranches. The synthetic CDO market is slowly recovering, as Figure 21.6 shows, but under new tougher capital rules banks no longer benefit as much from hedging their loans with CDS. Many remaining correlation traders are now involved in managing legacy synthetic CDOs. Correlation traders were some of the most active users of single-name CDS and with the decline in correlation trading liquidity in the single-name CDS market also shrank. An additional factor that depressed the single-name CDS activity was the EU's 2012 decision to ban buying naked sovereign CDS protection. One area where activity has been robust has been in the area of CDS on indices as some market participants use them for trading and hedging purposes. Index CDS has been used as hedges against macroeconomic uncertainty and as imperfect hedges against firm specific credit risk.

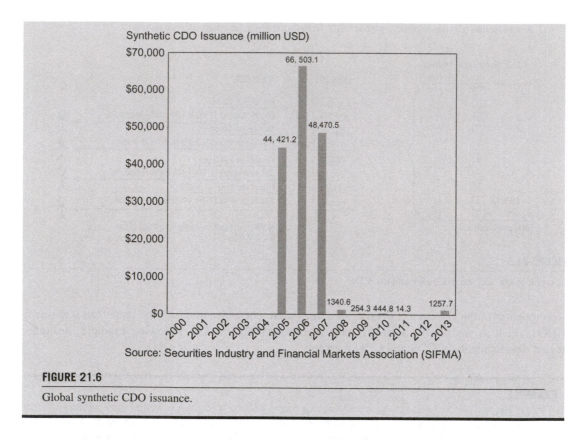

FIGURE 21.6

Global synthetic CDO issuance.

As the example above illustrates the single-name CDS and synthetic CDO market have shrunk compared to pre-GFC levels, but the products continue to trade. Figure 21.6 shows that global issuance in the synthetic CDO market has rebounded slightly in 2013 although it remains at <2% of the pre-GFC annual issuance level. There are reasons to believe that securitization as well as CDO markets will continue to exist. The reason is that following a period of critical evaluation of the role of these markets in the GFC following by regulation, governments realize that securitization and CDOs play an important role in providing funding to the economy. As stricter regulation and capital requirements have seen banks reduce lending, securitization may provide one of the avenues to increase lending to businesses and stimulate the economy. The reading above also highlights the role of correlation trading based on CDOs which we will examine in further detail in the next chapter.

21.4 A SETUP FOR CREDIT INDICES

We saw in Chapter 18 that a single-name CDS is a contract that provides protection against a default event on the part of a single issuer or *reference name*. The protection seller pays zero and receives a constant premium if no default event occurs. The CDS premium is cds_{t_0}. If a default event does occur the protection seller pays the difference between the promised (face) value of the underlying issue (100) and the market value of the defaulted bond. The recovery rate is denoted by R. Consider the single-name CDS in Figure 21.7. The maturity T is assumed to be 1 year for simplicity and the notional amount N is 100. The net payment by the protection seller in case of default is $(1 - R)100$.

A credit index is obtained by selecting n such reference entities indexed by $j = 1, \ldots, n$. This pool is called the reference portfolio. The associated CDS rates at time t, denoted by $\{cds_t^j\}$ are assumed to be arbitrage-free. A tradeable CDS index is formed by putting these names in a single contract, where if N dollars insurance is sold on the index, then this would correspond to a sale of insurance on *each* name by an amount $\frac{1}{n} N$.

Let I_t represent the spread on the credit index for the preselected n names. The "index" would then trade as a *separate* security from the underlying single-name CDSs. It should be regarded as a standalone security with a known maturity, coupon, and standardized documentation. Trading the index is equivalent to buying or selling protection on the reference portfolio names with *equal weights*.[7] The spread of this portfolio, i.e., I_t, is quoted separately from the underlying CDSs.[8]

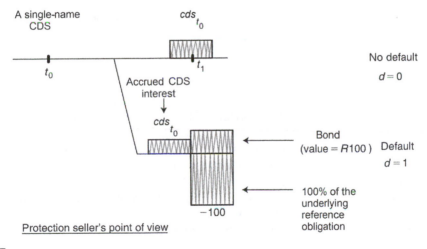

FIGURE 21.7

A single-name CDS.

[7]Trading the series 21 *iTraxx* Europe index requires an upfront payment based on a fixed coupon of 100 bps. The 5-year maturity version of the series was launched in 2014 and matures in 2019. Position in the series corresponds to buying or selling a security that pays a fixed coupon for 5 years with quarterly settlements. When a default occurs, the protection seller compensates the protection buyer by an amount equal to $\frac{1}{n} N$, with $n = 125$.

[8]Actual indices trade somewhat differently than the underlying CDSs. For example the single-name CDSs trade in first short coupon, whereas the indices trade in accrued. There are other differences concerning the recovery, restructuring, settlement, and other aspects, one of which is the constant contract spread (coupon) in the indices. See www.markit.com.

At the outset, one may think that I_t would (approximately) equal the simple average of the underlying CDSs. This turns out not to be the case, except in exceptional circumstances when all the CDS rates and their volatilities are the same. In general, we will have

$$I_t \leq \frac{1}{n} \sum_{j=1}^{n} \text{cds}_t^j \tag{21.1}$$

In fact, why should a traded credit index trade as if all credits are weighted equally? It is more reasonable that the pricing of a reference portfolio would weigh the underlying names using their survival probabilities as well as the level of the corresponding CDS rates. In other words, the index *spread* would be *DV01-weighted* even though the composition of the index is weighted equally.

EXAMPLE

The index has two names:

$$\text{cds}^1 = 20 \text{ bp} \tag{21.2}$$

$$\text{cds}^2 = 4500 \text{ bp} \tag{21.3}$$

reminiscent of the spreads during the credit crisis. The average spread will be:

$$\frac{4500 + 20}{2} = 2260 \tag{21.4}$$

According to these spreads it is much less likely that the first name defaults in the near future compared to the second name. This means that if an investor sold protection with notional $N = 50$ on each of the individual CDSs, the average receipt during the next five years is unlikely to be 2260. This is the case since the default of the second name appears to be imminent. Assuming that default occurs immediately after the transaction, the average return of the remaining 50 invested in the first credit would be 20 bp for the next five years.

On the other hand, if the index traded at 2260 bp, the index protection seller will continue to receive this spread on the remaining $50.

According to this, the index spread will deviate more from the simple average of the underlying CDS spreads, the more dispersed the latter are. This is due to the DV01 weighing mentioned above.

Thus, the credit indices are fundamentally different from the better-known equity indices such as S&P500 or Dow. The latter are supposed to equal some average of the price of the underlying stocks, otherwise there would be an (index) arbitrage opportunity. In the credit sector this difference is far from zero and the traders trade this difference if it deviates from a calculated fair value.

For the sake of presentation, the discussion will continue using the iTraxx IG index as representing the I_t. The next example will help to understand the cash flow structure of such an index.

EXAMPLE

Suppose $n = 3$. The underlying names are A, B, and C. We consider a 1-year maturity with settlement in arrears at the end of the year.

Suppose $N = 3$ is invested in this index. All names are equally weighted and all probabilities of default p^i are assumed to be the same. This makes the position similar to putting \$1 on each name in a reference portfolio of three single-name CDSs. However, as mentioned above, the spread on the portfolio as a whole may deviate significantly from the average of the three independent single-name CDSs.

Essentially, there will be four possibilities or scenarios at the end of the year. There may be no defaults, one default, two defaults or three defaults at time t_1. The structure will be as shown in Figure 21.8.

On the other hand, if the position was for more than 1 year the default possibilities would be more complicated for the second year. This is discussed later in this chapter.

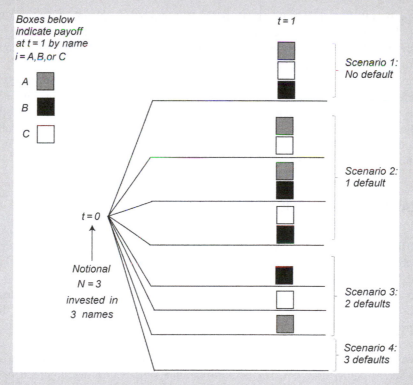

FIGURE 21.8

Index with three names.

What happens when an entity that belongs to the underlying reference names defaults? Consider the case of the *iTraxx* index with $n = 125$ names. The resulting process for this default is similar to that for a single-name CDS. The protection seller pays to the protection buyer a compensation equal to $\frac{1}{125} N$. In return, the protection buyer delivers deliverable bonds with the same face value.

After the default, trading will continue in the case of a credit index, albeit with the notional amount reduced by $\frac{1}{n} N$. The index will then have $n - 1$ names and any contracts written on it will continue as if the notional amount has become:

$$N - \frac{1}{n}N = \frac{n-1}{n}N \tag{21.5}$$

Further defaults would lower this notional in a similar way.

21.5 INDEX ARBITRAGE

Index arbitrage is well known in equity markets. A stock index is made of a given and known number of stocks. The Index itself can be traded in the futures markets. The carry strategy applies and by buying (short selling) the underlying stocks, one can create a synthetic futures contract which can be used to take arbitrage positions on the stock index.[9]

By analogy one can define index arbitrage in credit as positioning in the index itself versus its underlying reference names using the single-name CDSs. In equity markets, this strategy is reasonable. In credit markets, it faces several complications even though theoretically it sounds similar. One issue is the liquidity of single-name CDSs. In an index consisting of 125 names, not all underlying CDSs may be liquid. The bid—ask spreads for these individual CDSs may be too wide in many cases and trading the underlying against the index may become too costly.

21.5.1 REAL-WORLD COMPLICATIONS

There are two major issues that lead to different valuations in the index versus its constituents. The first is the differential treatment of *restructuring*. The second is a technical issue and deals with the *convexity* of the index versus the underlying CDSs. Credit index arbitrage is possible, but only after taking into account such divergences explicitly.

First note that as mentioned in Chapter 18, CDX indices trade with no restructuring. Yet, the CDSs for most of the IG credits treat restructuring as a credit event.[10] To correct for this valuation difference, the value of the restructuring risk must be subtracted from the individual CDSs. This value, however, is not observed separately and has to be "estimated."

The convexity issue is more technical. Fixed-income instruments have convexity as discussed previously, whereas equity does not. This also differentiates index arbitrage in credit from the index arbitrage in equity. Consider the 5-year CDS with a CDS rate cds_t^j bps for the jth name in the

---

[9]Specifically, this is done by selling the index and buying the underlying stocks to carry to the expiration date if the index is too high relative to the reference portfolio, or vice versa.

[10]A restructuring event may trigger a payment associated with the credit name on an individual CDS, yet this will not affect the corresponding index.

reference portfolio. As the underlying CDS rate changes, the market value of the future CDS coupon payments will change nonlinearly since the fixed coupons will be discounted using discount factors that will be a function of *survival probabilities*.[11] According to this, the CDS value would be a nonlinear function of the cds_t^j, the CDS rate, which leads to convexity gains.

On the other hand, an investor to the credit index receives a single coupon. The convexity of this cash flow will be different from the convexity of the portfolio of underlying CDSs, since the convexity of the average is different from the simple average of convexities. The effects of convexity and of restructuring should be taken out explicitly in order to come up with a *fair value* for the index.

21.5.2 CREDIT SKEW TRADE

We had seen earlier that volumes in single-name CDS trading have declined since the GFC. The following reading illustrates the interaction between regulation related to banks' capital requirements, index arbitrage activity, and liquidity in the single-name CDS market. Following the GFC capital requirements for banks have increased and this has increased the funding costs for certain trades. The index skew in the credit markets refers to the fair value or market value of the index versus its theoretical value. The *skew* is the difference between the value of the index using all of the underlying credit names and where the index is actually trading. In general, a positive skew implies that the skew is trading wider than its fair value.

EXAMPLE

Credit index arb trades under threat

> Dealers look to sell legacy books ahead of introduction of leverage rule
>
> Dealers are concerned that credit index arbitrage activity could be largely killed off by the coming introduction of a leverage ratio restriction on banks, further draining liquidity from the single-name CDS market as well as punishing banks with large legacy books.
>
> Arbitraging the difference between CDS indices and their underlying single-name components—often referred to as "skew"—is a favourite trade of market makers and hedge funds.
>
> But this activity becomes hugely punitive under the proposed leverage ratio rules as they currently stand. Dealers will only be able to net out long and short positions if they match precisely when executed with the same counterparty or if they face a clearing house, effectively ruling out skew trading.
>
> "Skew is a big problem for the banks, and it could be a big problem for liquidity. Roughly a third of liquidity in investment-grade CDS comes from skew activity. Funds are worried where the liquidity on single names comes from if that goes away," said the head of European credit trading at a major bank.
>
> Single-name CDS tend to lag credit spread movements on indices, which are investors' first port of call to hedge macro exposures. By lining up all the single-name CDS constituents in iTraxx Main versus the index itself, traders can pick up a handful of basis points for ironing out the discrepancy.
>
> It is a regular trade for sophisticated hedge funds, but dealers print skew packages as well to keep on their own books as a trading position, while also providing liquidity to the market. This type of activity has already become harder to carry off as liquidity in single-name CDS has dwindled since the financial crisis. These days, it is usually necessary to include multiple dealers in a package in order for a trader to line up all of the 125 single-name CDS in iTraxx Main in sufficient size, as banks' balance sheets have come under increasing pressure.

---

[11]Remember that a default event means the coupons expected by the protection seller would stop. Hence, the cash flow characteristic of the CDS changes with the default event and the discounting should take this into account.

> *Legacy issues*
>   *Many dealers have large legacy skew books sitting on their balance sheets, which look set to get clobbered by the leverage ratio. The current proposals focus on gross notional derivatives exposures and have strict rules about netting down positions. Even though a skew package is market risk neutral—buying the single names versus selling the index—such positions do not net.*
>
> **(Thomson Reuters IFR, 10 January 2014, "Credit index arb trades under threat", by Christopher Whittall)**

The above reading illustrates the importance of index arbitrage or skew trading for single-name CDS liquidity. The liquidity in single-name CDS is driven not only by regulation and capital requirements related to CDS markets but also to CDO and securitized products. The example shows the important role that regulation plays in the application of financial engineering principles and real-world arbitrage strategies.

21.6 TRANCHES: STANDARD AND BESPOKE

The popularity of the indices is mostly due to the existence of *standard* index tranches that permit trading credit risk at different levels of subordination. At the present time, *default correlation* can be traded only by using index tranches. The index itself is used to hedge the sensitivity of tranches to changes in the probability of default.[12] We discuss the formal aspects of pricing default correlation in the next chapter. In this section, we introduce standard index tranches and then introduce their relationship to default correlation.

The *equity tranche* is the first loss piece. By convention equity tranche bears the highest default risk.[13] A protection seller on the standard equity tranche bears the risk of the first 0−3% of defaults on the reference portfolio. The equity tranche spread is quoted in two components. The first, which is quoted by the market maker, is paid *upfront*. It is for the investor to "keep." The second is the 500 bps *running* fee which is paid depending on how much time passes between relevant events.[14]

The *mezzanine tranche* bears the second highest risk. A protection seller on the tranche is responsible, by convention, for 3−6% of the defaults.[15] This tranche is also upfront with a 500 bps spread.

The *senior* and *super senior iTraxx* tranches bear the default risk of 6−9% and 9−12% of names respectively. The 6−9% tranche also has a 300 bps fixed spread.

The numbers such as 0−3% are called the lower and upper *attachment points*. The ones introduced above are the attachment points for *standard tranches*. If attachment points are different from those and negotiated individually with the market maker they become *bespoke tranches*.[16] A bespoke trance will not naturally have the same liquidity as a standard tranche.

---

[12]Hence, the more popular correlation trading becomes, the higher will be the liquidity of the indices.

[13]Also, the *delta* of the equity tranche is the highest with respect to the underlying index. It has a higher sensitivity to index spread changes. This is another risk.

[14]For example, suppose all 3% of the defaults occur in exactly 6 months, then the running fee will be paid by the protection buyer only for 6 months. The upfront fee will not be affected by such timing issues.

[15]For the CDX index the attachment points of the Mezzanine tranche are different and they equal 3−7%.

[16]However, a more important characteristic of most bespoke tranches is that the selection of the reference portfolio may be different than the reference portfolio used in tradeable credit indices.

The value of the tranches depends on *two* important factors: The first is the risk of a change in the average probability of default; the second is the change in *default correlation*. This is a complex and important idea and leads to the market known as correlation trading. It turns out that from a typical bank's point of view, one of the biggest risks that may lead to bankruptcy is the event of defaults of its clients *at the same time*. Banks normally make provisions for "expected" defaults. It is part of the business. If individual defaults occur now, then it will not be very harmful. Yet, joint defaults of the borrowers can be fatal. This explains the importance of default correlation for the banking sector. Although most financial crises are characterized by, to most participants, surprising increases in correlations associated with declines in risky assets, it was the unprecedented increase in correlations in many of the underlyings in ABS and CDOs that led to outsized losses in this sector during the GFC. Investor and structures did not consider scenarios with historically unseen correlations.

In fact, the reference names in a credit index are affected by the same macroeconomic and financial conditions that prevail in an economy (sector) and hence are likely to be quite highly correlated at times, and the level of the correlation would change depending on the prevailing conditions. In an environment where credit conditions are benign and liquidity is ample, default correlation is likely to be low and any defaults occur mostly due to *idiosyncratic* reasons, that is to say, effects that relate to the defaulting company only, rather than the overall negative economic and financial conditions. During periods of stress this changes, and defaults occur in bunches due to the underlying adverse macroeconomic conditions.

Thus default correlation is a stochastic process itself. During the last few years, market professionals have learned how to strip, price, hedge, and trade the default correlation. We will study this more formally in the next chapter. Here we note that the value of the equity tranche depends *positively* on the level of default correlation. The higher the default correlation in the reference portfolio, the higher the value of an investment in the equity tranche, which means that the spread associated with it will be lower.[17] On the other hand, the value of the senior and super senior tranches depends *negatively* on the default correlation. As default correlation increases, the investment in a super senior tranche will become less valuable and its spread will increase. This is called the *correlation smile* and is discussed in the next chapter.

21.7 TRANCHE MODELING AND PRICING

Markets trade the indices and the index tranches actively. As a result, the spreads associated with these instruments should be considered to be arbitrage-free. Still, we would like to understand the price formation, and this requires a modeling effort. The market has over the past few years developed a market standard for this purpose. The specifics of this market standard model are discussed in the next chapter. Here we discuss the heuristics of CDO tranche valuation.

What determines the tranche values is of course the receipts due to spreads and the potential payoffs due to defaults. The general idea is the same as in any swap. The expected value of the properly discounted cash inflows should equal the expected value of the properly discounted cash outflows. The arbitrage-free spread is that number which makes the expected value of the two streams equal. Clearly,

---

[17]This is similar to the relationship between the value of a bond and its return.

in order to accomplish this we need to find a proper probability distribution to work with. We discuss this issue using tranche pricing. Tranche values depend on the probabilities that are associated with the payoffs the tranche protection seller will have to make. These probabilities are the ones associated with the number of defaulting companies during a particular time period and their correlation.

We limit ourselves to one period tranche contracts on an index with $n = 3$ names.

21.7.1 A MECHANICAL VIEW OF THE TRANCHES

Consider a reference portfolio of $n = 3$ names. Call them A, B, C, respectively. Limit the time frame to 1-year maturities. Let D represent, as usual, the total number of defaults in a year. The first step in discussing tranche valuation is to obtain the distribution of D. How many possible values can D have? It is clear that with $n = 3$, there are only *four* scenarios as shown in Figure 21.9.

At this moment ignore how such probabilities can be obtained and take them as given. Assume that the probability of default is the same for each name and that the recovery is R. Thus,

$$p^A = p^B = p^C = p \qquad (21.6)$$

We now show how tranche values depend on this probability and on the correlation between the defaults of the three names. We now assume p to be 5%. According to the figure the probability is quite high that there will be no defaults at all, i.e., $D = 0$. The probability associated with more defaults D then gets smaller.

Now, consider the *equity tranche*. We have three names, and the equity tranche is bearing the risk of *any* first name to default. Mezzanine is the protection for the second name, and senior tranche sells protection on the third name. Hence for a protection seller on the senior tranche to pay, all three names need to default. Suppose a market maker now sells protection on the equity tranche. In other words, the market maker will compensate the counterparty as soon as *any one* of A, B, or C defaults. What is the probability of this event? Let D denote the random variable representing the number of defaults. Then the probability that there will be at least one default is given by

$$P(D = 1) + P(D = 2) + P(D = 3) = 1 - P(D = 0) \qquad (21.7)$$

Going back to Figure 21.9, we see that this probability is $1 - 0.857375 = 14.2625\%$ in that particular case. This is much higher than the assumed 5% probability that any name defaults individually. Thus writing insurance on a first-to-default contract is much riskier than writing insurance on a particular name in the reference portfolio.[18] This is the risk associated with the equity tranche—the tranche that will get hit first in case of a default.

An investor may not be willing to take such a risk. He or she may want to write insurance *only* on the *second* default. This is the investment in the *mezzanine tranche*. The tranche has *subordination*, in the sense that there is a cushion between the defaults and the protection seller's loss. The first default will hit the equity tranche.

We can calculate the probability that the mezzanine tranche will lose money as,

---

[18]This is understandable. Consider the analogy. You go to school and you have 50 classmates. It is winter. The probability that tomorrow you come in with a cold is small. But the probability that *someone* in your class will have a cold is much higher. In fact, during a typical winter day this probability is quite close to one.

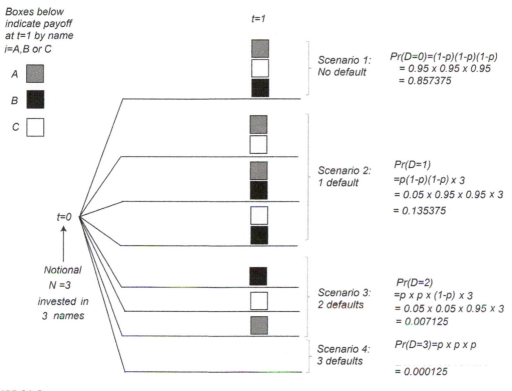

Boxes below
indicate payoff
at t=1 by name
i=A,B or C

A

B

C

t=1

Scenario 1:
No default

$Pr(D=0)=(1-p)(1-p)(1-p)$
$= 0.95 \times 0.95 \times 0.95$
$= 0.857375$

Scenario 2:
1 default

$Pr(D=1)$
$=p(1-p)(1-p) \times 3$
$= 0.05 \times 0.95 \times 0.95 \times 3$
$= 0.135375$

t=0

Notional
N =3
invested in
3 names

Scenario 3:
2 defaults

$Pr(D=2)$
$=p \times p \times (1-p) \times 3$
$= 0.05 \times 0.05 \times 0.95 \times 3$
$= 0.007125$

Scenario 4:
3 defaults

$Pr(D=3)=p \times p \times p$

$= 0.000125$

FIGURE 21.9

Default probability with three names.

$$P(D = 2) + P(D = 3) = 0.007125 + 0.000125 = 0.725\% \qquad (21.8)$$

One can also write protection for the third default. Here there is even *more* subordination. Before the insurer suffers any losses, three names must default. In this case, the investment will represent a *senior tranche*. The probability of making a payoff is simply

$$P(D = 3) = p \times p \times p = 0.05^3 = 0.000125 = 0.0125\% \qquad (21.9)$$

This simple case can be generalized easily to *iTraxx* indices.

21.7.2 TRANCHE VALUES AND THE DEFAULT DISTRIBUTION

We use Figure 21.10 to discuss the important relation between the area under the loss density function and the tranche values. First, note that the *iTraxx* attachment points slice the distribution of D into five separate pieces. Each tranche is associated with a different area under the density.

Consider the 3−6% mezzanine tranche as an example. The tranche has two *attachment points*, the lower attachment point is 3% and the upper attachment point is 6%. In heuristic terms, the lower attachment point represents the subordination, i.e., the cushion the investor has. Defaults up

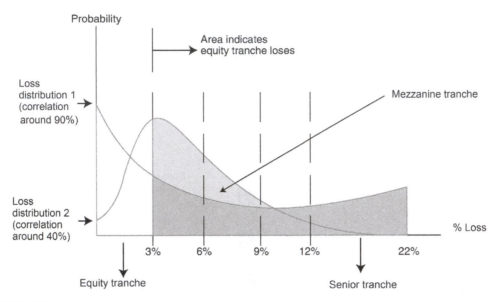

FIGURE 21.10

Loss distribution.

to this point do not result in payments of default insurance.[19] The upper attachment point represents a threshold of defaults after which the mezzanine protection seller has *exhausted* all the notional amount invested. Any defaults beyond this point do not hurt the mezzanine investor, simply because the investment does not exist anymore. Thus the area to the *right* of the upper attachment point is the probability of losing all the investment for that particular tranche.

Once this point about attachment points is understood, we can now show the relationship between default correlation and tranche values.

Consider Figure 21.11, giving the distribution of D, the total number of defaults. Note the difference between Figure 21.11 and Figure 21.10. Figure 21.11 shows the default density function while Figure 21.10 shows a loss distribution for a portfolio. Both distributions depend on the average probability of default p and on the default correlation ρ. Suppose in Figure 21.11 the correlation originally was low at $\rho = 0.1$. This leads to the default distribution 1. Then, keep the p the *same* and move the correlation up to, say, $\rho = 90\%$. This leads to loss distribution 2. The distribution will shift from 1 to 2 as shown in Figure 21.11. Consider the implications.

The first implication is that as correlation goes up, the distribution is being pressed downward from the middle. However, the area needs to equal one. So, as the middle is compressed, the weight goes to the two endpoints. Note that the figure does not plot parts of the distribution in the middle and somewhat exaggerates the right tail for illustration.

Second, in terms of defaults this means that as correlation increases the probability of no defaults increases compared to the low correlation case. This is because the probability of credits

---

[19]The spread on mezzanine tranche would go up since the cushion would be getting smaller.

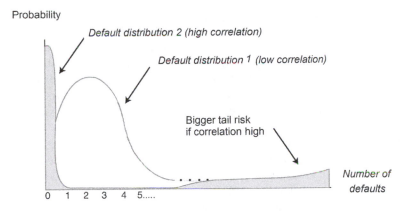

FIGURE 21.11

Default distribution.

surviving together increases. Second, the probability of a very large number of defaults also increases compared to the low correlation case. At the extreme as $\rho \rightarrow 1$ the distribution that is approximately bell-shaped starts looking more like a binomial distribution. As correlation goes to one, the default probabilities of the different assets start to behave more and more similarly and at the end, they become the same random variable. Hence the value assumed by the random variable D will be either n or 0, i.e., either every name will default or no name will default.

How do we translate the implications from Figure 21.11 to portfolio losses and tranche values? First consider another distribution, the *default loss distribution*. Consider a portfolio of defaultable assets issued by n different debtors. For simplicity, assume that the recovery rates are the same and are known at R. Also, the exposure to each name is $1, meaning that the total invest-ment in the portfolio is $n. Then, if one name defaults, the investor loses $(1 - R)$ dollars. The total default loss during a horizon of length T will depend on how many names default during this time interval. In other words, default loss depends on the *distribution* of the random variable D defined earlier. Figure 21.11 showed the distribution of defaults D.

We won't discuss here how, but let's assume it is possible to choose a corresponding distribu-tion for portfolio losses and plot the probability of losses as in Figure 21.10. Let's assume that the loss distribution 1 in Figure 21.10 corresponds to a default correlation $\rho = 40$, while the loss distri-bution 2 in Figure 21.10 corresponds to a default correlation $\rho = 90\%$. Thus we have again a low- and high-correlation scenario as in Figure 21.11 but with slightly different values which reflect the approximate shapes of the distributions. It turns out that the effect of an increase of default correla-tion has important insights for tranche valuation. First, note that the probability of the area near 0−3% goes up. The same is true for the probability associated with the other tail. The area near the 12−22% tranche also goes up.

But this has a second implication. The area on the right of the equity upper attachment point gets smaller. Implying the probability that the equity investor will lose all his investment has gone *down*. The equity tranche investor will benefit from this change in the distribution function. The equity tranche spread will go down and the investor who sold protection earlier at the higher rate will have mark-to-market gains. Hence the equity tranche protection seller is *long* default correlation. In other words, the equity tranche investor will benefit if the correlation increases.

Third, the probability associated with the cushion of the super senior tranche is also getting smaller. This implies that the probability is higher and the super senior protection writer will suffer some losses. The spread on the super senior tranche will go up and the investor who sold protection at a lower spread will have mark-to-market losses. Hence the super senior protection seller is *short* the default correlation. In other words, the super senior tranche investor will lose if default correlation decreases.

Finally, note that the effect of these movements is mixed on the mezzanine tranche. The area on the right of the upper attachment point has not changed that much. Hence the probabilities associated with mezzanine tranche losses are approximately the same and the mezzanine investor is more or less neutral toward the correlation changes. In the next chapter, we will discuss the sensitivity of tranche values to default correlation in further detail and explain how default correlation can be traded.

21.8 THE ROLL AND THE IMPLICATIONS

Every 6 months, on March and September 20th, the *iTraxx* and CDX indices *roll* and a new *series* starts trading. Some credit names may have defaulted, changed sector, merged, or been downgraded. Other CDSs may have become less liquid. These are considered as no longer eligible to be part of the index. Every such name is replaced by the next most liquid name available in its class. The "old series" continue to trade as long as there are open positions, but they are off-the-run. The new roll will be the *on-the-run* liquid index.

The *Roll* is an important characteristic of the indices from a financial engineering point of view since its presence leads to several strategies and complications that a financial engineer must be aware of. An obvious strategy is to guess the names that will leave the index and the names that will come in. But this happens in index revisions in the equity sector as well. Credit rolls have some novel additional strategies. Note that dropping lower-rated issuers from the index and replacing them with higher-rated ones normally means that the new index should start trading at a narrower spread than the old index, everything else being the same. But, surprisingly, this may not happen, as we will see below.

There is some empirical evidence that because the *buyers* of protection[20] would like to stay with the new, on-the-run index due to its liquidity, right before the new series, they will close their positions. This means they will *sell* protection and this will lower the spreads. Of course, as the new index starts trading, the same names will buy protection and this will widen the spreads.

There are also structured credit products that are partially based on the fact that the index will roll every 6 months. The constant proportion debt obligation (CPDO) is one good example. During a 6-month period, everything else being the same, a 5-year maturity *iTraxx* index will become a 4.5-year

---

[20]The buyers of protection are not necessarily the ones who desire insurance against default. Indices are mostly used by players who hedge the default risk in tranche positions that they have taken. Hence, historically hedge funds were often such buyers of protection.

maturity index. This means that the index will *roll down* the curve and, if the curve is upward sloping, spreads will tighten automatically. This is due to the shorter maturity and nothing else.

Another classic technical roll is the change in the *basis*. This is the difference between the index spread and the intrinsic value of the underlying CDSs of the referenced names in the index. The basis exists because normally there are more clients that *buy* protection than *sell* them.[21] In past rolls, the basis between the underlying CDSs and the index has narrowed significantly. This happens because more investors are selling protection than buying protection in order to move to the new, more liquid index. The example below summarizes the mechanics of the roll.

EXAMPLE

New York (Dow Jones)—As summer draws to a close, credit derivatives investors are eyeing September's changeover in the credit default swap indexes for trading opportunities. This time around though, trading patterns surrounding the change in the indexes' composition could differ from past ones as more issuers than ever before—a total of eight—will be dropping out of the investment-grade index family.

Credit default swaps allow investors to protect their holdings should issuers default on their debt. Since the inception of the credit default swap indexes—liquid benchmarks that allow investors to take positions on corporate debt issuers without having to buy or sell the underlying, often illiquid cash instruments—there have been [nine rolls until March 2008] which take place every six months.

At each roll date, issuers who have lost their investment-grade ranking by either Moody's Investors Service or Standard & Poor's are dropped from the Dow Jones CDX investment-grade credit default swap index and replaced by other issuers—chosen by a poll of the dealers who belong to the index consortium. At the first index roll, six issuers were replaced, at the second five and at the last roll in March 2006, just three issuers were dropped.

It is important to understand that at each roll the newly introduced series will have a new coupon and will be trading near par initially. The traders who buy and sell index protection are in fact buying and selling this newly introduced standardized contract.

21.8.1 ROLL AND DEFAULT RISK

Default risk in the indices is somewhat different than it looks at the outset due to the existence of the roll. In fact, at each roll the credit quality of the indices improves. In general, before a company defaults, its credit quality deteriorates. By the time the company is about to default, it is quite likely that it was dropped from the index during some previous roll. This brings up an important distinction. A 5-year position that stays with the *same* index will face the default risk of the underlying reference names, and these names will remain the same during 5 years. Some of these names will deteriorate and some may even default. This is what is meant by *default risk*.

---

[21]The market makers will hedge this discrepancy on their books using the underlying CDSs.

A position that always rolls to the new index faces a somewhat different default risk. The main default risk faced by such a position is when default occurs *all of a sudden*, without any indications. In this case a good corporation may go from a rating of AA to default, without the rating agencies having time to downgrade it, and before the roll date arrives. This is called *jump to default risk*. Note that its probability is significantly lower than staying with the same names during the 5 years and then seeing some of them default.

21.9 REGULATION, CREDIT RISK MANAGEMENT, AND TRANCHE PRICING

This section discusses the relationship between tranche pricing and credit risk management. Basel III is the framework associated with current credit risk management practices and indirectly it has a close relationship with tranche pricing.[22]

21.9.1 CREDIT RISK MANAGEMENT, DEFAULT LOSSES, AND MARK-TO-MARKET LOSSES

Suppose we would like to adapt a Value at Risk (VaR)-type (or Expected Shortfall) risk management approach to portfolios with credit risk.[23] This means that we would like to set enough capital to cover losses $0 < \alpha$ percent of the time. It turns out that there are *two* quite different empirical loss distributions that determine the calculation of possible losses.

In this setting, in order to calculate possible losses due to default, one first needs to obtain a distribution for this random variable L. This is the *default loss distribution* and is illustrated again in Figure 21.10. Using this distribution one can calculate the *expected default loss* and the threshold L_α which determines the extreme losses that occur with a probability of (at most) α, during the interval with length T.

The bank would then put aside enough capital to cover default losses up to the point L_α. These have a probability $1 - \alpha$ of occurring. Default losses may be greater than L_α only in α percent of the time. The bank is not obligated to cover these more extreme losses with additional capital.

But this is only *one* way of looking at credit risk. It involves the risk associated with the *default events* only. This way of managing risks is perhaps appropriate if the instruments under consideration are held until maturity and if the *mark-to-market* is not relevant.

On the other hand, if the portfolio under consideration is a trading portfolio where marking-to-market is important or if the instruments may not be held until maturity, then the bank or the hedge fund faces *another* risk.[24] This risk comes from credit quality changes, which incidentally also

---

[22]In this section, we ignore the market risk and concentrate on the *new* aspects of risk management brought about due to the existence of the default event.

[23]As discussed in Chapter 3, the latest Basel III proposals recommend the use of expected shortfall over VaR-type risk measures.

[24]A hedge fund position is a credit instrument and is usually financed by borrowed funds. This will be similar to repo, where the hedge fund buys the instrument and repos it to secure the funds to pay for it. The repo dealer would then mark this instrument to the market. Even when there is no default, the credit spreads may change and may create losses for the hedge fund.

involves defaults. If credit deteriorates and needs to be sold or marked-to-market, then the bank or the hedge fund will still suffer credit losses although no default has occurred. The *default* loss distribution cannot take such potential losses into account, because it is only directed toward measuring loss due to default. The loss due to credit quality changes requires an *additional* effort. The change in the market value of the portfolio due to credit quality changes can be calculated in a way similar to that of market risk.

The calculation of the *default* loss distribution requires the default probabilities for each name and the default correlations. The calculation of the *credit* return distribution on the other hand requires, in addition, the *conditional transition probabilities* concerning the rating migrations of each name in the portfolio as well as their correlations. Hence the calculation of this second distribution is much more involved. Note that default is only one of the states where credits can migrate. Hence default risk is included in the credit loss distributions due to credit deterioration.

21.9.2 CAPITAL REQUIREMENTS AND CDO ACTIVITY

As we saw at the beginning of this chapter, securitization activity has dropped significantly since pre-GFC levels. One of the reasons why securitization and CDO activity was so high before the GFC was that it represented a way for banks to reduce regulatory capital. As with many previous booms, this led to excesses and abuses. However, securitization activity continues and even under the current more stringent capital requirements banks find it possible to reduce capital requirements as the following example indicates:

EXAMPLE

In one example Chenavari provides to investors, a $1 billion portfolio of small business loans has a risk-weighting of 75% under Basel III's standardised approach, generating a risk-weighted asset (RWA) total of $750 million. That implies a bank would need around $75 million of equity capital to support it. Securitising the pool and transferring the 1.5–8% second-loss tranche—representing $65 million of exposure—to a third party, while retaining the first loss and senior tranches, could bring a bank's capital requirement down to around $30 million, a 60% saving. The bank also gets to keep the loans on its balance sheet and maintain its client relationships.

Fery refuses to mention counterparty names, but describes some deals he says are representative of the business the fund engages in. One involves a €1 billion portfolio of around 1,500 loans to German small and medium-sized enterprises (SMEs). The pool had a weighted average credit rating of BB+ and an average duration of two-and-a-half years. More than half the loans were investment-grade. Chenavari had a positive view of Germany's Mittelstand sector—the legion of mostly family-owned small businesses that power the country's export machine—so agreed to absorb the second loss on the pool via a €65 million note referencing the 1.5–8% tranche of the loan portfolio, paying 13.5% over three-month Euribor per annum. The bank retained the first loss exposure and the entire senior tranche.

The firm cut a similar deal with a UK bank that was seeking regulatory capital relief on a £1 billion portfolio of residential mortgage loans. The pool comprised nearly 8,000 loans originated before 2007, and included a large portion of interest-only and buy-to-let mortgages. This time, Chenavari acquired an eight-year note referencing the 2–8.5% second-loss tranche, paying a coupon of 12.5% over three-month LIBOR.

Fery and his team have structured a host of similar deals involving different types of underlying assets—everything from emerging market corporate loans and European mortgages to asset-backed securities, trade finance assets and derivatives counterparty risk. The average deal size is around $80 million, with estimated returns in the 10–20% range, he says.

> *One of the difficulties for rivals seeking to replicate the strategy is that deals happen infrequently and often require months of work. That means building relationships with a large number of banks, and having a solid grasp of credit analysis and structuring—on both counts, it helps that Fery used to run the global credit trading business at Calyon.*
>
> **(Risk magazine, 13 January 2014,** *"Hedge fund of the year: Chenavari Investment Managers"***, www.risk.net/2317763)**

21.9.3 LESSONS FROM THE GFC, POST-GFC CAPITAL REGULATION AND THE SECURITIZATION MARKET

Securitization and the issuance of CDOs was so actively pursued by banks since it allowed banks to remove loans from their balance sheets and thus improve their capital ratio. Commercial banks must have sufficient capital to cover potential depositors' withdrawals. The ability to lend is affected by capital requirements and if a bank does not have enough capital it cannot lend further unless it finds a way to remove the assets from its balance sheet. As we saw, CDOs are a form of securitization. Many of the junior CDO tranches issued by banks were backed by risky (*subprime*) loans. Investigations following the GFC revealed malpractice in the loan origination process which saw loans approved without proper vetting of borrowers and the violation of proper procedures. As a result many of the CDOs that were sold to investors were very risky but this did not become apparent to most investors until the US housing market boom slowed in 2006. It is estimated that of the $500 billion worth of CDOs outstanding at that point, about 50% were based on MBSs and around 75% of these contained subprime loans. Driven by the prospect of commissions, risky loan origination and CDO issuance continued nevertheless. As selling these increasingly risky assets became more difficult traders found ways to make the banks hold some of the riskier tranches of the CDOs in order to facilitate sales, achieve favorable credit ratings for the securities, and pay traders commissions. One of the factors that exacerbated the GFC was the prevalence of ratings arbitrage which involved pooling low-rated tranches to create CDOs. All these factors not only perpetuated the excesses but also made the eventual correction more painful as banks themselves were saddled with large losses and the financial system was crippled until banks were bailed out, wound down, or recapitalized. Housing bubbles occur regularly around the world, but the one preceding the GFC was exacerbated by abuses of securitization and misaligned interests.

In December 2013, the Basel Committee on Banking Supervision issued a second consultative document regarding revisions to the existing securitization framework which was found to be deficient. According to the Basle Committee, the GFC revealed a number of shortcomings in the current securitization framework which it seeks to address with the latest changes. First, practices in the run-up to the GFC showed an almost mechanistic reliance on external credit ratings. Second, there was a tendency to assign too low risk weights for highly rated securitization exposures. The reason for this was that the calibration assumptions typically used turned out to be questionable and there was a lack of sufficient risk drivers across approaches in determining risk weights. Third, risk weights for low-rated senior securization positions were found to be too *high*. Fourth, the securitization framework leads to undesirably procyclical economic dynamics or *cliff effects*.

Vis-à-vis initial post-GFC proposals by the committee the document suggests lower capital requirements but these are still more stringent than existing capital requirements. Thus, if the future Basel Committee securitization framework was based on the latest proposal this leaves the possibility that the securitization and CDO market would continue to be an important part of financial markets.

21.9.4 CAN SECURITIZATION CURE CANCER?

In this chapter, we discussed the economic rationale for securitization. We have seen that securitization and financial engineering can play a useful economic role if correctly implemented. As in all areas of economic activity regulation exists and is required to ensure that all market participants act legally and responsibly. Financial engineering concepts have, however, also applications outside finance and may provide not just economic benefits but other social welfare improvements. In recent work, Fagnan et al. (2013) pose the question of whether financial engineering can cure cancer, a leading cause of death in many industrialized countries. The authors observe that investment in the pharmaceutical industry has declined in recent years and make proposals on how securitization can help increase investment including key important cancer research. The authors hypothesize that one of the reasons for low private sector investment is the skewed distribution of drug trials. The authors provide a stylized example that assumes a hypothetical drug-development programme that requires $200 million development costs in present value terms, a 10-year development period during which no revenues are generated and which has a 5% chance of successful drug approval at the end of 10 years. If one assumes that such a programme generates $2 billion in annual revenues in the subsequent 10 years (i.e. years 11−20), many investors could be expected to be scared off by the return standard deviation of 423%.

One proposed solution to this problem is to create a fund that pools various drug-development programmes and securitizes the resulting revenue streams. The proposed name for the structured securities that give ownership interest to experimental drug compounds is *research backed obligations* or RBOs. Fagnan et al. (2013) suggest how techniques that were used in the creation of CDOs and ABSs can be used to provide guarantees and alter the capital structure of these securities in such a way that they represent an attractive investment opportunity for investors. To date no such RBO has been issued but the research shows that financial engineering techniques, if appropriately applied, may potentially have wider implications and social benefits beyond the narrowly defined domain of financial markets.

21.10 NEW INDEX MARKETS

The credit sector plays an important role in financial markets for several reasons. One of these is the methodical way new sectors are introduced by the major players. Markets are normally created endogenously with the interaction of thousands of traders, market makers, and risk managers. In the credit sector, this effort has been consciously directed by broker-dealers and has come after years of experience in new derivatives markets and trading strategies. As a result, broker-dealers have been quite successful in creating platforms for trading new risks and offering broad numbers of instruments for hedging a range of credit risk exposures. The Markit website provides an overview of a wide range of indices including iTraxx, CDX, ABX, and LCDX.

The ABX and LCDX indices are the most prominent of the new tradeable indices that permit trading of new risks. The ABX index is a carefully constituted index that permits trading and hedging of mortgage debt exposures. A player who is short mortgage debt will hedge this position by buying the ABX index. The LCDX is similar to the *iTraxx* or CDX, except that the underlying securities are loans instead of being bonds. Hence, this index helps hedging loan exposures. Levx is similar. It is an index of leveraged loans.

A player who is long will hedge the position by selling the index. These indices are traded in the form of series and offer a fixed coupon at each roll.

21.10.1 THE ABX INDEX

The ABX index is made of obligations issued by 20 issuers of *residential* MBSs. Altogether there are five subindices. These subindices are each made of one security from each one of these issuers. These securities and hence the subindices have ratings that range from AAA to BBB −. Similar to the roll in the *iTraxx* and CDX indices, the new series of ABX indices roll on January 19 and July 19. Each series has a new set of mortgage loans behind it. Thus, unlike the *iTraxx* and CDX index, the ABX indices have "vintages."

The securities included in the index are essentially debt securities entitling the investors to receive cash flows that depend on residential mortgages of one-to-four family residences. Hence variations in these cash flows, defaults, and delinquencies affect the value of the ABX indices. An investor in the ABX indices will receive (pay) a fixed coupon set at the roll date and will make floating payments (receipts) if there is an interest or principal shortfall in the cash flows.

21.10.2 THE LCDS, LCDX

The so-called LCDS (Loan Credit Default Swaps) contracts and their corresponding index LCDX are relatively new tools in the credit sector. Besides being important tools they provide a good opportunity to summarize the Credit indexing technology from a somewhat different angle. LCDS is similar to the single-name CDS that we saw in Chapter 18. In a CDS, the deliverable debts are bonds. With LCDS the deliverable is syndicated *secured* debt in the United States that was originally issued by a syndicate. LCDX is the corresponding tradeable index and it is similar to the *iTraxx* and CDX indices. The underlying for the LCDX is $n = 100$ equally weighted single-name LCDS. The loans in question trade in the secondary *leveraged loan* market. The index launches with a fixed coupon, paid quarterly. The index trades on a clean price basis. If the price goes down, the corresponding spread goes up.

The protection seller will receive a coupon fixed at the roll date and will make a compensating payment during a credit event. Also during a credit event the protection seller pays the notional amount N and gains possession of the loan. Cash settlement is also permitted. The amount of this cash settlement will be determined in an auction.

21.10.2.1 Cancelability

An issuer can repay the debt and may not issue new debt afterward. If there is no deliverable obligation then the LCDS cannot continue. Thus LCDS contracts are cancelable unlike the standard CDS. An issuer may be upgraded. When this occurs, the issuer can repay the original debt and issue new debt with lower interest cost.

A loan repayment starts a 30 business day period, and during the entity can initiate a new loan. If after 30 days no new loans are made, the name is removed from the index after a dealer vote.

This cancelability affects the valuation. If the LCDS is cancelable upon repayment of debt, then the final maturity for a given LCDS will be unknown. The calculation of the present values should then take into account the probability that the loan will be repaid early. This is similar to the use of the credit-risky DV01.

21.10.2.2 Quoting conventions

The iTraxx and CDS index families are quoted on a spread basis, except for the equity tranche which is quoted with an upfront.[25]

On the other hand, the LCDX and ABX indices are quoted on a price basis and this forms another difference.

21.11 STRUCTURED CREDIT PRODUCTS

CDS is the basic building block of the credit sector. Using CDS engineering, one can immediately create credit-market equivalents of risk-free fixed-income instruments. Some of these instruments are discussed in this section.

21.11.1 CREDIT OPTIONS

It is natural for options to be the first derivative to be written on credit indices. After all, there are liquid indices and these could serve as an attractive underlying for those who would like to hedge their credit volatility or for investors who would like to take positions in them. Yet, such options turn out to be much more complicated to structure and market than visualized at the outset. Although there is decent liquidity in the market, with daily references to iTraxx and CDX implied volatility, some difficulties remain.

There are essentially three problems associated with options on credit indices. First, the credit sector is heavily influenced by monetary policy and has a long credit cycle. The practitioners would need long-dated options. Second, although stocks are liquid, a large majority of corporate bonds have very little liquidity; but the third and main problem is the *index roll*.

---

[25]For updates on conventions, see http://www.markit.com/product/indices.

Essentially these indices change every 6 months and one cannot price a long-term option against such an unstable benchmark.

The credit option trader is then forced to operate in the shortest maturities and the exercise dates controlled by the roll dates.[26] This is not sufficient for traders since their needs are really 5-year options because these could be used to arbitrage the structured credit market.[27]

There is another important difference between credit index options and options from other markets. In the credit sector the longer dated an option, the more it becomes a correlation product. With longer-dated options, the underlying risk is to what extent referenced credits will move jointly. Shorter-dated options, on the other hand, are more like volatility products.

21.11.2 FORWARD START CDOs

Using forward start CDOs, one can take leveraged positions on the outlook for credit spreads in the distant future, say from 2015 to 2020. Such forward start products may be useful for some investors that want to hedge their positions on take exposure during the *credit cycle*.

The product can be structured by selling CDSs maturing, say, in 10 years, and buying CDSs maturing in a shorter maturity, say, 5 years. Such forward start instruments are marketed as bespoken deals on the iTraxx or CDX default swap indexes. The most common reference is the mezzanine tranche insurance against the 3−6% of defaults in a credit portfolio.

The net position of buying 5-year protection and selling 10-year protection is selling 5-year protection 5 years from now. Note that such a position will have positive carry. Due to this, such trades become popular if the iTraxx curve is relatively steep.

21.11.3 THE CMDS

The constant maturity default swap (CMDS) is an important component of the structured credit sector. A CMDS can be structured as follows.

Fix the maturity of the CDS at, say, 5 years. Consider a series of T-maturity CDSs starting at times t, $t+1$, $t+2$, $t+3$, $t+4$, and $t+5$. Note that the spread of the current CDS is known at time t, whereas future CDS spreads c_{t+i} will be known only in the future as time passes. Also note that these CDSs all have the same 5-year constant maturity. Let their spread be denoted by c_{t+i}. Then the CMDS will be the 5-year CDS that pays the floating spreads c_{t+i}.

Essentially this is an extension of the CMS swaps to the case where the underlying risk incorporates default risk. There are several uses of this crucial component. A market example follows.

---

[26]Two months before the next roll, the expirations can be at most 2 months.

[27]One proxy for long-dated options is the range accrual that has been marketed to players who want to arbitrage the structured credit sector. These notes pay a coupon, or *accrue*, depending on the number of days the index I_t remains within a prespecified range $[L^{min}, L^{max}]$. Note the roll problem can be solved here by resetting the range $[L^{min}, L^{max}]$ at every roll.

EXAMPLE

Investors want to sell protection today with the potential to take advantage of expected future spread widening; this is exactly what the CMDS product provides. Client interest in constant maturity CDOs (CM-CDOs) also helps establish the CMDS market by drawing clients' attention to what the product can achieve.

Nearly all new CDOs in the pipeline today come with the option of constant maturity technology embedded in them.

There are concerns about the possibility that profit and loss volatility may be injected into synthetic CDOs because of International Accounting Standards (IAS) 39 which requires all derivatives to be marked-to-market through the income statement. These concerns drive a lot of interest in CM-CDOs. Using CMDS to mitigate mark-to-market volatility is a legitimate reason for employing them. Market participants do not expect CMDS to outshine the market for tranched credit index products or credit options. But having standardized fixings would be useful both for closing CDO deals and for developing the market for cash-settled credit spread options and other structured credit and volatility products.

CMDS are generally viewed as a building block for other structured credit products. The challenge is that fixings for credit derivatives are not as straightforward as they are for interest rate derivatives. "The amount of information that you need is so much greater than it is in rates. And where do you stop with fixings? The universe of names in the credit default swap market is very large. How many do you fix, and do you fix for the five-year or the whole curve for each name?" asked a trader.

Fixings on the credit indices are relevant for CMDS contracts because if CDSs are written on indices, which most are, an independent fix is needed. Quoting an index, such as Dow Jones iTraxx, is not easier as the coupon must be linked to 125 names, but the liquidity in credit indices is greater than it is in the underlying single names.

Standardized fixings. About 18 dealers have been working with Credited, the electronic dealer-broker, and Mark-It Partners, the OCT derivatives valuation firm, to standardize fixings for credit indices. Six weekly fixing test runs have been completed so far (Thomson Reuters IFR 1552). No runs on single-name credit default swaps have been tested yet, though.

According to dealers, the standard resets that are being developed are essential for the continued growth of the product as opposed to relying on just dealer polls as documented in current contracts.

"Fixings are worthwhile for the credit derivative market not just for CMDS but for all other second- and third-generation credit derivative products. For example, if fixings are done for iTraxx index tranches, this will help breed a further range of derivatives on the tranches. So, in this way, fixings are essential just like LIBOR fixes every day for the swap market," one dealer added.

For the CMDS market to evolve, common documentation must also be forthcoming, dealers say. In CMDS, two documentation issues to overcome are whether the coupon is quarterly or semiannual and if the fixing is $T + 1$ or $T + 2$.

The CMDS can be used to take an exposure to the movements of the credit curve. If one expects the credit curve to steepen then one could, for example, buy a 5-year protection on the CMDS that references a 10-year CDS and sell protection on the 5-year CMDS that references a 3-year CDS. The reverse could be done if the credit curve is expected to flatten. One could also put together a swap of the CDSs: e.g., paying 10-year and receiving 3-year reference spreads.

21.11.4 LEVERAGED SUPER SENIOR NOTES

The spreads on super senior tranches are very tight—say, around 10 bps. This is not very attractive to the investors. Hence, the demand for super senior tranches is relatively low. Yet, banks have issued many mezzanine tranches on CDOs and have kept the super senior and equity tranches on their books. They need a way to generate a demand for these tranches. Leveraged super senior notes is one method that was devised for selling the super senior risk to others.

With the note investors, take the additional exposure to the mark-to-market value of the tranche, and the trigger protects the bank against the investor's credit risk. Given the leverage, super senior investment may lose an amount greater than the original investment. With the trigger, this risk is reduced.

In this sense, one can say that a leveraged super senior note is a modification of the super senior tranche. This modification occurs first in the leverage. N is collected from the investor, but λN with $1 < \lambda$ invested in the super senior tranche. Hence the return is also multiplied by λ. Second, there is a trigger on the market value of the note. If this trigger is reached the issuer of the note can unwind the position and return the mark-to-market value of the note to the investors.

EXAMPLE

Consider a USD1 billion portfolio. Let the senior tranche have attachment points of 12% and 30%. This gives a thickness of USD180 million.

Suppose we select a leverage ratio of 10. The leveraged super senior note will consist of an issue of USD18 million. This amount is multiplied by 10 and invested as a notional amount of 180 million in the super senior tranche. Assuming that the quoted spread for this tranche is 8 bp, the note will pay LIBOR + 100.

The market value trigger could be defined, for instance, as 70% of the issue amount of USD18 million. If the market value of the note falls below this limit, say becomes 12, then the 12 is returned to the investors instead of the original investment of 18.

There are two general tendencies in structured credit. The first is to introduce *leveraged transactions*; the second is the introduction of *market risk* in addition to default risk. The leveraged super senior notes are examples of the first tendency. The CPPI techniques applied to credit are an example of the second and they will be discussed in Chapter 23.

Where does risk lie in leveraged super senior notes? Owing to the substantial credit protection inherent in a super senior structure, the default risk itself is very limited. The risk borne by investors mainly lies in the behavior of the *market value* of the CDS tranche, which depends on spreads and correlations. Most transactions are actually structured so as to ignore correlation variations as market value parameters; they introduce instead a *trigger* on the portfolio average spread.

In such a trade, the investor is long the super senior risk, while the dealer goes short that risk. To boost the return on these investments, dealers have been constructing products for their clients using borrowed funds.

During the year 2000, dealers purchased significant amounts of mezzanine protection and, as a result, were left exposed to the senior and equity components of the capital structure. This is because they had marketed mezzanine products to clients and kept the senior and equity tranches on their books. Note that leveraged super senior trades is one way of *buying* protection on the senior and super senior tranches. Hence, the structured product is in fact useful to both the client and the dealer.

If dealers did not have such long senior and super senior positions, after buying protection from a client with a leveraged super senior issuance, they would hedge themselves through the iTraxx index or other individual CDSs, although there may be a significant basis risk between the two risks.

21.11.5 CoCos

CoCos are an example of post-GFC financial innovation. The structuring and valuation of CoCos represent an interesting application of the financial engineering principles that we discussed in this book, including the modeling of credit and default events. For this reason and since CoCos are a relatively recent and untested product, we discuss CoCos in some detail. A CoCo is a hybrid product. It is a debt instrument that automatically converts into equity or suffers a write down when a certain trigger event occurs. CoCos can therefore be viewed as an example of convertible debt which we first encountered in Chapter 19. However, the contingent conversion of CoCos into equity means that they have more in common with structured products such as reverse convertibles discussed in Chapter 20 than with convertible bonds. What CoCos have in common with reverse convertibles is that they expose the holder to a *limited upside* but a potentially *large downside*. However, in contrast to reverse convertibles the trigger for CoCos recently issued by banks is an accounting trigger such as the banks' Tier-1 capital ratios (typically 5−7% of risk-weighted assets) and not their share price. The first CoCo was issued by Lloyds banking group in 2009 and labeled *Enhanced Capital Note (ECN)*. The £7bn Lloyds ECNs convert into equity if the bank's tier one capital falls below 5%. The minimum level for banks to pass the European stress test is 5.5%.

There is another important difference compared to convertible bonds and reverse convertibles which is that the trigger may sometimes depend on regulatory intervention, as was the case with a CoCo launched by Credit Suisse in February 2011 under the name *Buffer Capital Notes*. The Swiss regulator has the right to intervene and force a conversion into shares of the Credit Suisse CoCo. Of course, such intervention that is external to the company makes modeling the risk significantly more complicated than when the trigger is just related to the share price.

21.11.5.1 Cocos versus convertible bonds

Figure 21.12 shows the difference in price behavior between a CoCo and a convertible bond. For simplicity here we assume that the underlying debt is riskless. This is reflected in the horizontal line that represents the flat bond floor. This is in contrast to Chapter 19 where convertible bond value for low stock price levels converged to zero. As the share price falls, the convertible bond value approaches the bond floor from above. The CoCo behaves differently. We assume that the

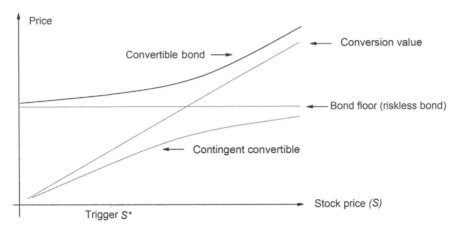

FIGURE 21.12

Contingent convertible versus convertible bond.

CoCo is converted into shares when the share price S falls below a market trigger S^*. As the stock price falls, the line representing the CoCo approaches the conversion value line from below. As the stock price rises the CoCo approaches the bond floor from below while the convertible bond approaches the conversion value from above. The slope of the line representing the conversion value depends on the conversion ratio (CR), discussed in Chapter 19.

21.11.5.2 Valuation of CoCos

CoCos provide a nice application of the financial engineering principles discussed in this book. Since CoCos are a hybrid security, it is not surprising that there are different valuation methods that can be used to price a CoCo. In fact there are three different approaches: the credit derivative approach, the equity derivative approach, and the structural credit approach. From a credit derivative perspective or reduced form approach, we price the CoCo as a fixed-income instrument with an enhanced yield as compensation for potential losses upon conversion. The risk of losses can be modeled using the default intensity or reduced form approach outlined in Chapter 18. One can model the trigger event that leads to the conversion of the CoCo into shares, an event similar to the way default is modeled statistically in the valuation of corporate debt or CDS. This approach could lead to an equation that expresses the credit spread on CoCos as a function of the default, or in this case trigger, intensity (λ), and the recovery rate. As a variation of Eq. (18.38) we would obtain the following relationship:

$$\text{spread}_{\text{CoCo}} = (1 - R_{\text{CoCo}}) \times \lambda_{\text{CoCo}} \tag{21.10}$$

From an equity derivatives perspective, we can apply the principles outlined in Chapter 20 and see the CoCo as a long position in CR shares, where CR is the conversion ratio. The knock-in feature is dependent on the trigger event and barrier option valuation approaches can be used to price it. Third, in Chapter 19 we saw structural models of default and credit valuation. CoCos are just one element of the issuer's capital structure and therefore we can apply the principles underlying the Merton model and its extensions to value CoCos. For this purpose, we would specify a stochastic process for the firm and model the value of different continent claims, including those of CoCo holders, as a function of the value of the assets and the trigger price. From an equity derivatives one can decompose the CoCo into a risky bond, a knock-in forward on the underlying share and a short position in a digital down-and-in option, which reflects the fact that the coupons on the bond will only be received as long as the trigger event does not occur. This leads to the following contractual equation:

$$
\boxed{\text{CoCo}} = \boxed{\begin{array}{l}\text{Corporate}\\\text{bond}\end{array}} + \boxed{\begin{array}{l}\text{Knock-in}\\\text{forward(s) on}\\\text{the shares}\end{array}} + \boxed{\begin{array}{l}\text{Short digital}\\\text{down-and-in}\\\text{option}\end{array}} \qquad (21.11)
$$

The end of chapter exercises provide CoCo valuation examples using the credit reduced form approach and the equity derivatives approach. Further details are available in the references listed at the end of the chapter.

21.11.5.3 The outlook for CoCos

The rationale for CoCos arose because of banks' desire to shore up their capital base in response to regulatory changes that led to more stringent capital requirements. The implicit hope on the part of regulators is that CoCos will help protect big banks from having to be bailed out by taxpayers again. Therefore, CoCos are also sometimes referred to as *bail-in bonds*. Bond holders were largely unaffected by bank losses during the GFC because taxpayers bailed out banks. The idea behind a bail-in bond is not to bail out, but to bail-in bond holders before taxpayers are asked to rescue a bank. Bail-in clauses are different from traditional bankruptcy procedures which have strict rules and court-supervised procedures. As Figure 21.12 shows, CoCos have a significant downside potential which has to be compensated by a higher yield. Thus, these instruments may be of interest to yield-hungry investors, provided that they understand the risks.

However as the following reading illustrates there is some concern about the hidden risk and future performance of CoCos.

EXAMPLE

Potential conflict of interest between regulators and investor can arise as a result of significant discretion that regulators have over the mandatory conversion feature embedded in CoCos. From an investor point of view, a CoCo is a hybrid product that is a deeply subordinated investment with uncertain income and significant capital risk. In a 2012 Financial Times article Satyajit Das warned that as a result of the dependence of conversion on regulatory decisions as well as credit risks, it is rather difficult for investors to quantify the probability of conversion. In addition to this risk, investors also face liquidity risk when they buy CoCos. The potential downside is made worse by the possibility of "death spirals". Such negative feedback loop spirals may arise when holders of a bank's shares try to sell shares to hedge a feared decline in the value of their holdings in the face of a mandatory conversion and potential dilution of their shares. This can potentially worsen the bank's financial position and lead to increased fear among other banks, counterparties, investors, and depositors. The probability of such spirals is increased by the fact that hybrids such as CoCos are marketed to private investors instead of institutional investors who face restrictions regarding purchases of hybrid securities. Of course, it is possible that investors are compensated for these risks in the form of generous coupons, but the recent example of CoCos issued by Credit Suisse and UBS which paid about 8% per annum suggests otherwise. Double-digit percentage coupons may be required to compensate for the missed couponsor mandatory conversion in times of distress. History contains several examples that should serve as a warning regarding the purchase of hybrids without careful due diligence. Perpetual Floating Rate Notes were a popular form of hybrid capital in the late 1980s but saddled their investors with large capital losses when they fell sharply in price. Another example is the case of some hybrid capital securities issued in the late 1990s in Australia which in 2012 were trading well below issue price. More recently, during the GFC several banks deferred coupon payments on hybrids which caused losses for investors.

As the above example illustrates, there are serious concerns about future performance of CoCos and several design flaws are apparent in the existing offerings.

The sensitivity of financial engineering and structured product innovation to regulatory changes was highlighted again in February 2014 when both Lloyds and Credit Suisse discussed plans to buy back their CoCos issued in 2009 after their respective regulators changed the way they viewed the debt and which instruments will count towards capital in bank stress tests. The February 2014 episode illustrates the significant uncertainty related to CoCos, their trigger events, their treatment by regulators for stress testing purposes, potential downward spirals and the wider effects on the legal protection of bond holders and the economy.

21.12 CONCLUSIONS

This chapter discussed securitization and the role that it has played before and after the GFC. We discussed one example of securitization in the form of CDOs and their tranches. We saw that CDO tranche values depend on default correlations which has led us to a discussion of correlation trading. The next chapter will provide a detailed example that shows the effect of correlation on tranche prices. This will lead us to a discussion of how to risk manage positions in CDO tranches and take views on increasing and decreasing correlations.

This chapter also provided interesting applications of financial engineering principles including option valuation and reduced form and structural models which we encountered in earlier chapters. We discussed credit structured products include post-GFC developments such as CoCos.

SUGGESTED READING

Brigo et al. (2010) provide a good account of credit models and CDOs during and after the GFC while **Schonbucher** (2004) and **Duffie and Singleton** (2002) are earlier academic approaches.

Some of the most useful references are in handbooks published by banks. We recommend the Handbook of Credit Derivatives by Merrill Lynch and one by JP Morgan. The latter is the closest approach to the market standard in this sector. For more details on and recent developments in the credit indices, see www.markit.com. **Francis et al.** (2003) is a good reference for credit correlation trading. **De Spiegeleer and Schoutens** (2012) provide an in-depth analysis of the valuation and structuring of CoCos.

EXERCISES

1.   What is the difference between an ABS and a CDO?

2.   What is the effect of default probabilities on CDO tranches? What is the effect of default correlations on CDO tranches? Explain.

3.   Consider the following news from Reuters:

     1008 GMT [Dow Jones] LONDON—SG recommends selling 7-year 0—3% tranche protection versus buying 5-year and 10-year 0—3% protection. 7-year equity correlation tightened versus 5-year and 10-year last year. SG's barbell plays a steepening of the 7-year bucket, as well as offering positive roll down, time decay, and jump to default.

     SG also thinks Alstom's (1022047.FR) 3—5-year curve is too steep, and recommends buying its 6.25% March 2010 bonds versus 3-year CDS.

     a.   What is a barbell? What is positive roll down, time decay?
     b.   What is jump to default?
     c.   Explain the logic behind SG's strategy.

4.   Consider the following quote:

     It is only when portfolios are tranched that the relative value of default correlation becomes meaningful.

     So, for subordinate tranches, the risk and spreads decrease as correlation between defaults increases, while for senior tranches the risk and spreads increase as default correlation increases.

     a.   Explain the first sentence carefully.
     b.   Explain the second paragraph.
     c.   Suppose you think that credit correlation would *decrease* in the near future. What type of trade would you put on?

5.   Consider the following quote:

     Until last year, this correlation pricing of single-tranche CDOs and first-to-default baskets was dependent on each bank or hedge fund's assessment of correlation. However, in 2003 the banks behind iBoxx and Trac-x started trading tranched versions of the indexes. This standardization in tranches has created a market where bank desks and hedge funds are assessing value and placing prices on the same products rather than on portfolios bespoke single-tranche CDOs and first-to-default baskets. Rather than the price of correlation being based on a model, it is now being set by the market.

   a. What is the *iTraxx* index?
   b. What is a *standard* tranche?
   c. Explain the differences between trading standardized tranches and the tranches of CDOs issued in the market place.

6. (Reduced form approach to CoCos valuation). The text mentions that CoCos can be valued from three different perspectives. One of the approaches is based on the reduced form or default intensity approach outlined in Chapter 18. Assume that a CoCo has a 5-year maturity. The underlying share price (S) is £100, the equity volatility is 20% and there are no dividend payments. The continuously compounded interest rate is 4%. Assume that the trigger event occurs then the share price (S) reaches the trigger price of $S^* = £50$. Calculate the credit spread using the trigger intensity model. Consult the end of chapter references for a derivation of the required probability of hitting the trigger, the trigger intensity and the recovery rate.

7. Consider the following reading, which deals with collateralized debt obligations (CDOs).
   Despite the deluge of downgrades in the collateralized debt obligation (CDO) market, banks are not focusing on the effect of interest rate swaps on arbitrage cash flow CDOs, Fitch Ratings said in a report released last week.
   Ineffective interest rate hedging strategies inflicted the hardest blows to the performance of high-yield bond CDOs completed during 1997–1999, the report noted.
   This combination of events caused some CDOs to become significantly over-hedged and out-of-the-money on their swap positions at the same time, the report found. For its report, Fitch used a random sample of 18 cash-flow deals that recently experienced downgrades.
   While half the CDOs benefited from falling rates, half did not. All nine of the over-hedged CDOs were high-yield bond transactions that closed before 1999.
   With the benefit of hindsight, a balanced guaranteed or customized swap would have mitigated the over-hedged CDO's risks. Plain vanilla swaps, which were economically advantageous during 1997–1999, ended up costing money in the long run because the notional balance of the swap is set at the deal's closing date and does not change over time, the report said. CDOs tend to use a plain vanilla swap instead of a customized swap because they are cheaper. (Thomson Reuters IFR Issue 1433, May 2002)
   a. Show the cash flows generated by a simple CDO on a graph. Suppose you are *short* the CDO.
   b. What are your risks and how would you hedge them?
   c. Show the cash flows of the CDO together with a hedge obtained using a plain vanilla swap.
   d. As time passes, default rates increase and interest rates decline, what happens to the CDO and to the hedge?
   e. What does the reading refer to with buying a customized swap?

CASE STUDY: CREDIT-LINKED NOTES

8. Read the following case study and answer the questions below. Overall, this case study deals with CDSs, synthetic corporate bonds, and, more interestingly, credit-linked notes.
   The case study highlights two issues.
   a. Cash flows and the risks associated with these instruments, and the reasons why these instruments are issued.
   b. The arbitrage opportunity that was created as a result of some of the recent issuance activity in credit-linked notes. Focus on these aspects when answering the questions that follow the readings.

Reading

Default swap quotes in key corporates have collapsed, as a rush to offset huge synthetic credit-linked notes has coincided with a shortage of bonds in the secondary market, and with a change in sentiment about the global credit outlook. The scramble to cover short derivatives positions has resulted in windfall arbitrage opportunities for dealers who chanced to be flat, and for their favored customers.

At least €5bn, and possibly as much as €15bn, equivalent of credit-linked note issuance has been seen in the last month. The resulting offsetting of short-credit default swap positions has caused a sharp widening in the negative basis between default swaps and the asset swap value of the underlying debt in the secondary bond market.

Dealers with access to corporate bonds have been able to buy default swaps at levels as much as 20 bp under the asset swap value of the debt, and to create synthetic packages for their clients where, in effect, the only risk is to the counterparty on the swap. Credit derivatives dealers who chanced to be flat have been turning huge profits by proprietary dealing—and from sales of these packages to their favored insurance company customers.

Deutsche Bank, Merrill Lynch, Bear Stearns, and Citigroup have been among the most aggressive sellers of default swaps in recent weeks, according to dealers at rival houses, and their crossing of bid/offer spreads has driven the negative default-swap basis to bonds ever wider.

A €2.25bn credit-linked note issued by Deutsche Bank is typical of the deals that have been fuelling this movement. The deal, Deutsche Bank Repon 2001–2014, offered exposure to 150 separate corporate credits, 51% from the US and 49% from Europe. Because DB had the deal rated, the terms of the issue spread across trading desks in London and New York, and rival dealers pulled back their bids on default swaps in the relevant corporates.

Other banks were selling similar unrated (and therefore, private) credit-linked notes at the same time, which led to a scramble to offset swap positions. Faced with a shortage of bonds in the secondary market, and repo rates at 0% for some corporate issues, dealers were forced to hit whatever bid was available in the default swap market, pushing the negative basis for many investment grade five-year default swaps from an 8 bp–16 bp negative basis to a 12 bp–20 bp basis last week.

This produced wild diversity between default swaps for corporates that had seen their debt used for credit-linked notes, and similar companies that had not. Lufthansa five-year default swaps were offered at 29 bp late last week, while British Airways offers in the same maturity were no lower than 50 bp, for example.

Many default swaps were also very low on an absolute basis. Single A-rated French pharmaceuticals company Aventis was quoted at 16 bp/20 bp for a five-year default swap at the close of dealing on Friday, for example. Other corporate default swaps were also at extremely tight levels, with Rolls-Royce offered as low as 27 bp in the five-year, Volkswagen at 26 bp, BMW offered at least as low as 26b, and Unilever at 21 bp.

Run for the Door

The movement was not limited to European credits. Offsetting of default swaps led to the sale of negative-basis packages in US names including Sears, Bank of America, and Philip Morris, with Bank of America trading at levels below 40 bp in the five-year, or less than half its trade point when fears about US bank credit quality were at their height earlier this year.

General market sentiment that the worst of the current downturn in credit quality has passed has amplified the effect of the default swap selling. Investors are happy to hold corporate bonds, which has left dealers struggling to buy paper to cover their positions as an alternative to selling default swaps.

"Everyone tried to run for the door at the same time," said one head dealer, describing trading in recent weeks. He predicted that the wide negative basis between swaps and bonds will be a trading feature for some time. Dealers worry that the banks which are selling default swaps most aggressively are doing so because they are lining up still more synthetic credit-linked notes. As long as they can maintain a margin between the notes and the level at which they can offset their exposure, they will keep hitting swap bids.

This collision of default swap-offset needs, a bond shortage, and improved credit sentiment is working in favor of corporate treasurers. WorldCom managed to sell the biggest deal yet from a US corporate last week, and saw spread talk on what proved to be a US$ 11.83bn equivalent deal tighten ahead of pricing. An issue of this size would normally prompt a sharp widening in default swaps on the relevant corporate, but WorldCom saw its five-year mid quotes fall from 150 bp two weeks ago to below 140 bp last week.

The decline in default swap quotes, and widening basis-to-asset swap levels for bonds, has been restricted to Europe and the US so far. If sentiment about the credit quality of Asian corporates improves there could be note issuance and spread movement. The dealers who have been struggling to cover their positions in the supposedly liquid US and European bond-and-swap markets may be reluctant to try the same approach in Asia, however.

With the prospect of more issuance of credit-linked notes on US and European corporates, and maintenance of the wide negative swap-to-bond basis, dealers who are allowed to run proprietary positions—and their insurance company clients—should reap further windfall arbitrage profits. The traders forced to offset deals issued by their structured note departments face further weeks of anxious hedging, however. (Thomson Reuters IFR, May 12, 2001)

Questions

a. What is a credit-linked note (CLN)? Why would investors buy credit-linked notes instead of, say, corporate bonds? Analyze the risks and the cash flows generated by these two instruments to see in what sense CLNs are preferable.

b. Suppose you issue a CLN. How would you hedge your position? Mention at least *two ways* of doing this. By the way, why do you need to hedge your position? Be specific.

c. As a continuation of the previous question, why is whether or not the investors sell their corporate bonds important in this situation?

d. Now we come to the arbitrage issue. What is the basis of the arbitrage argument mentioned in this reading? Be specific and explain in detail. Show your reasoning using cash flow diagrams.

e. What does a 0% repo rate for some corporate paper mean? Why is the rate zero?

f. Finally, why would this create an opportunity for corporate treasurers?

MATLAB EXERCISE

9. (Equity derivatives approach to the valuation of CoCos). Consider a CoCo, denominated in US dollars, with a maturity of 5 years and a face value of $1000. Assume that the current share price S is $100 and the equity volatility is 20%. The risk-free continuously compounded interest rate is 4% and there are no dividend payments. At the trigger moment a certain fraction (α) of the face value (N), that is αN is up for conversion. We assume that the conversion ($\alpha = 80$) will be triggered when the share price S falls below $50. The conversion price is equal to the price when the CoCo is issued. Therefore the conversion ratio CR equals $\alpha * 1000/100 = 8$. If the bank would like to issue the CoCo at par, what should be the coupon rate offered to the investors in order for the initial price of the CoCo to be $1000? Refer to the end of chapter references for details about the equity derivatives approach to the valuation of CoCos.

DEFAULT CORRELATION PRICING AND TRADING

CHAPTER OUTLINE

22.1 INTRODUCTION

There are three major issues with the credit sector. First there is the understanding and engineering of the credit risk itself. In other words, how does one strip the default risk component of a bond or a loan and trade it separately? The engineering of a CDS serves this purpose and was done in Chapter 18.

The second dimension in studying credit risk is the modeling aspect. Without modeling one cannot implement pricing, hedging, and risk management problems. Modeling helps to go from descriptive or graphical discussion to numbers. In credit risk, the modeling has a "novel" component. The risk in question is an *event*, the default. They are zero-one type random variables and are different than risk factors such as interest rates and stock prices which can be approximated by continuous state stochastic processes.[1] Credit risk, being a zero-one event, introduces a new dimensions in modeling.

The third major dimension of the credit sector has to do with the *tranching* of default risks. The main point of Chapter 21 was that tranching credit risk can eventually lead to stripping the *default correlation*. In this chapter, we talk about modeling default correlation and the resulting financial engineering of default correlation. We learn how to strip, hedge, and trade it. The recent correlation market provides the real-world background.

As we discussed in the previous chapter, banks extended the securitization process to CDOs, CLOs, and other products. The intermediate or mezzanine tranches were sold to investors. The very high-quality tranches, called *senior* and *super senior*, were also kept on banks' books because their implied return was too small for many investors. As a result, the banks found themselves *long equity tranche* and *long senior and super senior tranches*. They were short the *mezzanine* tranche which paid a good return and was rated investment grade. It turns out, as we will see in this chapter, that the equity tranche value is *positively* affected by the default correlation while the super senior tranche value is *negatively* affected. The sum of these positions formed the correlation books of the banks. Banks had to learn *correlation trading*, *hedging*, and *pricing* as a result.

The connection between default correlation and tranche values is a complex one and needs to be clarified step by step. We discuss market examples of how this idea can be exploited in setting up *correlation trades* and can be exploited in setting up new structured products. The first issue that we need to introduce is the dependence of the tranche pricing on the default correlation.

22.2 TWO SIMPLE EXAMPLES

We first discuss two simple cases to illustrate the logic of how default correlation movements affect tranche prices. It is through this logic that observed tranche trading can be used to back out the default correlation. This quantity will be called *implied correlation*.

---

[1]Remember that real-life indicators are actually not continuous state random variables. Instead the state space is countable. Markets have conventions in terms of decimal points. For example, the EUR/USD rate is quoted up to four decimal points. Thus the minimum tick is 0.0001 a pips. In a typical trading day the total number of plausible true states of the world is no more than, say, 200–300.

22.2.1 PORTFOLIO WITH THREE CREDIT NAMES

Let $n = 3$ so that there are only three credit names in the portfolio. With such a portfolio, we can consider only three tranches: equity, mezzanine, and senior. In this simple example, the equity tranche bears the risk of the first default (0−33%), the mezzanine tranche bears the risk of the second name defaulting (33−66%), and the senior tranche investors bear the default risk of the third default (66−100%).

In general, we follow the same notation as in the previous chapter. As a new parameter, we let $\rho_t^{i,j}$ be the default correlation between ith and jth names in the portfolio at time t. p_t^i is the probability of default and cds_t^i is the liquid CDS spread for the ith name, respectively. We make the following assumptions without any loss of generality. The default probabilities for the period $[t_0, T]$ are the same for all names, and they are constant

$$p_t^1 = p_t^2 = p_t^3 = p \quad \text{for all } t \in [t_0, T] \tag{22.1}$$

We also let the recovery rate be constant and be given by R.

We consider two extreme settings. In the first, default correlation is zero

$$\rho^{i,j} = 0 \tag{22.2}$$

The second is perfect correlation,

$$\rho^{i,j} = 1 \tag{22.3}$$

for all $t \in [t_0, T]$, i, j. These two extreme cases will convey the basic idea involved in correlation trading. We study a number of important concepts under these two assumptions. In particular, we obtain (i) default loss distributions, (ii) default correlations, and (iii) tranche pricing in this context.

Start with the independence case. In general, with n credit names, the probability distribution of D will be given by the binomial probability distribution. Letting p denote the constant probability of default for each name *and* assuming that defaults are independent, the number of defaults during a period $[t_0, T]$ will be distributed as

$$P(D = k) = \frac{n!}{k!(n-k)!} p^k (1-p)^{n-k} \tag{22.4}$$

Note that there is no ρ parameter in this distribution since the correlation is zero. Now consider the first numerical example.

22.2.1.1 Case 1: Independence

With $n = 3$ there are only four possibilities, $D = \{0, 1, 2, 3\}$. For zero default, $D = 0$ we have

$$P(D = 0) = (1 - p)^3 \tag{22.5}$$

For the remaining probabilities we obtain

$$P(D = 1) = p(1 - p)^2 + (1 - p)p(1 - p) + (1 - p)^2 p \tag{22.6}$$

$$P(D = 2) = p^2(1 - p) + p(1 - p)p + (1 - p)p^2$$

$$P(D = 3) = p^3 \tag{22.7}$$

Suppose p = 0.05. *Plugging in the formulas above we obtain first the probability that no default occurs*

$$P(D=0) = (1-0.05)^3 = 0.857375 \tag{22.8}$$

One default can occur in three different ways:

$$P(D=1) = (0.05)(1-0.05)(1-0.05) + (1-0.05)(0.05)(1-0.05) + (1-0.05)(1-0.05)(0.05)$$
$$= 0.135375 \tag{22.9}$$

Two defaults can occur again in three ways:

$$P(D=2) = (0.05)(0.05)(1-0.05) + (0.05)(1-0.05)(0.05) + (1-0.05)(0.05)(0.05)$$
$$= 0.007125 \tag{22.10}$$

For three defaults there is only one possibility and the probability is

$$P(D=3) = (0.05)(0.05)(0.05) = 0.000125 \tag{22.11}$$

As required, these probabilities sum to one.

The spreads associated with each tranche can be obtained easily from these numbers. Assume for simplicity that defaults occur only at the end of the year. In order to calculate the tranche spreads we first calculate the expected loss for each tranche under a proper working probability. Then the spread is set so that expected cash inflows equal this expected loss. The expected loss for the three tranches is given by

$$\text{Equity} = 0[P(D=0)] + 1[P(D=1) + P(D=2) + P(D=3)] = 0.142625 \tag{22.12}$$

$$\text{Mez} = 0[P(D=0) + P(D=1)] + 1[P(D=2) + P(D=3)] = 0.00725 \tag{22.13}$$

$$\text{Sen} = 0[P(D=0) + P(D=1) + P(D=2)] + 1[P(D=3)] = 0.000125 \tag{22.14}$$

In the case of 1-year maturity contracts it is easy to generalize these tranche spread calculations. Let $B(t, T)$ denote the appropriate time-t discount factor for \$1 to be received at time T. The recovery rate is given by R. Assuming that $N = 1$ and that defaults can occur only on date T, the tranche spreads at time t_0 denoted by $\text{cds}_{t_0}^j$, $j = e, m, s$ are then given by the following equation,

$$0 = B(t_0, T)\left[\text{cds}_{t_0}^e P(D=0) - \left[(1-R) - \text{cds}_{t_0}^e\right] P(D \geq 1)\right] \tag{22.15}$$

for the equity tranche which is hit with the first default. This can be written as

$$\text{cds}_{t_0}^e = (1-R) \times P(D \geq 1) \tag{22.16}$$

For the mezzanine tranche we have

$$0 = B(t_0, T)\left[\text{cds}_{t_0}^m (P(D=0) + P(D=1)) - \left[(1-R) - \text{cds}_{t_0}^m\right](P(D=2) + P(D=3))\right] \tag{22.17}$$

which gives

$$\text{cds}_{t_0}^m = (1-R) \times (P(D=2) + P(D=3)) \tag{22.18}$$

Finally, for the senior tranche

$$0 = B(t_0, T)\left[\text{cds}_{t_0}^m (P(D=0) + P(D=1) + P(D=2)) - \left[(1-R) - \text{cds}_{t_0}^m\right] \times P(D=3)\right] \tag{22.19a}$$

And we obtain

$$\text{cds}_{t_0}^s = (1-R) \times P(D=3) \tag{22.19b}$$

Note that, in general, with $n > 3$ and the number of tranches being less than n, there will be more than one possible value for R. One can obtain numerical values for these spreads by plugging in $p = 0.05$ and the recovery value $R = 40\%$.

If we plug in the above values, we obtain the following credit spreads for the different tranches: $\text{cds}_{t_0}^e = 0.085575$, $\text{cds}_{t_0}^m = 0.00435$, and $\text{cds}_{t_0}^s = 0.000075$. As expected the equity tranche is the riskiest and is compensated for this with the highest spread.

It is interesting to note that for each tranche, the relationship between *spreads* and *probabilities* of making floating payments is similar to the relation between *cds* and *p* we obtained for a single-name CDS in Chapter 18

$$\text{cds} = (1 - R) \times p. \tag{22.20}$$

22.2.1.2 Case 2: Perfect correlation

If default correlation increases and $\rho \to 1$ then all credit names become essentially the same, restricting default probability to be identical. So let

$$p_t^i = 0.05 \tag{22.21}$$

for all names and all times $t \in [t_0, T]$. Under these conditions, the probability distribution for D will be trivially given by the distribution.

$$P(D = 0) = (1 - 0.05) = 0.95 \tag{22.22}$$
$$P(D = 1) \to 0 \tag{22.23}$$
$$P(D = 2) \to 0 \tag{22.24}$$
$$P(D = 3) = 0.05 \tag{22.25}$$

The corresponding expected losses on each tranche can be calculated trivially. The loss is the same for all tranches and is equal to 0.05 if we assume same values for p and R as in Case 1. For the equity trance obtain the following credit spread:

$$0 = c_{t_0}^e P(D = 0) - \left[(1 - R) - c_{t_0}^e \right] (D \geq 1) \tag{22.26}$$
$$\text{cds}_{t_0}^e = (1 - R) \times P(D \geq 1) = (1 - 0.4)(0.05) = 0.03$$

The mezzanine tranche spread will be

$$\text{cds}_{t_0}^m = (1 - R) \times [P(D = 2) + P(D = 3)] = (1 - 0.4)(0.05) = 0.03 \tag{22.27}$$

Finally for the senior tranche we have

$$\text{cds}_{t_0}^s = (1 - R) \times [P(D = 3)] = (1 - 0.4)(0.05) = 0.03 \tag{22.28}$$

We can extract some general conclusions from the examples in cases 1 and 2 above. First of all, as $\rho \to 1$, all three tranche spreads become similar. This is to be expected since under these conditions all names start looking more and more alike. At the limit $\rho = 1$ there are only two possibilities, either nobody defaults or everybody defaults. There is no risk to "tranche" and sell separately. There is only *one* risk.

Second, we see that as correlation goes from zero to one, the expected loss of equity tranche decreases and the credit spread decreases. The expected loss of the senior tranche and the resulting credit spread, on the other hand, goes up. The mezzanine tranche is somewhere in between: the expected loss and credit spread goes up as correlation increases, but not as much as the senior tranche. In general, the sensitivity of the mezzanine tranche to correlation is dependent on its attachment points and size as well as the current credit spreads.

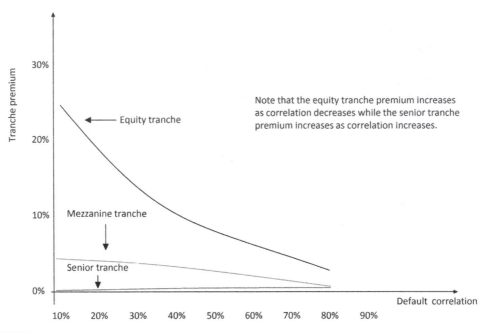

FIGURE 22.1

Tranche premiums as a function of default correlation.

Fourth, the expected loss of the portfolio is independent of the correlation. The loss is 0.15 in both cases if we add up the losses across the three tranches. However, the correlation determines how the expected losses are allocated across the different parts of the capital structure.

22.2.2 SENSITIVITY OF TRANCHE SPREADS AND BASKET DEFAULT SWAPS TO DEFAULT CORRELATION

Finally, note that as default correlation went up the distribution became more skewed with the "two" tails becoming heavier at both ends. In Chapter 21, we saw how the default distribution and portfolio loss distribution depend on the default correlation. Figure 22.1 is based on a hypothetical portfolio and illustrates that the tranche premium is a function of correlation and tranche subordination. The tranche premium is highest for the equity tranche for all levels of correlation. This reflects the fact that the equity tranche has no subordination and is the first to suffer losses. The senior tranche is protected by the subordination of the equity and mezzanine tranches and therefore it has the lowest risk. The line representing the equity tranche shows that as default correlation increases the equity tranche premium falls. The opposite is true if default correlation increases. For the senior tranche, the premium increases as the default correlation increases. For the mezzanine tranche, the relationship between default correlation and tranche premium is generally sensitivity to several parameters but in this illustrative example we assume that it decreases with default correlation. This is also what a comparison of cases 1 and 2 in the numerical example above revealed.

Note that the above example can also be interpreted as a basket default swap instead of an application of tranche pricing. A *basket default swap* differs from a single-name default swap in that the underlying is a basket of entities rather than one single entity. Examples of popular basket default swaps are *first-to-default, n-th-to-default, n-out-of-m-to-default*, and *all-to-default* swaps. In a single-name credit default swap, a credit event is usually a default of the entity. In the example there were three assets or names, and we calculated the spreads on the first-to-default, second-to-default, and third-to-default swaps. Thus the price of synthetic CDO tranches depends on default correlations in a similar way as a basket default swap. In a synthetic CDO, the underlying assets are CDS while in a *cash* CDO the underlying assets are bonds. As the CDO market evolved, *single-tranche trading* was developed. This refers to the trading of CDO tranches without the underlying portfolio of short CDS positions being created.

22.2.3 RECENT QUOTATION CONVENTION FOR CDO TRANCHES AND EVOLUTION OF TRANCHE SPREADS

The simplified example illustrated how the spreads of different tranches depend on the default correlation. Figure 22.1 illustrated this point using a hypothetical portfolio. Before we look at a real-world example of how (iTraxx Europe) tranche spreads have moved recently, we need to note that after the CDS *Big Bang* in 2009 CDS contracts were standardized and the quotation of the tranches changed. Until April 2009, all *iTraxx* tranches except for the equity tranche (0−3%) quoted in *basis points* with no fixed running spread. In other words, no money was exchanged upfront. The equity tranche, however, consisted of an upfront fee with a fixed 500 bps spread and was quoted in terms of *percentage of notional*. After April 2009, the quotation of tranches changed and since then the equity (0−3%) and mezzanine tranches (3−6%, 6−9%) consist of upfront fees with a fixed 500 bps spread and are quoted in terms of *percentage of notional*. The 9−12% and the senior 12−22% tranche continue to be quoted in basis points with no fixed running spread.

Therefore we will examine the years from 2007 until 2009 in Figure 22.2a and b first before discussing the years 2010−2013 which are reported in Figure 22.2c and d. As we see in Figure 22.2a and b, all tranche spreads increased significantly during the GFC from 2007 until 2009. The mezzanine 6−9% tranche spread rose from 11.95 bps on January 31, 2007 to 606.69 bps on January 30, 2009. This represents a more than 60-fold increase. All other tranches also increased significantly. There are many factors that can explain these changes, but two of them are related to the default probability and the default correlation. As market participants revised upwards their view of the default probability all tranche spreads should increase. In crisis times, correlations also increase which, however, as we saw in the examples in the previous section have a different impact on different tranches. As we would expect there is some evidence that proportionately spreads on mezzanine and senior tranches have gone up more than the spread on the equity tranche.

Some commentators have laid partial blame for the severity of the GFC on the Gaussian copula model use in the pricing of CDO backed by MBS. However, it is important to note that the limitations of the model were pointed out by practitioners and academics before the crisis. Moreover, laying blame on the model for mispricing the tranches and CDOs is akin to a trader laying blame for option losses that he or she suffered after using historical implied volatilities to price options before being surprised by an unexpected increase in implied volatility. Option prices and tranche prices

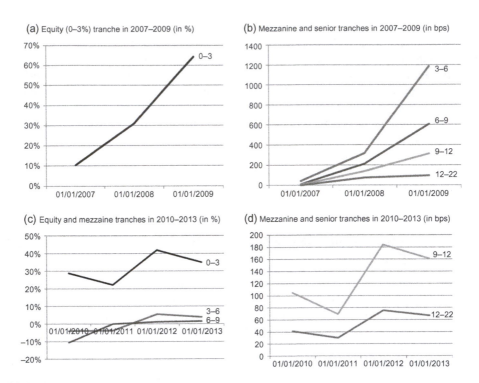

FIGURE 22.2

Tranche spreads during 2007–2013.

can move unexpectedly as market participants revise their expectations. The cause of the losses can be found with the users of the models. Even the brief discussion here highlights the limitations of the models. More refined models are constantly being developed but as we pointed out repeatedly in the book, models are just an approximation of reality and often serve as a communication tool or benchmark, but reality is significantly more complex than the models' assumptions make it out to be. Buyers and sellers of derivatives must understand the products that they trade and structurers of products must take care to explain products to clients and make sure that the products are suitable for their clients. This applies to all products and services and not just derivatives or structured products.

Figure 22.2c and d shows that tranche spread volatility decreased in the years 2010–2013. Figure 22.2c reports the new quotation conventions.[2] For the 0–3% equity tranche, the quotation on January 29, 2010, was 28.81% (in terms of percentage of notional). Since the tranche has a fixed coupon of 500 bps, there is an upfront payment at initiation. The amount paid upfront is equal to the present value of the difference between the current market spread and the fixed coupon. Since the contract has a fixed coupon of 500 bps but is trading at 2881 bps (28.81%), the protection buyer would make an upfront payment equal to the present value of the difference between 2881 and 500 bps. The figure shows that for some of the spread the market spread is lower than the fixed

---

[2]The data source is Creditex and Markit. See www.creditfixings.com for more details.

coupon in which case the protection *seller* would have to pay the difference between the two spreads upfront. In 2010–2013, tranche spreads have increased somewhat but the relative magnitude of the increase was smallest for the equity tranche and most pronounced for the senior and mezzanine 9–12% tranches. If we ignore technical factors and changes in liquidity, then we could interpret the movements in tranche prices over this period as being consistent with an increase in the perception of the default probability and the default correlation. The comovement of the different tranches suggests that there is some reversion to the mean. If a trader observed a divergence in tranche spreads, he or she could set up a position that bets on convergence due to the observed comovement of tranches over longer periods of time.

Our discussion related to Figure 22.2 here is heuristic and does not represent a rigorous analysis of tranche spread drivers since the main goal is to illustrate the dependence of tranches on default correlation.

22.3 STANDARD TRANCHE VALUATION MODEL

We now show how the distribution of D can be calculated under correlations different than 0 and 1. We discuss the standard market model for pricing standard tranches for a portfolio of n names.

22.3.1 THE GAUSSIAN COPULA MODEL

This model is equivalent to the so-called Gaussian copula model, in a one factor setting.

Let

$$\{S^j\}, \quad j = 1, \ldots, n \tag{22.29}$$

be a sequence of latent variables. Their role is to generate *statistically dependent* zero-one variables.[3] There is no model of a random variable that can generate dependent zero-one random variables with a *closed-form* density or distribution function. $\{S^j\}$ are used in a Monte Carlo approach to do this.

It is assumed that S^j follows a normal distribution and that the default of the ith name occurs the first time S^j falls below a threshold denoted by L_α, where a is the jth name's default probability. The important step is the way S^j is structured. Consider the following *one factor* case,

$$S^j = \rho^j F + \sqrt{1 - (\rho^j)^2} \, \epsilon^j \tag{22.30}$$

where F is a *common* latent variable independent of ϵ^j, ρ^j is a constant parameter, and ϵ^j is the *idiosyncratic* component. The random variables in this setup have some special characteristics. F has no superscript, so it is common to all S^j, $j = 1, \ldots, n$ and it has a standard normal distribution. ϵ^j are specific to each i and are also distributed as standard normal,

$$\epsilon^j \sim N(0, 1) \tag{22.31}$$

$$F \sim N(0, 1) \tag{22.32}$$

---

[3]The previous example dealt with an independent default case. Default is a zero-one variable, which made the random variable D, representing total number of defaults during $[t_0, T]$, follow a standard binomial distribution.

Finally, the common factor and the idiosyncratic components are uncorrelated.

$$E[\epsilon^j F] = 0 \qquad (22.33)$$

It turns out that ϵ^j and F may have any desired distribution.[4] If this distribution is assumed to be normal, then the model described becomes similar to a Gaussian copula model. Note this case is in fact a one factor latent variable model.

We obtain the *mean* and *variance* of a typical S^j as follows

$$E[S^j] = \rho E[F] + E\left[\sqrt{1 - (\rho^2}\,\epsilon^j\right]$$
$$= 0 \qquad (22.34)$$

and

$$E[(S^j)^2] = \rho^2 E[F^2] + E[(1 - \rho^2)(\epsilon^j)^2]$$
$$= 1 + \rho^2 - \rho^2 = 1 \qquad (22.35)$$

since both random variables on the right side have zero mean and unit second moment by assumption and since F is uncorrelated with ϵ^j. The model above has an important characteristic that we will use in stripping default correlation. It turns out that the correlation between defaults can be conveniently modeled using the common factor variable and the associated ρ^i. Calculate the correlation

$$E[S^j S^k] = E\left[\rho F + \sqrt{1 - \rho^2}\,\epsilon^j\right]\left[\rho F + \sqrt{1 - \rho^2}\,\epsilon^k\right]$$
$$= \rho^2 E[F^2] + E[2(1 - \rho^2)\epsilon^j \epsilon^i] + E\left[\rho\sqrt{(1 - \rho^2)}\epsilon^j F\right] + E\left[\rho\sqrt{(1 - \rho^2)}\epsilon^i F\right] \qquad (22.36)$$

This gives

$$\text{Corr}[S^i S^j] = \rho^i \rho^j \qquad (22.37)$$

The case where

$$\rho^i \rho^j \qquad (22.38)$$

for all i, j, is called the *compound correlation* and is the market convention. The compound default correlation coefficients is then given by ρ^2. The *ith* name default probability is defined as

$$p^i = P\left(\left(\rho F + \sqrt{1 - \rho^2}\,\epsilon^i\right) \le L_\alpha\right) \qquad (22.39)$$

The probability is that the value of S^i will fall below the level L_α. Remember that in Chapter 19 we introduced Merton's (1974) structural model of default where the default occurs when the value of the firm falls below the debt issued by the firm. Hence, at first glance, it appears that the market convention here is similar to Merton's model. This is somewhat misleading, however, since S^i has no structural interpretation here. It will be mainly used to generate correlated binomial variables.

---

[4]One caveat here is the following. The sum of two normal distributions is normal; yet, the sum of two arbitrary distributions may not necessarily belong to the same family. In fact, any closed-form distribution to model such a sum may not exist. One example is the sum of a normally distributed variable and a student's t distribution variable which cannot be represented by a closed-form distribution formula. Such exercises require the use of Monte Carlo and will generate the distributions numerically.

This above setup provides an agenda one can follow to price and hedge the tranches. It will also help us to back out an implied default correlation once we observe liquid tranche spreads in the market.

The agenda is as follows: First, observe the CDS rate cds_t^i for each name in the markets. Using this, calculate the risk-neutral default probability p^i. Using this find the corresponding L_α^i. Generate pseudo-random numbers for F and $\in^i$. Next, assume a value for ρ and calculate the implied S^i. Check to see if the value of S^i obtained this way is less than L_α^i. This way, obtain a simulated default process:

$$d^i = \begin{cases} 1 & \text{if } S^i < L_\alpha \\ 0 & \text{otherwise} \end{cases} \tag{22.40}$$

Finally calculate the number of defaults for the trial as

$$D = \sum_{i=1}^{n} d^i \tag{22.41}$$

Repeating this procedure m times will yield m replicas of the random variable D. If m is large, we can use the resulting histogram as the default loss distribution on the n reference names and then calculate the spreads for each tranche.

22.3.2 IMPLIED CORRELATIONS

The valuation of synthetic CDO tranches follows similar principles as the valuation of CDS described in Chapter 18. The *break-even spread* on a tranche is defined as the spread that sets the present value of the payments equals the present value of the payoffs of the tranche. The present value depends on the cash flows, the probability of receiving the cash flows and a discount factor. This is again analogous to the valuation of the single-name CDS. What is different in CDO tranche valuation is that now we have multiple underlying assets and the probability of cash flows must take into account the correlation structure between cash flows. This is what the standard market model described above achieves by using a Gaussian copula model specification. The standard Gaussian copula market model can be used in a similar way as the Black–Scholes model which is the standard market model in option markets. We used the Black–Scholes model to either calculate theoretical option prices based on given input parameters or to back out implied volatilities given observed market option prices. In the standard (Gaussian copula) market model, we can use default probabilities as input parameters and determine tranche spreads or we can start with market observed tranche spreads and calculate the implied probabilities.[5] Since we are using the standard market model, the default loss distribution will depend on a certain level of default correlation due to the choice of ρ. The *implied correlation* is that level of ρ which yields a calculated tranche spread that equals the observed spread in the markets. The coefficient ρ is the only unknown variable if we observe tranche spreads.

---

[5]If default probabilities are used as input parameters these can be based on bond yield data for example.

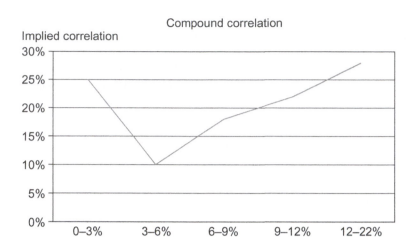

FIGURE 22.3

Compound correlation.

The reason why market participants prefer to quote implied correlation instead of the market spread directly is that correlation is independent of the overall spread level of the underlyings. This is similar to the market practice in option markets where implied volatilities instead of option prices are quoted.

It is common in the market to imply a correlation from market quotes for tranches. The correlation that if plugged into the standard model results in the spreads observed in the market is called the *compound correlation* or *tranche correlation*. In other words, the compound correlation is the correlation found by calibrating the Gaussian copula model to the price of a CDO tranche. The implied default correlation ρ is the correlation that, given the observed market spread, sets the present value of the two legs of a CDO tranche equal to zero. Figure 22.3 shows a typical compound correlation plot. It shows the compound correlation for different tranches. Similar to equity markets, where the Black–Scholes model leads to implied volatility skews, smiles, or frowns, the Gaussian copula approach to model the expected loss in a given tranche leads to different correlations. The compound correlations exhibit a typical *correlation skew* or *correlation smile*. With increasing tranche seniority, the implied correlation first decreases and then increases.

We observe that the implied compound correlation for the mezzanine tranche is lower than for the senior and equity tranches. This *dip* in correlations reflects the fact that the market assigns a higher price to the risk of losses in senior tranches than predicted by the Gaussian copula model compared to the equity tranche. In other words, the market view is that defaults tend to cluster as they become more frequent.

One of the limitations of compound correlations is that for some (mezzanine) tranches, the tranche correlation is not uniquely defined and may not even exist. These computation issues imply that the implied correlation is conceptually different from implied volatility. We will see in Section 22.6 how so-called base correlations address this issue.

Below we have a simple example that shows this process.

EXAMPLE

Let $n = 3$. Assume that the default probability is 1%; and let the default correlation be $\rho^2 = 0.36$. We generate 10,000 replicas for each $\{S^1, S^2, S^3\}$ using

$$S^1 = 0.25F + \sqrt{64}\,\epsilon^1 \tag{22.42}$$

$$S^2 = 0.25F + \sqrt{64}\,\epsilon^2 \tag{22.43}$$

$$S^3 = 0.25F + \sqrt{64}\,\epsilon^3 \tag{22.44}$$

For example, we select four standard pseudo-random variables

$$F = -1.12615 \tag{22.45}$$

$$\epsilon^1 = -2.17236 \tag{22.46}$$

$$\epsilon^2 = 0.64374 \tag{22.47}$$

$$\epsilon^3 = -0.326163 \tag{22.48}$$

Using these we obtain S^i as

$$S^1 = -2.41358 \tag{22.49}$$

$$S^2 = -0.160698 \tag{22.50}$$

$$S^3 = -0.936621 \tag{22.51}$$

We then calculate the corresponding S^j and see if they are less than $L_{.01} = -2.32$, where the latter is the threshold that gives a 1% tail probability in a standard normal distribution.

If $S^1 < -2.32$, $S^2 < -2.32$, $S^3 < -2.32$, then we let the corresponding $d^i = 1$. Otherwise they are zero. We then add the three d^i to get the D for that Monte Carlo run.

In this particular case, $d^1 = 1$, $d^2 = 0$, $d^3 = 0$. So the first Monte Carlo run gives $D = 1$.

Next we would like to show an example dealing with implied correlation calculations. The example again starts with a Monte Carlo sample on the D, obtains the tranche spreads, then extends this to three functions, calculated numerically, that show the mapping between tranche spreads and correlation.

EXAMPLE: *IMPLIED CORRELATION*

Suppose we are given a Monte Carlo sample of correlated defaults from a reference portfolio,

$$\text{Sample} = \{D_1, \ldots, D_m\} \tag{22.52}$$

Then, we can obtain the histogram of the number of defaults.

Assume $\rho = 0.3$, $n = 100$, and $m = 1000$. Running the procedure above we obtain 1000 replicas of D_i. We can do at least two analyses with the distribution of D. First we can calculate the fair value of the tranches of interest. For example, with recovery 40% and zero interest rates, we can compute the value of the equity tranche in this case as

$$\text{cds}^e = [0.05(0) + 0.12(33.33)0.6 + 0.15(66)0.6 + (1 - 0.05 - 0.12 - 0.15)]0.6 = 0.68 \quad (22.53)$$

This 1-year spread is based on the fact that the party that sells protection with nominal $N = 100$ on the first three defaults will lose nothing if there are no defaults, will lose a third of the investment if there is one default, will lose two-thirds if there are two defaults, and finally will lose all investments if there are three defaults.

Repeating this for all values of $\rho^2 \in [0, 1]$ we can get three surfaces. These graphs plot the value of the equity, mezzanine, and senior tranches against the fair value of the tranche.

Note that the value of the equity tranche goes up as correlation increases. The value of the mezzanine tranche is a U-shaped curve, whereas the value of the senior tranche is again monotonic. The end of chapter exercises provide another numerical example.

22.3.3 THE CENTRAL LIMIT EFFECT

Why does the default loss distribution change as a function of the correlation? This is an important technical problem that has to do with the central limit theorem and the assumptions behind it. Now define the *correlated* zero-one stochastic process z_i:

$$z_j = \begin{cases} 1 & \text{if } S^j \leq L_\alpha \\ 0 & \text{otherwise} \end{cases} \quad (22.54)$$

Next calculate a *sum* of these random variables as

$$Z^n = \sum_{j=1}^{n} z^j \quad (22.55)$$

According to the central limit theorem, even if z^j are individually very far from being normally distributed *and* are correlated, then the distribution of the sum will approach a normal distribution if the underlying processes have finite means and variances as in $n \to \infty$.

$$\frac{\sum_i^n z_i - \sum_i^n \mu_i}{\sqrt{\sum_i^n \sigma_i^2}} \to N(\mu^d, \sigma^d) \quad (22.56)$$

Thus if we had a very high number of names in the reference portfolio, the distribution of D will look like normal. On the other hand, with finite n *and* highly correlated z^j, this convergence effect will be slow. The higher the "correlation" ρ, the slower will the convergence be. In particular, with n around 100, the distribution of the D will be heavily dependent on the size of ρ and will be far from normal. This is where the relationship between Index tranches and correlation comes in. In the extreme case when correlation is perfect, the *sum* will involve the *same* z^j.

22.4 DEFAULT CORRELATION AND TRADING

Correlation impacts risk assessment and valuation of CDO tranches. Each tranche value reflects correlation in a different way. Perhaps the most interesting correlation relationship is shown in the equity (0−3%) tranche. This may first appear counterintuitive. It turns out that the higher the correlation, the lower the equity tranche risk, and the higher the value of the tranche. The reasoning behind this is as follows. Higher correlation makes extreme cases of very few defaults more likely. Everything else being the same, the more correlation, the lower the risk the investor takes on and the lower premium the investor is going to receive over the lifetime of the tranche.[6]

The influence of default correlation on the mezzanine tranche is not as clear. In fact, the value of the mezzanine tranche is less sensitive to default correlation. For the senior tranches, higher correlation of default implies a higher probability that losses will wipe out the equity and mezzanine tranches and inflict losses on the senior tranche as well. Thus, as default correlation rises, the value of the senior tranche falls. A similar reasoning applies to the super senior tranche.

Correlation trading is based on this different dependence of the tranche spreads on default correlation. One of the most popular trades of 2004 and the first half of 2005 was to sell the equity tranche and hedge the default probability movements (i.e., the market risk) by going long the mezzanine index.

This is a *long correlation* trade and it had significant positive carry.[7] The trade also had significant (positive) convexity exposure. In fact, as the position holder adjusts the *delta* hedge, the hedging gains would lead to a gain directly tied to index volatility. Finally, if the carry and convexity are higher, the position's exposure to correlation changes will be higher.

A long correlation position has two major risks that are in fact related. First, if a single-name credit event occurs, long correlation positions would realize a loss. This loss will depend on the way the position is structured and will be around 5−15%. The recovery value of the defaulted bonds will also affect this number.

Second, a change in correlation will lead to mark-to-market gains and losses. In fact, the change in correlation is equivalent to markets changing their view on an idiosyncratic event happening. A rule of thumb that can be used is that a 100 basis point drop in correlation will lead to a change in the value of a *delta*-hedged equity tranche by around 1%. For short equity tranche protection, long mezzanine tranche protection position this loss would be even larger. This means that the position may suffer significant mark-to-market losses if expected correlation declines. Sometimes these positions need to be liquidated.

A market participant can go

- *long correlation* by going either *long the equity* tranche or *short the senior* tranche.
- *short* correlation by going either *short* the equity tranche or *long* the senior tranche.

Note that the sensitivity to correlation can change through time since it depends on several variables, underlying spreads, number of defaults, and time decay.

---

[6]Note that when we talk about lower correlation, we keep the default probabilities the same. The sensitivity to correlation changes is conditional on this assumption. Otherwise, when default probabilities increase, all tranches would lose value.

[7]Around 300 bps on the average during 2005−2007.

22.5 *DELTA* HEDGING AND CORRELATION TRADING

The *delta* is the sensitivity of the individual tranche spreads toward movements in the underlying index, I_t.[8]

Let the tranche spreads at time t be denoted by cds_t^j, $j = e, m, s, sup$. The superscript represents the equity, mezzanine, senior, and super senior tranches, respectively. There will then be (at least) two variables affecting the tranche spread: the probability of default and the default correlation. Let the average probability of default be denoted by p_t and the compound default correlation be ρ_t. We can write the index spread as a function of these two variables.

$$cds^i = f^i(p_t, \rho_t) \tag{22.57}$$

It is important to remember that with changes in p_t the *sign* of the sensitivity is the same for *all* tranches. As probability of default goes up, the tranche spreads will all go up, albeit in different degrees. The sensitivities with respect to the index (or probability of default) will be given by

$$\Delta^i = \frac{\partial cds_t^i}{\partial I_t} > 0 \tag{22.58}$$

for all i. These constitute the tranche *deltas* with respect to the index itself. It is natural, given the level of subordination in higher seniority tranches, to find that

$$\Delta^e > \Delta^m > \Delta^s > \Delta^{sup} \tag{22.59}$$

Yet, the *correlation sensitivity* of the tranches is very different. As discussed earlier, even the sign changes.

$$\frac{\partial cds^e}{\partial \rho} < 0 \tag{22.60}$$

$$\frac{\partial cds^m}{\partial \rho} \cong 0 \tag{22.61}$$

$$\frac{\partial cds^s}{\partial \rho} > 0 \tag{22.62}$$

$$\frac{\partial cds^{sup}}{\partial \rho} > 0 \tag{22.63}$$

According to this, the equity protection seller benefits if the compound default correlation goes up. The super senior protection seller loses under these circumstances. The middle tranches are more or less neutral.

Suppose the market participant desires to take a position positively responsive to ρ_t but more or less neutral toward changes in p_t. This is clearly a *long correlation* exposure discussed earlier. How would one put such a correlation trade on in practice? To do this the market practitioner would use the *delta* of the tranches with respect to the index. The position will consist of *two* hedging efforts.

---

[8]*Deltas* can be calculated with respect to each other and reported individually. But the procedure outlined below comes to the same thing. We report the *deltas* with respect to the index for two reasons. First, this is what the market reports, and second, the positions are hedged with respect to the index and not by buying and selling equity or mezzanine tranches with direct hedging.

First, the right amounts of the equity and mezzanine tranches have to be purchased so that the portfolio is immune to changes in the default correlation. Second, each tranche should be hedged separately with respect to the changes in the index, as time passes.[9]

Let N^e and N^m be the two notional amounts. N^e is exposure on equity, while N^m is the exposure on the mezzanine tranche. What we want is a change in the index to not lead to a change in the total value of the position on these two tranches.

Thus the portfolio P_t will consist of selling default protection by the new amount

$$P_t = N^e - N^m \tag{22.64}$$

we let

$$N^m = \lambda N^e \tag{22.65}$$

where λ is the hedge ratio to be selected. Substituting we obtain

$$P_t = N^e - \lambda_t N^e \tag{22.66}$$

λ_t is the hedge ratio selected as

$$\lambda_t = \frac{\Delta_t^e}{\Delta_t^m} \tag{22.67}$$

With this selection, the portfolio value will be immune to changes in the index value at time t:

$$\frac{\partial}{\partial I_t} P_t = \frac{\partial}{\partial I_t} N^e - \frac{\partial}{\partial I_t} N^m \tag{22.68}$$

Replacing from Eq. (22.67)

$$\frac{\partial}{\partial I_t} P_t = \Delta^e - \frac{\Delta^e}{\Delta^m} \Delta^m = 0. \tag{22.69}$$

Hence the portfolio of selling N^E units of equity protection and buying simultaneously λN^E units of the mezzanine protection is *delta hedged*. As the underlying index moves, the portfolio would be neutral to the first order of approximation. Also, as the market moves, the hedge ratio λ_t would change and the *delta* hedge would need to be adjusted. This means that there will be *gamma* gains (losses).

22.5.1 **HOW TO CALCULATE DELTAS**

In the Black—Scholes world, *deltas* and other sensitivity factors are calculated by taking the appropriate derivatives of a function. This is often not possible in the credit analysis. In the case of tranche *deltas*, the approach is one of numerical calculation of sensitivity factors or of obtaining closed-form solutions by approximations.

One way is to use the Monte Carlo approach and one factor latent variable model to generate a sample $\{S_i^j\}$ and then determine the tranche spread. These results would depend on an initially

---

[9]*Deltas* can be calculated with respect to changes in other tranche spreads as well. But the market reports the *deltas* with respect to the index mainly because tranche positions are hedged with respect to the index and not by buying and selling equity or mezzanine tranches and doing direct hedging.

assumed index spread or default probability. To obtain *delta* one would divide the original value of the index by an amount ΔI and then repeat the valuation exercise. The difference in tranche spread divided by ΔI will provide a numerical estimate for *delta*.

22.5.2 GAMMA SENSITIVITY

Volatility in the spread movements is an important factor. Actively managing *delta* positions suggest that there may be *gamma gains*. The *gamma* effect seems to be most pronounced in the equity and senior tranches. There exist differences in *gamma* exposures depending on the particular tranche.

Unlike options, one can distinguish three different types of *gamma* in the credit sector.

- *Gamma* is defined as the portfolio convexity corresponding to a uniform relative shift in all the underlying CDS spreads.
- *iGamma* is the individual *gamma* defined as the portfolio convexity resulting from one CDS spread moving independently of the others; i.e., one spread moves and the others remain constant.
- *nGamma* is the negative *gamma* defined as the portfolio convexity corresponding to a uniform relative shift in underlying CDS spreads, with half of the credits widening and half of the credits tightening by a uniform amount.

Delta-hedged investors who are *long* correlation, benefiting if the correlation increases, will be *long gamma* while being *short iGamma* and *nGamma*.

22.5.3 CORRELATION TRADE AND GAMMA GAINS

Suppose one expects the compound default correlation to go up. This would be advantageous to the equity tranche protection seller and more or less neutral toward the mezzanine protection seller.[10] Thus, one would take a long correlation exposure by selling N^e units of equity protection while simultaneously buying N^m units of mezzanine exposure. The latter notional amount is determined according to

$$N^m = \lambda N^e \tag{22.70}$$

What would be the gains from such a position? What would be the risks? First, consider the gains.

It turns out that such long correlation positions have positive carry, meaning that, even if the anticipated change in correlation does not materialize and the status quo continues, the position holder will make money. In fact, historically this positive carry was significant at around $200-350$ bp depending on the prevailing levels of the index I_t and of the correlation ρ_t.

The position will also have positive *gamma* gains. As the *delta* hedge is adjusted, the hedge adjustments will monetize the I_t volatility because the long correlation position has, in fact, positive convexity with respect to I_t. In addition, the position will gain if the correlation goes up as expected.

---

[10]Depending on the level of the index and the level of correlation at time t_0.

The risks are related to *iGamma* effects and to the declining correlation. In both cases the position will suffer some losses, although these may not be big enough to make the overall return negative.

Below we see an example of the *gamma* gains and dynamic *delta* hedging.

EXAMPLE: DELTA *HEDGING AND* GAMMA *GAINS*

We are given the following information concerning iTraxx Index quotes (I_t) and the deltas of the equity and mezzanine tranches at various levels of *I*. For example if the index is at 30 bps, the equity tranche delta is 18.5, whereas the mezzanine tranche delta is 8.9.

I_t	Delta of 0−3 Tranche	Delta of 0−6 Tranche
30 bps	18.5	8.9
35 bps	17.6	9.4
40 bps	15.1	10.1

Now we use these to generate a dynamically adjusted series of delta hedges. Suppose we observe the following index movements across 3 days

$$I_1 = 30, I_2 = 40, I_3 = 30 \tag{22.71}$$

And suppose our notional is $N = 100$. Then a long correlation position will sell 100 units of equity protection and hedge this with $18.5/8.9 \times 100 = 207.865$ units of mezzanine protection.

The resulting position will be delta neutral with respect to changes in *I*.

During day 2, initially the position is not delta neutral since I_t has moved to 40 bps. The correct hedge ratio is smaller at $15.1/10.1 = 4.6$. The investor needs to reduce the long protection position on I_t by $[(18.5-8.9)-(15.1-10.1)] \times 100$.

During day 3, this position reverts back to the original position on the index.

Look what happened as a result of dynamic *delta* hedging of the tranche portfolio using the index. The investor sold protection when I_t widened and bought it back when I_t tightened. Clearly, these will lead to convexity gains similar to the case of options or long bonds.

22.6 REAL-WORLD COMPLICATIONS

The discussion of how to model and value tranche spreads has been a very simplified one. Real-world trading has several complications and also requires further modeling effort. Several additional questions also need to be addressed. We briefly discuss them below.

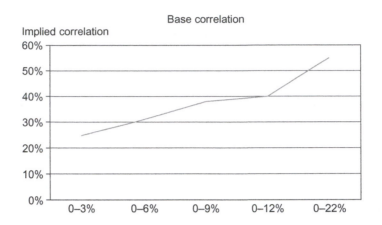

FIGURE 22.4

Base correlation.

22.6.1 BASE CORRELATIONS

The calculation of the implied compound correlation leads to the correlation smile in Figure 22.3. One of the issues is that the implied (compound) correlations can sometimes not be unique or may not even exist in mezzanine tranches. To address the issue of the nonmonotonicity of the mezzanine tranche, it became market practice to quote in another correlation, the base correlation, in addition to the compound correlation.

The intuition behind the base correlation approach is relatively simple. The approach consists of transforming tranche quotes into quotes for equity tranches with increasing attachment points.[11] The procedure to calculate base correlations in this way is referred to as a bootstrapping process and is outlined by McGinty et al. (2004). Base correlation will also give a skew, but this skew will always exist and be unique (see Figure 22.4). One of the additional advantages of base correlations is that one can interpolate correlations to detachment points for which no market tranches exist. For these reasons, Bloomberg quotes iTraxx tranches in terms of base correlations and tranche spreads and not compound correlations.

22.6.2 THE DISPERSION EFFECT

The *dispersion* effect refers to how different individual spreads are from the index spread. For example, if individual CDS spreads are more or less the same as the overall index spread, then we say that there is low dispersion.

---

[11]A base correlation first calculates the market implied spreads for successive equity tranches with attachment points, 0–3%, 0–6%, 0–9%, and so on. In fact, using the 0–3% equity tranche and the 3-6% mezzanine tranche spreads one can calculate an arbitrage-free 0–6% equity tranche spread.

The dispersion of individual CDS spreads in the portfolio versus the spread of the portfolio itself is an important factor influencing the tranche valuation. The effect varies with individual tranches. For the equity tranche, the higher the dispersion of spreads the higher the number of basis points. This is due to the increased riskiness of the tranche. In the mezzanine tranche, the lower the dispersion, the higher the number of bps indicating increased risk. The results of senior and super senior tranches are similar to the mezzanine.

22.6.3 THE TIME EFFECT

Time is also a variable affecting the tranche spreads. As the time to maturity gets shorter, tranche values tend to decline. In other words, the tranche spread curves are, in general, upward sloping.

22.6.4 DO DELTAS ADD UP TO ONE?

Suppose there are two tranches and only two names in the reference portfolio. We can define two tranches: the equity tranche and, say, the senior tranche. Can we say that the *deltas* of the entire reference portfolio weighted by tranche size should sum to one?

Suppose one default occurs. If we sold protection on the index itself, the index spread we would receive in the future would stay the same. However, a weighted average of tranche spreads would be significantly affected, since with the first default equity protection the seller would receive no further premiums. And, it turns out that the equity tranche spreads would be the highest.

This effect is due to the fact that tranche spreads are unevenly distributed whereas the index has a single spread. With the first default, the high spread tranche is triggered and the average spread of the remaining tranche portfolio decreases significantly. Clearly, if all tranches and the index traded on an upfront basis this effect would disappear.

22.7 DEFAULT CORRELATION CASE STUDY: MAY 2005

During May 2005, credit markets witnessed an intriguing event which looked like a puzzle and created a significant amount of discussion as to the correctness of the credit markets' standard models. The events are worth reviewing briefly since they are a good example of the way cash derivatives markets influence each other in the newly developing credit markets.

Essentially, the May 2005 events were triggered by General Motors and Ford being downgraded by rating agencies. These credits that had massive amounts of debt were investment grade and were downgraded to speculative grade, which meant that many institutional players would be prevented from holding them in their portfolios. The ensuing sales led to significant credit volatility in credit markets.

The biggest volatility happened in CDX and iTraxx tranche markets, since several players anticipating these events had put long correlation trades in place. The following IFR report shows the series of events as they happened in May 2005.

EXAMPLE: *MAY 2005 EVENTS*

Speculation about losses in the hedge fund industry sent a shudder through credit markets earlier this month. Fund managers were wrong-footed by going long on the equity piece of synthetic CDOs, while funding that position by shorting the mezzanine tranche on the view that spreads would move in the way predicted by their correlation assumptions.

This positive carry trade was popular with banks and hedge funds because the trade made money if spreads of the related pieces moved in a parallel fashion, as well as creating a hedge against large losses.

The risk was exposure to unexpected default in the portfolio, but with default rates at historic lows, investors were happy to assume this risk.

The ratings downgrade on GM and Ford, which are often included in these CDO structures because they are such large issuers of debt, made the banks change their outlook on default rates and prompted them to begin unwinding massive correlation books.

While equity spreads jumped higher, those on the mezzanine tranches headed in the other direction, ensuring these trades significantly underperformed. As a result, investors suffered on both legs of the trade.

Initially, the underperformance was most acutely felt in the United States, however, during the last couple of weeks the European market has been hit by hedge fund unwinding of iTraxx CDO tranche trades. It is estimated that between 20 and 30 hedge funds executed correlation trades in Europe.

"Selling is a result of the negative returns that these types of trades have generated in the US recently," said a European credit strategist.

The iTraxx spreads should not be affected by the sell-off in correlation positions. "Fundamentally, a risk re-distribution between different tranches, as reflected by a change in correlation, should have a very limited effect on the underlying market," said a credit strategist. However, from a sentimental point of view the impact was much more significant and the rush to unwind the same trade caused massive volatility in the DJ iTraxx crossover index last week.

The index ballooned to 475 bp in 5 years from 330 bp the previous week.

However, hedge funds putting on large shorts in a widening market coupled with a rally in US treasuries caused a violent snap back in the index to 380 bp towards the end of the week.

Attention also turned last week to the banking sector and the size of potential losses. Most of the fall-out will be felt in London and the United States where the majority of international banks keep their correlation books, whereas Asia is less exposed.

That is partly because synthetic tranching of purely Asian credit has not really taken off. Facilitating the interest in structured credit products in Europe and the United States has been the development and liquidity of the credit default swap indices; they have given transparency to implied correlation in those markets. Not so in Asia, where the regional indices have not proved hugely popular, there has been even less take-up of index tranching.

The end of June is critical for the high-yield market because most funds have quarterly liquidity, and there is the threat of huge redemptions. Many are unwinding positions in expectation that clients will redeem their cash.

We can now study this case in more detail. According to the series of events, the status quo before the volatility was as follows.

Several hedge funds and bank proprietary trading desks were long correlation, meaning that they had sold protection on the equity tranche by a notional amount N^E and bought protection on the mezzanine tranche by a notional amount N^M. These two notional amounts were related to each other according to

$$N_t^M = \lambda_t N^E \tag{22.72}$$

where λ_t is the hedge ratio selected so that the sensitivity of the portfolio to movements in the underlying index, or, in average default probability P_t is approximately zero. Letting V^E and V^M represent the value of the \$1 invested in the equity and mezzanine tranches, respectively, we can write the value

$$\frac{\partial}{\partial p_t} \left[N^E V_t^E - \lambda N^E V_t^M \right] = 0 \tag{22.73}$$

These positions were taken with the view that the compound default correlation coefficient denoted by ρ_t would go up. Since we had

$$\frac{\partial}{\partial \rho_t} V_t^E > 0 \tag{22.74}$$

and

$$\frac{\partial}{\partial \rho_t} V_t^M \cong 0 \tag{22.75}$$

The position would benefit as correlation increased,

$$\frac{\partial}{\partial \rho_t} V_t^E - \frac{\partial}{\partial \rho_t} V_t^M > 0 \tag{22.76}$$

in fact, as the reading above indicates, with the GM and Ford downgrades the reverse happened. One possible explanation of the May 2005 events is as follows.[12] As GM and Ford were downgraded, the equity spreads increased and V_t^E decreased, which led to a sudden collapse of the default correlation. Meanwhile, the mezzanine spreads decreased and V_t^M increased. This meant severe mark-to-market losses of the long correlation trades

$$\frac{\partial}{\partial \rho_t} V_t^E - \frac{\partial}{\partial \rho_t} V_t^M < 0 \tag{22.77}$$

since both legs of the trades started to have severe mark-to-market losses.

The dynamics of the tranche spreads observed during May 2005 contradicted what the theoretical models implied. The observed real-world relationships had the opposite sign of what the market standard models would imply. What was the reason behind this puzzle?

One explanation is, of course, the models themselves. If the market convention used the "wrong" model then it is natural that sometimes the real-world development would contradict the

---

[12]Remember that the Gaussian copula model is not a structural model that can provide a true "explanation" for these events. It is a mathematical construction that is used to calibrate various parameters we need during pricing and hedging. Hence, the discussion of May 2005 events can only be heuristic here.

model predictions. This is possible, but we know of no satisfactory new theoretical model that would explain the observed discrepancy at that time.

There is, on the other hand, a plausible structural explanation of the puzzle and it relates to the cash CDO market. Over the years, starting from the mid-1990s, banks had sold mezzanine tranches of cash CDOs (the banks were paying mezzanine spreads). Some of these positions could have been unhedged. As GM and Ford were downgraded, and as these names were heavily used in cash CDOs, the banks who were short the mezzanine tranche tried to cover these positions in the index market. This means that they had to sell protection on the mezzanine index tranche.[13] It is plausible that this dynamic created a downward spiral where the attempt to hedge the cash mezzanine position via selling mezzanine protection in the index market resulted in even further mezzanine spread tightening.

At the same time, since the correlation trade had gone in the opposite direction, hedge funds tried to close their position in the equity tranche. They were selling equity protection before, but closing the position meant buying equity protection and such a rush by institutions would create exactly the type of dynamic observed at that time. The equity spreads increased and the mezzanine spreads declined and, at the end, correlation declined. The downgrade by GM and Ford had created the opposite effect.

22.8 CONCLUSIONS

In this chapter, we have discussed the dependence of tranche spreads on default correlation in detail and explained how default correlation can be traded. We have applied the standard market model in the form of the Gaussian copula model and highlighted real-world complications. Similar to other model-based trading strategies that we discussed in previous chapters, we explained how correlation trading positions can be delta hedged and how gamma gains arise for such strategies.

SUGGESTED READING

The best references to continue with the discussion given in this chapter are handbooks and reports prepared by broker-dealers themselves. We recommend the **JP Morgan's Credit Derivatives Handbook** (2007), the two-volume Merrill Lynch set published in 2006 and the Credit Suisse (2007) Handbook on Credit Risk Modeling. **Francis et al.** (2003) is a good reference for correlation trading. The 2004 JP Morgan credit correlation guide is a good introduction to correlation, the difference between compound and base correlations and their relative merits. **Brigo et al.** (2010) provide a good account of credit models and CDOs during and after the GFC while **Schonbucher** (2004) and **Duffie and Singleton** (2002) are earlier academic approaches.

---

[13]This means that the banks who were paying the mezzanine spreads would now be receiving the mezzanine spreads. Obviously as mezzanine spreads declined, this would create mark-to-market losses and even negative P&L for those banks that were short the mezzanine tranche.

APPENDIX 22.1: SOME BASIC STATISTICAL CONCEPTS

Consider again the table discussed in the text, shown below, which gives the joint distribution of two random variables.

This simple table is an example of marginal and joint distribution functions associated with the two random variables d^A, d^B representing the default possibilities for the two references named A, B, respectively.

It is best to start the discussion with a small credit portfolio, then generalize to the, say, iTraxx index. Suppose we have an equally weighted CDS portfolio of $n = 2$ names denoted by $j = A$, B. We are interested in a horizon (i.e., maturity) of 1 year. According to this, there are two random variables d^A and d^B, each representing the credits $j = A$ and $j = B$. The d^j assumes the value of 1 if the jth name defaults, otherwise it is equal to 0.

Suppose the probabilities of default by A and B are given as in the table below.

	A Defaults ($d^A = 1$)	A Survives ($d^A = 0$)	Sum of Rows
B Defaults ($d^B = 1$)	0.02	0.03	0.05
B Survives ($d^B = 0$)	0.01	0.94	0.95
Sum of columns	0.03	0.97	1

Now we can discuss the most relevant point concerning this table. If the default-related random variables d^A and d^B were *independent*, then we would have

$$p^{11} = (1 - p^A)(1 - p^B) \tag{22.78}$$

$$p^{22} = (p^A)(p^B) \tag{22.79}$$

$$p^{12} = (1 - p^A)(p^B) \tag{22.80}$$

$$p^{21} = (p^A)(1 - p^B) \tag{22.81}$$

Otherwise, if any of these conditions were not satisfied, then the default random variables would be correlated. In this particular case, a quick calculation would reveal that none of these conditions are satisfied. For example for p^{11} we have

$$p^{11} = 0.02 \neq p^A p^B = 0.0015 \tag{22.82}$$

As a matter of fact, the joint probability of default is more than 10 times greater than the simple-minded (and in this case, wrong) calculation by simple multiplication of individual default probabilities.

Let us calculate the expected values, variances, and covariances of the two random variables d^A and d^B. For the expected values

$$E[d^A] = 1p^A + 0(1 - p^A) = p^A$$
$$= 0.03 \tag{22.83}$$

$$E[d^B] = 1p^B + 0(1 - p^A) = p^B$$
$$= 0.05 \tag{22.84}$$

For the variances, we have

$$E[(d^A - E[d^A])^2] = (1-p^A)^2 p^A + (0-p^A)^2(1-p^A) = (1-p^A)p^A$$
$$= (0.03)(0.97) \tag{22.85}$$

$$E[(d^B - E[d^B])^2] = (1-p^B)^2 p^B + (0-p^B)^2(1-p^B) = (1-p^B)p^B$$
$$= (0.05)(0.95) \tag{22.86}$$

Finally, the more important moment, the covariance between d^A and d^B is given by

$$E[(d^B - E[d^B])(d^A - E[d^A])] = (1-p^B)(1-p^A)p^{11} + (0-p^B)(0-p^A)p^{22}$$
$$+ (1-p^B)(0-p^A)p^{12} + (1-p^A)(0-p^B)p^{21} \tag{22.87}$$

where p^{ik} is the joint probability that the corresponding events happen jointly. Remember that we must have

$$p^{11} + p^{12} + p^{21} + p^{22} = 1 \tag{22.88}$$

To convert this into a correlation, we need to divide this by the square root of the variances of d^A and d^B.

EXERCISES

1. Explain the difference between a cash CDO and a synthetic CDO?

2. What is the difference between compound correlation and base correlation?

3. We consider a reference portfolio of three investment grade names with the following 1-year CDS rates:
   $c(1) = 116$
   $c(2) = 193$
   $c(3) = 140$
   The recovery rate is the same for all names at $R = 40$.
   The notional amount invested in every CDO tranche is $1.50. Consider the questions:
   a. What are the corresponding default probabilities?
   b. How would you use this information in predicting actual defaults?
   c. Suppose the defaults are uncorrelated. What is the distribution of the number of defaults during 1 year?
   d. How much would a 0−66% tranche lose under these conditions?
   e. Suppose there are two tranches: 0−50% and 50−100%. How much would each tranche pay over a year if you sell protection?
   f. Suppose all CDS rates are now equal and that we have $c(1) = c(2) = c(3) = 100$. Also, all defaults are correlated with a correlation of one. What is the loss distribution? What is the spread of the 0−50% tranche?

4.   We consider a reference portfolio of four investment grade names with the following 1-year CDS rates:

$c(1) = 14$
$c(2) = 7$
$c(3) = 895$
$c(4) = 33$

The recovery rate is the same for all names at $R = 30\%$.
The notional amount invested in every CDO tranche is \$1.00. Consider the questions:

a. What are the corresponding annual default probabilities?

b. Suppose the defaults are *uncorrelated*, what is the distribution of the number of defaults during 1 year?

c. Suppose there are three tranches:
- 0−50%
- 50−75%
- 75−100%

How much would each tranche pay over a year?

d. Suppose the default correlation becomes 1, and all CDS rates are equal at 60 bp, answer questions (a)−(c) again.

e. How do you hedge the risk that the probability of default will go up in the equity tranche?

5.   We consider a reference portfolio of three investment grade names with the following 1-year CDS rates:

$c(1) = 15$
$c(2) = 11$
$c(3) = 330$

The recovery rate is the same for all names at $R = 40$.
The notional amount invested in every CDO tranche is \$1.50. Consider the questions:

a. What are the corresponding default probabilities?

b. How would you use this information in predicting actual defaults?

c. Suppose the defaults are uncorrelated. What is the distribution of the number of defaults during 1 year?

d. How much would a 0−66% tranche lose under these conditions?

e. Suppose there are two tranches: 0−50% and 50−100%. How much would each tranche pay over a year if you sell protection?

f. Suppose all CDS rates are now equal and that we have $c(1) = c(2) = c(3) = 100$. Also, all defaults are correlated with a correlation of one. What is the loss distribution? What is the spread of the 0−50% tranche?

6.   Explain the following position using appropriate graphs. In particular, make sure that you define a barbell in credit sector. Finally, in what sense is this a convexity position?

   1008 GMT [Dow Jones] LONDON—SG recommends selling 7-year 0−3% tranche protection versus 5-year and 10-year 0−3% protection. 7-year equity correlation tightened versus 5-year and 10-year last year.

   SG's barbell plays a steepening of the 7-year bucket, as well as offering positive roll down, time decay, and jump to default.

7. (May 2005 Crisis) What was the effect of the 2005 downgrade of General Motors and Ford on the credit market. What was the effect on default correlations and correlation trading positions?

MATLAB EXERCISES

8. (Default Loss Distribution) Building on the tranche valuation model outlined in the text, write a MATLAB program to model the sensitivity of the default loss distribution of a portfolio consisting of 100 names/underlyings to changes in default correlation (ρ). Assume that the default probability (p) of each of the portfolio constituents is the same and equal to 5%. Take 100 numbers of latent variables to carry out the simulations and interpret the results.

9. (Correlation swaps and equity option implied correlation) In this chapter, we drew an analogy between tranche implied default correlations and implied volatilities and implied correlations in equity option markets. Equity correlation trading was discussed in Chapter 16. One approach to trading correlations is by means of a correlation swap.
   a. Explain how a synthetic equity correlation swap on the S&P500 or S&P100, for example, can be created. What are the similarities and differences between trading default correlation and equity option implied correlation?
   b. Download stock return and option data and calculate the implied and realized correlation for the index over the 2004–2013 period.
   c. Comment on the evolution of the so-called correlation risk premium, that is the difference between the implied and realized correlation.
   d. Are correlation swaps an attractive investment? What are the caveats?

PRINCIPAL PROTECTION TECHNIQUES

23.1 INTRODUCTION

Investment products, where the principal is protected, have always been popular in financial markets. However, until recently the so-called guaranteed products sector has relied mainly on *static* principal protection which consists of a static portfolio of a default-free bond plus a basket of options. This is also sometimes referred to as option-based portfolio insurance (OBPI). However, recent advances in financial engineering techniques have made it possible to create guaranteed products and offer protection in other ways. As structurers understood dynamic replication better, they realized that dynamic replication could synthetically create options on risks where no traded options exist. This led to the creation of dynamic rebalancing techniques, the best known being the constant proportion portfolio insurance (CPPI). CPPI products on a variety of assets have been sold by banks, including equity indices and credit default swap indices. With the development of credit markets, CPPI was applied to credit indices and the implied tranches as well. As expertise on the dynamic replication techniques grew, market activity led to new innovations such as the dynamic proportion portfolio insurance (DPPI). The heyday of CPPI product development and innovation was during the years before the GFC, but CPPI applications continue to exist in various markets.

A major reason for the popularity of the guaranteed product sector is regulatory behavior. Several countries do not let investment banks issue structured products involving "exotic" risks unless the product provides some principal protection guarantee. CPPI is a protected note and is not subject to these restrictions. In addition, many funds are not allowed, by law, to invest in securities that are speculative grade. Other dynamic proportion techniques are not principal protected, but both the principal and the coupon can be rated AAA by the main rating agencies, although they offer unusually high coupons. These make them attractive to conservative funds as well.[1]

In this chapter, we discuss the classical *static* principal protection methodology followed by an introduction to the CPPI as an extension of this classical methodology. We show the dynamics of the portfolio value that applies this technique under some simple conditions and provide the results of some simulations as well. We then deal with the application of the methodology to standard credit index tranches and discuss the complications that may arise when this is implemented. Finally, we introduce some modeling aspects and discuss the so-called *gap risk*, and deal briefly with the DPPI.

23.2 THE CLASSICAL CASE

At the simplest level a guaranteed product consists of either a zero-coupon bond and a static call option position or protective put option position. The former was discussed in detail in Chapter 20. Both of these approaches date back to Leland (1980) and are nowadays referred to as option-based portfolio insurance (OBPI). The protective put option strategy consists of protecting a risky investment from downside market movements using a put option, while allowing participation in the growth of a risky asset. The put option could be either based on a traded index option such as options on the S&P500 index or a synthetic put option based on dynamic replication using index futures. One of the complications in such guaranteed products is that the options may be too expensive, depending on the level of volatility. Out-of-the-money put options, for example, are typically very expensive. In Chapter 20 we discussed several limitations of guaranteed products based on a combination of a zero-coupon bond and call option.

These problems have led to several modifications of the traditional principal protection methods. Dynamically adjusted principal protection methods (versions of CPPI) and the dynamically adjusted methods that yield triple-A products have been developed as a result. We study the CPPI techniques first and discuss their application to standard credit tranches.

---

[1]We call these guaranteed products. It is important to realize that this guarantee was initially toward market risk; the classical guaranteed products could still carry a credit risk. The (institutional) investor could conceivably lose part of the investment if default occurs.

In practice structurers and portfolio managers also use other principal protection techniques such as risk control, volatility capping, and risk budgeting, but since they are not based on financial engineering principles in a way that OBPI and CPPI are, we do not discuss them further in this book.[2]

23.3 THE CPPI

The main advantage of the CPPI as a principal protection technique is that it gives a higher participation in the underlying asset than one can get from traditional capital protection. It also can be applied when interest rates are "too" low, or when options do not trade for some underlying risk. Before we discuss the CPPI algorithm and the associated risks we consider some market examples.

> **EXAMPLE**
>
> The CPPI investment is an alternative to standard tranche products, which offer limited upside (fixed premium) in exchange for unlimited downside (potential total loss of principal). CPPI offers limited downside (because of principal protection) and unlimited upside, but exposes investors to the market risk of the underlying default swaps contracts that comprise the coupon.
>
> With yields hovering near record lows until recently, credit investors are increasingly moving to structures that contain some element of market exposure. That has posed a problem for rating agencies, which are default oriented, and prompted a move to a more valuation-based approach for some products. Leveraged super senior tranches, also subject to market volatility, were the market risk product of choice last year, but in recent months have ceded popularity to CPPI.
>
> Banks profit from ratings on CPPI coupons because the regulatory capital treatment of rated products under Basel II is much kinder than on unrated holdings.

The CPPI is a structure which has constant *leverage*. Dynamic PPI (proportion portfolio insurance) is the name for strategies where the leverage ratio changes during the investment period. CPPI works by dynamically moving the investment between a safe asset and a risky asset, depending on the performance of the risky asset and depending on how much cash one has in hand. The main criterion in doing this is to protect the principal, while at the same time getting the highest participation rate.

The idea of CPPI can be related to the classical principal protection methods and is summarized as follows. In the classical principal protection, the investor buys a zero-coupon bond and invests the remaining funds to options. CPPI relaxes this with a clever modification. If the idea is to be long a bond with value 100 at T, then one can invest *any* carefully selected sum to the risky asset as long as one makes sure that the total value of the portfolio remains above the value

---

[2]Risk control or volatility capping controls the level of volatility of the risky asset by establishing a specific volatility target and dynamically adjusting the exposure to the risky asset based on its observed historical volatility.

of the zero-coupon bond during the investment period. Then, if the portfolio value is above the value of the zero-coupon bond, at any desired time the risky investment can be liquidated and the bond bought. This will still guarantee the protection of the principal, N. This way, the structurer is not limited to investing just the leftover funds. Instead, the procedure makes possible an investment of funds of any size, as long as risk management and risk preference constraints are met.

Let the initial investment of N be the principal. The principal is also the initial net asset value of the positions—call it V_{t_0}. Next, calculate the present value of N to be received in T years. Call this the *floor*, F_{t_0}.[3]

$$F_{t_0} = \frac{N}{\left(1 + r_{t_0}\right)^{\mathrm{T}}} \tag{23.1}$$

Let the increment Cu_t be called the *cushion*:

$$Cu_{t_0} = V_{t_0} - F_{t_0} \tag{23.2}$$

Then, select a *leverage ratio* λ in general satisfying $1 < \lambda$. This parameter has no time subscript and is constant during the life of the structured note. Using the λ calculate the amount to be invested in the *risky asset* R_{t_0} as:

$$R_{t_0} = \lambda \left[V_{t_0} - F_{t_0} \right] \tag{23.3}$$

This gives the *initial* exposure to the risky asset S_t. Invest R_{t_0} in the risky asset and deposit the remaining $V_{t_0} - R_{t_0}$ into a risk-free deposit account.[4]

$$D_{t_0} = V_{t_0} - R_{t_0} \tag{23.4}$$

Note that the cash deposit D_{t_0} is less than the time t_0 value of a risk-free zero-coupon bond that matures at time T, denoted by $B(t_0, T)N$, where $B(t_0, T)$ is the time t_0 price of a default-free discount bond with par value \$1. Hence, in case of a sudden and sizable downward *jump* in S_t, the note will not have enough cash in hand to switch to a zero-coupon bond. This is especially true if the jump in S_t leads to flight to quality and increases the $B(t_i, T)$ at some future date t_i. This is called the *gap risk* by the structurers and is studied later in the chapter.

Apply this algorithm at every rebalancing date t_i,

$$t_i - t_{i-1} = \delta_i \tag{23.5}$$

as long as $F_{t_i} < V_{t_i}$. This algorithm would increase the investment in the risky asset if things go well (i.e., if V_{t_i} increases), and decrease the investment in case markets decline (i.e., if V_{t_i} decreases).

Finally, if at some time τ^* V_{τ^*} falls and becomes equal to F_{τ^*}, liquidate the risky investment position and switch all investment to cash. Since the floor F_{t_i} is the present value at t_i, of 100 to be received at T, this will guarantee that principal N can be returned to the investors at T.

---

[3]If there are fees, then they should be deducted at this point from F_{t_0}. In reality, there are always such fees, but in this discussion we assume that they are zero to simplify the exposition.

[4]If $R_{t_0} < V_{t_0}$, then the risky asset investment does not require any additional borrowing. If, on the other hand, $R_{t_0} > V_{t_0}$, then invest R_{t_0} in the risky asset and borrow the remaining $R_{t_0} - V_{t_0}$.

23.4 MODELING THE CPPI DYNAMICS

We now obtain the equations that give the dynamics of V_t under the typical CPPI scheme. The equations are obtained from a relatively simple setting to highlight the important aspects of the methodology. The two points on which we focus are the following: first, we will see that from a single portfolio point of view, CPPI algorithms may be much more stable than they appear from the outside if there are no jumps. However, in reality there are jumps which lead to the gap risk. Second, we show that the CPPI methodology may have a more fragile structure with respect to yield curve movements than anticipated. This may be especially the case if sharp downward jumps in S_t are *correlated* with a sudden steepening of the curve—exactly what happens during periods of excessive market stress.[5]

Let us place ourselves in a Black–Scholes-type environment, with constant interest rates r and constant volatility σ. Further, the S_t follows the Wiener process-driven SDE,

$$dS_t = \mu S_t \, dt + \sigma S_t \, dW_t \tag{23.6}$$

We know from the previous discussion that the investment in risky assets is

$$R_t = \lambda Cu_t \tag{23.7}$$

and that the changes in the cushion are given by

$$dCu_t = d(V_t - F_t) \tag{23.8}$$

This means

$$dCu_t = (V_t - R_t)\frac{dB_t}{B_t} + R_t\frac{dS_t}{S_t} - dF_t \tag{23.9}$$

where B_t is the value of the zero-coupon bond at time t.[6] In this expression, we can replace V_t, R_t with their respective values to obtain

$$dCu_t = (Cu_t + F_t - \lambda Cu_t)\frac{dB_t}{B_t} + \lambda Cu_t\frac{dS_t}{S_t} - dF_t \tag{23.10}$$

But the value of the floor increases according to

$$dF_t = F_t\frac{dB_t}{B_t} \tag{23.11}$$

This means

$$dCu_t = (1 - \lambda)Cu_t\frac{dB_t}{B_t} + \lambda Cu_t\frac{dS_t}{S_t} \tag{23.12}$$

Substituting further,

$$dCu_t = (\lambda(\mu - r) + r)Cu_t \, dt + \lambda\sigma c \, dW_t \tag{23.13}$$

This is a *geometric* stochastic differential equation whose solution is given by

$$Cu_t = Cu_{t_0}e^{\left(\lambda(\mu-r)+r-((\lambda^2\sigma^2)/2)\right)(t-t_0)+\lambda\sigma W_t} \tag{23.14}$$

---

[5]As CPPI-type protection techniques became more popular, academic interest also increased. The work by Cont and Tankov (2009) provides an excellent view of the academic approach to this issue and is also quite accessible and practical. This section is based on this research.

[6]In this case, this value is given by the simple formula due to constant r, $B_t = B_{t_0}e^{r(t-t_0)}$.

We can combine this with F_{t_0} to get the behavior of the portfolio net asset value for all $t \in [t_0, T]$

$$V_t = F_{t_0} e^{r(t-t_0)} + Cu_t \qquad (23.15)$$

Using standard results from stochastic calculus, we can calculate the expected portfolio value as of time t

$$E_{t_0}^P[V_t] = F_{t_0} + \left(100 - F_{t_0} e^{r(T-t_0)}\right) e^{r(t-t_0) + \lambda(\mu - r)(t-t_0)} \qquad (23.16)$$

Note that the probability measure we use in this expectation is the real-world probability and not the risk-adjusted measure. This is the case since the drift of the S_t process was taken to be the μ and not the risk-free rate r.

23.4.1 INTERPRETATION

There are two equations in the above derivation that are very suggestive. The first relates to the dynamics of the cushion over time,

$$dCu_t = (\lambda(\mu - r) + r)Cu_t \, dt + \lambda Cu_t \frac{dS_t}{S_t} \qquad (23.17)$$

Note that with such a dynamic the cushion itself can never go below zero over time. In fact, suppose Cu_t becomes very small at time t. The first term on the right-hand side of the equation will be positive. Also, being a Wiener-driven system, S_t cannot exhibit jumps and over infinitesimal periods must change infinitesimally. The second term on the right-hand side then shows that the changes in S_t affect the cushion with a coefficient of λCu_t, which goes to zero as Cu_t approaches zero. Under these conditions, as the cushion Cu_t goes toward zero, the dCu_t will go to zero as well.

This leads to an interesting conclusion. The CPPI method will always "work" in the sense that as the market goes against the investor, the cushion will never become negative. In the worst case, the cushion will be zero which means that the risky investment is liquidated. This leaves the investor with the zero-coupon bond which matures at a value of 100. Hence, the principal is "always" protected.

Before we continue with the implications of this result consider the second interesting equation. The expected value of the portfolio was calculated as

$$E_{t_0}^P[V_t] = F_{t_0} + \left(100 - F_{t_0} e^{r(T-t_0)}\right) e^{r(t-t_0) + \lambda(\mu - r)(t-t_0)} \qquad (23.18)$$

This is also very suggestive because, as long as $r < \mu$, which means that the risky asset expected return is higher than the risk-free rate, the expected value of the portfolio can be increased indefinitely by picking higher and higher leverage factors, λ.

Thus we have reached an unrealistic result. The CPPI strategy will always work, in the sense that the investor's initial investment is always protected, while at the same time the higher the leverage the higher the expected gains. This implies picking the highest possible leverage factor that is available to the structurer. Yet, it is clear that in the real world this is not the prudent approach. This, in turn, suggests that the model we discussed above may be missing some critical features of real-world investment.

There are at least two possibilities, the first being the limits on borrowing. There are credit limits and the leverage cannot be increased indefinitely in the real world. The second possibility is more interesting.

In the real world, S_t process may not follow a geometric process and may contain jump factors. If this is the case, as $Cu_t \to 0$, a downward jump in S_t can make Cu_t negative. The portfolio value will fall *below* the floor and the investor's initial investment is lost,

$$V_t < F_t \tag{23.19}$$

This is the *gap risk*. Note that, if there are such jumps in S_t, then the presence of the leverage factor $1 < \lambda$ will magnify them. The higher the λ the higher will be the effect of a downward jump.

23.4.2 HOW TO PICK λ

The discussion involving the gap risk suggests a methodology for selecting a numerical value for the critical leverage parameter λ. The structurer would first determine an acceptable threshold for the gap probability using the investor's risk preferences. Then, using the observed volatility and jump parameters, the structurer would work backward and determine λ that makes the gap probability equal to this desired amount. This could be done with Monte Carlo or with semiparametric methods as in Cont and Tankov (2009).

Clearly this determination of λ will be model dependent. A structurer would have a number of other ways to deal with this gap risk, some of which are discussed below.

23.5 AN APPLICATION: CPPI AND EQUITY TRANCHES

We will use structured credit as an application of the CPPI technique. The credit sector has experienced significant growth and innovation in recent years. The GFC did not leave it unaffected but the sector has recovered since and its products are likely to remain a part of the financial landscape. Two paradigms are observed in the structured credit sector. The first tracks some credit derivatives index, and the second is managed credit derivatives funds. CPPI techniques can be useful in both of these trends as shown in the example from the markets.

EXAMPLE

Retail credit CPPI first ABN AMRO and AXA last week announced that they had closed the first principal protected credit derivatives fund targeted at retail investors. AXA persuaded its home regulator in France to permit the leveraged fund, and similar deals are expected to be launched in other regimes.

The fund is structured like an actively managed synthetic collateralized debt obligation. It will have a minimum of 100 credit default swap references and is starting with just under 120 names.

It uses CPPI from ABN AMRO to provide the capital protection that was necessary to obtain regulatory approval for its sale to retail investors in France. The insurance is provided on a binary basis if the basket of default swaps managed by AXA performs so poorly that a zero-coupon bond would have to be bought to guarantee investor capital, then there would be an effective wind-up.

In this section, we discuss the application of the CPPI technique to standard credit index tranches. It turns out that combining CPPI and iTraxx tranches is quite simple in terms of financial engineering, although there are some practical issues that need to be resolved in practice. Here is the algorithm.

1. Receive cash of $N = 100$ from the investor.
2. Calculate the difference between par and the cost of a zero-coupon bond F_t, and as before let this term be Cu_t, the cushion.
3. Multiply the cushion Cu_t by the leverage factor λ. This is the leveraged amount.
4. Subtract the leveraged amount, $Cu_t \times \lambda$, from the 100 total cash N and invest, $Cu_t \times \lambda$, in the *risky asset*, which in this case is assumed to be the iTraxx 0−3% equity tranche. Keep the remaining cash or reserve cash, $100 - Cu_t \times \lambda$, in a deposit account or as collateral.[7]
5. As the price of the iTraxx equity tranche goes up and down, adjust the allocation between risky asset and reserve cash to keep the leverage of the trade constant at λ.

Thus far this is a straightforward application of the previously discussed CPPI algorithm. The only major complication is in the definition of the risky asset. Standard equity tranches are quoted as upfront percentages and this will lead to a modification of the algorithm. In fact, suppose the amount to be invested in the risky asset is R_t, and suppose also that the iTraxx equity tranche quote is given by q_t where the latter is a pure number denoting the upfront payment as a percentage.[8] Then the relationship between the amount allocated in the risky asset and the notional amount invested in the equity tranche N^{Eq} will be as follows:

$$N^{Eq}_{t_i} = \frac{R_{t_i}}{(1 - q_t)} \tag{23.20}$$

In terms of the amount invested in the risky asset, R_{t_i}, this can also be written as

$$R_{t_i} = N^{Eq}_{t_i} \times (1 - q_t) = N^{Eq}_{t_i} - \frac{R_{t_i}}{(1 - q_t)} q_t \tag{23.21}$$

It may be helpful to discuss a numerical example at this point.

23.5.1 A NUMERICAL EXAMPLE

The following example shows how CPPI can be combined with the ITraxx and has similarities to a real-world product CPPI product that was called SPRING and developed by the IXIS CDO group.

First we note that the equity tranche of the iTraxx Index is quoted as an upfront percentage of the notional plus 500 basis point *running-fee* paid quarterly. Assume that the upfront fee of the iTraxx equity tranche is $q_t = 20\%$ of the notional amount invested in the equity tranche N^{Eq} and that the annual LIBOR rate is $L_t = 5\%$. For simplicity, suppose the swap curve is flat at 5.095% as well. The CPPI is applied with daily adjustment periods denoted by t_i, $i = 0, 1, \ldots, n$. The leverage factor is assumed to be 2. Assume no bid−ask spreads. Figure 23.1 illustrates how the leveraged amount $Cu_t \times \lambda$ is invested in the risky asset. It also shows that the equity tranche of the iTraxx index is quoted as an upfront cash amount.

---

[7]Remember that the iTraxx indices are unfunded.
[8]For example, if a market maker quotes $q_t = 12\%/12.5\%$, then a protection seller will receive $12 upfront for each 100 dollars of insurance sold. For the protection seller, this money is to keep.

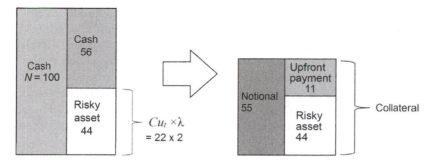

FIGURE 23.1

iTraxx CPPI example.

We apply the steps above in a straightforward fashion to a 5-year CPPI note where the underlying is the equity tranche.

23.5.1.1 The initial position

Initially the CPPI will have the following structure.

1. Receive $N = 100$.
2. The floor is

$$F_t = \frac{100}{(1+0.05095)^5}$$
$$= 78$$

(23.22)

3. The cushion is

$$Cu_t = 100 - 78$$
$$= 22$$

(23.23)

4. Assuming a leverage of

$$\lambda = 2$$

(23.24)

the amount to be invested in the risky asset is

$$22 \times 2 = 44$$

(23.25)

This is the investment to be allocated to the equity tranche.

5. The iTraxx equity tranche pays an upfront cash amount of $20\% \times N^{Eq}$. Therefore, if the risky asset exposure is $R_t = 44$, then the notional amount invested in the equity tranche N^{Eq} should be[9]

$$N^{Eq} = \frac{44}{(1-0.20)} = 55$$

(23.26)

---

[9]In other words, if we invest 55 in the equity tranche notional then our net exposure to the risky asset will be 44, since we did get 11 as upfront.

Thus, sell equity tranche protection with notional N^{Eq} of 55. In other words, an equity tranche notional amount N^{Eq} of 55 implies an uncovered exposure of 55 less 11 ($=q_t \times 55 = 20\% \times 55$) received upfront which leaves 44. The 44 is put into reserve cash.

6. When we buy the iTraxx equity tranche we enter into a swap on the iTraxx equity tranche. This implies that the cash allocated to the risky asset, in addition to the upfront cash received, is held as collateral for the swap. This is illustrated in Figure 23.1 in which we can see how the equity tranche notional amount N^{Eq} consists of the upfront payment (fee) of 11 and the risky asset. Note that the total amount of cash to be kept as collateral for the equity protection position is

$$q_t N^{Eq} + (100 - \lambda Cu_t) = 20\% \times 55 + 44 = 11 + 44 = 55 \tag{23.27}$$

Since the risky asset investment is 44 and the cash invested initially is $N = 100$, a balance of 56 remains. The balance of 56 is kept in reserve cash and receives LIBOR.

7. Does the trade indeed have a leverage of $\lambda = 2$ as desired? We invested 44 in the risky asset and 56 in reserve cash. The total portfolio value (PV) is 100 which corresponds to the 100 initially invested. The cost of the zero-coupon bond F_t was 78. Leverage can be calculated as follows:

$$\text{Leverage} = \frac{\text{risky asset}}{\text{portfolio value} - F_t} = \frac{44}{100 - 78} = 2 \tag{23.28}$$

8. As the equity tranche quote changes over the rebalancing periods $t_1, t_2, \ldots$ adjust the position dynamically, reducing the exposure to the risky asset as q_t increases, and increasing the exposure as q_t decreases.

Note that during this process the equity tranche position is actually taken as an unfunded investment. Still, the cash allocated to the risky asset, plus the upfront cash, is held as collateral for the position.

23.5.1.2 Dynamic adjustments

We can see from Eq. (23.28) that if the value of the risky asset increases or decreases, the leverage is going to change. Let, for example, $q_{t_1} = 15\%$, which can be interpreted as perceived risk decreasing (compared to the original value of $q_{t_1} = 15\%$), the index tightening, and the quoted price for protection falling. This corresponds to an increase in the value of the iTraxx equity tranche and the value of the risky asset investment. After all, one can buy protection at 15% and close the position with a profit. Yet, with the structured note the position is continued after an adjustment. We cover these steps below.

1. The value of the risky asset is

$$\begin{aligned} N(1 - q_{t_1}) \quad &= 44 \times (1 - 15\%) \\ &= 46.75 \end{aligned} \tag{23.29}$$

Due to unrealized gains, the value of the risky asset has gone up by

$$R_{t_1} - R_{t_0} = 2.75 \tag{23.30}$$

This is the case since we can close the position by buying equity protection at 15% upfront. Then we recover the 44 deposited as collateral for the equity tranche investment, and in addition we receive the realized gain

$$55 \times 0.05 = 2.75 \tag{23.31}$$

by not losing this position, it is as if we are investing 46.75 in the risky asset.

2. Calculate the V_{t_1} using

$$V_{t_1} = F_{t_0}\left(1 + L_{t_0}\right) + N_{t_0}\left(q_{t_1} - q_{t_0}\right) + N_{t_0}\left(1 + L_{t_0}\right) + N(0.05) \tag{23.32}$$

In this case, this amounts to

$$105.095 + 2.75 + (11)(0.05) + 55(0.05) \tag{23.33}$$

The first term is the interest on the 100 kept as cash or collateral. The second term is the capital gains from the q_t move, the third is the LIBOR earned from the upfront deposit. Finally, the fourth term is the 500 bp running fee on the 55.

3. What is the effect on leverage? The new leverage can be calculated as follows:

$$\text{Leverage} = \frac{\text{risky asset}}{\text{portfolio value} - F_t} = \frac{46.75}{102.75 - 78} = 1.8$$

To bring leverage back to the target level of 2, the portfolio needs to be rebalanced. The risky assets have a value of 46.75 and the reserve cash remains at 56. To obtain a leverage of 2 we need risky assets to increase to 49.5. This can be achieved by reallocating 2.75 from reserve cash to risky assets. As a result risky assets increase to 49.5 and reserve cash falls to 53.25. The increase in risky assets is used to buy more units of the iTraxx equity tranche. Since the current new upfront fee q_{t_1} is 15%, we can buy an equity notional amount of

$$N^{Eq} = \frac{2}{(1 - 0.15)} = 2.353 \tag{23.34}$$

Since the upfront fee q_{t_1} is 15%, this implies that we receive an upfront payment or fee that can be calculated as follows:

$$\text{fee} = 2.353 \times (1 - q_t) = 2.353 \times (1 - 15\%) = 0.353.$$

We can verify the cash collateral corresponding to the notional amount N^{Eq} of the equity tranche.

Collateral = risky asset cash + initial upfront cash + additional upfront cash = 49.5 + 11 + 0.353 = 60.853

Figure 23.2 illustrates the difference between the value of risky assets, which is 49.5, and the cash allocated to risky assets, which is 46.75. The difference of 2.75 is due to the unrealized capital gains. Here we have assumed that the zero-coupon bond value will remain unchanged, but in reality, changes in the value of the zero-coupon bond and income flows will also affect leverage.

The opposite adjustment will be implemented if q_{t_1} decreases. We leave the details of this case to the end-of-chapter exercises. Instead, we will consider the case of a default.

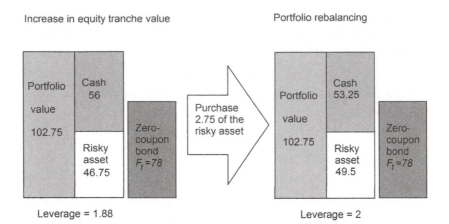

FIGURE 23.2

Dynamic adjustment.

23.5.1.3 Default and switching out of CPPI into zero-coupon bond

What is the liability of the bank that structures the CPPI if the iTraxx portfolio defaults? In the original numerical example, the investor received 100 and placed 56 into reserve cash and 44 into the risky asset. The notional amount of the equity tranche was 55 and it was backed by collateral of 44 in addition to the upfront fee of 11. In the scenario of a default of the iTraxx equity tranche, the equity tranche value would fall to zero. The structuring bank had bought principal protection for 78 which will be paid out to the purchase of the CPPI product. The principal protection costs the bank 78. Since the equity notional amount of the iTraxx tranch is 55, the bank needs to pay 55 to the counterparty, which implies that the total liability is 133 ($=55 + 78$). In terms of collateral, the bank holds cash of 44 related to the risky asset, reserve cash of 56 and the upfront fee payment of 11. The total cash assets are therefore 111. If we subtract assets from liabilities, we obtain a loss of 22. One possible strategy to avoid such a loss is to reduce the risk before default occurs. For example, if the value of the equity tranche deteriorates as credit quality drops, the portfolio could be switched from the CPPI structure to a zero-coupon bond when the value passes a certain threshold such as falling close to the cost of the zero-coupon bond.

23.6 DIFFERENCES BETWEEN CPDO AND CPPI

One of the more recent structured products that is sometimes described as sharing similarities with CPPI is called constant proportion debt obligation (CPDO). However, the two structures are very different and here we briefly contrast CPDO and CPPI to clear up any potential confusion. It is true that the CPDO approaches share some common features with CPPI products, but there are also important differences which imply that CPDOs are not really an example of principal protection products. What CPDOs have in common with CPPI structures is that they use a constant proportion approach to determining the leverage and they both rebalance the portfolio between a risky (credit) asset and the safe asset. As the name indicates, CPDOs are applied to credit indices such as iTraxx or CDX. There are two important differences between CPDO and CPPI structures. One of the selling points of CPDO is to try to achieve a high yield for investors. This is attempted by using a relatively high level of leverage. In contrast to CPPI, however, leverage in CPDO structures will be increased when the net asset value of the portfolio decreases and falls below the target level. If the net asset value of the portfolio rises and comes close to the target amount, leverage will be reduced. The way that CPPI products achieve principal protection is by guaranteeing the zero-coupon bond to the investor, while making the seller of the CPPI bear any losses. In a CPDO, on the other hand, there is no guaranteed principal protection. If the underlying risky asset that the CPDO invests in performs well, the investor receives promised coupon and principal payments, but if the asset does not perform well, only the remaining amount is paid to the investor and this may be less than the principal invested at the beginning. Another important difference between CPPI and CPDO is that different strategies are used to realize the intended performance. At inception the value of the CPDO is below the target portfolio value and the CPDO structure will at each time take risk exposure proportional to the amount the CPDO portfolio is lacking in order to reach the target value. This is in contrast to the CPPI structure which will take risk exposure positions depending on the amount of surplus the portfolio value has with respect to the floor value. The above discussion shows conceptually that CPDO structures do not offer the same protection as CPPI structures. This was also illustrated in real life as the GFC led to poor performance and closure of many CPDO structures.

23.7 A VARIANT: THE DPPI

Dynamic portfolio insurance (DPPI) methodology is a variation of the CPPI. The difference is that leverage is not constant but can change over time. Here is a brief example from the markets.

EXAMPLE

LONDON, June 14 (Reuters)—PIMCO, one of the world's biggest bond funds, has joined forces with Goldman Sachs to launch a range of derivative products. The investments include principal protected and leveraged structures aimed at institutional investors, high net worth individuals and private banks.

The main product, launched under PIMCO is a principle protected investment based on Goldman's Variable Proportion Portfolio Protection, similar to the better-known CPPI technology.

The leverage ratio λ_{t_i} which was constant during the CPPI adjustments can be made variable and becomes one of the unknowns to be determined. The structurer needs to provide an algorithm to do this. The idea is that the leverage ratio can be made to depend on some variables that one thinks are relevant to the problem under consideration in some optimal fashion. In particular, the exposure to the risky asset may depend on

1. The past behavior of the returns
2. The volatility of the returns
3. The liquidity observed in the market for the underlying asset, since the methodology is heavily dependent on the correct rebalancing
4. And upon the so-called *gap risk.*

Finally, another relevant variable may the dependence of λ_{t_i} on the swap curve parameters. The scenarios discussed above illustrated the importance of this. Note that this may be even more relevant for the credit market CPPI notes.

In DPPI, the allocation between the risky asset and cash is dynamically managed with a variable leverage ratio that will depend on one or more of these factors. Supposedly, the CPPI exposure to the risky asset increases when things "go well," and decreases when things "go badly." At the outset, a variable leverage ratio seems to be better able to handle changes in the yield curve environment than the classical CPPI procedures. For example, the leverage ratio λ_{t_i} may go down during *high*-volatility periods and may go *up* during low-volatility periods. The response to changes in the market liquidity could be similar.

23.8 APPLICATION OF CPPI IN THE INSURANCE SECTOR: ICPPI

CPPI techniques continue to be applied in different financial sectors despite the turmoil related to credit-linked CPPI products during the GFC. One example of the application of CPPI techniques is the insurance sector. The difficult economic climate following the GFC also affected the insurance sector and many insurance firm customers have become more cautious. There is still demand for guaranteed products that include upside participation. The UK's with profit or traditional variable annuity (VA) products typically included restrictions on protection levels and fund choices. Customers are now asking for less restrictive guaranteed products with upside participation. In response, insurance firms are exploring a variation on CPPI-type structures. Insurance companies compete with other financial services such as banks for long-term savings customers, but prefer not to offer products that add risks to their own balance sheets or require expensive hedges based on derivatives. CPPI solutions have been applied in several countries' insurance sectors. In Japan, insurers employ them to match guarantees embedded in VA policies. In Germany they were popular in 2006−2007 since they allowed insurers to circumvent restrictive investment rules for insurance companies that attempted to offer state-subsidised "Riester" pension products.

Traditional CPPI products however had several drawbacks. If the performance of the underlying risk asset resulted in the portfolio becoming cash-locked, investors missed out on future performance since the return reverted to the guaranteed level. This occurred in many instances during the GFC which implied that CPPI investors did not get the returns that they were expecting.

The market has responded to the problems related to the path-dependent nature of traditional CPPI structures by innovating and producing so-called individual constant proportion portfolio insurance (ICPPI). ICPPI products are customized for each policyholder and provide discretion on the level of the account floor, the duration of the contract and the choice of risky assets within the CPPI structure.

Another innovation found in some ICPPI products addresses the issue of rebalancing costs. As discussed earlier, traditional CPPI require regular rebalancing (between risky assets such as physical equities and safe assets) which incurs costs. To address this issue some ICPPI providers use derivatives to create quasi-synthetic portfolios that replicate equity movements without the need for purchases of physical equities. One drawback of this innovation is however that it introduces counterparty risk as well as market risk into ICPPI products. Insurers using such ICPPI structures consequently need to adapt their risk management practices related to counterparty risk, a topic that is discussed in the next chapter.

The following CPPI application example is from the insurance sector.

EXAMPLE

Asset management firm Allianz Global Investors is already using ICPPI structures. Kai Wallbaum, Frankfurt-based head of retail and life/asset solutions at the firm, says "We have installed an operational platform where we are able to manage on an individual account basis certain algorithmic strategies. One of the approaches clients often ask for is this individual CPPI, where we can manage a certain dynamic risk approach for each customer and provide a custom guarantee as well."

(Insurance Risk, 9 May 2013, "Insurers eye CPPI structures for next generation of savings products", www.risk.net/2266887.)

Another solution to the cash-lock scenario is to create an ICPPI product that is partly invested in a core insurance fund and partly invested in a third-party fund which contains its own guarantee level and allocations are changed between the two relatively infrequently, for example on a monthly instead of a daily basis. The guarantee from the third-party fund could be against decreases in the NAV by 80%, for example, and this provides protection against locking in a low return while reducing operational and rebalancing costs. The following example describes one such product that was launched in 2012.

EXAMPLE

Italian banking group Unicredit is one provider offering such a solution. The group pioneered an ICPPI product in 2012 called Green, aimed at its private banking clients, and achieved such success with this that it is now looking to partner with insurance providers to explore opportunities in other countries.

A further challenge for insurers is raising awareness around the products and dispelling the notion that long-term guarantees are not worth their price tag. "A lot of insurers are still at the concept stage when it comes to developing new products and customers are not yet at a place where they can understand the cost-benefit of ICPPIs," [...]

(Insurance Risk, 9 May 2013, "Insurers eye CPPI structures for next generation of savings products", www.risk.net/2266887.)

The reading illustrates that one of the motivations for the use of CPPI techniques by insurance companies is to avoid the costs of expensive derivatives such as options used in OBPI approaches.

As our review of CPPI structures showed in earlier sections, CPPI algorithms can be customized and depending on an investor's risk aversion, the cushion can be set at a particular level. Some insurers have implemented systems to offer tailor-made CPPI specifications. Some of the issues of CPPI are highlighted. As the risky asset in the CPPI structure declines, for example, the funds are placed in a safe deposit or bond permanently and this provides principal protection but also prevents pensioners from benefiting from future market recoveries. The reading also provides an illustration of the dynamic rebalancing mechanisms that we discussed earlier. Rebalancing incurs transaction costs and these costs are higher for physical assets than derivatives. Derivatives on the other hand introduce counterparty risk, which is something that we cover in detail in the subsequent chapter.

23.9 REAL-WORLD COMPLICATIONS

The idea behind CPPI techniques is simple. Actually, even the modeling is fairly straightforward. Yet, in practice, several difficulties arise. We will look at only some of them.

23.9.1 THE GAP RISK

If a structurer does not want exposure to gap risk then it could be sold to other investors through other structured products. For example, with structured products such as autocallables, a high coupon is paid to the investor, but the structure is called automatically if the underlying price hits a preselected level. Note that with autocallables the investor receives a high coupon but also assumes the risk of large downward movements in a basket. The extra coupon received by the investor can be visualized as the cost of insuring the *gap risk*.[10]

Another possibility frequently used in practice is to manage the gap risk using deep out-of-the-money puts. This is possible if the underlying is liquid. However, in the case of CPPI strategies, the underlying is often illiquid and this makes *delta*-hedging of the option positions difficult. Still, one can claim that during stress periods, correlations go toward one and liquid indices can become correlated with the illiquid underlying. Hence, carefully chosen deep out-of-the-money options on liquid indices can also hedge the gap risk. Banks generally charge a small "protection" or "gap" fee to cover gap risk, usually as a function of the notional leveraged exposure.

23.9.2 THE ISSUE OF LIQUIDITY

The issue of liquidity of the underlying is important to CPPI-type strategies for several reasons. First is the need to close the risky asset position when the cushion goes toward zero. If the underlying market is not liquid this may be very difficult to do, especially when markets are falling at a steep rate.[11]

---

[10]Alternatively, the gap risk can be insured by a reinsurance company.
[11]Note that the CPPI strategy will enhance the market direction. The structurer will sell (buy) when markets are falling (rising). Hence at the time when the risky asset position needs to be liquidated, other CPPI structurers may also be "selling."

Second, if the underlying is illiquid, then options on the underlying may not trade and hedging the gap risk through out-of-the-money options may be impossible.

The third issue is more technical. As mentioned in the previous section, gap risk can be modeled using the jump process augmented stochastic differential equations for S_t. In this setting, jump risk is the probability that Cu_t is negative. This determines the numerical value selected for the λ parameter. However, note that if the underlying is not liquid, then options on this underlying will not be liquid either. Yet, liquid option prices are needed to calibrate the parameters of the jump process. With illiquid option markets this may be impossible. Essentially, the selection of λ would depend on arbitrarily made assumptions and/or historical data.

23.9.3 WHICH PRINCIPAL PROTECTION TECHNIQUE IS BEST IN PRACTICE?

In the introduction to this chapter, we noted that different principal protection techniques exist including OBPI, CPPI, DPPI, and others. Which principal protection technique works best in practice? It turns out that this is an empirical question and no theoretical ranking of the techniques can be provided. Options can at certain times be very expensive, making OBPI unattractive. At first glance, CPPI may appear similar to standard option replication strategies. However, there are fundamental differences since standard CPPI does not make assumptions about future portfolio volatility, the distribution of portfolio returns, or the investor's time horizon. Because of the way that the cushion adjusts in CPPI, this approach will work well in a trending bull market with no reversals, since the hedger will keep increasing her exposure. There are some market environments when CPPI is less likely to do well. In a market without trends and oscillating prices, CPPI can be expected to perform poorly, since the hedge will buy on up-moves in prices only to see the market weaken subsequently and sell on down-moves in prices only to see the market rebound. Monte Carlo simulations and historical backtests that are specific to the underlying asset of interest and that incorporate assumptions about future market environments are required to decide which PPI technique generates the best risk-adjust performance for the hedger or investor. For the S&P500, for example, Lu (2010) provides a comparison of the performance of different portfolio insurance approaches including OBPI, CPPI, and DPPI.

23.10 CONCLUSIONS

There may be several other real-world complications. For example, consider the application of the CPPI to the credit sector. One very important question is what happens on a *roll?* Clearly the structurer would like to stay with on-the-run series, and during the roll there will be mark-to-market adjustments which may be infinitesimal and similar to jumps.

A second question is how to pick the leverage factor in some optimal fashion. It is clear that this will involve some Monte Carlo approach but the more difficult issue is how to *optimize* this.

SUGGESTED READING

The first academic papers on CPPI were by **Perold** (1986) who discusses CPPI for fixed-income instruments and **Black and Jones** (1987) who study CPPI for equity instruments. We also recommend the recent paper by **Cont and Tankov** (2009). **Joossens and Schoutens** (2010) discuss and compare CPPI and CPDO structures. **Jin and Whetten** (2005) summarize the evolution of Credit-Linked CPPI Variations.

EXERCISES

1. Consider the iTraxx CPPI example in this chapter. We assumed an increase in the value of the iTraxx tranche as q_{t_1} dropped from 20% to 15%. Now assume that spreads widen and credit quality decreases so that q_{t_1} increases from 20% to 25%. Calculate the resulting change in the leverage and explain the adjustment in detail required to return to the target leverage of 2.

2. The iTraxx crossover index followed the path given below during three successive time periods:

$$\{330, 360, 320\}$$

   Assume that there are 30 reference names in this portfolio.
   a. You decide to select a leverage ratio of 2 and structure a *five-year* CPPI note on iTraxx crossover index. LIBOR rates are 5%. Describe your general strategy and, more important, show your initial portfolio composition.
   b. Given the path above, calculate your portfolio adjustments for the three periods.
   c. In period four, iTraxx becomes 370 and one company defaults. Show your portfolio adjustments. (Assume a recovery of 40%. Reminder: Do not forget that there are 30 names in the portfolio.)

COUNTERPARTY RISK, MULTIPLE CURVES, CVA, DVA, AND FVA

CHAPTER OUTLINE

24.1 INTRODUCTION

In this chapter, we review recent innovations in financial engineering that attempt to better incorporate counterparty risk into derivatives pricing. The topics covered in this chapter are very much at the forefront of financial engineering practice and derivatives research. In previous chapters, we showed how basic principles such as swap cash flow engineering, option valuation, and dynamic replication can be used to construct and value more complex products. Our hope is that the reader now appreciates the commonality in the approaches to different asset classes and financial engineering problems. In this chapter, we will see how some of the tools that we studied earlier can be used to better account for counterparty risk. We will see that it is possible to model and incorporate counterparty risk in a derivatives transaction by applying several financial engineering tools that we encountered before including option valuation and CDS. In particular, credit valuation adjustment (CVA) is shown to resemble an option on the residual value of the portfolio with a random maturity given the default time of the counterparty. The default probability and default intensity for the counterparty can be backed out from CDS spreads. Thus some approaches to counterparty risk build on credit risk financial engineering tools such as CDS and structural models of default that we saw in Chapters 18 and 19. Some collateral agreements allow the choice of collateral and this choice leads to an embedded option. This optionality has recently been addressed by ISDA's so-called SCSA (standard credit support annex). Finally, some solutions that attempt to hedge counterparty risk build on CDO structuring ideas that we encountered in Chapter 21. Thus, as in previous chapters we can use basic financial engineering principles to address counterparty risk.

Incorporating counterparty risk-related adjustments significantly changes pricing formulae for derivatives and in practice it significantly changes valuations and behavior by market makers whose perspective is central to this book. The reason why we chose not to deal with counterparty risk by asset class in each chapter is that we first needed to develop the option valuation, CDS and CDO toolkit that we apply to address counterparty risk. Therefore, despite their importance, it is best to deal with more advanced counterparty risk issues at the end of the book so that the reader has had time to absorb the tools studied in the previous chapters.

Throughout this book, we have used financial engineering tools that price derivatives as if we lived in a world without counterparty risk. Counterparty risk affects many aspects of financial markets and banking including but not limited to derivatives pricing. First, incorporating counterparty risk has implications for financial derivatives pricing and the organization of banks. Many banks now make adjustments to the default-free derivatives that reflect the risk of their counterparty defaulting, so-called CVAs.[1] The principle of CVA is similar to that of a price discount for default risk assets in the sense that the value of a generic claim traded with a counterparty subject to default risk is always smaller than the value of the same claim traded with a counterparty having a null default probability. We discuss a case study that shows how CVA can be calculated for a portfolio of interest rates swaps. It is beyond the scope of this book to show how counterparty risk changes the pricing of derivatives for every asset class but we provide references at the end of the chapter that guide the interested reader to such applications.

The increasing attention that is being paid to counterparty risk led to an expansion in the role and workload of so-called *CVA desks* in banks, a functional unit that did not exist in many banks a few years ago. Banks also make adjustments arising from the bank's own default risk in the form of debit valuation adjustments (DVAs). Finally funding costs affect the valuation of derivatives and this is reflected in so-called funding valuation adjustments (FVAs). Therefore, the price of a derivative that incorporates counterparty risk can be decomposed in the default-free price plus an adjustment related to CVA, DVA, and FVA. The modeling choices related to CVA, DVA, and FVA are extensive and they imply that there is a wide dispersion in their application between banks. In previous chapters, we applied the overarching finance principle that the price of an asset should depend on the risk of the asset and be discounted by a discount rate that reflects this risk. However, in practice different borrowers have different funding costs and market practice is moving towards using different

---

[1]The Asian crisis and the bail out of LTCM led first-tier banks to pay more attention to counterparty risk and in the late 1990s large banks started using CVAs to assess the costs of counterparty risk.

discount rates that reflect differences in funding costs which represents a breach of the fundamental principle. As a result of counterparty risk adjustments there is no such thing as a fair value for a derivatives transaction since the value will depend on the counterparty. Moreover, counterparty risk introduces a new source of model risk. We had seen the prevalence of Black−Scholes−Merton option pricing model approaches, the Merton structural model of default, and the standard Gaussian copula market model. Our discussion will show that it is even conceptually not clear whether such a standard model can be found to reflect counterparty risk valuation in practice.

Second, counterparty risk affects capital requirements. In Basel II, there was a credit VaR charge related to counterparty risk. The most recent Basel III proposals require that capital requirements reflect risk that arises from mark-to-market losses related to CVA. However, these proposals do not allow for DVA which leads to a divergence between the treatment of counterparty risk from a capital charge regulation perspective and a derivatives pricing perspective.

Third, accounting standards (including IFRS and US GAAP) require credit risk to be reflected in the fair value measurement of derivatives. This has had multibillion dollar profit and loss implications for banks as they report the effects of DVA and CVA in their financial accounts.

Finally in this chapter, we will find that this discussion is everything but theoretical. These issues have transformed market practice, led to gains and losses for banks of the order of magnitude of several billion dollars over the last years and also created new professional opportunities and functions in the financial sector such as CVA desks.

The issue of counterparty risk also affects the choice of the discount rate that is used to discount cash flows under the risk-neutral measure. Typically LIBOR was used as a proxy for the riskless rate but after the GFC market practice has moved to the usage of OIS rates as a discount rate in collateralized transactions. This is partly due to the fact that LIBOR rates were significantly higher than OIS rates as a result of the counterparty risk that was priced into LIBOR rates. The move to OIS rates has three fundamental implications for derivatives pricing which we discuss in this chapter.

The topic of counterparty risk in financial engineering is a vast area and this chapter can only provide a brief summary. However, our focus is, as throughout the book, on showing how our existing financial engineering toolkit can be used to address these challenges and what the most important valuation and risk management issues are.

24.2 COUNTERPARTY RISK

Counterparty risk refers to the risk that a counterparty entering a transaction may default and not honor its payment obligations. In Chapter 2, we discussed how central clearing and CCPs can mitigate the effects of counterparty risk. Therefore, the counterparty risk associated with exchange-traded derivatives, which are always centrally cleared with CCPs, and with centrally cleared OTC transactions is remote since a clearing house guarantees the cash flows promised by the derivative to the counterparties.[2] However, the majority of derivatives transactions are *not* centrally cleared

---

[2]CCPs are normally highly capitalized and all clearing members post collateral which reduces but does not eliminate risk as the default of several CCPs around the world over the last decades proves (e.g., Caisse de Liquidation des Affaires en Marchandises in 1974, the Kuala Lumpur Commodity Clearing House in 1983, and the Hong Kong Futures Exchange in 1987 while the following were "near-misses": the CME and OCC in the US in 1987 and BM&F in Brazil in 1999). The increase in the volume of transactions via CCPs and the competition between an increasing number of CCPs raises the question whether CCPs might become more risky in the future.

which makes counterparty risk adjustments widespread and economically important. How large is the noncentrally cleared derivatives market that requires counterparty risk adjustments? According to recent BIS 2013 Triannual Survey, the OTC segment accounts for around 90% of the $750 trillion derivatives market in terms of notional amount outstanding.[3] Moreover, only approximately one-third of the OTC's market's notional is cleared via CCPs. Based on the above numbers, we estimate that roughly $440 trillion of the derivatives is not centrally cleared based on 2013 numbers.

Practitioners were aware of counterparty risk before the GFC, but, compared to other risks, such as market and credit risk, it was considered of second-order importance. There is however evidence that counterparty risk is of first-order importance. The following two motivating examples illustrate counterparty risk-related concepts that we will explain in detail in this chapter. First, mark-to-market losses due to counterparty risk can exceed mark-to-market losses due to market or credit risk at times as the following reading illustrates.

EXAMPLE

Under Basel II, the risk of counterparty default and credit migration risk were addressed but mark-to-market losses due to credit valuation adjustments (CVA) were not. During the financial crisis, however, roughly two-thirds of losses attributed to counterparty credit risk were due to CVA losses and only about one-third were due to actual defaults.

(Bank for International Settlements, 2011)

Second, recent profits reported by major banks reflected CVA and DVA adjustments that amount to several billion dollars. The following is just one recent example.

EXAMPLE

Goldman Sachs was the first large U.S. bank to report a third-quarter loss so far this earnings season, and its loss was due, in part, to its decision to hedge a potentially large accounting gain. [...]

The company hedged potential gains from its debit valuation adjustment, or DVA, by taking offsetting positions through insurance-like contracts called credit default swaps on a basket of other financials. [...]

Goldman's DVA gains in the third quarter totaled $450 million, about $300 million of which was recorded under its fixed income, currency and commodities trading segment and another $150 million recorded under equities trading.

That amount is comparatively smaller than the $1.9 billion in DVA gains that J.P. Morgan Chase and Citigroup each recorded for the third quarter. Bank of America reported $1.7 billion of DVA gains in its investment bank. All three banks were profitable and got a boost from the DVA gains, an accounting phenomenon that can reverse in subsequent quarters, given the direction of bond spreads.

(Wall Street Journal, Deal Journal blog, October 18, 2011, *"Goldman Sachs Hedges Its Way to Less Volatile Earnings"* by Liz Moyer and Katy Burne)

---

[3]The OTC market volume is equally divided between bilateral trading among market participants and multilateral trading.

Both of these examples illustrate how important CVA and DVA are in practice. We will discuss these concepts and how the derivatives industry has become progressively sophisticated in adjusting for counterparty risk in further detail below. Note that the formulae presented in this chapter are for uncollateralized derivatives transactions, but they can be generalized to collateralized deals too, where the CVA will be smaller and broadly speaking the option term will not be on the residual NPV at default but on the part of the residual NPV at default (if any) that will not be covered by collateral. This can be due to sudden profit and loss swings, possibly caused by the counterparty default itself or gap risk.

24.3 CREDIT VALUATION ADJUSTMENT

Throughout this book, we have assumed that there is no default risk in derivatives transactions. For example, in the Black−Scholes model we assumed that each counterparty will honor its obligations and that neither will default. The assumption that large financial institutions are default free is not realistic. For example, in 2008, eight credit events on financial institutions occurred in 1 month (Lehman Brothers, Fannie Mae, Freddie Mac, Washington Mutual, Landsbanki, Glitnir, Kaupthing, and Merrill Lynch).

24.3.1 COUNTERPARTY RISK EXAMPLE AND CVA

It was not just in the valuation of forwards, options, and credit default swaps that we assumed counterparty risk away. In our discussion of interest rate swaps (IRSs) in Chapter 3, we also did not consider the risk that a counterparty may default. How does counterparty risk affect the valuation of a simple IRS? Let's assume an IRS with a notional amount of N and maturity $T = t_4$, between bank B and a corporate counterparty C where B pays fixed and C pays floating. Figure 24.1a illustrates the cash flows if we assume that B and C cannot default. Let's continue to assume that B is risk free for now, but let's allow for the possibility of default by C at an unknown time τ. This is shown in Figure 24.1b. For illustration we assume that the default occurs at some point between dates t_3 and t_4. Apart from the trivial scenario of no default (that is $\tau > T$), we can distinguish two basic scenarios:

- Scenario 1 (default and B has positive exposure): C defaults at some point τ where $t_3 < \tau < t_4$, and the net present value (NPV) of the remaining payments from C to B is positive, that is, B has positive exposure. In this case, B receives only a recovery fraction REC of the net present value NPV.
- Scenario 2 (default and B has negative exposure): If C defaults at some point τ where $t_3 < \tau < t_4$, and the NPV of the remaining payments from C to B is negative, then B has negative exposure. In this case, B pays the liquidator of C a positive amount.

The example in Figure 24.1 illustrates the important insight that it is not just the stochastic default time that is important but also the sign and magnitude of the NPV of any remaining payments at default. These are features that any adjustment for counterparty risk must take into account.

There are several ways of incorporating counterparty risk into derivatives pricing. CVA is one such adjustment and it is simple and intuitive. The idea behind CVA is that if we enter into a portfolio derivative transaction with a counterparty, that counterparty might default and it might default at a time when we have an exposure, that is the value of the transaction to us is positive and negative to the counterparty. This has to be incorporated into the valuation of the transaction. First, we value the

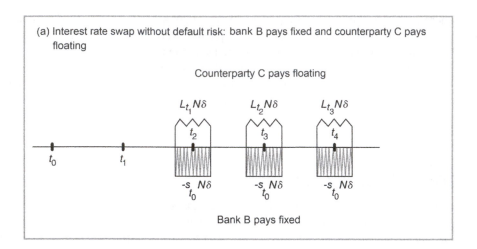

(a) Interest rate swap without default risk: bank B pays fixed and counterparty C pays floating

Counterparty C pays floating

$L_{t_1} N\delta$ $L_{t_2} N\delta$ $L_{t_3} N\delta$

t_2 t_3 t_4

t_0 t_1

$-s_{t_0} N\delta$ $-s_{t_0} N\delta$ $-s_{t_0} N\delta$

Bank B pays fixed

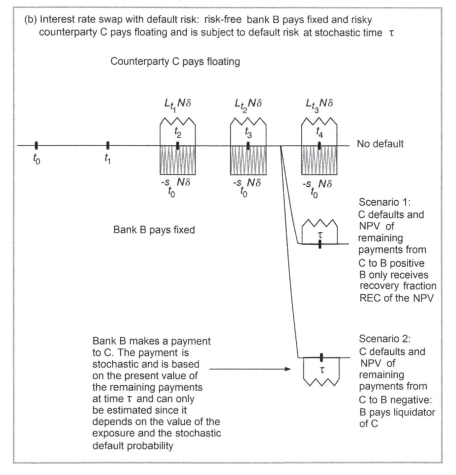

(b) Interest rate swap with default risk: risk-free bank B pays fixed and risky counterparty C pays floating and is subject to default risk at stochastic time τ

Counterparty C pays floating

$L_{t_1} N\delta$ $L_{t_2} N\delta$ $L_{t_3} N\delta$

t_2 t_3 t_4

No default

t_0 t_1

$-s_{t_0} N\delta$ $-s_{t_0} N\delta$ $-s_{t_0} N\delta$

Bank B pays fixed

Scenario 1:
C defaults and
NPV of
remaining
payments from
C to B positive
B only receives
recovery fraction
REC of the NPV

τ

Bank B makes a payment
to C. The payment is
stochastic and is based
on the present value of
the remaining payments
at time τ and can only
be estimated since it
depends on the value of the
exposure and the stochastic
default probability

Scenario 2:
C defaults and
NPV of
remaining
payments from
C to B negative:
B pays liquidator
of C

τ

FIGURE 24.1

IRS cash flows with and without counterparty risk.

transaction assuming that neither we nor the counterparty would default. Then we will subtract the CVA, or credit valuation adjustment, to reflect the possibility that the counterparty might default. Even before going into the details of CVA, it is clear that the above logic implies that another adjustment may be required for the fact that we could also default on the payment obligations to the counterparty. This is called DVA and we will discuss it in a subsequent section.

For illustration, we could assume that in the example above the counterparty C can be either the European Central Bank (ECB) or a Greek commercial bank in 2010. Clearly the ECB and the Greek bank had very different counterparty risks in 2010, as our discussion of sovereign default swaps in Chapter 18 showed. Incorporating counterparty risk into the evaluation of the payoffs from the transaction will necessarily imply that we will condition on whether a default by the bank's counterparty will occur or not. Intuitively, the inclusion of the risk of the bank's counterparty defaulting will add optionality into the payoff from the bank's perspective.

The CVA adjustment can be positive or negative. Consider an example involving bank B and its counterparty, a corporate client C. If C defaults and the present value of the portfolio at default is positive from the perspective of the surviving party (bank B), then B only gets a fraction of the portfolio equal to the recovery rate from C. It is also possible that the present value of the portfolio is negative from the perspective of the surviving party B. In this case, B has to pay the liquidators of the defaulted corporate C. This logic leads to a price of the derivatives transaction with counterparty risk ($P_{\text{with CR}}$) that is equal to the value of the deal without counterparty risk ($P_{\text{without CR}}$) minus a positive adjustment, labeled CVA, which captures the above option:

$$P_{\text{with CR}} = P_{\text{no CR}} - \text{CVA} \qquad (24.1)$$

This adjustment looks relatively straightforward, but several complications arise in practice. First, even for some simple derivatives incorporating counterparty risk makes the valuation model dependent, even if under default-free pricing it was model independent. One of the simplest derivatives that we have seen is an IRS as discussed in Chapter 4. With counterparty risk, the IRS valuation needs to include an option on the residual value of the portfolio and to price such an option we require an interest rate option pricing model.

Second, as we introduce a new source of risk in the form of counterparty risk, we need to also take into account correlations between interest rate risk and counterparty risk. This leads us to the concept of *wrong-way risk (WWR)*. WWR refers to the additional risk that arises when the underlying portfolio and the default of the counterparty are correlated in the worst possible way, i.e. when the portfolio value from bank B's perspective is positively correlated with the default probability of the counterparty. This could lead to a scenario when the counterparty C defaults in a state of the world when it owes bank B the most.

Third, to obtain volatility and correlation parameters we need to obtain such values under the risk-neutral probability. There are many counterparties that do not issue bonds or have CDS written on them. This makes it difficult to obtain default information about them. In practice, one can sometimes approximate the default probability under the risk-neutral measure with volatilities and correlations of default and other risks based on the real world measure.

24.3.2 CVA AS AN OPTION

The above discussion has been heuristic. Now we want to show rigorously that CVA can be interpreted as an option. For this purpose, we first make some assumptions. First, we assume that the counterparty risk valuation problem is *symmetric* for counterparties B and C. This means that the total value of the position for B as valued at a given time, including counterparty risk, is the opposite of the total value of the position valued by C at the time. Second, as in Figure 24.1 we continue to make the unrealistic *unilateral default assumption* (UDA) and assume that bank B is default free. We will allow for bilateral default and relax the UDA later in the chapter. The example with only one defaultable counterparty is already sufficient to explain the fundamental financial engineering issues that arise when dealing with counterparty risk. Third, we assume that there are no collateral agreements or other guarantees in place. Fourth, our framework assumes that the default risk is charged upfront to counterparty C and is therefore included in the risk-neutral valuation framework.

In this section, we will derive a general formula for unilateral counterparty risk valuation. We will see that the price in the presence of counterparty risk can be expressed as the default-free price minus a discounted option term in scenarios of early default multiplied by the loss given default. We will present a Monte Carlo simulation to calculate the CVA for a bank that has multiple IRSs with several different counterparties. The example will assume independence between interest rates and credit spreads (default intensities). This implies that, for simplicity, we are abstracting away from WWR.[4]

In what follows we present the calculations from the point of view of the investor, that is bank B. We continue to use the risk-neutral valuation framework which implies that expectations E_t are taken under the risk-neutral probability measure (Q). We follow Brigo et al. (2013) and use $\Pi(u,s)$ to denote the net cash flows of a generic claim as seen from the perspective of bank B and traded with the counterparty C between time u and time s, and discounted back to time u.[5] We define the default-free NPV at time t as $\text{NPV}(t) = E_t[\Pi(t,T)]$. In the context of the discussion of Figure 24.1 we used the term *exposure*. More formally, we can define exposure at time t for a position with a final maturity T and cash flows $\Pi(t,T)$ as

$$\mathbf{Ex}(t) = (E_t[\Pi(t,T)])^+ = (\mathbf{NPV}(t))^+. \tag{24.2}$$

To incorporate counterparty risk, we denote the payoff with a defaultable counterparty as $\overline{\Pi}(t,T)$ and define it as follows:

$$\overline{\Pi}(t,T) = 1_{\{\tau > T\}}\Pi(t,T) + 1_{\{t < \tau \le T\}}[\Pi(t,\tau) + D(t,\tau)(\mathbf{REC}(\mathbf{NPV}(\tau))^+ - (-\mathbf{NPV}(\tau))^+)]. \tag{24.3}$$

Equation (24.3) shows that if there is no default, that is $\tau > T$, then we obtain the risk-neutral valuation formula $\Pi(t,T)$ that we used throughout the book up until now. The remaining components of the right-hand side of Eq. (24.3) can be interpreted as consisting of the payments due before default occurs ($\Pi(t,\tau)$), the default-free stochastic discount factor at time t for maturity $\tau(D(t,\tau))$, and two terms with embedded options in the form of a positive recovery value ($\mathbf{REC}(\mathbf{NPV}(\tau))^+$) to be received by bank B if the net present value $\mathbf{NPV}(\tau)$ is positive and a payment from B to C if the NPV is negative.

24.3.2.1 Close-out proceedings

When counterparty C defaults, *close-out* proceedings begin. These are governed by the regulations and ISDA rules.[6] As part of the close-out process the residual value of the contract to bank B is

---

[4]The framework illustrated in the example can also be used to calculate counterparty risk for nonstandard swap contracts such as zero-coupon swaps and amortizing swaps.

[5]The alternative specification to Brigo et al. (2013) is that of Burgard and Kjaer (2011). Instead of "pricing via expectation" they follow the approach "pricing via hedging"—and replicating portfolio to derive CVA/DVA formulae.

[6]For details, see www.isda.org.

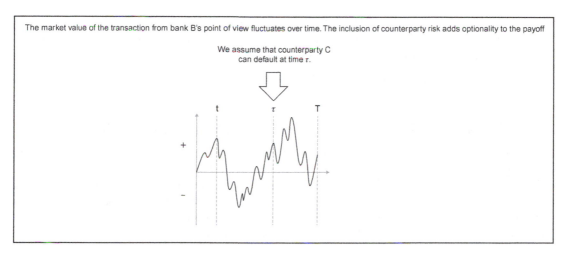

The market value of the transaction from bank B's point of view fluctuates over time. The inclusion of counterparty risk adds optionality to the payoff

We assume that counterparty C can default at time τ.

FIGURE 24.2

Counterparty risk adds optionality.

determined. If it is positive, the bank will receive a payment depending on the recovery rate and if it is negative, the bank will make a payment. When calculating the loss arising from the counterparty's default one assumes that the bank enters into a similar contract with another counterparty so that it maintains its market position. The banks' market position is unchanged after replacing the contract. The loss is a function of the contract's replacement cost at the time of default. There are therefore two scenarios:

- Scenario 1 (the contract value is positive for bank B at the time of default): In this case B closes out the position with C by paying it the market value of the contract. At the same time B enters into a similar contract with another counterparty and receives the market value of the contract, which implies that the net loss to B is zero.
- Scenario 2 (the contract value is positive for bank B at the time of default): In this case B closes out the position with C but does not receive any payment from the defaulting counterparty C. Since B enters into a corresponding contract with another counterparty, B suffers a net loss equal to the market value of the contract.

Figure 24.2 illustrates how the market value of the contract fluctuates over time for one assumed path. Scenarios 1 and 2 imply that the exposure of a bank that enters a derivative transaction with a counterparty is the maximum of contract i's market value or zero, which reflects the embedded optionality.

The above equation highlights the embedded optionality. Brigo et al. (2013) show that one can take expectations under the risk-neutral measure of Eq. (24.3) and obtain after some manipulations the following price of the payoff with maturity T under counterparty risk:

$$E_t(\overline{\Pi}(t,T)) = E_t[\Pi(t,T)] - E_t[\text{LGD} \times 1_{\{t < \tau \leq T\}} D(t,\tau)((\text{NPV}(\tau))^+)]. \tag{24.4}$$

The second term on the right-hand side of Eq. (24.4) can be interpreted as the counterparty risk adjustment. We can write Eq. (24.4) as the default-free net present value minus the unilateral CVA:

$$E_t(\overline{\Pi}(t,T)) = E_t[\Pi(t,T)] - U_{\text{CVA}}(t,T) \qquad (24.5)$$

Since the loss given default (**LGD**) is equal to 1 minus the recovery rate, we can also write the unilateral CVA in terms of the exposure $\mathbf{Ex}(\tau)$ at time t:

$$U_{\text{CVA}}(t,T) = E_t[(1 - \mathbf{REC}) \times D(t,\tau) \times 1_{\{t < \tau \le T\}} \times \mathbf{Ex}(\tau)] \qquad (24.6)$$

24.3.3 COUNTERPARTY RISK AND UNILATERAL CVA IN A SINGLE IRS

To provide an illustration of the application of the CVA concept, we will consider a simple IRS. As in Figure 24.1, we continue to assume that bank B pays a fixed rate s_{t_0} and receives a floating interest rate L from counterparty C. We further assume that the notional amount N is equal to 1. As discussed in Chapter 14, the fair (forward swap) rate s_{t_0} at time t_0 without counterparty risk is defined as the rate at which the present value of the fixed leg and floating legs payments is the same. For the 3-year paper swap in Figure 24.1, the discounted payoff IRS is

$$\sum_{i=2}^{4} D(t_0, t_i)\delta N\big(L(t_{i-1}, t_i) - s_{t_0}\big) \qquad (24.7)$$

We can denote the NPV at time t_0 of an IRS without default risk that starts at time t_1 and has a maturity of date of t_4 as $\Pi_{\text{IR}}(t_0, t_4)$. Since bank B is paying fixed and receiving floating payments from counterparty C, introducing the possibility that C may default implies that the correct swap rate to be paid as part of the fixed leg with counterparty risk should be lower than the rate without counterparty risk since the fixed rate payer is compensated for the additional risk in this way. If we apply Eq. (24.4) to the IRS example the net present value from the perspective of B with counterparty risk becomes:

$$\overline{\Pi}_{\text{IR}}(t_0, t_4) = \Pi_{\text{IR}}(t_0, t_4) - U_{\text{CVA}}(t_0, t_4) \qquad (24.8)$$

where $U_{\text{CVA}}(t_0, t_4)$ is again the unilateral credit valuation adjustment (UCVA) due to default. It can be shown that $U_{\text{CVA}}(t_0, t_4)$ can be rewritten as

$$U_{\text{CVA}}(t_0, t_4) = \mathbf{LGD} \times E_{t_0}\big[1_{\{t_0 < \tau \le t_4\}} D(t_0, \tau)((\mathbf{NPV}(\tau))^+)\big] \qquad (24.9)$$

or

$$U_{\text{CVA}}(t_0, t_4) = \mathbf{LGD} \times \int_{t_1}^{t_4} PS\big(t_0, u, t_4, s_{t_0}, \sigma_{u,t_4}\big) d_u P^Q\{\tau \le u\} \qquad (24.10)$$

In Eq. (24.10), $\int_{t_1}^{t_4} PS\big(t_0, u, t_4, K, s_{t_0}, \sigma_{u,t_4}\big)$ represents the price at time t_0 of a swaption with maturity u, strike price K, underlying forward swap rate s_{t_0}, volatility σ_{u,t_4}, and underlying swap with final maturity t_4.

The intuition for the result in Eq. (24.10) is straightforward. The key assumption is that the default time τ and interests rates are independent. Moreover, the residual net present value is a forward starting IRS starting at time τ. Therefore the option on the residual NPV can be interpreted as a sum of swaptions with maturities spanning the possible range of default times with weights equal to risk-neutral probabilities P^Q of defaulting around each time value. To value default-risky assets, it is essential to introduce the default times and default probabilities in the pricing framework. CDS are one liquid source of market risk-neutral default probabilities. An alternative would use credit spreads implied by corporate bond prices to calibrate the default probability. The resulting estimates may differ due to the CDS-bond basis discussed in Chapter 18. In Chapters 18 and 19, we have seen that different models can be used to calibrate CDS data to back out default probabilities. In this chapter, we will use reduced-form survival probability/hazard function models, but the end of chapter references show how structural models can be used to address counterparty risk.

24.3.4 NUMERICAL CVA EXAMPLE FOR IRS PORTFOLIO

We illustrate the calculation of unilateral CVA for IRSs in the context of a case study.[7]

EXAMPLE

Consider a bank B that holds a portfolio of vanilla interest rates swaps with five counterparties. The time t_0 is assumed to be December 14, 2007. We apply the unilateral CVA formula in Eq. (24.5) to calculate the CVA for bank B. Bank B enters into a total of 30 interest rates swaps with the five counterparties. The number of swaps and swap maturities per counterparty is as follows[8]:

- Counterparty 1: Six swaps with maturities ranging from June 15, 2009 to June 23, 2012.
- Counterparty 2: Two swaps with maturities ranging from May 12, 2012 to December 4, 2014.
- Counterparty 3: Three swaps with maturities ranging from July 12, 2010 to February 13, 2012.
- Counterparty 4: Seven swaps with maturities ranging from June 2, 2009 until July 20, 2014.
- Counterparty 5: Twelve swaps with maturities ranging from January 8, 2009 until June 5, 2014.

One can calculate the unilateral CVA by following the four steps:

- Step 1 (Scenario Generation): We simulate many future interest rate paths or scenarios which are the underlying risk factors for the swaps.
- Step 2 (Instrument Valuation): For each path in step 1 we value the underlying swap portfolio and derive future NPV(t) distributions for each trade within the portfolio.
- Step 3 (Portfolio Aggregation): For each counterparty and according to the netting agreement in place we aggregate the NPV(t) distributions of its trades to derive the NPV(t) distribution of the portfolio.
- Step 4 (Exposure Calculation): Then, for each counterparty we calculate the exposure using Equation (24.2) and we discount (scenario consistently) to t_0. We calculate the expected discounted exposure by simply taking the average of the discounted exposures at every t.
- Step 5 (Probability Weighting): We weight the expected exposures by the default probabilities. The default probabilities are extracted from CDS market quotes.

---

[7]The example uses various functions from Matlab's Financial Instruments toolbox. For details, please see the Matlab's Financial Instruments Toolbox Counterparty Credit Risk example.

[8]Further details about the most floating and fixed leg rate assumptions are available in Matlab's "cva-swap-portfolio.xls" file.

Table 24.1 Assumed Term Structured of CDS Quotes for Each Counterparty					
Date	CP1	CP2	CP3	CP4	CP5
March 20, 2008	140	85	115	170	140
March 20, 2009	185	120	150	205	175
March 20, 2010	215	170	195	245	210
March 20, 2011	275	215	240	285	265
March 20, 2012	340	255	290	320	310

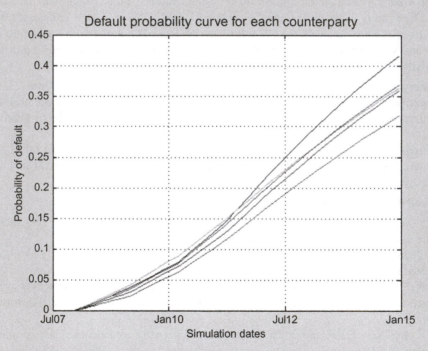

FIGURE 24.3

Default probability curves for each counterparty.

Table 24.1 shows the assumed term structure of CDS quotes for each counterparty (CP1, CP2, . . ., CP5) as of December 2007. Based on the CDS quotes in Table 24.1, we can extract default probabilities which are shown in Figure 24.3. At date t_0, each swap in the portfolio has a value close to zero. Figure 24.4 shows the assumed yield curve at settle date t_0 (December 14, 2007) for maturities of 3 months, 6 months, 1 year, 5 years, 7 years, 10 years,

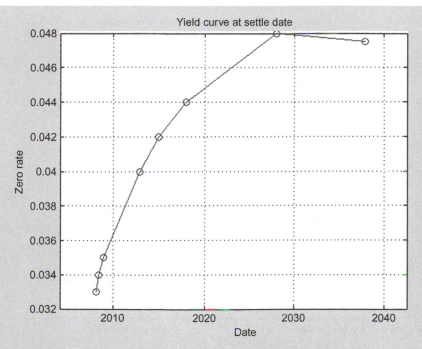

FIGURE 24.4

Yield curve at settle date.

20 years, and 30 years. As we see the yield curve is upward sloping for all maturities until 2027 and then becomes hump-shaped for the last 10 years.

To implement Step 2 described above, the example uses the Hull–White single factor model to simulate the value of the swap contracts. Figure 24.5 shows one scenario for the yield curve evolution from 2007 until 2015.

As part of Step 3, we calculate the mark-to-market value of each swap contract at each future simulation date. Figure 24.6 shows the mark-to-market values of the 30 swap contracts for one simulated scenario. As expected at time t_0, the swap contracts are valued close to zero.

If we aggregate across all 30 swap contracts at each point in time for a given simulated path, we obtain the simulated path of the entire portfolio. The exposure of a particular contract i at time t is the maximum of the contract value and 0. To better visualize the exposure profile, we can evaluate certain statistics of the exposure distribution at each simulation date. One can calculate the expected exposure profile, for example, by computing the expectation of each exposure at each simulation date. If we also discount the expected exposure, we can calculate the discounted expected exposure as reported in Figure 24.7.

It is clear from Figure 24.7 that the (discounted) expected exposure changes over time. There are two offsetting effects that determine the exposure profile over time: the diffusion effect and the amortization effect. The longer a simulation path is the greater the resulting

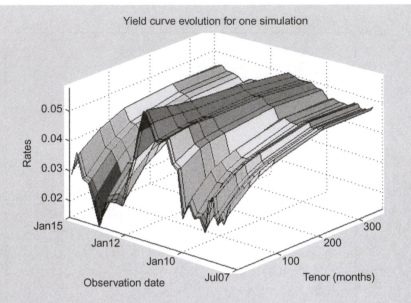

FIGURE 24.5

Yield curve evolution for one scenario.

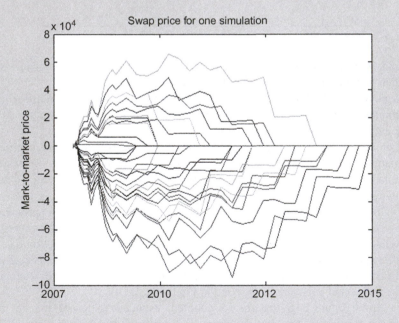

FIGURE 24.6

Swap prices for one scenario.

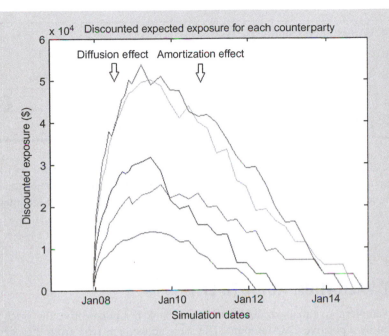

FIGURE 24.7

Discounted expected exposure for each counterparty.

variability in the risk factor (such as interest rates) and hence the further away the future value is from the current value. At the same time as the simulation path moves towards maturity of the contract, the value tends to decrease the exposure over time, since it reduces the remaining cash flows that are exposed to default. The two effects are offsetting since the diffusion effect increases the credit exposure and the amortization effect decreases it over time. On the one hand, for FX forwards and other single cash flow products, the potential exposures peak at the maturity of the transaction since the exposure is purely driven by the diffusion effect. For products with multiple cash flows, on the other hand, such as the IRS, the potential exposure usually peaks at one-third to one half of the way into the life of the transaction according to Pykthin and Zhu (2007). Figure 24.7 shows that this is also the case in our example with a portfolio of 30 IRSs.

The exposure profile does not just depend on the asset class and instrument but also on the underlying risk factors and the point in the market cycle. For example, when the yield curve is upward sloping, the exposure is greater for a payer swap than for a corresponding receiver swap, since the fixed payments in the early periods are greater than the floating payments. This leads to positive forward values on the payer swap. The opposite is true if the yield curve is downward sloping. We can see from Figure 24.4 that the yield curve was upward sloping in our example and the exposure profile is indeed as predicted. One of the uses of exposure profiles is for the purpose of calculating economic and regulatory capital, pricing and hedging counterparty risk as well as verifying compliance with credit limits.

Table 24.2 CVA for Each Counterparty

Counterparty	CP1	CP2	CP3	CP4	CP5
CVA	$2228.36	$2487.60	$920.39	$5478.50	$5859.30

Finally by using the default probabilities from Figure 24.3, we can calculate the unilateral CVA with respect to each counterparty. These are reported in Table 24.2.

The above example illustrates the calculation of unilateral CVA for a portfolio of simple interest rates swaps. For reasons of space we cannot cover the CVA examples for other asset classes such as credit, commodities, equities, or FX, but the references at the end of the chapter include such examples.

24.4 DEBIT VALUATION ADJUSTMENT

Consider the example above that we discussed. From bank B's perspective the price of the derivative was reduced by a positive CVA amount. The same adjustment seen from the perspective of the counterparty is called debit valuation adjustment (DVA). It is positive because the default of the counterparty C would lead to a discount on C's payment obligation and this can be interpreted as a gain. In this case, we continue to assume that the bank B itself is still default free. The DVA is *unilateral* since only the default risk of the client is included. For the two counterparties to agree, we require that the adjustment to the risk-free price be the same, but that it is added by counterparty C and subtracted by bank B. In other words we require:

$$\text{Unilateral CVA(Bank } B) = \text{Unilateral DVA(Corporate } C) \qquad (24.11)$$

Although the concept of DVA looks simple, there are several complications in practice. First, it is difficult to monetize DVA in practice. Consider what would be required to do this. One would need to unwind trades to the counterparty to monetize DVA since it is the reduced NPV of one's derivatives payment obligations to the counterparty. Prior to default this is practically impossible to achieve given that it would involve unwinding a complex derivatives portfolio. Second, it is important to note that the treatment of DVA for purposes of derivatives pricing is distinct from that for accounting purposes. As the second example in Section 24.2 has shown, DVA is a major element in financial reporting and there were recent instances when large banks such as Citibank saw their quarterly profit and loss reports change from a loss to a profit as the credit spreads of the banks widened. Although DVA is widely used to price new OTC derivatives, its accounting treatment is more ambiguous and many firms report DVA on a separate line and identify it as an accounting charge rather than a real profit and loss. Third, the treatment of DVA from a bank capital regulation perspective is also distinct from those for pricing and accounting purposes. Despite its use for accounting and derivatives pricing purposes, until recently the BIS explicitly excluded DVA from regulatory capital calculations. However, a recent Basel Committee on Banking Supervision consultative document discusses several options of recognizing DVA for capital purposes which implies the potential for a convergence between the accounting, regulatory, and derivatives pricing treatment of DVA.[9]

24.5 BILATERAL COUNTERPARTY RISK

So we have adopted the unilateral perspective. What would happen if both counterparties view each other as risky? For example, the corporate C could view bank B as risky. After the demise of Lehman Brothers in 2008 and multiple credit events involving financial institutions during the GFC this is a realistic scenario. The only way to deal with this situation is to allow both counterparties to include their own default besides the default of the counterparty into their valuation. This implies that each counterparty will subtract a positive CVA value from and add a positive DVA to the default risk-free price of the transaction. In other words, the CVA of one party will be the DVA of another party and vice versa. In the example above, we will obtain the following equation from bank B's perspective,

$$P^{\text{to bank B}}_{\text{with CR}} = P^{\text{to bank B}}_{\text{no CR}} + \text{DVA}_{\text{Bank B}} - \text{CVA}_{\text{Bank B}} \tag{24.12}$$

From the perspective of corporate C, the value of the deal has the opposite sign:

$$P^{\text{to bank B}}_{\text{no CR}} = - P^{\text{to corporate C}}_{\text{no CR}} \tag{24.13}$$

$$\text{DVA}_{\text{Bank B}} = \text{CVA}_{\text{Corporate C}} \tag{24.14}$$

$$\text{DVA}_{\text{Corporate C}} = - \text{CVA}_{\text{Bank B}} \tag{24.15}$$

Equations (24.12)−(24.15) imply that

$$P^{\text{to bank B}}_{\text{with CR}} = P^{\text{to corporate C}}_{\text{with CR}} \tag{24.16}$$

This means that both counterparties agree on the price. It is common to refer to the difference (DVA-CVA) as the bilateral valuation adjustment (BVA). One of the complications that arises with BVA is that it requires agreeing on which counterparty defaults first, since when one counterparty defaults its CVA is triggered and the other parties are not.

24.6 HEDGING COUNTERPARTY RISK

The main objectives of hedging CVA and DVA are threefold. First, hedging CVA and DVA allows one to reduce the sensitivity of a bank's profit and loss to changes in the credit spread of the counterparty or the banks' own creditworthiness changes. Second, hedging allows to lock in the CVA position for the duration of the trade with the counterparty. Third, hedging allows a reduction of a CVA exposure. In practice, CVA hedging is not that simple since hedging jump to default risk is difficult and WWR can make separating counterparty risk from other risks impossible.

[9]See http://www.bis.org/publ/bcbs214.pdf for details.

24.6.1 CVA AND DVA HEDGING IN PRACTICE

As we saw earlier, DVA and CVA adjustments can amount to billions of dollars in practice. The following reading illustrates this practice.

EXAMPLE

"Asked by an analyst to explain why the company's DVA gains were relatively muted, for a period in which its debt spreads also widened and which, in theory, should have led to a large DVA gain, Goldman Sachs Chief Financial Officer David Viniar said Tuesday that the company attempts to hedge using a basket of different financials. A Goldman spokesman confirmed that the company did this by selling credit default swaps on a range of financial firms. In selling CDS coverage, a firm is hoping to gain from an improvement in the credit quality of the underlying debt as evidenced by a narrowing in its risk spreads. Goldman wouldn't say what specific financials were in the basket, but Viniar confirmed to the analyst asking the question that the basket contained "a peer group."

Most would consider peers to Goldman to be other large banks with big investment-banking divisions, including Morgan Stanley, J.P. Morgan Chase, Bank of America, Citigroup, and others.

In Europe, a company could hedge against a range of banks at once through an index of credit default swaps called the iTraxx Europe Senior Financials, administered by Markit. But in the United States, there is no popularly traded index of CDS contracts on financials, so Goldman likely created a custom basket with another dealer."

(**Wall Street Journal, Deal Journal blog, October 18, 2011, "***Goldman Sachs Hedges Its Way to Less Volatile Earnings***" by Liz Moyer and Katy Burne**)

The example above also showcases the role of credit derivatives as hedges for DVA gains. What is at first sight counterintuitive is that a deterioration in a bank's credit quality can lead to accounting gains. Although Goldman Sachs denied using CDS on its own name this raised the issue of whether a bank could potentially increase its account profits by taking actions that would reduce its credit-worthiness.

According to a recent Ernst and Young survey[10] based on responses from 19 financial institutions, the most popular instruments to manage and hedge market and credit risk related to CVA are single-name CDS (used by 14 of the 19 respondents) and interest rate and FX products (14 users) to manage market and credit risk relating to CVA. The next most popular hedges are based on index-linked CDS (13 users) and volatility hedges (10 users), while bond-based hedging (8 users) and correlation products (7 users) as well as contingent CDS (CCDS) (6 users) are less popular. Some more recent and advanced approaches to deal with counterparty risk involve CVA-CDOs and margin lending.

Although CDS can be used for CVA hedging, there are some problems associated with static CDS hedging. The payout from a CDS is fixed and only depends on whether the reference name defaults and not on the amount of the exposure to the counterparty. This can be partly addressed by *dynamic hedging*, that is adjusting the CVA position over time and depending on the amount of exposure that one has today. There is also a tradeoff between single-name and index CDS hedges. Single-name hedging is more precise in case of bad news affecting a single firm rather than broad market moves, but

---

[10]See Ernst & Young (2012) survey "Reflecting credit and funding adjustments in fair value".

single-name CDS are not available for most counterparties as reference names. Index CDS hedges or other macro hedges are one practical alternative but they represent a rather imperfect hedge against bad news affecting a single firm.

24.6.2 CONTINGENT CDS

The above hedging limitations associated with vanilla CDS hedging have driven innovation in credit products such as CCDS. Whereas in a simple CDS, a credit event triggers the payment, in a CCDS, the trigger requires both a credit event and another specified event such as the level of a particular market or sector variable. A CCDS can be viewed as a derivative which contains an embedded knock-in option upon the default of the reference transaction. The economic rationale that leads to the development of the CCDS is that it eliminates the economic risk which arises from variations in the credit exposure of a counterparty as a result of deterioration in broader economic market forces. One of the caveats associated with CCDS is, however, that they are currently not standardized and this implies elevated contractual risks or deliverability risk compared to vanilla CDS. Moreover, in contrast to a vanilla CDS, a CCDS can currently only be used to create a static hedge and not a dynamic hedge, against the counterparty credit risk.

24.7 FUNDING VALUATION ADJUSTMENT

A final aspect of counterparty risk that needs to be taken into account relates to funding. When a trader manages a trading position, he or she must obtain cash in order to carry out different operations including (i) hedging the position and (ii) posting collateral. These operations have a cost and need to be funded. Accounting for this funding cost is called funding valuation adjustment. If we incorporate CVA, DVA, and FVA into the valuation of derivatives under counterparty risk, we obtain the following equation:

$$P^{\text{to bank B}}_{\text{with CR}} = P^{\text{to bank B}}_{\text{no CR}} + \text{DVA}_{\text{Bank B}} - \text{CVA}_{\text{Bank B}} - \text{FVA}_{\text{Bank B}} \qquad (24.17)$$

The practical implementation of FVA is not straightforward. The reason is that including funding costs leads to a recursive pricing problem that can be expressed as a backwards stochastic differential equation.

The pre-GFC market practice related to funding was that collateral and funding were second-order concerns when assessing the profit and loss of a trading desk. Banks were used to borrowing money by posting collateral via a variety of low-cost funding options. The borrowing cost was to a large extent offset by the interest paid on collateral by the receiving party. After the GFC, a fundamental shift occurred in the role of funding in the derivatives business since funding costs increased. What changed market practice is that the risk of credit lines being pulled if a firm is perceived as being at risk of default became very real. One practical example of the relevance of funding considerations is that a trade may appear to have a positive profit and loss, but it may also have high potential future exposure. If a credit support annex (CSA) is in place, then an adverse market could lead to significant additional funding needs in the form of a requirement to immediately post additional collateral.[11] Even if there is no CSA in place, expected future losses related to the trade may cause a loss of confidence in the bank and a withdrawal of credit lines to the firm. In practice, the primary role of FVA is to provide monetary incentives for trading desks to use less funding. FVA is currently not supported by accounting rules.

24.8 CVA DESK

The incorporation of counterparty risk adjustments into market practice has also affected the internal organization of banks as banks created *CVA desks*.[12] The rationale for their creation was to move counterparty risk management away from traditional asset classes trading desks by creating a specific counterparty risk trading desk. In principle, the creation of CVA desks could be interpreted as allowing traditional traders to work in a counterparty risk-free world in the same way as they did before the GFC. Some top-tier banks used to have CVA desks for many years before the GFC, but their importance has now grown and CVA desks can now be found in most large banks. One of the incentives to create CVA desks is that it reduces the workload and skill needs of traders. They are freed from the need to be familiar with advanced credit models that are linked to traditional derivatives trading models for different asset classes, as illustrated by the CVA study above. Moreover, the traders do not need to concern themselves with netting details. Finally the CVA desk aggregates counterparty risks across trades, thus enabling the bank to take a view of counterparty risk-related options on the whole portfolio of derivatives transactions.

The main job of a bank's CVA desks is to manage all counterparty risk for the bank. This implies several functions. First, CVA desks provide CVA/DVA and FVA pricing for trades with counterparties. Second, CVA desks can be involved in trade origination as they advise on structuring trades to minimize credit risk. Typically there is an upfront charge for the trade which contains CVA/DVA and it is allocated to the CVA desk. Thus the CVA desk provides protection to the origination/trading desk. Third, the CVA desk helps to monitor and dynamically hedge market, credit and cross partial risks as well as default risk. Fourth, in case a counterparty defaults, it is typically also the responsibility of the CVA desk to (i) pay the originating desk the exposure coverage, (ii) manage the workout with CDS protection providers, (iii) support trade termination activities, and (iv) maximize the recovery of the remaining trade value.

Although a CVA desk can in principle address some of these issues related to counterparty risk, one of the implications of incorporating counterparty risk is that there is no such thing as a fair value for derivative transaction since the deal value depends on the counterparty. It also implies that there is no such thing as mark-to-market at the deal level since the mark-to-market depends on the rest of the portfolio. Counterparty risk adjustments to some extent invalidate essential

---

[11]A CSA is a document which regulates credit support (collateral) for derivative transactions. See www.isda.org for further details. In 2014, ISDA revised the so-called standard CSA (SCSA) which had been introduced to address some issues such as the ability of counterparties to choose eligible collateral which leads to optionality under the CSA. The revision of the SCSA became necessary after three different regulatory changes made it too capital intensive for banks to use.

[12]Sometimes CVA desks are also referred to as XVA desks in the literature because they are managing different adjustments (CVA, FVA, DVA, KVA, etc.) and not only CVA.

assumptions behind risk-neutral valuation, that is market completeness and hedge availability to create a riskless portfolio. Moreover, model risk associated with CVA models is a new issue.

In practice, there are also some institutional and psychological aspects that affect the efficient functioning of CVA desks. Some traders may question the CVA fees that they have to pay to the CVA desk and be reluctant to cede authority to another unit such as the CVA desk. CVA desks may also give a false sense of security if traders incorrectly assume that counterparty risk can be perfectly hedged away by CVA desks. As we have seen in the previous section, CVA hedges are rather imperfect in practice.

24.9 CHOICE OF THE DISCOUNT RATE AND MULTIPLE CURVES

Throughout the book, we took the standard approach of pricing derivatives under the risk-neutral measure and using a risk-free discount rate. This was based on the fundamental financial engineering principle that a derivative can be valued by means of a replicating portfolio that leads to a risk-free portfolio. However, as will see below, one of the most common proxies for the risk-free rate before the GFC embeds a significant amount of counterparty risk which makes it an unsatisfactory proxy for the risk-free rate. We introduced LIBOR and the OIS rate in Chapter 3. Figure 24.8 shows the LIBOR—OIS spread over the December 2003—July 2014 period based on Bloomberg data. The solid line and the dashed line show the 1-month and 3-month LIBOR—OIS spreads, respectively. As the figure shows there were several episodes where the LIBOR and OIS rates diverged by several percentage points including the years during the GFC. The divergence between the two rates reflects counterparty risk and changes in the perception of creditworthiness of banks and their willingness to lend to each other.

Before the GFC, LIBOR was the most commonly used proxy for the riskless rate but after the GFC, market practice has moved to the usage of OIS rates as a discount rate in collateralized transactions. This is partly due to the fact that LIBOR rates were significantly higher than OIS rates as a result of the counterparty risk that was priced into LIBOR rates.

The following example illustrates that the move to OIS discounting affects hundreds of billions of dollars of derivatives transactions:

EXAMPLE

With an increasing proportion of interest rate swap trades now being valued using the overnight swap index (OIS) rather than Libor, LCH.Clearnet plans to begin using the measure to discount its US$218trn interest rate swap portfolio to ensure a more accurate valuation for risk management purposes. The move signals an ongoing trend that began during the financial crisis, and the shift could help to boost liquidity across the OIS curve.

Even prior to the crisis, many market participants argued that OIS was the most appropriate measure given that it represents an accurate measure of expectations of the federal funds rate over the term of the swap. With the Libor—OIS spread historically stuck around 10 bp, there was little impetus for change. But after peaking at more than 300 bp in late 2008, market participants were forced into reconsidering their valuation methods. Traders believe that the ongoing shift towards OIS discounting has already improved liquidity in longer maturities and more exotic currencies, and the shift by LCH.Clearnet should further boost liquidity.

(Thomson Reuters IFR 1838, June 19 to June 25, 2010, "LCH adopts OIS discounting")

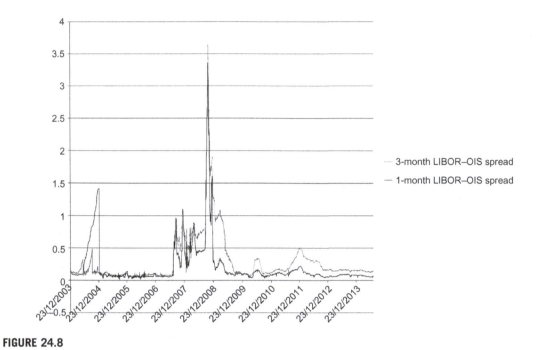

FIGURE 24.8

LIBOR−OIS spread.

The move to OIS rates has three fundamental implications for derivatives pricing. First, it is not possible anymore to use the same LIBOR/swap zero curves to price certain derivatives such as IRSs that we discussed in Chapter 4. Payoffs continue to depend on LIBOR rates, but discount rates will be based on OIS rates. Second, multiple curves need to be modeled to price fixed-income derivatives such as IRSs as well as caps and swaptions discussed in Chapter 17. The reason is not just due to the fact that OIS and LIBOR rates differ, but also because practitioners now use *multiple* LIBOR curves which reflect different levels of counterparty risk. Third, if one assumes that banks can risklessly borrow and lend at LIBOR and OIS rates then an implicit arbitrage arises.

The move to the OIS rate as the riskless rate also has implications for fixed-income financial engineering discussed in Chapter 14. We saw that one of the widely used models was the so-called Forward LIBOR or BGM (Brace Gatarek Musiela) Model. In principle, this model can be modified easily by replacing the LIBOR rate with the OIS rate, which would lead to a Forward OIS model. For some fixed-income derivatives, it is necessary to model the LIBOR and OIS curves simultaneously. In practice, many banks continue to use several models such as the 3-factor HJM and the Forward LIBOR model. One of the first-order effects in fixed-income modeling is stochastic volatility. It is important to incorporate stochastic volatility but the Forward LIBOR model is not very suitable for this purpose when large trading books are concerned since most Monte Carlo methods that would allow this tend to be too time consuming.

OIS discounting also brings with it some practical computational challenges as the following reading illustrates.

Finally there is an ongoing debate among practioners and among academics about the correct discount rate to use and this discussion is also related to a debate about the rationale for FVA.

24.10 CONCLUSIONS

In this chapter, we have reviewed recent innovations in financial engineering and derivatives pricing related to counterparty risk. The discussion has shown that many of the tools from the toolkit developed in previous chapters can be applied in corporate counterparty risk. CVA can be viewed as an option on the residual value of a portfolio upon default. Default probabilities for the calculation of CVA can be backed out from CDS market quotes. Finally, the option to choose the type of eligible collateral also. At the same time, we have seen that this is a very new and complex area and it is likely that practitioners, accountants, regulators, and academics will debate the best way of incorporating counterparty risk into accounting statements, pricing approaches, and bank capital charges for years to come.

SUGGESTED READING

This chapter could only provide a brief summary of the treatment of counterparty risk in financial engineering and derivatives pricing. **Brigo et al.** (2013) provide one of the best reviews of counterparty risk including detailed examples for each major asset class and a discussion of CVA-CDOs. The discussion in this chapter is partly based on this text. The best source for recent developments in counterparty risk market practice and thinking is in the form of articles in **Risk** magazine, which we strongly recommend. For details about the recent debate about the choice of the riskless rate proxy as well as the merits of FVA, for example, we refer the reader to Risk magazine as well as **Hull and White** (2014). For recent developments related to model risk in counterparty risk adjustments see **Kenyon and Stamm** (2012) and **Anfuso, Karyampas and Nawroth** (2014).

EXERCISES

1. What do the terms CVA, DVA, and FVA refer to?

2. What is the difference between unilateral and bilateral CVA?

3. Consider the reading below and explain how it is possible that a widening of credit spreads can lead to accounting gains? What is the difference between the treatment of DVA from an accounting and a bank's regulatory capital perspective?

    > On May 10, Jamie Dimon, chairman and chief executive of JP Morgan, announced that the bank would book a multi-billion dollar loss from "poorly executed" and "poorly monitored" derivatives trades made at its London-based Chief Investment Office.
    >
    > JP Morgan's credit spreads have steadily worsened since news of the trading loss broke. The cost of insuring against a JP Morgan default on five-year debt, using a CDS contract, increased from 110 basis points on May 9 to 165 bps on June 5, according to data from Markit.
    >
    > This means that the cost of protection on an annual basis over five years has jumped from $110,000 per $10m of debt at the start of the year, to $165,000.
    >
    > Pugachevsky worked out his approximate $1bn gain prediction from JP Morgan's previous quarterly results. At the end of the first quarter, JP Morgan reported a $900m DVA loss on the back of tightening spreads on its debt. Credit spreads had fallen from 144 bps on December 30 to 93 bps on March 30.
    >
    > Kinner Lakhani, a banks analyst at Citigroup, said, however, that while widening spreads generally convert to DVA gains, these gains are accounting "noise" that does nothing to "help boost a bank's regulatory capital".
    >
    > **Efinancialnews, June 6, 2012, "JP Morgan DVA gains could hit $1bn in second quarter"**

4. How can counterparty risk be hedged?

MATLAB EXERCISE

5. (CVA for interest rate portfolio) Consider the CVA case study for a portfolio of interest swaps in the text. The case study is based on a Matlab example. Update the example by collecting data on (i) a portfolio of IRSs between a bank B and five counterparties (making sure that the floating and fixed leg rates are chosen so that current swap values are close to zero), (ii) a recent yield curve data, and (iii) CDS data for the counterparties. Back out the implied probabilities of default from the CDS spreads by applying a reduced-form hazard function model. Now redo Steps 1−4 in the case study and plot
    a. the evolution of the yield curve for a given scenario
    b. the mark-to-market value of the swap portfolio for one simulation
    c. the mark-to-market value of the portfolio for all simulations
    d. the exposure of the portfolio for all simulations
    e. the expected exposure
    f. the discounted expected exposure for each counterparty.
    Finally calculate the CVA for each counterparty from bank B's perspective.

References

Adrian, T., Begalle, B., Copeland, A., Martin, A., 2013. Repo and Securities Lending, Federal Reserve Bank of New York Staff Reports.

Aït-Sahalia, Y., 1996. Testing continuous-time models of the spot interest rate. Rev. Financ. Stud. 9 (2), 385−426.

Allen, P., Granger, N., 2005. Correlation Vehicles. JP Morgan, European Equity Derivatives Strategy, London.

Andersen, L., Andreasen, J., 2000. Volatility skews and extensions of the LIBOR market model. Appl. Math. Finance 7 (1), 1−32.

Andersen, L.B.G., Piterbarg, V.V., 2010. Interest Rate Modeling, vol. 3. Products and Risk Management. Atlantic Financial Press.

Anfuso, F.K., Nawroth, A., 2014. Credit Exposure Backtesting for Basel III, RISK, September.

Avellaneda, M., Buff, R., Friedman, C., Grandchamp, N., Kruk, L., Newman, J., 2001. Weighted Monte Carlo: a new technique for calibrating asset-pricing models. Int. J. Theor. Appl. Finance 4 (1), 91−119.

Ayache, E., Forsyth, P.A., Vetzal, K.R., 2003. The valuation of convertible bonds with credit risk. Wilmott Magazine.

Baba, N., Packer, F., Nagano, T., 2008. The spillover of money market turbulence to FX swap and cross-currency swap markets. BIS Q. Rev. March.

Barkbu, B.B., Ong, L.L., 2012. FX Swaps: Implications for Financial and Economic Stability. IMF Working Paper.

Barone-Adesi, G., Whaley, R.E., 1987. Efficient analytic approximation of American option values. J. Finance 42 (2), 301−320.

Bielecki, T.R., Rutkowski, M., 2001. Credit Risk. Springer-Verlag, Berlin, Heidelberg, New York.

Black, F., Jones, R., 1987. Simplifying portfolio insurance. J. Portf. Manage 14 (1), 48−51.

Black, F., Derman, E., Toy, W., 1990. A one-factor model of interest rates and its application to treasury bond options. Financ. Anal. J. 46 (1), 33−39.

Blair, B.J., Poon, S.-H., Taylor, S.J., 2001. Forecasting S&P 100 volatility: the incremental information content of implied volatilities and high frequency index returns. J. Econ. 105 (1), 5−26.

Blanco, C., Pierce, M., 2013. OTC Commodity Swaps Valuation, Hedging and Trading, Risk.net/energy-risk.

Bodie, Z., Kane, A., Marcus, A.J., 2014. Investments, tenth ed. McGraw-Hill.

Bossu, S., Strasser, E., Guichard, R., 2005. Just What You Need to Know About Variance Swaps. JP Morgan.

Brace, A., Gatarek, D., Musiela, M., 1997. The market model of interest rate dynamics. Math. Finance 7, 127−154.

Brealey, R., Myers, S., 2013. Principles of Corporate Finance, eleventh ed. McGraw-Hill, New York, NY.

Brigo, D., Mercurio, F., 2006. Interest Rate Models—Theory and Practice: With Smile, Inflation and Credit (Springer Finance). Springer-Verlag, Berlin, Heidelberg, New York.

Brigo, D., Pallavicini, A., Torresetti, R., 2010. Credit Models and the Crisis: A Journey into CDOs, Copulas Correlations and Dynamic Models. The Wiley Finance Series.

Brigo, D., Morini, M., Pallavicini, A., 2013. Counterparty Credit Risk, Collateral and Funding: With Pricing Cases For All Asset Classes. Wiley Finance.

Buraschi, A., Kosowski, R.L., Trojani, F., 2014. When there is no place to hide—correlation risk and the cross-section of hedge fund. Rev. Financ. Stud. 27 (2), 581−616.

Burgard, C., Kjaer, M., 2011. Partial differential equation representations of derivatives with bilateral counter-party risk and funding costs. J. Credit Risk 7, 75−93.

Butler, K., 2012. Multinational Finance: Evaluating Opportunities, Costs, and Risks of Operations, fifth ed. John Wiley & Sons.

Cheng, I.-H., Xiong, W., 2013. The Financialization of Commodity Markets. NBER Working Paper No. 19642. National Bureau of Economic Research.

Choi, D., Getmansky Sherman, M., Hederson, B.J., Tookes, H., 2010. Convertible bond arbitrageurs as suppliers of capital. Rev. Financ. Stud. 23 (6), 2492−2522.

Choudhry, M., 2010. The Repo Handbook, second ed. Butterworth-Heinemann.

Clewlow, L., Strickland, C., 1998. Implementing Derivative Models. John Wiley & Sons, New Jersey.

Cloyle, B., Graham, A., 2000. Currency Swaps. Currency Risk Management Series. AMACOM.

Cont, R., Tankov, P., 2009. Constant proportion portfolio insurance in presence of jumps in asset prices. Math. Finance 19 (3), 379−401.

Copeland, A., Martin, A., Walker, M.W., 2010. The Tri-party Repo Market Before the 2010 Reforms, Federal Reserve Bank of New York Staff Report 477.

Copeland, A., Duffie, D., Martin, A., McLaughlin, S., 2012. Key mechanisms of the U.S. tri-party repo market. Econ. Policy Rev. 18, 3.

Cox, J.C., Ross, S.A., 1976a. The valuation of options for alternative stochastic processes. J. Financ. Econ. 3 (1), 145−166.

Cox, J.C., Ross, S.A., 1976b. A survey of some new results in financial option pricing theory. J. Finance 31 (2), 383−402.

Damadoran, A., 2012. Investment Valuation: Tools and Techniques for Determining the Value of Any Asset. Wiley.

Das, S., 1994. Swaps and Financial Derivatives: The Global Reference to Products, Pricing, Applications and Markets, second ed. Law Book Co., Sydney.

Das, S., 2000. Structured Products and Hybrid Securities, second ed. John Wiley & Sons, New Jersey.

Das, S., 2003. Swaps and Financial Derivatives: Products, Pricing, Applications and Risk Management, third ed. John Wiley & Sons, New Jersey.

Das, S., Sundaram, R., 2010. Derivatives. McGraw-Hill.

De Spiegeleer, J., Schoutens, W., 2012. Pricing contingent convertibles: a derivatives approach. J. Derivatives 20 (2), 27−36.

Demeterfi, K., Derman, E., Kamal, M., Zou, J., 1999. A guide to volatility and variance swaps. J. Derivatives 6 (4), 9−32.

Derman, E., Kani, I., 1994. The volatility smile and its implied tree. Quantitative Strategies Research Notes. Goldman Sachs.

Derman, E., Kani, I., Chriss, N., 1996. Implied trinomial trees of the volatility smile. Quantitative Strategies Research Notes. Goldman Sachs.

Duffie, D., 2001. Dynamic Asset Pricing Theory, third ed. Princeton University Press, Princeton, NJ.

Duffie, D., Singleton, K.J., 2002. Credit Risk: Pricing, Measurement, and Management. Princeton University Press, Princeton, NJ.

Duffie, D., Singleton, K.J., 2003. Credit Risk: Pricing, Measurement, and Management. Princeton University Press, Princeton, NJ.

Dupire, B., 1992. Arbitrage Pricing with Stochastic Volatility. Working Paper, Société Générale, Paris.

Edwards, D.W., 2010. Energy Trading and Investing. McGraw-Hill.

El Karoui, N., Jeanblanc-Picque, M., Shreve, S.E., 1998. Robustness of the Black and Scholes formula. Math. Finance.

FX week, April 16, 2014, article by Michael Watt, article title 'EU set to exempt FX swaps and forwards from initial margin requirements'. URL: <http://www.fxweek.com/fx-week/news/2340006/esa-exempts-fx-swaps-and-forwards-from-initial-margin-requirements>.

Fagnan, D.E., Fernandez, J.M., Lo, A.W., Stein, R.M., 2013. Can financial engineering cure cancer. Am. Econ. Rev. Pap. Proc. 103 (3), 406−411.

Fabozzi, F.J. (Ed.), 1998. Handbook of Structured Financial Products. John Wiley & Sons, New Jersey.

Flavell, R., 2009. Swaps and Other Derivatives, second ed. John Wiley & Sons, New Jersey.

Francis, C., Kakodkar, A., Martin, B., 2003. Credit Derivative Handbook. Merrill Lynch.

Gatheral, J., 2010. The Volatility Surface: A Practitioner's Guide. Wiley Finance.

Giesecke, K., 2002. Credit Risk Modeling and Valuation: An Introduction. Unpublished Manuscript.

Glasserman, P., Zhao, X., 2000. Arbitrage-free discretization of lognormal forward Libor and swap rate model. Finance Stoch. 4, 35−68.

Gorton, G., Metrick, A., 2012. Securitized banking and the run on repo. J. Financ. Econ. 104 (3), 425−451.

Gorton, G.B., Hayashi, F., Rouwenhorst, K.G., 2012. Review of Finance.

Gromb, D., Vayanos, D., 2010. Limits of arbitrage: the state of the theory. Annu. Rev. Financ. Econ. 2010 (2), 251−275.

Hirsa, A., Neftci, S.N., 2013. Introduction to the Mathematics of Financial Derivatives, third ed. Academic Press, New York, NY.

Hull, J., White, A., 2013. LIBOR vs. OIS: the derivatives discounting dilemma. J. Invest. Manage. 11 (3), 14−27.

Hull, J.C., 2014. Options, Futures and Other Derivatives, ninth ed. Prentice-Hall, New Jersey.

Hull, J.C., Predescu, M., White, A., 2005. Bond prices, default probabilities, and risk premiums. J. Credit Risk 1 (2), 53−60.

Jäckel, P., 2002. Monte Carlo Methods in Finance. Wiley.

Jacque, L., 2010. Global Derivatives Debacles—From Theory to Malpractice. World Scientific.

James, P., 2003. Option Theory. Wiley.

Jamshidian, F., 1997. LIBOR and swap market models and measures. Finance Stoch. 1, 293−330.

Jarrow, R.A., 2002. Modelling Fixed Income Securities and Interest Rate Options, second ed. Stanford University Press.

Jarrow, R.A., Turnbull, S., 1999. Derivative Securities: The Complete Investor's Guide, second ed. South-Western College Publishing.

Jegadeesh, N., Tuckman, B., 2000. Advanced Fixed-Income Valuation Tools. John Wiley & Sons, New Jersey.

Jin, W., Whetten, M., 2005. Anatomy of Credit CPPI. Nomura Fixed Income Research.

Joenväärä, J., Kosowski, R., 2013. An Analysis of the Convergence between Mainstream and Alternative Asset Management. EDHEC-Risk Institute Publication, February. Available at < http://www.edhec-risk. com/features/RISKArticle.2013-05-15.0924/attachments/EDHEC%20Publication%20An%20analysis%20of %20the%20convergence%20F.pdf > .

Joenväärä, J., Kosowski, R., Tolonen, P., 2014. The Effect of Investment Constraints on Hedge Fund Investor Returns. June 10. Available at SSRN: <http://ssrn.com/abstract=2362430>.

Johnson, S., Lee, H., 2003. Capturing the smile. Risk, March, 89−93.

Joossens, E., Schoutens, W., 2010. An Overview of Portfolio Insurances: CPPI and CPDO. JRC Scientific and Technical Reports.

Kat, H., 2001. Structured Equity Derivatives. Wiley.

Kenyon, C., Stamm, R., 2012. Discounting, Libor, CVA and Funding: Interest Rate and Credit Pricing. Palgrave Macmillan.

Kleinman, G., 2013. Trading Commodities and Financial Futures. Pearson Education, Inc., New Jersey.

Kloeden, P.E., Platen, E., 2011. Numerical Solution of Stochastic Differential Equations, third ed. Springer-Verlag, Berlin, Heidelberg, New York.

Kolb, R.W., 2007. Futures, Options, and Swaps, fifth ed. Blackwell Publishers.

Krishnamurthy, A., Nagel, S., Orlov, D., 2014. Sizing up repo. J. Finance, forthcoming.

Lea, M., 2010. International Comparison of Mortgage Product Offerings. Research Institute for Housing America.

Leland, H.E., 1980. Who should buy portfolio insurance? J. Finance 35, 2.

Lhabitant, F.-S., 2009. Handbook of Hedge Funds. Wiley.

Lipton, A., 2002. Assets with jumps. Risk September, 149−153.

Lu, J., 2010. Indexing Principal Protection. S&P Indices, Research Insights.

Mancini, L., Ranaldo, A., Wrampelmeyer, J., 2014. The Euro Interbank Repo Market. SSRN Working Paper. Available at SSRN: <http://ssrn.com/abstract=233135>.

Mancini-Griffoli, T., Ranaldo, A., 2013. Limits to Arbitrage During the Crisis: Funding Liquidity Constraints and Covered Interest Parity. SSRN Working Paper. Available at SSRN: <http://ssrn.com/abstract=1569504>.

McDougall, A., 1999. Mastering Swaps Markets: A Step-by-Step Guide to the Products, Applications and Risks. Financial Times Prentice-Hall.

McNeil, A.J., Frey, R., Embrechts, P., 2005. Quantitative Risk Management: Concepts, Techniques and Tools. Princeton University Press.

Merton, R.C., 1974. On the pricing of corporate debt: the risk structure of interest rates. J. Finance 29 (3), 449−470.

Merton, R.C., 1976. Option pricing when the underlying stock returns are discontinuous. J. Financ. Econ. 3 (1), 125−144.

Miltersen, K.R., Sandmann, K., Sondermann, D., 1997. Closed form solutions for term structure derivatives with log-normal interest rates. J. Finance 52 (1), 409−430.

Musiela, M., Rutkowski, M., 2007. Martingale Methods in Financial Modelling, second ed. Springer-Verlag, Berlin, Heidelberg, New York.

Natenberg, S., 2014. Option Volatility and Pricing: Advanced Trading Strategies and Techniques, second ed. McGraw-Hill Trade, New York.

Øksendal, B., 2010. Stochastic Differential Equations: An Introduction with Applications, sixth ed. Springer-Verlag, Berlin, Heidelberg, New York.

O'kane, D., 2008. Modelling Single-Name and Multi-Name Credit Derivatives. Wiley.

Pedersen, M., 1999. Bermudan Swaptions in the LIBOR Market Model. Manuscript.

Perold, A.F., 1986. Constant Proportion Portfolio Insurance. Harvard Business School.

Piros, C., 1998. Perfect hedge: to Quanto or not to Quanto. In: DeRosa, D.F. (Ed.), Currency Derivatives: Pricing Theory, Exotic Options, and Hedging Applications. John Wiley & Sons, New Jersey.

Piterbarg, V.V., 2005. Pricing and hedging callable libor exotics in forward Libor models. J. Comput. Finance 8 (2), 65−119 (Winter 2004/05).

Questa, G.S., 1999. Fixed Income Analysis for the Global Financial Market: Money Market, Foreign Exchange, Securities, and Derivatives. John Wiley & Sons, New Jersey.

Rebonato, R., 2000. Volatility and Correlation: In the Pricing of Equity, FX and Interest-Rate Options. John Wiley & Sons, New Jersey.

Rebonato, R., 2002. Modern Pricing of Interest-Rate Derivatives: The LIBOR Market Model and Beyond. Princeton University Press, Princeton, NJ.

Risknet/Asia Risk, April 2, 2014, Aaron Woolner and Xiao Wang "Dalian iron ore contract boosts overall market liquidity" <http://www.risk.net/asia-risk/feature/2337545/dalian-iron-ore-swaps-contract-boosts-overall-market-liquidity>.

Ross, S.A., Westerfield, R.W., Jaffe, J., 2012. Corporate Finance, tenth ed. McGraw-Hill College Div., New York, NY.

Roth, P., 1996. Mastering Foreign Exchange and Money Markets. Financial Times Market Editions.

Schofield, N.C., 2007. Commodity Derivatives: Markets and Applications. The Wiley Finance Series (Book 543).

Schönbucher, P.J., 2003. Credit Derivatives Pricing Models: Models, Pricing and Implementation. The Wiley Finance Series.

Serrat, A., Tuckman, B., 2011. Fixed Income Securities: Tools for Today's Markets, third ed. John Wiley & Sons, New Jersey.

Shleifer, A., Vishny, R.W., 1997. The limits of arbitrage. J. Finance 52 (1), 35–55.

Stefanini, F., 2010. Investment Strategies of Hedge Funds. Wiley Finance Series.

Steiner, R., 2012. Mastering Financial Calculations: A Step-by-Step Guide to the Mathematics of Financial Market Instruments. Financial Times Prentice-Hall.

Stojanovic, S., 2003. Computational Financial Mathematics Using MATHEMATICA: Optimal Trading in Stocks and Options. Birkhauser, Boston.

Sundaresan, S., 2013. A review of Merton's model of the firm's capital structure with its wide applications. Annu. Rev. Financ. Econ. 5.

Taleb, N.N., 1996. Dynamic Hedging: Managing Vanilla and Exotic Options. John Wiley & Sons, New Jersey.

Tavakoli, J.M., 2001. Credit Derivatives and Synthetic Structures: A Guide to Instruments and Applications, second ed. John Wiley & Sons, New Jersey.

Tepla, L., 2004a. KBC Alternative Investment Management (A): Convertible Bond Arbitrage. Case Study.

Tepla, L., 2004b. KBC Alternative Investment Management (B): Capital Structure Arbitrage INSEAD. Case Study.

Vasicek, O., 1977. An equilibrium characterisation of the term structure. J. Financ. Econ. 5, 177–188.

Weithers, T., 2006. Foreign Exchange: A Practical Guide to the FX Markets, first ed. John Wiley & Sons.

White, R., 2013. The Pricing and Risk Management of Credit Default Swaps, With a Focus on the ISDA Model. OpenGamma Quantitative Research.

Wilmott, P., 2006. Paul Wilmott on Quantitative Finance, vol. 3, second ed. John Wiley & Sons, New Jersey.

Wystup, U., 2006. FX Options and Structured Products. The Wiley Finance Series.

Index

CPSIA information can be obtained at www.ICGtesting.com
Printed in the USA
BVOW11*0055051214

378020BV00004B/9/P

9 780123 869685